T0358593

Quantum Foundations
and
Open Quantum Systems

Lecture Notes of the Advanced School

Quantum Foundations
and
Open Quantum Systems

Lecture Notes of the Advanced School

editors

Theo M Nieuwenhuizen
University of Amsterdam, The Netherlands

Claudia Pombo
Amsterdam, The Netherlands

Claudio Furtado
Federal University of Paraíba, Brazil

Andrei Yu Khrennikov
Linnaeus University, Sweden

Inácio A Pedrosa
Federal University of Paraíba, Brazil

Václav Špička
Academy of Sciences of the Czech Republic

 **World Scientific**

NEW JERSEY • LONDON • SINGAPORE • BEIJING • SHANGHAI • HONG KONG • TAIPEI • CHENNAI

Published by

World Scientific Publishing Co. Pte. Ltd.

5 Toh Tuck Link, Singapore 596224

USA office: 27 Warren Street, Suite 401-402, Hackensack, NJ 07601

UK office: 57 Shelton Street, Covent Garden, London WC2H 9HE

Library of Congress Cataloging-in-Publication Data

Quantum foundations and open quantum systems : lecture notes of the Advanced School / edited by Theo M. Nieuwenhuizen, University of Amsterdam, The Netherlands [and five others].

pages cm

Includes bibliographical references.

ISBN 978-9814616720

1. Quantum theory--Congresses. 2. Quantum logic--Congresses. 3. Quantum systems. 4. Wave-particle duality--Congresses. I. Nieuwenhuizen, Theo M., editor.

QC173.96.Q3615 2015

530.12--dc23

2014034482

British Library Cataloguing-in-Publication Data

A catalogue record for this book is available from the British Library.

Cover image: © David Costa (Navalha Photography)

Printed in Singapore

Preface

The Advanced School on Quantum Foundations and Open Quantum Systems took place in the city of Joao Pessoa in Northeast Brazil in July of 2012. The Advanced School served as a complement to the ongoing series of conferences: the Vaxjo, Sweden series "Foundations of Quantum Physics", and the Prague, Czech Republic series "Frontiers of Quantum and Mesoscopic Thermodynamics".

The Advanced School comprised an exceptional combination of lectures on several inter-related topics in quantum physics and quantum foundations. The program was specially designed to offer a broad and accessible outlook on current and near future fundamental research in theoretical physics, while at the same time offered a great opportunity to reunite researchers in the topical areas, who enjoyed discussions and exchange of ideas.

The topics and the lecturers were as follows.

- Quantum Solid State, presented by Knut Bakke, Ilde Guedes, Valdir Bezerra, Fernando Moraes, Antonio Murilo and Václav Špička;
- Dynamics of Quantum Open Systems, presented by Amir Caldeira, Giuseppe Falci, and Elisabetta Palladino;
- Quantum Information and Computation, presented by David Simon;
- Quantum Thermodynamics, presented by Karen Hovhannisyan;
- Quantum Measurements, presented by Theo Nieuwenhuizen and Martí Perarnau Llobet;
- Foundations of Quantum Mechanics, presented by Alexander Burinskii, Ana Maria Cetto, Luis de La Peña, Gerhard Groessing, Andrei Khrennikov, Kristel Michielsen, and Hans De Raedt; and
- History and Philosophy of Quantum Physics, presented by Carlos Baladrón and Arkady Plotnitsky.

This book contains a collection of nearly all the lectures given at the Advanced School, aimed primarily for students at the master level.

We wish to thank the scientific committee and the organizers, the lecturers and all the participants, who made this meeting possible. Our very special thanks to Edvaldo Nogueira, Alexandre Rosas, Ilde Guedes, Alexandre Carvalho and José Roberto do Nascimento for providing local organization. We kindly thank Isabela P. Geertsma and Xander Giphart for their secretarial and editorial support. This meeting could not have happened without the funding support of the Universidade Federal da Paraíba, CAPES, CNPq and FAPESQPB.

The Editors,
Theo Nieuwenhuizen, Claudia Pombo, Claudio Furtado,
Andrei Khrennikov, Inácio Pedrosa and Václav Špička

Contents

Chapter 1

The Physics of Quantum Computation

Giuseppe Falci* and Elisabette Paladino†

*Dipartimento di Fisica e Astronomia, Università degli Studi Catania &
CNR-IMM MATIS, Consiglio Nazionale delle Ricerche &
Istituto Nazionale di Fisica Nucleare, Sezione di Catania
Via Santa Sofia 64, 95123 Catania, Italy*

Quantum Computation has emerged in the past decades as a consequence of down-scaling of electronic devices to the mesoscopic regime and of advances in the ability of controlling and measuring microscopic quantum systems. QC has many interdisciplinary aspects, ranging from physics and chemistry to mathematics and computer science. In these lecture notes we focus on physical hardware, present day challenges and future directions for design of quantum architectures.

1. Preliminary concepts

The field of quantum computation[1] has emerged in the past decades as a consequence of down-scaling of electronic devices to the mesoscopic regime and of advances in the ability of controlling and measuring microscopic quantum systems. The question emerges whether there is any fundamental gain in resorting to new computational paradigms where quantum mechanical properties of physical objects are fully exploited. Positive answers came from Richard Feynman[2] who showed that quantum systems may be efficiently simulated only by other quantum systems, i.e. an analog Quantum Computer (QC), and from the Shor's breakthrough discovery of an efficient factoring algorithm on a digital QC.[3] This has motivated theoretical and experimental investigation on the implementation of Digital QCs, which will be mainly addressed in this article, although the scenario of Quantum Information is much wider, and present efforts are mainly focused on analog QCs.[4]

QC has many interdisciplinary aspects, ranging from physics and chemistry to mathematics and computer science. It also plays, in our opinion, a central role in the field of foundations of Quantum Mechanics (QM). However for the sake of clarity

---

*gfalci@dmfci.unict.it
†elisabetta.paladino@dmfci.unict.it

we will adopt here a totally different point of view, presenting QC as an application of QM and addressing the topic of quantum hardware. On the other hand, quite often people use the Copenhagen interpretation as an efficient tool to teach the how-to of QM. Despite of our choice it is worth stressing that ultimate answers on fundamental questions may critically depend on subtle hardware problems, as in the case of the violation of Bell inequalities.[5] Improvement in hardware as sources and detectors, and in the ability of faithful production of highly entangled states, belong indeed to the realm of Quantum Information and certainly will trigger new insight and provide new tools for fundamental physics studies.

In these lecture notes we focus on physical hardware, present day challenges and future directions for design of quantum architectures. An exhaustive but condensed review on the subject has been published by Ladd *et al.* few years ago.[6] We will first describe the main issues of digital QC[a] from which naturally emerges a set of requirements a quantum system should satisfy to be a good candidate for the implementation of a quantum computer, known as Di Vincenzo criteria.[7] Recent experimental progresses have dealt with the central problem of decoherence, showing that it can be challenged to a certain extent. Also it emerged that a key concept for quantum computation is *scaling*, which is referred to several aspects of the problem as quantum algorithms, fabrication of "fault tolerant" (protected from decoherence/defects and reliable) hardware, number of elementary control pulses required for preparation, manipulation and interaction of quantum bits. This has lead to a more mature reformulation of the physical criteria for the implementation of quantum architectures.[6,8]

1.1. *From miniaturization to quantum technologies*

Although microscopic systems are a natural hardware for Quantum Information, the strong motivation triggering this field comes from the quest for faster and low consuming "classical" computers (ClC), which is achieved by using smaller and smaller devices in the mesoscopic regime.[9] As for the technological roadmap to down-scaling, also fundamental studies in Condensed Matter Physics follow a "top-down" approach for the description of solid-state phenomena. This approach is sort of "adaptive": The underlying quantum nature is introduced judiciously, starting from phenomena related to quantization and then to coherence, which appears as long as macroscopic (semiclassical) physical systems shrink to become mesoscopic (see Fig. 1).

1.1.1. *Semiclassical electrons in electronic devices*

The golden era of semiconductor electronics started about sixty years ago with the invention of the bipolar transistor, and then with the discovery that also resistors

---

[a] Analog quantum computation, originally introduced by Feynman, is experiencing at present a boost of scientific interest. It will be briefly addressed in Sec. 2.3.

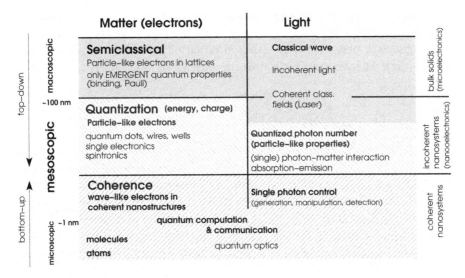

Fig. 1. Synoptic scheme of the roadmap from semiclassical to quantum coherent regimes in solid-state systems. New phenomena related to the quantum nature, in particular quantization and coherence, appear as long as macroscopic (semiclassical) physical systems shrink to become microscopic. In the intermediate mesoscopic regime, explored by Nanophysics, effects as quantum coherence in atoms survive on much larger time and space scales.

and capacitors can be made out of silicon, yielding the integrated circuit. The development of planar technology has allowed to reach integration levels from several thousands of transistor per unit chip area since 1970 to several billions nowadays. This trend has been empirically described by the Moore's law[b] (1965), which states that the number of transistor on a chip doubles roughly every two years. Scaling down the size of integrated circuits, keeping a much slower increase of costs per unit area, has been the main subject of microelectronics technologies in the last two decades, and has required enormous progresses in manufacturing.[9] Ideally it would allow to build larger and faster computers, with the same volume and price. However even in this scenario there are fundamental limitations to miniaturization. One of these is related to the so called Landauer principle[10] (see Sec. 1.2), other limitations being due to stray quantum effects coming into play, leading to malfunctioning of standard logic devices.[9] As we will see quantum computation leverages on these two issues to propose a new paradigm for information processing.

From the point of view of transport the regime of microelectronics is referred as semiclassical since, even if QM properties are crucial in explaining thermodynamics and even the very existence of solids, electrons move as classical particles, with a number of peculiarities which do not prevent the use of classical Boltzmann equation

---

[b]From Gordon Moore Intel.

as the main investigation tool for computational electronics. Also electromagnetic fields used to address optoelectronic microdevices are semiclassical. In particular laser light, is a phase coherent *classical* radiation, i.e. containing many many *photons*, even if phenomena involved in the physics of laser sources is exquisitely "quantum incoherent".

1.1.2. *Incoherent nanophysics*

The quantum signature appearing when devices are down-scaled towards nanosize is quantization, or better said discrete spectra. Quantization of energy appears in nanosystems having one or more "confined" dimension (size $L \sim 10 - 100\,\text{nm}$) along which the motion is completely frozen. Examples are quantum wells, wires and dots, being elementary building blocks of generic nanoarchitectures.[9]

When electrons are confined their electrostatic interaction becomes more important. Together with quantization of charge, electrostatic interaction allows to control the motion of single electrons. This happens in Single Electron Tunneling (SET) devices, which are architectures containing islands of conductors, isolated from the rest of the circuit apart for the possibility of electron tunneling. Moving even a single electron requires charging such islands, and this has a typical energy cost $E_C \sim e^2/(2C)$, C being a typical capacitance. Devices with $C \sim 10^{-15} - 10^{-17}$ F allow nowadays the usage of charging effects at $T \lesssim E_C/k_B$, thereby from cryostat to room temperatures.

Is it worth noticing that quantization is the only relevant phenomenon in most of the applications under the umbrella of nanoelectronics. Physically electrons either move as classical ballistic particles, as in point contact, or hop incoherently between different energy levels and positions, as in nanograin flash memories. Also for spintronic devices, where a discrete degree of freedom of exquisitely quantum origin is manipulated, the dynamics is incoherent. In the same scenario we can encompass most of the physics of new materials, from graphene, to conventional molecular electronics. The main goal of this research is to produce the functionality of a switch, which is the basis for conventional digital electronics. In a dual scheme nanophotonics deals with incoherent particle-like properties of light.

1.1.3. *Coherent nanoelectronics*

Further miniaturization of conventional devices as MOSFETs leads to the *mesoscopic regime*, where electron transport fully exhibits it's wave-like nature.[11] Despite of the fact that still many degrees of freedom are involved, the dynamics starts to show the most peculiar feature of microscopic quantum systems, namely coherence. For instance interference effects determine universal conductance fluctuations. Apart from their fundamental importance, these fluctuations would lead to scaling problems in a systems of MOS doped channels, since the standard characteristics of the device will be not be reproducible. On the other hand quantum effects can be turned in new functionalities. Historically the first device based on electron coherence was the Resonant Tunneling Diode, where negative differential conductance has been observed. Quantum Computation fully exploit these new possibilities offered by QM.

The physics of coherent nanosystems approaches that of electrons in atoms, in that long lived superposition of energy eigenstates come into play. Controlled dynamics may be achieved if first it is possible to prepare (write) and measure (read) a set of N observables $\{Q_i\}$, defining the basis of the *computational states* $|\{q_i\}\rangle$. If moreover it is possible to tune the Hamiltonian of the system, then the dynamics of a generic state

$$|\psi(t)\rangle = \sum_{q_1 \ldots q_N} c_{q_1 \ldots q_N}(t) |q_1, \ldots, q_N\rangle \tag{1}$$

may be controlled. In general $|\psi(t)\rangle$ is a *superposition* of computational states, which in a sense means that it may share the properties of each of them at the same time ("in parallel"). If $c_{\{q_i\}}(t)$ cannot be factorized in individual functions $c_{q_i}(t)$, then $|\psi(t)\rangle$ is said to be *entangled*,[1] which means that different variables, typically describing spatially separated parts of the system, exhibit *quantum correlations*, "much stronger than classical" ones[c]. The word *Coherence* denotes the ability to maintain a superposition of states, i.e. a well defined relation between the components $c_{\{q_i\}}(t)$ (in particular their phases) of the state Eq. (1) exists at the given time. Only in this case we can use *pure states* as Eq. (1), the relation between the $c_{\{q_i\}}(t)$ being the solution of the Schrödinger equation.

The natural question arises if coherence, in artificial atoms or in their natural counterpart could be of some help to break limits of ClCs and algorithms and to approach fundamental limits set by physics. This is achieved by quantum algorithms, which leverage on entanglement and on the synergistic use of interference and measurement.

Preserving superpositions and entanglement is a key requirement for a quantum device outperforming a classical one (at least for certain tasks). However in most of the cases physical systems undergo *decoherence* (see Sec. 3): As long as the phase relation is lost the system is not any any more described by a pure state. When eventually decoherence is complete the dynamics reduces to a (classical) stochastic process, the system no longer displaying interference and quantum correlations.

Phase coherent dynamics and control has been demonstrated in several physical systems from the atomic realm to solid state "macroscopic quantum coherent" devices.[6,15,16] Current implementations of quantum hardware (see Fig. 2) are discussed in Sec. 4. Many other proposals with coherent nanodevices, e.g. based on graphene or on topological excitations,[17] enrich an already wide scenario. In addition quantum control of photons, besides allowing a possible implementation of

---

[c]The Bell inequalities are commonly used to make this statement more quantitative.

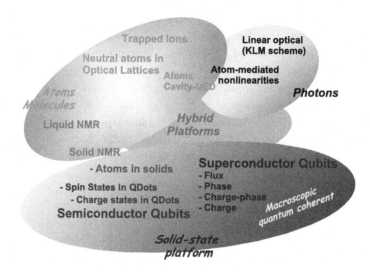

Fig. 2. Synoptic scheme of the current implementations of quantum hardware. They include: Localized atomic system (in cavities, or confined in ion traps and in optical lattices, or hosted by a solid matrix) and molecular systems addressed by NMR; Photons in linear optical circuits or photonic chips; Artificial atoms, which are nanodevices based on semiconductor quantum dots and on systems of superconducting tunnel junctions, where coherent dynamics has been demonstrated. Multiplatform hybrid systems are the new frontier of quantum computing.

quantum algorithms, provides a quite natural platform for Quantum Communication.[1] We will not touch this wide topic but still we point out that integrating electronic and phonic (or phononic) elements in distributed architectures is nowadays believed to be the new frontier for design of scalable quantum hardware (See Sec. 5).

1.2. *From bits to qubits*

1.2.1. *Classical computation*

The theory of classical computation developed in the effort of defining problems which could be in principle solved within certain computational schemes and sort them according to their *complexity class*, which measures how efficiently the solution could be achieved in practice. The theory is based on abstract mathematical models of computers which make no reference to the specific hardware and software. The best known example is the Turing Machine (TM),[1] a model where according to a set of instructions (transition rules) a head moves, reads and writes onto a infinitely long tape marked off into a sequence of cells. Independent mathematical models were proposed by A. Church, K. Gödel and E. Post. Despite the large apparent differences, these models are equivalent: Actually the TM provides an operative definition of computable functions, coinciding with the tasks identified

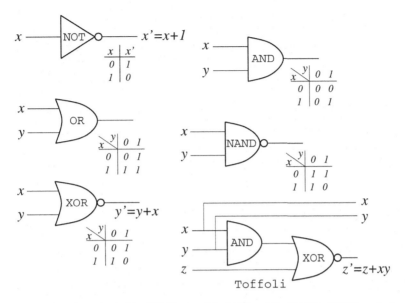

Fig. 3. Classical gates: single-bit (NOT), two-bit (AND, OR, NAND, XOR) and three-qubit (Toffoli) gates. Gates perform algebraic Boolean operations (e.g. NOT is $x \to x + 1$ modulo 2), represented by truth tables. NAND is part of a universal set of classical gates but it is not reversible; XOR can be made reversible (if a two-bit output is kept) but it is not universal. The Toffoli gate is both reversible and universal.

by the "competing" models. Moreover despite of their limitations, such models do encompass any conceivable ClC. Allowing for random choices the TM is extended to a Probabilistic TM (PTM), which calculates the same functions as TMs, possibly in a more efficient way but at the expenses of a finite probability to obtain an incorrect answer.

A different representation of the TM is based on the fact that transition rules are represented by classical *gates* acting on some input. Some of them are listed in Fig. 3 which are connected by wires to execute sequences of instructions. While gates may be implemented by devices, the resulting Classical Circuit Model (CCM) is again an abstract model equivalent to the TM. Moreover all gates can be obtained by combining a small number of elementary gates belonging to a *universal set*. An example is the two-bit NAND gate, supplemented by operations as swapping two bits (CROSSOVER) or making a copy (FANOUT). The Toffoli gate is another important example of universal set.

While classical models aim at a purely mathematical description of computation, physics creeps into the problem. Indeed a completely opposite perspective was raised by Rolf Landauer[10] who pointed out that the physical nature of hardware poses intrinsic limitations on the possibility of processing information (thereby "information is physical"). In particular heat is necessarily dissipated during irreversible transformations, and it is ultimately unavoidable for the erasure process

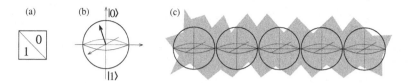

Fig. 4. (a) A classical bit is schematized by a cell storing a bistable variable $x \in \{0, 1\}$. (b) A qubit encodes such states as $\{|0\rangle, |1\rangle\}$ and all their superpositions, geometrically represented by the Bloch sphere. (c) No such geometric representation is available for a register; a set of Bloch spheres represents only a subset of states (factorized individual qubits, as in the "Benioff computer") and does not represent entanglement.

(CLEAR) in a ClC[d]. Erasure of a single bit determines a *minimum* increase of the environment entropy of $k_B \ln 2$ or equivalently a minimum heat generation of $k_B T \ln 2$. In fact dissipation is much larger in practice, also because many of the classical gates are irreversible, in particular the NAND. As a consequence a large scale of integration would produce heating and even melt parts of the computer.

1.2.2. *Reversibility and quantum computers*

In the 1970s, Charles Bennett of IBM pointed out the relation between logical and thermodynamical reversibility[19] arguing that if computation could be done reversibly dissipation could be limited to the effects of erasure only. He proposed that in principle this could be achieved using adiabatic dynamics. However elementary reversible gates (e.g. NOT and XOR) do not form a universal set in the CCM. A universal set of reversible gates can be found using the three-bit Toffoli gate, which can be used to perform both NAND and FANOUT operations. However physical models of reversible gates show large sensitivity to errors. Therefore they do not implement *fault tolerant* computation, this latter being the other key requirement besides universality.

The first step towards the QC is due to P. Benioff,[20] who introduced in 1980 a quantum Hamiltonian modeling a ClC, where classical gates are implemented by quantum dynamics, which is intrinsically reversible. In this ancestor of the Quantum Turing Machine (QMT) the unitary time-evolution of quantum bits (qubits) mimics a reversible TM. Therefore the computer deals with classical input and output states, even if it admits (to some extent) superposition at intermediate times. Soon afterward two different proposals aimed to exploit the full power of quantum coherence. In 1982 R. Feynman showed that no classical TM could simulate certain quantum phenomena without incurring in exponential slowdown,[2] but a *universal quantum simulator* (analog computer) could do the job. In 1985 Deutsch proposed that a QTM could fully use quantum dynamics (Fig. 4). All computational trajectories are allowed leading to arbitrary superpositions of classical states. Besides single

---

[d]Erasure in the quantum regime was studied in Ref. 18.

qubit non-classical states, already playing a role in the Benioff computer, the important new feature of the Deutsch QTM is that general many-qubit superpositions are as a rule entangled states, which are a key ingredient of quantum computation. Few years later Yao[21] proposed a Quantum Circuit Model (QCM) equivalent to the Deutsch QTM, formulated in terms of concatenated *quantum gates* which is an invaluable tool for the analysis of algorithms and protocols. Within the QCM sets of universal gates can be found, formed by the T ($\pi/8$-gate) and the H (Hadamard) single-qubit gates, by a two-qubit gate called CNOT (see Fig. 5), and supplemented by $S = T^2$ (phase gate), needed for fault-tolerant computation.

1.3. *Superpositions, entanglement, quantum parallelism*

Qubits can be introduced mathematically starting from a classical bit, $x \in \mathcal{B} = \{0, 1\}$, encoding it in a quantum two-state system (TSS) and allowing states defined by superpositions

$$|\psi\rangle = c_0|0\rangle + c_1|1\rangle \tag{2}$$

Several physical systems implement a TSS, exactly or approximately, as nuclear or electronic spins, two-level natural or artificial atoms, photon polarization or other degrees of freedom (see Fig. 2). New aspects relevant for computation emerge already from the representations of the computational states, namely encoding and parallelism.

1.3.1. *The Bloch sphere*

An efficient parametrization for the qubit follows from the general representation Eq. (2), by writing

$$|\psi\rangle = |c_0|\, e^{i\zeta_0}\, |0\rangle + |c_1|\, e^{i\zeta_1}\, |1\rangle = e^{i\zeta_0} \left[\cos(\theta/2)\, |0\rangle + \sin(\theta/2)\, e^{i\phi}\, |1\rangle\right]$$

where the irrelevant global phase factor ζ_0 is singled out, the normalization constraint $|c_0|^2 + |c_1|^2 = 1$ has been solved by introducing the angle $\theta \in [0, \pi]$ and $\phi = \zeta_1 - \zeta_0 \in [0, 2\pi]$ is the relative phase. This parametrization allows to draw a correspondence between each state and a point belonging to a unitary sphere, or equivalently a unitary vector **n** of the 3D space $\mathbb{R}^3$ (θ, ϕ) being its polar angles. It must be kept in mind that the Bloch sphere is in general a mathematical object, namely *a geometric representation of an abstract space* (in our case the projective Hilbert space $P(\mathrm{H})$, associated with physical states), as the complex plane represents C.

Alternatively the Bloch sphere emerges when we represent states by operators $|\psi\rangle \to \rho = |\psi\rangle\langle\psi|$ where the density matrix ρ is a Hermitian operator with unit trace. It can be represented uniquely as $\rho = \frac{1}{2}(I + \sum_{\alpha=1}^{3} s_\alpha \sigma_\alpha)$ in terms of the Pauli matrices,

$$\sigma_1 = \begin{bmatrix} 0 & 1 \\ 1 & 0 \end{bmatrix} \quad ; \quad \sigma_2 = \begin{bmatrix} 0 & -i \\ i & 0 \end{bmatrix} \quad ; \quad \sigma_3 = \begin{bmatrix} 1 & 0 \\ 0 & -1 \end{bmatrix} \tag{3}$$

which are Hermitian, unitary and have unit trace. As a consequence the coefficients s_α are real. It can be shown that they coincide with the Cartesian coordinates of $\mathbf{n} \leftrightarrow |\psi\rangle$, and also give the average of Pauli operators, $s_\alpha = \langle \psi | \sigma_\alpha | \psi \rangle$.

1.3.2. Encoding a qubit

It is useful to understand how to encode a system of qubits on a ClC. The binary encoding of a real number with an error smaller than ϵ requires $M = \log_{10}(1/\epsilon)$ digits, i.e. $\log_2(1/\epsilon)$ binary digits. Therefore a single qubit state $|\psi\rangle = \sum_{x \in \mathcal{B}} c_x |x\rangle \in P(\mathrm{C}^2)$ represented by a point of the Bloch 3D-sphere, requires $I_1 = 2M \log_2 10 \sim M$ bits. The generic state of a $N-$qubit computer can be written as

$$|\psi\rangle = \sum_{\mathbf{x} \in \mathcal{B}^N} c_\mathbf{x} |\mathbf{x}\rangle \tag{4}$$

where $|\mathbf{x}\rangle$ is a string of N binary digits representing a classical logic state of the N-qubit QC. The generic state belongs to the tensor product $\mathrm{H} = \bigotimes \mathrm{C}^2 = \mathrm{C}^{2^N}$ and requires $\sim 2^N$ coefficients (actually $P(\mathrm{H})$ is the unit sphere in $2^{N+1} - 1$ dimensions). Therefore it is encoded in $I_N \sim M \times 2^N$ bits. This is a huge number even for a small QC of $N = 100$ spins. On passing this shows that encoding a quantum system as a macromolecule on al ClC is already an impossible task. From the opposite point of view a suitably nanofabricated or nanoassembled quantum system could *emulate* the dynamics of other quantum systems, if sufficient control is available.

We stress here that the exponential blowup of I_N is due to increasing N, whereas the fact that possibly the Bloch sphere can be seen as a M-state system is not relevant.

1.3.3. Entanglement

Notice that Benioff deals with *factorizable* states having the form $|\psi_B\rangle = \prod_i (\sum_{x \in \mathcal{B}} c_x^i |x\rangle_i) = \sum_\mathbf{x} (\prod_i c_{x_i}^i) |\mathbf{x}\rangle$, which belong to the vector product of the individual qubit Hilbert space. Information is stored in $N I_1 \sim N \times M$ bits. Exponential blowup is due to the fact that a QC may process non factorizable states, called *entangled*, which represent the the great majority of the states in $P(\mathrm{C}^{2^N})$.

The characteristics of entangled states is that they encode correlations between different qubits. Well known examples of two-qubit entangled states are the Einstein-Podolski-Rosen (EPR) or Bell states

$$|\Psi^\pm\rangle = \frac{|01\rangle \pm |10\rangle}{\sqrt{2}} \quad , \quad |\Phi^\pm\rangle = \frac{|00\rangle \pm |11\rangle}{\sqrt{2}} \tag{5}$$

Such states describe correlations between measurements carried on two-qubits. For instance if the system is described by $|\Psi^\pm\rangle$ a measurement of the first qubit yielding $x \in \mathcal{B}$ implies that a measurement on the second qubit would yield $\bar{x}$. Actually correlations are much stronger: If for the first qubit yields x *in a arbitrary basis* the second will yield $\bar{x}$ in the same basis, i.e. it holds for noncommuting variables.

This observation, together with the fact that measurements can be carried out with the qubits being space-like separated (in relativistic language) lead EPR to formulate in 1935 the well celebrated paradox known under their names.[22] In in 1964 John Bell[23] found an inequality about correlations which is violated by QM[e]. This allows to assess quantitatively that quantum correlations due to entanglement are "stronger than classical ones".

1.3.4. *Quantum parallelism*

Dealing with superpositions has a very appealing feature for computation. Indeed processing a qubit generic input state is physically implemented by time evolution of the computer, which QM postulates to be linear

$$|\psi(t)\rangle = U(t)|\psi(0)\rangle = c_0 \, U(t)|0\rangle + c_1 \, U(t)|1\rangle$$

This suggests that the computer is able to process at once superpositions of two classical states. This property called *quantum parallelism*[24] becomes spectacular if a collection of N qubits is considered, since

$$|\psi(0)\rangle = \sum_{\mathbf{x} \in \mathcal{B}^N} c_{\mathbf{x}}|\mathbf{x}\rangle \; \rightarrow \; |\psi(t)\rangle = \sum_{\mathbf{x} \in \mathcal{B}^N} c_{\mathbf{x}} \, U(t)|\mathbf{x}\rangle$$

where $|\mathbf{x}\rangle$ is a string of N binary digits representing one of the classical logic states of the N-qubit computer. Therefore quantum parallelism means that 2^N classical inputs, stored as an entangled state, can be processed at once. Notice that in a classical parallel computer the input is divided in M parts which can be more efficiently processed on M processors at once, this yielding at most a polynomial speedup of the calculation. Instead processing entangled states may yield an exponential speedup, making clear how entanglement can be considered as the main resource for QC.

1.3.5. *Quantum gates*

Quantum gates are general transformations of H onto itself, and for a N-qubit quantum computer they are unitary operations belonging to $SU(2^N)$. Some of them can be obtained by generalizing truth tables to superpositions. For instance NOT, acting on classical states as $|x\rangle \rightarrow |\bar{x}\rangle$, is generalized to superpositions as $U_{\mathrm{NOT}}\left(c_0|0\rangle + c_1|1\rangle\right) = c_0|1\rangle + c_1|0\rangle$. Therefore in the representation of σ_3 we can write

$$U_{\mathrm{NOT}} = \begin{bmatrix} 0 & 1 \\ 1 & 0 \end{bmatrix} = \sigma_1$$

---

[e]Violation of Bell inequalities have been observed more and more clearly last few years with modern quantum technologies, since the observations at the beginning of the eighties.[25] Notice that they at most disprove the so called local realistic theories. Criticism has been raised against this point of view[12–14] and in any case violation of Bell inequalities does not imply, as sometimes stated, that quantum mechanics is the fundamental theory of nature.

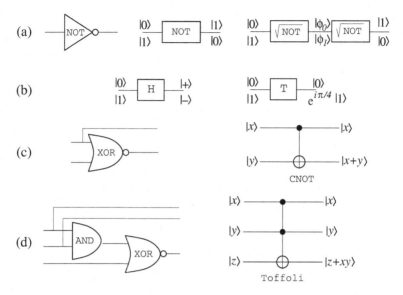

Fig. 5. (a) Quantum gates are obtained by generalizing truth tables to superpositions, e.g. NOT$\to \sigma_1$. Other quantum gates perform new tasks, as $\sqrt{\text{NOT}}$. (b) Examples of new single-qubit gates are the Hadamard (H), the T-gate and $S = T^2$. (c) The CNOT two-qubit gate is obtained by generalizing the classical XOR. It belongs to a universal set together with H, T and S. (d) The Toffoli gate can be also generalized to a three-qubit gate.

its action being a *bit flip*. More in general a classical gate is first written for a Benioff computer, then generalized by assuming that it operates linearly on superpositions.

While NOT is the only nontrivial classical single-bit gate, there are many non-trivial quantum single-qubit gates, which may perform new tasks. An example is the $\sqrt{\text{NOT}}$ which in the NMR notation[f] is defined as

$$X = \tfrac{1}{\sqrt{2}} \begin{bmatrix} 1 & -i \\ -i & 1 \end{bmatrix} \quad ; \qquad \begin{matrix} |0\rangle \\ |1\rangle \end{matrix} \boxed{X} \begin{matrix} |0\rangle - i\,|1\rangle \\ -i\,(|0\rangle + i\,|1\rangle) \end{matrix}$$

Acting on a classical state, X produces a superposition, e.g. $X|0\rangle = \frac{1}{\sqrt{2}} \left[|0\rangle - i|1\rangle \right]$. The double application yields $X^2 = -i\,\sigma_1$, therefore it is equivalent to NOT within an irrelevant global phase factor, and we may conclude that $X \equiv \sqrt{\text{NOT}}$.

An important class of single-qubit gates is given by H, T and $S = T^2$, since they belong to a universal set of gates. They are defined in Fig. 5, where $|\pm\rangle = (|0\rangle \pm |1\rangle)/\sqrt{2}$ are the eigenstates of σ_1. Since gates as H generate superpositions from classical states, they may be the key of new computational paradigms.

1.4. *Quantum speedup*

May quantum mechanical properties and new computational paradigms lead to a fundamental gain in efficiency? The concept of computational complexity classes

---

[f]X denotes a $\pi/2$ rotation about the $\mathbf{e}_1$ axis of the Bloch sphere.

make more quantitative this question. In classical computation efficiency is determined by the relation between the running time t_r of the computer to the size of the input I, expressed in bit and leads to a classification of problems in *complexity classes*. The growth rate of the computational time is used for classical deterministic algorithms. Problems solvable in polynomial time, $t_r \sim I^\alpha$ are considered tractable and form a class called **P**. Otherwise the problem is considered intractable, as in the case of exponential growth $t_r \sim a^I$, and belongs to one of several other classes of complexity. The class **NP** includes problems whose solutions can be quickly checked on a classical computer. This is the case of the problem of factoring composite integers, whose solution can of course be easily checked. Factoring is by itself a hard problem, and the best known algorithms (Multiple Polynomial Quadratic Sieve and Number Field Sieve) require $\exp[\alpha\, I^{1/3}(\ln I)^{2/3}]$ steps for an input integer n consisting of $I = [\log_2 n] + 1$ bits, thereby the problem is super-polynomial (but sub-exponential). For the **NP** class in principle it may exist a nondeterministic algorithm allowing to reasonably guess the solution and then easily check if it is correct. This strategy is also central in quantum algorithms.

Coming back to the question of fundamental gain, in 1992 Deutsch and Jozsa[26] found a toy-problem solved by a QC in poly-log time whereas linear time is required in a deterministic ClC, thereby the QC is exponentially faster. Later Berthiaume and Brassard found a problem solved in poly-time by a QC, beating a probabilistic ClC (at least exponentially).[27]

The 1994 Shor's[3] discovery of the first *significant* problems - number factoring and discrete logarithms - that a QC can solve better than any ClC, boosted the research in the field. They are both based on the quantum fast Fourier transform, which is also at the heart of other problems as phase estimation. A different class of algorithms is based on the quantum search problem solved by Grover.[1] This has boosted interest in finding quantum hardware, implementing such algorithms.

1.5. *Ingredients of quantum algorithms:*
Interference and measurement

Although quantum parallelism is a promising ingredient for computation, it must be kept in mind that its existence is not enough[28] to make it useful for QC, since it is impossible to read at once all the 2^N classical outputs resulting from quantum state processing. Indeed a projective measurements in the logic basis would yield only one of the classical outputs. For a single qubit, measurement allows to retrieve only a single bit of information: In order to reconstruct the state a large number of identically prepared qubits should be measured in the logic basis, using moreover special procedures of *quantum tomography* to determine the phase. Resources needed for the tomography of a quantum register grow exponentially with N making exponentially hard the problem of general time evolution.

However a class of functions exist which are computable by quantum parallelism as shown by R. Jozsa.[28] In this cases the answer is some collective property of the output, which can be measured in a proper basis. Therefore quantum algorithms must be arranged in a way such that in the output register constructive interference towards a "correct state" (or few) is obtained. Summing up, measurement puts constraint on what a quantum algorithm could calculate efficiently, providing implicit prescriptions for their design.

An important requirement of quantum algorithms is that they must be stable against imperfections, physically due to environmental noise. Encoding quantum states in more than one qubit allows to perform *quantum error correction*[29,30] (QEC). The general theory of fault-tolerant quantum computation[31] shows moreover how to deal with errors in encoding and on logical operations performed on encoded states, and provides an important threshold theorem[32,33] for quantum computation: provided the error rate per gate operation is below a certain threshold, it is possible to efficiently perform an arbitrarily large quantum computation. Early studies gave error thresholds of $10^{-4} - 10^{-5}$ but specific 2-D architectures may require a threshold $\lesssim 10^{-3}$.

2. Quantum physics and quantum information

2.1. *Quantum dynamics of qubits and gates*

2.1.1. *Single-qubit gates*

Single-qubit operations are general SU(2) rotations

$$R_{\mathbf{n}}(\alpha) = e^{-\frac{i}{2}\alpha\sigma_{\mathbf{n}}} \tag{6}$$

which represents a *counterclockwise* rotation of the Bloch vector, about $\mathbf{n}$ by an angle α. Special examples are rotations about the coordinate axes

$$R_1(\alpha) = e^{-\frac{i}{2}\alpha\sigma_1} = \cos\frac{\alpha}{2} - i\sin\frac{\alpha}{2}\sigma_1 - \begin{bmatrix} \cos\frac{\alpha}{2} & -i\sin\frac{\alpha}{2} \\ -i\sin\frac{\alpha}{2} & \cos\frac{\alpha}{2} \end{bmatrix}$$

$$R_2(\alpha) = e^{-\frac{i}{2}\alpha\sigma_2} = \cos\frac{\alpha}{2} - i\sin\frac{\alpha}{2}\sigma_2 = \begin{bmatrix} \cos\frac{\alpha}{2} & -\sin\frac{\alpha}{2} \\ \sin\frac{\alpha}{2} & \cos\frac{\alpha}{2} \end{bmatrix}$$

$$R_3(\alpha) = e^{-\frac{i}{2}\alpha\sigma_3} = \begin{bmatrix} e^{-\frac{i}{2}\alpha} & 0 \\ 0 & e^{\frac{i}{2}\alpha} \end{bmatrix}$$

Rotations map the Bloch sphere onto itself. A arbitrary unitary operator can be expressed uniquely as $U = e^{i\zeta}R_{\mathbf{n}}(\alpha)$.

Arbitrary rotations can be realized using a sequence of three rotations about two orthogonal axes, thereby any unitary operator (single-qubit gate) can be decomposed as

$$U = e^{i\alpha}R_{\mathbf{n}}(\beta)R_{\mathbf{m}}(\gamma)R_{\mathbf{n}}(\delta) \tag{7}$$

where $\mathbf{n} \perp \mathbf{m}$, for appropriate choice of the real numbers α, β, γ, δ. Therefore a arbitrary state transformation is achieved if control is available to operate at least two different kind of orthogonal rotations. This is what in practice quantum state processing requires.

2.1.2. *How to perform single-qubit gates*

The most common scheme to obtain unitary transformations is by controlled time evolution, operated by a suitably designed Hamiltonian. For a two-state system it reads

$$H = -\frac{1}{2}\epsilon\sigma_3 - \frac{1}{2}\Delta\sigma_1 \tag{8}$$

A pulse acts for a time t yielding the transformation $U(t) = \mathrm{e}^{-iHt}$. In order to perform standard gates one needs to have switchable couplings ϵ and/or Δ. Transformations changing the computational states (eigenstates of σ_3) are performed at $\epsilon = 0$ by switching on the "amplitude" Δ for a fixed t. For instance $t = \pi/\Delta$ yields the NOT gate. Alternatively one may operate at fixed Δ starting with a large "detuning" $\epsilon \gg \Delta$, then letting $\epsilon = 0$ for a fixed time t and then again $\epsilon \gg \Delta$. This option allows spin flips even if a modulable Δ is not available.

The problem of available control is very important in this context. For instance notice that although the Hamiltonian (8) allows in principle to perform rotations about the orthogonal axes $\mathbf{e}_1$ and $\mathbf{e}_3$, using dc pulses with $\Delta = 0$ is feasible in few implementations of qubits, and even in this cases decoherence prevents the use of both axes for rotations.

A more flexible control is provided by ac pulses. The Hamiltonian is

$$H = -\frac{1}{2}\Omega\sigma_3 + H_c(t) \quad ; \quad H_c(t) = \Omega_R\cos(\omega t + \phi)\,\sigma_1$$

where the control part $H_c(t)$ represents quasi resonant pulses, i.e. the detuning $\delta = \Omega - \omega$ is small compared to the natural frequency Ω, of amplitude Ω_R. The field is parametrized by (Ω_R, δ, ϕ) which can be easily controlled. This oscillating field can be decomposed into two counter-rotating components, one of which rotates in the same direction as the spin and it is in near resonance with it. The other component rotates in the opposite direction and it is very far off-resonance and can be neglected (Rotating Wave Approximation). The reduced problem can be conveniently seen from a rotating frame, defined by $|\psi\rangle_{rf} = U_{rf}^\dagger(t)|\psi\rangle$, where $U_{rf}(t) = \mathrm{e}^{\frac{i}{2}\omega t\sigma_3}$ and the corresponding effective Hamiltonian takes the form

$$\tilde{H} = U_{rf}^\dagger\, H\, U_{rf} - iU_{rf}^\dagger\,(dU_{rf}/dt) = -\frac{\delta}{2}\,\sigma_3 - \frac{\Omega_R}{2}\,(\cos\phi\,\sigma_1 + \sin\phi\,\sigma_2) \tag{9}$$

This is a remarkable result from various points of view. First it can be seen that population histories are unchanged by U_{rf}, which means that for most purposes one can even forget to transform back to the laboratory frame at the end of the protocol, and simply study the dynamics of $|\psi(t)\rangle_{rf}$. Second, the Hamiltonian (9), besides allowing to switch on and off rotations by modulations of the amplitude $\Omega_R(t)$ or of the detuning $\delta(t)$, also allows to change at will in the $\mathbf{e}_1 - \mathbf{e}_2$ plane the direction of the rotation by simply changing the phase ϕ of the driving field. In this way arbitrary rotations can be obtained from $\mathbf{e}_1 - \mathbf{e}_2$ standard rotations used in NMR,[36] in particular the H, T and S gates.

2.1.3. Two-qubit gates

As long as two-qubit gates are concerned, one should first notice that CNOT is an "entangling" gate, i.e. it turns a product state of two qubits in an entangled state. Not all gates are entangling, for instance the product of single qubit gates $U^{(2)} \otimes U^{(1)}$ is not. A generic entangling gate can be produced by an interaction Hamiltonian

$$H_{int} = \frac{1}{2} \sum_{\alpha,\beta=1}^{3} J_{\alpha\beta}\, \sigma_\alpha^{(2)} \otimes \sigma_\beta^{(1)}$$

In general quantum hardware allows only specific couplings to be present, however combining the two-qubit transformation $U_{int}(t)$ generated by a nonvanishing H_{int} with single qubit rotations the CNOT-gate has been implemented.

For gate design purposes it is important to define equivalent gates as two-qubit gates which can be obtained from each other supplementing the protocol with sequences of single-qubit gates. It has been proven that any entangling two-bit gate is universal for quantum computation,[34] when assisted by one-qubit gates, allowing in principle to find useful two-qubit gates for any interacting qubit system. A practical problem is however to achieve sufficient control to switch on and off the interaction H_{int} as will. Control of interaction can be done in several ways: (a) by direct tunable hardware, which poses important technological challenges; (b) by using extra quantum degrees of freedom as quantum buses, their coupling being switched dynamically, which requires non-trivial architectures but allows to perform such gates also between distant qubits; (c) by using fixed coupling and sequences of pulses dynamically canceling the interaction and compensating extra phases, which require large control resources to operate a gate.

2.2. Requirements for quantum hardware

Identifying candidates for quantum hardware is the first step towards the implementation of a quantum computer. On the basis of the structure of quantum algorithms, and of the experience gained by the first experiments, Di Vincenzo[7,35] identified a set of functional requirements which constitutes the paradigm of the Quantum Computer designer, namely[7,8]

(1) A scalable system of well defined qubits.
(2) A universal set of quantum gates of interacting qubits.
(3) An efficient procedure to initialize the quantum system to a known state.
(4) The ability to perform measurements on specific qubits.

(5) Coherence times much larger than the average time for an elementary logical operation (larger than the threshold for QEC).

Several proposals of implementation of quantum bits have been studied in the last years (see Fig. 2). This effort has yielded several important proof of principle of fundamental phenomena, from the manipulation of individual quantum systems to the production and detection of entanglement and important technological achievements, as for the ultrasensitive detection of individual excitations (spin, quantized charge, energy and magnetic flux of electronic systems, individual photons). Moreover invaluable information has been accumulated on the physical processes, in particular decoherence, which is the main obstacle to the implementation of working quantum architectures.

While all the proposals have been seen to meet specific requirements, satisfying all the Di Vincenzo criteria is a formidable experimental task. Moreover this roadmap has been updated and nowadays expressed in terms of favorable scaling of resources, universal logic and fault tolerance (see Sec. 5).

2.3. State of the art

Some of the goals of basic research in the physics of quantum hardware are illustrated in the following table.

Table 1. Order of magnitude of the number of qubits required to perform the indicated tasks.

Few	Tests on predictions of QM
~ 10	implement quantum coding schemes
$\rightarrow 100$	quantum simulators
~ 100	repeater in a noisy quantum cryptographic channel
$10^4 - 10^5$	QC factoring large integers

In the last decade systems of few (two or three) qubits were studied yielding important proof of principle and testing on different platforms predictions of quantum mechanics. A seven-qubit computer has implemented a demonstration of Shor's algorithm.[36] The present state of coherence in ion traps is 14 qubits.[37] The table shows that a quantum algorithm overcoming a ClC, requires a digital QC with number of qubits orders of magnitude smaller than what we are used to consider in ULSI devices. Moreover quantum architectures of ~ 100 qubits are enough to perform quantum simulations. Besides simulating quantum models, the analog QC would have important applications in the design of moderately large coherent systems, from nanostructures to drugs. Recently, variable-range Ising-like interactions have been demonstrated on a 300 qubit trapped-ion quantum simulator.[38] In the last few years QC based on adiabatic quantum computation (quantum annealing) have been also been implemented, and an algorithm on a 84-qubit system has been tested.[39]

3. Decoherence

Controlled dynamics of a coherent devices may be achieved if it is possible to prepare (write) and measure (read) a set of N observables $\{\hat{q}_i\}$,[1] denoting the basis of the computational states. If it is possible to tune the Hamiltonian then the dynamics of the system may be controlled, being ideally described by the pure state $|\psi, t\rangle = \sum_{q_1,\ldots,q_N} c_{q_1,\ldots,q_N}(t) |q_1,\ldots,q_N\rangle$. Coherence means that there is a well defined relation between the components $c_{\{q_i\}}(t)$. However the Hilbert space of the device is much larger than the computational space and this causes decoherence. The system (defined by the set $\{\hat{q}_i\}$) interacts with the environment (defined by all the other observables needed to complete the set), and this destroys the phase relation between $c_{\{q_i\}}(t)$. The system should be described by a reduced density matrix (RDM) $\rho_Q(t)$ rather than by a pure state. Considering a larger Hilbert space is needed because the coherent device is a many-body object and $\{\hat{q}_i\}$ is only a small set of collective variables. To appreciate the importance of the external environment in design considerations, notice that tunability requires to open ports to the external world which also couple noise sources to the device. *External* sources of decoherence may physically represent the apparata for preparation, measurement, and control of the Hamiltonian, which are themselves quantum systems enlarging the overall Hilbert space. *Internal* decoherence sources are instead specific to each implementation. For instance, in superconducting devices (see Sec. 4.1) internal decoherence is due to quasiparticles, and can be suppressed by working at temperatures much smaller than the BCS gap.[60] Moreover impurities in the substrates and in the oxides may act as a source of stray polarization of the qubit.

External degrees of freedom are accounted for by introducing the environment. In general, the Hamiltonian for a coherent nanosystem can be cast in the form[40]

$$\hat{H}_{\text{tot}} = \hat{H}_0 + \hat{H}_c(t) + \hat{H}_n(t) + \hat{H}_R + \hat{H}_I , \tag{10}$$

where $\hat{H}_0 + \hat{H}_c(t)$ describes the driven closed system, classical noise affecting the system is included in $\hat{H}_n(t)$ and $\hat{H}_R + \hat{H}_I$ represents quantum environment and its interaction with the system. When the system operates as a qubit, $\hat{H}_0$ reduces to Eq. (8), which can be conveniently written as

$$\hat{H}_q = \frac{\hbar\Omega}{2} \left(\cos\theta\,\sigma_3 + \sin\theta\,\sigma_1 \right) \tag{11}$$

where both the level splitting Ω and the polar angle θ shown in Fig. 6 depend on the parameters $\{q_i\}$. For simplicity, we suppose that a single control parameter, q, is used, then $\hat{H}_{\text{tot}}$ can be written as

$$\hat{H}_{\text{tot}} = \frac{\hbar}{2}\,\vec{\Omega}[q + \delta q(t)] \cdot \vec{\sigma} + \hat{H}_c(t) + \hat{H}_I + \hat{H}_R . \tag{12}$$

Expanding $\vec{\Omega}[q + \delta q(t)]$ about fixed value of the control parameter q, we obtain

$$\hat{H}_{\text{tot}} = \frac{\hbar}{2}\,\vec{\Omega}(q) \cdot \vec{\sigma} + \delta q(t)\frac{\partial \hbar\vec{\Omega}}{\partial q} \cdot \vec{\sigma} + \hat{H}_c(t) + \hat{H}_I + \hat{H}_R , \tag{13}$$

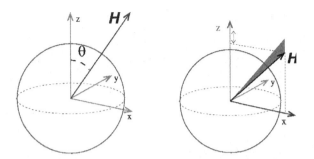

Fig. 6.   Left panel: Representation of the TSS Hamiltonian Eq. (11) in the Bloch sphere. Right panel: Pictorial description of asymmetric fluctuations of the TSS Hamiltonian Eq. (14).

and if q controls only one qubit component, for instance $\Omega_3(q)$, then Eq. (13) reduces to the commonly used form

$$\hat{H}_{\text{tot}} = \frac{\hbar\Omega_1}{2}\sigma_1 + \frac{\hbar\Omega_3(q)}{2}\sigma_3 + \frac{\hbar}{2}\left[E(t) + \hat{E}\right]\sigma_3 + \hat{H}_c(t) + \hat{H}_R. \qquad (14)$$

Here the qubit "working point" is parametrized by the angle θ_q which is tunable via the bias q, $\tan\theta_q = \Omega_1/\Omega_3(q)$, the classical noise term is $E(t) = 2\,\delta q(t)(\partial\Omega_3/\partial q)$, and $\hat{H}_I = \hbar\hat{E}\,\sigma_3/2$. With reference to the Bloch sphere (Fig. 6), when $\theta_q = \pi/2$ the qubit-environment interaction is transverse with respect to the free qubit Hamiltonian. It is usual to refer to this condition as "transversal noise". The operating condition $\theta_q = 0$ is instead referred to as "longitudinal noise".

This general Hamiltonian sometimes can be derived from a microscopic description of the quantum device implementing the qubit. More often it results from a phenomenological description of the environment.[40] Depending on the implementation, the system may prevalently be affected by classical noise sources, quantum environments or both. The ensuing decoherence crucially depends on the most relevant decoherence channel. For instance, the coherent dynamics of atoms in cavity QED is limited prevalently by spontaneous decay and by cavity losses. These processes are due to the interaction with the quantum environment formed by the cavity modes. Nanodevices instead suffer from the variety of noise sources typical of the solid-state. This requires considering both classical fluctuations of the control parameters and quantum fluctuations of the electromagnetic environment. The ensuing decoherence processes critically depend on the environment characteristics (for instance its correlation time or power spectrum) and on the coupling conditions with the quantum system. In the following we will summarize the most common approaches developed to predict decoherence and introduce the typical time scales which are routinely used to quantify the performance of single qubit gates. We will also briefly mention extensions of these concepts to universal two-qubit gates.

3.1. Weak coupling theory

For a weakly coupled environment the dynamics of the reduced density matrix $\rho_Q(t)$ of the qubit can be obtained by solving a Markovian Master equation[41] and the relevant decay times can be read by working in the basis of the eigenstates of $\hat{H}_q$. For the asymmetric coupling Eq. (14) the relaxation rate, $1/T_1$, governing the decay of the populations $\langle i|\rho_Q(t)|i\rangle$ towards equilibrium, and the decoherence rate $1/T_2$, which describes the vanishing of the coherences ij, are readily found

$$1/T_1 = \pi \sin^2 \theta\, S(\Omega)\,, \tag{15}$$

$$1/T_2 = 1/2T_1 + 1/T_2^* \tag{16}$$

where $1/T_2^* = \pi \cos^2 \theta S(0)$ and $S(\omega)$ is the power spectrum of the operator $\hat{E}$, calculated with the equilibrium density matrix $\hat{w} = exp[-\hat{H}_R/(k_B T)]$ of the isolated environment,

$$S(\omega) = \int_{-\infty}^{\infty} dt\, \frac{1}{2}\, \langle \hat{E}(t)\hat{E}(0) + \hat{E}(0)\hat{E}(t)\rangle\, e^{i\omega t}\,. \tag{17}$$

These results are valid if the environment is markovian, i.e. if its the correlation time τ_c is the smallest time scale in the problem. The time scales T_1 and T_2^* describe two different physical processes. Noise at frequencies of the order of the system's splittings, Ω, may induce incoherent energy exchanges between qubit and environment. These processes occur also in the zero temperature limit, where they induce spontaneous decay (*quantum noise*). Relaxation processes occur only if the qubit-environment interaction induces spin flips in the qubit eigenbasis, i.e. $\theta_q \neq 0$. Relaxation processes are the only environmental effect in the presence markovian transversal noise. Longitudinal noise instead does not induce spin flips, but it is responsible for pure dephasing with the "adiabatic" decay-time T_2^*.[42] This is often the major problems in solid state coherent nanodevices. We observe that when adiabatic fluctuations dominate, $1/T_2^* \sim \langle E^2 \rangle \tau_c > 1/\tau_c$, then Eqs. (15)–(16) are not valid. However they suggest that the transversal noise condition, $\theta = \pi/2$, represents an "optimal operating point" where low-frequency noise is canceled in lowest order. In order to predict the effect of adiabatic noise it is necessary to adopt a different approach which we briefly sketch below.

3.2. Approximate approaches for decoherence due to classical low-frequency noise

In the context of quantum computation, the effects of stochastic processes with long-time correlations depend on the quantum operation performed and/or on the measurement protocol. Since quantum measurements require averages of measurements runs, the main effect of low-frequency fluctuations is defocusing, similarly to inhomogeneous broadening in NMR.[42] This scenario is typical of solid-state implementations where the noise spectrum of some observable acting on the system's relevant degrees of freedom often displays a $1/f^\alpha$ behavior with $\alpha \sim 1$. Fluctuations with large spectral components at low frequencies can be treated as stochastic processes in the adiabatic approximation (*adiabatic noise*). Approximate approaches proposed to predict dephasing due to $1/f$ noise and its interplay with quantum noise have been described in the recent review.[40] In simplest cases, the effects of the two noise components add up independently in the coherences time-dependence.

An approach based on the adiabatic approximation[43,44] allows simple explanations of peculiar non-exponential decay reported in different experiments with various setups. In particular defocusing originated from inhomogeneous broadening (from measurement's repetitions) is obtained in the "static-path approximation" (SPA)[43] or "static-noise" approximation,[44] where the instantaneous level splitting, $\Omega(E(t))$, is replaced by a random variable, $\Omega(E)$, with standard deviation $\Sigma = \sqrt{\langle \delta\Omega^2 \rangle - \langle \delta\Omega \rangle^2}$, where $\delta\Omega = \Omega(E) - \Omega$. The qubit coherences in the SPA read

$$\rho_{ij}(t) \approx \rho_{ij}(0) \int dE P(E)\, e^{-i\Omega(E)t} \equiv \rho_{ij}(0) \langle e^{-i\Omega(E)t} \rangle, \qquad (18)$$

where the probability density, in relevant cases, can be taken of Gaussian form, $P(E) = \exp[-E^2/2\Sigma^2]/\sqrt{2\pi}\Sigma$. The splittings $\Omega(E)$ come both from longitudinal and from transverse terms in (14). Thus the short-times decay of qubit coherences depends on the symmetry of the qubit-environment coupling Hamiltonian. The decay factor turns from a $\propto \exp[-(bt)^2/2]$ behavior at $\theta \approx 0$, with $b = \cos\theta\Sigma$, (longitudinal noise), to a power law $\propto [1+(at)^2]^{-1/4}$ with $a = \sin\theta\Sigma/\Omega$, at $\theta \approx \pi/2$ (transverse noise).[43] Thus defocusing is minimized when noise is transverse with respect to the qubit Hamiltonian. The qubit is said to operate at an "optimal point" characterized by algebraic short-times behavior followed by exponential decay on a scale $2T_1$ due to quantum noise.[43–45] The characteristic time scales of entangling gates can to be identified as well. In the presence of both quantum and adiabatic noise this has been done in Ref. 46.

4. Current implementations of quantum bits

Quantum information science has by now provided us with a positive answer to the question whether one can expect to gain some advantage by storing, transmitting and processing information encoded in systems that exhibit quantum properties. The central, and still open, question is what form of quantum computing "hardware" will it take. In the past decade there's been a tremendous progress in the experimental development of a quantum computer and different hardware implementations are currently under investigation.

The easiest quantum bit one can imagine comes from the smallest form of matter: an isolated atom. "Natural atoms" occurs in cavity electrodynamics, in ion traps and optical lattices. However qubits may likewise be made far larger than routine electronic components, as in some superconducting systems. In solid-state implementations (based semiconducting and superconducting materials) qubits are effective bi-dimensional "artificial atoms" which can be addressed by voltages, currents and magnetic fields.

In this section we will describe some of the implementations where major efforts concentrated. We advise the reader that a large number of other technologies exhibiting quantum coherence, beside the ones considered, have been proposed and tested for quantum computation. For extended discussions we recommend the reviews.[6,47] These lecture notes cannot cover with proper details such a broad field. We will mainly focus on the current status of superconducting implementations, illustrating how the field evolved over the past two decades. For the other considered quantum implementations we will mention the working principles and latest achievements.

4.1. *Superconducting qubits*

Superconducting circuits are typically μm-scale circuits operated at mK temperatures. Although macroscopic, they can still exhibit quantum behavior, which can be harnessed for quantum computing purposes. Different superconducting nano-circuits have been fabricated, which can be classified according to their computational variable and operating regime and are commonly referred as charge, phase and flux qubits. Single qubit oscillations have been observed in a variety of cases and few examples of two-qubit gates have been experimentally realized as well. Up to now the Di Vincenzo criteria seem to be satisfied. Within the newly framework named *circuit quantum electrodynamics* (cQED), based on the analogy of superconducting circuits and the fields of atomic physics and quantum optics, important steps further have been done. Among the newest we mention the achievement of three-qubit entanglement with superconducting nano-circuits[48,49] which, in combination with longer qubit coherence, illustrate a potentially viable approach to factoring numbers,[50] implementing quantum algorithms[51,52] and simple quantum error correction codes.[53-55]

There are already several excellent reviews on superconducting implementation[56-59] and we do not attempt to write another review on the subject. Rather we will try to give a flavor of the flexibility of such systems and how this flexibility can be used advantageously.

4.1.1. *Superconductivity*

The reason why superconductors enable atomic-scale phenomena to be observed at the macroscopic level is explained elegantly by the theory of Bardeen, Cooper and Schrieffer.[60] In a given superconductor all of the Cooper pairs of electrons (which have charge $2e$, mass $2m_e$ and spin zero, and are responsible for carrying a supercurrent) are condensed into a single macroscopic state described by a wavefunction $\Psi(r,t)$ (where r is the spatial variable and t is time.) Like all quantum-mechanical wavefunctions, $\Psi(r,t)$ can be written as $\Psi(r,t) = |\Psi(r,t)| \exp\{i\Phi(r,t)\}$: that is, as the product of an amplitude and a factor involving the phase $\Phi(r, t)$. Furthermore, in "conventional" superconductors such as Nb, Pb and Al, the quasiparticles (electron-like and hole-like excitations) are separated in energy from the condensate by an energy gap $\Delta(T) = 1.76 k_B T_c$ (where T_c is the superconducting transition temperature). Thus, at temperatures $T \ll T_c$, the density of quasiparticles becomes exponentially small, as does the intrinsic dissipation for frequencies of less than $2\Delta(0)/h$, roughly 10^{11} Hz for Al. The macroscopic wavefunction leads to two phenomena that are essential for qubits: flux quantization and Josephson tunneling. The first one occurs when a closed ring is cooled through its superconducting transition temperature in a magnetic field. When the field is then switched off, the magnetic flux Φ in the ring, maintained by a circulating supercurrent, is quantized in integer values of the flux quantum $\Phi_0 = h/2e \approx 2.07 \times 10^{-15}$ T m^2. This quantization arises from the requirement that $\Psi(r, t)$ be single valued.[61]

4.1.2. *Josephson effect*

A Josephson element is a tunnel junction between two superconductors.[61] Its simplest realization is by deposition of two superconducting thin films forming the tunnel barrier with the oxidation of the bottom layer. In addition to quasi-particles current which can flow when a voltage bias is applied through the junction, Cooper pairs can tunnel across the barrier, this latter being a dissipationless process. As first predicted by Josephson even in the absence of a voltage drop a supercurrent can flow, depending on the phase difference of the order parameters of the two superconductors. The existence of an equilibrium current can be traced back to fact that in addition to the superconducting condensation energy of the two superconductors there is another phase dependent energy associated to their coupling, the Josephson potential energy

$$\mathcal{V}_J(\varphi) = -E_J \left(\cos \varphi - 1 \right). \tag{19}$$

The Josephson energy E_J is related to the critical current I_J (the maximum current that can flow through the junction without dissipation) by the relation $I_J = 2e E_J / \hbar$. It can be estimated from the Ambegaokar-Baratoff formula. At zero temperature $E_J \approx h\Delta(0)/(8e^2 R_N)$, where R_N is the normal state resistance.[61] Eq. (19) is not gauge invariant. The proper gauge invariant combination that should enter the energy is obtained through the replacement $\varphi \to \gamma = \varphi - \frac{2\pi}{\Phi_0} \int_1^2 \vec{A} \cdot d\vec{l}$, where $\vec{A}$ is the electromagnetic vector potential. The line integral in is calculated along a path connecting the two superconducting electrodes. The gauge invariant phase is relevant when loops are present in the superconducting circuit. The prototype example is the SQUID,[61] a superconducting loop interrupted by two Josephson junctions and pierced by a magnetic flux Φ_x. A simple analysis shows that in this case the maximum supercurrent can be modulated by changing Φ_x, the loop behaving as a single junction with $E_J^{eff}(\Phi_x) = 2E_J \cos(2\pi \Phi_x / \Phi_0)$.

Quantum mechanics enters into play in submicron junctions with very large normal state resistance, $R_N \gtrsim h/(4e^2) \approx 6.5\,k\Omega$. In this regime dissipation due to normal tunneling can be ignored but important effects come from the electrostatic energy of the charge $\hat{Q}$ imbalanced at the junction. For a single junction the Hamiltonian reads

$$H = \frac{\hat{Q}^2}{2C} - E_J \cos \hat{\varphi} \tag{20}$$

where C is the capacitance of the junction. The number of Cooper pairs $\hat{q} = \hat{Q}/2e$ can be considered as the momentum conjugated to the phase. Quantization implies that they do not commute, $[\hat{\varphi}, \hat{q}] = i$, this leading to the Heisenberg equations of motion:

$$\frac{\hbar}{2e}\frac{d\hat{\varphi}}{dt} = \frac{\hat{Q}}{C} = \hat{V} \quad ; \quad \frac{d\hat{Q}}{dt} = \frac{2e}{\hbar} E_J \sin \hat{\varphi} \tag{21}$$

The first is the celebrated Josephson relation, relating the phase evolution to the voltage at the junction. The second gives the relation between the current and the phase difference at the junction.

The Josephson junction is a unique *quantum non-linear* electrical circuit. Although many other non-linear electrical systems exist, their internal degrees of freedom all get frozen by the time one reaches the temperatures at which quantum effects implying collective variables (like charge) can be observed. By suitably inserting the Josephson element as the basic building block in a given nanocircuit it is possible to realize several type of qubits as described below.

4.1.3. Cooper-pair-box based qubits

An elementary unit for superconducting qubits is the Cooper pair box (CPB),[62] whose basic design is shown in Fig. 7 (a). It is described by the Hamiltonian

$$\mathcal{H}_{\text{CPB}} = 4E_C (\hat{q} - q_x)^2 - E_J \cos \hat{\varphi} \tag{22}$$

where the charging energy $E_C = e^2/(2C)$ is fixed by the geometry, whereas the other parameters are tunable, allowing external control. The gate voltage fixes the gate charge, $2eq_x = C_g V_x$ and improvements of the basic design allow to control also E_J. In the basis $|q\rangle$ of the eigenstates of the charge in the island the phase acts as $e^{\pm i\hat{\varphi}}|q\rangle = |q \pm 1\rangle$. Therefore the Josephson term changes the number of Cooper pairs in units.

The Cooper pair box implements a qubit if its dynamics is restricted to the two lowest energy eigenstates. The **charge qubit**[63–65] is obtained in the limit $E_J \ll E_C$, where states $|q\rangle$ are weakly mixed, unless near the "optimal" operating points, $q_x = 2 + q_0$, with q_0 integer, where the average electrostatic energy is the same for the charge states $|q_0\rangle$ and $|q_0 + 1\rangle$ (charge degeneracy point). If $q_x \in [0, 1]$ the dynamics at low temperatures, $k_B T \ll E_C$ is essentially limited to at most the two charge states, $\{|0\rangle, |1\rangle\}$. Projection of the box Hamiltonian onto this subspace is presented in terms of the pseudospin operators, $\sigma_z = |0\rangle\langle 0| - |1\rangle\langle 1|$ and $\sigma_x = |0\rangle\langle 1| + |1\rangle\langle 0|$ and it is obtained from the projections $P\hat{q}P = (1 - \sigma_z)$ and $P\cos\hat{\varphi}P = \sigma_x/2$

$$\mathcal{H}_Q = P\mathcal{H}_{\text{CPB}}P = -\frac{1}{2}\epsilon\sigma_z - \frac{1}{2}\Delta\sigma_x \tag{23}$$

where $\epsilon = 4E_C(1 - 2q_x)$ and $\Delta = E_J$. Important parameters are the level splitting $\Omega = \sqrt{\epsilon^2 + \Delta^2}$ and the mixing angle $\tan\theta = \Delta/\epsilon$. Typical clock frequencies are $\Omega/(2\pi) \sim 10$ GHz. Parameters are tunable and allow manipulation of the qubit (q_x tunes ϵ and Φ_x tunes Δ, see Fig. 7 (a)). Fast control in charge qubits is possible only via the electrostatic port $\epsilon(t)$ because coupling to the magnetic field is weak and large amplitude $\vec{B}$ pulses would require much too large ramping times. On the other hand sensitivity to charge bias also implies sensitivity to charge noise. Indeed as shown in Fig. 7 (b) the spectrum of a charge qubit may substantially change in response to a stray charge bias $\delta q_x(t)$. Cooper-pair boxes are particularly sensitive to low-frequency noise from electrons moving among defects and can show sudden large jumps in q_x. The development of more advanced charge qubits such as the **quantronium**[66] and the **transmon**[67] has greatly ameliorated this problem.

The **quantronium**[66] is based on a Cooper pair box implemented by a SQUID geometry, see Fig. 7 (c). The main difference in the design with respect to the charge qubit is the presence of an extra large Josephson junction in the loop, whose phase φ_t is in principle an extra dynamical quantum variable, but in practice it behaves classically. This provides a second control port for the qubit, since a current bias I_x fixes φ_t and consequently the total phase across the two smaller junctions connecting the island to the rest of the circuit. The effective Hamiltonian is again Eq. (23)

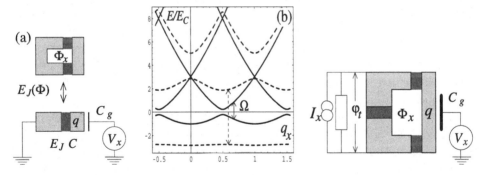

Fig. 7. (a) Basic design of the Cooper pair box. It is a superconducting island storing the variable $\hat{q}$ and connected to a circuit via a Josephson junction and a capacitor C_g. Control is operated via V_x. Replacing the junction by a SQUID it is possible to tune also $E_J(\Phi_x)$ by changing the magnetic flux. (b) Spectrum of the box for fixed $E_J/E_C = 0.44$ (solid lines, corresponding to the charge qubit of[63]). In charge qubits the spectrum is very sensitive to fluctuations of q_x. Instead for $E_J/E_C > 1$ (dashed lines, $EJ/EC = 5.04$, correspond to the quantronium[66]) the splitting is less sensitive to charge fluctuations. (c) The two-port Cooper pair box implementing the quantronium. External control is operated via ac pulses $V_x(t)$ during processing and via step pulses $I_x(t)$ for measurement.

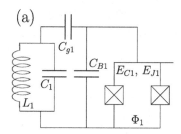

Fig. 8. Effective circuit diagram of the transmon qubit. The two Josephson junctions E_{J1} with capacitance and Josephson energy C_{J1} and E_{J1} are shunted by an additional large capacitance C_{B1}, matched by a comparably large gate capacitance C_{g1}. The CPB is coupled to a transmission line used for control and readout and modeled as a resonator, $L_1 C_1$ circuit.[67]

where the Josephson energy terms takes the form $\Delta(\Phi_x) = E_J(\Phi_x) \cos \varphi_t \cos \hat{\varphi}$. In the limit $E_J \ll E_C$ the Hamiltonian of the quantronium reduces to the charge qubit Hamiltonian Eq. (22). The important difference between the two regimes come from the availability of different sets of controls and on the convenient computational basis. In charge qubits the charge in the island couples more strongly to the external world. Therefore charge states are the natural computational basis even if sensitivity to charge noise determines much larger decoherence rates than in the quantronium.

The quantronium two-port design gives the possibility of using two knobs to tune optimally the device when processing. The two control parameters are in fact used to maintain the qubit at the double-degeneracy point where the two qubit state are (to first order) insensitive to low frequency fluctuations both of the charge and of the magnetic flux. This insensitivity however implies that the two states of the qubit cannot be distinguished at the optimal point. To measure the qubit state, a current pulse is applied that moves the qubit away from the flux degeneracy point. This produces a clockwise or anti-clockwise current in the loop, depending on the qubit state. The large junction is used to measure the circulating current rather than the charge in the island. More recently, a much faster detection scheme has been implemented by the introduction of a Josephson bifurcation amplifier which has been shown to approach the quantum non-demolition limit.[68]

The device is operated by an additional ac gate voltage coupling to the charge in the island. By operating at the charge degeneracy point, the quantronium implements the usual Hamiltonian of an NMR system, where the rf field is orthogonal to the static external magnetic field determining the Zeeman splitting, or the Hamiltonian of a two-level atom driven by a classical field coupled via the dipole moment, as discussed in Sec. 2.1.2. This allows to implement several standard protocols for NMR.[69,70]

The **transmon** is a small CPB that is made relatively insensitive to charge by shunting the Josephson junction with a large external capacitor to increase E_C and by increasing the gate capacitor to the same size, see Fig. 8. Consequently, the energy bands are of the same type shown in Fig. 7 (b), but become almost flat, and the eigenstates are a combination of many Cooper-Pair-Box charge states. The transmon is thus insensitive to low-frequency charge noise at all operating points. At the same time, the large gate capacitor provides strong coupling to external microwaves even at the level of a single photon. Embedding the transmon in a superconducting transmission line resonator opens up the possibility of control and readout of the qubit state, a scenario that has been termed *circuit QED*.[71,72]

4.1.4. *Phase qubits*

Phase qubits are fabricated with larger Josephson junctions, where $E_J \gg E_C$. Eigenstates of the Hamiltonian (22) are superpositions of charge states. This key property allows substantial protection against charge noise because of the reduced sensitivity of the energy spectrum to fluctuations of q_x (see Fig. 7 (b)). Consequently driving the qubit operating with V_x requires larger signal amplitudes.

On the other hand since fluctuations of the phase $\hat{\varphi}$ are smaller than in charge qubits, special design solutions allow to shape and tune the Josephson potential. For instance Josephson junctions can be also biased by a constant external current I_x accounted by an extra term $(\hbar/2e)I_x\hat{\varphi}$ in the Hamiltonian which represents a fictitious particle in washboard potential.[61] In the quantum regime and if the external current is close to but below the critical current, the system has a number of quasi-bound states. Although there are more level inside the well, thanks to the anharmonicity of the potential the levels are not equidistant and therefore it is possible to address selectively only the pair chosen for the qubit. Operations are implemented by $I_x(t)$ pulses. As in the quantronium the single Josephson junction qubit is quite insensitive to charge or flux noise, but couples to fluctuations of $I_x(t)$. Moreover the larger junction area opens the possibility that E_J also fluctuates because of impurities in the tunnel oxide. The readout is naturally built in since when I_x approaches the critical current I_J, level broadening due to macroscopic quantum tunneling starts to play a role. Since the tunneling rate is exponentially sensitive to the energy barrier, it is possible to make likely the escape if the system is in the state $|1\rangle$, whereas $|0\rangle$ stays trapped. Several experimental groups have realized the Josephson Junction qubit.[73]

4.1.5. *Flux qubits*

A qubit can also be realized with superconducting nano-circuits in the limit $E_J \gg E_C$.[74,75] An rf-SQUID (a superconducting loop interrupted by a Josephson junction) for example provides the prototype of such a device. In order to fulfill various operational requirements the flux qubit proposed by Mooij *et al.*,[74] is based on a asymmetric three-junction SQUID design one of the junction having area smaller by a factor α (Fig. 9 (a)). Since junctions are large, $E_J/E_C = 40$, the classical Josephson energy of the three junctions is the dominant energy. It depends on two gauge invariant phases γ_i, since the phase at the smaller junction is fixed by

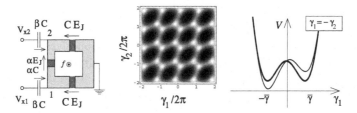

Fig. 9. (a) The flux qubit unit is a SQUID ring interrupted by three Josephson junctions (the crossed boxes) and it is manipulated by ac flux pulses, $f_x = f_x(t)$. (b) Making one of the junctions slightly smaller the Josephson potential $V(\gamma_1, \gamma_2)$ has pairs of neighboring minima. (c) The potential for $\gamma_1 = -\gamma_2$ has a double well shape at the symmetry point $f_x = 1/2$ (thick line). Deviations from the symmetry point correspond to a bias for the double well (thin lines).

the fluxoid quantization relation. The two-dimensional periodic potential $V(\gamma_1, \gamma_2)$ can be shaped with α and f_x. For $0.5 < \alpha$ and $f_x = 1/2$ the minima form a two-dimensional cell pattern containing two neighboring minima, see Fig. 9(b). Cells are repeated in a two-dimensional square lattice. For $\alpha < 1$ the potential barrier separating such pair of minima is much larger than the intercell barriers. Thus V is a double well potential which distinguishes the computational states, each carrying a persistent current I_p of opposite sign (Fig. 9 (c)). Deviation from $f_x = 1/2$ corresponds to a bias. The finite electrostatic energy allows tunneling between minima. The spectrum can be obtained from the full Hamiltonian and shows a pair of well separated lowest energy levels, which are rather insensitive to charge fluctuations, i.e. the device is protected from charge noise. The device can be driven with magnetic microwave pulses. Spectroscopy has confirmed the validity of this model.[76] The measuring device is a big SQUID loop with two large Josephson junctions attached to the qubit. During state processing the SQUID is decoupled. At the end of the protocol a current pulse makes the SQUID meter sensitive to the qubit current circulating in a certain direction, which triggers switching events of the SQUID to a finite voltage state. In principle the optimal point to operate is the flux degeneracy point, where flux noise vanishes in first order and the drive is orthogonal to the static Hamiltonian. However the expectation values of the circulating current are very small and difficult to detect. Therefore coherent dynamics has been initially demonstrated by working at a non optimal point.[77] Shortly later measurements at the optimal point have been performed with a more strongly coupled measuring SQUID which in the idle state behaves as a coupled quantum harmonic oscillator.[78]

4.1.6. *Time-domain measurements and estimates of T_1 and T_2*

The first experimental evidence that a given qubit is a functional device consists in making spectroscopy, i.e. measuring its energy-level splitting as a function of relevant control parameters. In all the mentioned implementations spectroscopic evidence of quantum two-level system behavior has been performed initially.

Measurements in the time domain are also necessary to determine the dynamical behavior of a qubit and the feasibility of quantum gates. Time measurements involve manipulating the state of the qubit by using either dc-pulses or appropriate microwave pulses, often combinations of dc and ac pulses are employed. Various pulse measurements can be visualized in the Bloch sphere description. Well-known quantum mechanical effects like Rabi oscillations are are routinely reproduced in all types of superconducting qubits.

Moreover, by measuring the qubit time evolution it is possible to extract the decay times T_1, T_2 and T_2^* introduced in Sec. 3. These time scales provide a quantitative indication of the efficiency of single-qubit gates. More generally, since different decay laws result from noise sources having different statistical properties, time-domain measurements also provide indications on the most detrimental decoherence channels in each implementation.

The simplest way to measure the relaxation time T_1 is to irradiate the qubit with microwaves at the frequency corresponding to the energy-level splitting between the ground and first excited states for a time much greater than T_1. After the pulse has been turned off, the qubit has an equal probability of being in either state; the probability $p_1(t)$ of its being in the excited state $|1\rangle$ subsequently decays with time as $\exp(-t/T1)$. Measurements of $p_1(t)$ as a function of t yield the value of T_1. It should be emphasized that each measurement of p_1 at a given time delay involves a large number of measurements, typically 10^4 or 10^5. T_1 can vary from values of the order of 1 ns to many microseconds.

In measuring the dephasing time, it is crucial to distinguish the two times T_2 and the pure dephasing time T_2^* which essentially results of an ensemble measurement in the presence of low-frequency noise. The ensemble is formed because experiments on a single qubit need to be carried out repeatedly so that sufficiently precise data are acquired. Even though the different measurements are nominally identical, slow fluctuations on the timescale of a single run result in a change in the operating conditions between runs. This reduces the observed coherence time to T_2^* (which is smaller that T_2). We remind that the leading effect of low-frequency fluctuations is to induce non-exponential decay of the qubit coherences. The estimate of T_2^* is obtained by inverting $\rho_{01}(T_2^*) = \exp(-1)$. T_2^* and T_2 can be measured separately. T_2^*, which includes the effects of low-frequency noise, can be measured by using Ramsey fringes, while T_2 is usually estimated by using a spin-echo technique, which eliminates certain low-frequency contributions.[42,69]

Measuring the times T_1, T_2 and T_2^* provides an important initial characterization of qubit coherence. In Table 2 (from Ref. 47) we report recent values of these time scales in the different charge, flux and phase qubits. We remark that other factors such as pulse inaccuracy, relaxation during measurement and more complex decoherence effects result in measurement errors. A more complete measure of a qubit is fidelity, a single number that represents the difference between the ideal and the actual outcome of the experiment. Determining the fidelity involves quantum-

Table 2.   Improvement of coherence times of superconducting qubits until 2011, table taken from Ref. 47.

Year	T_1	T_2 (echo)	Qubit	Ref.
1999	1 ns	-	Charge	63
2002	580 ns	2 ns	Charge	80
2002	100 ns	100 ns	Phase	81
2002	1.8 μs	500 ns	Hybrid (charge/phase)	66
2003	0.9μs	30 ns	Flux	77
2006	1.9μs	3.5μs	Flux	82
2008	1.87μs	2.22μs	Hybrid (charge/phase)	83
2009	350 ns	-	Flux	84
2010	1.6μs	1.3μs	Hybrid (phase/flux)	85
2011	12 μs	23μs	Flux	86
2011	0.2 ms	-	Charge	87

process tomography (a repeated set of state tomographies), which characterizes a quantum mechanical process for all possible initial states. In a Ramsey-fringe tomography experiment, Matthias Steffen et al.[79] found a fidelity of 80%, where 10% of the loss was attributed to read-out errors and another 10% to pulse-timing uncertainty.

4.1.7. *Coupling qubits: towards the implementation of quantum algorithms*

An unique feature of solid-state qubits in general and superconducting qubits in particular is that schemes can be implemented that both couple them strongly to each other and turn off their interaction in situ by purely electronic means. Because the coupling of qubits is central to the architecture of quantum computers, this subject has attracted much attention, in terms of both theory and experiment.

Implementing a universal two-qubit gate involving an entanglement operation on two quantum bits represents a necessary step toward the construction of a scalable quantum computer.[1] The best way to achieve controllable coupling and a universal two-qubit gate are still open questions. A number of theoretical proposals[88–92] and experimental attempts have been put forward.[93,94]

The most natural way to realize two-qubit entanglement is via a fixed, capacitive or inductive, coupling scheme.[88,91] With tunable single-qubit energy spacing, fixed coupling has been used to demonstrate two-qubit logic gates.[93] In order to achieve mutual resonance during the gate operation at least one qubit has to be moved away from the working point of minimal sensitivity to parameters variations, the "optimal point".[63,66,77] This is so far the main drawback of fixed-coupling schemes for most of the Josephson implementations. Recent proposals have attempted to solve this problem by tunable coupling schemes.[89,90] Most of them rely on additional circuit elements, such as switchable Josephson junctions, inductors, lumped or cavity-type resonators or further qubits. Some of them gain their tunability from ac-driving,[90] others from "adiabatic" couplers.[92] Some of these schemes have been experimentally tested and are potentially scalable.[72,94,95] However, none of these implementations is totally immune from imperfections. Some ac-schemes require strong driving or result in slow operations and non-adiabatic corrections may prevent complete switching-off of adiabatic couplers. Cross-talk due to always-on coupling and to measurement is present also in most of the ac-driven operational modes. Moreover, any additional circuit element is a new port to noise. A strategy to identify two-qubit optimal points within fixed coupling schemes has been proposed in[96] where it has been demonstrated that for selected *optimal couplings* a high-efficiency universal two-qubit gate can be implemented even in the presence of the detrimental $1/f$ noise sources.

On the basis of these coupling schemes, several architectures have been proposed for scaling up from two qubits to a quantum computer. The central idea of most proposals is to couple all qubits to a long central coupling element, a *quantum bus*, and to use frequency selection to determine which qubits can be coupled.[65,88,91] This scheme has been experimentally demonstrated. As couplers become longer, they become transmission lines that have electromagnetic modes. For example, qubits have been coupled by placing them at the anti nodes of a standing wave on a stripline.[72,95] Coupling between specific pairs of qubits can result in a scalable architecture. By first coupling a qubit to the standing-wave mode using frequency selection, a photon is excited and then stored after decoupling. Subsequently, a second qubit is coupled to the mode, and the photon transfers the quantum state to the second qubit.

4.2. Quantum dots and atomic systems in solid-state hosts

Quantum Dots (QD) are "artificial atoms" formed when a small semiconductor nanostructure, impurity or impurity complex, binds one or more electrons or holes (empty valence-band states) into a localized potential with discrete energy levels, which is analogous to an electron bound to an atomic nucleus. Quantum dots come in many varieties. Some are electrostatically defined quantum dots, where the confinement is created by controlled voltages on lithographically defined metallic gates. Others are self-assembled quantum dots, where a random semiconductor growth process creates the potential for confining electrons or holes. The first ones require operating temperatures of $< 1K$ and are primarily controlled electrostatically, the second ones operate at higher temperatures $\sim 4K$ and are controlled optically.

One of the earliest proposals for quantum computation in semiconductors, that of Loss and DiVincenzo,[97] envisioned arrays of electrostatically defined dots, each containing a single electron whose two spin states provide a qubit. Quantum logic would be accomplished by changing voltages on the electrostatic gates to move electrons closer and further from each other, activating and deactivating the exchange interaction.[36] The detection of a single electron spin on the dot can be performed via "spin-to-charge-conversion". The ability of the single electron to tunnel in and out of the dot in fact depends on its spin state (because of the Pauli exclusion principle), for this reason the spin state can be inferred from the measurement of the charge on the dot. The electron charge can be detected from the change of the conductance of a nearby single-electron-transistor or quantum point contact.

Qubits may be also defined by clusters of exchange-coupled spins, with effective single-qubit logic controlled by the pairwise exchange interaction. For instance, the collective spin state of two electrons trapped in a double-QD can be used as a qubit. The control of individual spins via microwave magnetic and electric fields, or of exchange-coupled double-QD qubit via voltage pulses, allowed to estimate the T_2 and T_2^* times.

A critical issue of most of these nanostructures, which are based on GaAs, is the presence of nuclear spins in the semiconductor substrate. The nuclear spins create an inhomogeneous magnetic field which couples to the electron spins and result in dephasing with $T_2^* \approx 10$ ns, whereas T_2 is few microseconds. One way to eliminate the effect of nuclear spins is to use Si and Ge-based quantum dots where many of the accomplishments demonstrated in GaAs have been demonstrated.[98]

Similarly to the above silicon-based QD proposals,[99] the quantum dot can be replaced by a single impurity, in particular a single phosphorus atom, which binds a donor electron at low temperature. Quantum information may then be stored in either the donor electron, or in the state of the single ^{31}P nuclear spin, accessed via the electron-nuclear hyperfine coupling.

Alternative spin qubits can be formed by nitrogen-vacancy (NV) centers in diamond. NV-centers are point defect in the diamond lattice, consisting of a nearest-neighbor pair made of a nitrogen atom, substituting a carbon atom, and a lattice vacancy. The negatively charged state of the NV-center forms a triplet spin system. Under optical illumination, spin-selective relaxations enable efficient optical pumping of the system into a single spin state, allowing fast (250 ns) initialization of the spin qubit and coherent manipulation via resonant microwave fields.[100] Due to a nearly nuclear-spin-free carbon lattice and low spin-orbit coupling, NV-centers show longer spin coherence times than GaAs quantum dots, even at room temperature.[101,102]

4.3. *Liquid NMR*

In 1996, methods were proposed[103] for building small quantum computers using nuclear spins in molecules in liquid solutions based on the existing magnetic resonance technology. Immersed in a strong magnetic field, nuclear spins can be identified through their Larmor frequency. In a molecule, nuclear Larmor frequencies vary from atom to atom because of shielding effects from electrons in molecular bonds. Irradiating the nuclei with resonant radio-frequency pulses allows manipulation of nuclei of a distinct frequency, giving generic one-qubit gates. Two-qubit interactions arise from the indirect coupling mediated through molecular electrons. Measurement is achieved by observing the induced current in a coil surrounding the sample of an ensemble of such qubits.

Liquid-state nuclear magnetic resonance has allowed the manipulation of quantum processors with up to a dozen qubits,[104] and the implementation of algorithms[104] and QEC protocols. Initialization is an important challenge for nuclear magnetic resonance quantum computers. Moreover some of the proposed techniques face the problem of scalability. One way to address the scalability limitation is to move to solid-state nuclear magnetic resonance, for which a variety of dynamic nuclear polarization techniques exist.

To date, no bulk nuclear magnetic resonance technique has shown sufficient initialization or measurement capabilities for effective correctability, but nuclear magnetic resonance has led the way in many qubit quantum control. The many-second T_2 times are comparable to gate times in liquids and much longer than the sub-millisecond gate times in solids, but are still short in comparison to timescales for initialization and measurement. The many lessons learned in NMR quantum computation research are most likely to be relevant for advancing the development of other quantum technologies.

4.4. *Trapped ions and cold atoms*

Ions can be trapped by electrical (or magnetic) fields, laser-cooled and manipulated with high precision.[105] Quantum information can be encoded either in the internal (hyperfine or Zeeman sublevels, or the ground and excited states of an optical transition), or in the motional states (the collective motion of the ions). While the internal states exhibit very long coherence times (hyperfine transitions > 20s and optical transitions > 1s) the motional states have typical lifetimes of < 100 ms.

Trapped ion qubits can be entangled through a laser-induced coupling of the spins mediated by a collective mode of harmonic motion in the trap. Cirac and Zoller in 1995[106] made the first proposal of a simple realization of this interaction to form entangling quantum gates and Wineland in the same year[107] demonstrated it in the laboratory. Recently, up to eight trapped ion qubits have been entangled by extending this approach employing optical spin-dependent forces.[108] The scaling of trapped-ion Coulomb gates becomes difficult when large numbers of ions participate in the collective motion. One method for scaling ion trap qubits is to couple small collections of Coulomb-coupled ions through photonic interactions, offering the advantage of having a communication channel that can easily traverse large distances. Recently, atomic ions have been entangled over macroscopic distances in this way.[109]

Neutral atoms provide qubits similar to trapped ions. An array of cold neutral atoms may be confined in free space by a pattern of crossed laser beams, forming an optical lattice.[110]

For trapped atoms and ions, coherence times are many orders of magnitude longer than initialization, multi-qubit control, and measurement times. The critical challenge for the future of trapped atom quantum computers will be to preserve the high-fidelity control already demonstrated in small systems while scaling to larger, more complex architectures.

4.5. *Atoms in cavity QED*

Cavity-QED in its simplest form consists of a single atom with two relevant quantum states, coupled to a single mode of the electromagnetic field, defined for instance by a pair of mirrors.[111] A photon in the cavity, bouncing back and forth between the mirrors, can be absorbed by the atom; conversely, if the atom is excited, it can decay by emitting a photon into the cavity. The rate of this atom-light interaction is proportional both to the dipole moment of the atom and to the electric field of the photon at the atom's location.

The undesired processes in this case are photons losses from the cavity resulting from imperfect mirrors, and the decay of the atom into other channels. The strong coupling regime of cavity QED is obtained when the rate of absorption or emission of a single photon by the atom is more rapid than any of the rates of loss. In this case, an excited atom in an initially empty cavity will emit one (and only one) photon, which can then be trapped and reabsorbed again (vacuum Rabi oscillations). Because of the cavity, the spontaneous emission from the atom which is usually an irreversible process, results in a coherent and reversible oscillation. In the language of quantum computation, strong coupling means that quantum information can be exchanged back and forth between the atom and the photon many times before it is lost for ever.

The challenge for strong-coupling cavity QED is to maximize the vacuum Rabi frequency simultaneously minimizing the decay. A further difficulty is the placement and trapping of a single atom at a desired location in the cavity. Despite the obvious technical challenges, there are several examples of strong-coupling cavity QED using real atoms. For optical photons trapped between mirrors, the vacuum Rabi splitting was first observed in 1992.[112] Another approach uses Rydberg atoms,[113] which are highly excited atomic states that have very large dipole moments and low energy transitions at ~ 50 GHz. In this case, the photons are in the microwave domain and the cavity consists of a superconducting metallic box a few wavelengths accross (several centimeters).

Other efforts have focused on strong coupling with semiconductor quantum dots as the emitters. These have the potential advantage of emitting at infrared wavelengths, close to those used for telecommunication.[114] The coupling of internal states of an atomic ion to its motion in a trap[115] can also be seen as a realization of strong-coupling cavity QED. In fact, it contains the same essential ingredients of a two-level system (the ion) interacting with a harmonic oscillator (the quantized motion of the ion, or phonons).[116]

Many beautiful experiments have been done in the last two decades using these strongly coupled cavity QED systems, performing textbook demonstrations of fundamental quantum phenomena, such as decoherence and entanglement. The ability to control the interactions of atoms and single photons in a quantum mechanical way has intriguing implications for quantum computation and communication. Photons in a cavity,[117] or phonons in an ion trap,[106] can be used to generate entanglement and make a quantum bus to communicate quantum information between multiple atoms. The technical difficulties of achieving sufficiently strong coupling, trapping many atoms, and individually addressing and controlling them, make it difficult at present to build large-scale quantum systems. Circuit QED is a more recent attempt to bring about strong coupling within an integrated superconducting circuit. Circuit QED has also been used for quantum communication and coupling between qubits.

4.6. *Photons*

Realizing a qubit as the polarization state of a photon is appealing. In fact photons are relatively free of the decoherence that plagues other quantum systems. Polarization rotations (one-qubit gates) can easily be done using "waveplates" made of birefringent material. Achieving the needed interactions between photons for universal multi-qubit control presents a major difficulty. The necessary interactions appear to require optical nonlinearities stronger than those available in conventional nonlinear media.

In 2001, Knill-Laflamme and Milburn[118] demonstrated that scalable quantum computing is possible using only single-photon sources and detectors, and linear optical circuits. This scheme relies on quantum interference with auxiliary photons at a beamsplitter and single-photon detection to induce interactions nondeterministically. Since this proposal, several experiments have been put forward.[6] Recently, impressive progresses have been achieved with sources and detectors of single photons and with the introduction of the so-called photon chips. Today, efforts are also focused on devices that would enable a deterministic interaction between photons and chip-scale waveguide quantum circuits. Regardless of the approach used for photon sources, detectors and nonlinearities, photon loss remains a significant challenge, and provides the closest comparison to T_2 decoherence in solid-state qubits.

5. Conclusions and perspectives

We conclude these lecture notes with the observation that, in the course of about two decades, the ideas for implementing quantum computing have diversified. It has been observed that qubits are not a prerequisite; quantum d-state systems (qudits) or quantum continuous variables may also enable QC. Similarly, quantum computers need not be made with gates. In adiabatic quantum computation the answer to a computational problem is represented by the ground state of a complex network of interactions between qubits.[119] The DiVincenzo criteria as originally stated are difficult to apply to many of the emerging concepts. Today, DiVincenzo's original considerations can be rephrased into three more general requirements which we summarise here following Ladd *et al.* in Ref. 6. The first condition to be fulfilled is "**Scalability**: The computer must operate in a Hilbert space whose dimensions can grow exponentially without an exponential cost in resources (such as time, space or energy)."[6] Declaring a technology "scalable" is quite tricky because it requires that also the classical resources employed for control, detection etc must be made scalable. This demands complex engineering issues and the infrastructures for large-scale technologies. The **Universal logic** criterion can be rewritten as follows: "The large Hilbert space must be accessible using finite set of control operations; the resources for this set must also not grow exponentially."[6] Finally, the last requirement involving quantum measurement and quantum-error correction can be rephrased as "**Correctability**: It must be possible to extract the entropy of the computer to maintain the computer's quantum state."[6] These are the requirements that any quantum technology must fulfill in order to implement a quantum computer.

A number of other technologies exhibiting quantum coherence is presently being considered beyond the ones we illustrated in these lecture notes. The question about which quantum hardware has the most promise to achieve a large-scale quantum computer nowadays is still open. At present, an exciting prospect is combining advantages of different implementations to form hybrid systems.[47] Natural atoms, with their long decoherence times, are envisaged by many as quantum memories, while tunable artificial atoms may be used for the "quantum processing unit". Both natural and artificial atoms may be coupled with photons via a cavity. Such cavities could be used as input/output interfaces and for long-distance communication.

Acknowledgments

We Acknowledge partial support from the Centro Siciliano di Fisica Nucleare e di Struttura della Materia (CSFNSM), Catania (Italy) and by MIUR through Grant No. PON02-00355-3391233, Tecnologie per l'ENERGia e l'Efficienza energETICa - ENERGETIC.

References

1. M. Nielsen and I. Chuang, *Quantum Computation and Quantum Information* (Cambridge University Press, Cambridge, 2000).
2. R. P. Feynman, Simulating physics with computers *Int. J. Theor. Phys.* **21**, 467-488 (1982).
3. P. Shor, Polynomial-time algorithms for prime factorization and discrete logarithms on a quantum computer, *SIAM Journal on Computing* **26**, 1484-1509 (1997).
4. Andreas Trabesinger, Quantum Simulation *Nature Physics* **8**, 263-263 (2012).
5. J. S. Bell, *Foundations of Physics* **12** 989 (1982).
6. T. D. Ladd *et al.*, Quantum computers, *Nature* **464**, 45 (2010)
7. D. P. DiVincenzo, Quantum computation *Science* **270**, 255 (1995),
8. M. Steffen, D. DiVincenzo, J. M. Chow, T. N. Theis and M. B. Ketchen, Quantum computing: An IBM perspective, *IBM J. Res. & Dev.* **55**, 13 (2011).
9. Nanoelectronics and Information Technology, Ed. by R. Waser, Wiley-VCH GmbH & Co. (2003).
10. R. Landauer, Irreversibility and Heat Generation in the Computing Process, *IBM J. Res. Dev.* **5**, 183 (1961).

11. Yu. V. Nazarov and Y. Blanter, *Quantum Transport: Introduction to Nanoscience*, Cambridge University Press, Cambridge, UK (2009).
12. A. Y. Khrennikov, *Contextual Approach to Quantum Formalism* (Springer, Berlin, 2009).
13. T. M. Nieuwenhuizen, Is the contextuality loophole fatal for the derivation of Bell inequalities?, *Found. Phys.* **41**, 580, (2011).
14. K. Hess, H. De Raedt, and K. Michielsen, Hidden assumptions in the derivation of the theorem of Bell, *Phys. Scr.* **T151**, 014002, (2012).
15. A. E. D. Bouwmeester and A. Z. (Eds.) (eds.), *The Physics of Quantum Information* (Springer Verlag, Berlin, 2000).
16. G. Chen *et al.*, *Quantum Computing Devices: Principles, Designs and Analysis* (Chapman and Hall/CRC, 2007).
17. A. Stern and N. H. Lindner, Topological quantum computation — From basic concepts to first experiments, *Science* **339** 1179 (2013).
18. A. E. Allahverdyan and Th. M. Nieuwenhuizen, Breakdown of the Landauer bound for information erasure in the quantum regime, Phys. Rev. E 64, 056117, 1-9 (2001).
19. C. H. Bennett, On constructing a molecular computer, *IBM J. Res. & Dev.* **17**, 525-532 (1973).
20. P. Benioff, The computer as a physical system: A microscopic quantum mechanical Hamiltonian model of computers as represented by Turing machines, *Jour. Stat. Phys.* **22**, 563 (1980).
21. A. Yao, Proc. of the 34th IEEE Symposium of Computer Science, IEEE Computer Society Press, Los Alemitos, CA, pp. 352-360 (1993).
22. A. Einstein, B. Podolsky, and N. Rosen, Can quantum-mechanical description of physical reality be considered complete?, *Phys. Rev.* **47**, 777 (1935).
23. J. S. Bell, On the Einstein-Poldolsky-Rosen paradox, *Physics* **1**, 195 (1964).
24. D. Deutsch, Quantum Theory, the Church-Turing Principle and the Universal Quantum Computer, *Proc. Roy. Soc. London* **400**, 97-117 (1985).
25. A. Aspect, J. D'Alibard and G. Roger, Experimental Test of Bell's Inequalities Using Time- Varying Analyzers, *Phys. Rev. Lett.* **49**, 1804-1807 (1982).
26. D. Deutsch and R. Jozsa, Rapid Solution of Problems by Quantum Computation, *Proc. Roy. Soc. London* **439A**, 553 (1992).
27. A. Berthiaume and G. Brassard, Oracle quantum computing, in Proc. Workshop on Physics of. Computation, *IEEE Press*, 195 (1992).
28. R. Josza, Characterizing Classes of Functions Computable by Quantum Parallelism, *Proc. Roy. Soc. London* **435A**, 563-574 (1991).
29. P. Shor, Scheme for reducing decoherence in quantum computer memory, *Phys. Rev. A* **52**, 2493-2496(R) (1995).
30. A. M. Steane, Error Correcting Codes in Quantum Theory *Phys. Rev. Lett.* **77**, 793-797 (1996).
31. D. Gottesman, *Stabilizer codes and quantum error correction*, Ph.D. dissertation, California Inst. Technol., Pasadena, CA, (1997); A. Y. Kitaev, Fault-tolerant quantum computation by anyons, *Ann. Phys.* **303**, 2 (2003); A. Y. Kitaev, Quantum computations: Algorithms and error correction, *Russ. Math. Surv.* **52**, 1191 (1997).
32. D. Aharonov and M. Ben-Or, Fault-tolerant quantum computation with constant error, in *Proc. 29th Annu. ACM Symp. Theory Comput., El Paso, TX, May 4-6*, 176 (1997).
33. D. Aharonov and M. Ben-Or, Fault-Tolerant Quantum Computation with Constant Error Rate, *SIAM J. Comput.* **38**(4), 1207-1282 (2008).

34. M. J. Bremner *et al.*, Practical Scheme for Quantum Computation with Any Two-Qubit Entangling Gate, *Phys. Rev. Lett.* **89**, 247902 (2002).
35. D. P. DiVincenzo, The Physical Implementation of Quantum Computation *Fortschr. Phys.* **48**, 771-783 (2000).
36. Hanson, R. *et al.*, Spins in few-electron quantum dots, Rev. Mod. Phys. **79**, 1217 (2007).
37. T. Monz *et al.*, 14-Qubit Entanglement: Creation and Coherence, *Phys. Rev. Lett.* **106**, 130506 (2011).
38. J. W. Britton *et al.*, Engineered two-dimensional Ising interactions in a trapped-ion quantum simulator with hundreds of spins, *Nature* **484**, 489 (2012).
39. Z. Bian *et al.*, Experimental Determination of Ramsey Numbers, *Phys. Rev. Lett.* **111**, 130505 (2013).
40. E. Paladino, Y. M. Galperin, G. Falci and B. L. Altushuler 1/f *noise: implications for solid-state quantum information*, Rev. Mod. Phys. **86**, 361 (2014).
41. C. Cohen-Tannoudji, J. Dupont-Roc and G. Grynberg, *Atom-Photon Interactions: Basic Processes and Applications* (Whiley, New York, 1992).
42. C. P. Schlichter, *Principles of Magnetic Resonance 3ED (Springer Series in Solid-State Sciences)* (Springer Verlag, 1992).
43. G. Falci, A. D'Arrigo, A. Mastellone and E. Paladino, Initial Decoherence in Solid State Qubits, *Phys. Rev. Lett.* **94**, 167002 (2005).
44. G. Ithier *et al.*, Decoherence in a superconducting quantum bit circuit, *Phys. Rev. B* **72**, 134519 (2005).
45. F. Chiarello *et al.*, Superconducting qubit manipulated by fast pulses: experimental observation of distinct decoherence regimes *New J. Phys.* **14** 023031 (2012).
46. A. D'Arrigo and E. Paladino, Optimal operating conditions of an entangling two-transmon gate, *New J. Phys.* **14**, 053035 (2012); E. Paladino, A. D'Arrigo, A. Mastellone and G. Falci, Decoherence times of universal two-qubit gates in the presence of broad-band noise, *New J. Phys.* **13**, 093037 (2011); B. Bellomo *et al.*, Entanglement degradation in the solid state: Interplay of adiabatic and quantum noise, *Phys. Rev. A* **81**, 062309 (2010); A. D'Arrigo, A. Mastellone, E. Paladino, G. Falci, Effects of low-frequency noise cross-correlations in coupled superconducting qubits, *New J. Phys.* **10**, 115006 (2008); E. Paladino, M. Sassetti, G. Falci, Modulation of dephasing due to a spin-boson environment, *Chem. Phys.* **296**, 325 (2004).
47. I. Buluta, S. Ashhab and F. Nori, Natural an artificial atoms for quantum computation, *Rep. Prog. Phys.* **74**, 104401 (2011).
48. L. Di Carlo *et al.*, Preparation and Measurement of Three-Qubit Entanglement in a Superconducting Circuit, *Nature* **467**, 574 (2010).
49. M. Neeley *et al.*, Generation of Three-Qubit Entangled States using Superconducting Phase Qubits, *Nature* **467**, 570 (2010).
50. E. Lucero *et al.*, Computing prime factors with a Josephson phase qubit quantum processor, *Nat. Physics* **8**, 719 (2012).
51. M. Mariantoni *et al.*, Implementing the Quantum von Neumann Architecture with Superconducting Circuits, *Science* **334**, 61 (2011).
52. A. Fedorov *et al.*, Implementation of a Toffoli gate with superconducting circuits, *Nature* **481**, 170 (2012).
53. M. D. Reed *et al.*, Realization of three-qubit quantum error correction with super-conducting circuits *Nature* **482**, 382 (2012).
54. J. M. Chow *et al.*, Universal Quantum Gate Set Approaching Fault-Tolerant Thresholds with Superconducting Qubits, *Phys. Rev. Lett.* **109**, 060501 (2012).

55. C. Rigetti *et al.*, Superconducting qubit in a waveguide cavity with a coherence time approaching 0.1 ms, *Phys. Rev. B* **86**, 100506(R) (2012).

56. Y. Makhlin, G. Schön, and A. Shnirman, Quantum-state engineering with Josephson-junction devices, *Rev. Mod. Phys.* **73**, 357 (2001); D. Esteve and D. Vion, in Les Houches School, Session LXXXI, 2004, H. Bouchiat, Y. Gefen, S. Gueron, G. Montambaux, and J. Dalibard Eds.; G. Falci and E. Paladino and R. Fazio in *Quantum Phenomena in Mesoscopic Systems*, 173, IOS Press, (Amsterdam) (2003).

57. J. Clarke and F. K. Wilhelm, Superconducting quantum bits, *Nature*, **453**, 1031 (2008).

58. J. Q. You and F. Nori, Atomic physics and quantum optics using superconducting circuits, *Nature* **474** 589 (2011).

59. Z.-L. Xiang, S. Ashhab, J. Q. You and F. Nori, Hybrid quantum circuits: Superconducting circuits interacting with other quantum systems, *Rev. Mod. Phys.*, **85**, 623 (2013).

60. J. Bardeen, L. N. Cooper and J. R. Schrieffer, Theory of Superconductivity, *Phys. Rev.* **108**, 1175 (1957).

61. M. Tinkham, *Introduction to Superconductivity* (McGraw-Hill, New York, 1996).

62. V. Bouchiat, *et al.*, Quantum coherence with a single Cooper pair, *Physica Scripta* **T76**, 165 (1998).

63. Y. Nakamura, Y. A. Pashkin and J. S. Tsai, Coherent control of macroscopic quantum states in a single-Cooper-pair box, *Nature* **398**, 786 (1999).

64. A. Shnirman, G. Schön and Z. Hermon, Quantum Manipulations of Small Josephson Junctions, *Phys. Rev. Lett.* **79**, 2371 (1997).

65. Y. Makhlin, G. Schön and A. Shnirman, Josephson-junction qubits with controlled couplings *Nature* **398**, 305 (1999).

66. D. Vion *et al.*, Manipulating the quantum state of an electrical circuit *Science* **296**, 886 (2002).

67. Jens Koch *et al.*, Charge-insensitive qubit design derived from the Cooper pair box, *Phys. Rev. A* **76**, 042319 (2007).

68. I. Siddiqi *et al.*, Direct Observation of Dynamical Bifurcation between Two Driven Oscillation States of a Josephson Junction *Phys. Rev. Lett.* **94**, 027005 (2005).

69. L. M. K. Vandersypen and I. L. Chuang, NMR techniques for quantum control and computation *Rev. Mod. Phys.* **76**, 1037 (2004).

70. G. Falci *et al.*, Design of a Lambda system for population transfer in superconducting nanocircuits, *Phys. Rev. B* **87**, 214515 (2013); G. Falci *et al.*, Effects of low-frequency noise in driven coherent nanodevices, *Phys. Scr.* **T151**, 014020 (2012).

71. A. Wallraff *et al.*, Strong coupling of a single photon to a superconducting qubit using circuit quantum electrodynamics *Nature* **431**, 162 (2004).

72. Blais A. *et al.*, Cavity quantum electrodynamics for superconducting electrical circuits: An architecture for quantum computation *Phys. Rev. A* **69**, 062320 (2004).

73. Y. Yu *et al.*, Coherent Temporal Oscillations of Macroscopic Quantum States in a Josephson Junction, *Science* **296**, 889 (2002); J. Martinis *et al.*, Rabi Oscillations in a Large Josephson-Junction Qubit, *Phys. Rev. Lett.* **89**, 117901 (2002); J. M. Martinis *et al.* , Decoherence of a superconducting qubit due to bias noise, *Phys. Rev. B* **67**, 094510 (2003); R. W. Simmonds *et al.*, Decoherence in Josephson Qubits from Junction Resonances, *Phys. Rev. Lett.* **93**, 077003 (2004); K. B. Cooper *et al.* Observation of quantum oscillations between a Josephson phase qubit and a microscopic resonator using fast readout, *Phys. Rev. Lett.* **93**, 180401 (2004); J. Claudon *et al.*, Coherent Oscillations in a Superconducting Multilevel Quantum System, *Phys. Rev. Lett.* **93**, 187003 (2004).

74. J. E. Mooij *et al.*, Josephson persistent-current qubit, *Science* **285**, 1036 (1999)
75. L. B. Ioffe *et al.*, Environmentally decoupled sds -wave Josephson junctions for quantum computing, *Nature* **398**, 679 (1999).
76. C. van der Wal *et al.*, Quantum Superposition of Macroscopic Persistent-Current States, *Science* **290**, 773 (2000).
77. I. Chiorescu, Y. Nakamura, C. J. P. M. Harmans, J. E. Mooij, Coherent quantum dynamics of a superconducting flux qubit, *Science* **299**, 1869 (2003).
78. I. Chiorescu *et al.*, Coherent dynamics of a flux qubit coupled to a harmonic oscillator, *Nature* **431**, 159 (2004).
79. M. Steffen *et al.*, Measurement of the Entanglement of Two Superconducting Qubits via State Tomography, *Science* **313**, 1423 (2006).
80. Y. Nakamura, Y. A. Pashkin, J. S. Yamamoto and T. Tsai, Charge echo in a Cooper-pair box, *Phys. Rev. Lett.* **88**, 047901 (2002).
81. J. M. Martinis, S. Nam, J. Aumentado and C. Urbina, Rabi Oscillations in a Large Josephson-Junction Qubit, *Phys. Rev. Lett.* **89**, 117901 (2002).
82. F. Yoshihara *et al.*, Decoherence of Flux Qubits due to 1/f Flux Noise, *Phys. Rev. Lett.* **97**, 167001 (2006).
83. J. A. Schreier *et al.*, Suppressing charge noise decoherence in superconducting charge qubits, *Phys. Rev. B* **77**, 180502 (2008).
84. V. E. Manucharyan, J. Koch, L. I. Glazman and M. H. Devoret, Fluxonium: Single Cooper-Pair Circuit Free of Charge Offsets, *Science* **326**, 113 (2009).
85. Steffen M *et al.*, High-Coherence Hybrid Superconducting Qubit, *Phys. Rev. Lett.* **105** 100502 (2010).
86. J. Bylander *et al.*, Noise spectroscopy through dynamical decoupling with a superconducting flux qubit, *Nature Physics* **7**, 565 (2011).
87. Z. Kim *et al.*, Decoupling a Cooper-pair box to enhance the lifetime to 0.2 ms, *Phys. Rev. Lett.* **106** 120501 (2011).
88. J. Q. You, J. S. Tsai, and F. Nori, Scalable Quantum Computing with Josephson Charge Qubits, *Phys. Rev. Lett.* **89**, 197902 (2002).
89. D. V. Averin, C. Bruder, Variable Electrostatic Transformer: Controllable Coupling of Two Charge Qubits, *Phys. Rev. Lett.* **91**, 057003 (2003); A. Blais, A. Maassen van den Brink, and A. M. Zagoskin, Tunable Coupling of Superconducting Qubits *Phys. Rev. Lett.* **90**, 127901 (2003); B. Plourde *et al.*, Entangling flux qubits with a bipolar dynamic inductance, *Phys. Rev. B* **70**, 140501(R) (2004); A. O. Niskanen, Y. Nakamura, and J. S. Tsai, Tunable coupling scheme for flux qubits at the optimal point *Phys. Rev. B* **73**, 094506 (2006); P. Bertet, C. J. Harmans, and J. E. Mooij, Parametric coupling for superconducting qubits, *Phys. Rev. B* **73**, 064512 (2006); Y.-D. Wang, A. Kemp, and K. Semba, Coupling superconducting flux qubits at optimal point via dynamic decoupling with the quantum bus, *Phys. Rev. B* **79**, 024502 (2009).
90. C. Rigetti, A. Blais, and M. Devoret, Protocol for Universal Gates in Optimally Biased Superconducting Qubits, *Phys. Rev. Lett.* **94**, 240502 (2005); Yu-xi Liu, L. F. Wei, J. S. Tsai and Franco Nori Controllable Coupling between Flux Qubits, *Phys. Rev. Lett.* **96**, 067003 (2006); G. S. Paraoanu, Microwave-induced coupling of superconducting qubits, *Phys. Rev. B* **74**, 140504(R) (2006).
91. F. Plastina, G. Falci, Communicating Josephson qubits, *Phys. Rev. B* **67**, 224514 (2003).
92. T. V. Filippov *et al.*, Tunable Transformer for Qubits Based on Flux States, *IEEE Trans. Appl. Supercond.* **13**, 1005 (2003); A. Maassen van den Brink, A. J. Berkley, and M. Yalowsky, Mediated tunable coupling of flux qubits, *New J. Phys.* **7**, 230 (2005).

93. Yu. A. Pashkin *et al.*, Quantum oscillations in two coupled charge qubits, *Nature* **421**, 823 (2003); A. J. Berkley *et al.*, Entangled Macroscopic Quantum States in Two Superconducting Qubits *Science* **300**, 1548 (2003); T. Yamamoto *et al.*, Demonstration of conditional gate operation using superconducting charge qubits, *Nature* **425**, 941 (2003); J. B. Majer *et al.*, Spectroscopy on Two Coupled Superconducting Flux Qubits *Phys. Rev. Lett.* **94**, 090501 (2005); J. H. Plantenberg, P. C. de Groot, C. J. P. M. Harmans, and J. E. Mooij, Demonstration of controlled-NOT quantum gates on a pair of superconducting quantum bits, *Nature* **447**, 836 (2007).

94. A. Izmalkov *et al.*, *Phys. Rev. Lett.* Evidence for Entangled States of Two Coupled Flux Qubits, **93**, 037003 (2004); T. Hime *et al.*, Solid-State Qubits with Current-Controlled Coupling, *Science* **314**, 1427 (2006); A. O. Niskanen *et al.*, Quantum Coherent Tunable Coupling of Superconducting Qubits, *Science* **316** 723 (2007).

95. J. Majer *et al.*, Coupling superconducting qubits via a cavity bus, *Nature* **449**, 443 (2007); S. H. W. van der Ploeg *et al.*, Controllable Coupling of Superconducting Flux Qubits, *Phys. Rev. Lett.* **98**, 057004 (2007); A. Fay *et al.*, Strong Tunable Coupling between a Superconducting Charge and Phase Qubit, *Phys. Rev. Lett.* **100**, 187003 (2008); Helmer, F. *et al.*, Two-dimensional cavity grid for scalable quantum computation with superconducting circuits, *EPL* **85** 50007 (2009); Sillanpaa, M. A., Park, J. I., and Simmonds, R. W., Coherent quantum state storage and transfer between two phase qubits via a resonant cavity. *Nature* **449**, 438-442 (2007)

96. E. Paladino, A. Mastellone, A. D'Arrigo, G. Falci, Optimal tuning of solid-state quantum gates: A universal two-qubit gate, *Phys. Rev. B*, **81**, 052502 (2010).

97. D. Loss and D. P. DiVincenzo, Quantum computation with quantum dots, *Phys. Rev. A* **57**, 120 (1998).

98. Liu, H. W. *et al.*, A gate-defined silicon quantum dot molecule, *Appl. Phys. Lett.* **92**, 222104 (2008); Simmons, C. B. *et al.*, Charge sensing and controllable tunnel coupling in a Si/SiGe double quantum dot, *Nano Lett.* **9**, 3234-3238 (2009).

99. Kane, B. E. A silicon-based nuclear spin quantum computer, *Nature* **393**, 133 (1998); Vrijen, R. *et al.* Electron-spin-resonance transistors for quantum computing in silicon-germanium heterostructures, *Phys. Rev. A* **62**, 012306 (2000).

100. Berezovsky, J. *et al.*, Nondestructive optical measurements of a single electron spin in a quantum dot, *Science* **314**, 1916 (2006); Harrison, J., Sellars, M. J. and Manson, N. B. Measurement of the optically induced spin polarisation of N-V centres in diamond, *Diamond Related Mater.* **15**, 586 (2006).

101. Dutt, M. V. G. *et al.* Quantum register based on individual electronic and nuclear spin qubits in diamond, *Science* **316**, 1312 (2007).

102. Balasubramanian, G. *et al.* Ultralong spin coherence time in isotopically engineered diamond, *Nature Mater.* **8**, 383 (2009).

103. Cory, D. G. Fahmy, A. F. and Havel, T. F. Ensemble quantum computing by NMRspectroscopy, *Proc. Natl Acad. Sci. USA* **94**, 1634 (1997). Gershenfeld, N. A. and Chuang, I. L. Bulk spin resonance quantum computation, *Science* **275**, 350 (1997).

104. Negrevergne, C. *et al.*, Benchmarking quantum control methods on a 12-qubit system, *Phys. Rev. Lett.* **96**, 170501 (2006).

105. Blatt R. and Wineland D. J., Entangled states of trapped atomic ions, *Nature* **453** 1008 (2008).

106. Cirac, J. I. and Zoller, P. Quantum Computations with Cold Trapped Ions, *Phys. Rev. Lett.* **74**, 4091 (1995).

107. Wineland, D. J. *et al.*, Experimental issues in coherent quantum-state manipulation of trapped atomic ions, *J. Res. Natl. Inst. Stand. Technol.* **103**, 259 (1998).

108. Wineland, D. and Blatt, R. Entangled states of trapped atomic ions *Nature* **453**, 1008 (2008)
109. Olmschenk, S. *et al.* Quantum teleportation between distant matter qubits *Science* **323**, 486 (2009).
110. Morsch, O. and Oberthaler, M. Dynamics of Bose-Einstein condensates in optical lattices, *Rev. Mod. Phys.* **78**, 179 (2006).
111. S. Haroche and J. M. Raimond, *Exploring the Quantum: Atoms, Cavities, and Photons* (Oxford Univ. Press, 2006); Walther, H. *et al.*, Cavity quantum electrodynamics, *Rep. Prog. Phys.* **69**, 1325 (2006); Miller, T. E. *et al.*, Trapped atoms in cavity QED: coupling quantized light and matter, *J. Phys. B* **38**, S551 (2005).
112. Thompson, R. J., Rempe, G. and Kimble, H. J. Observation of normal-mode splitting for an atom in an optical cavity, *Phys. Rev. Lett.* **68**, 1132 (1992).
113. Raimond, J. M., Brune, M. and Haroche, S. Manipulating quantum entanglement with atoms and photons in a cavity, *Rev. Mod Phys.* **73**, 565 (2001); Meschede, D., Walther, H. and Muller, G. One-Atom Maser, *Phys. Rev. Lett.* **54**, 551 (1985); Rempe, G., Walther, H. and Klein, N. Observation of quantum collapse and revival in a one-atom maser, *Phys. Rev. Lett.* **58**, 353 (1987); Brune, M. *et al.*, Quantum Rabi Oscillation: A Direct Test of Field Quantization in a Cavity, *Phys. Rev. Lett.* **76**, 1800 (1996).
114. Vahala, K. J. Optical microcavities, *Nature* **424**, 839 (2003).
115. Leibfried, D. *et al.*, Quantum dynamics of single trapped ions *Rev. Mod. Phys.* **75**, 281 (2003); Gabrielse, G. and Dehmelt, H. Observation of inhibited spontaneous emission, *Phys. Rev. Lett.* **55**, 67 (1985).
116. Cavaliere, F. *et al.*, Phonon distributions of a single-bath mode coupled to a quantum dot, *New J. Phys.* **10**, 115004 (2008).
117. Osnaghi, S. *et al.*, Coherent Control of an Atomic Collision in a Cavity, *Phys. Rev. Lett.* **87**, 037902 (2001); Pellizari, T. , Gardiner, S. A., Cirac, J. I. and Zoller, P. Decoherence, Continuous Observation, and Quantum Computing: A Cavity QED Model, *Phys. Rev. Lett.* **75**, 3788 (1995).
118. Knill, E., Laflamme, R. and Milburn, G. J. A scheme for efficient quantum computation with linear optics, *Nature* **409**, 46 (2001).
119. A. Mizel, D. Lidar and M. Mitchell, Simple proof of equivalence between adiabatic quantum computation and the circuit model, *Phys. Rev. Lett.* **99**, 070502 (2007).

Chapter 2

Quantum Information in Communication and Imaging

David S. Simon[*,†,**], Gregg Jaeger[†,‡,§,††] and Alexander V. Sergienko[†,‡,‡‡]

*Department of Physics and Astronomy, Stonehill College,
320 Washington Street, Easton, MA 02357 USA*

†*Department of Electrical and Computer Engineering, Boston University,
8 Saint Mary's Street, Boston, MA 02215, USA*

‡*Department of Physics, Boston University, 590 Commonwealth Avenue,
Boston, MA 02215, USA*

§*Division of Natural Sciences and Mathematics, Boston University,
Boston, MA 02215, USA*

A brief introduction to quantum information theory in the context of quantum optics is presented. After presenting the fundamental theoretical basis of the subject, experimental evaluation of entanglement measures are discussed, followed by applications to communication and imaging.

1. From bit to qubit

1.1. *Introduction*

Quantum information science is a rapidly developing area of interdisciplinary investigation, which plays a significant role in a number of sub-disciplines of physics and engineering. Quantum communication (including quantum key distribution for cryptography) and quantum imaging are currently two of the most exciting applications of quantum information science. For this reason, we focus here on quantum optical systems, a natural choice because communication and imaging are typically optical in the current era. Further, interferometry is central to quantum information processing and interferometry has primarily progressed through optical physics. Quantum theory was developed by Einstein, Bohr, Schrödinger, Heisenberg, Dirac, and others, and given a unified formalization first by Dirac[a] and later

---

**simond@bu.edu

††jaeger@math.bu.edu

‡‡alexserg@bu.edu

[a]In his *The Principles of Quantum Mechanics*.

by von Neumann[b] in the first third of the 20th century. It serves as a basis for understanding quantum field theory, wherein Dirac again played a key role. By the end of the twentieth century, quantum information science, which was developed entirely within this formalism, became a subject in its own right. In practice, it can be best understood as a range of interferometric systems acting as realizations of specifically quantum mechanical physical communications layers, protocols, and algorithms. These are primarily based on the use of the quantum information unit, the "qubit." The term "qubit" originated with Benjamin Schumacher, who replaced "the classical idea of a binary digit with a quantum two-state system... These quantum bits, or 'qubits,' are the fundamental units of quantum information."[1]

The quantum difference from classical information arises from the *superposition principle* of quantum mechanics. This means that, despite its being two-valued in the chosen computational basis, a qubit system can be in one of an infinite number of physically significant states: while a bit is capable of being in only one of two significant states at a given moment, a qubit system in general can be considered as potentially being in one measurable state and the other opposite state at the same time. In further contrast to classical states, a single unknown state of a qubit system cannot generally be found by a single measurement, but rather requires an ensemble of them to be determined. It is precisely the superposition of individual qubits that provides the possibility of secure quantum key distribution, for example.

One striking consequence of superposition in quantum mechanics, is the possibility of *entanglement*, in which the state of a composite system can not be factored into a product of states describing each of its subsystems separately. To be more specific, consider a bipartite composite system, formed from the pair of subsystems A and B; for example, A and B may be labels for two photons, two atoms, or any other pair of quantum systems. These two systems may be separated by arbitrarily large distances. We may then form the composite system $A \otimes B$, whose Hilbert space is the product of the two individual Hilbert spaces: $\mathcal{H}_{AB} = \mathcal{H}_A \times \mathcal{H}_B$. Consider a pure state $|\psi_{AB}\rangle$ of the composite system. Then the state is said to be *separable* if it can be written in some basis in the form $|\psi_{AB}\rangle = |\psi_A\rangle \otimes |\psi_B\rangle$, where $|\psi_A\rangle \in \mathcal{H}_A$ nd $|\psi_B\rangle \in \mathcal{H}_B$. The state is then defined to be entangled if it is not separable. We will describe the consequences of entanglement and how it may be quantified in more detail below. As we are primarily concerned here with optical systems, we will also describe in detail the process of spontaneous parametric downconversion (SPDC), which provides a convenient and versatile means of producing entangled pairs of photons. As we introduce measures of information and of entanglement, we will apply them to the downconversion process and its applications.

Our main focus is on quantum communication. The most thoroughly studied application of quantum communications is quantum cryptography, also known as quantum key distribution. After describing the basic ideas in this area, we move on to a topic which has been less well studied from a quantum information theoret-

---

[b]In his *Mathematische Grundlagen der Quantenmechanik*.

ical viewpoint, namely quantum imaging. We will take a broad view of the word communication in order to include the reconstruction of images over a distance. In order to quantify our ability to communicate, it will be necessary to investigate the amount of information extracted per photon and the amount of entanglement per pair of photons, as well as the amount of mutual information carried per pair. We will exclusively discuss *bipartite* systems, i.e. those with two subsystems.

One related topic we will not discuss in detail is quantum computing, which applies the quantum superposition principle to a collection of *stored* qubits, which can be thought of as forming a compound quantum system. The space of possible quantum states available to such multiple-qubit systems grows more rapidly than does the space of states available to multiple-bit systems. The size of the parameter space describing a quantum system for information encoding and computing grows exponentially in the number of qubits—in a classical system it grows only linearly in the number of bits. This also provides a unique sort of computational parallelism, which can be harnessed to make tractable some important computational tasks that are thought intractable using classical means only. This improvement in efficiency is known as "quantum speedup." Multiple-qubit states however are also very fragile, being susceptible to decoherence effects.[2,3] After a short period of time, the initially pure quantum states described are inevitably altered by interactions with their environments and must then be described instead by a mixed quantum states.

In the remainder of this section, we introduce basic notions of quantum communication theory in the context of quantum optics, including a detailed discussion of spontaneous parametric down conversion, which is the principle source of entangled optical states for experiments. Section 2 moves on to applications with a discussion of quantum cryptography, followed by a discussion of how the main ideas generalize from the context of qubits to so-called qudits. Section 3 follows up by discussing a specific realization of qudits in the form of orbital angular momentum states. These same states are then applied in the context of imaging, leading to the idea that the mutual information shared by entangled states may serve as probes of geometric symmetries.

1.2. *From bits to qubits*

The properties of a qubit system are two-valued and can be probabilistically predicted like a classical system that randomly takes one of two computationally relevant values. But unlike the classical system, which can only be in one of the two states at any time irrespective of how it may be measured, a qubit can be in both states simultaneously. The unit of classical information is sometimes referred in quantum information science as *c-bits*.[4] A putative inherently probabilistic bit can be called a *probabit*.[5] The probabilities of the outcomes of measurements of any classical system are due only to *ignorance* of the actual state of the system. In the quantum case, it arises also from a *fundamental* indeterminacy of properties, entirely so in the case of the pure quantum states, defined below. The quantum bit is, therefore, not reducible to the probabilistic bit.[c]

Let us begin by considering various representations of qubit states. Recall that quantum states are associated with a complex Hilbert vector space, $\mathcal{H}$, via a special class of linear operators acting in it, the *statistical operators*, $\hat{\rho}$, constituting the quantum state-space. For pure qubit states, the statistical operators are projectors onto one-dimensional subspaces. The projective operators $P(|\psi\rangle)$ can be uniquely associated with points on the boundary of the Bloch ball, known as the Poincaré–Bloch sphere. These states can also be, and typically are, represented by the state-vectors

$$|\psi\rangle \in \mathcal{H} \tag{1}$$

spanning them. The remaining states of the Bloch ball are essentially statistical or mixed states, defined as those which are not pure but still satisfy the definition of a density operator. The mixed states can be formed from these projectors by appropriate linear combinations and lie in the interior of the Bloch ball. However, the mixed states *cannot* be written as linear combinations of state vectors.

The set of statistical states available to a qubit system is representable by the 2×2 complex Hermitian trace-one matrices

$$[\hat{\rho}_{ij}] \in H(2). \tag{2}$$

By contrast, for the full physical state description of a quantum system *in spacetime*, an infinite-dimensional spatial representation is required in which the state-vectors are called *wavefunctions*. Quantum information theory is based on the behavior of qubits and has thus far overwhelmingly dealt with quantities with discrete eigenvalue spectra in the non-relativistic regime, the state-vectors considered here are usually taken to lie within finite-dimensional Hilbert spaces constructed by taking the tensor product of multiple copies of *two-dimensional* complex Hilbert space, or other finite-dimensional spaces. The Hilbert spaces considered here are only finite-dimensional *subspaces* of the larger full physical state-spaces of particles, the other subspaces of which are rarely taken into account in the study of quantum information processing. In many cases we consider the polarization states or orbital angular momentum states of photons, without considering the corresponding full photon wavefunctions.

The *purity*, $\mathcal{P}$, of a quantum state specified by the statistical operator $\hat{\rho}$ is the trace of its square,

$$\mathcal{P}(\hat{\rho}) = \operatorname{tr} \hat{\rho}^2 \,, \tag{3}$$

where $\frac{1}{d} \leq \mathcal{P}(\hat{\rho}) \leq 1$ and d is the dimension of the Hilbert space, $\mathcal{H}$, attributed to the system it describes. The quantum state is *pure* if $\mathcal{P}(\hat{\rho}) = 1$, that is, if it spans

---

[c]Note that we will here use "qubit" and "quantum bit" to refer to both physical systems on which quantum information can be encoded, as well as to the quantum bit of information in the sense of information theory, depending on context to make clear which is intended in any given instance.

a one-dimensional subspace of $\mathcal{H}$. One can then naturally define state *mixedness* as the complement of purity, $\mathcal{M}(\hat{\rho}) \equiv 1 - \mathcal{P}(\hat{\rho})$. The *Unitary* linear operators, U, are those for which $U^{\dagger}U = UU^{\dagger} = \mathbb{I}$, where "$\dagger$" indicates Hermitian conjugation (see Sec. A.3). Here, the time-evolution is prescribed by the Schrödinger equation, assuming a time-independent Hamiltonian. (In general, temporal evolution in quantum mechanics is not always so simple; *cf.* Sec. 2.1 of Sakurai.[6]) The purity and mixedness of a quantum state are invariant under transformations of the form $\hat{\rho} \rightarrow U\hat{\rho}\,U^{\dagger}$, where U is unitary, most importantly under the *dynamical mapping* $U(t, t_0) = e^{-\frac{i}{\hbar}H(t-t_0)}$, where H is the Hamiltonian operator, which can readily be seen upon recalling that the trace operation $\mathrm{tr}(\cdot)$ is cyclic. Pure states are those which are *maximally specified* within quantum mechanics. A quantum state is pure if and only if the statistical operator $\hat{\rho}$ is *idempotent*, that is,

$$\hat{\rho}^2 = \hat{\rho}\,, \tag{4}$$

providing a convenient test for maximal state purity. It is then also a *projector*, $P(|\psi_i\rangle)$, where $|\psi_i\rangle$ is the normalized vector representative of the corresponding one-dimensional subspace of its Hilbert space. Rays cannot be added, whereas vectors $|\psi_i\rangle$ can be, making the latter better for use in calculations involving pure states, where superpositions are formed by addition. A Hermitian operator P acting in a Hilbert space $\mathcal{H}$ is a *projector* if and only if $P^2 = P$. It follows from this definition that $P^{\perp} \equiv \mathbb{I} - P$, where $\mathbb{I}$ is the identity operator, is also a projector. The projectors P and $P^{\perp}$ project onto orthogonal subspaces within $\mathcal{H}$, $\mathcal{H}_s$, and $\mathcal{H}_s^{\perp}$, respectively, thereby providing a decomposition of $\mathcal{H}$ as $\mathcal{H}_s \oplus \mathcal{H}_s^{\perp}$; two subspaces are said to be *orthogonal* if every vector in one is orthogonal to every vector in the other. In the case of a general state of a single qubit, one may write $\hat{\rho} = p_1 P(|\psi\rangle) + p_2 P(|\psi^{\perp}\rangle)$, where the weights p_i are the eigenvalues of the statistical operator $\hat{\rho}$.

A quantum state is thus mixed if it is *not* a pure state, that is, if $\mathcal{P}(\hat{\rho}) < 1$. In the Dirac notation, projectors are written

$$P(|\psi_i\rangle) \equiv |\psi_i\rangle\langle\psi_i|\,. \tag{5}$$

Consider a finite set, $\{P(|\psi_i\rangle)\}$, of projectors corresponding to distinct, orthogonal pure states $|\psi_i\rangle$. Any state $\hat{\rho}'$ that can be written

$$\hat{\rho}' = \sum_i p_i P(|\psi_i\rangle)\,, \tag{6}$$

with $0 < p_i < 1$ and $\sum_i p_i = 1$, is then a normalized mixed state. The superposition principle implies that any (complex) linear combination of qubit basis states, such as $|0\rangle$ and $|1\rangle$, that is,

$$|\psi\rangle = a_0|0\rangle + a_1|1\rangle \tag{7}$$

with $a_i \in \mathbb{C}$ and $|a_0|^2 + |a_1|^2 = 1$, is *also* a physical state of the qubit and is, as we have seen, also a pure state. The scalar coefficients a_0 and a_1 are called quantum probability amplitudes, because their square magnitudes, $|a_0|^2$ and $|a_1|^2$, are the probabilities p_0 and p_1, respectively, of the qubit described by state $|\psi\rangle$ being found in these basis states $|0\rangle$ and $|1\rangle$, respectively, upon measurement.

The superposition principle is ultimately the source of many of the quantum phenomena that we will use in the forthcoming sections; in particular it underlies entanglement, interference phenomena, and the inability to distinguish nonorthogonal states, all of which will be used for applications in Secs. 2 and 3. Consider the normalized sums

$$| \nearrow \rangle \equiv \frac{1}{\sqrt{2}}(|0\rangle + |1\rangle) \text{ and } | \searrow \rangle \equiv \frac{1}{\sqrt{2}}(|0\rangle - |1\rangle) . \tag{8}$$

of two orthogonal pure state-vectors $|0\rangle \doteq (1\ 0)^{\mathrm{T}}$ and $|1\rangle \doteq (0\ 1)^{\mathrm{T}}$ of a qubit, the r.h.s.'s being given in the matrix representation and $(\cdots)^{\mathrm{T}}$ indicating matrix transposition. The superpositions in Eq. (8) are pure states, as can be immediately verified by taking their square moduli. The corresponding projectors are $P(| \nearrow \rangle) = | \nearrow \rangle\langle \nearrow |$, $P(| \searrow \rangle) = | \searrow \rangle\langle \searrow |$. However, the normalized sum of a pair of *projectors*, for example, $P(|0\rangle)$ and $P(|1\rangle)$ corresponding to pure states $|0\rangle$ and $|1\rangle$, namely,

$$\hat{\rho}_+ = \frac{1}{2}\big(P(|0\rangle) + P(|1\rangle)\big) , \tag{9}$$

is a *mixed* state that can also be written

$$\hat{\rho}_+ = \frac{1}{2}\big(P(| \nearrow \rangle) + P(| \searrow \rangle)\big) . \tag{10}$$

Finally, note that he statistical operator corresponding to the normalized sum of $| \nearrow \rangle$ and $| \searrow \rangle$ is $P(|0\rangle) \neq \hat{\rho}_+$.

The pure states of the qubit can be represented by vectors in the two-dimensional complex Hilbert space, $\mathcal{H} = \mathbb{C}^2$. Any orthonormal basis for this space can be put in correspondence with two bit values, 0 and 1, in order to act as the single-qubit *computational basis*, sometimes also called the *rectilinear basis*, and written $\{|0\rangle, |1\rangle\}$. The vectors of the computational basis can be represented in matrix form as

$$|0\rangle \doteq \begin{pmatrix} 1 \\ 0 \end{pmatrix}, \qquad |1\rangle \doteq \begin{pmatrix} 0 \\ 1 \end{pmatrix}. \tag{11}$$

Other commonly used bases are the *diagonal basis*, $\{| \nearrow \rangle, | \searrow \rangle\}$, sometimes also written $\{|+\rangle, |-\rangle\}$, and the *circular basis* $\{|r\rangle, |l\rangle\}$:

$$|r\rangle \equiv \frac{1}{\sqrt{2}}(|0\rangle + i|1\rangle) , \qquad |l\rangle \equiv \frac{1}{\sqrt{2}}(|0\rangle - i|1\rangle) , \tag{12}$$

sometimes also written $\{| \circlearrowleft \rangle, | \circlearrowright \rangle\}$, is also useful for quantum cryptography, being conjugate to both the computational and diagonal bases.

All three of the above bases are mutually conjugate and are used in protocols for quantum key distribution (Sec. 2.2); the probabilities of qubits in the states $|r\rangle$ and $|l\rangle$ being found in the states $|0\rangle$, $|1\rangle$, $| \nearrow \rangle$, and $| \searrow \rangle$ are all $\frac{1}{2}$, and vice-versa.

The generic mixed state, $\hat{\rho}$, lies in the interior of the Bloch ball, can be written as a convex combination of basis-element projectors corresponding to the pure-state bases described above. The effect of a general operation on a qubit can be viewed as a (possibly stochastic) transformation within this ball; for illustrations of this in practical context, see.[7] The parametrization required to adequately describe mixed states is now discussed in detail.

The density matrix and the *Stokes four-vector*, S_μ, are related by

$$\hat{\rho} = \frac{1}{2} \sum_{\mu=0}^{3} S_\mu \sigma_\mu, \tag{13}$$

where σ_μ ($\mu = 1, 2, 3$) are the *Pauli operators* which, together with the identity $\sigma_0 = \mathbb{I}_2$, are represented in the matrix space $H(2)$ by the Pauli matrices. The Pauli matrices form a basis for $H(2)$, which contains the qubit density matrices. The qubit density matrices themselves are the positive-definite, trace-class elements of the set of 2×2 complex Hermitian matrices $H(2)$ of unit trace, that is, for which the total probability S_0 is one, as prescribed by the Born rule for quantum probabilities and the well-definedness of quantum probabilities as such. Density matrices are similarly defined for systems of countable dimension. The nontrivial products of the four Pauli matrices—those between the σ_i for $i = 1, 2, 3$—are given by

$$\sigma_i \sigma_j = \delta_{ij} \sigma_0 + i \epsilon_{ijk} \sigma_k , \tag{14}$$

which defines their algebra. Appropriately exponentiating the Pauli matrices provides the rotation operators, $R_i(\xi) = e^{-i\xi \sigma_i / 2}$, for Stokes vectors about the corresponding directions i;[6] these rotations realize the group $SO(3)$.

The Stokes parameters S_μ ($\mu = 0, 1, 2, 3$) also allow one to directly visualize the qubit state geometrically in the Bloch ball via S_1, S_2, S_3. The Euclidean length of this three-vector (also known as the *Stokes vector*, or *Bloch vector*) is the radius $r = (S_1^2 + S_2^2 + S_3^2)^{1/2}$ of the sphere produced by rotations of this vector. With the matrix vector $\vec{\sigma} = (\sigma_1, \sigma_2, \sigma_3)$ and the three-vector $\vec{S} = (S_1, S_2, S_3)$, one has

$$\hat{\rho} = \frac{1}{2}(S_0 \mathbb{I}_2 + S_1 \sigma_1 + S_2 \sigma_2 + S_3 \sigma_3) , \tag{15}$$

known as the *Bloch-vector representation* of the statistical operator, in accord with Eq. (15). In optical situations, where $\vec{S}$ describes a polarization state of a photon, the degree of polarization is given by $P = r/S_0$, where S_0 is positive. For the qubit, when the state is normalized so that $S_0 = 1$, S_0 corresponds to total quantum probability. The density matrix of a single qubit is then of the form

$$\hat{\rho} \doteq \begin{pmatrix} \rho_{00} & \rho_{01} \\ \rho_{10} & \rho_{11} \end{pmatrix} , \tag{16}$$

where $\rho_{00} + \rho_{11} = 1$, $\rho_{ii} = \rho_{ii}^*$ with ($i = 0, 1$), and $\rho_{10} = \rho_{10}^*$, where * indicates complex conjugation. One can write the Pauli matrices for $\mu = 1, 2, 3$ in terms of

outer products of computational basis vectors, as follows. The Stokes parameters
are expressed in terms of the density matrix as

$$S_\mu = \text{tr}(\hat{\rho}\sigma_\mu) , \qquad (17)$$

which are probabilities corresponding to ideal normalized counting rates of measurements in the standard eigenbases.

1.3. *Optical qubits*

For specificity, let us now take the system in question to be a photon. Light is
easy to produce and to detect, and has properties that are both well understood
and easily controlled. As a result, most experiments in quantum information and
communication are carried out on optical systems. Consequently, we will focus
henceforth exclusively on quantum optical systems. We begin by describing how
optical qubits can be created.

Consider the beam splitter shown in Fig. 1(a). A beam splitter (BS) is a device
for splitting a single optical beam into two: a portion of the beam is transmitted
through the BS, while a portion is reflected. Throughout, we assume that all beam
splitters used are 50-50, i.e. that equal amounts of light are reflected and transmitted. We also consider only nonpolarizing beam splitters. A beam splitter is a linear,
passive four-port device, with two input ports (a and b) and two output ports (c
and d). To describe its action, we form the operator-valued column vectors

$$\begin{pmatrix} \hat{a}^\dagger \\ \hat{b}^\dagger \end{pmatrix} \qquad \text{and} \qquad \begin{pmatrix} \hat{c}^\dagger \\ \hat{d}^\dagger \end{pmatrix}, \qquad (18)$$

where $\hat{a}^\dagger$, $\hat{b}^\dagger$, $\hat{c}^\dagger$, $\hat{d}^\dagger$ are the creation operators for photon states at the corresponding ports. Then we may denote the action of the beam splitter by a matrix T
relating ingoing and outgoing operators, $\begin{pmatrix} \hat{c}^\dagger \\ \hat{d}^\dagger \end{pmatrix} = T \begin{pmatrix} \hat{a}^\dagger \\ \hat{b}^\dagger \end{pmatrix}$. The form of this matrix is easy to determine: the photon is unchanged when it is transmitted and picks
up a phase of $\frac{\pi}{2}$ when reflected, so the BS matrix is

$$T = \frac{1}{\sqrt{2}} \begin{pmatrix} 1 & i \\ i & 1 \end{pmatrix} . \qquad (19)$$

We may now think of photons entering or leaving from above the BS (i.e. ports
a and d) as representing state $|0\rangle$, while those entering or leaving below the beam
splitter (i.e. b and c) represent $|1\rangle$ states. This provides a representation of physical
qubits as spatial modes, and then allows us to think of the BS matrix T as taking
combinations of input bits to combinations of output bits. In particular, if a *bit* 0 is
input, the resulting output is the *qubit* $\frac{1}{\sqrt{2}}(|0\rangle + |1\rangle)$. Thus, we have a simple way
of producing spatial qubits from classical bits.

We may form more general spatial qubits with the Mach-Zehnder interferometer
(Fig. 1(b)). This is equivalent to a double-slit-like arrangement where only two

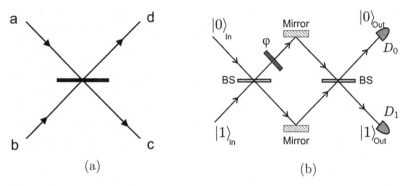

(a)   (b)

Fig. 1. (a) A 50/50 beamsplitter. A photon entering either input port, **a** or **b**, has equal probability of being transmitted or reflected out either output port, **c** or **d**. (b) The Mach–Zehnder interferometer providing a range of qubit states as the input qubit amplitudes a_i and phases ϕ_i are changed. The detectors provide count rates proportional to the probability of lying in the output computational-basis states described by state-projectors $P(|0\rangle)$ and $P(|1\rangle)$, for input amplitudes $a_0 = 0$, $a_1 = 1$, namely, $p(0) = \sin^2[(\phi_0 - \phi_1)/2]$ and $p(1) = \cos^2[(\phi_0 - \phi_1)/2]$.

directions are available to the self-interfering system, so that the exit ports of a beam-splitter act as "slits". In this interferometer, a photon enters from the left into a beam-splitter, with two exit paths on the right. It provides a spatial qubit, consisting of occupation of one and/or the other interior beam path. Each path then encounters a mirror, a phase shifter, a second beam-splitter, and finally a particle detector. Since only the relative phase between arms matters, the phase shift in one path can be set to zero without loss of generality. One can also use this interferometer to prepare a *phase qubit* by selecting only those systems entering a single initial input port and exiting a single final output port.

The action of the interferometer may be described by the matrix $B = T\Phi T$, where T is the BS matrix above and the phase shift is described by the matrix

$$\Phi = \begin{pmatrix} e^{i\phi} & 0 \\ 0 & 1 \end{pmatrix}. \tag{20}$$

Multiplying out the matrices, we find that the action on an incoming bit $|0\rangle$ is:

$$|0\rangle \rightarrow \frac{1}{2}\left[\left(e^{i\phi} - 1\right)|0\rangle + i\left(e^{i\phi} + 1\right)|1\rangle\right], \tag{21}$$

allowing construction of a family of phase qubits.

Rather than spatial or phase qubits, we may consider superposition states of some other degree of polarization. A common choice is the polarization qubit, such as

$$|\psi\rangle = a_0|\uparrow\rangle + a_1|\rightarrow\rangle, \tag{22}$$

or in the diagonal basis:

$$|\psi\rangle = a_0'|\nearrow\rangle + a_1'|\searrow\rangle. \tag{23}$$

The next subsection shows one means of creating entangled polarization qubits. A further type of optical qubit, formed from superpositions of orbital angular momentum states is considered in Sec. 3.

1.4. *Spontaneous parametric down conversion*

The most reliable and versatile means of producing entangled photon pairs is via *spontaneous parametric down conversion* (SPDC) inside a nonlinear crystal, such as β-barium borate (BBO) or potassium titanyl phosphate (KTP). In this process, a high frequency incoming photon (the *pump*) is converted into a pair of lower frequency outgoing photons (known for historical reasons as the *signal* and *idler* photons). Although the signal and idler beams are individually spatially and temporally incoherent, the signal and idler are mutually coherent, in the sense that the two photons in a given pair always leave the interaction point with a stable relation between their phases. The resulting photons are entangled in a number of different variables: position, momentum, frequency, time, polarization, and orbital angular momentum. In fact, the eigenstates of these multiple variables for the two photons are intertwined through entanglement; for example, the joint signal polarization-idler momentum states are entangled, a phenomenon which is known as *hyperentanglement*.[8-10] Note that the output is entangled in both continuous *and* discrete degrees of freedom. In later sections, we will use the entangled state produced in down conversion for several communication and cryptography applications; this state has found a number of other uses in diverse areas such as dispersion and aberration cancelation, quantum optical coherence tomography, and precision measurement of polarization mode dispersion.[11-19]

When an electric field is applied to a material with a nonlinear response, the polarization may be expanded in powers of the field. Here we concentrate on the second-order term, $\hat{P}_i^{(2)} = \chi_{ijk}^{(2)} \hat{E}_j \hat{E}_k$, where the indices label spatial components and repeated indices are summed over. The corresponding interaction Hamiltonian is

$$\hat{H}_{\text{int}}(t) = \epsilon_0 \int d^3r \hat{P}^{(2)} \cdot \hat{E} = \epsilon_0 \int d^3r \chi_{jkl}^{(2)} \hat{E}_{\text{p}j} \hat{E}_{\text{s}k} \hat{E}_{\text{i}l}. \tag{24}$$

The labels p, s, i have been added to distinguish the pump, signal and idler fields. We may expand each field in terms of plane wave components,

$$\hat{E}_j(\boldsymbol{r}, t) = \int d^3k \left[\hat{E}^-(\boldsymbol{k}) e^{-i(\omega t - \boldsymbol{k} \cdot \boldsymbol{r})} + \hat{E}^+(\boldsymbol{k}) e^{i(\omega t - \boldsymbol{k} \cdot \boldsymbol{r})} \right], \tag{25}$$

where, for quantization volume V, the positive and negative frequency parts are given by

$$\hat{E}_j^{(-)}(\boldsymbol{k}) = i\sqrt{\frac{2\pi\hbar\omega}{V}} \hat{a}_j^\dagger(\boldsymbol{k}), \qquad \hat{E}_j^{(+)}(\boldsymbol{k}) = -i\sqrt{\frac{2\pi\hbar\omega}{V}} \hat{a}_j(\boldsymbol{k}). \tag{26}$$

Substituting Eqs. (25)–(26) into Eq. (24) and keeping only the terms that give a nonzero result when wedged between a one-photon incoming state and two-photon outgoing state, the result is:

$$\hat{H}_{\text{int}}(t) = C \int d^3k_{\text{s}} d^3k_{\text{i}} e^{i(\omega_{\text{p}} - \omega_{\text{s}} - \omega_{\text{i}})t} \int_0^L dz e^{i(k_{\text{p}z} - k_{\text{s}z} - k_{\text{i}z})z} \tag{27}$$

$$\times \int_A d^2r_\perp e^{i(k_{\text{s}\perp} + k_{\text{i}\perp})r_\perp} \hat{a}^\dagger(\mathbf{k}_{\text{s}}) \hat{a}^\dagger(\mathbf{k}_{\text{i}}) + h.c.$$

Here, we have assumed that the incoming intensity is high enough to treat the pump as a classical field, and we have swept all of the overall constants into a single constant, C. It has also been assumed that the pump is a plane wave aligned along the z-axis, with no transverse momentum. In addition, the $\sqrt{\omega}$ terms coming from Eq. (26) are very slowly varying compared to the exponentials, and so were treated as constants. L is the length of the crystal in the z direction, and A is the area of the interaction region, i.e. the region of the crystal where the pump is intense enough for significant downconversion to take place. Since the interaction area A is normally much larger than the wavelength, we may approximate by taking $A \to \infty$, making the transverse integral trivial:

$$\int_A d^2r_\perp e^{i(\mathbf{k}_{s\perp} + \mathbf{k}_{i\perp}) \cdot \mathbf{r}_\perp} = 2\pi\delta^{(2)}(\mathbf{k}_{s\perp} + \mathbf{k}_{i\perp}). \tag{28}$$

Defining the longitudinal momentum mismatch, $\Delta k = k_{\text{p}z} - \mathbf{k}_{\text{s}z} - \mathbf{k}_{\text{i}z}$, the longitudinal integration may also be carried out:

$$\Phi(\Delta k\, L) \equiv \int_0^L dz\, e^{i\Delta k\, z} \equiv 2e^{i\Delta k\, L} \frac{\sin \frac{\Delta k\, L}{2}}{\Delta k\, L} = e^{i\Delta k\, L} \text{sinc}\left(\frac{\Delta k\, L}{2}\right), \tag{29}$$

where the sinc function is defined by $\text{sinc}(x) = \frac{\sin x}{x}$. In the limit of a long crystal, $L \to \infty$, this phase-matching function becomes a delta function for the longitudinal momenta: $\lim_{L \to \infty} \Phi(\Delta k\, L) = \pi\delta(\Delta k)$.

The result, finally, is that the relevant part of the interaction Hamiltonian may be written as

$$H_{\text{int}}(t) = C' \int d^3k_{\text{s}}\, d^3k_{\text{i}}\, \Phi(\Delta k\, L) e^{i(\omega_{\text{p}} - \omega_{\text{s}} - \omega_{\text{i}})t} \delta^{(2)}(\mathbf{k}_{\text{s}\perp} + \mathbf{k}_{\text{i}\perp}) \hat{a}^\dagger(\mathbf{k}_{\text{s}}) \hat{a}^\dagger(\mathbf{k}_{\text{i}}) + h.c. \tag{30}$$

This, of course, must be supplemented by the appropriate dispersion relations connecting the frequencies to the wavevectors in the birefringent crystal. The resulting *phase matching conditions* (equivalent to energy-momentum conservation) that must be satisfied by the outgoing fields are thus dependent on the polarizations of the photons. The down conversion is called *Type I* if the signal and idler have the same polarization (opposite to the pump), and *Type II* if the signal and idler have opposite polarizations to each other. Henceforth, we assume Type II parametric down conversion, e $\to$ {e, o}, with o being the idler and the pump and signal both being e-polarized. (o and e denote ordinary and extraordinary polarizations.)

For a weak interaction Hamiltonian $\hat{H}_{\text{int}}$ which is only nonzero for times in the interval $-T < t < T$, perturbation theory tells us that $\hat{H}_{\text{int}}$ will transform an initial vacuum state (before the interaction) $|vac\rangle$ into a new state $|\Psi\rangle$ afterwards:

$$|\Psi\rangle = \int_{-T}^{T} dt \, e^{-\frac{i}{\hbar}\hat{H}_{int}t}|vac\rangle = \left(1 - \frac{i}{\hbar}\int_{-T}^{T} dt \, H_{int} + \ldots\right)|vac\rangle. \quad (31)$$

Taking $T \to \infty$, the time integration becomes $\int_{-\infty}^{\infty} dt \, e^{i(\omega_p - \omega_s - \omega_i)t} = 2\pi\delta(\omega_p - \omega_s - \omega_i)$. Using the Hamiltonian of Eq. (30), we have the biphoton state:

$$|\Psi\rangle = -\frac{iC'}{\hbar}\int d^3k_e \, d^3k_o \, \delta(\omega_e + \omega_o - \omega_p) \quad (32)$$

$$\times \, \delta^{(3)}(\boldsymbol{k}_{s\perp} + \boldsymbol{k}_{i\perp})\Phi(\Delta k \, L)\hat{a}_H^\dagger(\boldsymbol{k}_e)\hat{a}_V^\dagger(\boldsymbol{k}_o)|vac\rangle.$$

Using the dispersion relations the k integrations may be rewritten as frequency integrations, so:

$$|\Psi\rangle = \int d\omega_e \, d\omega_o e^{-i\Delta kL/2} E(\omega_e + \omega_o)\Phi(\omega_e, \omega_o)|\omega_e\rangle_H|\omega_o\rangle_V, \quad (33)$$

where we have generalized the situation to include a non-plane-wave pump with envelope $E(\omega_p) = E(\omega_e + \omega_o)$. The momentum mismatch is now written in terms of frequency:

$$\Delta k = \frac{1}{c}\left[n(\omega_p)\omega_p - n(\omega_e)\omega_e - n(\omega_o)\omega_o\right]. \quad (34)$$

Due to the nonfactorability of $\Phi(\omega_e, \omega_o)$ into a product of terms each involving only one of the frequencies, the state of Eq. (33) is clearly entangled in terms of the various frequency states. It is also entangled in polarization; in particular, if the frequencies are held fixed (by means of filters, for example), we have $|\Psi\rangle = \frac{1}{2}\left[|H\rangle_s|V\rangle_i + |V\rangle_s|H\rangle_i\right]$. The latter is a realization of the well-known Bell state $|\psi^+\rangle$, i.e. a maximally entangled bipartite state.

We now look at several ways of quantifying the entanglement of the biphoton state.

1.5. *Concurrence in down conversion*

If the state of a system is known, then one readily computable measure of entanglement is the concurrence. Given a two-qubit pure state $|\Psi\rangle$ on Hilbert space $\mathcal{H}_A \otimes \mathcal{H}_B$, define the *spin-flipped state* $|\tilde{\Psi}\rangle = \sigma_2^{(A)} \otimes \sigma_2^{(B)}|\Psi\rangle$. More generally, the spin-flipped state corresponding to two-qubit state $\hat{\rho}$ is $\tilde{\rho} = \sigma_2^{(A)} \otimes \sigma_2^{(B)}\hat{\rho}\sigma_2^{(A)} \otimes \sigma_2^{(B)}$. Let $\lambda_1 \geq \lambda_2 \geq \lambda_3 \geq \lambda_4$ denote the eigenvalues of the density operator $\sqrt{\tilde{\rho}\rho}$, in descending order. Then the *concurrence* of the bipartite system is defined to be

$$C = max\left\{0, \lambda_1 - \lambda_2 - \lambda_3 - \lambda_4\right\}. \quad (35)$$

For a pure state, this reduces to an inner product: $C = \langle\Psi|\tilde{\Psi}\rangle$. (A more general definition applying to bipartite systems of arbitrary dimension can be given.[5])

The concurrence in the frequency spectrum of Type II SPDC has been calculated by Grice and Walmsley[20] in approximate form and more exactly by Erenso.[21] Here we follow the latter.

If the spectral bandwidth of downconversion is relatively small, the phase matching function $\Phi(\omega_1, \omega_2)$ is approximately symmetric in the frequencies. However, due to the birefringence of the crystal, the function necessarily shows noticeable asymmetry at larger bandwidths. As a result, we write the Type II two-photon down conversion state as

$$|\Psi\rangle = C \int d\omega_e d\omega_o E(\omega_e + \omega_o) \left(\Phi(\omega_e, \omega_o) |\omega_e\rangle_H |\omega_o\rangle_V + \Phi(\omega_o, \omega_e) |\omega_o\rangle_H |\omega_e\rangle_V \right), \quad (36)$$

where

$$\Phi(\omega_e, \omega_o) = \frac{\sin\left([k_o(\omega_o) + k_e(\omega_e) - k_p(\omega_o + \omega_e)] L\right)}{[k_o(\omega_o) + k_e(\omega_e) - k_p(\omega_o + \omega_e)] L}, \quad (37)$$

and we take the spectral envelope function of the pump to be

$$E(\omega_e + \omega_o) = \exp\left\{ -\frac{[2\omega_0 - (\omega_o + \omega_e)]^2}{s\sigma_p^2} \right\}. \quad (38)$$

Here, $2\omega_0$ and σ_p are the central frequency and bandwidth of the pump. Expanding k_e and k_o about ω_0, and k_p about $2\omega_0$, we find that

$$\Phi(\omega_1, \omega_2) = \frac{\sin(\tau_1 \nu_1 + \tau_2 \nu_2)}{[\tau_1 \nu_1 + \tau_2 \nu_2] L}. \quad (39)$$

In the latter expression, $\nu_j = \omega_j - \omega_0$ (for $j = 1, 2$) are the frequency detunings of the two photons, while $\tau_j = \left(\frac{\partial k_p}{\partial \omega} |_{\omega=2\omega_0} - \frac{\partial k_j}{\partial \omega} |_{\omega=\omega_0} \right) L$ are the differences in time delay of the pump photon relative to photon j during transit through the crystal. So the state is

$$|\Psi\rangle = \mathcal{N} \int d\nu_e d\nu_o \left\{ f(\nu_e, \nu_o) |\omega_e\rangle_H |\omega_o\rangle_V + f(\nu_o, \nu_e) |\omega_o\rangle_H |\omega_e\rangle_V \right\}, \quad (40)$$

where

$$f(\nu_i, \nu_j) = \mathcal{N} e^{-\frac{(\nu_i + \nu_j)^2}{2\sigma_p^2}} \frac{\sin\left([\tau_1 \nu_1 + \tau_2 \nu_2]\right)}{[\tau_1 \nu_1 + \tau_2 \nu_2] L}. \quad (41)$$

Computing the density operator and its eigenvalues, the concurrence is then given by

$$C = \mathcal{N}^2 \int d\nu_e d\nu_o \, e^{-\frac{(\nu_e + \nu_o)^2}{2\sigma_p^2}} \frac{\sin\left([\tau_1 \nu_e + \tau_2 \nu_o]\right)}{[\tau_1 \nu_e + \tau_2 \nu_o] L} \frac{\sin\left([\tau_1 \nu_o + \tau_2 \nu_e]\right)}{[\tau_1 \nu_o + \tau_2 \nu_e] L}. \quad (42)$$

This expression may be readily plotted as a function of transit times for a given pump beam and crystal. Examples of such plots were constructed by Erenso,[21] in which it can be seen that when the transit time of the pump beam through the crystal is small compared to that of the signal and idler, the concurrence is close to one. As the pump transit time decreases relative to the others, the concurrence decays. Thus, one means to control the degree of spatial entanglement is to alter the frequency and polarization dependence of the index of refraction, thus altering the transit times.

1.6. *Schmidt number and von Neumann entropy*

One of the most useful tools in quantum information theory is the *Schmidt decomposition*.[22] In Schmidt form, a bipartite state vector is "diagonal", in the sense that the basis vectors of the first and second Hilbert spaces are matched up in one-to-one fashion,

$$|\Psi\rangle = \sum_{i=1}^{d_{min}} \sqrt{\lambda_i}|u_i\rangle|v_i\rangle, \tag{43}$$

where d_{min} is the dimension of the smaller of the two Hilbert spaces. λ_i is the ith eigenvalue of the density matrix, and so gives the probability of measuring the ith term in the expansion, $p_i = \lambda_i$. The quantum correlations present in entangled systems are now manifest in this form: whenever the first system is measured to be in state $|u_i\rangle$, the second system is guaranteed to be in state $|v_i\rangle$. The number of nonzero terms in the expansion is known as the *Schmidt number*, K, and serves as a simple measure of entanglement: $K = 1$ for an unentangled product state, and increasing with increasing number of entangled states in the sum. Interpreting λ_k as the probability of the kth state, the *average* probability per state in the sum is $\sum_k p(k)\lambda_k = \sum_k \lambda_k^2$, so the average effective number of nonzero components in the decomposition is $1/\sum \lambda_i^2$. Thus, if the Schmidt decomposition is known, then the Schmidt number can be computed from the coefficients:[23]

$$K = 1/\sum \lambda_i^2. \tag{44}$$

Being essentially a count of available states, the Schmidt number is bounded above by the number of states that can fit into the phase space volume accessible to the system. So K is finite, even for continuous degrees of freedom, as long as the available phase space volume is finite.

Once the system is put into Schmidt form, the *von Neumann entropy* can then be computed:

$$S(\hat{\rho}) = -\mathrm{tr}\,\hat{\rho}\log_2\hat{\rho} = -\sum_i \lambda_i \log_2 \lambda_i. \tag{45}$$

The von Neumann entropy is a measure of the mixedness of a state: $S(\hat{\rho}) = 0$ for a pure state $\hat{\rho} = |\psi\rangle\langle\psi|$ and attains a maximum of $\log_2 d$ for the maximally mixed state $\hat{\rho} = \frac{1}{d}\hat{I}$. The von Neumann entropy is the quantum analog of the Shannon entropy to be discussed in the next section, and is essentially a measure of the information gained by measurement of the state. An explicit recipe can be constructed for putting a state into Schmidt form. Consider some pure state

$|\Psi\rangle = \sum_{ij} C_{ij} |u'_i\rangle |v'_j\rangle$, so that the density operator is of the form

$$\hat{\rho} = \sum_{ijkl} C_{ij} C^*_{kl} |u'_i\rangle\langle u'_k| \otimes |v'_j\rangle\langle v'_l|. \tag{46}$$

We first rotate from the $|u'_i\rangle$ basis to the basis $|u_i\rangle$ in which $\hat{\rho}_u = \text{tr}_v \, \hat{\rho}$ is diagonal. $\sum_{ij} C_{ij} C^*_{kl} = \delta_{ik} |g_i|^2$ in this basis, for some constants g_i, and $\hat{\rho}_u = \sum_i |g_i|^2 |u_i\rangle\langle u_i|$. For each nonzero g_i, we also define a new basis for the second Hilbert space, $|v_i\rangle = \sum_j \frac{C_{ij}}{|g_i|} |v'_j\rangle$. We then find that

$$\hat{\rho} = \sum_{ik} |g_i| \, |g_k| \, |u_i\rangle\langle u_k| \otimes |v_i\rangle\langle v_k| = |\tilde{\Psi}\rangle\langle\tilde{\Psi}|, \tag{47}$$

with corresponding state vector $|\tilde{\Psi}\rangle = |g_i| \, |u_i\rangle |v_i\rangle$, which is of Schmidt form. Therefore, $\sqrt{\lambda_i} = |g_i|$.

The two-photon state in Type II SPDC can be written

$$|\psi\rangle = \int \Phi(\omega_1, \omega_2) a^\dagger_H(\omega_1) a^\dagger_V(\omega_2) |0\rangle_H |0\rangle_V \, d\omega_1 d\omega_2, \tag{48}$$

and the spectral amplitude $\mathcal{A}(\omega_1, \omega_2)$ then decomposed into Schmidt form:

$$\mathcal{A}(\omega_1, \omega_2) = \sum_n \sqrt{\lambda_n} \psi_n(\omega_1) \phi_n(\omega_2), \tag{49}$$

where the eigenvalues and eigenfunctions λ_n, ψ_n, and ϕ_n are solutions to the integral equations

$$\int K_1(\omega, \omega') \psi_n(\omega') d\omega' = \lambda_n \psi_n(\omega) \tag{50}$$

$$\int K_2(\omega, \omega') \phi_n(\omega') d\omega' = \lambda_n \psi_n(\omega). \tag{51}$$

The integral kernels in these equations are given by

$$K_1(\omega, \omega') = \int \mathcal{A}(\omega, \omega_2) \mathcal{A}^*(\omega', \omega_2) d\omega_2 \tag{52}$$

$$K_2(\omega, \omega') = \int \mathcal{A}(\omega_1, \omega) \mathcal{A}^*(\omega_1, \omega') d\omega_1. \tag{53}$$

The eigenfunctions ψ_n and ϕ_n can be used to define a new set of effective creation operators for horizontally and vertically polarized photons,

$$\hat{b}^\dagger_n = \int \psi_n(\omega_1) \hat{a}^\dagger_H(\omega_1) d\omega_1, \qquad \hat{c}^\dagger_n = \int \phi_n(\omega_2) \hat{a}^\dagger_V(\omega_2) d\omega_2. \tag{54}$$

In terms of these, we can rewrite the Schmidt decomposition of the biphoton state as

$$|\psi\rangle = \sum_n \sqrt{\lambda_n} \hat{b}^\dagger_n \hat{c}^\dagger_n |vac\rangle_H |vac\rangle_V. \tag{55}$$

For SPDC, we may split the amplitude into a pump envelope and a phase-matching function Φ: $\mathcal{A}(\omega_1, \omega_2) = \tilde{E}(\omega_1 + \omega_2)\Phi(\omega_1, \omega_2)$. Law, Walmsley, and Eberly[24] have calculated the eigenvalues for this case and found that the sizes of the terms in the sums of Eqs. (49) and (55) drop rapidly, leaving only a small number of eigenvalues of non-negligible size. As a result, the *effective Schmidt number K* of the spectrally-entangled system is in fact relatively small. In fact, for the parameter values they used, the authors found that 96% of the state could be accounted for by the first six eigenvalues. The von Neumann entropy computed from these first six eigenvalues gives a value $S = 1.4$, compared to the large K limit of 1.8. By narrowing the bandwidth, correlations between the spectral components increases; as a result, the von Neumann entropy and the effective Schmidt number both increase. Bandwidth therefore determines the level of entanglement present in the current situation.

Rather than frequency entanglement, we can take a similar approach to quantify the entanglement in some other degree of freedom, for example the spatial entanglement carried by the momentum vectors. This was investigated by Law and Eberley,[25] as follows. Let k and q be the transverse spatial momenta of the two photons. (Transverse here means perpendicular to the direction of the pump beam, taken to be along the z-axis.) As a simple model of downconversion that allows analytic calculation of the Schmidt number, take the biphoton amplitude in transverse momentum space to be of Gaussian form,

$$\mathcal{A}(k, q) = E(k + q)\Phi(k - q) = C_g e^{-\frac{|k+q|^2}{\sigma^2}} e^{-b^2|k-q|^2}, \tag{56}$$

where the two terms represent the pump envelope and the phase matching function in momentum space. For this form, the Schmidt number can be found exactly:[25] $K = \frac{1}{4}\left(b\sigma + \frac{1}{b\sigma}\right)^2$. The degree of entanglement thus depends only on $b\sigma$, the ratio of widths of the two exponentials. K increases whenever $b\sigma \gg 1$ or $b\sigma \ll 1$, with a minimum at $b\sigma = 1$.

The Gaussian form given above is unrealistic; a more realistic approximation for the amplitude is given by replacing the second exponential (the phase-matching term) by a sinc function, as we have seen in Sec. 1.4:

$$\mathcal{A}(k, q) = E_{\text{p}}(k + q)\Phi(k - q) = C_g e^{-\frac{|k+q|^2}{\sigma^2}} sinc\left(b^2|k - q|^2\right), \tag{57}$$

where $b^2 = L/4k_{pump}$. The Schmidt number now has to be calculated numerically, but the result is qualitatively similar to the Gaussian model, with K becoming large whenever $b\sigma$ is either much larger or much smaller than 1.[25] Thus, spatial entanglement can be increased by, for example, increasing the transverse momentum spread. For some parameter ranges, the effective number of states K can be in the hundreds, but not all of these states are necessarily accessible; we will return to this issue in Sec. 3.

The analysis of Law and Eberly[25] has been generalized by van Exter *et al.*[26] Among other things, these authors showed that a one-dimensional Schmidt number K_{1d} can be calculated for photon pairs confined to propagate in a single plane, and that the full two-dimensional Schmidt number is simply $K_{2d} = K_{1d}^2$. They also added in the effect of a finite-sized detection aperture (diameter a), showing that in this case

$$K_{2d} = \frac{(1/\sigma^2 + b^2 + a^2)^2}{(1/\sigma^2 + b^2 + a^2)^2 - (\frac{1}{\sigma^2} - b^2)^2}. \tag{58}$$

This decreases asymptotically to $K_{2d} = 1$ as $a \to \infty$, demonstrating the role of spatial filtering by the detector and reminding us that the degree of entanglement, as well as the information content, will be dependent on our measuring devices and is not entirely intrinsic to the system being measured.

How is the Schmidt number measured experimentally? It can be shown[27] that for transverse spatial modes in the quasi-homogeneous approximation the Schmidt number can be written in a form analogous the étendue[28] of an optical system:

$$K = \frac{1}{\lambda^2} \frac{\left[\int I_S(\boldsymbol{x}) d\boldsymbol{x}\right]^2}{\int I_S^2(\boldsymbol{x}) d\boldsymbol{x}} \times \frac{\left[\int I_{FF}(\boldsymbol{\theta}) d\boldsymbol{\theta}\right]^2}{\int I_{FF}^2(\boldsymbol{\theta}) d\boldsymbol{\theta}}, \tag{59}$$

where I_S and I_{FF} are the near-field (source) and far-field intensities. Thus intensity measurements in two planes suffice to determine the Schmidt number.

The Schmidt number for the output of SPDC depends strongly on the properties of both the pump beam and the crystal. For some parameter ranges, it can be extremely large; for example, in the experiment of Dixon *et al.*,[29] the number of product states superposed in the outgoing spatially-entangled biphoton state was $K \sim 1400$! In contrast, we have seen that for the parameter values considered by Law, Walmsley, and Eberly,[24] the effective number of polarization-entangled terms was very small, on the order of $K \approx 2$. This is one of the reasons that down conversion is such an important source for optical experiments: by appropriately tuning the input parameters or measuring different variables we can exert a great deal of control over the output state and can vary its properties over a very wide range.

1.7. *Other measures of entanglement*

Many other measures of entanglement have been defined. (See Plenio and Virmani[30] for comprehensive reviews.) Here, we briefly mention a couple of these.

For a bipartite system on a Hilbert space $\mathcal{H}_A \times \mathcal{H}_B$, the *partial transpose* operations T_A or T_B consist of taking the transpose of the part of an operator's action only on one of the two subsystems. Thus, for example, the partial transpose of a density operator relative to the A subsystem is defined by $\langle i_A j_B | \hat{\rho}^{T_A} | k_A l_B \rangle = \langle k_A j_B | \hat{\rho} | i_A l_B \rangle$. According to the *Peres-Horodečki criterion*,[31,32] a system is entangled if (either) partial transpose of the density matrix is negative. This can be associated with a numerical measure by defining the *negativity*:

$$\mathcal{N}(\hat{\rho}) = \frac{1}{2} \left(||\hat{\rho}^{T_A}||_1 - 1 \right), \tag{60}$$

where $||\hat{\mathcal{O}}||_1 \equiv \mathrm{tr}\sqrt{\mathcal{O}^\dagger \mathcal{O}}$ is the trace-norm of Hermitian operator $\mathcal{O}$. This may also be expressed as $\mathcal{N}(\hat{\rho}) = |\sum_i \lambda_i|$, where λ_i represent the *negative* eigenvalues of $\hat{\rho}^{T_A}$. The negativity is bounded by the concurrence, $\mathcal{N}(\hat{\rho}) \le c(\hat{\rho})$

A further fundamental entanglement measure that can be related to the concurrence is the entanglement of formation: $E_f(\hat{\rho}) = h(c(\hat{\rho}))$, where $h(x) = -x \log_2 x - (1 - x) \log_2(1 - x)$; see[5] for more information.

2. Communication and cryptography

2.1. *Information and channel capacity*

In the 1940's, the major problems in telecommunications included the questions of how to quantify the amount of information being carried on a communication channel, how to determine what the maximum information a given channel could carry, and how to understand the effect of noise on information capacity. These questions were largely answered by Shannon and his contemporaries. Here we briefly discuss these questions and their generalizations to quantum theory. Then in the following subsections, we look at some communication phenomena that exist only in the quantum case.

The most basic quantity in classical information theory is the Shannon entropy. Given random variable X, we will denote the possible values that it can take by $x_1, x_2, x_2, \ldots$; x will be used to denote a generic value. These values occur according to some probability distribution $p(X)$. We will restrict ourselves here to discrete distributions for simplicity. Then the *Shannon entropy* associated with variable X is

$$H(X) = -\sum_i p(x_i) \log_2 p(x_i) = -E[\log_2(X)], \tag{61}$$

where E denotes expectation value or mean.

The significance of $H(X)$ is that it tells you the "surprise value" or average amount of new knowledge you gain from a measurement of X. For example, consider a variable X which can take on two values x_1 and x_2. If $p(x_1) = 1$ and $p(x_2) = 0$ (a state of maximal a priori knowledge), then the Shannon information vanishes; this is as expected from the fact that we know X will always take the value x_1, so a measurement tells us nothing new. In contrast, if $p(x_1) = p(x_2) = \frac{1}{2}$ (the state of maximal a priori uncertainty) the entropy reaches its maximum value ($H(X) = \log_2 2 = 1$), since in this case we learn the most from each measurement.

The entropy depends only on the probability distribution associated with the random variable, $H(X) = H(p(X))$, is concave, and is nonnegative: $H(X) \geq 0$ for all X, with equality if and only if only a single value of X has nonzero probability. Conceptually, the Shannon entropy is a measure of how much redundancy is occurring in a message, or equivalently how much the message can be compressed. This is the content of the Shannon noiseless coding theorem: a message of length n can be coded by a string of only nH bits, as $n \to \infty$.

Similarly, the *Shannon noisy coding theorem* tells how much additional redundancy must be encoded into a message transmitted over a noisy channel in order to allow for error correction. For the simplest case, a binary symmetric channel with error probability q per bit, the theorem says that each binary digit may carry no more than $1 - h(q)$ bits of information, where $h(q) = q \log_2(q) + (1-q) \log_2(1-q)$ is the entropy of the error probability distribution. $h(q)$ serves as a measure of the amount of redundancy that must be built into a message to enable error correction.

The *von Neumann entropy* was introduced in the last section, and can be viewed as the quantum analog of the Shannon entropy for a quantum state $\hat{\rho}$:

$$S(\hat{\rho}) = -\text{tr}\hat{\rho}\log_2\hat{\rho} = -\sum_i \lambda_i \log_2 \lambda_i, \tag{62}$$

where the λ_i are the Schmidt coefficients. $S(\hat{\rho})$ is a measure of the mixedness of the state: for a system of dimension n, the von Neumann entropy is bounded by $0 \leq S(\hat{\rho}) \leq \log_2 n$, with the lower limit reached by pure states and the upper limit achieved for the maximally mixed states $\hat{\rho} = \frac{1}{n}\hat{I}$.

For a statistical mixture of states $\hat{\rho} = \sum_i p_i\hat{\rho}_i$ it can be shown that

$$\sum_i p_i S(\hat{\rho}_i) \leq S(\hat{\rho}) \leq H(p_1, \ldots, p_n) + \sum_i p_i S(\hat{\rho}_i). \tag{63}$$

The left-hand inequality simply expresses the concavity of the von Neumann entropy; as for the right hand inequality, the first term on the right describes the classical uncertainty due to the statistical mixture of the states, while the second term describes the uncertainty inherent in the quantum states themselves. If the $\hat{\rho}_i$ are pure states, the latter terms vanish, so that $S(\hat{\rho}) \leq H(p_1, \ldots, p_n)$; thus the quantum uncertainty is less than the uncertainty of the corresponding classical system. This is a reflection of the fact that quantum systems can contain correlations stronger than are possible classically, as reflected by the well-known Bell inequalities.[33,34]

The von Neumann entropy to a large extent plays a role in quantum systems similar to that of the Shannon entropy in classical systems. For example, there is a theorem (the Schumacher theorem[1]) for quantum systems analogous to that of the Shannon noiseless coding theorem, with S replacing H.

Rather than investigating the formal properties of entropy and information in detail, we move on in the next section to discuss attempts to communicate secretly by means of encryption keys shared between two parties. We will see that the laws of quantum mechanics will prevents an eavesdropper from gaining information about the key without causing disturbances that can be detected by the communicating parties.

2.2. *Quantum key distribution*

The goal is to generate a secret key for encrypting and decrypting messages that is shared between two legitimate users, usually known as Alice and Bob, and which cannot be broken by an eavesdropper, usually called Eve. The only truly unbreakable code is the *one-time pad* or *Vernam cipher* in which the secret key k is a random string of binary digits which is used only once, and then discarded. If the text to be encoded is given as a binary string m, then the encoded message is given by $m \oplus k$, where $\oplus$ is base-two addition. To decode the message, Bob simply adds the same key to the encoded message: since $k \oplus k = 0$, it follows that $m \oplus k \oplus k = m$. The randomness of the key means that there are no patterns that can be used to break the code: the key has the maximum possible entropy and carries no information. However, if the same key is used multiple times, detectable patterns in the messages themselves will cause correlations in the sum $m \oplus k$, which in principle can leak information about the messages. Therefore, it is essential that each key not be reused. Although the key itself is unbreakable, there is still the problem of distribution: Alice and Bob must use the *same* key, so Eve may be able to intercept the passing of the key from one to the other, destroying the security of the message.

The Vernam cipher solves the problem of encrypting a message in an unbreakable manner, once the participants share a random key. However, this does not solve the problem of *distributing* the key among the legitimate users without it being intercepted. Classically there is no foolproof means for completely secure key distribution; this is where quantum mechanics becomes essential. In *quantum cryptography* or *quantum key distribution* (QKD), the goal is to generate a one-time encryption key and to share it between the two legitimate users, Alice and Bob, while using the laws of quantum mechanics to prevent illegitimate eavesdroppers from obtaining the key undetected. We will see that, although the eavesdropping itself is not preventable, it will always be possible to detect it if it is occurring, so that it will be ineffective; if Eve is detected, Alice and Bob know their communication line has been compromised, so they must stop using it and seek another communication channel.

What makes QKD possible is the existence of noncommuting operators in quantum mechanics. Suppose we have two Hermitian operators $\hat{\mathcal{O}}$ and $\hat{\mathcal{O}}'$ which fail to commute: $\left[\hat{\mathcal{O}}, \hat{\mathcal{O}}'\right] \neq 0$. We assume that either (i) Alice prepares a state, makes a measurement on it and sends it to Bob, or else (ii) a third party sends an entangled pair of states (half of the pair to Alice, the other to Bob), in which the values of the relevant operators are either correlated or anticorrelated between the two states. Alice chooses randomly to measure the value of either $\hat{\mathcal{O}}$ or $\hat{\mathcal{O}}'$ on the state, obtaining some value o_A which is an eigenvalue of whichever operator was used. This determines the value o_B Bob will measure *if* he measures the same operator. However, if he measures the other operator, the value he obtains is random (indeterminate), due to the fact that $\hat{\mathcal{O}}$ and $\hat{\mathcal{O}}'$ are incompatible observables. The operators should be chosen so that application of one operator makes all possible eigenvalues of the other equally likely; such operators are called *mutually unbiased, conjugate,* or *incompatible*. The communication procedure then consists schematically of the following steps: (i) Alice generates a sequence of states. (ii) For each state, she randomly chooses $\hat{\mathcal{O}}$ or $\hat{\mathcal{O}}'$ and makes a measurement. (iii) Bob then randomly chooses $\hat{\mathcal{O}}$ or $\hat{\mathcal{O}}'$ for each state and also makes a measurement. (iv) Alice and Bob then communicate over a classical communication channel. This channel can be completely public. They tell each other which measurement operator they chose for each state, but *not* the result of the measurement. (v) They keep only those values for which they made the *same* choice, discarding the rest. This process is called *sifting*. (vi) They randomly select a subset of the sifted trials to subject to a security check. They compare the values obtained on these trials, and check to see if the these values have the correlation (or anticorrelation) expected. Unexpected drops in correlation signal the activity of an eavesdropper. (vii) If the security trials have the expected level of correlation, then they can be certain that no eavesdropping occurred. They can therefore use the values they measured on the remaining trials (after sifting and security trials) as the digits of the one-time key. Although they have not told each other their values, the fact that they measured the same operator on these correlated or anticorrelated states guarantees that each can deduce the other's value from their own.

This procedure is safe, because if Eve is intercepting the states and making her own measurements, she has no way of knowing whether Alice chose to measure $\hat{\mathcal{O}}$ or $\hat{\mathcal{O}}'$ on each trial. She has to guess, and has only a 50% probability per trial of guessing correctly. Suppose on a given trial Alice measures $\hat{\mathcal{O}}$. Then if Eve also measures $\hat{\mathcal{O}}$ she will measure the correct value o_A and can generate a copy of the state to send to Bob. Bob (if he also measures $\hat{\mathcal{O}}$) will also determine the value o_A, and so the tampering will not be detected. But when Eve chooses to measure the wrong operator $\hat{\mathcal{O}}'$, she will sometimes (let us say with probability p, where $p < 1$) measure the correct value o_A, but will also sometimes measure an incorrect value o_A' with probability $1 - p$. When this happens, it will show up during the security check: instead of Alice and Bob agreeing 100% of the time when they used the same operator, they will find that they now only agree on a fraction $\frac{p+1}{2}$ of the trials. This drop in correlation between their values immediately signals the presence of an eavesdropper. (The strategy used here by Eve is called the *intercept-resend strategy*. It is the simplest type of eavesdropping attack. More sophisticated attacks are also possible, which may require additional safeguards.[35–39])

To make the protocol as safe as possible, we want the overlap between each eigenstate of one operator with all of the eigenstates of the other to be as uniform as possible, i.e. we want mutually unbiased operators. In the most common case, the different operators represent projections onto polarization states measured along the axes of different bases; for example, $\hat{\mathcal{O}}_j = |\psi_j\rangle\langle\psi_j|$ and $\hat{\mathcal{O}}_k' = |\psi_k'\rangle\langle\psi_k'|$. In that case, we specify the different operators by specifying the bases, and the eigenvectors represent the basis vectors. For the case of two incompatible bases, the best possible choice is $\langle\psi_i|\psi_j'\rangle = \frac{1}{\sqrt{2}}$ for all $i \in \{1,2\}$ and $j \in \{1,2\}$ (so that $p = 1 - p = \frac{1}{2}$). Here, $|\psi_i\rangle$ and $|\psi_j'\rangle$ are respectively the eigenvectors of $\hat{\mathcal{O}}$ and $\hat{\mathcal{O}}'$. More generally, we may use m incompatible operators $\hat{\mathcal{O}}_1, \ldots, \hat{\mathcal{O}}_m$, such that

$$\langle\psi_i^{(\mu)}|\psi_j^{(\nu)\prime}\rangle = \frac{1}{\sqrt{m}}, \tag{64}$$

for all $i, j \in \{1, 2\}$ and all $\mu, \nu \in \{1, 2, \ldots m\}$. (The superscripts μ, ν label the operator, while subscripts i, j label the states within the set of eigenstates of each operator.) Bases satisfying the conditions Eq. (64) are called *mutually unbiased* or *conjugate* bases. Mutually unbiased bases are such that a measurement in one basis gives no information about the value in the other basis: a measurement in one basis completely randomizes values in the other, with a uniform probability distribution. (For a review of mutually unbiased bases and their construction, Durt *et al.*[40])

To make this more concrete, let us consider the first successful QKD method, invented by Bennet and Brassard[41] and known as the BB84 protocol. Here, we take the states to be polarization states of a photon, and the operator $\hat{O}$ to be the polarization operator in a coordinate system defined by a pair of perpendicular axis, the horizontal (H) and vertical (V) axes. We take $o_A = 0$ if the polarization is horizontal and $o_A = 1$ if it is vertical, with corresponding eigenvectors $|0\rangle = |H\rangle$ and $|1\rangle = |V\rangle$. The second operator $\hat{O}'$, incompatible with $\hat{O}$, is the polarization operator in a system defined by two axes ($| \nearrow \rangle$ and $| \searrow \rangle$) at $\pm 45°$ to the horizontal. We will denote the eigenvectors $|0'\rangle = | \nearrow \rangle$ and $|1'\rangle = | \searrow \rangle$, and the eigenvalues $o'_A = \{0, 1\}$. These two bases are clearly mutually unbiased.

To generate a secure key, Alice randomly selects one of the two bases for each photon and makes a measurement of the polarization in that basis. She then sends the photon on to Bob, who similarly makes a random choice among the two bases and measures polarization. If they both chose the same basis, then they should always measure the same value for polarization. However, if they make different choices, then (due to the incompatibility of the bases), the result of Bob's measurement should be completely random and independent of the basis. This is the key to the security. Alice and Bob select a random subset $\mathcal{S}$ of photons to use for a security check, and tell each other (over a classical and potentially public channel) both their basis choices and the results of their measurements. The trials on which they used different bases are discarded. For the rest, they compare their measurements. Assuming ideal conditions (negligible noise, perfect detectors, etc.), their measurements should match 100% of the time if there is no eavesdropping, but only $(\frac{100-\eta}{2})\%$ of the time if Eve has intercepted and resent a fraction η of the photons. The presence of eavesdropping is therefore immediately detectable, unless the eavesdropping rate η is so small that Eve cannot obtain significant information anyway. If no eavesdropping has been detected, then for the remaining photons (those not in $\mathcal{S}$) classical information is again exchanged between Alice and Bob, but only concerning the choice of bases, *not* the actual polarization values in those bases. The photons for which the choices disagreed are again discarded. For the remainder, the polarizations are guaranteed to match; these polarizations then form a random sequence which is then used as the key.

A common variation on the BB84 idea is the E91 protocol. Here, rather than Alice sending a photon to Bob, a third party sends out a pair of *entangled* photons, one to Alice and one to Bob. Usually these photons were produced in Type II SPDC, so their polarizations are perfectly anticorrelated. Now Alice and Bob proceed as before, choosing bases, discarding trials on which the choices differ, checking for security by comparing measurement results on $\mathcal{S}$, and using the random sequence determined by the remaining trials as a key. (In this variation, one possible means of verifying the absence of eavesdropping is to verify that there is no decrease in Bell inequality violations.[33,34])

Other protocols are possible as well, including one that only requires the use of two nonorthogonal states.[42] A slightly different approach is to use the visibility of interference patterns instead of correlations between polarizations.[43] (For an interference pattern that oscillates between intensity values I_{min} and I_{max}, the visibility is defined by $V = (I_{max} - I_{min})/(I_{max} + I_{min}.)$ The interference used here is not the familiar interference between *amplitudes* that occurs, for example, in the Young two-slit experiment; rather it is interference between *intensities*, involving the fourth order correlation function between the two fields or second order correlation between intensities. Two independent detectors measure intensities at different output ports of the interferometer, then each detector feeds its signal into a computer which measures the correlation function between the signals. At low intensities, when only a single photon at a time is likely to be striking the detectors, this becomes a coincidence counting setup, in which an event is registered only when both detectors see a photon simultaneously (i.e. within a very short coincidence time window). Such coincidence counting or intensity correlation experiments are common for investigating entanglement effects in quantum optics. Quantum correlations can lead to very high visibility in these experiments, close to 100%, whereas the presence of background terms reduce the maximum visibility to 70.7% in the classical case. This classical visibility limit is directly analogous to (and stems from the same source as) the Bell inequality for correlations. When an eavesdropper interferes with the photon traveling to Bob, it is detectable by a sudden drop in the visibility of the interference pattern. For more details on this approach, see Sergienko *et al.*[43]

2.3. *Quantum ghost imaging and secure image distribution*

In order to prepare for applications in the next section, we now discuss that idea of forming images through spatial correlations between pairs of photons. This is two-photon imaging process is known as *ghost imaging* or *correlated imaging*. The spatial correlations involved may be either classical correlations or quantum mechanical correlations due to entanglement. Although ghost imaging has been achieved using classically-correlated light sources,[44–50] we focus here on the original version of ghost imaging (*quantum ghost imaging*),[51,52] which relies on pairs of entangled photons produced by SPDC. The essential idea, shown schematically in Fig. 2(a) is that one photon encounters the object to be imaged, then passes on to a single-pixel detector, known as a *bucket detector*, D_A, which has no spatial resolution. This detector only registers the presence or absence of a photon, recording no information about the spatial location or momentum. The other photon does not encounter the object at all, but proceeds directly to a second, spatially-resolving detector, D_B. Clearly, neither detector by itself is capable of imaging the object: one detector gains no spatial information, the other detector only sees photons that never interact with the object. But when the detectors are connected to a coincidence circuit, the image reappears in the coincidence rate between the detectors (i.e in the intensity correlations). The photon interrogating the object essentially acts as a gate, which opens the detection window for the second photon only when the first photon has not been blocked by the object; the second photon then provides the spatial information needed to reconstruct the image.

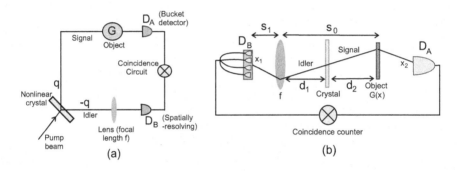

Fig. 2. (a) Quantum ghost imaging with entangled photons. (b) Klyshko backward-wave picture, in which the signal and idler are treated as a single ray passing through the crystal from one detector to the other.

The signal is transmitted from the object, then detected by bucket detector D_A. D_A should be large enough to collect all of the signal photons arriving at the right end of the apparatus. It only registers whether the photons passed through the object or were blocked. Detector D_B on the other hand has high spatial resolution: it can be a CCD camera, an array of avalanche photodiodes, or a single small detector scanned over the imaging region. The lens in branch 2 has focal length f. Let d_1 and d_2 be the distances from the source to object and source to lens, and let $s_0 = d_1 + d_2$. s_1 is the distance from the lens to detector D_B. The distances s_0 and s_1 satisfy the imaging condition $\frac{1}{s_0} + \frac{1}{s_1} = \frac{1}{f}$ When the information from the two detectors is combined via coincidence counting, the image reappears if the coincidence rate is plotted versus position in D_B. The imaging process is therefore highly nonlocal; in fact, the original motivation of this line of inquiry was to investigate the nonlocal causal structure of quantum mechanics and the Einstein-Podolsky-Rosen (EPR) "paradox". Although it is now clear that only classical correlation between the spatial degrees of freedom is required for the information between the two photons to be correctly integrated, we will look below at a variation in which true entanglement is needed.

The imaging property of the apparatus is more clearly shown by displaying a schematic version drawn in the Klyshko "backward wave" picture[53,54] (Fig. 2(b)). Here, we view the signal and idler as a single photon passing through the crystal. The signal is viewed as traveling backward from the object, into the crystal, where it converts into the forward-moving idler, then travels onward to the detector D_B. The detector D_A acts like the source in this view. Alternatively, we could fold the picture over, so the signal appears to reflect off the crystal in order to form the idler. In this latter version, the crystal acts as a mirror, and the pump determines the properties of that mirror. We will assume that the pump beam is approximately a plane wave, so the crystal acts as a planar mirror.

The ghost imaging apparatus has improved resolution compared to the image formed in ordinary imaging with a comparable single lens, and in fact beats the usual diffraction limit by a factor of 2. Effectively, the resolution is determined by the shorter pump frequency, rather than the longer signal or idler frequencies of the detected photons. This fact has formed the basis for the process of *quantum lithography*,[55,56] in which two-photon imaging is used to write subdiffraction-sized structures onto a semiconductor surface. The idea has been extended to N-photon imaging with $N > 2$, although the prospects for this to become practical seem limited, due to the difficulty of producing sufficiently entangled states of more than two photons.

One question that could be asked at this point is whether we can use quantum mechanics to securely transmit images from Alice to Bob. Of course, the answer is obviously yes, since we can encode an image digitally and then encode with a quantum key, as in Sec. 2.2. But can we use some variation on ghost imaging to accomplish secure quantum image distribution in an analog manner, without digitizing? Suppose that the object whose image is to be transmitted is in Alice's lab, along with the bucket detector, D_A. The spatially-resolving detector D_B is in Bob's lab, and Alice wishes to send the object's image to him, while keeping it safe from eavesdroppers. Note that if Alice and Bob have detectors with sufficient time resolution and which have been well-synchronized (taking the transit time from A to B into account), then in place of using a coincidence counter they can simply compare the times at which they have detected photons and discard those times at which they did not make simultaneous detections. So Alice may send a list of her detection times (via a classical channel) to Bob, who then compares it with his list of detection times, thus determining the coincidence times. Since Bob also knows the spatial locations (the specific pixels) of each detection event, he can now reconstruct the image. To anyone eavesdropping on the classical channel, the list of random detection times is meaningless, unless they have also intercepted the quantum channel (Bob's photon), which contains the spatial information. To prevent this, Alice and Bob can use the same means as in the E91 protocol: they place polarizers, randomly switching between two bases, in front of the detectors. Alice then sends the choice of polarization basis along with the detection times.

They keep only events on which they chose the same basis. By comparing the polarizations measurements on a random subset, eavesdropping may be detected, exactly as before.

The procedure is essentially a three-dimensional version of the E91 protocol, where the two transverse spatial dimensions plus time replace the two-dimensional qubit space of the conventional E91 case. This indicates that it might be advantageous to investigate more generally what happens when we replace our two-dimensional qubits with quantum degrees of freedom belonging to higher-dimensional Hilbert spaces. This will be the topic of the next section.

3. Qudits and imaging

The generalization from qubits built from a two-dimensional effective Hilbert space spanned by states $|0\rangle$ and $|1\rangle$ to a d-dimensional *qudit* on a space spanned by states $|0\rangle, \ldots, |d-1\rangle$ is obvious:

$$|\psi\rangle = a_0|0\rangle + a_1|1\rangle \cdots + a_{d-1}|d-1\rangle, \tag{65}$$

with $\sum_{i=0}^{d-1} |a_i|^2 = 1$. These are known as *qutrits* for $d = 3$ and *ququats* for $d = 4$.

Since we will again be looking at QKD, we need to find sets of mutually unbiased (or mutually complementary) bases for these states. Let m be the number of bases we seek, so we seek sets of basis vectors $|\psi_i^{(\mu)}\rangle$, where $\mu = 1, \ldots, m$ labels the basis, while $i = 0, 1, \ldots, d-1$ labels the vector in that basis. We then require orthonormality,

$$|\langle \psi_i^{(\mu)}|\psi_j^{(\mu)}\rangle| = \delta_{ij}, \text{ for all } \mu, i, j \tag{66}$$

and mutual complementarity (mutually unbiasedness),

$$|\langle \psi_i^{(\mu)}|\psi_j^{(\nu)}\rangle| = \frac{1}{\sqrt{d}}, \text{ for all } i, j, \mu \neq \nu. \tag{67}$$

Wootters and Fields[57] showed that there exist $m = d + 1$ mutually unbiased bases whenever $d = p^k$, with p prime and k non-negative integer.

For $m = 2$ bases and d dimensions, we can take one basis arbitrarily, $|\psi_0^{(1)}\rangle, |\psi_1^{(1)}\rangle, \ldots, |\psi_{d-1}^{(1)}\rangle$, then construct the second basis according to

$$|\psi_k^{(2)}\rangle = \frac{1}{\sqrt{d}} \sum_{n=0}^{d-1} e^{2\pi i k n/d}|\psi_n^{(1)}\rangle \tag{68}$$

It has been shown that higher values of d and m can lead to both higher capacity and improvements in security against eavesdropping.[58–62] In addition, they maintain their security in the face of greater amounts of noise. Consider a pure state $|\psi\rangle$, and add some some admixture of noise F ($0 \leq F \leq 1$) by defining the density operator $\hat{\rho} = (1 - F)|\psi\rangle\langle\psi| + F\hat{\rho}_{noise}$, where $\hat{\rho}_{noise} = \frac{1}{9}I$ is the density matrix for a completely chaotic system. Einstein's conditions on a physical theory, represented in the EPR assumptions, have come to be known in the physics literature as *local realism.* These preconditions have turned out to be too strong but do not preclude either locality or realism.[5,63,64] For F too large, apparent Bell inequality violations can be due to noise-induced errors, and so the security of quantum cryptography breaks down. The value F for which this occurs is $\frac{2-\sqrt{2}}{2} = .293$ for maximally entangled qubits ($d = 2$); in contrast, this value increases to $\frac{11-6\sqrt{3}}{2} = .304$[65] for maximally entangled qutrits ($d = 3$), and to .309 for $d = 4$.[66] Thus, QKD can be carried out in the presence of larger amounts of noise as the dimension of the Hilbert space increases.

A number of realizations of qudits have been carried out experimentally, including polarization entangled four-photon states,[67] time-energy entangled qutrits using single photon in a three-arm interferometer,[68] and time-bin-entangled photons is produced by a train of laser pulses.[69] Here, however, we will concentrate one specific realization, optical orbital angular momentum, which we introduce in the next subsection.

3.1. *Orbital angular momentum*

In addition to the intrinsic or spin angular momentum that leads to the existence of polarization states, it is somewhat less well known that photons can also carry orbital angular momentum (OAM). This OAM is due to the possibility of the photon state having nontrivial spatial structure. It wasn't until the 1990's that a thorough investigation of optical OAM began to be carried out and that a simple way was found to produce it controllably. After the seminal paper of Allen *et al.*,[70] a flood of papers began which continues to grow today. A number of excellent reviews of the subject exist.[71–73]

The key observation is that if an approximate plane wave is given an azimuthally-dependent phase shift of the form $e^{il\phi}$, where ϕ is the angle about the propagation axis, z, the resulting wave has angular momentum about the z-axis given by $L_z = l\hbar$. (Note that single-valuedness of the field forces the *topological charge l* to be quantized to integer values.) This phase factor has the effect of tilting the wavefronts by an increasing amount as the axis is circumnavigated, so that the wavefronts have a corkscrew shape. The Poynting vector $\boldsymbol{S}$ must be perpendicular to the wavefront, so it is at an angle to the propagation axis. $\boldsymbol{S}$ therefore rotates about the axis as the wave propagates, leading to the existence of nonzero orbital angular momentum.

A number of different beam modes can carry OAM, including higher-order Bessel or Hermite-Gauss modes. Here, we focus on Laguerre-Gauss (LG) modes. The LG wavefunction with OAM $l\hbar$ and with p radial nodes is[74]

$$u_{lp}(r, z, \phi) = \frac{C_{\mathrm{p}}^{|l|}}{w(z)} \left(\frac{\sqrt{2}r}{w(z)} \right)^{|l|} e^{-r^2/w^2(r)} L_{\mathrm{p}}^{|l|} \left(\frac{2r^2}{w^2(r)} \right)$$

$$\times e^{-ikr^2 z/\left(2(z^2+z_R^2)\right)} e^{-i\phi l + i(2p+|l|+1)\arctan(z/z_R)}, \tag{69}$$

with normalization $C_p^{|l|} = \sqrt{\frac{2p!}{\pi(p+|l|)!}}$ and beam radius $w(z) = w_0\sqrt{1 + \frac{z}{z_R}}$ at z. $z_r = \frac{\pi w_0^2}{\lambda}$ is the Rayleigh range and the arctangent term is the Gouy phase.

There are a number of ways to generate optical OAM states, the most common being the use of spiral phase plates (plates whose optical thickness varies azimuthally according to $\frac{l\phi}{k(n-1)}$[75]), computer generated holograms of forked diffraction gratings,[76] which convert Guassian modes into OAM modes in first-order diffraction, or spatial light modulators (SLM).

3.2. *Entangled OAM pairs*

The SPDC-generated biphoton state is most often written as an expansion in the space of transverse linear momenta of the outgoing signal and idler, as was done in Sec. 1. Now, though, we instead wish to expand in the space of orbital angular momentum. Consider a pump beam of spatial profile $E(\boldsymbol{r}) = u_{l_0 p_0}(\boldsymbol{r})$ encountering a χ^2 nonlinear crystal, producing two outgoing beams via SPDC. For fixed beam waist, the range of OAM values produced by the crystal is roughly inversely proportional to the square root of the crystal thickness L.[77] We wish a broad OAM bandwidth, so we assume a thin crystal located at the beam waist ($z = 0$). The output is an entangled state,[78] with a superposition of terms of form $u_{l_1',p_1'} u_{l_2',p_2'}$, angular momentum conservation requiring $l_0 = l_1' + l_2'$. We will take the pump to have $l_0 = 0$, so that the OAM values just after the crystal are equal and opposite: $l_1' = -l_2' \equiv l$. The p_1', p_2' values are unconstrained, although the amplitudes drop rapidly with increasing p' values (see Eq. (82) below). The output of the crystal may be expanded as a superposition of signal and idler LG states:

$$|\Psi\rangle = \sum_{l_1',l_2'=-\infty}^{\infty} \sum_{p_1',p_2'=0}^{\infty} C_{p_1'p_2'}^{l_1',l_2'} |l_1',p_1';l_2',p_2'\rangle \delta(l_0 - l_1' - l_2'), \qquad (70)$$

where the coupling coefficients are given by

$$C_{p_1'p_2'}^{l_1',l_2'} = \int d^2r \; \Phi(\boldsymbol{r}) \left[u_{l_1'p_1'}(\boldsymbol{r})u_{l_2'p_2'}(\boldsymbol{r})\right]^*. \qquad (71)$$

Explicit expressions for the $C_{p_1'p_2'}^{l_1',l_2'}$ coefficients have been calculated Torres, Alexandrescu, and Torner.[77]

How entangled are the angular momenta of the beams? One way to answer this is to again compute the Schmidt number. Taking $p_1 = p_2 = 0$ for simplicity, Eq. (70) reduces to

$$|\Psi\rangle = \sum_{l=0}^{\infty} C_{00}^{l,-l} |l,0\rangle |-l,0\rangle \equiv \sum_{l=0}^{\infty} \sqrt{\lambda_l}|l\rangle|-l\rangle, \qquad (72)$$

from this, $\lambda_l = \left(C_{00}^{l,-l}\right)^2$ can be calculated explicitly.[77] The state is already in Schmidt form. From the λ_l, the Schmidt number and von Neumann entropy can then be found.

Salakhutdinov *et al.*[79] examine the Schmidt number for parametric down conversion in detail. Looking at the case of a Gaussian pump ($l = p = 0$) and vanishing radial quantum numbers for both signal and idler ($p_s = p_i = 0$), they found that a pump beam of waist $w = 325 \ \mu m$ and wavelength $\lambda_p = 413 \ nm$ on a crystal of thickness $L = 2 \ mm$, the total Schmidt number was $K \approx 350$. However, those associated with entangled azimuthal degrees of freedom (OAM) only accounted for roughly $K_{az} \approx 2\sqrt{K} \approx 37$ of them. A more detailed analysis found that pure radially entangled modes (p_s and p_i values entangled) account for a further $K_r \approx \sqrt{K} \approx 18$ modes. The remainder are modes of radial-azimuthal cross-correlation, with p and l values jointly entangled.

3.3. *Quantum cryptography with OAM*

The first successful demonstration of QKD with orbital angular momentum was achieved by Gröblacher *et al.*,[62] using qutrits formed by superpositions of $l = 0, \pm 1$ states. A Gaussian beam ($l = p = 0$) was used to pump a nonlinear crystal. Parametric downconversion then produced photon pairs with opposite momenta $\pm l$. Only pairs with $l = 0, 1$ were used. A transmission hologram was placed in each outgoing beam. When the beam strikes on-axis, the hologram changes l by one unit, so for example $l_{initial} = 0$ changes to $l_{final} = 1$. However, when the beam strikes the hologram off-center, the result is a superposition of $l_{initial}$ and l_{final}. By changing the displacement of the hologram within the beam, we may control the relative weights in this superposition. By this means, it is possible to produce a maximally-entangled state $|\psi\rangle = \frac{1}{\sqrt{3}} (|00\rangle + |1, -1\rangle + |-1, 1\rangle)$ where the two numbers in each ket represent the OAM in arms A and B, respectively. In this manner, the successful construction of a quantum key shared between two experimenters was demonstrated.[62] 150 qutrits were sent in, and security was maintained by checking the parity of 3-qutrit blocks and discarding those with parity mismatch. The result was a final key of 72 qutrits, which was used to code and decode a 72-bit message without error. The Bell parameter (a measure of supposed violation of local realism) for this experiment was $S_3 = 2.688$, well above the classical upper limit of $S_3 = 2$.

3.4. *Digital spiral imaging*

Digital spiral imaging[80,81] is a form of angular momentum spectroscopy in which properties of an object are reconstructed based on how it alters the OAM spectrum of light used to illuminate it (Fig. 3). The input and output light may be expanded in LG functions, with the object acting by transforming the coefficients of the ingoing expansion into those of the outgoing expansion. Information about the transmission profiles of both phase and amplitude objects may be retrieved.[77,80]

The idea naturally arises of trying to use the measured OAM spectrum to reconstruct an image of the object. But, since only intensities are measured, the lack

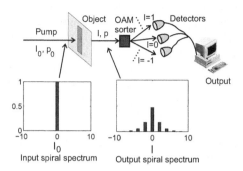

Fig. 3. Digital spiral imaging: the presence of an object in the light beam alters the distribution of angular momentum values in the outgoing light.

of phase information prevents this. One way to extract the necessary phase is to use pairs of beams, in some sort of interferometric arrangement. This leads naturally to the idea of using pairs of photons with entangled OAM states, so we next investigate OAM entanglement in SPDC.

3.5. *Joint OAM spectra*

We now investigate the use of two beams, rather than one, in combination with spiral imaging. The full benefits of doing this will emerge in Sec. 3.7. In the current section, we focus on examination of the OAM correlations. We begin with an entangled version, where the light source is parametric downconversion in a nonlinear crystal such as β-barium borate (BBO). Imagine an object in the signal beam (Fig. 4). Since OAM conservation holds exactly only in the paraxial case, we assume the signal and idler are produced in *collinear* downconversion, then directed into separate branches by a beam splitter. (We assume all beam splitters are 50-50.) Assume perfect detectors for simplicity (imperfect detectors can be accounted for by the method in[82]).

Let $P(l_1, p_1; l_2, p_2)$ be the joint probability for detecting the signal with quantum numbers l_1, p_1 and the idler with values l_2, p_2. The marginal probabilities at the two detectors (probabilities for detection of a single photon, rather than for coincidence detection) are

$$P_s(l_1, p_1) = \sum_{l_2, p_2} P(l_1, p_1; l_2, p_2), \qquad P_i(l_2, p_2) = \sum_{l_1, p_1} P(l_1, p_1; l_2, p_2). \qquad (73)$$

Then the mutual information for the pair is

$$I(s, i) = \sum_{l_1, l_2 = l_{min}}^{l_{max}} \sum_{p_1, p_2 = 0}^{p_{max}} P(l_1, p_1; l_2, p_2) \log_2 \left(\frac{P(l_1, p_1; l_2, p_2)}{P_s(l_1, p_1) P_i(l_2, p_2)} \right) \qquad (74)$$

The most common experimental cases are when (i) $p_{max} = \infty$ (p_1 and p_2 are not measured, so all possible values must be summed), or (ii) $p_{max} = 0$. Except when stated otherwise, we will use $l_{max} = -l_{min} = 10$ and $p_{max} = 0$.

Suppose the transmission profile for the object is $T(\boldsymbol{x})$, where $\boldsymbol{x}$ is position in the plane transverse to the beam axis. The goal is determine the function $T(\boldsymbol{x})$ from measurements of orbital angular momentum correlations *only*. The coincidence probabilities $P(l_1, p_1; l_2, p_2) = |A_{p_1 p_2}^{l_1 l_2}|^2$ have amplitudes

$$A_{p_1 p_2}^{l_1 l_2} = C_0 \sum_{p_1'} C_{p_1' p_2}^{-l_2, l_2} a_{p_1' p_1}^{-l_2, l_1}(z), \tag{75}$$

$$a_{p_1' p_1}^{l_1' l_1}(z) = \int u_{l_1' p_1'}(\boldsymbol{x}, z) \left[u_{l_1 p_1}(\boldsymbol{x}, z)\right]^* T(\boldsymbol{x}) d^2 x \tag{76}$$

where C_0 is a normalization constant. Here it is assumed that the total distance in each branch is $2z$ (see Fig. 4). We define an operator $\hat{T}$ to represent the effect of the object on the beam. We may expand this operator in the position basis,

$$\hat{T} = \int d^2r \, d^2r' |\boldsymbol{r}'\rangle T(\boldsymbol{r}, \boldsymbol{r}') \langle \boldsymbol{r}| = \int d^2r \, |\boldsymbol{r}\rangle T(\boldsymbol{r}) \langle \boldsymbol{r}|, \tag{77}$$

where the last line assumes that the operator is local, i.e. diagonal in the position space basis. So the function $T(\boldsymbol{r})$ is then given by

$$T(\boldsymbol{r}) = \langle \boldsymbol{r}|\hat{T}|\boldsymbol{r}\rangle. \tag{78}$$

Alternately, the object operator may be expanded in the Laguerre-Gauss basis,

$$\hat{T} = \sum_{ll'} \sum_{pp'} d_{p'p}^{l'l} |l'p'\rangle \langle lp|. \tag{79}$$

Making use of these definitions and of Eq. (76), it follows immediately that

$$d_{p_1', p_1}^{l_1', l_1} = \langle l_1' p_1'|\hat{T}|l_1 p_1\rangle = a_{p_1, p_1'}^{l_1, l_1'}. \tag{80}$$

Using this result in Eq. (79), then applying Eq. (78) and the fact that $u_{lp}(r) = \langle r|lp\rangle$, we find that determination of the $a_{p_1', p_1}^{l_1', l_1}$ coefficients is equivalent to reconstructing the object, since

$$T(\boldsymbol{r}) = \langle \boldsymbol{r}|\hat{T}|\boldsymbol{r}\rangle = \sum_{ll'} \sum_{pp'} a_{p_1', p_1}^{l_1', l_1} u_{l_1 p_1}(\boldsymbol{r}) \left[u_{l_1' p_1'}(\boldsymbol{r})\right]^* \tag{81}$$

That the object's size and shape affect the coincidence rate is easy to see. For example, Fig. 5 shows the calculated spectrum when a single opaque strip of width d is placed in the beam. We see a clear effect from changing an object parameter (the strip width). Similarly, if the corresponding mutual information is calculated it is found to vary with width, exihibiting a minimum at $d = w_0$.

The central peak of the spectrum (Fig. 5) broadens as d increases from zero, reducing the correlation between l_1 and l_2; the mutual information between them thus declines over the range $d/w_0 < 1$. But at $d/w_0 \approx 1$, the central peak in $\{l_1, l_2\}$ space bifurcates into two narrower peaks (right side of Fig. 5); the information thus goes back up as the peaks separate in the region $d/w_0 > 1$, after passing through the minimum at $d = w_0$. If we continue to sufficiently large d, the two peaks once

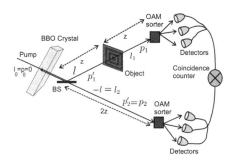

Fig. 4. Setup for analyzing object via orbital angular momentum of entangled photon pairs.

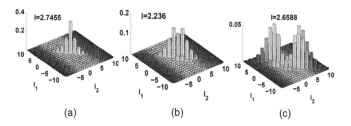

Fig. 5. An opaque strip of width d placed in the signal path. The widths are (a) $d = .1w_0$, (b) $d = .9w_0$, (c) $d = 2.5w_0$. The outgoing joint angular momentum spectra are plotted. As the width increases, the peak in the spectrum broadens, then (at $d = w_0$) splits into two peaks.

again broaden and the mutual information decays gradually to zero. In addition, the total intensity getting past the opaque strip will continue to drop, so coincidence counts decay rapidly.

3.6. Mutual information and symmetry

Figure 6 shows the computed mutual information for several simple shapes. It can be seen that I depends strongly on the size and shape of the object, so that for object identification from among a small set a comparison of the I values rather than of the full probability distribution may suffice.

If the object has rotational symmetry about the pump axis, then its transmission function $T(r)$ depends only on radial distance r, not on azimuthal angle ϕ. The angular integral in Eq. (76) is then $\int_0^{2\pi} e^{-i\phi(l-l')} d\phi = 2\pi\delta_{l,l'}$. So the joint probabilities reduce to the form $P(l_1, l_2) = f(l_1)\delta_{l_1,l_2}$ (assuming $p_1 = p_2 = 0$) for some function f. The marginal probabilities for each arm reduce to $P_1(l_1) = f(l_1)$ and $P_2(l_2) = f(l_2)$. The mutual information $I(L_1, L_2) = S_1(L_1)$ where $S_1(L_1) = -\sum_{l_1} f(l_1) \ln f(l_1)$ is the Shannon information of the object arm OAM spectrum. Thus in the case of rotational symmetry, the second arm becomes irrelevant from an information standpoint. In this sense, the quantity $\mu(L_1, L_2) \equiv |I(L_1, L_2) - S_1(L_1)|$ is an order parameter, capable of detecting breaking of rotational symmetry.

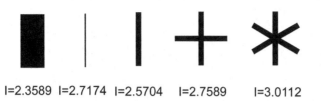

$l=2.3589 \quad l=2.7174 \quad l=2.5704 \quad l=2.7589 \quad\quad l=3.0112$

Fig. 6. The mutual information depends strongly on size and shape of the object. Here, the two objects on the left have widths $1.5w_0$ and $.2w_0$; all other widths are $.4w_0$.

More generally, suppose that the object has a rotational symmetry group of order N; i.e., it is invariant under $\phi \to \phi + \frac{2\pi}{N}$. From Eqs. (69) and (76) it follows that the coefficients must then satisfy $a^{l'_1 l_1}_{p'_1 p_1} = e^{\frac{2\pi i}{N}(l'_1 - l_1)} a^{l'_1 l_1}_{p'_1 p_1}$, which implies $a^{l'_1 l_1}_{p'_1 p_1} = 0$ except when $\frac{l'_1 - l_1}{N}$ is integer. When N goes up (enlarged symmetry group), the number of nonzero $a^{l'_1 l_1}_{p'_1 p_1}$ goes down; with the probability concentrated in a smaller number of configurations, correlations increase and mutual information goes up. This may be seen in the three right-most objects of Fig. 6, for example.

The first experimental use of this correlated spiral spectrum method has recently been carried out for the purpose of object identification.[83] Figure 7 shows a simple example: if the coincidence rate is plotted versus the OAM of two entangled photons, normally angular momentum conservation forces the coincidence rate to vanish off the diagonal. However, as seen in the plots when objects with four-fold and six-fold rotational symmetry are placed in one beam off-diagonal terms appear, shifted respectively by three or four units from the diagonal. This allows the symmetry structure of the object to be easily determined, opening up possible applications such as rapid recognition of defective items on an assembly lines or irregular and diseased cells in a tissue sample.

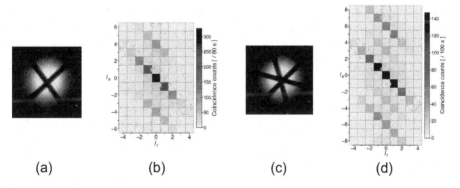

Fig. 7. An object with four-fold rotational symmetry (a) is placed in one output beam of a down conversion crystal. The joint OAM spectrum (b) of the two outgoing beams shows, in addition to the OAM-conserving main diagonal, a pair of secondary bands displaced from the diagonal by four units. Similarly a six-fold symmetric object (c) has joint OAM spectrum with bands displaced by six units (d).

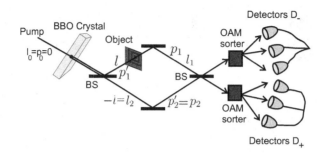

Fig. 8.　A configuration allowing image reconstruction via phase-sensitive measurement of entangled OAM content.

3.7. *Imaging with entangled OAM*

The inability of digital spiral imaging to produce images due to loss of phase information has been pointed out. But a variation on the entangled OAM setup can be used to find the expansion coefficients *including phase*.

Assume that the beam waist for the OAM expansion (which is determined by the size and location of the detector aperture) is equal to the pump waist. Then, for the case of a Gaussian pump ($l_0 = p_0 = 0$) the expansion coefficients of Subsec. 3.2 reduce to:[82]

$$C_{p_1,p_2}^{l,-l} = \sum_{m=0}^{p_1} \sum_{n=0}^{p_2} \frac{\left(\frac{2}{3}\right)^{m+n+l} (-1)^{m+n} \sqrt{p_1! p_2! (l+p_1)! (l+p_2)!} \, (l+m+n)!}{(p_1-m)!(p_2-n)!(l+m)!(l+n)! \, m! \, n!}. \tag{82}$$

Using the latter expression for the coefficients, it can be shown[84] that determining the coincidence amplitudes $A_{p_1 p_2}^{l_1 l_2}$ is sufficient to determine the $a_{p_1' 0}$ coefficients, including phase. The measurement of the $A_{p_1 p_2}^{l_1 l_2}$ is accomplished by inserting a beam splitter to mix the signal and idler beams before detection, as in Fig. 8, erasing information about which photon followed which path. We then count singles rates in the two detection stages, rather than the coincidence rate. If value l is detected at a given detector it could have arrive by two different paths, so interference occurs between these two possibilities. The detection amplitudes in the two sets of detectors D_+ and D_- involve factors $A_+ \sim \left(1 + i a_{00}^{l_0 - l_2, l_1}\right)$ and $A_- \sim \left(i + a_{00}^{l_0 - l_2, l_1}\right)$, with detection rates $R_\pm \sim 1 + |a_{00}^{l_0 - l_2, l_1}|^2 \pm 2i \, \text{Im} \, a_{00}^{l_0 - l_2, l_1}$. From these counting rates, both the amplitudes and the relative phases of all coefficients can be found, allowing full image reconstruction, by inversion of Eq. (76) to find the object transmission function $T(\boldsymbol{r})$.

3.8. *Pixel entanglement*

In the previous sections, we have discussed entanglement between orbital angular momenta. Another avenue for investigating the spatial entanglement in imaging situations is via *pixel entanglement*.[85] Here, a spatially-resolving detector is used in each branch of a correlated-imaging setup, and spatial correlations are found by measuring coincidence counts between corresponding pixels in the two detectors. If the detectors are in the imaging plane, the result is the position-space correlation function, while if the detectors are in the Fourier plane of the imaging system, then the momentum space correlations are measured instead. The mutual information carried by the spatial correlations has been studied[29] using a Gaussian model similar that used in Subsec. 1.6: in the notation of Dixon *et al.*,[29] the biphoton state is

$$|\psi\rangle = \int d^2x_a \, d^2x_b \, f(\boldsymbol{x}_a, \boldsymbol{x}_b) \hat{a}_a^\dagger \, \hat{a}_b^\dagger |vac\rangle = \int d^2k_a \, d^2k_b \, \tilde{f}(\boldsymbol{k}_a, \boldsymbol{k}_b) \, \hat{a}_a^\dagger \hat{a}_b^\dagger |vac\rangle, \quad (83)$$

where the position- and momentum-state amplitudes are

$$f(\boldsymbol{x}_a, \boldsymbol{x}_b) = \mathcal{N} e^{-\frac{|\boldsymbol{x}_a - \boldsymbol{x}_b|^2}{4\sigma_c^2}} e^{-\frac{|\boldsymbol{x}_a + \boldsymbol{x}_b|^2}{16\sigma_p^2}} \quad (84)$$

$$\tilde{f}(\boldsymbol{k}_a, \boldsymbol{k}_b) = (4\sigma_p\sigma_c)^2 \mathcal{N} e^{-\sigma_c^2 |\boldsymbol{k}_a - \boldsymbol{k}_b|^2} e^{-4\sigma_p^2 |\boldsymbol{k}_a + \boldsymbol{k}_b|^2}, \quad (85)$$

and $\mathcal{N} = \frac{1}{2\pi\sigma_c\sigma_p}$. Defining joint and marginal probability densities $p(\boldsymbol{x}_a, \boldsymbol{x}_b) = |f(\boldsymbol{x}_a, \boldsymbol{x}_b)|^2$, $p(\boldsymbol{x}_a) = \int p(\boldsymbol{x}_a, \boldsymbol{x}_b) d^2x_b$, and $p(\boldsymbol{x}_b) = \int p(\boldsymbol{x}_a, \boldsymbol{x}_b) d^2x_a$, the mutual information (the same for either the position or momentum basis) is

$$I(A:B) = -\int p(\boldsymbol{x}_a, \boldsymbol{x}_b) \log_2 \frac{p(\boldsymbol{x}_a, \boldsymbol{x}_b)}{p(\boldsymbol{x}_a) p(\boldsymbol{x}_b)} \, d^2x_a \, d^2x_b = \log_2 \left(\frac{4\sigma_p^2 + \sigma_c^2}{4\sigma_c\sigma_p} \right)^2. \quad (86)$$

For the limiting case $\frac{\sigma_p}{\sigma_c} \gg 1$, we find the simple result $I(A:B) = \log\left(\frac{\sigma_p}{\sigma_c}\right)^2$. For the values $\sigma_p = 1500 \ \mu m$ and $\sigma_c = 40 \ \mu m$, this predicts $I(A:B) \approx 10.5$ bits per photon; experimentally, the value was found to be $7.1 \pm .7$ in the position basis and $7.2 \pm .3$ in the momentum basis.[29] For these approximate Gaussian states, the Schmidt number is approximately $K \approx \left(\frac{\sigma_p}{\sigma_c}\right)^2 = 1400$, indicating a highly entangled state.

3.9. *Conclusions*

We have seen that the unique properties of quantum systems, superposition and entanglement in particular, allow for a number of interesting and useful phenomena in communication that are not possible in classical systems. Focusing on cryptography and imaging, we have seen that not only does the noncommutativity of quantum operators allow truly secret communication, but that many of the applications examined have involved extraction of multiple bits of information in a single photon. Thus, further developments along these lines offer the promise of someday being able to securely transmit enormous amounts of information with a handful of photons. It is impossible to say what further interesting phenomena may be uncovered in the course of these investigations.

References

1. B. W. Schumacher, Quantum coding, *Phys. Rev. A* **51**, 2738 (1995).
2. R. Omnès, General theory of the decoherence effect in quantum mechanics, *Phys. Rev. A* **56**, 3383 (1997).
3. H. D. Zeh, On the interpretation of measurement in quantum theory, *Found. Phys.* **1**, 69 (1970).
4. C. Caves, and C. A. Fuchs, Quantum information: How much information is in a state vector? In *Sixty years of EPR*, Ann. Phys. Soc., Israel, A. Mann and M. Revzen, Eds. (1996).
5. G. Jaeger, *Quantum Information - An Overview* (Springer, Berlin, 2007).
6. J. J. Sakurai, *Modern Quantum Mechanics, Revised edition* (Addison-Wesley, Reading, MA, 1994).
7. N. Peters, J. Altepeter, E. Jeffrey, D. Branning, and P. Kwiat, Precise creation, characterization and manipulation of single optical qubits, *Q. Info. Comp.* **3**, 503 (2003).
8. M. Atatüre, G. Di Giuseppe, M. D. Shaw, A. V. Sergienko, B. E. A. Saleh, M. C. Teich. Multiparameter entanglement in quantum interferometry, *Phys. Rev. A* **66**, 023822 (2002).
9. A. V. Sergienko, G. S.Jaeger, G. Giuseppe, B. E. A. Saleh, M. C. Teich, Quantum Metrology and Quantum Information Processing with Hyper-Entangled Quantum States. In *Quantum Communication and Information Technologies (NATO Science Series)*, A. S. Shumovsky, V. I. Rupasov, Eds., Springer, Dordrecht, The Netherlands (2003).
10. C. Bonato, D. Simon, P. Villoresi, and A. V. Sergienko, Multiparameter entangled-state engineering using adaptive optics, *Phys. Rev. A* **79**, 062304 (2009).
11. A. M. Steinberg, P. G. Kwiat, R. Y. Chiao, Dispersion cancellation in a measurement of the single-photon propagation velocity in glass, *Phys. Rev. Lett.* **68**, 2421 (1992).
12. J. D. Franson, Nonlocal cancellation of dispersion, *Phys. Rev. A* **45**, 3126 (1992).
13. A. Abouraddy, M. B. Nasr, B. E. A. Saleh, A. V. Sergienko, and M. C. Teich, Quantum-optical coherence tomography with dispersion cancellation, *Phys. Rev. A* **65**, 053817 (2002).
14. D. S. Simon and A. V. Sergienko, Spatial-Dispersion Cancellation in Quantum Interferometry, *Phys. Rev. A*, **80**, 053813 (2009).
15. O. Minaeva, C. Bonato, B. E. Saleh, D. S. Simon, A. V. Sergienko, Odd- and Even-Order Dispersion Cancellation in Quantum Interferometry, *Phys. Rev. Lett.* **102**, 100504 (2009).
16. D. S. Simon and A. V. Sergienko. Odd-Order Aberration Cancellation in Correlated-Photon Imaging, *Physical Review A* **82**, 023819 (2010).
17. D. S. Simon and A. V. Sergienko, Correlated-Photon Imaging with Cancellation of Object-Induced Aberration, *J. Optical Society of America B* **28**, 247 (2011).
18. A. Fraine, D. S. Simon, O. Minaeva, R. Egorov, and A. V. Sergienko, Evaluation of Polarization Mode Dispersion and Chromatic Dispersion using Quantum Interferometry, *Optics Express* **19**, 22820-22836 (2011).
19. A. M. Fraine, O. M. Minaeva, D. S. Simon, R. Egorov, A. V. Sergienko, Evaluation of Polarization Mode Dispersion in a Wavelength Selective Switch using Quantum Interferometry, *Optics Express* **20**, 2025 (2012).
20. W. P. Grice, I. A. and Walmsley, Spectral information and distinguishability in Type-II down-conversion with a broadband pump, *Physical Review A*, **56**, 1627 (1997).
21. D. Erenso, Pump Spectral Bandwidth, Birefringence, and Entanglement in Type-II Parametric Down Conversion, *Research Letters in Optics,* Article ID 387580 (2009).

22. A. Ekert, P. L. Knight, Entangled quantum systems and the Schmidt decomposition, *Am. J. Phys.*, **63**, 415 (1995).
23. B. Grobe, K. Rzazewski, J. H. Eberly, Measure of electron-electron correlation in atomic physics, *J. Phys. B: At. Mol. Opt. Phys.* **27**, L503, 1994).
24. C. K. Law, I. A. Walmsley, J. H. Eberly, Continuous Frequency Entanglement: Effective Finite Hilbert Space and Entropy Control, *Phys. Rev. Lett.* **84**, 5304 (2000).
25. I. A. Law, J. H. Eberly, Analysis and Interpretation of High Transverse Entanglement in Opticical Parametric Down Conversion, *Phys. Rev. Lett.* **92**, 127903 (2004).
26. M. P. van Exter, A. R. Aiello, S. S. Oemrawsingh, G. Nienhuis, J. P. Woerdman, Effect of spatial filtering on the Schmidt decomposition of entangled photons, *Phys. Rev. A* **74**, 012309 (2006).
27. H. Di Lorenso Pires, C. H. Monken, M. P. van Exter, Direct measurement of transverse-mode entanglement in two-photon states, *Phys. Rev. A* **80**, 022307 (2009).
28. J. Mertz, *Introduction to Optical Microscopy*, Robert and Co. Publishers (2010).
29. P. B. Dixon, G. A. Howland, J. Schneeloch, J. C. Howell, Quantum Mutual Information Capacity for High-Dimensional Entangled States, *Phys. Rev. Lett.* **108**, 143603 (2012).
30. M. B. Plenio, S. Virmani, An introduction to entanglement measures, *Quant. Inf. and Comp.* **7** 001 (2001).
31. A. Peres, Separability Criterion for Density Matrices, *Phys. Rev. Lett.* **77**, 1413 (1996).
32. M. Horodečki, P. Horodečki, R. Horodečki, Separability of mixed states: necessary and sufficient conditions, *Phys. Lett. A* **223**, 1 (1996).
33. J. S. Bell, On the problem of hidden variables in quantum mechanics. *Rev. Mod. Phys.* **38**, 447 (1966).
34. J. S. Bell, On the Einstein Podolsky Rosen Paradox. *Physics* **1**, 195 (1964).
35. G. Brassard, N. Lütkenhaus, T. Mor, and B. C. Sanders, Limitations on Practical Quantum Cryptography, *Phys. Rev. Lett.*, **85**, 1330 (2000).
36. A. Vakhitov, V. Makarov and D. R. Hjelme, Large pulse attack as a method of conventional optical eavesdropping in quantum cryptography, *J. Mod. Opt.* **48**, 2023 (2001).
37. V. Makarov and D. R. Hjelme, Faked states attack on quantum cryptosystems, *J. Mod. Opt.* **52**, 691. (2005).
38. C.-H. F. Fung, B. Qi, K. Tamaki, and H.-K. Lo. Phase-remapping attack in practical quantum-key-distribution systems, *Phys. Rev. A* **75**, 032314 (2007).
39. B. Qi, C.-H. F. Fung, H.-K. Lo, and X. Ma, Time-shift attack in practical quantum cryptosystems, *Quant. Info. Compu.* **7**, 73 (2007).
40. T. Durt, B. G. Englert, I. Bengtsson K. Zyczkowski, On mutually unbiased bases, *Int. J. Quant. Info.*, **8**, 535-640 (2010).
41. C. H. Bennett, G. Brassard, Quantum cryptography: Public key distribution and coin tossing. In *Proceedings of the IEEE International Conference on Computers, Systems, and Signal Processing*, 175, Bangalore (1984).
42. C. Bennett, Quantum cryptography using any two nonorthogonal states, *Phys. Rev. Lett.* **68**, 3121 (1992).
43. A. V. Sergienko, M. Atatüre, Z. Walton, G. Jaeger, B. E. A. Saleh, M. C. Teich, Quantum cryptography using femtosecond-pulsed parametric down-conversion, *Phys. Rev. A* **60**, R2622 (1999).
44. R. S. Bennink, J. Bentley, R. W. Boyd, J. C. Howell, Quantum and Classical Coincidence Imaging, *Phys. Rev. Lett.* **92**, 033601 (2004).
45. A. Gatti, E. Brambilla, M. Bache, L. A. Lugiato, Correlated imaging, quantum and classical, *Phys. Rev. A* **70**, 013802 (2004).

46. A. Gatti, E. Brambilla, M. Bache, L. A. Lugiato, Ghost Imaging with Thermal Light: Comparing Entanglement and Classical Correlation, *Phys. Rev. Lett.* **93**, 093602 (2004).

47. Y. J. Cai, S. Y. Zhu, Ghost interference with partially coherent radiation, *Opt. Lett.* **29**, 2716 (2004).

48. Y. J. Cai, S. Y. Zhu, Ghost imaging with incoherent and partially coherent light radiation, *Phys. Rev. E* **71**, 056607 (2005).

49. A. Valencia, G. Scarcelli, M. D'Angelo, Y. H. Shih, Two-Photon Imaging with Thermal Light, *Phys. Rev. Lett.* **94**, 063601 (2005).

50. F. Ferri, D. Magatti, A. Gatti, M. Bache, E. Brambilla, L. A. Lugiato, High-Resolution Ghost Image and Ghost Diffraction Experiments with Thermal Light, *Phys. Rev. Lett.* **94**, 183602 (2005).

51. A. V. Belinskii and D. N. Klyshko, Two-photon optics: diffractlon, holography, and transformation of two-dimensional signals, *Sov. Phys. JETP* **78**, 259 (1994).

52. T. B. Pittman, Y. H. Shih, D. V. Strekalov, A. V. Sergienko, Optical imaging by means of two-photon quantum entanglement, *Phys. Rev. A* **52**, R3429 (1995).

53. D. Klyshko, Transverse photon bunching and twophoton processes in the field of parametrically scattered light, *hurnal Eksperimental'noi i Teoreticheskoi Fiziki*, 83: 1313-1323 (1982) [*Soviet Journal of Experimental and Theoretical Physics* 56 753 (1982)].

54. D. Klyshko, Effect of focusing on photon correlation in parametric light scattering, *Zhurnal Eksperimental'noi i Teoreticheskoi Fiziki* **94**, 82 (1988).

55. A. Boto, P. Kok, D. Abrams, S. Braunstein, C. Williams, J. and Dowling, Quantum interferometric optical lithography: Exploiting entanglement to beat the diffraction limit, *Phys. Rev. Lett.* **85**, 2733 (2000).

56. P. Kok, A. Boto, D. Abrams, C. Williams, S. Braunstein, and J. Dowling, Quantum-interferometric optical lithography: Towards arbitrary two-dimensional patterns, *Phys. Rev. A* **63**, 063407 (2001).

57. W. K. Wootters, B. D. Fields, Optimal state-determination by mutually unbiased measurements, *Ann. Phys. (N.Y.)* **191**, 363 (1989).

58. D. Bruß, Optimal Eavesdropping in Quantum Cryptography with Six States, Quantum Cryptography with 3-State Systems, *Phys. Rev. Lett.* **81**, 3018 (1998).

59. H. Bechmann-Pasquinucci, A. Peres, Quantum Cryptography with 3-State Systems, *Phys. Rev. Lett.* **85**, 3313 (2000).

60. M. Bourennane, A. Karlsson, G. Björk, Quantum key distribution using multilevel encoding, *Phys. Rev. A* **64**, 012306 (2001).

61. N. J. Cerf, M. Bourennane, A. Karlsson, N. Gisin, Security of Quantum Key Distribution Using d -Level Systems, *Phys. Rev. Lett.* **88**, 127902 (2002).

62. S. Groblacher, T. Jennewein, A. Vaziri, G. Weihs, A. Zeilinger, Experimental quantum cryptography with qutrits, *New J. Phys.* **8**, 75 (2006).

63. Th. M. Nieuwenhuizen, Is the Contextuality Loophole Fatal for the Derivation of Bell Inequalities? *Found. Phys.* **41**, 580 (2010).

64. K. Hess, H. De Raedt and K. Michielsen, Hidden assumptions in the derivation of the theorem of Bell, *Phys. Scr.* **T151**, 014002 (2012).

65. J. L. Chen, D. Kaszlikowski, L. C. Kwek, C. H. Oh, M. Zukowski, Entangled three-state systems violate local realism more strongly than qubits: An analytical proof, *Phys. Rev. A* **64**, 052109 (2001).

66. M. Genovese, P Traina, Review on Qudits Production and Their Application to Quantum Communication and Studies on Local Realism, *Adv. Sci. Lett.* **1**, 153 (2008).

67. J. C. Howell, A. Lamas-Linares, D. Bouwmeester, Experimental Violation of a Spin-1

Bell Inequality Using Maximally Entangled Four-Photon States, *Phys. Rev. Lett.* **88**, 030401 (2002).

68. R. Thew, A. Acin, H. Zbinden, and N. Gisin, Experimental realization of entangled qutrits for quantum communication, *Quant. Inf. and Comp.* **4**, 93 (2004).
69. D. Stucki, H. Zbinden, N. Gisin, A FabryPerot-like two-photon interferometer for high-dimensional time-bin entanglement, *J. Mod. Opt.* **52**, 2637 (2005).
70. L. Allen, M. W. Beijersbergen, R. J. C. Spreeuw, and J. P. Woerdman, Orbital angular momentum of light and the transformation of Laguerre-Gaussian laser modes, *Phys. Rev. A*, **45**, 8185 (1992).
71. A. M. Yao and M. J. Padgett, Orbital angular momentum: origins, behavior and applications, *Adv. in Opt. and Phot.* **3**, 161 (2011).
72. J. P. Torres and L. Torner, Eds., *Twisted Photons: Applications of Light with Orbital Angular Momentum* (Wiley, Hoboken, 2011).
73. S. Franke-Arnold, L. Allen, M. Padgett, Advances in optical angular momentum, *Laser and Photonics Reviews* **2**, 299 (2008).
74. L. Allen, M. Padgett, M. Babiker, The Orbital Angular Momentum of Light, *Prog. Opt.* **39**, 291 (1999).
75. M. W. Beijersbergen, R. Coerwinkel, M. Kristensen, and J. P. Woerdman, Helical-wavefront laser beams produced with a spiral phaseplate, *Opt. Commun.* **112**, 321(1994).
76. V. Yu. Bazhenov, M. V. Vasnetsov, and M. S. Soskin, Laser beams with screw dislocations in their wavefronts, *JETP Lett.* **52**, 429 (1990).
77. J. P. Torres, A. Alexandrescu, L. Torner, Quantum spiral bandwidth of entangled two-photon states, *Phys. Rev. A* **68**, 050301(R) (2003).
78. A. Mair, A. Vaziri, G. Weihs, A. Zeilinger, Entanglement of the orbital angular momentum states of photons, *Nature* **412**, 313 (2001).
79. V. D. Salakhutdinov, E. R. Eliel, W. Löffler, Full-Field Quantum Correlations of Spatially Entangled Photons, *Phys. Rev. Lett.* **108**, 173604 (2012).
80. L. Torner, J. P. Torres, S. Carrasco, Digital spiral imaging, Opt. Exp. **13**, 873 (2005).
81. G. Molina-Terriza, L. Rebane, J. P. Torres, L. Torner, S. Carrasco, Probing canonical geometrical objects by digital spiral imaging, *J. Eur. Opt. Soc.* **2**, 07014 (2007).
82. X. F. Ren, G. P. Guo, B. Yu, J. Li, and. G. C. Guo, The orbital angular momentum of down-converted photons, *J. Opt. B: Quantum Semicluss. Opt.* **6**, 243 (2004).
83. N. Uribe-Patarroyo, A. M. Fraine, D. S. Simon, O. M. Minaeva, A. V. Sergienko, Object Identification Using Correlated Orbital Angular Momentum States, *Phys. Rev. Lett.* **110**, 043601 (2013).
84. D. S. Simon, A. V. Sergienko, Two-photon spiral imaging with correlated orbital angular momentum states, *Phys. Rev. A* **85**, 043825 (2012).
85. M. N. O'Sullivan-Hale, I. Ali Khan, R. W. Boyd, J. C. Howell, Pixel Entanglement: Experimental Realization of Optically Entangled d=3 and d=6 Qudits, *Phys. Rev. Lett.* **94**, 220501 (2005).

Chapter 3

Electron Systems Out of Equilibrium: Nonequilibrium Green's Function Approach

Václav Špička*,‡, Bedřich Velický*,† and Anděla Kalvová*

*Institute of Physics, Academy of Sciences of the Czech Republic,
Na Slovance 2, 182 21 Praha 8, Czech Republic
†Charles University, Faculty of Mathematics and Physics, DCMP,
Ke Karlovu 5, 121 16 Praha 2, Czech Republic

This review deals with the state of the art and perspectives of description of non-equilibrium many body systems using the non-equilibrium Green's function (NGF) method. The basic aim is to describe time evolution of the many-body system from its initial state over its transient dynamics to its long time asymptotic evolution. First, we discuss basic aims of transport theories to motivate the introduction of the NGF techniques. Second, this article summarizes the present view on construction of the electron transport equations formulated within the NGF approach to non-equilibrium. We discuss incorporation of complex initial conditions to the NGF formalism, and the NGF reconstruction theorem, which serves as a tool to derive simplified kinetic equations. Three stages of evolution of the non-equilibrium, the first described by the full NGF description, the second by a Non-Markovian Generalized Master Equation and the third by a Markovian Master Equation will be related to each other.

1. Introduction

The aim of this article is to show, how to describe time evolution of one particle observables of many-body electron systems out of equilibrium within the Non-equilibrium Green's function (NGF) approach. This article will be orientated on the non-equilibrium quantum field theory on the real time Schwinger-Keldysh contour. We will demonstrate that NGF provide useful tools how to deal with several open questions of non-equilibrium statistical physics, and enables to formulate the consistent quantum field theory of description of non-equilibrium quantum systems.

Recently developed experimental techniques enable us to observe details of time evolution of various electron systems far from equilibrium and to perform many

---

‡spicka@fzu.cz

interesting measurements on various natural or artificially prepared structures including mesoscopic (nanoscopic) systems,[1-3] where quantum processes, like quantum coherences, play decisive role. The growing area of non-equilibrium mesoscopic systems is naturally pervaded by open questions. Some of the problem open up newly during the research, some other have been resolved already, but in a provisional or an incomplete fashion. In general, the possibilities of the description of non-equilibrium many body systems, not only mesoscopic ones, is far from being satisfactory due to the complexity of problems involved.

To understand complex behavior of many-body systems out of equilibrium, and to interpret results of various recent experiments on mesoscopic systems, it is necessary to combine and to further improve methods of quantum field theory,[4-7] many body physics,[8-14] statistical physics,[15-40] and quantum transport theory.[38-58] Before going to details of the NGF approach, which uses knowledge of all these fields, we will mention problems, which every candidate on a successful theory of non-equilibrium systems has to tackle.

There are several key problems of non-equilibrium statistical physics to be understood on the way to adequate methods of the description of many body systems out of equilibrium. Here we will mention some of them.

1.1. *Challenges, open questions, techniques*

- Proper description of many-body character of systems, which represents a real challenge already in equilibrium. In addition there is a problem with consistency of used approximations: to ensure this consistency we have to check conservation laws, which is often not an easy task;
- Formulation and incorporation of non-trivial initial conditions into the formalism;
- Understanding of different non-equilibrium regimes and their description from short to long times evolution;
- Influence of external fields on time evolution;
- Time evolution of open systems: formulation, what is the system and its surrounding representing reservoirs, and proper treatment of interactions between these two parts, loss of quantum coherences, and dissipation processes.

These problems are still far from their complete solutions. Due to complexity of problems and technical difficulties involved, several complementary approaches have been developed, which are dealing with various aspects of the above problems in more or less details. These are the following techniques: Time Dependent Density Functional Theory (TDDFT),[59-69] Time Dependent Dynamical Mean-field Theory (TDMFT),[70-74] and various versions of Density Matrix Renormalization Group (DMRG) techniques.[75-80] We will not follow these "competitive" techniques to solve above mentioned key problems of non-equilibrium dynamics here.

1.2. *Non-equilibrium Green's functions*

In this review we will address problems of non-equilibrium statistical physics via Nonequilibrium Green's functions techniques.[81–131]

As we will see later on this approach is based on methods of quantum field theory, and is able to deal with many important problems of statistical physics[17–27] and generalization of equilibrium many body techniques[8–14] to non-equilibrium systems.

This approach has been used for such diverse non-equilibrium systems as particles in plasmas,[119,120] electrons, spins and phonons in various condensed matter systems like metals, semiconductors, superconductors and mesoscopic systems[132–175] or nuclei[106,128,176–181] as it is well documented also in the special volume dedicated to workshops and conferences about the NGF techniques.[111–115] It enables to describe non-equilibrium extended systems as well as mesoscopic (nanoscopic) systems, which have to be treated like open systems. The time evolution of observables can be calculated in various non-equilibrium regimes. In particular, the NGF formalism can be also conveniently used for the description of various steady state and equilibrium situations. NGF have been used not only for calculations of non-equilibrium occupation numbers, currents, current densities, but they have been also generalized to provide noise characteristics.[182–186] They have been able to describe such different processes like decay of initial correlations, dynamics of formation of correlations or quasiparticles, various transient and transport regimes, fast electron and spin dynamics, quantum coherence and decoherence processes, thermalization, physics of non-equilibrium cold atoms, e.g. dynamics of bosonic and fermionic systems in traps. Nowadays computers enable to solve complicated NGF equations for simplified, but often quite realistic models. Over the recent years many numerical calculations of the NGF equations have been performed.[111–115,120,129,187–192]

1.2.1. *Note on the NGF history*

The text of the article will not follow the historical developments of the NGF techniques. It is, however, useful to mention several key figures and lines of early developments of the NGF technique. The beginning of the NGF technique is related to Julian Schwinger at the end of the forties followed by works of his school represented e.g. by P. Martin, L.P. Kadanoff, G. Baym, V. Korenman, L.A. Craig, N. Horing. Later on two streams of developments have been related to the influential articles and the book written by Kadanoff and Baym[85] on the one side and Keldysh[89] and his followers on the other side.

The reader can find many interesting details about early developments of the NGF techniques in the following references: Refs. 81–101.

1.3. *Topics*

We may now formulate more precisely the subjects of this review:
to overview the possibilities of the NGF techniques to describe non-equilibrium

behavior of many body systems either of bulk (very large) or small sizes for the whole time evolution of the system: from short to long times.

The general problems of non-equilibrium statistical physics mentioned above are mirrored in the following topics, which will be discussed in this article within the NGF frame.

1.3.1. *Formulation of transport theory and NGF*

The first topic of this paper is to formulate demands on transport theory to motivate the NGF approach to non-equilibrium systems, see Section 2.

We will introduce NGF in more details in the following Section 3. An important aspect of the theory are controlled approximations. We will briefly mention related conservation laws like particle number and energy conservation, Ward identities and their generalization for non-equilibrium situations at the end of Section.

1.3.2. *Initial conditions*

The second topic deals with the task how to incorporate a nontrivial initial condition to the NGF formalism and implementation of the NGF techniques in the case of a fast transient process starting at a finite initial time and induced by a non-stationary initial condition and/or by an external field turned on at the initial time. The ways of incorporating complex initial conditions in the NGF methods are already known in several variants.[92,102,110,118,125,129,131,154,167,168,171,193–210]

We will discuss various approaches to initial conditions in Section 4.

1.3.3. *Transport in open systems*

There is still another aspect especially concerning electronic systems. Originally, the NGF based transport theory has been formulated for extended systems. The attention is presently shifting more and more towards the nano-structures and nano-devices. We will not deal with this aspect separately, but the presentation will be broad enough to encompass these systems.

The important point is that, in addition to the internal non-equilibrium dynamics of electrons driven perhaps by external fields, there may be expected important effects of a *changing* environment. Consider a suddenly opened thermal link between the system and a phonon bath. This will have an immediate influence on the electron Green's function, which must adapt to the new decay channels which will cause a loss of coherence of the propagation, reduce the relaxation time, etc. This process of adaptation will be gradual, having the character of a transient.

It turns out that there is a possibility of a uniform treatment of transient processes induced by external fields and by changes in the environment of the system. This is achieved by reformulating the complex initial conditions for the many-particle state in terms of the history of its single-particle Green's function, and a consistently constructed description of the sub-sequent transient evolution.

To achieve this task we will introduce the time-partitioning formalism[168] which is suited for for description of such process: the preparation stage is without the thermal contact and it defines the initial state for the follow-up, the process of relaxation of the electron sub-system induced by the suddenly attached thermal bath. We have thus at our disposal a Green's function formalism parallel to the Nakajima-Zwanzig projection method or to the path integral formulation, both developed within the density matrix technique.

1.3.4. *Reconstruction theorems*

The fourth topic of the review, see Section 5, is a general scheme, based on the so-called Reconstruction theorem, which provides a scheme how to calculate NGF and in the same time reduction of the complete NGF machinery to kinetic equations, which will be discussed in more details in the following Section 6.

1.3.5. *NGF and kinetic equations*

The last topic of this review, see Section 6, will close the circle: we will return to the formulation of the transport theory and derivation of quantum kinetic equations based on NGF techniques.

The initial period of a transient process is characterized by a complicated evolution known under the name of decay of initial correlations. If the external conditions are not very irregular, the system tends to a kinetic stage as the intermediate and late period of the process. The evolution during the kinetic stage is characterized by availability of simplified dynamic equations, quantum kinetic equations, generally speaking. This is the contents of the famous Bogolyubov conjecture.[38-40]

This conjecture brings us to the question of a meaning of the simplified description of dynamics with the help of kinetic equations. Nowadays, due to powerful computers, it is already possible to solve directly equations for NGF in many cases. In the same time, however, it is important to have simplified description for studies of more complicated systems. As we will see the NGF method enables us to construct quantum kinetic equations in a well controlled way. These equations are easier to solve than the full NGF equations. This approach and its applications has been a flourishing and successful research field over the past 30 years. There are at least three reasons why to address the current state of this problem.

First, we have to ask, which advantages are offered by this approach as compared with the direct use of NGF – at the time of an important breakthrough in the area of NGF solvers and the increasing power of contemporary computers.

Second, this theme is not really new: the basic concept of an "Ansatz", an approximate truncation reducing the NGF equations to transport equations for a quantum distribution, has been introduced by Kadanoff and Baym more than 50 years ago.[85,109] Yet the procedure is still not quite routine, as witnessed by the continuing research activity, and the question Why? is fully justified.

A standard tool for this task has been one type of approximate decoupling of the particle correlation function going by name of Kadanoff-Baym Ansatz.[85,109] This Ansatz has been modified to a class of the so-called causal Ansatzes including a particularly successful Generalized Kadanoff-Baym Ansatz.[109,116,118,125,129,211–217] We will describe an improved version we call Quasi-Particle Kadanoff-Baym Ansatz.[125] It is based on the notion of non-equilibrium quasi-particle.[218]

The third reason is specific for electrons in atomic systems (we will have in mind the electrons as a specific example) (... solids). Just as it happened before for systems in equilibrium and the Kohn-Sham Density Functional Theory, the systematic quantum field treatment of many body systems out of equilibrium seems to face emerging competitors in the Time dependent density functional theory (TDDFT), Time dependent dynamical mean-field theory (TDMFT) and various versions of Density matrix renormalization group (DMRG) techniques.

1.4. *Advantages of NGF*

Problems indicated above are very often formulated within various reduced density matrix approaches.[219–239] It makes now sense to indicate the space open for the use of the largely complementary approach based on NGF.

First, they are congruous with tasks of finding one-particle observables, like currents and current densities, occupation numbers, spin densities.
Second, they accentuate a "holistic" look on the system, permitting any strength of the coupling between the system and the environment and easily providing the true observable quantities, like the currents within the leads, for example.
Third, they capture, thanks to their two-time nature, the coherence and decoherence in the system in a natural way. As a consequence, NGF appear to permit a controlled transition to simplified theories suitable for moderately fast transient processes and describing the time evolution by a quantum transport equation, in this case having the form of a generalized master equation (GME, a master equation with memory). Conditions for a further reduction to a plain master equation are also at hand.

It should be stated that the problems discussed in this review are posed on a general level. All results are equally valid for both, extended and small (mesoscopic) systems. Even if we have in mind mainly electrons out of equilibrium, these techniques can be easily adapted also to other fermions, and are widely used not only to describe fermions, but also to describe dynamics of bosons.

2. Transport theory: Prototype description of nonequilibrium systems

To define the "transport theory", we may start from the more general nonequilibrium quantum dynamics of a system described by the many-body statistical operator (density matrix) $\mathcal{P}(t)$. The dynamics is driven by the full Hamiltonian $\mathcal{H}$ which can contain also an additive external time dependent disturbance $\mathcal{U}(t)$.

An initial state $\mathcal{P}_I$ at $t = t_I$ has also to be specified, and it may be an arbitrary equilibrium or out-of-equilibrium state. For any observable $\mathcal{X}$, the average value is $\langle \mathcal{X} \rangle_t = \text{Tr} \mathcal{X} \mathcal{P}(t)$. A transport theory can be derived, if the observables are restricted to those, which are relevant to the observed sub-system and additive. This permits a description in terms of reduced quantities, namely the single-particle distributions. There are two possibilities.

Generalizing the Landau theory of Fermi liquids,[8-14] one makes use of the quasiparticle distribution $f(\vec{r}, \vec{k}, t)$. To obtain a closed theory, generalizing famous Boltzmann approach for classical systems, a time-local Quantum Boltzmann equation is usually constructed.

The other possibility is to work with the single-particle density matrix $\rho(t)$ defined by the correspondence

$$\begin{array}{ll} \langle \mathcal{X} \rangle_t = \text{Tr} \mathcal{X} \mathcal{P}(t) & \xrightarrow[\text{REDUCTION}]{\text{additive} \mathcal{X}} & \langle \mathcal{X} \rangle_t \equiv \overline{X}(t) = \text{tr} X \rho(t), \\ \mathcal{P}(t) & \longrightarrow & \rho(t). \end{array} \tag{2.1}$$

In other words, the average values are, for all additive (single-particle) observables A, given by the single particle reduced density matrix:

$$A_{\text{Av}}(t) = \text{tr}(\rho(t)A), \tag{2.2}$$

$$\rho(x, y, t) = \text{Tr}\left(\mathcal{P}(t)\psi^\dagger(y)\psi(x)\right), \tag{2.3}$$

Here $\mathcal{P}$ is the full many-body statistical operator, and $\psi^\dagger$ and ψ are field operators for particles (often electrons in this article).

In particular, the local particle density n of electrons is given by

$$n(x, t) = \rho(x, x, t). \tag{2.4}$$

For a single band $\epsilon(k)$, the local current density j equals

$$j(x, t) = \frac{1}{2m}\left(\frac{\partial \epsilon}{\partial k}(i\nabla_x) + \frac{\partial \epsilon}{\partial k}(-i\nabla_y)\right)\rho(x, y, t)\bigg|_{y=x}. \tag{2.5}$$

We will now discuss these two approaches starting from the Boltzmann equation.

2.1. *Prototype transport equation: Boltzmann equation*

A prototype of all kinetic equations, the Boltzmann equation of classical physics[15,16] reads

$$\frac{\partial f}{\partial t} - \text{drift}[f(t)] = I_{\text{in}}[f(t)] - I_{\text{out}}[f(t)], \tag{2.6}$$

where the drift term is the classical Poisson bracket

$$\mathsf{drift}[f(t)] = \frac{\partial \varepsilon}{\partial k}\frac{\partial f}{\partial r} - \frac{\partial \varepsilon}{\partial r}\frac{\partial f}{\partial k} . \tag{2.7}$$

This is, in principle, an equation for a single particle distribution function $f(k, r, t)$ in phase space, representing balance between the drift of particles (with energy $\varepsilon = \epsilon_k + U(r, t)$ and velocity $\frac{\partial}{\partial k}\varepsilon = \frac{\partial}{\partial k}\epsilon_k$ in the external force field $-\frac{\partial}{\partial r}\varepsilon = -\frac{\partial}{\partial r}U$), as given by the left hand side of the BE, and the irreversible evolution due to collisions described by the scattering integrals I_{in} and I_{out} on the right hand side.

2.2. Transport theory: Physical concepts

To understand physics beyond the Boltzmann equation, it will be important to recall several fundamental ideas, which run through the development of the transport theory. As will be seen, the physical principles of the contemporary NGF approach are not different from those which were in the minds of the founders of non-equilibrium statistical physics. The issues remain, the understanding and attitudes change. We pick up just the following topics.

I. The central idea of the transport theory has always been a **reduced description** of the system. The relevant information about a gas of particles was contained in the one-particle distribution function $f(x, p, t)$. The essential step was to search for a closed equation governing this function, that is for the transport equation. This was only possible, if, say, the description of binary collisions in the gas was given in terms of the function f. As a systematic program, this seems to appear first in the work of Chapman and Enskog[15,16] but the principle dates back to Boltzmann himself with his Stosszahlansatz for the binary distribution f_{12} expressed as $f_{12} \sim f_1 \times f_2$. This type of factorization, or, more generally, decoupling of higher correlation functions was systematically developed in the BBGKY technique. We have to say more about Bogolyubov, to whom we owe the postulate[38–40] that *in a chaotized many-body system all higher particle correlation functions become a functional of the distribution function.*

Thus

$$f_{12} \xrightarrow{\mathsf{chaot.}} \Phi[f_1, f_2] . \tag{2.8}$$

The functional has yet to be specified, but this is of secondary importance. As will be described later, the KB Ansatz also singles out the one-particle distribution as the determining characteristic of the non-equilibrium system. This brings us to the **reconstruction problem**: under which conditions the full description of the many-body interacting system can be built up from the knowledge of single-particle characteristics? This seemingly outrageous question was seriously treated in several contexts. We will discuss later two of them: the time-dependent density functional theory (TDDFT), and the Generalized Kadanoff-Baym Ansatz related techniques.

II.   The second crucial notion is the **hierarchy of characteristic times**. There are three intrinsic times related to a many-particle interacting system. In a reminiscence of a non-dilute gas, they are often identified as the *collision duration time* $\tau_c = a/\bar{v}$, the *collision time* $\tau_r = \ell/\bar{v}$ and the *hydrodynamic time* $\tau_h = L/\bar{v}$. Here, $\bar{v}$ is the average thermal velocity for a classical gas, the Fermi velocity v_F for a degenerate Fermi gas. The characteristic lengths are: a is the interaction potential range (particle size), ℓ is the mean free path (mean inter-particle distance), L is the characteristic length of spatial inhomogeneities in the system. In the modern interpretation, the three times are: $\tau_c \cdots$ the chaotization time characterizing the decay of correlations, $\tau_r \cdots$ the relaxation time characterizing the thermalization of the system (local relaxation) and, finally, $\tau_h \cdots$ characterizes the process of relaxation of spatial inhomogeneities. In "normal" situations, the three times obey the inequality

$$\tau_c \ll \tau_r \ll \tau_h, \tag{2.9}$$

separating the chaotization stage, the kinetic stage, and the hydrodynamic stage. This structuring of the spontaneous return to equilibrium was also introduced by Bogolyubov; he postulated equation (2.8) only for times later than the chaotization time τ_c. The Boltzmann equation proper corresponds to the limit

$$0 \leftarrow \tau_c \lll \tau_r \lll \tau_h \to \infty \,. \tag{2.10}$$

We will be mostly concerned with the opposite, more realistic case, when the distinction between τ_c and τ_r will be less sharp. Further, the time range of interest will be specified by the external fields. For example, an optical pulse is characterized by its duration, the ground period of the signal and its Rabi period measuring the pulse strength. These times should be compared with the intrinsic times of the system. This will, at last, specify the situation and the necessary version of transport theory used.

III.   The last basic concept is that of the **quasi-particles**.[8–14] With that, we, as a matter of fact, move over to the quantum realm. There are several streams merging into the generalized notion of a quasi-particle. Firstly, the polaron, an electron dragging along a cloud of lattice polarization. In the quantum field language, the electron is dressed by virtual phonons. This compound object has some features characteristic of a particle, like a dispersion law (renormalized by the self-energy), which has an operational meaning in experiments. The self-energy is typically complex, and this leads to the finite life-time τ of the quasi-particles, closely related to the transport relaxation time τ_r. Secondly, and even more to the point, in a non-dilute system of interacting particles, their individuality is suppressed by mutual correlations and the use of a transport equation seems to be hopeless. However, the weakly excited states may appear to mimic a gas of weakly interacting quasi-particles. This was at the bottom of the Landau theory of the Fermi liquid, whose part was a proper adaptation of the Boltzmann equation. This approximate, but highly precise description of the Fermion systems had, of course, a number of predecessors, like the Sommerfeld electrons in simple metals, and parallels, like the quasi-particles in nuclei, where the self-energy has been originally introduced under the name of the optical potential. It might seem that the quantum transport equations for strongly interacting systems will all deal with the quasi-particles. This is, to some extent, true. However, the quasi-particles are vulnerable and elusive objects and cease to exist under some harsher conditions, like beyond the quasi-classical regime, for strong and/or transient disturbances, or if the system itself is not favoring their existence. This can sometime be judged by the Landau-Peierls criterion. Take the gas of quasi-particles with energies around the Fermi level E_F and a lifetime τ due to impurity scattering. The criterion reads: if $\tau \simeq \hbar/E_F$, the BE-like transport theory is not applicable. Now $\hbar/E_F$ can be interpreted as the "quasi-particle formation time" τ_Q. If $\tau \simeq \tau_Q$, the quasi-particle decays before having formed. On the other hand, τ is closely related to τ_r, whereas τ_Q appears to play the role of τ_c. The Landau-Peierls criterion thus says that if the quasi-particles do not form, then τ_r becomes comparable to τ_c and the condition (2.10) is not obeyed.

2.3. Boltzmann equation for quantum systems

After overviewing the basic concepts beyond the Boltzmann equation and the transport theory in general we can return to the properties of the Boltzmann equations to discuss possibilities how to generalize it towards quantum systems.

First of all we note that in the scattering integrals of the Boltzmann equation, collisions are approximated as instant randomizing events, so that the BE is Markovian. This is in agreement with the first inequality in (2.10). The other inequality demands all inhomogeneities in the system, including external fields, to be smooth enough to allow a sufficient time for local equilibration.

The BE has been used for many classical systems and it was extended very early also to the transport by quantized particles, in particular by electrons in metals. The quantum effects were incorporated mainly in the scattering integrals where the collision rates were calculated by the Fermi Golden Rule and the exclusion principle was taken into account by means of the Pauli blocking factors.

The average values of additive observables were calculated from formulas taken over from classical kinetic theory of gases. Thus, the particle density and the particle current density were

$$n(r,t) = \int \frac{dk}{(2\pi)^3} f(k,r,t), \qquad (2.11)$$

$$j(r,t) = \int \frac{dk}{(2\pi)^3} \frac{\partial \epsilon}{\partial k} f(k,r,t). \qquad (2.12)$$

Now, we depart from the BE in the direction towards quantum dynamic equations far from equilibrium. There are several extensions of the BE we will first explore separately:

(1) From particles to quasi-particles. The interaction energy will no longer be negligible, but will be renormalized in the quasi-particle transformation.
(2) From near-equilibrium to "arbitrary" non-equilibrium. This will be reflected in the memory of the system and taken into account by the non-Markovian GME

2.3.1. *Quasi-particles and transport*

In dense systems, the particle interactions cannot be reduced to rare collisions randomizing the motion of otherwise free particles. Quantum statistics for such systems at near-equilibrium is often well described by the Landau theory of quasi-particles. In this theory, weakly excited states of the interacting particles are described as a gas of quasi-particles with energy

$$\varepsilon(k, r, t) = \epsilon_k + U_{\text{eff}}(r, t) + \sigma(\varepsilon, k, r, t), \qquad (2.13)$$

where U_{eff} may include the mean-field part of the interaction and the self-energy σ describes the mass renormalization. The quasi-particles are coupled by a weak residual interaction.

2.3.2. *Quantum kinetic equation for quasi-particles*

These quasi-particle features have consequences for transport properties, which are described by the Landau kinetic equation. It has exactly the structure of the BE (2.6). However, the energy entering the drift term is now the quasi-particle energy ε. Scattering integrals I_{in} and I_{out} contain scattering rates calculated again by the Fermi Golden Rule, but with the residual interactions reflecting that they are reduced by the many-particle wave–function renormalization. Finally, the function $f(k, r, t)$ is the *quasi-particle distribution function*.

Because of its intuitive character, the "Boltzmann" equation for quasi-particles provided many insights into the behavior of non-equilibrium many-particle systems. In particular, it offered a retrospective explanation of the over-successful Sommerfeld model of metals with non-interacting electrons responding to external fields. While various modifications of the BE-like approach have been applied successfully to many systems and situations, there are also many physical cases, in which it is bound to fail. The limitations of the Boltzmann equation include: 1. The BE well describes quantum systems if the quasi-particle picture is justified. This may no longer be valid in highly non-equilibrium quantum systems. 2. It is based on the intuitive idea of instant collisions between quasi-particles, which requires the collision duration time to be very short. 3. At the same time, the BE will not be well suited to describe systems with abrupt changes in space, including small structures, where quantum effects are essential and the quasi-classical approach hidden behind the BE is far from being sufficient.

2.3.3. Observables and quantum distribution function

Relations between the quasi-particle distribution function and the expectation values of observables are more complicated than (2.11), (2.12). For example, the expression for current density must incorporate the back-flow accompanying the motion of quasi-particles. Rules for evaluating observables have been from the outset an integral part of the Landau equilibrium theory. A natural question emerges: which is the relation between the quasi-particle distribution function f and the expectation values out of equilibrium?

Our task is then to find a functional relation

$$\{f,\, A\} \mapsto A_{\mathsf{Av}} = \mathrm{tr}(\rho A),$$

which is equivalent to finding a functional $\rho[f]$ associating the single-particle density matrix with any given quasi-particle distribution. To relate the reduced density matrix ρ with the distribution function $f(k, r, t)$ it is convenient to express ρ in the (k, r) representation. This can be done in different ways, but we choose the so-called Wigner representation

$$\widetilde{\rho}(k, r, t) = \int dx e^{-ikx} \rho\left(r + \frac{x}{2}, r - \frac{x}{2}, t\right). \tag{2.14}$$

The reduced density $\widetilde{\rho}(k, r, t)$ in the Wigner representation is commonly known under the name of Wigner distribution. It can be viewed as a distribution of electrons in the phase space and the expectation values

$$n(r, t) = \int \frac{dk}{(2\pi)^3} \widetilde{\rho}(k, r, t), \tag{2.15}$$

$$j(r, t) = \int \frac{dk}{(2\pi)^3} \frac{\partial \epsilon}{\partial k} \widetilde{\rho}(k, r, t), \tag{2.16}$$

have the same form as in the Boltzmann theory.

2.3.4. Wigner and quasiparticle distributions

These suggestive properties of the Wigner distribution should not lead us to the conclude that this function is the "right" quantum generalization of the classical distribution function, and that the proper quantum generalization of the Boltzmann equation will be a kinetic equation of the Boltzmann form (2.6) for the Wigner distribution function. We can see that this is not the case from the following argument given already by Landau.

Consider a homogeneous system. In equilibrium, the BE-like kinetic equation for electrons is solved by the Fermi-Dirac function regardless of the interactions in the system. Thus, at zero temperature, the distribution f jumps from 1 to 0 at the Fermi level. In contrast, the Wigner distribution describes the occupation numbers of true particles and it differs from the Fermi-Dirac function by depletion of the momentum states below the Fermi level. These missing states emerge as states above the Fermi level, where they form the so-called high-momenta tails of the Wigner function. The step of the distribution $\widetilde{\rho}$ is accordingly reduced to $1 - 2z$, where z is the renormalization constant for quasi-particles at the Fermi level.

We may summarize that the quasi-particle kinetic theory consists of two steps. First, the quantum kinetic equation is solved for the quasi-particle distribution. Second, the true quantum particle distribution is constructed by means of the functional $\rho[f]$ and the expectation values of the observables are calculated.

2.4. *Generalized Master Equations: Beyond the quantum kinetic equations*

In this setion, we will not follow the Boltzmann like direction, which is the description using the concept of a quasiparticle distribution function and is suited for "slow" processes, like a stationary transport. Instead of this approach, we will now discuss the second possibilty how to develop a quantum transport theory: we will deal with description based on the single particle density matrix ρ. So, the aim will be to introduce (and later on to derived from the full set of the NGF equations) a General Master Equation (GME) which governs ρ.

A closed Quantum Transport Equation for ρ has the general form

$$\frac{\partial \rho}{\partial t} - \text{drift} = \Phi_t[\rho(\tau); \tau < t], \tag{2.17}$$

where "drift" means the bare one-particle dynamics and the effect of all interactions is contained in the generalized collision term on the r.h.s. The functional Φ_t has a form parametrically dependent on time and is functionally dependent on the full history of the distribution function itself. Thus, the equation is a one-particle version of the so-called Generalized Master Equation. This is very formal. We should now address several questions, whose answers will depend on the physical nature of the system under consideration:

⋄ Proof of the existence of the Quantum Transport Equation.
⋄ Explicit construction of the generalized collision term Φ_t.
⋄ Introduction of the initial conditions at t_0, both explicitly and also through the form of Φ_t.

We intend to analyze these points from the angle of the Green's functions. It will be seen, however, that physical principles of the contemporary NGF approach are very close to those which were formulated by the founders of non-equilibrium statistical physics.

As already mentioned, quantum kinetic equations of the Boltzmann type have a restricted range of validity and these limits can be transgressed only by resorting to a more general framework, permitting, at least in principle, to start from a fully

quantum description and work directly with an equation for the reduced density
matrix (or the equivalent Wigner function), the so called quantum Generalized
Master Equation (GME). There are several approaches and approximations leading
to the GME. Its general form (2.17) can be rewritten as

$$\frac{\partial \rho}{\partial t} - \quad \text{drift}[\rho(t)] \qquad = \text{interaction term} ,$$

$$\frac{\partial \rho}{\partial t} + \mathrm{i}\,[\, \underbrace{T + U_{\mathrm{eff}}(t)}_{H_0}, \rho(t)\,]_- = \int_{-\infty}^{t} d\bar{t} F[\rho(\bar{t})] . \tag{2.18}$$

This equation is shown in an entirely symbolic form on the first line, and it is not
really explicit on the second one either. We will discuss the detailed structure of
this type of equation below. Here, we only touch its most salient features. The
GME is a closed equation for ρ. It is non-markovian, because the interaction term
is non-local in time.

At the end of Sec. 2.3.2, we pointed out three important limiting factors for the
use of the quantum kinetic equations. These factors will now be discussed from the
point of view of the GME.

1. GME is an equation for ρ

Equation (2.18) is a dynamical equation for the distribution of particles instead of
quasi-particles as was the case of the BE. Therefore, it does not hinge on the use
of quasi-particles, and it is free of the physical limitations necessary for introducing
the gas of quasi-particles. In particular, the system may evolve under conditions
which do not permit the quasi-particle states to consolidate. This does not preclude
using some of the quasi-particle features where applicable. Thus, the left hand side
of the equation describes bare particles drifting under the influence of the effective
field. All other features involving interactions are included in the right hand side. It
may be possible to transfer parts of the r.h.s to the drift term and achieve evolution
in re-normalized bands.

2. Interaction term

We prefer to call the right hand side of (2.18) interaction term rather than scattering integrals. In principle, the GME is exact, so that the interaction term must
incorporate all of the sub-dynamics of ρ reflecting not only particle collisions, but
also the short time dynamics, off-shell propagation and coherence between the collisions and with the external fields, multiparticle correlations, gradual saturation of
the scattering rates after the onset of a non-equilibrium process, etc.

It is remarkable that all this rich physics can be absorbed in an interaction term
depending only on the one-particle density matrix. This is made possible by the
subtle memory effect reflected in the time integration over the full depth of the past
in (2.18). Clearly, this form of the interaction term agrees with the Bogolyubov
postulate quoted above in Sec. 2.2, at least for times beyond τ_c after the onset of

the process. We will return to the related Reconstruction Theorems later on.

3. Quasi-classical expansion

In the GME (2.18), neither the drift term, nor the interaction integral, are restricted to smooth variation of the fields and distributions in space. The drift term is given by the quantum Poisson bracket rather than by the classical one appearing in (2.7). For smooth functions, it can be quasiclassically expanded; the formal expansion parameter is $\hbar^2$. For the Wigner distribution $\widetilde{\rho}$ defined in Eq. (2.14), we obtain, writing, by exception, the Planck constant explicitly:

$$\text{drift}[\rho(t)] = (i\hbar)^{-1}[T + U_{\text{eff}}(t), \rho(t)]_- \longrightarrow$$

$$\text{(2.19)}$$

$$\text{drift}[\widetilde{\rho}(t)] = \frac{\partial \epsilon}{\partial p}\frac{\partial \widetilde{\rho}}{\partial r} - \frac{\partial U_{\text{eff}}}{\partial r}\frac{\partial \widetilde{\rho}}{\partial p} - \frac{\hbar^2}{3!}\left(\frac{\partial^3 \epsilon}{\partial p^3}\frac{\partial^3 \widetilde{\rho}}{\partial r^3} - \frac{\partial^3 U_{\text{eff}}}{\partial r^3}\frac{\partial^3 \widetilde{\rho}}{\partial p^3}\right) + \cdots$$

This expansion, suited for comparison of the GME in the quasiclassical limit with the corresponding kinetic equation is, in fact, a Fourier transformed expansion around the space diagonal $x_1 = x_2$.

3. NGF approach to nonequilibrium systems: Basic concepts

This section is the first of the central parts of the paper, in which we introduce and formally elaborate on the Nonequilibrium Green's functions (NGF).[102–131] Later on we will discuss the NGF method from two points of view:
1. To describe properly the dynamics of non-equilibrium systems by directly solving full set of equations for the NGF;
2. To derive and analyze the simplified non-equilibrium dynamic equations of either the BE or the GME type as we discussed them in previous sections of this article.

3.1. Definition of the system

The NGF theory is formulated for closed systems, which have a well defined Hamiltonian and undergo a strictly unitary evolution. This does not preclude irreversible evolution, if at least some parts of the whole system are extended and possess a continuous spectrum. In fact, a finite isolated system possessing bound states would be a difficult case for the present approach. The system as a whole has to incorporate both the "relevant" sub-system and all other components, like thermal baths an particle reservoirs. The relevant sub-system then appears as open and its evolution has the signatures of irreversibility. The baths are typically taken as ideal, that is inert, not participating in the dynamics, but acting to impose temperature and/or chemical potential. The state of the system is then specified by the density matrix of the relevant sub-system and the parameters of its environment, its dynamics is governed by the corresponding Hamiltonian.

We have already given practical reasons, why the systems we will specifically consider will be electrons under various external conditions and driven by classical external fields. They will be interacting by instantaneous pair forces, typically of Coulomb origin.

The Hamiltonian of the system will be denoted $\mathcal{H}(t)$. Calligraphic letters relate to many-body quantities. The Hamiltonian has several parts of different nature:

$$\mathcal{H} = \mathcal{H}_0(t) + \mathcal{W}, \quad \mathcal{H}_0(t) = \mathcal{T} + \mathcal{V} + \mathcal{H}'_e(t). \tag{3.1}$$

$\mathcal{H}_0(t)$ is the one-particle part of the Hamiltonian. For a quiescent spatially homogeneous system, this would be just the kinetic energy operator $\mathcal{T}$. In general, it also incorporates $\mathcal{V}$ which accounts for static internal fields specifying the geometry, atomic composition and other characteristics. $\mathcal{V}$ thus uniquely defines the system under consideration. The departure from the equilibrium state is driven by $\mathcal{H}'_e(t)$ which for simplicity is assumed to include only scalar local external fields. Finally, $\mathcal{W}$ is the pair interaction term, as characterized above.

In a second-quantized form written for fermions, we have

$$\mathcal{H}_0(t) = \int dx \psi^\dagger(x)(-\frac{1}{2m}\Delta + V(x))\psi(x) + \int dx \psi^\dagger(x)V_e(x,t)\psi(x),$$

$$= \int dx \psi^\dagger(x)h_0(t)\psi(x), \tag{3.2}$$

$$\mathcal{W} = \frac{1}{2}\iint dxdy \psi^\dagger(x)\psi^\dagger(y)w(x,y)\psi(y)\psi(x), \tag{3.3}$$

$$x = \{r,\sigma\}, \quad \int dx = \int dr \sum_\sigma,$$

$$[\psi(x),\psi^\dagger(x')]_+ = \delta(x-x') \equiv \delta(r-r')\delta_{\sigma\sigma'},$$
$$[\psi(x),\psi(x')]_+ = 0, \quad [\psi^\dagger(x),\psi^\dagger(x')]_+ = 0. \tag{3.4}$$

3.2. *Evolution operator*

In the non-equilibrium physics, the Hamiltonian determines primarily the time evolution of the system from given initial conditions. For example, for an initial state given by a (many-body) wave function $|\Psi\rangle_{t=t_I} = |\Psi_I\rangle$ at an initial time t_I, we have to solve the Schrödinger equation

$$i\frac{\partial}{\partial t}|\Psi\rangle_t = \mathcal{H}(t)|\Psi\rangle_t, \tag{3.5}$$

This solution can be formally written in a form universal for all initial conditions employing the evolution operator $\mathcal{S}$:

$$|\Psi\rangle_t = \mathcal{S}(t,t_I)|\Psi_I\rangle, \tag{3.6}$$

The evolution operator will serve as a universal tool in this paper and we shall discuss some of its properties now.

3.2.1. *Properties of the evolution operator*

Given the Hamiltonian, its evolution operator $\mathcal{S}(t, t')$ is a function of two time arguments. It is determined by the Schrödinger equation and the initial condition at equal time arguments, consistently with Eq. (3.6):

$$i\frac{\partial}{\partial t}\mathcal{S}(t, t') = \mathcal{H}(t)\mathcal{S}(t, t'), \quad \mathcal{S}(t', t') = \mathbf{1}. \tag{3.7}$$

Equivalently, it is given by an analogous initial value problem for the other time variable:

$$i\frac{\partial}{\partial t'}\mathcal{S}(t, t') = -\mathcal{S}(t, t')\mathcal{H}(t'), \quad \mathcal{S}(t, t) = \mathbf{1}. \tag{3.8}$$

The evolution operator obeys two basic rules, it is unitary and has the group property:

$$\mathcal{S}(t, t')\mathcal{S}^{\dagger}(t, t') = \mathcal{S}^{\dagger}(t, t')\mathcal{S}(t, t') = \mathbf{1}, \tag{3.9}$$

$$\mathcal{S}(t, t') = \mathcal{S}(t, t'')\mathcal{S}(t'', t'). \tag{3.10}$$

The so-called group property expresses a composition rule for two subsequent time segments of evolution. Because the evolution is unitary, there is no restriction on the values of the times involved, the intermediate time may in fact precede both terminal times, etc. As a consequence, the following identities are obtained in particular:

$$\mathcal{S}^{\dagger}(t, t') = \mathcal{S}^{-1}(t, t') = \mathcal{S}(t', t). \tag{3.11}$$

3.2.2. *Schrödinger and Heisenberg pictures*

In the customary representation of quantum dynamics, the Schrödinger picture, the quantum states evolve in time. For pure states, this evolution is described by Eq. (3.6). Similarly, the evolution operator allows to express the evolution of a general state of the many-body system by its state operator $\mathcal{P}$ (many-body density matrix) from an initial time t_I to the time t as

$$\mathcal{P}_S(t) = \mathcal{S}(t, t_I)\mathcal{P}(t_I)\mathcal{S}^{\dagger}(t, t_I). \tag{3.12}$$

The vertically displaced calligraphic capital $\mathcal{P}$ is used to denote capital Greek Rho rather than Roman P. The label indicates the Schrödinger picture.

In Green's function theory, it is preferable to work in the Heisenberg picture, in which the state operator is time independent and the time evolution is transferred to the operators of observables such that the resulting average values are the same in both pictures:

$$\langle \mathcal{X} \rangle_t = \mathrm{Tr}(\mathcal{P}_S(t)\mathcal{X}) = \mathrm{Tr}(\mathcal{S}(t, t_I)\mathcal{P}(t_I)\mathcal{S}^{\dagger}(t, t_I)\mathcal{X}(t))$$

$$= \mathrm{Tr}(\mathcal{P}(t_I)\underbrace{\mathcal{S}^{\dagger}(t, t_I)\mathcal{X}(t)\mathcal{S}(t, t_I)}_{\mathcal{X}_H(t)}). \tag{3.13}$$

Clearly, the Heisenberg operator may combine two distinct time dependences, the explicit one, and the other one reflecting the evolution of the system. We get the following equation of motion for the Heisenberg operator of any observable and the associated initial value problem:

$$i\frac{\partial \mathcal{X}_H(t)}{\partial t} = [\mathcal{X}_H(t), \mathcal{H}_H(t)]_- + i\left(\frac{\partial}{\partial t}\mathcal{X}_S(t)\right)_H, \qquad \mathcal{X}_H(t_I) = \mathcal{X}_S(t_I). \tag{3.14}$$

The operators in the Heisenberg picture can be written in the form derived in (3.13), or in its modification employing (3.11):

$$\mathcal{X}_H(t) = \mathcal{S}^\dagger(t, t_I)\mathcal{X}(t)\mathcal{S}(t, t_I) = \mathcal{S}(t_I, t)\mathcal{X}(t)\mathcal{S}(t, t_I). \tag{3.15}$$

The second form is suggestive of evolution from t_I to t and back. This idea is at the heart of the Schwinger-Keldysh NGF formalism, as will become clear soon.

The Heisenberg picture may be seen to offer two important advantages. First, the averaging at all times is performed over the same time-independent many-body state. Second, in this manner, it is possible to obtain averages of any number of observables detected at their individual times and in this way to study space-time correlations of an arbitrary order.

3.2.3. Explicit expressions for the evolution operator

Finding the evolution operator is an immensely difficult task, except for simple special cases. One problem is an admitted time dependence of the Hamiltonian. If the Hamiltonian is time-independent, the equation (3.7) is easily solved by

$$\mathcal{S}(t, \hat{t}') = e^{-i\mathcal{H}\cdot(t-t')}. \tag{3.16}$$

For a time-dependent Hamiltonian, Eq. (3.16) is easily generalized in the rather exceptional case that $\mathcal{H}(t)$ can be diagonalized for all times in the same basis:

$$\mathcal{S}(t, t') = e^{-i\int_{t'}^t d\tau\, \mathcal{H}(\tau)}, \qquad [\mathcal{H}(t_1), \mathcal{H}(t_2)]_- = 0, \tag{3.17}$$

In the general case, a solution extending the last expression can be given with the use of the time-ordering (chronological) operator T:

$$\mathcal{S}(t, t') = Te^{-i\int_{t'}^t d\tau\, \mathcal{H}(\tau)}, \qquad [\mathcal{H}(t_1), \mathcal{H}(t_2)]_- \neq 0, \quad t > t' \tag{3.18}$$

To understand the structure of this deceivingly simple formula, we have to expand the exponential into the power series. The T "operator" acts on operators. It is linear and in a product of several operators, it rearranges them in the order of their time arguments, the latest time coming first on the left. For two operators this means

$$T\{A(t)B(t')\} = A(t)B(t') \quad t > t', \tag{3.19}$$

$$T\{A(t)B(t')\} = B(t')A(t) \quad t' > t. \tag{3.20}$$

The expression (3.18) for $\mathcal{S}$ becomes

$$\mathcal{S}(t,t') = \sum_n \frac{1}{n!}(-\mathrm{i})^n \int_{t'}^{t} \cdots \int_{t'}^{t} \mathrm{d}t_1 \cdots \mathrm{d}t_n \, T\{\mathcal{H}(t_1)\cdots\mathcal{H}(t_n)\}. \tag{3.21}$$

Consider now the n-th order term of the series. The time-ordering operator permutes the n factors in all possible manners and yields a non-zero result when the permuted times have their values ordered. Thus the integral splits into $n!$ contributions mutually equal because of the symmetric structure of the integral and it is enough to keep just one multiplied by $n!$. With the notation for permutations

$$P: \{1,\cdots,n\} \;\rightarrow\; \{1_P,\cdots,n_P\},$$

we have

$$\mathcal{S}(t,t') = \sum_n \frac{1}{n!}(-\mathrm{i})^n \sum_P \int_{t'}^{t} \mathrm{d}t_{1_P} \; \cdots \int_{t'}^{t_{(n-1)_P}} \mathrm{d}t_{n_P} \mathcal{H}(t_{1_P})\cdots\mathcal{H}(t_{n_P})$$

$$= \sum_n (-\mathrm{i})^n \int_{t'}^{t} \mathrm{d}t_1 \; \cdots \int_{t'}^{t_{(n-1)}} \mathrm{d}t_n \mathcal{H}(t_1)\cdots\mathcal{H}(t_n). \tag{3.22}$$

The last series is the well known iterative solution of the initial value problem (3.7) which is conveniently recast into a Volterra integral equation for that purpose:

$$\mathcal{S}(t,t') = \mathbf{1} - \mathrm{i} \int_{t'}^{t} \mathrm{d}t_1 \mathcal{H}(t_1)\mathcal{S}(t_1,t'). \tag{3.23}$$

In a similar fashion, (3.8) has the integral form

$$\mathcal{S}(t,t') = \mathbf{1} + \mathrm{i} \int_{t}^{t'} \mathrm{d}t_1 \mathcal{S}(t',t_1)\mathcal{H}(t_1). \tag{3.24}$$

and its solution for $t' > t$ is given by

$$\mathcal{S}(t,t') = \widetilde{T}e^{+\mathrm{i}\int_{t}^{t'} \mathrm{d}\tau\, \mathcal{H}(\tau)}, \qquad t' > t. \tag{3.25}$$

This is of the same structure as Eq. (3.18), but involving the anti-chronological time-ordering operator $\widetilde{T}$ which orders the operators in the order of time arguments increasing from the right to the left.

3.2.4. *Dirac picture*

For an actual work, often neither of the two pictures, Schrödinger or Heisenberg, is suited, and an intermediate Dirac picture, encompassing both as limiting cases, has to be introduced. The general scheme of the Dirac picture starts from decomposing the Hamiltonian in question into two parts,

$$\mathcal{H} = \mathcal{H}_\mathsf{F} + \mathcal{H}_\mathsf{P}. \tag{3.26}$$

Although it is not indicated explicitly, all components of the Hamiltonian may depend on time explicitly. The F ("free") component of the Hamiltonian is considered as a reference, while the P ("perturbation") component has to be included as an additional perturbation. This division is, of course, possible in different ways. For example, the pair interaction is taken as the perturbation, and the Dirac picture then serves to develop the many body perturbation expansion. Alternatively, the external field enters as the perturbation, and the Dirac picture leads to the general non-linear response theory.

The average values of observables should not depend on the picture used, and this will lead us directly to the Dirac picture definition by extending the procedure outlined in Eq. (3.13). The idea is to let the observable $\mathcal{X}$ undergo the "free" evolution and the state $\mathcal{P}$ to compensate for the difference between the free and the full evolution. Starting again from the Schrödinger picture, we get

$$\langle \mathcal{X} \rangle_t = \mathrm{Tr}(\mathcal{P}_\mathsf{S}(t)\mathcal{X}) = \mathrm{Tr}(\mathcal{S}(t,t_\mathrm{I})\mathcal{P}(t_\mathrm{I})\mathcal{S}^\dagger(t,t_\mathrm{I})\mathcal{X})$$

$$= \mathrm{Tr}(\mathcal{S}(t,t_\mathrm{I})\mathcal{P}(t_\mathrm{I})\mathcal{S}^\dagger(t,t_\mathrm{I})\mathcal{S}_\mathsf{F}(t,t_\mathrm{I})\underbrace{\mathcal{S}_\mathsf{F}^\dagger(t,t_\mathrm{I})\mathcal{X}\mathcal{S}_\mathsf{F}(t,t_\mathrm{I})}_{\mathcal{X}_\mathrm{D}(t)}\mathcal{S}_\mathsf{F}^\dagger(t,t_\mathrm{I}))$$

$$= \mathrm{Tr}(\underbrace{\overbrace{\mathcal{S}_\mathsf{F}^\dagger(t,t_\mathrm{I})\mathcal{S}(t,t_\mathrm{I})}^{\mathcal{S}_D(t,t_\mathrm{I})}\mathcal{P}(t_\mathrm{I})\overbrace{\mathcal{S}^\dagger(t,t_\mathrm{I})\mathcal{S}_\mathsf{F}(t,t_\mathrm{I})}^{\mathcal{S}_D^\dagger(t,t_\mathrm{I})}}_{\mathcal{P}_\mathrm{D}(t)}\underbrace{\mathcal{S}_\mathsf{F}^\dagger(t,t_\mathrm{I})\mathcal{X}\mathcal{S}_\mathsf{F}(t,t_\mathrm{I})}_{\mathcal{X}_\mathrm{D}(t)})$$

$$\langle \mathcal{X} \rangle_t = \mathrm{Tr}(\mathcal{P}_\mathrm{D}(t)\mathcal{X}_\mathrm{D}(t)). \tag{3.27}$$

With the evolution operator in the Dirac picture thus defined, pure states evolve in time according to

$$|\Psi_\mathrm{D}\rangle(t) = \mathcal{S}_D(t,t_\mathrm{I})|\Psi_\mathrm{D}\rangle(t_\mathrm{I}),$$
$$\langle\Psi_\mathrm{D}|(t) = \langle\Psi_\mathrm{D}|(t_\mathrm{I})\mathcal{S}_D^\dagger(t,t_\mathrm{I}) = \langle\Psi_\mathrm{D}|(t_\mathrm{I})\mathcal{S}_D(t_\mathrm{I},t). \tag{3.28}$$

The extension of the Dirac evolution operator as defined in (3.27) to an arbitrary pair of times is

$$\mathcal{S}_D(t,t') = \mathcal{S}_\mathsf{F}^\dagger(t,t_\mathrm{I})\mathcal{S}(t,t')\mathcal{S}_\mathsf{F}(t,t_\mathrm{I}). \tag{3.29}$$

It has the properties generalizing Eqs. (3.9), (3.10) and (3.11)

$$\mathcal{S}_D(t,t')\mathcal{S}_D^\dagger(t,t') = \mathcal{S}_D^\dagger(t,t')\mathcal{S}_D(t,t') = \mathbf{1}, \tag{3.30}$$

$$\mathcal{S}_D(t,t') = \mathcal{S}_D(t,t'')\mathcal{S}_D(t'',t'). \tag{3.31}$$

$$\mathcal{S}_D^\dagger(t,t') = \mathcal{S}_D^{-1}(t,t') = \mathcal{S}_D(t',t). \tag{3.32}$$

The equation of motion for the evolution operator in the Dirac picture is

$$i\frac{\partial}{\partial t}\mathcal{S}_D(t,t') = \mathcal{H}_D(t)\mathcal{S}_D(t,t'), \qquad \mathcal{S}_D(t',t') = \mathbf{1}. \tag{3.33}$$

The Hamiltonian in the Dirac picture has the form

$$\mathcal{H}_D(t) = \mathcal{S}_F^\dagger(t,t_I)\mathcal{H}_P\mathcal{S}_F(t,t_I). \tag{3.34}$$

Now it is straightforward to repeat the procedure of Sec. 3.2.3. The integral form of the equation of motion for $\mathcal{S}_D$ is

$$\mathcal{S}_D(t,t') = \mathbf{1} - i\int_{t'}^{t}dt_1 \mathcal{H}_D(t_1)\mathcal{S}_D(t_1,t'). \tag{3.35}$$

Its solution results as

$$\mathcal{S}_D(t,t') = T e^{-i\int_{t'}^{t}d\tau\,\mathcal{H}_D(\tau)}, \qquad t > t'. \tag{3.36}$$

Similarly, for $t' > t$, $\mathcal{S}_D$ is given by

$$\mathcal{S}_D(t,t') = \widetilde{T} e^{+i\int_{t}^{t'}d\tau\,\mathcal{H}_D(\tau)}, \qquad t' > t. \tag{3.37}$$

The last two relations find numerous applications in many body theory.

3.3. *Motivation for introducing NGF*

It has been said that the Green's or correlation functions play a central role in quantum statistical physics, because they provide a link between experimentally relevant quantities and easily calculable quantities.[104,105,109,110,129,131] In this paragraph we look first at the physical relevance of NGF.

3.3.1. *Particle and hole correlation functions*

There are two guiding principles here: to use reduced quantities, as exemplified by the whole transport theory, but to avoid an excessive loss of physical content and flexibility in the process. This may be shown for the single particle density matrix ρ introduced in (2.2). In Heisenberg picture,

$$\mathcal{X}(t) = \int dx_1 dx_2 \psi^\dagger(x_2, t) X(x_1, x_2) \psi(x_1, t),$$

$$\langle \mathcal{X} \rangle_t = \mathrm{Tr}(\mathcal{P}_\mathrm{I} \mathcal{X}(t)) \equiv \mathrm{tr}(\rho(t)X),$$

$$\rho(x_1, x_2; t) = \mathrm{Tr}\left(\mathcal{P}_\mathrm{I} \psi^\dagger(x_2, t)\psi(x_1, t)\right)$$

$$\equiv \langle \psi^\dagger(x_2, t)\psi(x_1, t)\rangle.$$

(3.38)

Here $\mathcal{P}_\mathrm{I}$ is the full many-body statistical operator and $\psi^\dagger$ and ψ are the Heisenberg field operators. To find the density matrix ρ we have to solve the corresponding equation of motion,

$$\frac{\partial \rho}{\partial t} = \mathrm{Tr}\mathcal{P}_\mathrm{I}\left(\frac{\partial \psi^\dagger(x_2, t)}{\partial t}\psi(x_1, t) + \psi^\dagger(x_2, t)\frac{\partial \psi(x_1, t)}{\partial t}\right).$$

(3.39)

In view of Eqs. (3.2), (3.3), the r.h.s. time derivatives lead to expressions involving four field operators, and this appears as the beginning of a BBGKY-like hierarchy of equations. It is rather difficult to formulate well defined approximations terminating or self-consistently closing this hierarchy. The difficulty is two-fold: *both* field operators have the same time argument, and, as a consequence, they are differentiated with respect to this single time argument simultaneously.

This is conveniently overcome by introducing a two time generalization of ρ, the so-called particle correlation function $g^<$, defined as

$$g^<(1, 2) = \mathrm{Tr}\left(\mathcal{P}_\mathrm{I}\psi^\dagger(2)\psi(1)\right)$$

$$\equiv \langle \psi^\dagger(2)\psi(1)\rangle.$$

(3.40)

Following convention, we introduce cumulative variables $1 \equiv x_1, t_1$, etc., denoted by numbers, instead of x, t.

The single particle density matrix is the time diagonal part of $g^<$:

$$\rho(x_1, x_2; t) = g^<(1, 2)\big|_{t_{1,2}=t}.$$

(3.41)

The technical advantage of $g^<$ is that it depends on two time arguments, so that the dynamical behavior of each of the field operators can be treated independently, as will be shown in detail is Sec. 3.4. The two time structure of $g^<$ is also rich in physical content. It describes quantum coherences in a natural way as the time off-diagonal elements, and captures the related memory effects. A simple interpretation of $g^<$ refers to the general notion of a "survival amplitude". Let $t_1 < t_2$ (notice the $<$ sign in $g^<$). Then the meaning of $g^<$ is roughly as follows: an electron is extracted from the system at the worldpoint $1 \equiv x_1, t_1$, the resulting one-particle (in fact, a hole) excitation propagates, until an electron is injected back at $2 \equiv x_2, t_2$. The correlation function measures the amplitude of such an elementary process. Clearly, this amplitude depends on the dynamical behavior of the system, but also on the particle distribution: an electron can be extracted only from occupied places.

To probe in a similar fashion the unoccupied states, a complementary correlation function $g^>$ is introduced in an obvious manner:

$$g^>(1,2) = \text{Tr}\left(\mathcal{P}\psi(1)\psi^\dagger(2)\right)$$
$$\equiv \langle \psi(1)\psi^\dagger(2)\rangle. \tag{3.42}$$

Its interpretation seems even more intuitive, this time for $t_1 > t_2$. An electron is injected at t_2 and later, at t_1, extracted again.

An important step further is to define the spectral density operator,

$$a(1,2) = g^>(1,2) + g^<(1,2). \tag{3.43}$$

By the anticommutation relation (3.4), the time diagonal part of the spectral density is a unit operator in the space of single-particle functions:

$$a(x_1, t, x_2, t) = \delta(x_1 - x_2) \tag{3.44}$$
$$= g^>(x_1, t, x_2, t) + g^<(x_1, t, x_2, t) \tag{3.45}$$
$$\equiv \rho_h(x_1, x_2; t) + \rho(x_1, x_2; t), \tag{3.46}$$
$$\rho_h = 1 - \rho. \tag{3.47}$$

This appears in the first line. On the second line, the unit operator is decomposed into the time diagonal parts of the two correlation functions. As shown on the third line, one is the single-particle density matrix. In analogy, a single hole density matrix is introduced. The last line shows, in an operator form, that the two density matrices complement each other to the unit operator.

3.3.2. *Equilibrium as a special case. Fluctuation-dissipation theorem*

The state of the system in thermal equilibrium has some very particular properties. We will explore them for their own sake and also to shed some light back at the non-equilibrium.

First, in equilibrium, as in any stationary state, the system is homogeneous in time, so that the correlation functions depend only on the difference of both times:

$$g^<(1,2) = g^<(x_1, t_1, x_2, t_2) = g^<(x_1, t_1 - t_2, x_2, 0) \equiv g^<(x_1, x_2; t = t_1 - t_2). \tag{3.48}$$

This permits to go over to the spectral representation:

$$g^<(x_1, x_2; E) = \int dt \exp(iEt) g^<(x_1, x_2; t). \tag{3.49}$$

These relations are written for $g^<$ for definiteness, but the same holds for the other two correlation functions $g^>$ and a.

The equilibrium state of the many-body system may be conveniently taken as the grand-canonical density matrix,

$$\mathcal{P}_{\text{eq}} = Z^{-1}e^{-\beta(\mathcal{H}-\mu\mathcal{N})}, \quad Z = \text{Tr}e^{-\beta(\mathcal{H}-\mu\mathcal{N})}. \tag{3.50}$$

Similarly, the evolution operator is simply

$$\mathcal{S}_{\text{eq}}(t_1, t_2) = e^{-i\mathcal{H}\cdot(t_1-t_2)}. \tag{3.51}$$

These two operators mutually commute. This, together with the anticommutation relations (3.4), permits to derive the relation

$$g^<(x_1, x_2; E) = e^{-\beta(E-\mu)} g^>(x_1, x_2; E). \tag{3.52}$$

With the above interpretation of the correlation functions as transition amplitudes, this relation may be said to reflect the principle of detailed balancing. An even more striking result is obtained, if the spectral density is introduced into the last equation:

$$g^<(x_1, x_2; E) = f_{\mathrm{FD}}(E) a(x_1, x_2; E), \tag{3.53}$$

$$g^>(x_1, x_2; E) = (1 - f_{\mathrm{FD}}(E)) a(x_1, x_2; E). \tag{3.54}$$

The last identities bear the name "Green's function fluctuation-dissipation theorem" in analogy to the FDT known from the linear response theory. In the GF approach, this equilibrium identity mirrors a deep relation between fluctuations (described generally by correlation functions) and dissipation (described by the spectral function). Alternatively, it may be said that the spectral properties of an equilibrium system determine also the statistical information contained in the particle distribution function.

To summarize, the consequence of either of the relations (3.52), (3.54) is that there is only one independent correlation (or Green's) function in equilibrium systems, which is an essential difference from the non-equilibrium situations, for which the knowledge of two independent correlation functions is

3.3.3. Orbital representation

The correlation functions $g^<$, $g^>$ were defined in the coordinate representation. This is natural in a local field theory and, given the structure of the Hamiltonian (3.2), (3.3), this representation is best suited for the general theory, as it will be developed below. In applications, some other orbital representation may be suitable. Consider an orthonormal basis of spinorbitals $\{\varphi_{\lambda\sigma}(\boldsymbol{r})\}$. The associated annihilation and creation operators are related to the field operators ψ, $\psi^\dagger$ by

$$
\begin{aligned}
\psi_\sigma(\boldsymbol{r}) &= \sum_\lambda \varphi_{\lambda\sigma}(\boldsymbol{r}) \cdot c_{\lambda\sigma} & c_{\lambda\sigma} &= \int \mathrm{d}^3 r \, \varphi_{\lambda\sigma}^*(\boldsymbol{r}) \psi_\sigma(\boldsymbol{r}), \\
\psi_\sigma^\dagger(\boldsymbol{r}) &= \sum_\lambda \varphi_{\lambda\sigma}^*(\boldsymbol{r}) \cdot c_{\lambda\sigma}^\dagger & c_{\lambda\sigma}^\dagger &= \int \mathrm{d}^3 r \, \varphi_{\lambda\sigma}(\boldsymbol{r}) \psi_\sigma^\dagger(\boldsymbol{r}).
\end{aligned} \tag{3.55}
$$

These relations in the Schrödinger picture are equally valid in the Heisenberg picture. Substituted into the defining relation (3.40) for the particle correlation function $g^<$, they yield

$$g^<(1,2) = \sum_{\lambda_1 \lambda_2} \varphi_{\lambda_1\sigma_1}(\boldsymbol{r}_1) g^<_{\lambda_1\sigma_1;\lambda_2\sigma_2}(t_1, t_2) \varphi^*_{\lambda_2\sigma_2}(\boldsymbol{r}_2),$$

$$g^<_{\lambda_1\sigma_1;\lambda_2\sigma_2}(t_1, t_2) = \langle c^\dagger_{\lambda_2\sigma_2}(t_2) c_{\lambda_1\sigma_1}(t_1) \rangle. \tag{3.56}$$

These results mean two things. First, $g^<$ may be regarded as an operator in the space of single particle states, whose matrix representation transforms according to the usual rules. Second, the matrix elements of $g^<$ in an arbitrary representation can be obtained directly in a mechanical fashion, if the corresponding annihilation and creation operators are known. The same is true for $g^>$, therefore also for the spectral density and for other Green's functions, which are all formed from the two correlation functions $g^<$, $g^>$ by linear operations, as will be seen shortly.

The choice of representation will often be suggested by the symmetry of the system. The best known example is a homogeneous electron gas, for which the translational symmetry points to the momentum representation. In equilibrium, the Green's functions are k-diagonal and spin-diagonal. For example, the spectral density becomes

$$A_{\sigma_1\sigma_2}(\boldsymbol{k}_1,t_1;\boldsymbol{k}_2,t_2) = A_{\sigma_1}(\boldsymbol{k}_1;t_1,t_2)\delta_{\boldsymbol{k}_1\boldsymbol{k}_2}\delta_{\sigma_1\sigma_2}. \qquad (3.57)$$

Notice the discrete Kronecker δ for the wave vector. It results from the use of periodic boundary conditions making the system finite, although perhaps large, so that the k-vectors form a quasicontinuum.

When setting up a model, the orbital basis is more or less dictated by its structure, without reference to the coordinate representation. Consider, for example, the simplest nanostructure consisting of a molecular island connected by tunneling junctions between two leads. The molecule may be described in a minimum basis set of atomic orbitals, while the leads will be mesoscopic, that is possessing a band of conducting states filled each by a Fermi sea of electrons. The molecular orbitals (MO) may be considered "relevant" and then we will be interested in just the projection $P_{\mathrm{MO}} \cdots P_{\mathrm{MO}}$, where P_{MO} is the corresponding projector in the single particle state space and $\cdots$ dots stand for the Green's function in question.[123]

3.4. *Keldysh time contour*

In this section, the basic formalism of the non-equilibrium Green's Function's will be developed. It should be made clear right here that there is no unique "canonical" NGF machinery, although the core of the theory is common to all variants at stock. This core includes the closed time loop as the time range, construction of the Dyson equations, and the transformation of the loop NGF to a matrix Green's function of real time. The main differences concern the perturbation theoretical *vs.* non-perturbative treatment and the ways of properly respecting the initial conditions. Our present choice will be to start with the Keldysh theory of NGF, which employs simple, the so-called "uncorrelated", initial conditions. This permits to develop the theory in an easy manner. We shall proceed in a non-perturbative way, at variance with Keldysh. This way is less laborious and more physical than that based on the traditional many-body perturbation theory (MBPT). Only in the introductory subsection, we shall recapitulate the original Keldysh line of reasoning, and in Sec. 3.4.6, the diagrammatic perturbation expansion will be recovered using the functional derivative technique.

3.4.1. *From equilibrium MBPT to Keldysh non-equilibrium GF technique*

L. V. Keldysh entitled his fundamental paper "Diagram technique for nonequilibrium processes". His aim was to modify the Feynman diagrams of the MBPT for the equilibrium Green's function in order to allow for processes reaching arbitrarily far from equilibrium. We shall paraphrase in a heuristic manner the ideas of Keldysh.

Let us contrast the two physical situations for the simplest case of zero temperature, starting from the equilibrium case. The external fields are then switched off, $\mathcal{H}'_e(t) = 0$ and the Hamiltonian $\mathcal{H}$ is time-independent. It is assumed that the system in question is normal, that is, it possesses a non-degenerate ground state $|\Psi\rangle$ with the ground state energy E_g which can be found as an eigensolution of the Schrödinger equation

$$\mathcal{H}|\Psi\rangle = E_g|\Psi\rangle, \quad \mathcal{H} = \mathcal{H}_0 + \mathcal{W}. \tag{3.58}$$

The statistical operator of the system in this ground state is

$$\mathcal{P} = |\Psi\rangle\langle\Psi|, \tag{3.59}$$

so that the $g^>$, $g^<$ functions defined by (3.40), (3.42) are equal to

$$\begin{aligned} g^>(1,2) &= \langle\Psi|\psi(1)\psi^\dagger(2)|\Psi\rangle, \\ g^<(1,2) &= \langle\Psi|\psi^\dagger(2)\psi(1)|\Psi\rangle. \end{aligned} \tag{3.60}$$

There is only one independent Green's function in equilibrium, and for the purpose of the MBPT expansion, the proper one is the *real-time causal GF* formed of $g^>$, $g^<$ as follows:

$$G^c(1,2) = -i(\theta(t_1 - t_2)g^>(1,2) - \theta(t_2 - t_1)g^<(1,2)) \tag{3.61}$$

$$= -i\langle\Psi|\,T(\psi(1)\psi^\dagger(2))|\Psi\rangle. \tag{3.62}$$

In the first line, $\theta(t)$ is the Heaviside function, i.e., the unit step from 0 to 1 at $t = 0$. The $>$, $<$ superscripts were historically introduced precisely to indicate that G^c is given by $g^>$, $g^<$ for the time order $t_1 > t_2$, $t_1 < t_2$, respectively. A shorthand for this time ordering prescription is given in the second line. Although we use the same notation T for the chronological operator as before, its definition (3.20) is extended: If there are also fermion field operators among the operators to be time-ordered, an exchange of their time-order is accompanied by a sign change. As is apparent from the first line of (3.61), for any pair of fermion field operators, the new definition reads

$$\begin{aligned} T\{A(t)B(t')\} &= A(t)B(t') \quad t > t', \\ T\{A(t)B(t')\} &= -B(t')A(t) \quad t' > t. \end{aligned} \tag{3.63}$$

This reduces to the previous definition (3.20) for operators forming the Hamiltonians like (3.2) or (3.3), because the fermion field operators enter them always in pairs.

Let $t_1 > t_2$. Then the Green's function (3.61) is represented by

$$G^c(1, 2) = -i\langle \Psi | \psi(1) \psi^\dagger(2)) | \Psi \rangle, \qquad t_1 > t_2. \tag{3.64}$$

This relation will be rewritten in the Dirac picture (properly called an interaction picture in this case) introduced in Sec. 3.2.4. The Hamiltonian is split according to Eq. (3.26) with the correspondence

$$\begin{array}{ccc} \mathcal{H} = & \mathcal{H}_\mathsf{F} & + & \mathcal{H}_\mathsf{P} \\ & \downarrow & & \downarrow \\ \mathcal{H} = & \mathcal{H}_0 & + & \mathcal{W}. \end{array} \tag{3.65}$$

To make the equation visually close to the notation in the literature, we use the caret for the operators in the Dirac picture and plain S for the Dirac evolution operator:

$$\mathcal{X}_D(t) \to \hat{\mathcal{X}}(t), \qquad \mathcal{S}_D(t, t') \to S(t, t'). \tag{3.66}$$

With Eqs. (3.65), (3.66), the Dirac evolution operator (3.29) is written as

$$S(t, t') = T e^{-i \int_{t'}^t d\tau \, \hat{\mathcal{W}}(\tau)}, \qquad t > t'. \tag{3.67}$$

Let us continue with Eq. (3.64). It becomes

$$G^c(1, 2) = -i\langle \Psi | S(t_\mathrm{I}, t_1) \hat{\psi}(1) S(t_1, t_2) \hat{\psi}^\dagger(2) S(t_2, t_\mathrm{I}) | \Psi \rangle, \qquad t_1 > t_2. \tag{3.68}$$

This equation is still not suitable for the perturbation expansion, because the average is performed over the eigenstates of the full Hamiltonian. This can be overcome (following the inspiration by the quantum field theory) using the adiabatic turning on and off of the perturbation. The underlying *adiabatic hypothesis* or adiabatic theorem is a delicate mathematical problem,[240] but here we present the essence of it following.[9,241] Starting at $t \to -\infty$ from the ground state $|\Phi\rangle$ of the unperturbed, i.e. non-interacting, system, the interaction is adiabatically turned on, until the fully dressed ground state $|\Psi\rangle$ of the interacting system is reached at t_I. After that, the interaction is switched off adiabatically again, until at $t \to \infty$ the system returns to the non-interacting ground state $|\Phi\rangle$. Of course, during the process, the wave functions acquire phase factors, so that the three stages of the adiabatic process may be formalized as

$$\begin{array}{ccc} \gamma_2 |\Phi\rangle = S(+\infty, t_\mathrm{I})|\Psi\rangle \longleftarrow & \gamma_1 |\Psi\rangle = S(t_\mathrm{I}, -\infty)|\Phi\rangle \longleftarrow & |\Phi\rangle \\ t = +\infty & t = t_\mathrm{I} & t = -\infty. \end{array} \tag{3.69}$$

While the phase factors γ_1, γ_2 are not known individually, if follows from (3.69) immediately that

$$\gamma_1 \gamma_2 = \langle \Phi | S(+\infty, -\infty) | \Phi \rangle. \tag{3.70}$$

Putting all this together, the expression (3.68) for G^c can be brought to

$$G^c(1,2) = \frac{-i\langle\Phi|S(+\infty,t_1)\hat{\psi}(1)S(t_1,t_2)\hat{\psi}^\dagger(2)S(t_2,-\infty)|\Phi\rangle}{\langle\Phi|S(+\infty,-\infty)|\Phi\rangle}, \qquad t_1 > t_2. \quad (3.71)$$

An analogous result is obtained, if we proceed the same way for $t_2 > t_1$. Together, the two expressions yield the celebrated formula

$$G^c(1,2) = \frac{-i\langle T\{\hat{\psi}(1)\hat{\psi}^\dagger(2)S(+\infty,-\infty)\}\rangle_0}{\langle S(+\infty,-\infty)\rangle_0} \qquad (3.72)$$

as the final result.

We introduce the symbol $\langle\cdots\rangle_0$ for the average over an unperturbed stationary state in general, $\langle\cdots\rangle_0 = \langle\Phi|\cdots|\Phi\rangle$ in Eq. (3.71). This form of G^c is suitable for the perturbation expansion, which also gives a clear interpretation to the symbolic expressions for both the numerator and the denominator. By expanding the evolution operator (3.67) into a power series, which is without problems about commutativity under the T sign, the numerator becomes

$$-i\sum_n \frac{1}{n!}(-i)^n \int\limits_{-\infty}^{+\infty}\cdots\int\limits_{-\infty}^{+\infty}d\bar{t}_1\cdots d\bar{t}_n \langle T\{\hat{\psi}(1)\hat{\psi}^\dagger(2)\hat{W}(\bar{t}_1)\cdots\hat{W}(\bar{t}_n)\}\rangle_0. \quad (3.73)$$

The averages of the time ordered products of operators can be disentangled using the Wick theorem and the bookkeeping of the resulting expressions is best achieved using the Feynman diagrams; these have disconnected parts which are exactly cancelled by the denominator of (3.71) and the result is G^c as a sum over all connected Feynman diagrams. We are not going to follow the details of this classical subject-matter, see, for example, Refs. 9, 17 for details.

Now we are ready to look into the non-equilibrium process driven by an external field V_e. The reference time t_I will be selected such that $\mathcal{H}'_e(t) \neq 0$ for $t > t_I$. Prior to t_I, the system is identical with the equilibrium one, so that we may go over to the interaction picture and start from the equation (3.68). The external field term is included in the unperturbed Hamiltonian. The evolution operator is thus given by (3.67) as before, only we have to remember that it incorporates the forces driving the system out of equilibrium through the unperturbed evolution operator which is hidden in the perturbation $\hat{W}(\tau)$ in the Dirac picture. The whole scenario of the particular Keldysh process we consider starts at $t = -\infty$ from the ground state of the unperturbed system. First, the adiabatic switching on of the interaction leads to the fully dressed ground state, which is reached at $t = t_I$. After that, the system is already driven out of equilibrium. The adiabatic mode of evolution holds no more, so the meaning of the state reached by $t = +\infty$ is not clear. It will not be needed, however, if we exactly retrace the evolution back first to $t = t_I$ and then further up to $t = -\infty$. This brings the system back to the unperturbed ground state. All this may be summarized by a formula written in two equivalent forms:

$$G^c(1,2) = -i\langle S(-\infty, +\infty)\, T\{\hat{\psi}(1)\hat{\psi}^\dagger(2)S(+\infty, -\infty)\}\rangle_0 \qquad (3.74)$$
$$= -i\langle\Phi|S^{-1}(+\infty, -\infty)\, T\{\hat{\psi}(1)\hat{\psi}^\dagger(2)S(+\infty, -\infty)\}|\Phi\rangle. \qquad (3.75)$$

The form (3.74) corresponds precisely to the verbal description just given. The time order of the operators seems to be corrupted by the trip in the reverse direction. The factors are ordered correctly, however, if we apply a new rule that the factors coming later into play stand more to the left. This is just the idea of the Keldysh closed time contour. The other form (3.75) is shown as a link to Eq. (3.72). In the equilibrium case, the adiabatic passage permits to take the S^{-1} factor out of the average as a number, a phase factor, into the denominator and the previous expression Eq. (3.72) is recovered.

3.4.2. *Green's functions on the time contour*

In the previous section, it was shown, how the goal of extending the diagrammatic methods of the many body perturbation theory to non-equilibrium leads to the Keldysh contour in a natural way. This contour is a special limiting case of the contour introduced by Schwinger.[83] The non-equilibrium Green's function may be defined on the Schwinger contour without a direct reference to the perturbation expansion. It then serves to describe general non-equilibrium processes in a non-perturbative manner. We shall follow this way of reasoning, turning to the perturbation-theoretical point of view as appropriate.

The Schwinger contour $\mathfrak{C}$, commonly but not quite precisely called Schwinger-Keldysh contour today, is sketched in Fig. 1. It has two branches labeled by $+$ and $-$. The $+$ branch starts at t_I and extends to some very large time, $t_\infty \to +\infty$. There the contour makes a U-turn and returns as the $-$ branch back to t_I. The two branches are depicted as parallel for clarity, but they both lie strictly on the (real) time axis.

The Keldysh contour is drawn in the conventional manner, as simply stretching from $-\infty$ to $+\infty$ and back. As indicated, this means $t_I \to -\infty$. To specify the

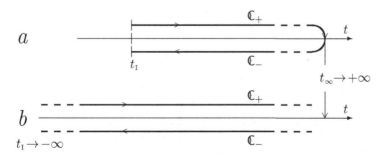

Fig. 1.   NGF contour by (*a*) Schwinger; (*b*) Keldysh.

position of a time variable on either contour, two things are needed: its numerical value and the branch label. Then, two time arguments may be ordered along the contour in an understandable way: If $t < t'$, then the ordering along the contour indicated by the precedence or succession symbols $\prec$, $\succ$ is

$$
\begin{aligned}
t \prec t' && \text{if } t \text{ is at } \ \mathbb{C}_+ \ \text{ and } t' \text{ is at } \ \mathbb{C}_+ \\
t \prec t' && \text{if } t \text{ is at } \ \mathbb{C}_+ \ \text{ and } t' \text{ is at } \ \mathbb{C}_- \\
t \succ t' && \text{if } t \text{ is at } \ \mathbb{C}_- \ \text{ and } t' \text{ is at } \ \mathbb{C}_+ \\
t \succ t' && \text{if } t \text{ is at } \ \mathbb{C}_- \ \text{ and } t' \text{ is at } \ \mathbb{C}_-.
\end{aligned}
\tag{3.76}
$$

The non-equilibrium Green's Function, causal on the contour, is defined by

$$
G(1,2) = -i\mathrm{Tr}\left(\mathcal{P}\,\mathcal{T}_{\mathrm{c}}\{\psi(1|t_{\mathrm{I}})\psi^{\dagger}(2|t_{\mathrm{I}})\}\right)
\tag{3.77}
$$

with the Heisenberg field operators $\psi(1)$, $\psi^{\dagger}(2)$ anchored at t_{I}, as explicitly shown here and tacitly understood in the following. We shall use this convention also for the Keldysh contour. The time-ordering (or chronological) operator $\mathcal{T}_{\mathrm{c}}$ is acting along the contour $\mathbb{C}$ of Fig. 1 according to the ordering rules (3.76) in a manner similar to the usual time ordering operator. If A, B denotes field operators,

$$
\mathcal{T}_{\mathrm{c}}\{A(t)B(t')\} = A(t)B(t') \quad t \succ t',
\tag{3.78}
$$

$$
\mathcal{T}_{\mathrm{c}}\{A(t)B(t')\} = \mp B(t')A(t) \quad t' \succ t
\tag{3.79}
$$

where the $-$ sign holds for fermions, $+$ for bosons.

The Green's function $G(1,2)$ defined in Eq. (3.77) contains all necessary ingredients to fully specify a general non-equilibrium process: the initial time t_{I}, the initial state $\mathcal{P}$ of the system, and the dynamical process driven by the Hamiltonian $\mathcal{H}(t)$ and captured by the Heisenberg field operators. Its physical content is made clear from its decomposition into four real time functions:

The NGF (3.77) defined on the contour comprises in fact four functions of real time (... defined on time axis $-\infty < t < +\infty$) according to the position of the time arguments on the $\mathbb{C}$ contour, as shown in Fig. 2:

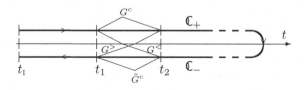

Fig. 2.   NGF time contour.

$$G^{++}(1,2) \equiv G^c(1,2) = -\mathrm{iTr}\left(\rho\, T\{\psi(1)\psi^\dagger(2)\}\right) \quad t_1, t_2 \text{ at } \mathbb{C}_+, \qquad (3.80)$$

$$G^{+-}(1,2) \equiv G^<(1,2) = +\mathrm{iTr}\left(\rho\psi^\dagger(2)\psi(1)\right) \quad t_1 \text{ at } \mathbb{C}_+, \ t_2 \text{ at } \mathbb{C}_- \quad (3.81)$$

$$G^{-+}(1,2) \equiv G^>(1,2) = -\mathrm{iTr}\left(\rho\psi(1)\psi^\dagger(2)\right) \quad t_1 \text{ at } \mathbb{C}_-, \ t_2 \text{ at } \mathbb{C}_+ \quad (3.82)$$

$$G^{--}(1,2) \equiv \widetilde{G}^c(1,2) = +\mathrm{iTr}\left(\rho\, \widetilde{T}\{\psi(1)\psi^\dagger(2)\}\right) \quad t_2, t_1 \text{ at } \mathbb{C}_-. \qquad (3.83)$$

We know two of these functions already, as $G^>$ and $G^<$ differ from $g^>$ and $g^<$ introduced in Eqs. (3.40) and (3.42) just by a prefactor $\pm\mathrm{i}$. The other two quantities are the causal and anti-causal Green's functions. The time-ordering operators $T, \widetilde{T}$ act in accordance with T_c, see Eq. (3.79), but for real times. T was introduced in Sec. 3.4.1 by Eq. (3.63), $\widetilde{T}$ acts similarly. In an algebraic form,

$$
\begin{aligned}
T\{\psi(1)\psi^\dagger(2)\} &= \theta(t_1 - t_2)\psi(1)\psi^\dagger(2) - \theta(t_2 - t_1)\psi^\dagger(2)\psi(1), \\
\widetilde{T}\{\psi(1)\psi^\dagger(2)\} &= \theta(t_1 - t_2)\psi^\dagger(2)\psi(1) - \theta(t_2 - t_1)\psi(1)\psi^\dagger(2).
\end{aligned}
\qquad (3.84)
$$

It would be possible to work directly with the quadruplet (3.80-3.83) already now, but the contour Green's function has important advantageous properties for the formal derivations of the subsequent paragraphs. We shall analyze the decomposition (3.80-3.83) in detail in Sec. 3.5.

3.4.3. *Equation of motion for NGF*

The non-perturbative treatment of NGF is based on the method of the equations of motion. This requires a proper treatment of the initial conditions. The reason we are developing the theory for the Keldysh Green's functions first, is that the uncorrelated initial condition in the distant past can be readily satisfied and this permits to proceed with the basic structure of the equations of motion and the ways of their solution. We shall extend the theory to the general case of arbitrary finite time initial conditions in Section 4.

The Non-equilibrium Green's function (3.77) on the $\mathbb{C}$ contour can be written with the chronological T_c operator expressed in the manner resembling Eq. (3.84):

$$G(t_1, t_2) = \theta_c(t_1, t_2)G^>(t_1, t_2) + \theta_c(t_2, t_1)G^<(t_1, t_2). \qquad (3.85)$$

Here, $\theta_c(t_1, t_2)$ is the step function defined on the path $\mathbb{C}$, with reference to (3.76), the definition reads

$$\theta_c(t_1, t_2) = 1 \quad t_1 \succ t_2, \qquad \theta_c(t_1, t_2) = 0 \quad t_1 \prec t_2. \qquad (3.86)$$

The equation of motion for the Green's function follows from the equation (3.14) for Heisenberg operators. Applied to the field operator, this yields (with the H

subscripts suppressed again):

$$i\frac{\partial \psi(1)}{\partial t_1} = [\psi(1), \mathcal{H}(t_1)]_-$$
$$= h_0(1)\psi(1) + \int d3w(1,3)\psi^\dagger(3)\psi(3)\psi(1) \qquad (3.87)$$
$$w(1,3) \leftarrow w(x_1, x_3)\delta(t_1 - t_3).$$

As indicated, the notation for the interaction w has been changed so as to incorporate its time dependence, i.e., an instantaneous action. The integral then means

$$\int d3 = \int_{\mathbb{C}} dt_3 \int dx_3. \qquad (3.88)$$

Taking the time derivative of the expression (3.85) and using (3.87), we obtain an equation for the NGF on the $\mathbb{C}$ contour:

$$\left(i\frac{\partial}{\partial t_1} - h_0(1)\right) G(1,2) = \delta_c(t_1, t_2) - i\int d3w(1^+, 3)G_2(1,3,2,3^+). \qquad (3.89)$$

A similar "conjugate" equation is derived in the same way and the result is

$$\left(-i\frac{\partial}{\partial t_2} - h_0(2)\right) G(1,2) = \delta_c(t_1, t_2) - i\int d3w(2^+, 3)G_2(1,3,2,3^+). \qquad (3.90)$$

Two new functions enter this result. One is the contour δ-function appearing as the time derivative of the contour step function (3.86):

$$\delta_c(t_1, t_2) = \frac{d}{dt}\theta_c(t_1, t_2) = \begin{cases} +\delta(t_1 - t_2) & \text{if } t_1, t_2 \text{ are both at } \mathbb{C}^+, \\ -\delta(t_1 - t_2) & \text{if } t_1, t_2 \text{ are both at } \mathbb{C}^-, \\ 0 & \text{otherwise.} \end{cases} \qquad (3.91)$$

The other new function is the two particle NGF G_2 on the $\mathbb{C}$ contour, defined by

$$G_2(1,3,2,4) = (-i)^2 \text{Tr}\left(\mathcal{P}_\text{I}\, T_c\{\psi(1|t_1)\psi(3|t_1)\psi^\dagger(4|t_1)\psi^\dagger(2|t_1)\}\right). \qquad (3.92)$$

Except for t_2, all time arguments of G_2 in Eq. (3.89) are equal, $t_1 = t_3$ because of the instantaneous interaction w, $t_3 = t_4$ as follows from the equation of motion. In order to preserve the correct order of multiplications, $\psi^\dagger(3)\psi(3)\psi(1)$ under the action of the time-ordering operator, the two arguments are infinitesimally shifted: $t_4 \equiv t_3^+ = t_3 + 0$, $t_3 = t_1^+ \equiv t_1 + 0$. The shifts are along the $\mathbb{C}$ contour, that is, ± 0 in the algebraic sense on $\mathbb{C}_+$, $\mathbb{C}_-$ resp. Similarly for Eq. (3.90).

The equation (3.89) has the desired form of an equation of motion, but it is not closed, as it contains an unknown Green's function of a higher order. For this GF, there could be obtained an analogous equation containing a three-particle Green's function, etc., and in this way an infinite chain of equations would be developed, the so-called Martin-Schwinger hierarchy. Instead, we turn to methods of converting Eq. (3.89) to a closed equation for the single-particle Green's function.

3.4.4. *Auxiliary fields and the technique of functional derivatives*

The method for the derivation of a closed equation of motion for the non-equilibrium Green's Function on the contour is based on a formal device of subjecting the system to an additional fictitious external field traditionally denoted by $U(t)$. The U field is defined along the contour and it may assume different values on its both branches, $U_+(t)$ and $U_-(t)$. The Green's function incorporating this field is defined as

$$G(1,2;U) = -i \frac{\text{Tr}(\mathcal{P}_I \mathcal{T}_c\{S_U \psi(1)\psi^\dagger(2)\})}{\text{Tr}(\mathcal{P}_I \mathcal{T}_c S_U)}. \tag{3.93}$$

Here,

$$S_U = \mathcal{T}_c \, e^{-i \int d\bar{1} \, \psi^\dagger(\bar{1}^+) U(\bar{1})\psi(\bar{1})}, \tag{3.94}$$

has the form reminiscent of the S matrix in the interaction representation and the whole Green's function (3.93) may be compared with Eq. (3.72). Presently, the perturbation is the additional external field, so that in the transition to the Dirac picture all field operators have the full Heisenberg time dependence including the interactions. Therefore, they have no hats (carets). The integral in (3.94) is given by (3.88), that is, the time integration extends over the whole $\mathfrak{C}$. In the limit of a "physical" disturbance $U_+(t) = U_-(t)$, the denominator of (3.93) would become equal to unity and the whole GF would reduce to the GF with an additional external field written in the Dirac picture. We continue with the general U defined on the contour. The Green's function $G(1,2;U)$ obeys equations of motion extending the equations (3.89), (3.90) for $G(1,2)$:

$$\left(i\frac{\partial}{\partial t_1} - h_0(1) - U(1)\right)G(1,2;U),$$

$$= \delta_c(t_1, t_2) - i\int d3 w(1^+,3)G_2(1,3,2,3^+;U) \tag{3.95}$$

$$\left(-i\frac{\partial}{\partial t_2} - h_0(2) - U(2)\right)G(1,2;U),$$

$$= \delta_c(t_1, t_2) - i\int d3 w(2^+,3)G_2(1,3,2,3^+;U). \tag{3.96}$$

$$G_2(1,3,2,4;U) = (-i)^2 \frac{\text{Tr}\left(\mathcal{P}_I \mathcal{T}_c\{S_U \psi(1)\psi(3)\psi^\dagger(4)\psi^\dagger(2)\}\right)}{\text{Tr}(\mathcal{P}_I \mathcal{T}_c S_U)}. \tag{3.97}$$

The functional derivative with respect to the functional variable U is defined by a relation generalizing the notion of a full differential of a function of many variables expressed in terms of partial derivatives: U is changed by a small variation δU. A functional $\Phi[U]$ changes by $\delta\Phi$ and the linear part of the variation corresponds to the total differential. It must have the form of a linear functional of δU:

$$\delta\Phi[U] = \int d\bar{1}\, \frac{\delta\Phi}{\delta U(\bar{1})} \cdot \delta U(\bar{1}) + \text{higher order terms} \qquad (3.98)$$

and the coefficients at $\delta U(\bar{1})$ define the functional derivative as a function of the variable $\bar{1} = \{x_1, \bar{t}_1\}$. For example, writing $U(1) = \int d2\delta(1-2)U(2)$, we get from (3.98) the useful identity

$$\frac{\delta U(1)}{\delta U(2)} = \delta(1-2). \qquad (3.99)$$

The functional derivative introduced by (3.98) is the physicist's conception of the *Fréchet derivative*. On the same level of mathematical rigor, it may be transformed to the *Volterra derivative*, more convenient in some respects. Its definition is local: let $\delta U(\bar{1}) \neq 0$ only in a small neigborhood of the point 2. Then, by Eq. (3.98),

$$\frac{\Phi[U+\delta U] - \Phi[U]}{\int d\bar{1}\, \delta U(\bar{1})} = \frac{\delta\Phi}{\delta U(2)} + \text{higher order terms} \qquad (3.100)$$

in case that $\delta\Phi/\delta U(1)$ is continuous in a neighborhood of the point 2. Thus the Volterra definition is not suitable for obtaining (3.99).

The main task of this subsection is to deduce the functional derivative of $G(1,2;U)$. By (3.93), G is a ratio of two functionals. The rules for functional derivatives are no different from the usual ones, schematically $(u/v)' = u'/v - uv'/v^2$. Let us sketch the calculation for the denominator of (3.93). It has the explicit form (3.94). The derivative under the chronological operator may be performed ignoring non-commutativity problems:

$$\frac{\delta}{\delta U(3)} \text{Tr}(\mathcal{P}_1 T_c S_U) = \frac{\delta}{\delta U(3)} \text{Tr}(\mathcal{P}_1 T_c\, e^{-i\int d\bar{1}\, \psi^\dagger(\bar{1}^+)U(\bar{1})\psi(\bar{1})})$$

$$= -i\text{Tr}(\mathcal{P}_1 T_c\, e^{-i\int d\bar{1}\, \psi^\dagger(\bar{1}^+)U(\bar{1})\psi(\bar{1})}\psi^\dagger(3^+)\psi^\dagger(3))$$

$$= -i\text{Tr}(\mathcal{P}_1 T_c S_U \psi^\dagger(3^+)\psi^\dagger(3)). \qquad (3.101)$$

The derivative of the whole GF is obtained by similar steps:

$$\frac{\delta G(1,2;U)}{\delta U(3)} = G(3,3^+;U)G(1,2;U) - G_2(1,3;2,3^+;U). \qquad (3.102)$$

By this identity, the equation of motion (3.95) becomes a closed differential equation:

$$\left(i\frac{\partial}{\partial t_1} - h_0(1) - U(1)\right)G(1,2;U)$$

$$= \delta_c(t_1, t_2) - i\int d3\, w(1^+, 3)\left(G(3,3^+;U)G(1,2;U) - \frac{\delta G(1,2;U)}{\delta U(3)}\right). \qquad (3.103)$$

This equation is satisfactory in many respects. It is a non-perturbative equation which closes the Martin-Schwinger hierarchy at the one-particle level. The mean field (Hartree) potential is separated out. For a local spin-independent interaction, like the Coulomb force, we get explicitly:

$$-i\int d3w(1^+, 3)G(3, 3^+; U) = \int dx_3 w(\boldsymbol{r}_1 - \boldsymbol{r}_3)\langle n(\boldsymbol{r}_3, t_1; U)\rangle \equiv V_{\mathrm{H}}(1; U). \quad (3.104)$$

Returning to the physical case $U \to 0$, we may compare the starting equation (3.89) with the final form

$$\left(i\frac{\partial}{\partial t_1} - h_0(1)\right)G(1, 2)$$

$$= \delta_c(t_1, t_2) - i\int d3w(1^+, 3)\left(G(3, 3^+)G(1, 2) - \frac{\delta G(1, 2; U = 0)}{\delta U(3)}\right). \quad (3.105)$$

While Eq. (3.89) was linear – being a mere first link in the chain of linear Martin-Schwinger equations, the autonomous Eq. (3.105) is non-linear, as is already seen from the self-consistent nature of the mean field term. The other term involving the functional derivative incorporates everything beyond the mean field, that is all exchange and correlations. This inner many-particle dynamical structure of the GF is seen to be given by a response function probing the reaction of the system to the U field.

There are two serious technical obstacles on the way to the solution of the equation (3.105). Symptomatically, they are the essential physical constituents of the problem at the same time. One is just $\delta G/\delta U$, the pair correlation function by (3.102), for which unfortunately no methods of direct handling are available. The other technical issue is a proper inclusion of the initial/boundary conditions. Here, we are going to employ the Keldysh initial conditions, as motivated in the introduction to the whole section 3.4: with these initial conditions, several methods how to solve Eq. (3.105) can be developed in a close parallel to analogous procedures known for equilibrium systems.

3.4.5. *Keldysh initial condition*

While the finite time initial condition envisaged for the GF (3.77) on the Schwinger contour is explicit, but requires a complex technical treatment in general, as will be described in Section 4, the initial condition for the Keldysh choice is much easier to work with, but needs a precise explanation.

As has been discussed in Section 3.4.1, if the interactions are switched off adiabatically as $t_{\mathrm{I}} \to -\infty$, and the external fields do not act yet in that distant past, the Hamiltonian tends to

$$\mathcal{H}_0(t \to -\infty) = \mathcal{T} + \mathcal{V}, \quad (3.106)$$

and the system assumes a stationary state $\mathcal{P}_{\mathrm{I}}$ of the isolated non-interacting system asymptotically:

$$[\mathcal{P}_I, \mathcal{H}_0(t \to -\infty)]_- = 0. \tag{3.107}$$

The reference time for both the Heisenberg and Dirac operators coincides with $t_I \to -\infty$. An actual value of this elusive time is not critical, because $\mathcal{P}_I$ is stationary in the asymptotic region and serves as the Heisenberg state of the system throughout the whole process. All averages similar to (3.77) have the meaning

$$\langle \cdots \rangle = \mathrm{Tr}(\mathcal{P}_I \cdots). \tag{3.108}$$

In the forward time direction, $\mathcal{P}_I$ acts as the initial state of both the non-interacting state, and the complete state with interactions included, at $t_I \to -\infty$. This initial state often will, but need not, be a state of equilibrium. For example, it may describe a nanostructure + uncoupled leads with a mutual bias, or even with a temperature difference.[136]

This initial condition is imposed on the NGF through the unperturbed GF corresponding to $\mathcal{W} = 0$. It is introduced by relations analogous to Eqs. (3.93) and (3.94), but, naturally, with all operators in the interaction representation (*cf.* Eq. (3.66)):

$$G_0(1,2;U) = -i \frac{\mathrm{Tr}(\mathcal{P}_I \, T_c \{\hat{S}_U \hat{\psi}(1) \hat{\psi}^\dagger(2)\})}{\mathrm{Tr}(\mathcal{P}_I \, T_c \hat{S}_U)}, \tag{3.109}$$

$$\hat{S}_U = T_c \, e^{-i \int d\bar{1} \, \hat{\psi}^\dagger(\bar{1}^+) U(\bar{1}) \hat{\psi}(\bar{1})}. \tag{3.110}$$

With G_0 at hand, the differential equation (3.103) for G may be converted to an integral form:

$$G(1,2;U) = G_0(1,2;U) - i \iint d4d3 G_0(1,4;U)$$
$$\times w(4^+,3) \left(G(3,3^+;U)G(4,2;U) - \frac{\delta G(4,2;U)}{\delta U(3)} \right). \tag{3.111}$$

The function G_0 satisfies the "free" EOM

$$\left(i\frac{\partial}{\partial t_1} - h_0(1) - U(1) \right) G_0(1,2;U) = \delta_c(t_1,t_2), \tag{3.112}$$

$$\left(-i\frac{\partial}{\partial t_2} - h_0(2) - U(1) \right) G_0(1,2;U) = \delta_c(t_1,t_2). \tag{3.113}$$

It is then readily verified that the GF given by (3.111) satisfies the full equation of motion (3.103). The integral form (3.111) incorporates, in addition, the boundary condition set by the free GF G_0. This involves the asymptotic initial condition common to the free GF and to the full GF, and the external fields. The only viable method for solving the equation (3.111) is to iterate it starting from the zeroth order solution $G^{(0)} = G_0$. Let us try the first iteration:

$$G^{(1)}(1,2;U) = G_0(1,2;U) - i \iint d4d3 G_0(1,4;U)$$

$$\times w(4^+,3) \left(G_0(3,3^+;U)G_0(4,2;U) - \frac{\delta G_0(4,2;U)}{\delta U(3)} \right). \quad (3.114)$$

The key quantity is the functional derivative again. In analogy to Eq. (3.102), it is given by

$$\frac{\delta G_0(1,2;U)}{\delta U(3)} = G_0(3,3^+;U)G_0(1,2;U) - G_{02}(1,3;2,3^+;U) \quad (3.115)$$

with the obvious definition of G_{02}. In the next iteration, G_{03} would enter, etc., invoking gradually the full Martin-Schwinger hierarchy of the unperturbed n-particle Green's functions. This expansion depends entirely on the initial state $\mathcal{P}_I$ reflecting its inner correlations.

We concentrate on the particular class of initial states having no inner correlations, termed appropriately the *uncorrelated initial states*. We define them as those, for which the free two particle GF factorizes to an anti-symmetric product of a pair of single particle GF. Then the functional derivative of G_0 is expressed in terms of G_0 itself. By (3.115),

$$\frac{\delta G_0(1,2;U)}{\delta U(3)} = G_0(3,3^+;U)G_0(1,2;U)$$

$$\overbrace{-\left\{ G_0(1,2;U)G_0(3,3^+;U) - G_0(1,3;U)G_0(3^+,2;U) \right\}}^{\text{uncorrelated } G_{02}},$$

$$\frac{\delta G_0(1,2;U)}{\delta U(3)} = G_0(1,3;U)G_0(3,2;U). \quad (3.116)$$

As a result, the iteration of (3.111) leads to closed expressions for G in terms of G_0 and w in each iteration, and the usual many body perturbation expansion is possible. This will be shown in the next subsection. In view of the discussion in Sec. 3.4.1, it is then proper to identify the uncorrelated $\mathcal{P}_I$ states with those which obey the *Keldysh initial condition*, and to call the non-equilibrium processes unfolding from these initial conditions the *Keldysh processes*.

The deeper reason for this result is that the states uncorrelated according to the definition given are those, for which the Wick theorem, properly generalized, works. Let us sketch a simple *ad hoc* proof.

Let us consider the physical case $U \to 0$ for clarity. The field operator in the interaction representation is governed by a simple equation of motion (*cf.* (3.87))

$$i\frac{\partial \hat{\psi}(1)}{\partial t_1} = h_0(1)\hat{\psi}(1). \quad (3.117)$$

This equation is easily solved like a single particle Schrödinger equation. We introduce the corresponding evolution operator $s(t, t')$ by

$$i\frac{\partial}{\partial t}s(t, t') = h_0(t)\, s(t, t'), \qquad s(t', t') = 1_{\text{op}}, \tag{3.118}$$

and for the initial state we employ the decomposition into the representation of the Hamiltonian (3.106):

$$\mathcal{H}_0(t \to -\infty) = \mathcal{T} + \mathcal{V} = \sum_\alpha \epsilon_\alpha c_\alpha^\dagger c_\alpha. \tag{3.119}$$

We get

$$\hat{\psi}(1) = \sum_\alpha \langle x_1 | s(t_1, t_I) | \alpha \rangle c_\alpha, \qquad \hat{\psi}^\dagger(1) = \sum_\alpha c_\alpha^\dagger \langle \alpha | s(t_I, t_1) | x_1 \rangle. \tag{3.120}$$

The one-particle GF G_0 is obtained as follows. An initial time t_I is selected in the asymptotic region. By Eqs. (3.77), (3.86) and (3.108),

$$G_0(1, 2) = -i\, \text{Tr}\left(\mathcal{P}_I\, T_c\{\hat{\psi}(1)\hat{\psi}^\dagger(2)\}\right)$$
$$= -i\theta_c(t_1, t_2)\langle \hat{\psi}(1)\hat{\psi}^\dagger(2)\rangle + i\theta_c(t_2, t_1)\langle \hat{\psi}^\dagger(2)\hat{\psi}(1)\rangle. \tag{3.121}$$

Inserting the expressions (3.120) for the field operators, we obtain

$$G_0(1, 2) = -i\sum_\alpha \sum_\beta \langle x_1 | s(t_1, t_I) | \alpha \rangle \times$$
$$\left\{\, \theta_c(t_1, t_2)\langle c_\alpha c_\beta^\dagger \rangle - \theta_c(t_2, t_1)\langle c_\beta^\dagger c_\alpha \rangle \,\right\}$$
$$\times \langle \beta | s(t_I, t_2) | x_2 \rangle. \tag{3.122}$$

This may be written in a compact transparent form as

$$G_0(1, 2) = -i\theta_c(t_1, t_2)\langle x_1 | s(t_1, t_I) \left(1 - \rho_I\right) s(t_I, t_2) | x_2 \rangle$$
$$+ i\theta_c(t_2, t_1)\langle x_1 | s(t_1, t_I) \rho_I s(t_I, t_2) | x_2 \rangle, \tag{3.123}$$

with

$$\rho_I = \sum_\alpha \sum_\beta |\alpha\rangle \langle c_\beta^\dagger c_\alpha \rangle \langle \beta|. \tag{3.124}$$

This result seems to depend on t_I, but this dependence is eliminated by the asymptotic stationarity condition (3.107). It is usually assumed that the external fields are turned on at a finite time, say t_P, so that the stationarity condition holds for $t < t_P$. Then Eqs. (3.122), (3.123) are valid for any $t_I < t_P$, in particular for $t_I \to -\infty$.

The free two-particle Green's function can be obtained in the same way. Its structure is given by the formula (3.125), but with the field operators in the interaction representation again:

$$G_{02}(1,3,2,4) = (-\mathrm{i})^2 \mathrm{Tr}\left(\mathcal{P}_\mathrm{I} \mathcal{T}_\mathrm{c}\{\hat{\psi}(1)\hat{\psi}(3)\hat{\psi}^\dagger(4)\hat{\psi}^\dagger(2)\}\right). \qquad (3.125)$$

By substitution from Eqs. (3.120), the Green's function is shown to evolve by the action of four evolution operators $\langle x_i|s(t_i,t_\mathrm{I})|\alpha_i\rangle$ from an initial condition at t_I. The initial condition is given by a quadruple sum of averages of four c, $c^\dagger$ operators whose order is given by the order of the corresponding times. Consider the example of time order leading to the two particle density matrix:

$$t_1 \prec t_3 \prec t_4 \prec t_2 \qquad \mathrm{IC} \cdots \left\{ + \langle c^\dagger_{\alpha_2} c^\dagger_{\alpha_4} c_{\alpha_3} c_{\alpha_1}\rangle \right\}. \qquad (3.126)$$

Next, the requirement that G_{02} factorize (see (3.116)) into

$$G_{02}(1,3,2,4) \to G_0(1,2)G_0(3,4) - G_0(1,4)G_0(3,2), \qquad (3.127)$$

brings about a coincident requirement on the initial conditions. It is easy to verify that in our example (3.126) it is required that

$$t_1 \prec t_3 \prec t_4 \prec t_2 \qquad \langle c^\dagger_{\alpha_2} c^\dagger_{\alpha_4} c_{\alpha_3} c_{\alpha_1}\rangle \overset{!}{=} \langle c^\dagger_{\alpha_2} c_{\alpha_1}\rangle\langle c^\dagger_{\alpha_4} c_{\alpha_3}\rangle - \langle c^\dagger_{\alpha_2} c_{\alpha_3}\rangle\langle c^\dagger_{\alpha_4} c_{\alpha_1}\rangle. \qquad (3.128)$$

This decomposition is the content of the Wick theorem and it is seen that its validity is equivalent to the decomposition (3.127) of the Green's function. This equivalence has been studied in detail by van Leeuwen, see.[131] It seems to have been clearly stated for the first time by Danielewicz in.[102] This author also established the most general form of an uncorrelated initial density matrix,

$$\mathcal{P}_\mathrm{I} = \frac{e^{-\mathcal{A}}}{\mathrm{Tr}e^{-\mathcal{A}}}, \qquad \mathcal{A} = \sum_\alpha A_\alpha c^\dagger_\alpha c_\alpha. \qquad (3.129)$$

This statistical operator is fully specified by an arbitrary numerical sequence $\{A_\alpha\}$. The grand canonical ensemble is obtained as a special case for $A_\alpha = \epsilon_\alpha - \mu$. The result (3.129) thus crowns the previous work on the Wick theorem in statistical physics.[9,131,242–247] The one-particle density matrix for (3.129) has the form

$$\rho = \sum_\alpha |\alpha\rangle f_\alpha \langle\alpha|, \qquad f_\alpha = \left(1 + e^{A_\alpha}\right)^{-1}. \qquad (3.130)$$

The one-particle density matrix thus specifies the uncorrelated initial condition in full. To see the richness of the set of admissible uncorrelated initial conditions, it is enough to choose the various sequences f_α and to recalculate the corresponding $\{A_\alpha\}$. In particular, the N particle ground state is obtained in the limit $f_\alpha \to \theta(\mu - \epsilon_\alpha)$ leading to $A_\alpha \to \pm\infty$ for $\epsilon_\alpha \lessgtr \mu$. By a similar limiting process, any excited Slater determinant of unperturbed one-particle states can be created, etc. In conclusion of this discussion, we mention that Keldysh himself seems to have circumvented the question of uncorrelated initial states in Ref. 89. This he may have done, because the paper was only concerned with spatially homogeneous systems, for which the Wick theorem may be derived under much weaker assumptions, see[17] for example.

3.4.6. Perturbation expansion

Now we are ready to develop the perturbation expansion for the Green's function. The aim is to end up with the Dyson equation for G. By this procedure we depart, in this paragraph, from the general non-perturbative approach of this review. The benefit will be an insight into the relationship of various ways towards the Dyson equation. An outline of the classical method of perturbation expansion in terms of Feynman diagrams was sketched/reminded of in Sec. 3.4.1 at Eq. (3.73). Here, we generate the Feynman diagrams using the alternative technique of functional derivatives. We follow the book of Kadanoff and Baym.[85] Only two equations are needed: Eq. (3.111) slightly rearranged, and the functional derivative (3.116) of G_0:

$$G(1,2;U) = G_0(1,2;U) - \iint d4d3 G_0(1,4;U)[iw(4^+,3)]G(3,3^+;U)G(4,2;U)$$

$$+ \iint d4d3 G_0(1,4;U) \times [iw(4^+,3)]\frac{\delta G(4,2;U)}{\delta U(3)}, \qquad (3.131)$$

$$\frac{\delta G_0(1,2;U)}{\delta U(3)} = G_0(1,3;U)G_0(3,2;U). \qquad (3.132)$$

The expansion in powers of the interaction w,

$$\begin{aligned} G &= G^{(0)} + G^{(1)} + G^{(2)} + \cdots \\ G^{(0)} &= G_0, \end{aligned} \qquad (3.133)$$

is obtained successively from the recurrent relation

$$G^{(n+1)}(1,2;U) = -\sum_{s=0}^{n} \iint d4d3 G_0(1,4;U)[iw(4^+,3)]G^{(s)}(3,3^+;U)G^{(n-s)}(4,2;U)$$

$$+ \iint d4d3 G_0(1,4;U) \times [iw(4^+,3)]\frac{\delta G^{(n)}(4,2;U)}{\delta U(3)}. \qquad (3.134)$$

If all perturbation corrections up to $G^{(n)}$ are expressed in terms of G_0 and w, the same is true for $G^{(n+1)}$ because of Eq. (3.132). It follows by induction that the Green's function is a functional of G_0 and w to all orders of the perturbation expansion, which is called a $G_0 w$ expansion therefore. This result is generally taken to mean that simply $G = G[G_0, w]$.

The expansion (3.133) must coincide with the usual perturbation series. Then the individual terms forming together the nth order correction should be represented by standard Feynman diagrams. This will be illustrated by the diagrams of the first order. There are three elements of the diagrams, the propagator line, the interaction line and a vertex, shown in the table together with their analytical equivalents:

$G_0(1,2)$	1 2
$i\,w(1,2)$	1 2
$\int d3$	3

(3.135)

By (3.134) and (3.132), the first order correction is

$$G^{(1)}(1,2;U) = -\iint d4d3 G_0(1,4;U)[iw(4^+,3)]G_0(3,3^+;U)G_0(4,2;U)$$

$$+\iint d4d3 G_0(1,3;U)[iw(3,4)]G_0(3,4^+;U)G_0(4,2;U). \quad (3.136)$$

The two integrals correspond to the diagrams

$$\text{(3.137)}$$

Even these simplest examples show the basic overall features of the diagrammatic expansion based on Eqs. (3.134) and (3.132). ⋄ The diagrams of any order are obtained recurrently following purely mechanical rules which are easy to establish from the analytical equations; ⋄ a single representant is obtained for each set of topologically equivalent diagrams; ⋄ all diagrams are connected, i.e., diagrams with disconnected parts, like ⊶, are excluded automatically. Altogether, the present method of functional derivatives leads straight to the *connected diagram expansion* of the Green's function. This is a convenient alternative to the more common direct use of the Wick theorem referred to at the end of Sec. 3.4.1, see, for example, Ref. 131.

In order to construct the analytical expressions for the members of the perturbation series starting from the diagrams, it is enough to follow the *General rules for Feynman diagrams*, which are: ⋄a assign weight 1 to each diagram; ⋄b use the correspondence (3.135); ⋄c for each closed loop of Fermion lines add a prefactor of -1. Applied to the diagrams (3.137), these rules recover the expressions (3.136). Notice that the ⋄c rule has to be used for the first (Hartree) diagram.

The whole Green's function is a sum of all (...expressions corresponding to the) Feynman diagrams of the typical form

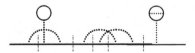

where the time arrows are left out for simplicity.

Each such diagram consists of a chain of bare (free) propagator lines joined together by inserts consisting of propagator lines and interaction lines and having exactly two terminal points. The whole diagram is said to be *reducible*, because it can be split into disconnected parts by cutting a single propagator line. The inserts are the *irreducible parts*, called so, because they cannot be split by such a single cut. This distinction is illustrated in the diagram by a few dot-dashed vertical lines. The diagrams may be of various lengths, i.e., have any number of irreducible inserts. Using the notation IR for these irreducible parts, we may write the whole Green's function as a symbolic sum

$$G = G_0 + G_0 \sum_{n=1}^{\infty} \sum_{\mathsf{IR}_1} \cdots \sum_{\mathsf{IR}_n} \mathsf{IR}_1 G_0 \cdots \mathsf{IR}_n G_0. \tag{3.138}$$

The sum may be rearranged to

$$G = G_0 + G_0 \sum_{\mathsf{IR}_1} \mathsf{IR}_1 \left\{ G_0 + G_0 \sum_{n=2}^{\infty} \sum_{\mathsf{IR}_2} \cdots \sum_{\mathsf{IR}_n} \mathsf{IR}_2 G_0 \cdots \mathsf{IR}_n G_0 \right\}. \tag{3.139}$$

All sums involved are infinite and we see that in the last formula $\{\cdots\} = G$ for any IR_1. The outer sum may then be executed with the result

$$G = G_0 + G_0 \Sigma G, \tag{3.140}$$

$$\Sigma = \sum_{\mathsf{IR}} \mathsf{IR}. \tag{3.141}$$

The first line is a symbolic shorthand for the Dyson equation well known from the equilibrium theory; the second line defines the

self-energy Σ = sum of all irreducible two terminal diagrams.

As an explicit example, we quote the lowest order approximation for the self-energy,

$$\Sigma^{(1)}(1, 2; U) = \qquad\qquad\qquad + \qquad\qquad \tag{3.142}$$

$$= -\mathrm{i}\delta_c(1, 2)\!\int\! \mathrm{d}3 w(1, 3) G_0(3, 3^+; U) + \mathrm{i} w(1, 2) G_0(1, 2^+; U).$$

It is easy to check that, had we started from Eq. (3.138) with a reverse order of the inner sums, $\mathsf{IR}_n \cdots \mathsf{IR}_1$, the same manipulations would have led to the Dyson equation $G = G_0 + G \Sigma G_0$ with the reverse order of factors, but with the same self-energy. This is an exceedingly important result.

In the standard notation, the final outcome of this analysis is the Dyson equation

$$G(1, 2; U) = G_0(1, 2; U) + \int\! \mathrm{d}\bar{3}\!\int\! \mathrm{d}\bar{4} G_0(1, \bar{3}; U) \Sigma(\bar{3}, \bar{4}; U) G(\bar{4}, 2; U) \tag{3.143}$$

and its conjugate

$$G(1,2;U) = G_0(1,2;U) + \int d\bar{3} \int d\bar{4} \, G(1,\bar{3};U)\Sigma(\bar{3},\bar{4};U)G_0(\bar{4},2;U). \tag{3.144}$$

To sum up,

◇ it was shown that the pair interaction can be incorporated into the non-equilibrium Green's function by means of a perturbation series, whose individual terms are classified by Feynman diagrams identical with those known from the equilibrium many body theory. This is an important illustration of the rule set by D. Langreth long ago:[95]

> "I will not give a set of diagrammatic rules and simply say: use your own rules. They will work here as well!"

which, as is apparent, should be supplemented by the proviso "... for an uncorrelated initial condition". We shall return to these matters in Sec. 3.4.1.

◇ In particular, the method of summation of infinite subsets of Feynman diagrams is applicable just as in the equilibrium theory, and it was employed to derive the Dyson equations. These equations have a non-perturbative nature, which will be used below. It should be remembered, however, that this result was obtained to all orders of the perturbation theory only, again just like in usual equilibrium statistical physics.

3.4.7. *Self-energy and Dyson equation*

The self-energy could have been introduced without recourse to the perturbation expansion starting from the equations of motion (3.95), (3.96), in which the term with the two-particle function G_2 would be replaced by the self-energy using the definitions

$$\begin{aligned} -\mathrm{i} \int d3w(1^+,3)G_2(1,3,2,3^+;U) &= \int d3\Sigma(1,3;U)G(3,2;U) \quad \text{in (3.89)}, \\ -\mathrm{i} \int d3w(2^+,3)G_2(1,3,2,3^+;U) &= \int d3G(1,3;U)\widetilde{\Sigma}(3,2;U) \quad \text{in (3.90)}, \end{aligned} \tag{3.145}$$

for the self-energy Σ and the conjugate self-energy $\widetilde{\Sigma}$. Then the first task would have been to prove that they must be equal, a fact which follows from the diagrammatic analysis directly:

$$\Sigma(1,2;U) = \widetilde{\Sigma}(1,2;U). \tag{3.146}$$

The proof can be completed, if the assumption of the uncorrelated initial condition is employed.

Substitution of Eq. (3.145) back in the equations of motion turns them into Dyson equations in differential form. It will be convenient to introduce the inverse Green's functions by the following steps. First, the inverse free Green's function is defined by

$$G_0^{-1}(1,2;U) = \left(\mathrm{i}\frac{\partial}{\partial t_1} - h_0(t_1) - U(t_1)\right)\delta_c(1,2), \qquad \delta_c(1,2) \equiv \delta_c(t_1,t_2)\delta(x_1,x_2). \tag{3.147}$$

It satisfies the equations

$$\int d\bar{3} G_0^{-1}(1,\bar{3};U) G_0(\bar{3},2;U) = \delta_c(1,2),$$

$$\int d\bar{3} G_0(1,\bar{3};U) G_0^{-1}(\bar{3},2;U) = \delta_c(1,2),$$

(3.148)

and in this sense is an inverse operator to G_0. The last identities considered as equations for G_0 are solved by any admissible G_0, of course, so that the formal "inverse" G_0^{-1} can only be inverted back in conjunction with a specific initial condition. With all these prerequisites, we can rewrite the equations of motion (3.95), (3.96) to the differential Dyson equations in an operator form:

$$\int d\bar{3} \left(G_0^{-1}(1,\bar{3};U) - \Sigma(1,\bar{3};U) \right) G(\bar{3},2;U) = \delta_c(1,2),$$

$$\int d\bar{3} G(1,\bar{3};U) \left(G_0^{-1}(\bar{3},2;U) - \Sigma(\bar{3},2;U) \right) = \delta_c(1,2).$$

(3.149)

These equations can also be considered a definition of the inverse full Green's function:

$$G^{-1}(1,2;U) = G_0^{-1}(1,2;U) - \Sigma(1,2;U),$$

$$\int d\bar{3} G(1,\bar{3};U) G^{-1}(\bar{3},2;U) = \int d\bar{3} G^{-1}(1,\bar{3};U) G(\bar{3},2;U) = \delta_c(1,2).$$

(3.150)

Now we turn to the functional equation (3.103) again, this time in order to generate self-consistent equations for the self-energy. We have

$$\int d3 \Sigma(1,3;U) G(3,2;U) = -i \int d3 w(1^+,3) \left(G(3,3^+;U) G(1,2;U) - \frac{\delta G(1,2;U)}{\delta U(3)} \right).$$

(3.151)

The identity $\delta G = \delta G(G^{-1}G) = (\delta G G^{-1})G = (-G\delta G^{-1})G$, which follows from (3.150), has the explicit form given by Eq. (3.149)

$$\frac{\delta G(1,2;U)}{\delta U(3)} = -\int d4 d5 G(1,4;U) \left(\delta_c(3,4)\delta_c(4,5) + \frac{\delta \Sigma(4,5;U)}{\delta U(3)} \right) G(5,2;U).$$

(3.152)

A closed equation for the self-energy then follows from (3.151). It can be written in two slightly different forms:

$$\Sigma(1,2;U) = \overbrace{-i\int d3w(1^+,3)G(3,3^+;U)\delta_c(1,2)}^{\Sigma_{\mathrm{HF}}(1,2;U) = V_{\mathrm{HF}}(x_1,x_2,t_1;U)\delta_c(t_1,t_2)} + iw(1,2)G(1,2;U)$$

$$+i\int d3d4w(1,4)G(1,3;U)\frac{\delta\Sigma(3,2;U)}{\delta U(4)}, \qquad (3.153)$$

$$\Sigma(1,2;U) = \overbrace{-i\int d3w(1^+,3)G(3,3^+;U)\delta_c(1,2)}^{\Sigma_{\mathrm{H}}(1,2;U)}$$

$$+i\int d3d4w(1,4)G(1,3;U)\Big(\overbrace{\delta_c(3,2)\delta_c(3,4) + \frac{\delta\Sigma(3,2;U)}{\delta U(4)}}^{\Gamma(3,2;4;U)}\Big). \quad (3.154)$$

The first form (3.153) collects the two terms which are 'singular', i.e., time-local, to the Hartree-Fock self-energy on the first line. The remaining term involves the vertex correction. This equation can be solved by iteration. The full GF is taken as given during the process, so that the iteration leads to a formal expansion of Σ in powers of the interaction w af fixed G. The resulting series can be represented by means of diagrams which are rather similar to those of the plain perturbation expansion of Σ, with two differences: ◇ Full lines in the diagrams correspond to G rather than to G_0, ◇ There are no self-energy insertions at these lines.

The important conclusion is that the self-energy is expressed solely in terms of G and w and its U dependence is thus mediated through G. In other words, the self-energy is a functional $\Sigma[G]$ of the Green's function G. This functional dependence complements the Dyson equation (3.149), which is, in fact, just an identity between G_0, Σ and G. A closed equation for G results,

$$\int d\bar{3}\left(G_0^{-1}(1,\bar{3};U) - \Sigma(1,\bar{3})[G]\right)G(\bar{3},2;U) = \delta_c(1,2),$$

$$\int d\bar{3}G(1,\bar{3};U)\left(G_0^{-1}(\bar{3},2;U) - \Sigma(\bar{3},2)[G]\right) = \delta_c(1,2). \qquad (3.155)$$

The equation (3.154) preserves the separation of the self-energy into the local mean-field part and the exchange-correlation rest. The expression for Σ may be compared with the lowest order iteration $\Sigma^{(1)}$ of the perturbation series (3.142). There, two diagrams represent the Hartree term and the Fock exchange term, both expressed in terms of the unperturbed Green's function and of the interaction. The exact self-energy (3.154) is obtained from (3.142) by renormalizing G_0 to G and by renormalizing *one* of the vertices in the exchange term from a simple point-like *bare vertex*, in which two G and one interaction w meet, to a three-point structure of the *full many body vertex*. The *vertex correction* $\delta\Sigma/\delta U$ is responsible for the correlation effects, everything "beyond the Hartree-Fock". This can be represented diagramatically as

$$\Sigma(1,2;U) = \qquad\qquad + \qquad\qquad \tag{3.156}$$

Once the self-energy functional $\Sigma[G]$ is given, exact or approximate, the auxiliary U field is not needed. A system of self-consistent equations for determination of the NGF can be derived. First, a closed equation for the Γ vertex is obtained. By (3.154) and (3.152),

$$\Gamma(1,2;3) = \delta_c(1,2)\delta_c(1,3) + \frac{\delta\Sigma(1,2;U)}{\delta U(3)}$$

$$= \delta_c(1,2)\delta_c(1,3) + \int d4d5 \frac{\delta\Sigma(1,2;U)}{\delta G(4,5;U)}\frac{\delta G(4,5;U)}{\delta U(3)}, \tag{3.157}$$

and setting $U \to 0$, we obtain the integral equation for Γ

$$\Gamma(1,2;3) = \delta_c(1,2)\delta_c(1,3) + \int d4d5d6d7 \frac{\delta\Sigma(1,2)}{\delta G(4,5)}G(4,6)G(7,5)\Gamma(6,7;3). \tag{3.158}$$

Together with the expression Eq. (3.154) for the self-energy and the Dyson equation (3.149) for G, collected here in a concise form,

$$\Sigma(1,2) = \Sigma_{\mathrm{H}}(1,2) + i\int d3d4 w(1,4)G(1,3)\Gamma(3,2;4), \tag{3.159}$$

$$\left(i\frac{\partial}{\partial t_1} - h_0(1)\right)G(1,2) = \delta_c(1,2) + \int d3\Sigma(1,3)G(3,2), \tag{3.160}$$

we have a self-consistent system of equations for the GF description of an arbitrary non-equilibrium process starting from a Keldysh initial condition. In order to obtain a solution of the system (3.158-3.160), a physical approximation for the four-point vertex $\delta\Sigma/\delta G$ has to be chosen. This is the true heart of the whole task, the rest is technical.

Comparing the results of this part with the perturbative approach of Sec. 3.4.6, we may summarize that there, the self-energy was generated as a functional $\Sigma[G_0]$. This has been superseded by $\Sigma[G]$ presently. The gains from this transition are manifold: the free GF G_0, whose meaning at finite times is spurious, has been eliminated, the relation $G \leftrightarrow \Sigma$ is non-perturbative and self-consistent, the many-body vertex structure is separated from single particle fields.

The transition to the self-consistent formalism may be symbolized simply as going from the G_0w formalism with bare both the GF and the interaction to the Gw formalism, in which the GF is dressed (renormalized), but the interaction remains bare. We sketch now one more step, to the GW formalism, in which the interaction is also renormalized. The procedure is universal. It is inevitable for the Coulomb interaction, where it accounts for the all-important screening effects.

We start from the equation of motion (3.103) and transfer the mean-field part of the r.h.s. to the left as the Hartree field $V_H(1;U)$ according to Eq. (3.104). We introduce the screened field $U_{eff}(1) = U(1) + (V_H(1;U) - V_H(1;U=0))$ and get

$$\left(i\frac{\partial}{\partial t_1} - h_0(1) - V_H(1;U=0) - U_{eff}(1) \right) G(1,2;U)$$

$$= \delta_c(t_1, t_2) + i\int d3\, w(1^+, 3)\frac{\delta G(1,2;U)}{\delta U(3)}. \tag{3.161}$$

U_{eff} will be substituted as the new variational variable instead of U. The integral on the r.h.s. equals to $\int d3(\Sigma - \Sigma_H)G$ by (3.151). We denote the difference of self-energies as Σ_{XC} for eXchange and Correlation:

$$\int d3\, \Sigma_{XC}(1,3;U)G(3,2;U) = -i\int d3\, d4\, w(1^+, 3)\frac{\delta U_{eff}(4)}{\delta U(3)}\frac{\delta G(1,2;U)}{\delta U_{eff}(4)}. \tag{3.162}$$

Proceeding as before, we obtain

$$\frac{\delta G(1,2;U)}{\delta U(3)} = -\int d4\, d5\, G(1,4;U)\Gamma_s(4,5;3)G(5,2;U), \tag{3.163}$$

$$\Sigma_{XC}(1,2) = i\int d3\, d4\, w_s(1,4)G(1,3)\Gamma_s(3,2;4). \tag{3.164}$$

with the renormalized interaction w_s and the vertex Γ_s defined by

$$w_s(1,2;U_{eff}) = \int d3\, w(1^+, 3)\frac{\delta U_{eff}(2)}{\delta U(3)}, \tag{3.165}$$

$$\Gamma_s(3,2;4;U_{eff}) = \delta_c(3,2)\delta_c(3,4) + \frac{\delta \Sigma_{XC}(3,2;U_{eff})}{\delta U_{eff}(4)}. \tag{3.166}$$

Eqs. (3.164) and (3.166) permit to obtain an expansion of Σ_{XC} in terms of G and w_s. In the diagrammatic representation of the resulting series, the diagrams representing insertions between interaction lines are absent and all diagrams correspond to vertex corrections of an ever increasing topological complexity.

The renormalized interaction is given by the equation

$$w_s(1,2) = w(1,2) + \int d3\, d4\, w(1,3)\Pi(3,4)w_s(4,2), \tag{3.167}$$

$$\Pi(3,4) = \int d5\, d6\, G(3,5)G(6,3)\Gamma_s(5,6;4). \tag{3.168}$$

The equation (3.167) has the structure of a Dyson equation. The quantity Π is the *polarization operator*. It is apparent now that the self-energy is a functional of G again, with no explicit U dependence, so that the final equation closing the self-consistent set for the case of a screened interaction is

$$\Gamma_s(1,2;3) = \delta_c(1,2)\delta_c(1,3) + \int d4 d5 d6 d7 \frac{\delta\Sigma_{\mathrm{xc}}(1,2)}{\delta G(4,5)} G(4,6)G(7,5)\Gamma_s(6,7;3).$$

(3.169)

The equations forming the self-consistent system were already written for $U_{\mathit{eff}} = 0$, that is $U = 0$. These are: (3.164), (3.167), (3.168), and (3.169). Finally, the Dyson equation is added, in the form which follows from Eqs. (3.161) and (3.162):

$$\left(i\frac{\partial}{\partial t_1} - h_0(1) - V_{\mathrm{H}}(1)\right) G(1,2;U) = \delta_c(t_1, t_2) + \int d3 \Sigma_{\mathrm{xc}}(1,3)G(3,2). \quad (3.170)$$

We made this excursion into the GW version of the GF formalism, because it is widely used in the area of electronic structure computations and it is gradually becoming standard also in the non-equilibrium problems. Its approximate formulations include the well known random phase approximation (RPA) and the popular so-called GW approximation consisting in the neglect of the vertex correction in Eq. (3.164). The reader may be referred to.[131] Here, this direction will not be pursued further.

3.4.8. *A note to approximate theories*

In this review, we are concerned primarily with the general structure of the NGF theory and do not analyze in detail properties of the inevitable approximations which make the whole formalism tractable. On the whole, two classes of approximations are in use, one class is formed by approximations of the decoupling type, which are usually of an ad hoc nature, based on physical motivation. The best known example is the Hartree-Fock theory. The other class encompasses the approximations of a systematic nature, usually based on the existence of small parameters, which serve for a systematic expansion and offering ways of testing and/or improving an approximation of certain degree. Here, we have to mention the RPA method as a classical example and the GW approximation as the technique on the rise and promising further improvement.

It is clear that very often there is no quantitative criterion to judge an approximation. It is then essential to use only (or at least as much as possible) approximate theories which are physically consistent, in other words, which are qualitatively correct and do not contain an inner contradiction.

Here, we only briefly mention several consistency requirements, which approximate theories have to fulfill. The reader can find a useful overview of these topics here.[129,131]

To ensure proper physical meaning of calculated observables, approximations must lead to conservation laws for observables like the number of particles and total energy, total momentum and total angular momentum. Within the framework of NGF approach this is closely related to the so called Ward identities and

ϕ derivable approximations for the self-energy. The conserving approximations for the NGF were thoroughly discussed by Kadanoff and Baym.[85,248,249] Ward identities, including their non-equilibrium variants, are discussed in the following articles.[125,162,250–253]

3.5. *Matrix Green's function of real time*

The formalism of the non-equilibrium Green's Function's defined on the time loop has an equivalent reformulation working with functions of real time, as we have mentioned preliminarily in Sec. 3.4.2. There are several variants of this matrix representation which have been used in the literature and a brief overview follows.

3.5.1. *Matrix representation for general functions on the contour*

In Sec. 3.4.2, Eqs. (3.80)–(3.83) and Fig. 2, it was shown on the example of the Green's function, how a single function of double times produces four real time functions. For further work, this quadruplet is conveniently arranged into a 2×2 matrix. In the case of the Green's function, we have

$$G \longleftrightarrow \overset{\supset}{\mathbf{G}} \equiv \begin{vmatrix} G^c & G^< \\ G^> & \tilde{G}^c \end{vmatrix} \equiv \begin{vmatrix} G^{++} & G^{+-} \\ G^{-+} & G^{--} \end{vmatrix} \equiv \begin{vmatrix} G^{11} & G^{12} \\ G^{21} & G^{22} \end{vmatrix}. \tag{3.171}$$

The first representation is descriptive, peculiar to the Green's function, and we shall return to it in Sec. 3.5.2. The $\pm$ notation originates from Keldysh and the signs refer to the two branches of the time contour. The numerical labels have the same meaning and we shall use the third variant for the formal developments now. In general, for a function $F(s,t)$ of two times s,t on the contour, the correspondence is

$$F(s,t) \longleftrightarrow \overset{\supset}{\mathbf{F}} \equiv \begin{vmatrix} F^{11}(s,t) & F^{12}(s,t) \\ F^{21}(s,t) & F^{22}(s,t) \end{vmatrix} \equiv \begin{vmatrix} F(s^+,t^+) & F(s^+,t^-) \\ F(s^-,t^+) & F(s^-,t^-) \end{vmatrix} \tag{3.172}$$

where on the right all times run from $-\infty$ to $+\infty$. These matrices are denoted by boldface characters and are — just for the present discussion — tagged by the $\supset$ symbol as a reminder of the time loop. The functions on the loop can be added and multiplied one with another and their assembly contains a neutral element for addition, zero, represented simply by the zero function, and also a neutral element for multiplication, unity, represented by the delta function $\delta_c(s,t)$ — if such functions, singular at $s = t$, are also admitted. Altogether, the functions on the loop form a (unitary) ring $\mathfrak{C}$. [a] On going to the matrix representation, it appears that addition is mapped on the matrices trivially, but for multiplication,

---

[a] Remaining axioms of a ring, commutativity of addition, associativity of multiplication, distributivity of multiplication, are all verified by inspection.

say $D = BC$, we obtain

$$D(s, u) = \int_{\mathbb{C}} dt\, B(s, t) C(t, u) \tag{3.173}$$

$$\longleftrightarrow D^{\alpha\beta}(s, u) = \int_{-\infty}^{+\infty} dt\, B^{\alpha 1}(s, t) C^{1\beta}(t, u) + \int_{+\infty}^{-\infty} dt\, B^{\alpha 2}(s, t) C^{2\beta}(t, u)$$

$$= \int_{-\infty}^{+\infty} dt\, B^{\alpha 1}(s, t) C^{1\beta}(t, u) - \int_{-\infty}^{+\infty} dt\, B^{\alpha 2}(s, t) C^{2\beta}(t, u) \tag{3.174}$$

$$= \int_{-\infty}^{+\infty} dt\, \left(\overset{\supset}{\mathbf{B}}(s, t) \tau_3 \overset{\supset}{\mathbf{C}}(t, u) \right)^{\alpha\beta}, \qquad \tau_3 = \begin{vmatrix} 1 & 0 \\ 0 & -1 \end{vmatrix}. \tag{3.175}$$

Here, τ_3 is one of the Pauli matrices; for the rest, we use a corresponding notation,

$$\tau_0 = \begin{vmatrix} 1 & 0 \\ 0 & 1 \end{vmatrix}, \quad \tau_1 = \begin{vmatrix} 0 & 1 \\ 1 & 0 \end{vmatrix}, \quad \tau_2 = \begin{vmatrix} 0 & -i \\ i & 0 \end{vmatrix}. \tag{3.176}$$

We are thus led to work with matrices

$$\overset{\rightrightarrows}{\mathbf{F}} = \tau_3 \overset{\supset}{\mathbf{F}} = \begin{vmatrix} F^{11} & F^{12} \\ -F^{21} & -F^{22} \end{vmatrix}. \tag{3.177}$$

Their set is closed with respect to addition and multiplication, i.e., Eq. (3.173) becomes

$$\overset{\rightrightarrows}{\mathbf{D}}(s, u) = \int_{-\infty}^{+\infty} dt\, \overset{\rightrightarrows}{\mathbf{B}}(s, t) \overset{\rightrightarrows}{\mathbf{C}}(t, u). \tag{3.178}$$

In other words, the set is algebraically isomorphic with the original ring of functions on the contour. It is possible to multiply more factors consecutively, so that, for example, the Dyson equation (3.143) reads

$$\overset{\rightrightarrows}{\mathbf{G}} = \overset{\rightrightarrows}{\mathbf{G}}_0 + \overset{\rightrightarrows}{\mathbf{G}}_0 \overset{\rightrightarrows}{\mathbf{\Sigma}} \overset{\rightrightarrows}{\mathbf{G}}. \tag{3.179}$$

The matrix representation of the delta-function follows from Eq. (3.91):

$$\delta_c(t_1, t_2) \quad \leftrightarrow \quad \overset{\supset}{\boldsymbol{\delta}}(t_1, t_2) = \tau_3 \delta(t_1 - t_2) \quad \leftrightarrow \quad \overset{\rightrightarrows}{\boldsymbol{\delta}}(t_1, t_2) = \tau_0 \delta(t_1 - t_2). \tag{3.180}$$

By the last transformation, a plain delta-function is thus obtained, so that, for example, the definition (3.147) of the inverse free Green's function has the form

$$\overset{\rightrightarrows}{\mathbf{G}}_0^{-1}(1, 2; U) = \left(i\frac{\partial}{\partial t_1} - h_0(1) - U(1) \right) \tau_0\, \delta(1, 2) \equiv G_0^{-1}(1, 2; U)\delta(1, 2) \tag{3.181}$$

$$\delta(1, 2) \equiv \delta(t_1 - t_2) \tau_0 \delta(x_1, x_2).$$

3.5.2. *G-like functions*

The definition (3.77) of the Green's function repeated here for convenience,

$$G(1,2) = -i\text{Tr}\left(\mathcal{p}\,T_c\{\psi(1|t_1)\psi^\dagger(2|t_1)\}\right).$$

implies several symmetry properties of G as a function on the contour discussed here following Danielewicz:[102]

- *Symmetry with respect to the branches of the contour*

We introduce the following notation. Let t_1 be at a branch of the contour. Then t_1^T lies oppositely on the other branch. Further, if $t_1, x_1 \equiv 1$, then $t_1^T, x_1 \equiv 1^T$. The symmetry of G reads:

$$\text{Let } t_1 > t_2 \text{ algebraically. Then } t_2 \prec t_1,\ t_1^T \Rightarrow G(1,2) = G(1^T,2),$$
$$\text{Let } t_1 < t_2 \text{ algebraically. Then } t_1 \prec t_2,\ t_2^T \Rightarrow G(1,2) = G(1,2^T). \tag{3.182}$$

These rules lead to the form (3.85) for G, which in turn is equivalent with the first matrix form of G in Eq. (3.171).

Other quantities also obey the symmetry rules (3.182), and together they constitute the *class of G-like functions* we shall denote as $\mathfrak{K}$. The general structure of functions from $\mathfrak{K}$ is

$$F(t_1, t_2) = \theta_c(t_1, t_2)F^>(t_1, t_2) + F^\delta + \theta_c(t_2, t_1)F^<(t_1, t_2) \tag{3.183}$$

where $F^>$, $F^<$ are arbitrary and

$$F^\delta(t_1, t_2) = f_0(t_1)\delta_c(t_1, t_2) + f_1(t_1)\delta_c'(t_1, t_2) + \cdots \tag{3.184}$$

is the singular component already introduced for a general F in the preceding paragraph.

It turns out that the G-like functions (3.183) form a subset of all functions on the contour, which is closed with respect to addition and multiplication, and contains unity δ_c. In other words, $\mathfrak{K}$ is a (unitary) sub-ring of $\mathfrak{C}$. If the inverse to an $F \in \mathfrak{K}$ exists, then it also belongs to this class of G-like functions. In particular, for the self-energy we get the important result that $\Sigma \in \mathfrak{K}$, explicitly

$$\Sigma(1,2) = \theta_c(t_1, t_2)\Sigma^>(1,2) + V_{\text{HF}}(1,2)\delta_c(t_1, t_2) + \theta_c(t_2, t_1)\Sigma^<(1,2). \tag{3.185}$$

NOTE The set of G-like functions is referred to as Keldysh space in Ref. 131 and other works of R. van Leeuwen and coworkers. In Ref. 129 even the whole $\mathfrak{C}$ ring is called the Keldysh space. It should be noted that this terminology is at variance with most other literature, where the term Keldysh space is reserved, probably starting from the influential review,[104] to the representation of contour functions by matrices of real time functions. Even this latter convention is not accepted universally, but we are going to adhere to it.

- *Symmetry with respect to complex conjugation*

Another symmetry of G which follows directly from the definition (3.77) or from Eq. (3.85) and Eqs. (3.81)–(3.82) can be written in two equivalent forms:

$$[G(1,2)]^* = -G(2^T, 1^T), \tag{3.186}$$

$$[G(1,2)]^\dagger = -G(1^T, 2^T). \tag{3.187}$$

The important functions from $\mathfrak{K}$, like G_0, G_0^{-1}, G^{-1} and Σ, all may be shown to obey the same symmetry condition

$$\begin{aligned} [F(1,2)]^* &= -F(2^T, 1^T), \\ [F(1,2)]^\dagger &= -F(1^T, 2^T). \end{aligned} \tag{3.188}$$

For an F function of the general form (3.183), the conditions (3.188) yield the explicit relationships

$$\begin{aligned} \left[F^\delta(1,2)\right]^* &= F^\delta(2,1) & \left[\, F^\gtrless(1,2)\,\right]^* &= -F^\gtrless(2,1), \\ \left[F^\delta(1,2)\right]^\dagger &= F^\delta(1,2) & \left[\, F^\gtrless(1,2)\,\right]^\dagger &= -F^\gtrless(1,2), \end{aligned} \tag{3.189}$$

where $F^\gtrless$ are functions of real time.

3.5.3. *G-like functions in the Keldysh space*

In analogy to (3.171), a function $F \in \mathfrak{K}$ is mapped on

$$F \longleftrightarrow \overset{\supset}{\mathbf{F}} \equiv \begin{vmatrix} F^c & F^< \\ F^> & \widetilde{F}^c \end{vmatrix} \equiv \begin{vmatrix} F^{++} & F^{+-} \\ F^{-+} & F^{--} \end{vmatrix} \equiv \begin{vmatrix} F^{11} & F^{12} \\ F^{21} & F^{22} \end{vmatrix}, \tag{3.190}$$

and the matrix elements, by Eqs. (3.183) and (3.184), are

$$\begin{aligned} F^{11}(1,2) &= F^\delta(1,2) + \theta(t_1 - t_2)F^>(t_1, t_2) + \theta(t_2 - t_1)F^<(t_1, t_2), \\ F^{12}(1,2) &= F^<(1,2), \\ F^{21}(1,2) &= F^>(1,2), \\ F^{22}(1,2) &= -F^\delta(1,2) + \theta(t_1 - t_2)F^<(t_1, t_2) + \theta(t_2 - t_1)F^>(t_1, t_2). \end{aligned} \tag{3.191}$$

Four matrix elements are expressed by three functions, which is reflected by the identity

$$F^{11} + F^{22} = F^{12} + F^{21}. \tag{3.192}$$

As a special quantity, the *Keldysh function* is defined by

$$F^K = F^{12} + F^{21}. \tag{3.193}$$

Other functions in common use are

$$F^R = F^{11} - F^{12} = F^{12} - F^{22}, \tag{3.194}$$

$$F^A = F^{11} - F^{21} = F^{21} - F^{22}. \tag{3.195}$$

These are the retarded component (3.194) and the advanced component (3.195) respectively. Their explicit form follows from the relations (3.191):

$$F^R(1,2) = F^\delta(1,2) + \theta(t_1 - t_2)\left(F^>(t_1, t_2) - F^<(t_1, t_2)\right),$$
$$F^A(1,2) = F^\delta(1,2) - \theta(t_2 - t_1)\left(F^>(t_1, t_2) - F^<(t_1, t_2)\right). \tag{3.196}$$

Finally, F^R, F^A satisfy the *spectral identity*

$$A^F \equiv \mathrm{i}(F^R - F^A) = \mathrm{i}(F^{21} - F^{12}), \tag{3.197}$$

by which another important quantity, the *spectral density* A^F, is introduced.

The action of complex conjugation may be demonstrated on the example of the Green's function for which $G^\delta = 0$. Combining Eq. (3.189) and the explicit expressions (3.191), we get

$$\left[\, G^c(1,2) \,\right]^* = -\widetilde{G}^c(2,1),$$
$$\left[\, G^<(1,2) \,\right]^* = -G^<(2,1), \qquad \left[\, G^>(1,2) \,\right]^* = -G^>(2,1),$$
$$\left[\, G^K(1,2) \,\right]^* = -G^K(2,1), \tag{3.198}$$
$$\left[\, G^R(1,2) \,\right]^* = -G^A(2,1), \qquad \left[\, G^A(1,2) \,\right]^* = -G^R(2,1),$$
$$\left[\, A(1,2) \,\right]^* = +A(2,1).$$

3.5.4. *Isomorphic transformations of the Keldysh space*

Now we return to Sec. 3.5.1. It was shown there that in order to obtain an isomorphic mapping of contour functions onto the matrix functions of real time, the correspondence should not be $F \mapsto \overset{\supset}{\mathbf{F}}$ but rather should involve matrices we denoted there by $\overset{\rightrightarrows}{\mathbf{F}}$:

$$F \mapsto \tau_3\, \overset{\supset}{\mathbf{F}} = \begin{vmatrix} F^{++} & F^{+-} \\ -F^{-+} & -F^{--} \end{vmatrix} \equiv \overset{\rightrightarrows}{\mathbf{F}}. \tag{3.199}$$

This simplest transformation is recommended e.g. by Kita,[126] and the matrix (3.199) is the representation of choice for general formal work.

Often, it turns out as more convenient to take into account the linear dependence of the four matrix elements, Eq. (3.192), and to eliminate one of them by means of a linear transformation. A popular variant was introduced by Keldysh himself in Ref. 89. By a unitary transformation (known as "*Keldysh rotation*") with the matrix

$$\mathsf{L} = \frac{1}{\sqrt{2}}(\tau_0 - \mathrm{i}\tau_2) = \frac{1}{\sqrt{2}}\begin{vmatrix} 1 & -1 \\ 1 & 1 \end{vmatrix}, \tag{3.200}$$

he obtained the GF matrix in the form

$$G \mapsto \mathsf{L}\,\overset{\supset}{\mathbf{G}}\mathsf{L}^{-1} = \begin{vmatrix} 0 & G^A \\ G^R & K \end{vmatrix}. \tag{3.201}$$

A modification proposed by Larkin and Ovchinikov[254] yields a more convenient upper triangular $\mathbf{F}$ matrix. The same unitary matrix (3.200) is employed, but the basic transformation (3.199) is performed first. The GF matrix, self-energy matrix, etc., all obtain as

$$F \mapsto \mathsf{L}\tau_3\, \overset{\supset}{\mathbf{F}}\mathsf{L}^{-1} = \begin{vmatrix} F^R & F^K \\ 0 & F^A \end{vmatrix} \equiv \mathbf{F_K}. \tag{3.202}$$

This "KLO" form of matrix functions is also widely used in the literature. We shall denote it by the $\mathbf{K}$ label. It should be warned that the Keldysh, and the Larkin Ovchinnikov, transforms are often mutually confused in the literature.

Yet another representation was proposed and developed in detail by Langreth and Wilkins[94] and popularized by Langreth in his famous lecture.[95] This variant employs a similarity transformation which is not unitary and this leads to an asymmetric result, in which the Keldysh function is replaced by $F^<$:

$$F \mapsto M\tau_3\, \overset{\supset}{\mathbf{F}}M^{-1} = \begin{vmatrix} F^R & F^< \\ 0 & F^A \end{vmatrix} \equiv \mathbf{F}, \tag{3.203}$$

with

$$M = \tau_0 + \tfrac{1}{2}(\tau_1 - i\tau_2) = \begin{vmatrix} 1 & 0 \\ 1 & 1 \end{vmatrix}. \tag{3.204}$$

This transformation was introduced, because it is particularly well suited for transport problems including the derivation of quantum transport equations. For the same reason, it will be employed in the rest of this review predominantly. The associated dialect of NGF will be referred to as LW for brevity and the Langreth-Wilkins matrices will be used without label.

The LW representation is intuitively appealing, because it operates with the propagators and the particle correlation function, and these quantities have an immediate physical meaning. The formal advantage of LW formalism emerges when two matrix functions are multiplied:

$$\mathbf{B} = \mathbf{CD} = \begin{vmatrix} C^R & C^< \\ 0 & C^A \end{vmatrix}\begin{vmatrix} D^R & D^< \\ 0 & D^A \end{vmatrix},$$

$$\mathbf{B} = \begin{vmatrix} B^R & B^< \\ 0 & B^A \end{vmatrix} = \begin{vmatrix} C^R D^R & C^R D^< + C^< D^A \\ 0 & C^A D^A \end{vmatrix}. \tag{3.205}$$

The resulting multiplication formulae are widely known as the so-called *Langreth rules*. They have been derived in a number of ways, originally by a distortion of the Keldysh trajectory to another shape with two U-turns.[95] For a purely analytical derivation see, e.g., Ref. 131. Here, they are seen as a corolary to the algebraic structure of the LW representation. The rules are easily extended to more factors:

$$(\mathbf{CDEF}\cdots)^R = C^R D^R E^R F^R \cdots$$
$$(\mathbf{CDEF}\cdots)^A = C^A D^A E^A F^A \cdots$$

For the *less*-component, the pattern already emerges for three factors:

$$(\mathbf{CDE})^< = C^R D^R E^< + C^R D^< E^A + C^< D^A E^A. \tag{3.206}$$

From a comparison of Eq. (3.202) with (3.203) it is apparent that the KLO multiplication rules are identically structured with the LW rules. We have, in particular:

$$
\begin{aligned}
B &= CD & &\text{on } \mathbb{C}, \\
B^< &= C^R D^< + C^< D^A & &\text{LW}, \\
B^K &= C^R D^K + C^K D^A & &\text{KLO}, \\
B^> &= C^R D^> + C^> D^A.
\end{aligned}
\tag{3.207}
$$

The fourth line is obtained by subtracting the second line from the third one.

3.5.5. Dyson equation — various representations

The Dyson equation in the basic representation (3.199) has already been given in Eq. (3.179). In the differential form, this equation and its conjugate read

$$\vec{\mathbf{G}}_0^{-1}\vec{\mathbf{G}} = \vec{\delta} + \vec{\Sigma}\vec{\mathbf{G}}, \tag{3.208}$$

$$\vec{\mathbf{G}}\vec{\mathbf{G}}_0^{-1} = \vec{\delta} + \vec{\mathbf{G}}\vec{\Sigma}. \tag{3.209}$$

The inverse $\vec{\mathbf{G}}_0^{-1}$ of the free GF is given by (3.182) with $U = 0$.

- **Kadanoff-Baym equations**

It is enough to consider two of the matrix components of either of the equations (3.208)–(3.209), because of the relations (3.191) valid both for $G \in \mathfrak{K}$ and $\Sigma \in \mathfrak{K}$. Choosing the off-diagonal elements and using the specific correspondence (3.85), (3.185), we are led to two equivalent pairs of equations for $G^>$ and $G^<$, one arising from Eq. (3.208),

$$\left(i\frac{\partial}{\partial t_1} - h(1)\right) G^>(1,2) - \int dx_3\, V_{\mathrm{HF}}(1,3^+)G^>(3,2)$$

$$= \int_{-\infty}^{t_1} d3\, [\Sigma^>(1,3) - \Sigma^<(1,3)]\, G^>(3,2) - \int_{-\infty}^{t_2} d3\, \Sigma^>(1,3)\, [G^>(3,2) - G^<(3,2)],$$

$$\tag{3.210}$$

$$\left(i\frac{\partial}{\partial t_1} - h(1)\right) G^<(1,2) - \int dx_3\, V_{\mathrm{HF}}(1,3^+)G^<(3,2)$$

$$= \int_{-\infty}^{t_1} d3\, [\Sigma^>(1,3) - \Sigma^<(1,3)]\, G^<(3,2) - \int_{-\infty}^{t_2} d3\, \Sigma^<(1,3)\, [G^>(3,2) - G^<(3,2)],$$

the other one arising from the conjugate Eq. (3.209):

$$\left(-i\frac{\partial}{\partial t_2} - h(2)\right) G^>(1,2) - \int dx_3\, G^>(1,3)V_{HF}(3,2^+)$$

$$= \int_{-\infty}^{t_1} d3\, [G^>(1,3) - G^<(1,3)]\, \Sigma^>(3,2) - \int_{-\infty}^{t_2} d3\, G^>(1,3)\, [\Sigma^>(3,2) - \Sigma^<(3,2)],$$
(3.211)

$$\left(-i\frac{\partial}{\partial t_2} - h(2)\right) G^<(1,2) - \int dx_3\, G^<(1,3)V_{HF}(3,2^+)$$

$$= \int_{-\infty}^{t_1} d3\, [G^>(1,3) - G^<(1,3)]\, \Sigma^<(3,2) - \int_{-\infty}^{t_2} d3\, G^<(1,3)\, [\Sigma^>(3,2) - \Sigma^<(3,2)].$$

These are the well-known *Kadanoff-Baym equations*, best represented in Ref. 85. We have a minimum set of two coupled integro-differential equations for the elemental quantities $G^>$, $G^<$ as the two unknowns. The discontinuities of G^c and $\widetilde{G}^c$ come out here as the finite upper integration limits taking care of causality. On the whole, the KB equations leave the field theoretic idiom and are written in a manner close to the transport equations. In Sec. 5, we shall combine Eqs. (3.210) and (3.211) to a proto-transport equation called the Generalized Kadanoff-Baym equation. A certain drawback of the KB equations is that the spectral and statistical aspects are not distinct and there is no direct way of separating both. This is better treated in the LW equations working explicitly with the set of propagators and the particle correlation function.

- *Dyson equation in the LW representation*

The Dyson equations in the KLO representation and in the LW representation are handled in the same manner, and we shall work out the LW case. Performing the transformation (3.203) on the "left" and "right" Dyson equations (3.208), (3.209), we get

$$\mathbf{G}_0^{-1}\mathbf{G} = \boldsymbol{\delta} + \boldsymbol{\Sigma}\mathbf{G},$$
(3.212)

$$\mathbf{G}\mathbf{G}_0^{-1} = \boldsymbol{\delta} + \mathbf{G}\boldsymbol{\Sigma}.$$
(3.213)

Here, $\mathbf{G}_0^{-1} = \vec{\mathbf{G}}_0^{-1}$ and $\boldsymbol{\delta} = \vec{\boldsymbol{\delta}}$, because these matrices are diagonal, see Eqs. (3.180) and (3.182). By matrix multiplication or by the Langreth rules (3.205), the Dyson equations for the R, A components are found as

$$G_0^{-1}G^R = \delta + \Sigma^R G^R, \qquad G_0^{-1}G^A = \delta + \Sigma^A G^A,$$
$$G^R G_0^{-1} = \delta + G^R \Sigma^R, \qquad G^A G_0^{-1} = \delta + G^A \Sigma^A.$$
(3.214)

It suffices to treat one of these equations explicitly. By left-multiplying by G_0^R, the left equation for G^R is transformed to an integral equation, assuming, of course, the self-energy Σ^R to be known:

$$G^R = G_0^R + G_0^R \Sigma^R G^R.$$
(3.215)

The integral equation incorporates the boundary condition through the free propagator. It is important to realize that these boundary conditions do not depend on the initial condition specific for the selected uncorrelated initial state. To see that, it is enough to inspect the equation of motion for G_0^R:

$$G_0^{-1} G_0^R = \delta \cdots \left(i \frac{\partial}{\partial t_1} - h_0(1) - U(1) \right) G_0^R(1,2) = \delta(1,2), \qquad (3.216)$$

with the boundary condition

$$G_0^R(1,2) = 0 \qquad \text{for } t_1 < t_2. \qquad (3.217)$$

The solution is fixed by the jump conditions at equal times and the initial condition does not enter at all. In fact, the free propagator depends on the internal and external fields, but as concerns various initial conditions, it is universal for all of them. Explicitly,

$$G_0^R(1,2) = -i\, s(1,2) \theta(t_1 - t_2). \qquad (3.218)$$

The evolution operator s has been defined by (3.118).

The *less*-component of **G** is governed by the equations

$$G_0^{-1} G^< = \Sigma^R G^< + \Sigma^< G^A, \qquad (3.219)$$

$$G^< G_0^{-1} = G^R \Sigma^< + G^< \Sigma^A. \qquad (3.220)$$

These equations are, in fact, identical with the second Kadanoff-Baym equation (3.210), as can be verified with the use of relations (3.196), and its conjugate. The form (3.219) is well suited for a numerical integration. More importantly, it permits a formal explicit solution in a closed form. For this, the right Dyson equation for G^R is needed. From (3.213) or (3.214), it follows that $G^R(G_0^{-1} - \Sigma^R) - \delta$. With this identity, the equation (3.219) multiplied by G^R from the left becomes

$$G^< = G^R \Sigma^< G^A. \qquad (3.221)$$

This exceedingly simple result is another one of the NGF relations most frequently quoted – and used. It was seemingly obtained without invoking the initial conditions in three steps: first – G_0^R from (3.217); second – G^R from the Dyson equation; finally, in the third step, G^R alone is enough to specify the solution of Eq. (3.219) uniquely. This was only made possible by the Keldysh initial conditions. These enter the solution implicitly through the self-energies Σ^R, $\Sigma^<$, which incorporate the uncorrelated initial condition in the present case. Returning to Eq. (3.221), the following features are apparent: ◇ The $<$ component of the *integral* Dyson equation is, in fact, a formula. ◇ Neither the initial conditions nor the unperturbed GF enter the result. ◇ The expression for $G^<$ has the causal structure $R \cdots < \cdots A$, and the finite integration limits, explicit in the KB equations, are imposed by the boundary conditions for both propagators. ◇ Finally, we quote the KLO equivalent of Eq. (3.221). As expected,

$$K = G^R \Omega G^A; \qquad K \equiv G^K, \quad \Omega \equiv \Sigma^K, \tag{3.222}$$

where we use the conventional Keldysh notation.

4. Finite time initial conditions

In the previous chapter, we were able to present in some detail ways of handling the non-equilibrium Green's functions of the Keldysh type, that is defined on the Keldysh contour extending to an infinitely remote past and obeying an uncorrelated initial condition there. In this chapter, we will be concerned with the same task, but for the general case of Green's functions defined on a Schwinger contour with an arbitrary, typically finite, initial time t_I, and starting from an arbitrary initial many-body state $\mathcal{P}_I$.

The problem of general initial conditions is often stated as a problem of *correlated* initial conditions. This may be understood in two ways: either as a requirement that at any finite time the description of the system take account of the particle correlations, or simply as an opposite to the Keldysh initial conditions. This second interpretation is more than a trivial logical figure, because it points to the basic formal difficulty that the Wick theorem does not hold in the correlated case. See the discussion in Sec. 3.4.5. This means in turn that neither the perturbative expansion based on Feynman diagrams of Sec. 3.4.6, nor its non-perturbative equivalents of Sec. 3.4.7 are valid, and the whole theory has to be reconsidered anew.

Historically, the problem of general initial conditions was not fully appreciated at first, partly because it has little importance for the steady state non-equilibrium quantum transport, the topic in focus of the early NGF work. We cannot review here the beginnings of the investigations on the finite time initial conditions started by Fujita,[193] Hall,[194] Craig[92] and Kukharenko, Tikhodeev.[195] The modern period in this field was opened by the work[102] of Danielewicz, which we had occasion to cite several times already. There is a vast literature devoted to the finite time initial conditions. Here are some of the important citations.[102,110,118,125,129,131,154,167,168,171,196–210]

There are two basic approaches in current use, as sketched in Fig. 3.

A. A direct construction of the NGF with arbitrary initial conditions on the Schwinger contour, Fig. 3a.

B. A perturbative treatment of the correlated initial condition on an extended contour, the so-called Kadanoff-Baym trajectory. This is the Schwinger countour extended by an imaginary time stretch beyond the final time t_I^-, as shown in Fig. 3b. This is, without doubt, the most widely used technique in the field.

C. An extension of the Schwinger contour to the full Keldysh contour, Fig. 3c, can be used for the same purpose, yielding simple and physically transparent results.

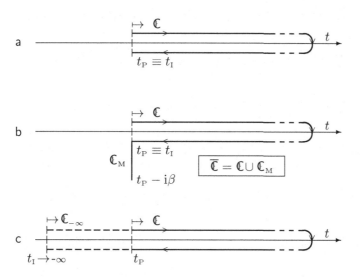

Fig. 3. The contours used to deal with finite time initial conditions. The time t_P marks the onset of the relevant non-equilibrium process. (a) *Schwinger trajectory* $\mathbb{C}$ of the process under study. The initial time t_I, at which the trajectory is starting and ending, coincides with t_P. (b) *Extended* (Kadanoff-Baym) trajectory $\overline{\mathbb{C}}$ consists of the Schwinger trajectory $\mathbb{C}$ and of its extension to imaginary times $\mathbb{C}_M$; the label M stands for "Matsubara". The trajectory is not closed, but still it involves the excursion to "$+\infty$" and back. (c) *Keldysh* trajectory $\mathbb{C}_{-\infty}$ for a "host" process starting at $t_I \to -\infty$ and partitioned at t_P marking the onset of the physical process in quetion, and the end of a "prelude" unfolding in the "past".

First, the case A. will be introduced briefly. It is remarkable that the NGF problem can be attacked directly from the definition, so to say, for an initial many-body state $\mathcal{P}_P$ which may be arbitrary, that is, no temperature, chemical potential, etc. can nor need be ascribed to it. The only condition is that it is normalized to a prescribed particle number,

$$\mathrm{Tr}(\mathcal{P}_P \mathcal{N}) = N. \tag{4.1}$$

The task of finding the Green's function from the definition (3.77) might seem hopeless, but, in fact, several practicable algorithms have been devised for it. A fully self-consistent approach based on the functional derivative method has been developed and partly applied with success.[197,201,207,208] A parallel treatment is the very basic procedure employing the perturbation expansion. It was outlined in[102] and worked out in detail in recent years[247] with the outcome of a generalized Wick theorem according to which the perturbation expansion gradually incorporates also the initial condition in the form of the correlation parts of the reduced density matrices of ever increasing order. An uncorrelated initial state then falls back to the standard Wick expansion in an automatic fashion. The self-consistent equations have a diagrammatic representation too, with the GF lines dressed, $G_0 \dashrightarrow G$, while the correlation inserts remain intact. All these matters are excluded from the present review, in which we concentrate rather on the cases B. and C. listed above. Even there, the presentation will be but brief.

4.1. *Extended (Kadanoff-Baym) contour*

This method was originally discovered for the initial condition given by the grand-canonical ensemble (3.51) with a prescribed temperature and chemical potential. We shall briefly outline this important case and mention possible extensions only at the end.

4.1.1. *Green's function on the extended contour*

For the grand canonical thermal average, the Green's function becomes

$$G(1,2) = -\mathrm{i}\, \frac{\mathrm{Tr}\left[e^{-\beta(\mathcal{H}_{\mathrm{eq}}-\mu\mathcal{N})}\, \mathcal{T}_{\mathfrak{c}}(\psi(1)\psi^\dagger(2))\right]}{\mathrm{Tr}\, e^{-\beta(\mathcal{H}-\mu\mathcal{N})}}, \tag{4.2}$$

where the times t_1, t_2 are on the real time Schwinger contour so far and

$$\mathcal{H}_{\mathrm{eq}} = \mathcal{H}(t_{\mathrm{P}}) = \mathcal{H}_0(t_{\mathrm{P}}) + \mathcal{W} = \mathcal{T} + \mathcal{V} + \mathcal{W}, \tag{4.3}$$

is the equilibrium Hamiltonian of the stand alone system before the external fields $\mathcal{H}'_{\mathrm{e}}(t)$ are turned on, see Eq. (3.1). The statistical operator obeys the Matsubara analogy with the evolution operator of a constant Hamiltonian and can be formally attached to the evolution operator on the loop as its extension beyond the end point t_{P}^-, that is for times $t_{\mathrm{P}}^- - \mathrm{i}\tau$, $0 \leq \tau \leq \beta$. The evolution operator is defined on the whole extended contour in Fig. 3b by

$$\begin{aligned}\mathcal{S}(t,t') &\equiv \mathcal{S}(t,t') && \text{for } t,t' \in \mathbb{C} \\ &= \mathcal{S}_{\mathrm{M}}(t,t_{\mathrm{P}})\mathcal{S}(t_{\mathrm{P}},t') && \text{for } t \in \mathbb{C}_{\mathrm{M}}, t' \in \mathbb{C} \\ &= \mathcal{S}_{\mathrm{M}}(t,t') && \text{for } t,t' \in \mathbb{C}_{\mathrm{M}} \\ &\text{etc.} \end{aligned} \tag{4.4}$$

$$\mathcal{S}_{\mathrm{M}}(t,t') = e^{-\mathrm{i}(t-t')\mathcal{H}_{\mathrm{M}}}, \qquad \mathcal{H}_{\mathrm{M}} = \mathcal{H}_{\mathrm{eq}} - \mu\mathcal{N} \equiv \mathcal{H}_{\mathrm{M0}} + \mathcal{W}.$$

It is important to define the GF for time arguments on the whole extended contour $\overline{\mathbb{C}} = \mathbb{C} \cup \mathbb{C}_{\mathrm{M}}$. To this end, we have to modify the definition of the Heisenberg operators by restricting it to the second form of Eq. (3.15) which is suitable for both real and imaginary times:

$$\mathcal{X}_{\mathrm{H}}(t) \equiv \mathcal{X}(t) = \mathcal{S}(t_{\mathrm{I}},t)\mathcal{X}(t)\mathcal{S}(t,t_{\mathrm{I}}). \tag{4.5}$$

With the definition (4.4), the GF (4.2) becomes

$$G(1,2) = -\mathrm{i}\, \frac{\mathrm{Tr}\left[\mathcal{S}(t_{\mathrm{I}} - \mathrm{i}\beta, t_{\mathrm{I}})\, \mathcal{T}_{\overline{\mathfrak{c}}}(\psi(1)\psi^\dagger(2))\right]}{\mathrm{Tr}\, \mathcal{S}(t_{\mathrm{I}} - \mathrm{i}\beta, t_{\mathrm{I}})}, \tag{4.6}$$

where the range of the time-ordering operator $\mathcal{T}_{\overline{\mathfrak{c}}}$ is now also extended to the whole $\overline{\mathbb{C}}$ contour and the times from the $\mathbb{C}_{\mathrm{M}}$ extension come later than all times on the basic real time contour.

4.1.2. *Kubo-Martin-Schwinger conditions*

The so-called Kubo-Martin-Schwinger (KMS) conditions are boundary conditions for the Green's function (4.6). They have been derived originally for the equilibrium Green's functions. Quite remarkably, they can be extended to the systems out of equilibrium. The boundary condition they represent links the values of the Green's function at the initial time t_I and at the final time $t_I - i\beta$:

$$
\begin{aligned}
G(x_1, t_I - i\beta, 2) &= -i \frac{\text{Tr}\left[\psi(x_1)\mathcal{S}(t_I - i\beta, t_2)\psi^\dagger(x_2)\mathcal{S}(t_2, t_I)\right]}{\text{Tr}\,\mathcal{S}(t_I - i\beta, t_I)} \\
&= -i \frac{\text{Tr}\left[\mathcal{S}(t_I - i\beta, t_2)\psi^\dagger(x_2)\mathcal{S}(t_2, t_I)\psi(x_1)\right]}{\text{Tr}\,\mathcal{S}(t_I - i\beta, t_I)} = -G(x_1, t_I, 2).
\end{aligned}
$$
(4.7)

In a similar fashion, the other KMS relation can be derived:

$$
G(1, x_2, t_I - i\beta) = -G(1, x_2, t_I). \tag{4.8}
$$

As is seen, these conditions express the anti-periodicity of the Green's function along the imaginary time axis.

4.1.3. *Perturbation expansion on the extended contour*

Our task will be to transform the Green's function (4.6) into the interaction picture in the literal sense, with the interaction $\mathcal{W}$ playing the role of the perturbation, like in Sec. 3, but on the extended contour $\overline{\mathfrak{C}}$. We recall the notation with carets over the operators in the interaction picture and plain S for the evolution operator in the interaction picture introduced in Sec. 3.4.1. Then

$$
\mathcal{S}(t, t_P) = \mathcal{S}_0(t, t_P)S(t, t_P). \tag{4.9}
$$

The free evolution operator is defined in analogy with Eq. (4.4), but for the free Hamiltonian. In particular,

$$
\mathcal{S}_0(t_P - i\beta, t_P) = e^{-\beta \mathcal{H}_{M0}}. \tag{4.10}
$$

The unperturbed statistical operator is thus equal to

$$
\mathcal{P}_0 = \frac{\mathcal{S}_0(t_P - i\beta, t_P)}{\text{Tr}\,\mathcal{S}_0(t_P - i\beta, t_P)}. \tag{4.11}
$$

Finally, we define

$$
S = \mathcal{T}_{\overline{c}}\, e^{-i\int d\tau\, \hat{\mathcal{W}}(\tau)}, \qquad \boxed{\int = \int_{\overline{\mathfrak{C}}}}. \tag{4.12}
$$

With all this notation, it is easy to bring (4.6) to

$$
G(1, 2) = -i \frac{\text{Tr}\left[\mathcal{P}_0 \mathcal{T}_{\overline{c}}(\psi(1)\psi^\dagger(2)S)\right]}{\text{Tr}\left[\mathcal{P}_0 S\right]}. \tag{4.13}
$$

The structure of this formula is the same as that of the groundstate expression (3.72). The average is now taken over the free grand canonical ensemble, which is clearly an uncorrelated state. Expanding the expression (4.12) for S in terms of $\mathcal{W}$, we obtain a power series both in the numerator and in the denominator, to which the Wick theorem applies. The resulting expansion is organized according to the standard connected Feynman diagrams, only the integrals involved now run over the whole extended contour.

4.1.4. Functional derivatives

It is illuminating to see, how the same result can be obtained by a procedure closely following Sec. 3.4.6. An auxiliary external scalar field is introduced into the system, so that the Green's function is U-dependent. The equation (3.131)

$$G(1,2;U) = G_0(1,2;U) - \iint d4d3G_0(1,4;U)[iw(4^+,3)]G(3,3^+;U)G(4,2;U)$$
$$+ \iint d4d3G_0(1,4;U) \times [iw(4^+,3)]\frac{\delta G(4,2;U)}{\delta U(3)}. \tag{4.14}$$

can be derived as before, because the initial conditions are not used in the derivation. Now comes the critical point. This equation should serve as a basis for the iterative procedure yielding G as series in powers of iw. In order to get the conventional Feynman diagrams, the simple expression for the functional derivative of G_0 is needed:

$$\frac{\delta G_0(1,2;U)}{\delta U(3)} = G_0(1,3;U)G_0(3,2;U). \tag{4.15}$$

In Sec. 3.4.6, the latter relation was valid as a consequence of the uncorrelated Keldysh initial conditions. Following Kadanoff and Baym,[85] we may try to derive it from the identity $G_0^{-1}G_0 = 1$:

$$\delta[G_0^{-1}G_0] = \delta(G_0^{-1})G_0 + G_0^{-1}\delta G_0 = 0. \tag{4.16}$$

We are tempted to conclude that

$$\delta G_0 = -G_0\delta(G_0^{-1})G_0. \tag{4.17}$$
$$\tag{4.18}$$

This conclusion is too rash, because to δG_0 in (4.16), there may be added an arbitrary solution of the homogeneous equation

$$\left(-i\frac{\partial}{\partial t_1} - h_0(1)\right)Q(1,2) = 0. \tag{4.19}$$

The result (4.18) follows also from the other relation, $G_0 G_0^{-1} = 1$, so that Q should also satisfy the conjugate equation. At this point enter the KMS conditions, which hold for $\delta G_0/\delta U$ as a corrolary to (4.7), (4.8) and finally determine that $Q = 0$. Thanks to that, the whole iterative procedure becomes possible, the self-energy can be uniquely defined as the sum of all irreducible two-point diagrams and the Dyson equations of the extended loop follow.

4.1.5. *Matrix Green's function*

Just like in Sec. 3.5.3, it is important for a practical work with the NGF to rewrite the formalism on the extended $\overline{\mathfrak{C}}$ contour into equations for the components. Now, there are three stretches of the contour, $+$, $-$ and M, so that we get nine combinations of the time arguments:

$$\mathbf{G} = \begin{vmatrix} G^{++} & G^{+-} & G^{+\mathrm{M}} \\ G^{-+} & G^{--} & G^{-\mathrm{M}} \\ G^{\mathrm{M}+} & G^{\mathrm{M}-} & G^{\mathrm{MM}} \end{vmatrix} = \begin{vmatrix} G^c & G^< & G^\rceil \\ G^> & \widetilde{G}^c & G^\rceil \\ G^\lceil & G^\lceil & G^{\mathrm{M}} \end{vmatrix}. \tag{4.20}$$

In the second form of the matrix, we use the more common notation nowadays, in particular the symbols $\lceil$ and $\rceil$ for the "mixed" components, whose one time argument is real, the other imaginary. This notation also makes explicit that $G^{\mathrm{M}+} = G^{\mathrm{M}-}$ and similarly the other pair; the reason is that the imaginary stretch follows in contour ordering both the $+$ branch and the $-$ branch of the contour. For $\mathbf{G}$, the multiplication rules have to be extended, but are still called the Langreth-Wilkins rules. For example, if $C = AB$ on the extended contour, then

$$C^< = A^R B^< + A^< B + A^\rceil B^\lceil, \tag{4.21}$$

$$C^R = A^R B^R, \tag{4.22}$$

$$C^\lceil = A^\lceil B^A + A^{\mathrm{M}} B^\lceil, \tag{4.23}$$

with the products meaning the appropriate integrations. Correspondingly enriched are the Kadanoff-Baym equations.

4.1.6. *Notes to the extended contour*

The first remark concerns terminology. The extended contour has been given various names, like Schwinger-Keldysh c., which appears as somewhat inappropriate; in the book,[131] the authors coin the name of Konstantinov and Perel, which would be a just tribute to the very early work of these authors,[255] where the extended contour is a central concept. We prefer to call the extended contour the Kadanoff-Baym contour. It is true that neither in the famous papers,[248,249] nor in the book,[85] the contour is introduced at all. Yet it can be construed from these works in the hindsight, as pointed out by D. Langreth in Ref. 95.

The method of Kadanoff and Baym is very different from the present understanding of the use of the extended contour. They start from a Matsubara-like Green's function of imaginary times, but with an external field U included, which depends on time analytically. This non-equilibrium function is found to satisfy the KMS boundary conditions just like in equilibrium, and these boundary conditions serve to fix uniquely the analytical continuation of both pieces of the causal function from the imaginary axis to real times; they are identified with $G^>$ and $G^<$, and the two functions are shown to be controlled by the Kadanoff-Baym equations. The initial value problem is obviated by sending t_{P} to $-\infty$, together with the imaginary time stretch. The evolution operator is, naturally, the same in the whole complex time plane, and this gives the left KMS condition the form

$$G(x_1, t_{\mathrm{I}} - \mathrm{i}\beta, 2) = -e^{\beta\mu}G(x_1, t_{\mathrm{I}}, 2)$$

and similarly for the right one. Reminiscences of this technique appear from time to time in the literature in phrases like "derive the LW rules by analytical continuation" used for contemporary quite remote techniques.

It has appeared clearly from the derivations in Sec. 4.1.1 that the $\mathbb{C}_{\mathrm{M}}$ extension of the Schwinger loop is simply an additional integration range contiguous with the return track of the loop, not necessarily lying in a complex time plane; in that sense, the idea of the extended contour is more general, admitting initial conditions having nothing in common with the dynamics of the system for real times. This has been brought to an extreme by M. Wagner,[196] who proposed the following construction. Take an arbitrary initial state $\mathcal{P}_{\mathrm{I}}$ which is positive definite, and a real positive number λ (the case of positive semidefinite states, like a pure state $|\Psi\rangle\langle\Psi|$, seems to be overstretched). Then a self-adjoint operator $\mathcal{B}$ exists such that

$$\mathcal{P}_{\mathrm{I}} = e^{-\lambda\mathcal{B}}. \tag{4.24}$$

Further a single particle "Hamiltonian" $\mathcal{B}_0$ can be introduced, and their difference is the "interaction" $\mathcal{Y}$. Making now the replacements

$$\begin{aligned}
\mathcal{H}_{\mathrm{M}} &\dashrightarrow \mathcal{B} \\
\mathcal{H}_{\mathrm{M0}} &\dashrightarrow \mathcal{B}_0 \\
\mathcal{W} &\dashrightarrow \mathcal{Y} \\
\beta &\dashrightarrow \lambda
\end{aligned} \tag{4.25}$$

in Eq. (4.4), we may define a generalized version of NGF on the extended contour and continue up to Eq. (4.13) without change. Then, of course, the excessive generality causes a problem. The formal interaction $\mathcal{Y}$ involves instantaneous collisions of arbitrarily large clusters in general, and an attempt to generate a perturbation expansion is bound to fail.

There is a restricted choice of $\mathcal{B}$, however, which is tractable with an effort similar to the basic case (4.4). Namely, as proposed by Danielewicz in Ref. 102, the initial state is taken to correspond to a Hamiltonian with pair interactions only, but the interaction term $\mathcal{Y}$ may be chosen at will, for example stronger than $\mathcal{W}$, or stronger in certain parts of the phase space. This will produce an overcorrelated initial state. This concept should be understood relative to the "true" dynamics with the interaction $\mathcal{W}$, according to which the system will relax. Similarly, the one-particle Hamiltonian defining the initial state may be selected different from that acting on the loop. For example, the system may be squeezed initially, and then released. We see that the extended contour permits a wide flexibility in the admissible initial conditions.

4.2. *Folding down the Keldysh contour (time partitioning)*

Finally, we briefly introduce a third method of respecting the finite time initial conditions, as it was defined in Fig. 3c. The goal is to construct a NGF with an initial time t_P, at which the correlated initial state is $\mathcal{P}_P$. This initial state is not arbitrary, but it coincides with the state at which an antecedent ("preparation") process has arrived at the initial time. Beyond that time, the evolution will continue as the dynamical process in question ("measurement" or "observation").

This interpretation of the two stages as the past and the future with respect to the initial time is entirely subjective and our task is, in fact, to compare two processes differing only in their time definition range, as shown in the figure. The process under study evolves along the Schwinger-Keldysh trajectory $\mathfrak{C}$ starting and ending at t_P and is described by the Green's function we denote $\mathbf{G}$ suppressing the subscript. This process is augmented by a preparatory stage running between $t_I \rightarrow -\infty$ and t_P. Together, an extended host process results with the $\mathfrak{C}_{-\infty}$ trajectory, and $\mathbf{G}_{-\infty}$ the associated NGF . The $\mathfrak{C}_P$ process is *embedded* in the $\mathfrak{C}_{-\infty}$ one. Both processes describe the same evolution beyond t_P. This coincidence permits to build up the NGF of the shorter process starting from that of the long process. Once this is done, the t_P process may be viewed as an autonomous "restart" process which is being restarted from a frozen initial state $\mathcal{P}_P = \mathcal{P}_{-\infty}(t_P)$ at t_P. The perturbation scheme is based on the time partitioning method. We will not go to details of its derivation here.[168] We will, however, mention at the beginning a principle, which lies in foundations of this method, namely the invariance of the NGF with respect to the restart time. In this section, we will work with the NGF of real time rather than on the contour. The advantages will be apparent.

4.2.1. *Invariance of the NGF with respect to the restart time*

Now we will demonstrate that the NGF is invariant with respect to the choice of the initial time. To this end, we compare two non-equilibrium Green's Functions differing by their initial times and, hence, by their definition ranges D_{t_0}, $D_{t_{-\infty}}$ (Fig. 4a). We consider the *less* correlation function; $G^>$ would be treated similarly and the R, A components are their combinations. The two correlation functions define also the whole GF on the contour by Eq. (3.85). We have

$$\begin{aligned}
G^<_{t_{-\infty}}(1,1') &= -i\operatorname{Tr}(\mathcal{P}_{t_{-\infty}}\,\psi^\dagger(1'|t_{-\infty})\psi(1|t_{-\infty})), \\
G^<_{t_0}(1,1') &= -i\operatorname{Tr}(\mathcal{P}_{t_0}\,\psi^\dagger(1'|t_0)\psi(1|t_0)),
\end{aligned} \tag{4.26}$$

in the respective definition ranges $D_{t_{-\infty}}\{t,t' \geq t_{-\infty}\}$ and $D_{t_0}\{t,t' \geq t_0\}$.
The Heisenberg field operators are evolving from the respective initial times according to the full many-particle unitary evolution operator $\mathcal{K}(t,t')$

$$\begin{aligned}
\psi(1|t_{-\infty}) &= \mathcal{K}(t_{-\infty},t)\psi(x)\mathcal{K}(t,t_{-\infty}), & \psi^\dagger(1'|t_{-\infty}) &= \cdots \\
\psi(1|t_0) &= \mathcal{K}(t_0,t)\psi(x)\mathcal{K}(t,t_0), & \psi^\dagger(1'|t_0) &= \cdots
\end{aligned} \tag{4.27}$$

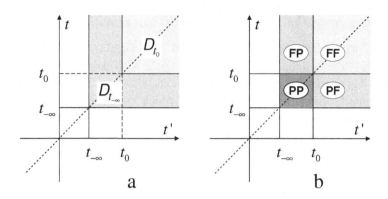

Fig. 4. (a) Definition ranges $D_{t_{-\infty}}$ of the full process and D_{t_0} of the restarted process are the first quadrants with the initial times specifying their lower left corners, see (4.26). (b) In the partitioning language the whole $D_{t_{-\infty}}$ range is cut into four partitions at the crossing point $[t_0, t_0]$. The future-future partition coincides with the restart process time range D_{t_0}.

while the two initial states, over which the trace is performed, are mutually related by

$$\mathcal{P}_{t_0} = \mathcal{K}(t_0, t_{-\infty}) \mathcal{P}_{t_{-\infty}} \mathcal{K}(t_{-\infty}, t_0), \tag{4.28}$$

see Fig. 4.

Introducing all these relations into the definition (4.26), we find that, in fact, the values of the two GF's are identical over the common definition range D_{t_0} and can be denoted by $G^<$ without the time label:

$$\begin{aligned} G^<(1, 1') &\equiv G^<_{t_{-\infty}}(1, 1'), \quad t, t' \geq t_{-\infty}, \\ G^<(1, 1') &\equiv G^<_{t_0}(1, 1'), \quad t, t' \geq t_0. \end{aligned} \tag{4.29}$$

The restart time may be an arbitrary time later than $t_{-\infty}$ and the result (4.29) thus proves that the definition of the GF for an embedded process is invariant with respect to shifting its initial ("restart") time t_0.

4.2.2. *Dyson equation with initial conditions*

The NGF for a general finite time initial condition satisfies Dyson equations, but with a self-energy having additional terms singular at the initial time. In this section, we will arrive at this structure of the NGF and its self-energy on the basis of the relationship between the host process and the embedded process. For definiteness, we first summarize the expected structure of the Dyson equation working on general terms. The NGF satisfies the Dyson equation on the finite time contour, but it is convenient to use the real time matrix form. The new feature is that now the self-energy $\boldsymbol{\Sigma}$ is replaced by the self-energy $\boldsymbol{\Sigma}_{\mathbf{IC}}$ having additional terms, so that the Dyson equation reads

$$\mathbf{G} = \mathbf{G}_0 + \mathbf{G}_0 \boldsymbol{\Sigma}_{\mathbf{IC}} \mathbf{G}. \tag{4.30}$$

The properties of the self-energy are quite different for different components and we will consider them separately.

The Dyson equation for the propagators $G^{R,A}$ retains its structure. The mean field G_0 is not changed at all; the effect of the self-energy is gradual, so that we have

$$G^{R,A} = G_0^{R,A} + G_0^{R,A}\, \Sigma^{R,A} G^{R,A}, \text{ etc.}, \tag{4.31}$$

with the self-energies being "regular" two-time functions.

The *less*-component of the Dyson equation, in contrast, has the form

$$G^<_{t_{\mathrm{P}}}(t,t') = \int\limits_{t_{\mathrm{P}}}^{t} \mathrm{d}\bar{t} \int\limits_{t_{\mathrm{P}}}^{t'} \mathrm{d}\bar{\bar{t}}\, G^R \Xi^< G^A, \qquad \boxed{t \geq t_{\mathrm{P}},\, t' \geq t_{\mathrm{P}}}, \tag{4.32}$$

where

$$\Xi^< = {}_{\circ}\Sigma^<_{\circ} + \Sigma_{IC}, \qquad \Sigma_{IC} = {}_{\circ}\Sigma^<_{\bullet} + {}_{\bullet}\Sigma^<_{\circ} + {}_{\bullet}\Sigma^<_{\bullet}. \tag{4.33}$$

Note that the integrations in (4.32) start at $t_0 = t_{\mathrm{P}}$. The four terms have a varying degree of singularity at the initial time. The open circles indicate a time variable fixed at t_{P}, the filled ones a time variable continuous in (t_{P}, ∞). The regular term ${}_{\bullet}\Sigma^<_{\bullet}$ corresponds to the Dyson equation as it is usually written for $t_{\mathrm{P}} \to -\infty$, namely $G^< = G^R \Sigma^< G^A$. The other terms play each a specific role. In particular, ${}_{\circ}\Sigma^<_{\bullet}$ and ${}_{\bullet}\Sigma^<_{\circ}$ (the self-energies Σ^c, Σ_c of Ref. 102) are related to the initial correlations. They have the form

$$\begin{aligned}
{}_{\bullet}\Sigma^<_{\circ}(t,t') &= \Lambda^<_{\circ}(t,t_{\mathrm{P}})\delta(t'-t_{\mathrm{P}}^+), \\
{}_{\circ}\Sigma^<_{\bullet}(t,t') &= {}_{\circ}\Lambda^<(t_{\mathrm{P}},t')\delta(t+t_{\mathrm{P}}^+), \qquad \boxed{t_{\mathrm{P}}^+ = t_{\mathrm{P}} + 0},
\end{aligned} \tag{4.34}$$

and are thus equivalent to single-time continuous functions $\Lambda^<_{\circ}(t,t_{\mathrm{P}})$, ${}_{\circ}\Lambda^<(t_{\mathrm{P}},t')$ dependent on t_{P} as on a parameter. For the correlated initial conditions, these two functions must be determined in addition to the regular *less* self-energy.

The last term,

$$\begin{aligned}
{}_{\circ}\Sigma^<_{\circ}(t,t') &= \mathrm{i}\rho(t_{\mathrm{P}})\delta(t-t_{\mathrm{P}}^+)\delta(t'-t_{\mathrm{P}}^+), \\
\rho(t) &= -\mathrm{i}G^<(t,t),
\end{aligned} \tag{4.35}$$

represents the uncorrelated part of the initial conditions. This is the only part of $\Xi^<$ which enters the free particle correlation function

$$G_0^< = G_0^R \, {}_{\circ}\Sigma^<_{\circ} G_0^A. \tag{4.36}$$

To verify the uncorrelated IC limit of Eq. (4.32), let us write $\Sigma^<$ for ${}_{\bullet}\Sigma^<_{\bullet}$ and use the uncorrelated, that is unperturbed, $\rho(t_{\mathrm{P}}) \to \rho_0(t_{\mathrm{P}}) = \mathrm{i}G_0^<(t_{\mathrm{P}},t_{\mathrm{P}})$ to transform

$$\begin{aligned}
G^R \, {}_{\circ}\Sigma^<_{\circ} \, G^A &\to G^R(\mathrm{i}\rho_0(t_{\mathrm{P}}))G^A \\
&= G^R[G_0^R]^{-1}G_0^R(\mathrm{i}\rho_0(t_{\mathrm{P}}))G_0^A[G_0^A]^{-1}G^A \equiv f^<,
\end{aligned}$$

and finally set $_0\Sigma_\bullet^<$ and $_\bullet\Sigma_\circ^<$ to zero. Eq. (4.32) becomes

$$
\begin{aligned}
G^< &= f^< + G^R \Sigma^< G^A, \\
f^< &= (1 + G^R \Sigma^R) G_0^< (1 + G^A \Sigma^A).
\end{aligned}
\tag{4.37}
$$

This is identical with the famous form of the Dyson equation with uncorrelated initial conditions for $G^<$ given by Keldysh, see Ref. 109.

4.2.3. *Time partitioning and equations for NGF*

The details of the derivation of the time partitioning formulas will be skipped. The basic idea is simple: it is required that the host and the embedded Green's function coincide according to the invariance theorem (4.29). The host GF has its "usual" Dyson equation, while the embedded GF has the $<$ self-energy according to (4.33), whose correction singular terms serve to compensate for the left-out effect of the past. Combining all that, the partitioning expressions are obtained.

To present the partitioned form of $G^<$, it is convenient to introduce the decomposition (4.33) into the Dyson equation (4.32) and write the latter in an explicit form:

$$
\begin{aligned}
G_{t_P}^<(t, t') &= \mathrm{i} G^R(t, t_P) \rho(t_P) G^A(t_P, t') \\
&+ G^R(t, t_P) \times \int_{t_P}^{t'} \mathrm{d}u \, {}_0\Lambda^<(t_P, u) G^A(u, t') \\
&+ \int_{t_P}^{t} \mathrm{d}v G^R(t, v) \, \Lambda_\circ^<(v, t_P) \times G^A(t_P, t') \\
&+ \int_{t_P}^{t} \mathrm{d}v \int_{t_P}^{t'} \mathrm{d}u \, G^R(t, v) {}_\bullet\Sigma_\bullet^<(v, u) G^A(u, t'), \\
& \qquad\qquad\qquad\qquad t > t_P, \, t' > t_P.
\end{aligned}
\tag{4.38}
$$

Notice that the lower integration limit is t_P, while t_I has been shifted to the remote past at the onset of the preparatory stage. The whole host process is partitioned into the past prior to t_P and into the future after t_P. This *partitioning* is reflected in the form of the components of the less self-energy entering Eq. (4.38),

$$
\begin{aligned}
{}_0\Lambda^<(t_P, u) &= \mathrm{i} \int_{t_I}^{t_P} \mathrm{d}\bar{t} \left\{ G^R \Sigma^< + G^< \Sigma^A \right\}, \\
\Lambda_\circ^<(v, t_P) &= -\mathrm{i} \int_{t_I}^{t_P} \mathrm{d}\bar{\bar{t}} \left\{ \Sigma^< G^A + \Sigma^R G^< \right\}, \\
{}_\bullet\Sigma_\bullet^<(v, u) &= \Sigma^<(v, u) \\
&+ \int_{t_I}^{t_P} \mathrm{d}\bar{t} \int_{t_I}^{t_P} \mathrm{d}\bar{\bar{t}} \{ \Sigma^R G^R \Sigma^< + \Sigma^R G^< \Sigma^A + \Sigma^< G^A \Sigma^A \}
\end{aligned}
\tag{4.39}
$$

$$
\Sigma^R G^R \Sigma^< \mapsto \Sigma^R(u, \bar{t}) G^R(\bar{t}, \bar{\bar{t}}) \Sigma^<(\bar{\bar{t}}, v), \text{ etc.}
$$

By these relations, the self-energy of the embedded process is expressed by integrals involving time blocks of the Green's functions and the self-energies of the host process. The external arguments u, v refer to the process in the future and are always greater than t_P, while the integration variables, denoted by $\bar{t}, \bar{\bar{t}}$ for clarity, belong entirely to the past and are less than t_P. The propagation takes place entirely in the past. The history and the future are interconnected by the off-diagonal blocks of the self-energies. The singular components of the self-energy have no analogue in the host process, while the regular term $\,_\bullet\Sigma_\bullet^<$ has two parts

$$\,_\bullet\Sigma_\bullet^< = \Sigma^<(t, t') + \widehat{\Sigma}_{t_\mathrm{P}}^<(t, t'). \tag{4.40}$$

The first term comes from the host process without change (where it would enter the Dyson equation in the usual form, $G^< = G^R \Sigma^< G^A$). It is supplemented by the second term, $\widehat{\Sigma}_{t_\mathrm{P}}^<(t, t')$, which takes the finite time initial condition into account.

4.2.4. *Notes to the time partitioning method*

The infinite Schwinger-Keldysh contour can accommodate an extremely rich class of processes bearing the generic name of the *Keldysh switch-on processes*. These processes start from general uncorrelated states, become correlated as the interactions are switched on and are driven by an endless variety of external influences. It may be said that the intermediate states passed through by the system in the course of such processes form in their entirety the class of *all physically attainable states* of the system, including states out of equilibrium and incorporating correlations of widely different nature and strength. Any of these states may be used as an initial state for the transient process we wish to study. Thus, we can start the transient from "all" physically meaningful initial states. Still, the direct method of including finite time initial conditions[247] admits, in principle at least, quite arbitrary initial states, including those which are artificially overcorrelated, etc. Such states fall out of our scope by definition.

The formal tool for using an arbitrary Keldysh switch-on process as preparatory for the relevant process starting at a finite time t_P is the time partitioning. It is clear that the time partitioning is universal and does not depend on any assumptions, like equilibrium, about the "past". We may thus place the splitting time to any convenient time instant.

It is essential, however, that the "past" preparatory stage and the envisaged embedded "relevant" transient are a part of one uninterrupted host process. Just as in the method of the extended Kadanoff-Baym contour, the mixed components of the GF and the self-energy played a crucial role, the time partitioning expressions contain the coupling between the past and the future with respect to the dividing time t_P. Fortunately, the time depth of the coupling is typically quite restricted. According to the Bogolyubov principle, the mutual correlation will die out within a time of the order of the collision duration time.

Within the NGF formalism, the Bogolyubov principle is translated into the assumption about the behavior of the system expressed in terms of self-energies, as will be discussed further in Sec. 6.1: All components of the host self-energy should be concentrated to a strip

$$|t - t'| < \mathcal{O}(\tau^\star) \qquad \tau^\star = \mathcal{O}(\tau_c, \tau_Q).$$

The two times appearing at the r.h.s. are: τ_c, often called the collision duration time in transport theory, relevant for $\Sigma^<(t, t')$, and τ_Q, usually called the quasiparticle formation time, characteristic for $\Sigma^{R,A}(t, t')$. This is illustrated in Fig. 5. To conclude, we have presented two methods of incorporating the finite time initial conditions, both of which transform the inclusion of the complex many body initial condition at a single time to invoking the single particle GF for all times in the past with respect to the dividing time point; we have proposed to distinguish these methods as *diachronous* as opposed to the *synchronous* direct methods defined as case A at the beginning of the whole Sec. 4.

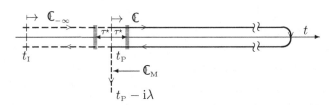

Fig. 5. A finite-time Schwinger-Keldysh contour and its diachronous extensions. $\mathbb{C}$: The closed time contour for a transient starts at the initial time t_P, goes to $+\infty$ and returns to t_P. $\mathbb{C}_M$: Extension by a Matsubara-like imaginary time interval according to Refs. 102, 107 or 66, 196. $\mathbb{C}_{-\infty}$: Extension of $\mathbb{C}$ along the time axis to the past. The resulting contour starts at t_I. The dashed segment preceding t_P accommodates the preparation process. The two adjoining processes form together the host process, in which the transient is embedded. For $t_I \to -\infty$, the host process with an uncorrelated initial condition becomes a Keldysh switch-on process. Two diffuse boundaries bracketing t_P at the distance $\sim \tau^\star$: If a finite correlation decay time $\tau^\star$ exists, the lower boundary indicates the depth of coupling between the preparation process and the transient, the upper one marks the extent of penetration of the initial correlations into the future, cf. Sec. 6.1.

5. Reconstruction theorems

The aim of this section is to introduce the so-called Reconstruction theorems. Their name derives from the fact that they are built over the better known reconstruction equations[109,110,116,118,125,129,154,171,211] and make them an integral part of an alternative system of NGF equations. These equations are exact, but one of their features is that they are suggestive of the approximations leading to quantum kinetic equations of the GME type. In fact, the way to the exact reconstruction technique was just the reverse: it was inspired and motivated by a simplified description of quantum dynamics based on the quantum transport theory, in which all dynamics is expressed in a kinetic equation governing a single-particle single-time density matrix.

Throughout the whole section, we confine the considerations to the Keldysh initial condition without correlations. Only in the last subsection, the modifications brought about by correlated initial conditions will be indicated.

5.1. *Generalized Kadanoff-Baym equation*

We will start the way towards Reconstruction theorems by aiming at a quantum kinetic equation for the single particle density matrix $\rho(t)$ from the NGF equations.

5.1.1. *Precursor kinetic equation*

The well known starting point on the way from the double time NGF to single time kinetic equations is the differential equation called the Generalized Kadanoff-Baym Equation (GKBE). This equation is obtained directly from the Dyson equation.

First, we subtract the Dyson equations (3.219), (3.220) one from another. Second, we use the Dyson equations (3.214) in the form

$$[G^{R,A}]^{-1} = [G_0^{R,A}]^{-1} - \Sigma^{R,A}. \tag{5.1}$$

After easy manipulations we get the identity

$$G_0^{-1}G^< - G^< G_0^{-1} = \Sigma^R G^< - G^< \Sigma^A$$
$$-G^R \Sigma^< + \Sigma^< G^A. \tag{5.2}$$

This (still exact) equation is the *Generalized Kadanoff-Baym equation* (GKBE). It has already a structure closely related to transport equations: its l.h.s. contains information about the drift of free particles, the four terms on the r.h.s. represent the generalized collision terms. The equation still has the double time structure, however.

In order to obtain an equation for $\rho(t)$, we make the limit $t_1 = t = t_2$ of Eq. (5.2) and recall that at equal times, $t_1 = t = t_2$, the one particle density matrix ρ is given by the time-diagonal of $G^<$

$$G^<(t,t) = i\rho(t). \tag{5.3}$$

Using the explicit expression (3.182) for G_0^{-1},

$$G_0^{-1}(t_1, t_2) = \{i\partial t_1 - h_0(t_1)\}\delta(t_1 - t_2) = \{-i\partial t_2 - h_0(t_2)\}\delta(t_1 - t_2) \tag{5.4}$$

the l.h.s. of the (5.2) is transformed to an unrenormalized drift of the one-particle density matrix ρ:

$$\text{l.h.s. of (5.2)} \xrightarrow{t_1 = t = t_2} \frac{\partial \rho}{\partial t} + i[\, h_0(t), \rho \,]_- \,. \tag{5.5}$$

This already has the form consistent with a QTE, in which $h_0(t)$ is the mean field one-particle Hamiltonian, and the whole equation (5.2) becomes the desired *Precursor Quantum Kinetic Equation* (PKE):

$$\frac{\partial \rho}{\partial t} + i\,[\,h_0(t), \rho\,]_- = (\Sigma^R G^< - G^< \Sigma^A)_{t_1=t=t_2}$$
$$-(G^R \Sigma^< - \Sigma^< G^A)_{t_1=t=t_2}\,. \qquad (5.6)$$

The "generalized collision" terms on the r.h.s. still involve double time *less* quantities. The related integrals preserve causality extending only to the past because of the presence of the propagator factors.

5.1.2. *On the way to the quantum kinetic equation*

To convert the precursor kinetic equation to a true closed kinetic equation for ρ, we first have to specify the *physical approximation for the self-energies* in the form

$$\mathbf{\Sigma} = \mathbf{\Sigma}[\mathbf{G}], \qquad (5.7)$$

that is

$$\Sigma^{R,A} = \Sigma^{R,A}[G^{R,A}, G^<], \quad \Sigma^< = \Sigma^<[G^{R,A}, G^<]\,. \qquad (5.8)$$

Next comes the crucial point, $G^<$ is expressed in the functional form

$$G^< = G^<[\rho, G^{R,A}]\,. \qquad (5.9)$$

By introducing (5.9) into (5.8) and directly into Eq. (5.6), the double time function $G^<$ is eliminated in favor of its time diagonal ρ, and by this, a true kinetic equation is obtained.

We will deal with the question of an actual implementation of this program in the following subsections. Here we only note that the first successful (approximate) attempt in this direction was connected with the famous Kadanoff-Baym Ansatz (Sec. 6.2.1), which was followed by other Ansatzes, a parallel derivation of the reconstruction equations, and culminated by formulation of the Reconstruction theorems.[116,118,125,154,168,171]

5.2. *Concept of reconstruction theorems*

The general plan of constructing quantum transport equations within the NGF scheme, as outlined at the end of the preceding subsection, is to transform the GKBE (5.6) (... precursor kinetic equation) to a closed kinetic equation for the single particle distribution function. We may assume a more abstract position and detach the construction of transport equations from the more fundamental issue formulated already in the Introduction as the *reconstruction problem*:

> Can the full description of a many-body interacting system be built up from its single-particle characteristics, and if yes, then under which conditions?

It appears, and has been documented in this paper, that virtually all relevant information about a non-equilibrium many body system can in principle be unfolded from its properly chosen reduced characteristics, such as the pair of double-point quantities, $G^<$ and $G^>$. Now we touch at the possibility that, actually, it may be enough to know or control even less: just a function of a single time variable, the one-particle density matrix.

5.2.1. *Various approaches to reconstruction problem*

Let us begin by a look at several alternative approaches *vis-à-vis* the reconstruction problem. This will help us to grasp the NGF reconstruction hypothesis properly.

First of these approaches is based on *the Bogoluybov postulate*:[38-40]

> In an autonomous system, the evolution is controlled by a hierarchy of times (2.9) and for times past the initial decay of correlations, a transport equation of the form (2.17) is valid.

which we properly recall here once more. It should be noted that this postulate has no universal proof, but its validity has been verified for a whole row of systems of a rather different nature. The postulate is related to the reconstruction problem in the following sense. After the decay of initial correlations the systems is expected to assume a state characterized by a set of robust observables, which can be re-constructed from the knowledge of the sole one-particle distribution function. The related reconstruction has a rather symbolic form

$$\mathcal{P} = \widetilde{\Phi}[\rho], \qquad t > t_0 + \tau_c. \tag{5.10}$$

A complementary stream of research closely related to the reconstruction questions, although distinct in some respect, is the so-called *Inversion Problem*. It stems from the fundamental paper by Schwinger[83] on the use of the generating functional in non-equilibrium physics. The key concept is the functional inversion (or substitution) based on the Legendre transformation. It has been amply used in the field theoretical studies of many-body problems and it found its best known application in the Density Functional Theory (DFT) first in equilibrium, then in the extension to the Time Dependent Density functional Theory (TDDFT).[59-69]

Schwinger introduced, as discussed already, the closed time path $\mathfrak{C}$ and the generating functional dependent on an external field $U^\pm(t)$ depending on the branch of the contour,

$$e^{iW(U^+,U^-)} = \text{Tr } \mathcal{P}(t_I) \, \widetilde{\mathbf{T}} \, e^{+i\int_{t_I}^\infty d\tau (\mathcal{H} - U^-(\tau)X)} \, \mathbf{T} \, e^{-i\int_{t_I}^\infty d\tau (\mathcal{H} - U^+(\tau)X)}. \tag{5.11}$$

For a local field U, the response of local density of particles could be obtained by functional derivatives, and this solved the related transport problem in a closed form:

$$\overline{n}(x,t) = \frac{\delta W}{\delta U^+(x,t)}\bigg|_{U^+=U^-=U} = -\frac{\delta W}{\delta U^-(x,t)}\bigg|_{U^+=U^-=U}. \tag{5.12}$$

It comes immediately to mind that the latter relations could be *inverted*, so that the field U would be expressed in terms of $\bar{n}$. Introducing this back into the time derivative of (5.12) should lead to a transport equation. This can be further formalized by working with a Legendre transform of the Schwinger functional. All this has been thoroughly investigated, e.g., by the Fukuda group.[256,257]

This is one of the classical cases of the inversion expressed symbolically by the relation

$$U(t), \ \{t_I \leq t < \infty\} \rightleftharpoons \bar{n}(t), \ \{t_I \leq t < \infty\}. \tag{5.13}$$

The importance of the inversion consists in the following: Starting from $\bar{n}$, an ultimately reduced data set, we may go, by (5.13), in the *inverse* direction to U. This, in the "forward" direction, implemented by means of any quantum transport formalism, defines uniquely, i.e. reconstructs, the behavior of the whole many-body system. Thus, whenever the inversion is possible, it proves and clarifies the Bogolyubov symbolic reconstruction loop (5.10).

This is similar to the well known development for equilibrium systems, where the famous Density Functional Theory is based on the Hohenberg-Kohn theorem stating that in equilibrium the bijection (5.13) is valid.

The foundations of the time dependent extension of DFT have long lagged behind its intuitive introduction and use, based on the simplest approximations for the effective potentials.

The TDDFT analogue to the Kohn-Hohenberg theorem, established relatively long ago, is the **Runge-Gross Theorem**: Let $U(t)$ be a local potential smooth in time. Then, for a fixed initial state $|\Psi_0\rangle$, the functional relation $\bar{n}[U]$ is bijective and can be inverted.

A consistent time dependent counterpart to the Kohn-Sham energy functional has only been found recently in the Schwinger functional for effective non-interacting particles moving in an effective potential local in space and time. We will not follow the details of this developing field, see Refs. 69, 131.

5.2.2. *NGF approach to reconstruction problem*

Now it is possible to sum up some lessons. Within the NGF context, there is an exact formal theory, which permits, in principle at least, to obtain a complete description of the evolution of a non-equilibrium system. Our present aim is to recast the theory to a form, where the decisive quantity, sufficient for a complete reconstruction of the NGF results, would be the single particle density matrix.

In Sec. 5.1, we have outlined one part of the program, namely the reduction of GKBE to a kinetic equation. There, the problem of initial conditions was avoided by using the Keldysh IC. The Bogolyubov postulate warns us that the simple method may work in the saturated kinetic regime only, while the early stage of the decay of correlations may be difficult. On the other hand, the Runge-Gross theorem should be valid in the whole time range starting from t_I, which is encouraging. Of course, this theorem is existential, not operational, so that it suggests no way to follow. It is well known that all density functional theories involve an introduction of ill-controlled approximations for the exchange and correlations. Besides, the reciprocity theorems of the density functional theories use pairs of variables like local potential – local particle density. The theories based on the use of the reduced density matrix (RDDMFT) also exist, but they offer no truly viable alternative. There should be no surprise that the technique which is really inspirational here, is the inversion technique of Schwinger as represented by (5.13).

Let us summarize the requirements on a reconstruction technique in the NGF theory:

◇ The central quantity will be the single particle density matrix ρ;

◇ The technique should represent a complete operational scheme;

◇ It should be exact in principle, not based on *ad hoc* approximations;

◇ It should incorporate general finite time initial conditions.

All these requirements are fulfilled by the reconstruction equations we are going to discuss in the next subsection.

5.3. *Reconstruction equations*

An integral part of the reconstruction formalism are the *Reconstruction Equations* (RE), which permit an exact reconstruction of $G^<$ from ρ.[109,110,116,118,125,154,168,171,211]

If the Keldysh initial condition is assumed, they easily follow from the Dyson equations. To derive the RE, we first split the exact $G^<$ as follows:

$$G^<(t_1, t_2) = G_R^<(t_1, t_2) - G_A^<(t_1, t_2), \tag{5.14}$$

$$G_R^<(t_1, t_2) = \theta(t_1 - t_2)G^<(t_1, t_2), \tag{5.15}$$

$$G_A^<(t_1, t_2) = -\theta(t_2 - t_1)G^<(t_1, t_2) - [G_R^<(t_2, t_1)]^\dagger . \tag{5.16}$$

It is convenient to calculate

$$\{G^R\}^{-1}G_R^< \big|_{t_1, t_2} = (G_0^{-1} - \Sigma^R)G_R^< \big|_{t_1, t_2}$$

$$= \delta(t_1 - t_2)\rho(t_2)$$

$$+\theta(t_1 - t_2)(\{G^R\}^{-1}G^< + \Sigma^R G^< - \Sigma^R G_R^<) \big|_{t_1, t_2} \tag{5.17}$$

$$= \delta(t_1 - t_2)\rho(t_2) + \theta(t_1 - t_2)(\Sigma^< G^A + \Sigma^R G_A^<) \big|_{t_1, t_2}$$

where Dyson equations (5.1, 3.221), and definition of G_0 were employed. Multiplication by G^R from the left yields the first equation for $G_R^<$, which coincides with $G^<$ in the $t_1 > t_2 > -\infty$ wedge. In the complementary time region $-\infty < t_1 < t_2$, an analogous equation for $G^< = G_A^<$ results. The two equations are conjugate, as it holds $G^<(t_1, t_2) = -[G^<(t_2, t_1)]^\dagger$.

In other words, we have obtained the *reconstruction equations* in the form

$$\boxed{t > t'}$$ $$\boxed{t < t'}$$

$$G^<(t,t') =$$

$$-G^R(t,t')\rho(t') \qquad\qquad\qquad +\rho(t)G^A(t,t')$$

$$+\int_{t'}^{t} d\bar t \int_{-\infty}^{t'} d\bar{\bar t} G^R(t,\bar t)\Sigma^<(\bar t,\bar{\bar t})G^A(\bar{\bar t},t') \ \bigg|\bigg| \ +\int_{t}^{t'} d\bar{\bar t} \int_{-\infty}^{t} d\bar t G^R(t,\bar t)\Sigma^<(\bar t,\bar{\bar t})G^A(\bar{\bar t},t')$$

$$+\int_{t'}^{t} d\bar t \int_{-\infty}^{t'} d\bar{\bar t} G^R(t,\bar t)\Sigma^R(\bar t,\bar{\bar t})G^<(\bar{\bar t},t') \ \bigg|\bigg| \ +\int_{t}^{t'} d\bar{\bar t} \int_{-\infty}^{t} d\bar t G^<(t,\bar t)\Sigma^A(\bar t,\bar{\bar t})G^A(\bar{\bar t},t')$$

$$(5.18)$$

These equations represent the most important step on the way to the NGF Reconstruction Theorem.

5.3.1. *Properties of reconstruction equations*

The reconstruction equations are inhomogeneous and their source terms contain ρ as an input. The unknown $G^<$ is also contained in both integral terms: in the second one explicitly, in the first one through the functional dependence $\Sigma^< = \Sigma^<[G^<]$.

By solving Eqs. (5.18), the double time correlation function $G^<(t_1, t_2 \neq t_1)$ is generated from the knowledge of its time-diagonal part, the one-particle single-time density $\rho(t) \propto G^<(t,t)$, and of the propagators taken also as known. The role of propagators will be discussed in detail below. The integrals are written down explicitly to show the complicated interplay of the integration limits leading to an integration range consisting of two off-diagonal blocks.

We have to warn that the Reconstruction Equations alone do not solve the reconstruction problem:

◇ For $t = t'$, they turn into the tautology $\rho = \rho$, thus an independent input of ρ is required.

◇ This input is not arbitrary: while the equations would lead to a formal solution $G^<$ for a wide range of input $\rho(t)$, the two functions need not be compatible as constituents of the same Green's function, however. This would lead to unphysical, invalid results. In particular, this would impair the conservation laws.

The compatibility is provided by the precursor kinetic equation (5.6). Recall now that also the PKE is a consequence of the Dyson equation for $G^<$. In fact, RE and PKE together are equivalent with the Dyson equation:

Dyson equation for $G^<$	$\Big\{$	Reconstruction equations for $G^<$, Eq. (5.18)
$\Longleftrightarrow$		
Eq. (3.221)		Precursor kinetic eq. for ρ, Eq. (5.6)

$$(5.19)$$

5.4. Reconstruction theorems

The preceding subsection set the stage for a compact formulation of the reconstruction theorem. It will be done in three consecutive steps, first for the correlation function $G^<$, then for the complete NGF assuming the Keldysh initial condition, and, finally, the finite time initial conditions will be incorporated.

5.4.1. Reconstruction of $G^<$

The use of the Dyson equation split into two parts according to the r.h.s. of the scheme (5.19) clearly hints at a cyclic solution for $G^<$, giving rise to the cycle

$$\boxed{\text{reconstruction eq.}} \rightleftharpoons \boxed{\text{precursor kinetic eq.}} \tag{5.20}$$

constituting a new alternative for generating the full $G^<$ correlation function.

The reconstruction equations are thus not stand-alone equations, but one part of a linked twin process whose other constitutive part is the precursor kinetic equation: $G^<$ is substituted from the reconstruction equations, the kinetic equation is solved for the density matrix $\rho(t)$ and this in turn enters the reconstruction equation as an input.

Before going to discuss general formulation of reconstruction theorems, we make a point, which will be important in the next section. The equations (5.18) can be used in two alternative ways:
1. they are perfectly suited for a direct numerical solution aiming at $G^<$ connected with a particular physical situation.
2. They provide an efficient tool for generating transport equations.

5.4.2. Reconstruction theorem

The new reformulation of the NGF method starts from putting ρ at the hub. The whole process of finding the complete NGF, including $G^<$ and the propagator pair G^R, G^A, is restructured to a dual task:

NGF RECONSTRUCTION SCHEME

$$
\begin{array}{|ccc|}
\hline
\rho & \longrightarrow & \boxed{\text{DE}}\;\boxed{\text{RE}} \\
\boxed{\text{PKE}} & \longleftarrow & G^{R,A}\;\;G^< \\
\hline
\end{array}
\tag{5.21}
$$

In one (the upper) stream, the single-particle distribution ρ is introduced into the reconstruction equations for $G^<$ and the Dyson equations for propagators so that the NGF is reconstructed from a known ρ. In the reverse stream, the NGF components are introduced into the PKE, which then acts as a quantum kinetic equation, and yields ρ. These two mutually coupled streams thus complete the reconstruction cycle.

This scheme gives a full description of the NGF reconstruction procedure in operational terms. On a more abstract level, this result may be summarized as a mathematical statement central in the present context:

NGF Reconstruction Theorem:

> For a non-equilibrium process starting from the Keldysh initial condition, the complete double-time NGF **G** and the single-time single-particle density matrix ρ are in a bijective relationship,

$$\{G^<, G^R, G^A\} \rightleftharpoons \rho. \tag{5.22}$$

This theorem satisfies all requirements set at the end of Sec. 5.2.2. It should be compared with the alternative schemes listed in the preceding section. Let us start with the Runge-Gross theorem. Firstly, on one side of our dual relation stands the full ρ rather than its space diagonal $\bar{n}$. This may appear as a weaker result, but, in fact, the Legendre duality behind the NGF theorem extends to arbitrary non-local fields U and thus covers a class of physical situations much wider than those limited to the local external fields. Secondly, in comparison with the Runge-Gross *existence* theorem, the present theorem has a constructive algorithmic nature. The validity of the theorem is thus established in each specific case by an actual reconstruction process. Thirdly, the dual relation (5.22) links ρ with the NGF rather than with the external field U in the vein of Eq. (5.13). It may be said that, in this sense, the NGF reconstruction theorem makes true the Bogolyubov conjecture (5.10), albeit in a realistically restricted form.

An important question remains about the feasibility of the reconstruction scheme (5.21). Several procedures offer themselves, all based on some type of successive approximations:

1. *NGF solver.* The equations have a structure suitable for a novel type of the NGF solver.[144,187] With all equations cast in the differential form, the solution proceeds in steps incremental in time. This may appear as abandoning the kinetic equation approach, but, in fact, this solver would permit to implement a still unemployed concept: an auto-adaptive scheme NGF/QKE, in which the full NGF solver would only be acting when necessary, like at the instant of a rapid transient, while it would downgrade to the quantum kinetic equation when possible, like over the long periods of autonomous relaxation.

2. *Interaction strength as small parameter.* Iteration of the reconstruction equations is less promising for computations than a solver, but it has a more basic context. It permits various interpretations. For non-interacting particles, the self-energies vanish and the particle correlation function is given by the absolute term of Eqs. (5.18) exactly. Thus, the iteration can be interpreted as a perturbation expansion in the particle interaction strength. This has the advantage that the reconstruction is turned into a systematic procedure and also brought close to the

direct methods of deriving the QKE. The drawback of this expansion is that it mixes two consecutive levels by respecting the many-particle correlations and the kinetic behavior simultaneously.

3. *Collision duration time as small parameter.*   This choice is fully consistent with the present approach, as it clearly separates the many-body level and the kinetic behavior. The theory may work then with dressed GF, so that it is fully renormalized. It is true that the collision duration time is not an *a priori* well defined quantity. It may be argued, however, that the Reconstruction Equations serve to offer an operational definition of the collision duration time in the course of their solution. For this, a special feature of the RE (5.18), namely their off-diagonal integration range, may be employed. This in turn reflects the time/spectral structure of the self-energies. This is sketched in an idealized representation in Fig. 6. The "small parameter" measuring the corrections to the GKBA in the RE is the triangular overlap region of the integration range and of the strip around the time diagonal in which the values of self-energies are significant. The effective width of the strip is identified with the quasi-particle formation time τ_Q for Σ^R and with the collision duration time τ_c for $\Sigma^<$. The two times may be different, but both should be "small". It is known that rather than the interaction strength alone, it is the whole inner dynamics of build-up processes induced in the system during a non-equilibrium evolution, whose subtle details are decisive for the magnitude of these characteristic times, and, thus, for the iteration procedure. Even if the strips are less sharply defined, the iteration provides clues as to the effective values of τ_Q and τ_c. To conclude, this approach parallels neatly other transport theories, in particular, it is a NGF implementation of the Bogolyubov principle.

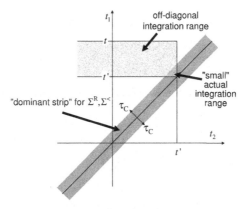

Fig. 6.   Integration range for the Reconstruction equations (retarded part).

5.5. Generalized reconstruction theorem: Correlated initial state

In the case of a *correlated initial state at a finite initial time*, the process is similar as in the case of the Keldysh initial condition already treated, but it is more involved. Both the precursor kinetic equation and the reconstruction equations are modified.

5.5.1. Precursor quantum kinetic equation with initial conditions

The starting point on the way from the double time NGF to single time transport equations is the Generalized Kadanoff-Baym Equation (GKBE). We have already introduced the GKBE for the special case of Keldysh initial conditions in Sec. 5.1. We will now generalize GKBE for the more general initial conditions. As in the previous case, we begin with a differential equation obtained from Eq. (4.32) combined with Eq. (4.31) written as $[G^{R,A}]^{-1} = [G_0^{R,A}]^{-1} - \Sigma^{R,A}$.

$$
G_0^{-1}G^< - G^<G_0^{-1} = \Sigma^R G^< - G^< \Sigma^A
$$
$$
-G^R \Xi^< + \Xi^< G^A . \tag{5.23}
$$

Again, this (exact) equation has already a structure closely related to transport equations. The four terms on the r.h.s. represent the generalized collision terms. The equation still has the double time structure, however, and, in contrast to common transport equations, it also incorporates the initial conditions through the self-energy $\Xi^<$. Using Eq. (4.32) and splitting the selfenergy $\Xi^<$ into the self-energy $\Sigma^<$ of the host process and the remainder $\Theta^<$ including all contributions from the initial correlations,

$$
\Theta_{t_0}^< = {}_\circ\Sigma_\circ^< + {}_\circ\Sigma_\bullet^< + {}_\bullet\Sigma_\circ^< + \widehat{\Sigma}_{t_\mathrm{P}}^<(t,t'), \quad \widehat{\Sigma}_{t_\mathrm{P}}^<(t,t') = {}_\bullet\Sigma_\bullet^< - \Sigma^<(t,t'),
$$

we can rewrite Eq. (5.23) into the following form:

$$
G_0^{-1}G^< - G^<G_0^{-1} = \Sigma^R G^< - G^< \Sigma^A
$$
$$
-G^R \Sigma^< + \Sigma^< G^A
$$
$$
-G^R \Theta_{t_0}^< + \Theta_{t_0}^< G^A. \tag{5.24}
$$

Finally, we arrive at the precursor kinetic equation generalizing Eq. (5.6),

$$
\frac{\partial \rho}{\partial t} + i\,[\,h_0(t), \rho\,]_- = (\Sigma^R G^< - G^< \Sigma^A)_{t_1=t=t_2}
$$
$$
-(G^R \Sigma^< - \Sigma^< G^A)_{t_1=t=t_2}
$$
$$
-(G^R \Theta_{t_0}^< + \Theta_{t_0}^< G^A)_{t_1=t=t_2}. \tag{5.25}
$$

Notice again that according to (4.40) the fully nonsingular term of selfenergy ${}_\bullet\Sigma_\bullet^<$ contains also the part $\widehat{\Sigma}_{t_\mathrm{P}}^<(t,t')$ which is dependent on initial correlations and is dying out with time.

5.5.2. *Reconstruction equations with finite time initial conditions*

We write down the exact reconstruction equations for $G^<$ extended so as to incorporate the finite time initial conditions right away. They are an exact consequence of the Dyson equation (4.32) like in the uncorrelated case before. They have the appearance of an integral equation serving to reconstruct full $G^<$ from the inhomogeneous term which coincides with the analogous term in Eq. (5.18)

$$
G^<(t,t') = -G^R(t,t')\rho(t') \qquad \boxed{t \geq t' \geq t_I}
$$
$$
+ \int_{t'}^{t} d\bar{t} \int_{t_I}^{t'} d\bar{\bar{t}}\, G^R(t,\bar{t}) \Sigma^R(\bar{t},\bar{\bar{t}}) G^<(\bar{\bar{t}},t')
$$
$$
+ \int_{t'}^{t} d\bar{t} \int_{t_I}^{t'} d\bar{\bar{t}}\, G^R(t,\bar{t}) \Xi^<(\bar{t},\bar{\bar{t}}) G^A(\bar{\bar{t}},t') , \tag{5.26}
$$
$$
G^<(t,t') = \rho(t')G^A(t,t') + \cdots \qquad \boxed{t' \geq t \geq t_I} .
$$

In other words, Eq. (5.26) serves to reconstruct a full double time function from the knowledge of its time diagonal $i\rho$ (and propagators). Unfortunately, the integral terms involve also the $\Xi^<$ self-energy, see (4.33), with all initial time corrections and this is a problem in general. In the uncorrelated case, these corrections vanish and the self energy can be obtained from Eq. (5.7). This defines the reconstruction procedure of the preceding section.

5.5.3. *Reconstruction scheme with initial conditions*

The generalized Reconstruction theorem states that the solution of the Dyson equation for $G^<$ can be obtained according to the scheme, which represents a substantial modification of the scheme for the Keldysh case, because of the necessity to include the initial conditions. We draw the scheme in a slightly different format from the diagram (5.21), but the core of the process remains the same and is easily identified:

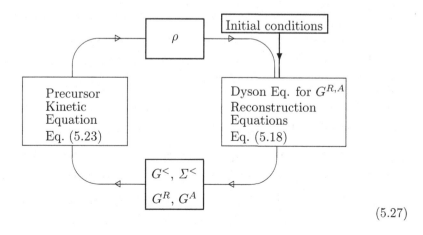

$$\tag{5.27}$$

The basic cycle is indicated by open arrows. The cycle follows the true physical process under investigation. It starts at t_I and is incremental in time. The initial conditions — correlated or uncorrelated alike — have to be fed in from outside, as shown symbolically by a box "Initial conditions" in the top right corner of Eq. (5.27).

We have to distinguish the uncorrelated initial condition from the correlated one, however. The uncorrelated initial conditions amount just to the initial one-particle density matrix, which is known. This would be enough to reconstruct the initial correlation function and its self-energy in terms of initial ρ (right hand box). These enter the precursor transport equation and the result would be ρ for the first incremental time (left hand box). This cycle would then proceed on and on. This is the practical use of the reconstruction theorem, by which the Dyson equation is solved for all times using the already known section of ρ as a function of time.

In the correlated case, the initial conditions which have to be input during each cycle in addition to ρ_I involve basically the initial time corrections to $\Xi^<$. The reconstruction cycle may become practicable, if it will be supplemented by the corresponding algorithm, which will depend on the method of incorporating the initial conditions. We are not aware of a practical attempt to make use of the scheme for correlated initial conditions, and do not enter into details for that reason.

6. Long time asymptotics: Quantum kinetic equations

This section will be devoted to the second main theme of this review, namely the use of the non-equilibrium Green's functions as a reliable tool for derivation and justification of simplified quantum kinetic equations, both of the type of transport equations, augmenting the Boltzmann equation for extended systems, and of the type of Generalized Master equation extending the common Pauli equation to systems with some quantum memory.

We will introduce two schemes how to construct quantum kinetic equations for a single time distribution function within the NGF approach: the first one is based on the quasiclassical expansion and leads to various versions of quantum Boltzmann equation, the second one is related to a short time expansion and it enables us to describe time dependent processes via the GME type of kinetic equations. both these scheme are approximate only, and are based on certain type of decoupling schemes which are commonly known under the name of "Ansatzes". The way towards the Boltzmann equation is based on the use of the *Kadanoff-Baym Ansatz*, while the derivation of the GME uses the *Generalized Kadanoff-Baym Ansatz*. Both schemes start from the precursor quantum kinetic equation (PKE) as introduced in Sec. 5.1.1 and 5.5.1.

6.1. *Assumptions about the time structure of the self-energy*

For convenience, we repeat here the equation (5.24):

$$G_0^{-1}G^< - G^<G_0^{-1} = \Sigma^R G^< - G^< \Sigma^A$$
$$-G^R \Sigma^< + \Sigma^< G^A$$
$$-G^R \Theta_{t_0}^< + \Theta_{t_0}^< G^A.$$

We intend to use it for times of steady external conditions. More precisely, for times which are sufficiently separated from the time t_{Last} of the last abrupt ("jerky") perturbation of the system. To understand how this can simplify the precursor equation, we have to review once more the time structure of the self-energies.

The assumption we make is that there exists a time $\tau^\star$, such that

$$\Sigma^\natural(t, t') \approx 0 \text{ for } |t - t'| > \tau^\star, \qquad \natural = R, A, > . \tag{6.1}$$

This condition should not be taken too literally, as, e.g., the band edges lead to weak power law tails. Still, it captures in a simple manner the essential feature of a typical self-energy, to be concentrated along the time diagonal, and is in agreement with the notion of the decay of correlations, as explained above. We will use (6.1) for convenience and simplicity. The finite range of the IC corrections is just a corollary.

To see, how the condition (6.1) works, we note that the time t_{Last} plays the role of the initial time for the smooth phase of the system evolution we consider. The system quickly loses the memory of the disturbance, which means that the last two generalized collision terms of Eq. (5.24), which take the initial conditions into account explicitly, become zero as the running time t exceeds $t_{\text{Last}} + 2\tau^\star$. This is in accord with the Bogolyubov conjecture. Thus, it is enough to wait for just $2\tau^\star$ past t_{Last}, and the precursor QTE becomes free of any IC terms:

$$G_0^{-1}G^< - G^<G_0^{-1} = \Sigma^R G^< - G^< \Sigma^A$$
$$+\Sigma^< G^A - G^R \Sigma^<, \qquad t_1, t_2 > t_{\text{Last}} + 2\tau^\star. \tag{6.2}$$

In the remaining collision terms, the time integration range is restricted by the condition (6.1) to a finite interval $(t - \tau^\star, t)$. This is the memory depth of the system. For a typical term in Eq. (6.2), $\Sigma^R G^<$, this means

$$\{\Sigma^R G^<\}(t, t) = \int_{t-\infty}^{t} d\bar{t}\, \Sigma^R(t, \bar{t}) G^<(\bar{t}, t)$$

$$\xrightarrow{\tau^\star} \int_{t-\tau^\star}^{t} d\bar{t}\, \Sigma^R(t, \bar{t}) G^<(\bar{t}, t) . \tag{6.3}$$

For $t > t_{\text{Last}} + 3\tau^\star$, the self-energy in (6.3) extends already over the steady time span, and while the turbulent past may be reflected in the values of $G^<$, it will in no case show in the form of the EOM (6.2). And this is all the Bogolyubov conjecture says.

The equation (6.2) is just the Generalized Kadanoff-Baym equation (5.6). This is the equation, which enabled Kadanoff and Baym to derive the quantum Boltzmann equation.

6.2. Kadanoff-Baym Ansatz and quantum Boltzmann equation

To arrive at a quantum generalization of the Boltzmann equation, it is natural to convert the transport equation (6.2) for correlation function $G^<$ to the form which contains explicitly spectral properties of the system and has a shape which is more suitable for quasi-classical approximation. To this end, we first introduce into (6.2) spectral functions A and Γ by separating the imaginary and real parts of retarded and advanced functions

$$A = i(G^R - G^A), \quad G = \frac{1}{2}(G^R + G^A), \tag{6.4}$$

$$\Gamma = i(\Sigma^R - \Sigma^A), \quad \Sigma = \frac{1}{2}(\Sigma^R + \Sigma^A). \tag{6.5}$$

In the second step with the help of the following identities

$$i\left(G^R\sigma^< - \sigma^< G^A\right) = i[G, \sigma^<]_- + \frac{1}{2}[A, \sigma^<]_+ , \tag{6.6}$$

$$i\left(\Sigma^R g^< - g^< \Sigma^A\right) = i[\Sigma, g^<]_- + \frac{1}{2}[\Gamma, g^<]_+ , \tag{6.7}$$

where $[A, B]_\pm = AB \pm BA$ are anticommutators or commutators, we rearrange (6.2) into the following equation

$$-i[G_0^{-1} - \Sigma, g^<]_- - i[G, \sigma^<]_- = \frac{1}{2}[A, \sigma^<]_+ - \frac{1}{2}[\Gamma, g^<]_+ . \tag{6.8}$$

Here we used instead of $G^<$ its counterpart $g^<$ given by (3.40), and correspondingly $\sigma^<$.

This equation is just different form of the GKBE equation and it is thus also often called the Generalized Kadanoff-Baym (GKB) Equation.

6.2.1. Kinetic equation for quasi-particle distribution function: Kadanoff-Baym approach

Our aim is to find a quantum generalization of the BE, in a controlled way, i.e., we want to find a kinetic equation for a distribution function $f(k, r, t)$ from the GKB Equation as an asymptotic equation. Naturally, the first step is to introduce the Wigner representation of the correlation function

$$g^<(\omega, k, r, t) = \int dx d\tau e^{i\omega\tau - ikx}$$
$$\times g^< \left(r + \frac{x}{2}, t + \frac{\tau}{2}, r - \frac{x}{2}, t - \frac{\tau}{2}\right), \tag{6.9}$$

where $(r + \frac{x}{2}, t + \frac{\tau}{2}) \equiv 1$ and $(r - \frac{x}{2}, t - \frac{\tau}{2}) \equiv 2$ in the cumulative variable notation. From now on, we will use Roman types for operators in the cumulative variable representation and Sans-serif types for operators in Wigner's representation.

The expression for Wigner's distribution function,

$$\rho(k, r, t) = \int\limits_{-\infty}^{\infty} \frac{d\omega}{2\pi} g^<(\omega, k, r, t), \tag{6.10}$$

has the essential property that it includes contributions from all independent energies ω. Thus, a perturbation scheme constructed for $g^<$ instead for ρ will enable us to keep the energy as an independent variable until we select how to determine the energy of a particle from its position in phase space. By this step, we will find a proper distribution function avoiding the problems with the high momenta tails of Wigner's function. The existence of the independent energy permits to distinguish two very different contributions in the transport equations for the correlation function $g^<$). First, the on-shell contributions, for which a dispersion relation between the energy and the position of the particle in phase space holds. Second, the off-shell contributions, for which no such relation exists. As we will see later, the possibility of this distinction leads to a formulation of the perturbation schemes which better suit the demands of the kinetic equations. To summarize this philosophy, we will close the perturbative expansion for $g^<$. A quantum kinetic equation will be derived as an asymptotic limit of the equation for $g^<$. This asymptotic equation is not closed for the Wigner distribution, but only for the on-shell part of $g^<$ which can be interpreted as the quasi-particle distribution. The observables will be obtained from $g^<$ in the second step via (6.10).

6.2.2. *Quasi-classical approximation*

To proceed on the way to a kinetic equation, we now need to convert the GKB Equation (6.8) to the Wigner representation. For matrix products, $C = AB$, it reads

$$c = \exp\left(\frac{i}{2}\left(\frac{\partial}{\partial\omega}\frac{\partial}{\partial t'} - \frac{\partial}{\partial t}\frac{\partial}{\partial\omega'} - \frac{\partial}{\partial k}\frac{\partial}{\partial r'} + \frac{\partial}{\partial r}\frac{\partial}{\partial k'}\right)\right)$$

$$a(\omega, k, r, t)b(\omega', k', r', t')_{\omega'=\omega, k'=k, r'=r, t'=t}. \tag{6.11}$$

The Boltzmann equation is known to be valid only for slowly varying fields. For this case we can expect that a dependence on the hydrodynamical variables r and t will be proportional to gradients of the external fields, i.e., small. In this situation, the gradient expansion (6.11) can be approximated by its lowest order terms. The anticommutator is even in gradients, thus it is approximated by a simple product, $C = \frac{1}{2}[A, B]_+$,

$$c = ab. \tag{6.12}$$

The commutator is odd in gradients and it thus turns to the Poisson bracket, $C = -i[A, B]_-$,

$$c = [a, b] \equiv \frac{\partial a}{\partial \omega} \frac{\partial b}{\partial t} - \frac{\partial a}{\partial t} \frac{\partial b}{\partial \omega} - \frac{\partial a}{\partial k} \frac{\partial b}{\partial r} + \frac{\partial a}{\partial r} \frac{\partial b}{\partial k}. \tag{6.13}$$

Using (6.12) and (6.13) we find the GKB Equation (6.8) in the gradient approximation

$$[\omega - \epsilon - U_{\text{eff}} - \sigma, g^<] + [g, \sigma^<] = a\sigma^< - \gamma g^<. \tag{6.14}$$

This equation is a good starting point for quantum generalizations of the BE.

6.2.3. Kadanoff-Baym Ansatz

Contrary to the KB equation, which is the equation for $g^<(\omega, k, r, t)$, any kinetic equation of the Boltzmann type is an equation for a distribution function $f(k, r, t)$, so that there is no independent energy variable in it. Therefore, any construction of an asymptotic kinetic equation from the KB equations means to find an auxiliary functional $g^<[f]$ by which the independent energy becomes fixed and related to the phase-space variables. Kadanoff and Baym suggested as a starting point for their construction of the auxiliary functional the relation

$$g^<(\omega, k, r, t) \approx \phi(\omega, k, r, t) a(\omega, k, r, t), \tag{6.15}$$

i.e. they assumed that the relation between the correlation function $g^<$, the spectral function a and the Fermi Dirac distribution f_{FD}, in the form

$$g^<(k, \omega, r, t) \overset{\text{equil.}}{\equiv} g^<(k, \omega)$$
$$= f_{\text{FD}}(\omega) \times A(k, \omega) , \tag{6.16}$$

which is exact in equilibrium, was approximately valid near to equilibrium also for a non-equilibrium function ϕ. In the following step they eliminated the energy argument from ϕ by the assumption that the scattering rate was small, so that the spectral function had a singularity

$$a(\omega, k, r, t) \approx 2\pi\delta(\omega - \epsilon - U_{\text{eff}}(r, t)) . \tag{6.17}$$

The energy argument of ϕ can be replaced by the mean-field energy $\epsilon + U_{\text{eff}}$. Kadanoff and Baym proposed to approximate $g^<$ as

$$g^<(\omega, k, r, t) \approx f(k, r, t) a(\omega, k, r, t) \tag{6.18}$$
$$\approx f(k, r, t) 2\pi\delta(\omega - \epsilon - U_{\text{eff}}(r, t)) . \tag{6.19}$$

Taking the approximation (6.18), one does not obtain the BE, since this approximation leads to the violation of energy conservation of individual scattering events. Kadanoff and Baym thus used the (mean-field) non-renormalized pole approximation (6.19), which is called the Kadanoff and Baym Ansatz (KBA).

6.2.4. *BE in the mean-field approximation*

From the KB equation (6.14) and the KB Ansatz (6.19) of $g^<$ we get

$$2\pi\delta(\omega - \epsilon - U_{\text{eff}})[\omega - \epsilon - U_{\text{eff}} - \sigma, f] + [g, \sigma^<]$$
$$= 2\pi\delta(\omega - \epsilon - U_{\text{eff}})(\sigma^< - \gamma f), \tag{6.20}$$

where we used $[\omega - \epsilon - U_{\text{eff}}, \delta(\omega - \epsilon - U_{\text{eff}}] = 0$, and the δ–function can be extracted from the first Poisson bracket.

At this point, Kadanoff and Baym adopted an idea that only the pole terms, in other words only the terms with the δ–function, contributed to the BE. They completely neglected the Poisson bracket $[g, \sigma^<]$. By this step, Kadanoff and Baym eliminated the ω–dependence using the pole value, $\omega = \epsilon + U_{\text{eff}}$ and they derived the following kinetic equation:

$$\frac{\partial f}{\partial t} + \frac{\partial \epsilon}{\partial k}\frac{\partial f}{\partial r} - \frac{\partial U_{\text{eff}}}{\partial r}\frac{\partial f}{\partial k} = I_{in}[f] - I_{out}[f], \tag{6.21}$$

where the scattering integrals are given by

$$I_{in}[f] = \sigma^<_{\omega=\epsilon+U_{\text{eff}}}, \tag{6.22}$$
$$I_{out}[f] = f\gamma_{\omega=\epsilon+U_{\text{eff}}}. \tag{6.23}$$

There are various generalization of the Kadanoff-Baym Ansatz, which go beyond the mean-field approximation, the most advanced of which is the so called extended quasiparticle approximation. It is out of scope of this article to discuss them.[106,258–267] Instead we will deal with its generalizations in different direction, in which the starting point is the Generalized Kadanoff-Baym Ansatz.

6.3. *Generalized Kadanoff-Baym Ansatz and generalized master equation*

In this subsection, the Generalized Kadanoff and Baym Ansatz (GKBA) will be defined, its salient properties will be analyzed, and the Ansatz will be finally used as a tool for deriving the GME. As mentioned the GKBA has originally been introduced as an alternative to the Kadanoff-Baym Ansatz. We shall follow this line of reasoning now. The KBA is given by

$$G^<(t_1, t_2) = f(\frac{t_1 + t_2}{2})[-G^R(t_1, t_2) + G^A(t_1, t_2)], \tag{6.24}$$

The *Generalized Kadanoff and Baym Ansatz* has been written as

$$G^<(t_1, t_2) = -G^R(t_1, t_2)\rho(t_2) + \rho(t_1)G^A(t_1, t_2). \tag{6.25}$$

In comparison with the KBA it differs in four constitutive properties:

Causal structure: The two propagators are now separated and each is multiplied
  by its own distribution function corresponding to the earlier time argument,
  depending on the retarded or advanced propagation

Equal time limit: The distribution function is the true one-particle density matrix, not the quasi-particle distribution. Thus, at equal times, the GKBA for $G^<$
  obeys identity (3.41). This implies an automatic particle number conservation.

General representation: The order of factors corresponds to the products interpreted as a matrix multiplication. Therefore, a diagonal representation for
  Green's functions is not assumed.

Arbitrary non-equilibrium: No assumptions about slowly varying disturbances,
  small deviations from equilibrium, or quasi-particle approximation are built in,
  the Ansatz is formally quite general.

In addition, there is a basic difference, which distinguishes the GKBA from the
KBA. Namely, while both Ansatzes are but approximate constructions, only the
expression (6.25) can serve as a starting approximation for a process leading to an
exact reconstruction of $G^<$. The GKBA thus has an important meaning, independent of and going beyond the restricted context of transport equations: it offers a
path to the Reconstruction Problem as discussed in the previous Section devoted
to the Reconstruction theorem.

The GKBA particle correlation function perfectly fits into the formal NGF
scheme. It was constructed in this manner in order to overcome the limitations
of the KBA. When inserted into the machinery of deriving quantum transport
equations from the NGF, it leads to the GME, rather than to the BE like quantum
kinetic equations.

6.3.1. Towards GME

To derive a GME, i.e. to obtain a single time kinetic equation, we proceed in three
steps:

1st *step:* We recall the *precursor quantum kinetic equation* (PKE) (5.6) . In other
words on the way to GME we start from

$$\frac{\partial \rho}{\partial t} - \text{drift} = \left[-G^R \Sigma^< + \Sigma^< G^A + \Sigma^R G^< - G^< \Sigma^A \right]_{\text{equal times}}. \qquad (6.26)$$

2nd *step:* A self-consistent *physical approximation* (5.7) is selected, that is, the
self-energy is expressed in terms of the GF,

$$\Sigma = \Sigma[\mathbf{G}].$$

A typical approximation could be RPA for interacting electrons, or the Migdal
approximation for an electron-phonon system.

3rd *step:* The l.h.s. of (6.26) already has the desired GME form. The r.h.s. contains
a number of double time quantities, which have to be eliminated. In this final step,

use is made of the Generalized Kadanoff-Baym Ansatz in order to eliminate $G^<$ in favor of its time diagonal ρ:

$$G^<(t,t') = \underset{t > t'}{-G^R(t,t')\rho(t')} + \underset{t < t'}{\rho(t)G^A(t,t')} . \tag{6.27}$$

When introduced into the PKE (6.26), this leads finally to the quantum kinetic equation represented by a closed GME for ρ:

$$\frac{\partial \rho}{\partial t} + \mathrm{i}[h_0, \rho]_- \qquad \boxed{\Sigma^< = \Sigma^<[\rho|G^R, G^A]} \tag{6.28}$$

$$= -\mathrm{i} \int_{-\infty}^{t} d\bar{t}(G^R(t,\bar{t})\Sigma^<(\bar{t},t) - \Sigma^<(t,\bar{t})G^A(\bar{t},t))$$

$$+ \mathrm{i} \int_{-\infty}^{t} d\bar{t}(\Sigma^R(t,\bar{t})\rho(\bar{t})G^A(\bar{t},t) + G^R(t,\bar{t})\rho(\bar{t})\Sigma^A(\bar{t},t)) .$$

The memory kernel at the r.h.s. is seen to be the generalized collision term; the role of the propagators is essential, and this will be further accented by the discussion in the following subsections.

The result (6.28) is still very general. If it is specified that h_0 corresponds to an extended system of electrons moving in a single band under the influence of a strong *dc* or *ac* homogeneous electric field, we find that an equivalent equation, usually referred to as the Levinson equation,[268] has originally been derived by the density matrix techniques. Its derivation from the NGF has been the first success of the GKBA. In fact, it re-established the Green functions as a serious contender in the everlasting competition with the more direct density matrix approach. Namely, the GME obtained previously with the help of the KB Ansatz failed to describe correctly even the *ac* linear response in the weak scattering limit, as it predicted spurious sub-harmonic frequencies in the induced current. This artifact of the approximations was not appearing for the original Levinson equation; it was eliminated in the NGF method by simply using the GKBA instead of the KBA.

To summarize, a closed transport equation for ρ was derived in three steps,

$$\frac{\partial \rho}{\partial t} - \mathrm{drift} = \Phi[\rho(\tau); \tau < t \,|G^R, G^A] , \tag{6.29}$$

having precisely the form (2.17), only here the functional dependence on the propagators is explicitly indicated.

With the aid of GKBA, the program for deriving GME like equations outlined in Sec. 5.1.2 has been successfully completed.

6.3.2. *GME and the quantum BE compared*

The GME resembles the BE since it has the drift term on the left hand side and interaction terms similar to the scattering integrals on the right hand side. In detail, however, there are three basic differences between both equations: the GME captures more of the quantum dynamics, it is not limited to the quasiclassical limit, and it is not conditional on the quasi-particle picture. This mirrors in some obvious features of the GME. First, it is a non-Markovian equation thanks to the time integration in interaction terms. Second, its drift term does not include the quasi-particle energy renormalization. Finally, it is an equation directly for the reduced density matrix, so in comparison to the BE it does not need any accompanying functional serving to calculate observables. The first r.h.s. integral describes the back scattering of electrons. In (6.28), it is still kept in the closed symbolic form. Its expanded form will depend on the specific approximation used for $\Sigma^<[G^<]$. We will not discuss here any specific interaction. The second, forward scattering integral, on the other hand, has a sufficiently explicit structure to be compared with the BE.

7. GKB Ansatz, its variants and GME

The Generalized Kadanoff-Baym Ansatz has been introduced in Sec. 6.3 in an intuitive manner on the basis of several physically motivated requirements. Then we concentrated on demonstrating its use in deriving the quantum kinetic equations. In this last section of the whole review, we intend to seek for a better justification of the Ansatz both from the formal and the physical point of view. It may be pointed out that GKBA has been applied to a number of problems with an excellent or at least a good success. In this sense, it has already been accepted as a reliable tested part of the inventory of the tools of condensed matter theory. A purely theoretical background of its use may still be important. In particular, we will find that a detailed study of the properties of GKBA suggests its possible modifications promising improved reconstruction schemes or offering a hindsight justification of the reconstruction schemes which have been employed in practice.

7.1. GKBA and the reconstruction equations

The GKB Ansatz and the reconstruction equations are intimately related, as is seen at a single glance; for convenience, we repeat here one half of the RE, Eq. (5.18). For $t_1 > t_2$ we have

$$
\begin{aligned}
G^<(t_1, t_2) = &-G^R(t_1, t_2)\rho(t_2) \\
&+ \int_{t_2}^{t_1} d\bar{\bar{t}} \int_{-\infty}^{t_2} d\bar{t}\, G^R(t_1, \bar{\bar{t}})\Sigma^<(\bar{\bar{t}}, \bar{t})G^A(\bar{t}, t_2) \\
&+ \int_{t_2}^{t_1} d\bar{\bar{t}} \int_{-\infty}^{t_2} d\bar{t}\, G^R(t_1, \bar{\bar{t}})\Sigma^R(\bar{\bar{t}}, \bar{t})G^<(\bar{t}, t_2).
\end{aligned}
\tag{7.1}
$$

The absolute term of this inhomogeneous equation is identical with the corresponding part of the GKBA formula (6.25). If the integral terms of (7.1) could be neglected, the Ansatz would be justified. An iteration of this equation would yield a corrected GKBA, but this still would represent $G^<$ as a functional of ρ and the propagators, which could be used for the same reconstruction mechanism as the Ansatz itself. There are two drawbacks of such procedure. One is practical. Computation of the double integrals may be truly time consuming. In a steady state, some simplifications would be possible, but for transient processes the computational effort would be prohibitive. The other difficulty is fundamental. The special feature of the Ansatz is the factorization of $G^<$ at one sharp instant of time, and, of course, any integral corrections would blur that sharply time local factorization over some range, most likely on the order of the collision duration time.

7.2. *GKBA and the multiplicative rule for propagators*

In this section the GKBA will be illuminated from a different angle, following.[138] The dynamic nature of the GKBA will be linked with the so-called semi-group character of the single-particle propagation.

7.2.1. *Non-interacting particles*

For independent particles, it holds

$$G_0^<(t_1, t_2) = iG_0^R(t_1, t_0)\rho(t_0)G_0^A(t_0, t_2) . \qquad (7.2)$$

The propagators $G_0^{R,A}$ are proportional to the unitary single-particle evolution operator. They obey a multiplication rule (have the *"semi-group property"*). For G_0^R, which obeys

$$G_0^R(t_1, t_0) = iG_0^R(t_1, t_2)G_0^R(t_2, t_0), \qquad t_1 > t_2 > t_0. \qquad (7.3)$$

Using this semi-group property in (7.2) we obtain, for $t_1 > t_2 > t_0$,

$$
\begin{aligned}
G_0^<(t_1, t_2) &= \quad G_0^R(t_1, t_2)G_0^R(t_2, t_0)\rho(t_0)G_0^A(t_0, t_2) \\
&= -G_0^R(t_1, t_2)\rho(t_2) , \qquad (7.4)
\end{aligned}
$$

which is just the retarded half of the GKB Ansatz. The advanced half for the converse order of times can be obtained by hermitian conjugation. The GKB Ansatz appears here as an exact identity based on the semi-group property of free particle propagators.

7.2.2. *Interacting particles*

In the case of interacting particles we follow the heuristic argument of[138] and start from the Dyson-Keldysh equation (3.221):

$$G^<(t_1, t_2) = \int\limits_{-\infty}^{t_1} dt_3 \int\limits_{-\infty}^{t_2} dt_4 G^R(t_1, t_3)\Sigma^<(t_3, t_4)G^A(t_4, t_2). \qquad (7.5)$$

Let $t_1 > t_2 > t_0 \to -\infty$. The formal integration region shown in Fig. 7 is a rectangle with corners at t_0, t_1, t_2. Following Sec. 6.1, it will be assumed that the actual integration involves only an intersection with the strip of a width $2\tau_c$ along the time diagonal $t_1 = t_2$, where the self-energy $\Sigma^<$ is significantly different from zero, cf. Eq. (6.1). If the "small" area C is neglected, the integral extends only over the square $t_3 < t_2, t_4 < t_2$. Notice that neglecting the contribution of C is identical with setting the first integral correction in the reconstruction equation (7.1) to zero.

In the remaining square integration range, let us assume an approximate validity of the semi-group rule generalized from (7.3) to

$$G^R(t_1, t_0) \approx iG^R(t_1, t_2)G^R(t_2, t_0), \qquad t_1 > t_2 > t_0. \tag{7.6}$$

Then the equation (7.5) is reduced to the form of the GKB Ansatz

$$G^<(t_1, t_2) = -G^R(t_1, t_2)\rho(t_2), \qquad t_1 > t_2, \tag{7.7}$$

where

$$\begin{aligned}
\rho(t_2) &= -iG^<(t_2, t_2) \\
&= -i \int_{-\infty}^{t_2} dt_3 \int_{-\infty}^{t_2} dt_4 G^R(t_2, t_3)\Sigma^<(t_3, t_4)G^A(t_4, t_2), \tag{7.8}
\end{aligned}$$

according to Eq. (7.5) for $t_1 = t_2$.

In deriving the Generalized Kadanoff-Baym Ansatz, three assumptions were invoked. Two have a general character: an uncorrelated initial condition (not essential), and a uniformly small particle correlation time. The third one, the semi-group rule, is specific for this consideration. We have found that the semi-group property (7.6) is equivalent to an elimination of the second integral term from the reconstruction equation (7.1). Due to the interactions, the semi-group condition for the propagators cannot be valid exactly. The conclusion is that the Ansatz for $G^<$ can be checked or improved by the study of the multiplicative properties of propagators $G^{R,A}$.

7.3. *Quasi-particle GKB Ansatz*

We have seen the link between the GKB Ansatz and the semigroup property of propagators. In interacting systems, validity of this semigroup property is doubtful. A modified multiplication rule and a related modification of the Ansatz can be motivated, however, if the quasi-particle picture is known to work.

7.3.1. *Quasi-particle multiplication rule for G^R*

Consider first an equilibrium quasi-particle. The propagator $G^R(t_1, t_2)$ is characterized by three quantities: the pole energy $E_W = E - i\tau^{-1}$, the renormalization constant z and the time of formation τ_Q, which corresponds to the time spread of the kernel of the retarded self-energy Σ^R, just like τ_c is related to $\Sigma^<$. In order to

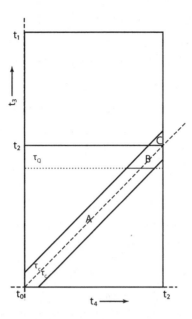

Fig. 7.   Schematic structure of integration regions for deriving the GKBA from (7.5). To obtain $G^<(t_1, t_2)$, the integration region is the big rectangle. The actual integration range is the strip of the width $2\tau_c$ flanking the time diagonal. The effective integration region for obtaining the GKBA is the diagonal square. The propagators should be factorized according to the semi-group rule at its top side. The initial time t_0 should be set to $-\infty$ at the end. Derivation of the QKBA runs along similar lines; for details, see the main text.

achieve the correct hierarchy of the characteristic times, the so-called quasi-particle condition, $\tau_Q \ll \tau_r$, is required, see Ref. 116. Here, τ_r is the transport relaxation time which is comparable with the quasi-particle life-time τ . Then

$$
G^R(t_1, t_2) = \begin{cases} \text{QP formation process } t_1 < t_2 + \tau_Q, \\[2mm] z\, G_W^R(t_1, t_2) \qquad\quad t_1 > t_2 + \tau_Q, \end{cases}
\tag{7.9}
$$

$$
G_W^R(t_1, t_2) = -\mathrm{i}\, \exp(-\mathrm{i}E(t_1 - t_2) - \tau^{-1}(t_1 - t_2)),
$$

with G_W^R the Weisskopf-Wigner propagator. This propagator obeys, as a defining property, the exact multiplication rule

$$
G_W^R(t_1, t_2) = \mathrm{i}\, G_W^R(t_1, \widetilde{t}) G_W^R(\widetilde{t}, t_2) .
\tag{7.10}
$$

For the following hierarchy of times,

$$
t_1 > \widetilde{t} > t_2 + \tau_Q > t_2 > t_0,
\tag{7.11}
$$

the modified multiplication rule for G^R is obvious:

$$
G^R(t_1, t_2) = \mathrm{i}\, G_W^R(t_1, \widetilde{t}) G^R(\widetilde{t}, t_2) .
\tag{7.12}
$$

In contrast, the original semigroup factorization (7.3) is not satisfactory: at $t_1 = \tilde{t}$, a spurious kink appears because of a repeated quasi-particle formation.[138,141,144] For times $t_1 > \tilde{t} + \tau_Q$, the factorized expression has the value $z^2 G_W^R(t_1, t_2)$ rather than the correct one as given by (7.9).

The rule may now be generalized to non-equilibrium by postulating that, to the true propagator G^R, a time of formation τ_Q and a construct called $G_W^R(t_1, t_2)$ characterized by (7.10) exists such that for times satisfying (7.11) the modified composition rule (7.12) holds. The conditions for validity of this rule, presently just postulated, are discussed in Ref. 218

7.3.2. Quasi-particle Ansatz for $G^<$

On the way to the new Ansatz we continue as above in Sec. 7.2.2. We start from the exact relation (7.5) and substitute everywhere G^R by its factorized form. Only now we use (7.12) instead of (7.6). Integration regions are sketched in Fig. 7 again. The newly invoked feature is the dashed horizontal line at $t_3 = t_2 - \tau_Q$. Below this line, the rule (7.12) is exact. The quasidiagonal integration strip is thus divided into three regions, A, B and C. In A, our transformation is exact. The top region C reaching above t_2 has to be neglected again and the region B remains as the principal source of error caused by (7.12). Ignoring this, we obtain

$$G^<(t_1, t_2) = -G_W^R(t_1, t_2)\rho(t_2), \qquad t_1 > t_2 , \qquad (7.13)$$

where ρ is given by the exact expression (7.8), just as before.

7.4. GKBA, QKBA, and family of causal Ansatzes

The main problem of the GKBA is the uncertain role of the involved unknown propagators. We have already seen that one of possibilities is to use quasiparticle propagators in the structure of the GKBA. We will now first summarize properties of the GKBA before we discuss variants of generalization of the GKBA and their role in generation of GME.

The GKBA fits well into the general formal and physical structure of the NGF theory. Its basic property is the correct causal structure. In addition, the GKBA has the particle-hole symmetry, the correct equal time limit, the correct asymptotic behavior for $|t_1 - t_2| \gg \max\{\tau_c, \tau_Q\}$. It is correct in the true Boltzmann limit. In the limit of non-interacting particles, it is exact for excitations arbitrarily far from equilibrium. Thus, it emerges as an interpolation scheme between two crucial exact limits. Furthermore, the GKBA does not depend on the quasi-classical expansion in space, and, in fact, it is not associated with any specific representation for the GF. All these qualitatively correct features do not guarantee that the GKBA, as a truncation, is satisfactory also quantitatively.

However, practical experience with this "standard approach" based on the GKBA has covered with good success several areas of physics including the hot electron transport (Levinson equation), and the response of electrons in semiconductors to sub-picosecond pulses (Quantum Optical Bloch Equations). Computed properties and processes were in an excellent qualitative and very good quantitative agreement with experimental data.

A closer look offers an explanation for these good results. It was naturally not practicable to compute the exact propagators entering the GKBA, and the propagators constructed on model grounds worked well, most likely compensating the error of the Ansatz itself.

Such approach, seemingly a pragmatic expedient, has, in fact, a deep meaning. In reality, the Ansatz used was not the GKBA proper, but its modification employing rather the quasi-particle propagators, the "QKBA". We meet here an instance of an Ansatz scheme possessing the same general properties as the GKBA. A whole family of such "Causal Ansatzes" already exists. Our goal in the next part of the paper will be to discuss the means of their systematic generation, comparison and assessment. This is also due to the general concept of the reconstruction procedure discussed above.

Clearly, neither the GKBA (7.7), nor the QKBA (7.13) have been proven, but firstly, both have been linked with the related multiplication rule for propagators. Secondly, their physical motivation has been put on the same footing. In fact, the QKBA appears as somewhat better justified than the GKBA proper once the quasi-particle behavior of the system can be assumed. Thirdly, the two Ansatzes appear as two instances of a general factorization $G_x^< = G_x^R \rho$ with the subscript $_x$ meaning an arbitrary choice of the effective propagator. If this propagator satisfies basic requirements, the corresponding "Ansatz" will possess the general properties listed for the GKBA like Causal Ansatzes.

7.4.1. *Causal Ansatzes*

We have already discussed that the GKBA Ansatz has many useful properties. On the whole, the GKBA has a number of important properties which make it consistent with the general NGF scheme. It would be natural to take this set of properties as requirements to be fulfilled by any other Ansatz approximation scheme. We will term any approximation a **Causal Ansatz**, if it fullfills the same following characteristics as the GKBA Ansatz:

◇ Its basic property is that of possessing the correct causal structure.

◇ In addition, the GKBA has the particle-hole symmetry,

◇ the correct equal time limit,

◇ the correct asymptotic behavior for $|t_1 - t_2| \gg \max\{\tau_c, \tau_Q\}$.

◇ It coincides with the KBA in the true Boltzmann limit. In the limit of non-interacting particles,

◇ it is exact for excitations arbitrarily far from equilibrium (cf. Sec. 7.2.1). ◇ Thus, it emerges as an interpolation scheme between two crucial exact limits.

◇ As said already, it fits not only into the physical framework, but also into the formal structure of the exact NGF equations. ⋄ Furthermore, the GKBA does not depend on the quasi-classical expansion in space, and, in fact, it is not associated with any specific representation for the GF.

A wide family of Causal Ansatzes already exists. The GKBA, in addition to its direct importance, is the primary member of this family. Still, all these qualitative properties do not guarantee that the GKB approximation, amounting to the first term of an asymptotic expansion, will be quantitatively satisfactory.

Finally, we have in this section outlined a simple and graphic way to the Causal Ansatz family, which offers new possibilities of devising, improving and analyzing its further members.

7.5. *Causal Ansatzes and various GME*

We will now shortly show how variants of the GKBA can influence the related form of corresponding GME.

To this end we first shortly summarize the general method of generating a quantum transport equation from the precursor transport equation :

⋄ We will start from the precursor quantum transport equation:

$$\frac{\partial \rho}{\partial t} + i\, [\, H_0, \rho\,]_- = -(\Sigma^R G^< - G^< \Sigma^A)$$
$$+(G^R \Sigma^< - \Sigma^< G^A)$$
$$+\Theta^<_{t_0} G^A - G^R \Theta^<_{t_0}. \tag{7.14}$$

This already has the form of a QTE, in which $H_0(t)$ is the mean field one-particle Hamiltonian. The "generalized collision" terms on the r.h.s. still involve double time *less* quantities. The related integrals extend only to the past because of the presence of the propagator factors.

⋄ We will apply the Keldysh initial conditions at $t_0 \rightarrow t_I$,

⋄ The double-time $G^<(t,t')$ is replaced by an approximate 'Ansatz' expression involving only its time diagonal $\rho(t) = -iG^<(t,t)$ and propagators.

⋄ This expression is also introduced into the self-energy, for which a self-consistent approximation specifying the self-energy in terms of the Green's function is assumed.

⋄ As a parallel input the knowledge of the propagator components of the Green's function in necessary. In general, the components of the NGF are interconnected, but often this is of a lesser importance and the propagators can be found beforehand once for ever.

7.5.1. *From an Ansatz to the GME*

The resulting transport equation depends on the approximate replacement of $G^<$.

As we already discussed, the historically first one is the famous Kadanoff-Baym Ansatz, schematically $G^< = \frac{1}{2}[\rho, G^A - G^R]_+$. This choice was connected with the subsequent quasi-classical expansion made by Kadanoff and Baym. This is irrelevant for systems with discrete levels.

We will concentrate on the so-called XKB Ansatzes, belonging to the general class of the causal Ansatzes, and exemplified by the original Generalized Kadanoff-Baym Ansatz — GKBA $G^< = -G^R \rho + \rho G^A$. This Ansatz is exact for independent particles governed by a one-particle Schrödinger equation. In general, it is an approximation and other XKBA variants have been considered:

$$G_{\mathsf{x}}^< = -G_{\mathsf{x}}^R \rho + \rho G_{\mathsf{x}}^A . \tag{7.15}$$

Some more important examples are summarized in the table:

Abbr.	Ansatz
GKBA	$G_{\mathsf{G}}^< = -G_{\mathsf{G}}^R \rho + \rho G_{\mathsf{G}}^A$ (7.16)
QKBA	$G_{\mathsf{Q}}^< = -G_{\mathsf{Q}}^R \rho + \rho G_{\mathsf{Q}}^A$ (7.17)
FKBA	$G_0^< = -G_0^R \rho + \rho G_0^A$ (7.18)

Each Ansatz is specified by a particular choice for the form of the propagators. This is indicated by the label. Here $G_{\mathsf{G}}^{R,A}$ denotes the true, fully renormalized propagators, $G_{\mathsf{Q}}^{R,A}$ the quasiparticle GF and $G_0^{R,A}$ the free particle propagators. When using the Ansatz, all participating GF have to be labeled. Thus, although the GKBA given by (7.16) is clearly identical with (6.25), we use the label 'G', so that $G_{\mathsf{G}}^R \equiv G^R$, etc. It may be asked which use one can expect from approximating the true propagators by another choice. This goes beyond the present discussion, but reasons have been given for using the Quasiparticle Kadanoff-Baym Ansatz (QKBA) in particular, in which the propagators are represented by the unrenormalized pole part of the true propagators. FKBA has been used inadvertently in derivations of the Boltzmann equation or the master equation in the second order weak coupling theory equivalent with the use of the plain Fermi Golden Rule.

Returning to our main task, it is enough to approximate the exact generalized scattering terms in the precursor Quantum Transport Equation (5.6) by a properly selected Ansatz approximate expression, and the GME for the true particle distribution ρ results:

$$\frac{\partial \rho}{\partial t} + \mathrm{i}\,[\,H_0, \rho\,]_- = -(\Sigma_{\mathsf{x}}^R G_{\mathsf{x}}^< - G_{\mathsf{x}}^< \Sigma_{\mathsf{x}}^A)$$
$$+ (G_{\mathsf{x}}^R \Sigma_{\mathsf{x}}^< - \Sigma_{\mathsf{x}}^< G_{\mathsf{x}}^A) . \tag{7.19}$$

We can clearly see that the quality with which the above GME describes the dynamics of the consider systems depends on the construction of suitable approximations for propagators.

Acknowledgments

This research was supported by the Czech Science Foundation within the grant project P204/12/0897.

List of abbreviations

BE Boltzmann equation
BSE Bethe-Salpeter equation
DE Dyson equation
DFT Density Functional Theory
FDT Fluctuation-Dissipation Theorem
GF Green's Function
GKBA Generalized Kadanoff-Baym Ansatz
GKBE Generalized Kadanoff-Baym equation
GME Generalized master equation
IC initial condition
IR irreducible parts
KB Kadanoff-Baym
KBA Kadanoff-Baym Ansatz
KMS Kubo-Martin-Schwinger
l.h.s. left-hand side
MO molecular orbitals
NGF non-equilibrium Green's Function
nE non-equilibrium
PE Phase Equation
QP Quasi-Particle
QKBA Quasi-Particle Kadanoff-Baym Ansatz
QRC Quasi-Particle Composition Rule
QTE Quantum Transport Equation
RE Reconstruction equations
r.h.s. right-hand side
SCBA Self-Consistent Born Approximation
TD Time Dependent
vC vertex correction
WI Ward Identity
WTI Ward-Takahashi Identity
WW Weisskopf-Wigner
XC exchange and Correlation

References

1. E. Akkermans, G. Montambaux, *Mesoscopic Physics of Electrons and Photons* (Cambridge University Press, Cambridge, 2007).
2. M. Di Ventra, *Electrical Transport in Nanoscale Systems*, (Oxford University Press, Oxford 2008).
3. J. Imry, *Introduction to mesoscopic physics* (Oxford University press, Oxford, Second ed., 2009).
4. N. Nagaosa, *Quantum Field Theory in Condensed Matter Physics* (Springer-Verlag, Berlin, 1999).
5. X.-G. Weng, *Quantum Field Theory of Many-Body Systems* (Oxford University Press, Oxford, 2004).
6. A. Altland, B. Simons, *Condensed Matter Field Theory* (Cambridge University Press, Cambridge, 2010).
7. E. Fradkin, *Field Theories of Condensed Matter Physics* (Cambridge University Press, Cambridge, second ediition, 2013).
8. P. Nozieres, *Theory of Interacting Fermi Systems* (Benjamin, New York, 1964).
9. A. L. Fetter, J. D. Walecka, *Quantum Theory of Many Particle Systems* (McGraw-Hill, New York, 1971).
10. J. W. Negele and H. Orland, *Quantum Many-Particle Systems* (Addison-Wesley, Redwood City, 1988).
11. G. Baym and Ch. Pethick, *Landau Fermi-Liquid Theory: Concepts and Applications* (Wiley, New York, 1991).
12. E. Lipparini, *Modern Many-Particle Physics: Atomic Gases, Quantum Dots and Quantum Fluids* (World Scientific, London 2003).
13. H. Bruus, K. Flensberg, *Many-body quantum theory in condensed matter physics* (Oxford University Press, Oxford, 2004).
14. G. F. Giuliani, G. Vignale, *Quantum Theory of the Electron Liquids* (Cambridge University Press, Cambridge, 2005).
15. J. O. Hirschfelder, C. F. Curtiss, R. H. Bird, *Molecular Theory of Gases and Liquids* (Wiley, New York, 1954).
16. S. Chapman, T. G. Cowling, *The Mathematical Theory of Non-uniform Gases* (Cambridge University Press, Cambridge, 1990).
17. A. A. Abrikosov, L. P. Gorkov, I. E. Dzyaloshinskii, *Methods of Quantum Field Theory in Statistical Physics* (Prentice Hall, 1963).
18. S. Fujita, *Introduction to Non-Equilibrium Quantum Statistical Mechanics* (W.B. Saunders Comp, Philadelphia, 1966).
19. D. N. Zubarev, *Nonequilibrium Statistical Thermodynamics* (Consultants, New York, 1974).
20. E. B. Davies, *Quantum Theory of Open Systems* (Academic Press, London, 1976).
21. H. Spohn, Kinetic equations from Hamiltonian dynamics: Markovian limits, Rev. Mod. Phys. **53**, 569 (1980).
22. S. W. Lovesey, *Condensed Matter Physics, Dynamic Correlations* (The Benjaming Publishing Company, Reading, 1980).
23. R. L. Liboff, *Kinetic Theory, Classical, Quantum and Relativistic Descriptions* (Prentice Hall, London, 1990, 3rd.ed, Springer 2003).
24. R. Balian, *From Microphysics to Macrophysics, Methods and Applications of Statistical Physics, Vol. I* (Springer-Verlag, Berlin, 1992).
25. R. Balian, *From Microphysics to Macrophysics, Methods and Applications of Statistical Physics, Vol. II* (Springer-Verlag, Berlin, 1992).

26. G. Mahler, V. A. Weberrus, *Quantum Networks, Dynamics of Open Nanostructures* (Springer, Berlin, 1995).

27. C. M. Van Vliet, *Equilibrium and Non-equilibrium Statistical Mechanics* (World Scientific, 2008).

28. P. Hänggi and H. Thomas, *Stochastic processes: Time evolution, symmetries, and linear response,* Phys. Rep. **88**, 207 (1982).

29. E. Nelson, *Quantum Fluctuations* (Princeton University Press, Princeton, 1985).

30. C. W. Gardiner, *Handbook of Stochastic Methods for Physics, Chemistry and the Natural Sciences* (Springer, Berlin, 2nd Edition, 1985, 1997).

31. P. Hänggi, P. Talkner and M. Borkovec, Reaction-Rate Theory: Fifty Years after Kramers, Rev. Mod. Phys. **62**, 251 (1990).

32. *Quantum Fluctuations in Mesoscopic and Macroscopic Systems,* H. A. Cerdeira, F. Guinea-Lopez and U. Weiss (Eds.) (World Scientific, Singapore, 1991).

33. N. G. van Kampen, *Stochastic Processes in Physics and Chemistry* (North-Holland, 1992).

34. C. W. Gardiner, P. Zoller, *Quantum Noise, A Handbook of Markovian and Non-markovian Quantum Stochastic Methods with Applications to Quantum Optics* (Springer, Berlin 2000).

35. A. Polkovnikov, K. Sengupta, A. Silva, M. Vengalattore, Colloquium: Nonequilibrium dynamics of closed interacting quantum systems, Rev. Mod. Phys. **83**, 863 (2011).

36. *Nonequilibrium Statistical Physics of Small Systems: Fluctuation Relations and Beyond,* R. Klages, W. Just, C. Jarzynski (Eds.) (Wiley-VCH, Weinheim, 2012).

37. Hao Gea, Min Qian, Hong Qian, Stochastic theory of nonequilibrium steady states. Part II: Applications in chemical biophysics, Phys. Rep. **510**, 87 (2012).

38. N. N. Bogolyubov, in G. Uhlenbeck and J. DeBoer (eds), *Studies in Statistical Mechanics, Vol. 1* (North-Holland, Amsterdam, 1962).

39. G. V. Chester, The Theory of Irreversible Process, Rep. Prog. Phys. **26**, 411 (1963).

40. A. I. Akhiezer and S. V. Peletminskij, *Methods of Statistical Physics* (Pergamon, N.Y. 1980).

41. W. Kohn, J. M. Luttinger, Quantum Theory of Electrical Transport Phenomena, Phys. Rev. **108**, 590 (1957).

42. J. M. Luttinger, W. Kohn, Quantum Theory of Electrical Transport Phenomena II, Phys. Rev. **109**, 1892 (1958).

43. H. Smith, H. H. Jensen, *Transport Phenomena* (Clarendon Press, Oxford, 1989).

44. *Quantum Transport in Semiconductors* D. K. Ferry, C. Jacoboni (Eds.) (Plenum Press, 1992).

45. *Quantum Tunnelling in Condensed Matter,* Y. U. Kagan and A. J. Leggett (Eds.) (Elsevier Publishers, New, York, 1993).

46. T. Dittrich, P. Hanggi, G.-L. Ingold, B. Kramer, G. Schon, W. Zwerger, *Quantum transport and dissipation* (Wiley-WCH, Weinheim, 1998).

47. S. Datta, *Quantum Transport: Atom to Transistor* (Cambridge University Press, Cambridge, 2005).

48. J. Freericks, *Transport in Multirayed Nanostructures* (Imperial College Press, London, 2006).

49. Y. V. Nazarov, Y. M. Blunter, *Quantum transport* (Cambridge University Press, Cambridge, 2009).

50. J. C. Cuevas, E. Scheer, *Molecular Electronics* (World Scientific, Singapore, 2010).

51. Y. Dubi, M. Di Ventra, Colloquium: Heat flow and thermoelectricity in atomic and molecular junctions, Rev. Mod. Phys. **83**, 131 (2011).

52. S. Datta, *Lessons from Nanoelectronics* (World Scientific, Canmbridge, Singapore, 2012).
53. S. Hershfield, Reformulation of Steady State Non-equilibrium Quantum Statistical Mechanics, Phys. Rev. Lett. **70**, 2134 (1993).
54. *Quantum Transport and Dissipation* T. Dittrich, P. Hänggi, G. L. Ingold, B. Kramer, G. Schon, and W. Zwerger (Eds.) (Wiley, New York, 1998).
55. G. Lang, U. Weiss, Nonlinear quantum transport and current noise, Ann. Phys. (Leipzig) 9 (2000) 804.
56. Coherent Evolution in Noisy Environments A. Buchleitner, K. Hornberger (Eds.), Lect. Notes Phys. 611, (Springer-Verlag, Berlin, 2002).
57. V. Pechakus and U. Weiss (Eds.), Chem. Phys. **268**, Issues 1-3 (2001).
58. U. Weiss, *Quantum dissipative systems* (World Scientific, 3rd. edition, 2008).
59. E. Runge and E. K. U. Gross, Density-Functional Theory for Time-Dependent Systems Phys. Rev. Lett. **52**, 997 (1984).
60. K. Burke, E. K. U. Gross, A Guided Tour of Time-Dependent Density Functional Theory, pp. 116 in D. Joubert(ed), *Density Functionals: Theory and Applications*, Lecture Notes in Phys. **500** (Springer, Berlin 1998).
61. R. Van Leeuwen, Key Concepts in Time-Dependent Density-Functional Theory, Int. J. Mod. Phys. B **15**, 1969 (2001).
62. G. Onida, L. Reining, and A. Rubio, Electronic excitations: density-functional versus many-body Green's functional approaches, Rev. Mod. Phys. **74**, 601 (2002).
63. G. Stefanucci, C.-O. Almbladh, Time-dependent quantum transport: an exact formulation based on TDDFT, Europhys. Lett. **67**, 14 (2004).
64. R. van Leeuwen, N. E. Dahlen, G. Stefanucci, C. O. Almbladh, and U. von Barth, Introduction to the Keldysh formalism and applications to time-dependent density-functional theory cond-mat/0506130 (2005).
65. S. Kurth, G. Stefanucci, C.-O. Almbladh, A. Rubio, E. K. U. Gross, Time-dependent quantum transport: A practical scheme using density functional theory, Phys. Rev. B **72**, 035308 (2005).
66. A. R. Rocha, V. M. Garcia-Suarez, S. Bailey, C. Lambert, J. Ferrer, and S. Sanvito, Spin and molecular electronics in atomically generated orbital landscapes, Phys. Rev. B **73**, 085414 (2006).
67. N. E. Dahlen, R. van Leeuwen, Solving the Kadanoff-Baym Equations for Inhomogeneous Systems: Application to Atoms and Molecules Phys. Rev. Lett. **98**, 153004 (2007).
68. C. Verdozzi, D. Karlsson, M. Puig von Friesen, C.-O. Almbladh, and U. von Barth, Some open questions in TDDFT: Clues from Lattice Models and Kadanoff-Baym Dynamics, Chem. Phys. **391**, 37 (2011).
69. C. Ullrich, *Time-Dependent Density-Functional Theory: Concepts and Applications* (Oxford University Press, Oxford, 2012).
70. J. K. Freericks, V. M. Turkowski, V. Zlatic, Nonequilibrium dynamical mean-field theory, Phys. Rev. Lett **97**, 266408 (2006).
71. V. M. Turkowski, J. K. Freericks, Nonequilibrium perturbation theory of the spinless Falicov-Kimball model: Second-order truncated expansion in U, Phys. Rev. B **75**, 125110 (2007).
72. O. P. Matveev, A. M. Shvaika, J. K. Freericks, Optical and dc-transport properties of a strongly correlated charge-density-wave system: Exact solution in the ordered phase of the spinless Falicov-Kimball model with dynamical mean-field theory, Phys. Rev. B **77**, 035102 (2008).

73. A. V. Joura, J. K. Freericks, T. Pruschke, Steady-State Nonequilibrium Density of States of Driven Strongly Correlated Lattice Models in Infinite Dimensions, Phys. Rev. Lett **101**, 196401 (2008).

74. N. Tsui, T. Oka, H. Aoki, Correlated electron systems periodically driven out of equilibrium: Floquet+DMFT formalism, Phys. Rev. B **78**, 235124 (2008).

75. T. A. Costi, Renormalization group approach to non-equilibrium Green's function, Phys. Rev. B **55**, 3003 (1997).

76. M. Keil, H. Schoeller, The real-time renormalization group approach for the spin-boson model in nonequilibrium, Chem. Phys. **268**, 11 (2001).

77. M. A. Cazalila, J. B. Marston, Time-dependent Density-Matrix Renormalization Group, Phys. Rev. Lett. **88**, 256403 (2002).

78. P. Schmitteckert, Nonequilibrium electron transport using the density matrix renormalization group, Phys. Rev. B **70**, 121302 (2004).

79. S. G. Jakobs, V. Meden, H. Schoeller, Nonequilibrium Functional Renormalization Group for Interacting Quantum Systems, Phys. Rev. B **70**, 121302 (2004).

80. F. B. Anders, Steady-State Currents through Nanodevices: A Scattering Staes Numerical Renormalization Group Approach to Open Quantum Systems, Phys. Rev. Lett. **101** 066804 (2008).

81. J. Schwinger, Quantum Electrodynamics III. The Electromagnetic Properties of the Electron-Radiative Corrections to Scattering, Phys. Rev. **76**, 796 (1949).

82. P. C. Martin, J. Schwinger, Theory of Many Particle Systems I Phys. Rev. **115**, 1342 (1959).

83. J. Schwinger, Brownian Motion of a Quantum Oscillator, J. Math. Phys. **2**, 407 (1961).

84. K. T. Mahanthappa, Multiple Production of Photons in Quantum Electrodynamics, Phys. Rev. **126**, 329 (1962).

85. L. P. Kadanoff and G. Baym, *Quantum Statistical Mechanics* (Benjamin, New York, 1962).

86. P. M. Bakshi, K. T. Mahanthappa, Expectation Value Formalism in Quantum Field Theory I, II J. Math. Phys. **4**, pp 1, ibid pp. 12 (1963).

87. S. Engelsberg, J. R. Schrieffer, Coupled Electron-Phonon System Phys. Rev. **131**, 993 (1963).

88. R. E. Prange and L. P. Kadanoff, Transport Theory for Electron-Phonon Interactions in Metals Phys. Rev. **134**, A566 (1964).

89. L. V. Keldysh, Real-time Nonequilibrium Green's Functions, Zh. Exp. Teor. Fiz. **47** 1515 (1964), [Sov. Phys. JETP **20**, (1965) 1018].

90. V. Korenman, Non-equilibrium Quantum Statistics: Applications to the Laser, Ann. Phys. **39**, 72 (1966).

91. R. E. Prange, A. Sachs, Transport Phenomena in the Simple Metals, Phys. Rev. **158**, 672 (1967).

92. R. A. Craig, Perturbation Expansion for Real Time Green's Functions, J. Math. Phys. **9**, 605 (1968).

93. R. Mills, *Propagators for many-particle systems* (Gordon and Branch, NY, 1969).

94. D. C. Langreth and G. Wilkins, Theory of spin resonance in dilute magnetic alloys, Phys. Rev. B **6**, 3189 (1972).

95. D. C. Langreth, *Linear and Non-Linear Response Theory with Applications* in Linear and Nonlinear Electron Transport in Solids (J. T. Devreese and E. van Boren, eds.) (Plenum, New York, 1976).

96. F. Cooper, Nonequilibrium Problems in Quantum Field Theory and Schwinger's Closed Time Path Formalism, arXiv:hep-th/9504073 (1995).

97. P. C. Martin, Quantum Kinetic Equations, pp. 2, in M. Bonitz (ed.), *Progress in Non-equilibrium Green's Functions* (World Scientific, Singapore, 2000).
98. G. Baym, Conservation Laws and the Quantum Theory of Transport: the Early Days, p. 17, in M. Bonitz (ed.), Progress in Non-equilibrium Green's Functions (World Scientific, Singapore, 2000).
99. L. V. Keldysh, Real-time Nonequilibrium Green's Functions, p. 2, in M. Bonitz, D. Semkat (eds.), *Progress in Non-equilibrium Green's Functions II*, World Scientific (Singapore, 2003).
100. S. S. Schweber, The sources of Schwinger's Green's functions, PNAS **102**, 7783 (2005).
101. E. M. Lifshitz, L. P. Pitaevskii, *Physical Kinetics, L. D. Landau and E. M. Lifshitz Course of Theoretical Physics, Vol. 10* (Pergamon Press 1981).
102. P. Danielewicz, Quantum theory of nonequilibrium processes, Ann. Phys. (N.Y.) **152**, 239 (1984).
103. K-chao. Chou, Z-bin. Su, B-lin Hao, L. Yu, Equilibrium and Nonequilibrium Formalism Made Unified, Phys. Rep. **118**, 1 (1985).
104. J. Rammer, H. Smith, Quantum field-theoretical methods in transport theory of metals, Rev. Mod. Phys. **58**, 323 (1986).
105. G. Strinati, Application of the Green's functions method to the study of the optical properties of semiconductors Riv. Il Nuovo Cim. 11, 1 (1988).
106. W. Botermans, R. Malfliet, Quantum Transport Theory of Nuclear Matter, Phys. Rep. **198**, 115 (1990).
107. *G. D. Mahan, Many Particle Physics*, (Plenum Press, New York, 1990, 3rd edition 2000).
108. H. Schoeller, A New Transport Equation for Single-time Greens Functions in an Arbitrary Quantum System. General Formalism, Ann. Phys. (N.Y.) **229**, 273 (1994).
109. H. Haug, A. P. Jauho, *Quantum Kinetics in Transport and Optics of Semiconductors* (Springer, Berlin, 1997, 2nd. ed. 2007).
110. M. Bonitz, *Quantum Kinetic Theory* (Teubner, Stuttgart, 1998).
111. M. Bonitz, R. Nareyka, D. Semkat (eds.), *Progress in Non-equilibrium Green's Functions I* (World Scientific, Singapore, 2000).
112. M. Bonitz, D. Semkat (eds.), *Progress in Non-equilibrium Green's Functions II* (World Scientific, Singapore, 2003).
113. M. Bonitz, A. Filinov (eds.), *Progress in Non-equilibrium Green's Functions III*, J. Phys: Conference Series **35**, 2006.
114. M. Bonitz, K. Balzer (eds.), *Progress in Non-equilibrium Green's Functions IV*, J. Phys: Conference Series **220**, 2010.
115. R. Tuovinen, R. van Leeuwen, M. Bonitz (eds.), *Progress in Non-equilibrium Green's Functions V* (J. Phys: Conference Series **427**, 2013).
116. V. Špička, B. Velický, A. Kalvová, Long and short time quantum dynamics: I. Between Green's functions and transport equations, Physica E **29**, 154 (2005).
117. V. Špička, B. Velický, A. Kalvová, Long and short time quantum dynamics: II. Kinetic regime, Physica E **29**, 175 (2005).
118. V. Špička, B. Velický, A. Kalvová, Long and short time quantum dynamics: III. Transients, Physica E **29**, 196 (2005).
119. D. Kremp, W. Schlanges, W. D. Kraeft, T. Bornath, *Quantum Statistics of Nonideal Plasmas* (Springer-Verlag, Berlin, 2005).
120. M. Bonitz, D. Semkat (eds.), *Introduction to Computational Methods in Many Body Physics* (Rinton Press, Inc., New Jersey 2006).

121. L. A. Banyai, *Non-equilibrium Theory of Condensed Matter* (World Scientific, Singapore, 2006).
122. G. Tatara, H. Kohno, J. Shibata, Microscopic approach to current driven domain wall dynamics, Phys. Reports **468**, 213 (2008).
123. D. A. Ryndyk, R. Gutierrez, B. Song, G. Cuniberti, Green function techniques in the transport at the molecular scale, p. 213 in I. Burghardt et al. (eds), *Energy transfer dynamics in biomaterial systems* (Springer-Verlag, Berlin, 2009).
124. H. Kleinert, *Path integrals in quantum mechanics, statistics, polymer physics, and financial markets* (World Scientific, 5th edition, Singapore, 2009).
125. V. Špička, A. Kalvová, B. Velický, Dynamics of mesoscopic systems:Non-equilibrium Green's functions approach, Physica E **42**, 525 (2010).
126. T. Kita, Introduction to Nonequilibrium Statistical Mechanics with Quantum Field Mechanics, Prog. Theor. Phys. **123**, 581 (2010).
127. A. Kamenev, *Field Theory of Non-Equilibrium Systems* (Cambridge University Press, Cambridge, 2011).
128. A. Rios, B. Barker, and P. Danielewicz, Towards a nonequilibrium Green's function description of nuclear matter, J. Phys.: Coference Series **427**, 012010 (2013).
129. K. Balzer, M. Bonitz, *Nonequilibrium Green's functions Approach to Inhomogenous Systems* (Springer, Berlin, 2013).
130. R. van Leeuwen, G. Stefanucci, Equilibrium and nonequilibrium many-body perturbation theory: a unified framework based on the Martin-Schwinger hierarchy, J. Phys.: Conference Series **427**, 012001, (2013).
131. G. Stefanucci, R. van Leeuwen, *Nonequilibrium Many-Body Theory of Quantum Systems: A Modern Introduction* (Cambridge University Press, Cambridge, 2013).
132. F. Sols, Scattering, Dissipation, and Transport in Mesoscopic Systems, Ann. Phys. **214**, 386 (1992).
133. R. Lake, S. Datta, Non-equilibrium Green's function method applied to double-barrier resonant-tunneling diodes, Phys. Rev. B **45**, 6670 (1992).
134. E. Runge, H. Ehrenreich, Non-equilibrium transport in alloy based resonant tunneling systems, Ann. Phys. **219**, 55 (1992).
135. Y. Meir, N. S. Wingreen, Landauer formula for the current through an interacting electron region, Phys. Rev. Lett. **68**, 2512 (1992).
136. A. P. Jauho, N. S. Wingreen and Y. Meir, Time-dependent transport in interacting and noninteracting resonant tunneling systems, Phys. Rev. B **50**, 5528 (1994).
137. H. Schoeller, G. Schon, Mesoscopic quantum transport: Resonant tunneling in the presence of a strong Coulomb interaction, Phys. Rev. B **50**, 18436 (1994).
138. B. Velický, A. Kalvová, Build-up and Decoherence of Optical Transients in Disordered Semiconductors, Phys. Stat. Sol. b **188**, 515 (1995).
139. S. M. Alamoudi, D. Boyanovsky, H.J. de Vega, and R. Holman, Quantum kinetics and thermalization in an exactly solvable model, Phys. Rev. D **59**, 025003 (1998).
140. S. M. Alamoudi, D. Boyanovsky, and H. J. de Vega, Quantum kinetics and thermalization in a particle bath model, Phys. Rev. E **60**, 94 (1999).
141. B. Velický, A. Kalvová, Transient Quasi-particle Dynamics, Phys. Stat. Sol. b **206**, 341 (1998).
142. A. Wacker, A. P. Jauho, S. Rott, A. Markus, P. Binder, and G. H. Dohler, Inelastic Quantum Transport in Superlattices: Success and Failrure of the Boltzmann Equation, Phys. Rev. Lett. **83**, 836 (1999).
143. W. Belzig, F. K. Wilhelm, Ch. Bruder, and G. Schon, Quasiclassical Green's function approach to mesoscopic superconductivity, Superlattices and Microstructures **25**, 1251 (1999).

144. A. Kalvová, B. Velický, Photoexcited transients in disordered semiconductors: Quantum coherence at very short to intermediate times, Phys. Rev. B **65**, 155329 (2002).
145. S. C. Lee, A. Wacker, Non-equilibrium Green's function theory for transport and gain properties of quantum cascade structures, Phys. Rev. B **66**, 245314 (2002).
146. A. Wacker, Semiconductor superlattices: A model system for nonlinear transport, Phys. Rep. **357**, 1 (2002).
147. Y. Xue, S. Datta, and M. A. Ratner, First-principles based matrix Green's function approach to molecular electronic devices: general formalism, Chem. Phys. **281**, 151 (2002).
148. Z. Y. Zeng, B. Li, F. Claro, Electronic transport in hybrid mesoscopic structures: A non-equilibrium Green function approach, Phys. Rev. B **68**, 115319 (2003).
149. F. Michael, M. D. Johnson, Replacing leads by self-energies using non-equilibrium Green's functions, Physica B **339**, 31 (2003).
150. R. Swirkowicz, J. Barnas, and M. Wilczynski, Non-equilibrium Kondo effect in quantum dots, Phys. Rev. B **68**, 195318 (2003).
151. P. Schwab and R. Raimondi Quasiclassical theory of charge transport in disordered interacting electron systems, Annalen der Physik 12, 471 2003.
152. A. P. Jauho, Nonequilibrium Green function modelling of transport in mesoscopic systems, p. 181 in *Progress in Non-equilibrium Green's Functions II*, M. Bonitz, D. Semkat (eds.), (World Scientifi, Singapore, 2003).
153. Zhu, Y; Maciejko, J; Ji, T; et al. Time-dependent quantum transport: Direct analysis in the time domain, Phys. Rev. B **71**, 075317 (2005).
154. B. Velický, A. Kalvová and V. Špička, Between Green's Functions and Transport Equations: Reconstruction Theorems and the Role of Initial Conditions, J. Phys: Conference Series **35**, 1 (2006).
155. J. Maciejko, Joseph, J. Wang, H. Guo, Time-dependent quantum transport far from equilibrium: An exact nonlinear response theory, Phys. Rev. B **74**, 085324 (2006).
156. A. P. Jauho, Modelling of inelastic effects in molecular electronics, J. Phys: Conference Series **35**, 313 (2006).
157. T. Frederiksen, M. Paulsson, M. Brandbyge, and A. P. Jauho, Inelastic transport theory from first principles: Methodology and application to nanoscale devices, Phys. Rev. B **75**, 205413 (2007).
158. M. Galperin, M. A. Ratner, and A. Nitzan, Molecular transport junctions: vibrational effects, J. Phys.: Cond. Matt. **19**, 103201 (2007).
159. G. Stefanucci, Bound states in ab initio approaches to quantum transport: A time-dependent formulation, Phys. Rev. B **75**, 195115 (2007).
160. Z. Feng, J. Maciejko, J. Wang, H. Guo, Current fluctuations in the transient regime: An exact formulation for mesoscopic systems, Phys. Rev. B **77**, 075302 (2008).
161. T. L. Schmidt, P. Werner, L. Muhlbacher, and A. Komnik, Transient dynamics of the Anderson impurity model out of equilibrium, Phys. Rev. B **78**, 235110 (2008).
162. B. Velický, A. Kalvová and V. Špička, Ward identity for nonequilibrium Fermi systems, Phys. Rev. B **77**, 041201(R) (2008).
163. A. P. Jauho, M. Engelund, T. Markussen, and M. Brandbyge, Ab initio vibrations in nonequilibrium nanowires, J. Phys.: Conference Series **220**, 012010 (2010).
164. D. F. Urban, R. Avriller, and A. Levy Yeyati, Nonlinear effects of phonon fluctuations on transport through nanoscale junctions, Phys. Rev. B **82**, 121414(R) (2010).
165. B. K. Nikolic, L. P. Zarbo, S. Souma, Spin currents in semiconductor nanostructures: A nonequilibrium Green function approach, Chapter 24, 814 in Volume I of *The Oxford Handbook on Nanoscience and Technology: Frontiers and Advances*, Eds. A. V. Narlikar and Y. Y. Fu, Oxford University Press, Oxford, 2010; arXiv:0907.4122.

166. D. A. Areshkin and B. K. Nikolic, Electron density and transport in top-gated graphene nanoribbon devices: First principles Green function algorithms for systems containing large number of atoms, Phys. Rev. B **81**, 155450 (2010).
167. B. Velický, A. Kalvová and V. Špička, Single molecule bridge as a testing ground for using NGF outside of the steady current regime, Physica E **42**, 539 (2010).
168. B. Velický, A. Kalvová and V. Špička, Correlated initial condition for an embedded process by time partitioning, Phys. Rev. B **81**, 235116 (2010).
169. N. A. Zimbovskaya, M. K. Pedersen, Electron transport through molecular junctions, Phys. Rep. **509**, 1 (2011).
170. A. Kalvová, V. Špička, B. Velický, Fast Transients in Mesoscopic Systems, Non-equilibrium Statistical Physics Today, AIP Conf. Proc. **1332**, 223 (2011).
171. V. Špička, A. Kalvová, B. Velický, Fast dynamics of molecular bridges, Physica Scripta T **151**, 014037 (2012).
172. T. Koch, H. Feshke, and J. Loss, Phonon-affected steady-state transport through molecular quantum dots, Phys. Scripta T **151**, 014039 (2012).
173. Tae-Ho Park, M. Galperin, A time-dependent response to optical excitation in molecular junctions, Phys. Scripta T **151**, 014038 (2012).
174. K. Kaasbjerg, T. Novotny, A. Nitzan, Charge-induced frequency renormalization, damping, and heating of vibrationa modes in nanoscale junctions Phys. Rev. B **88**, 201405(R) (2013).
175. A. Kalvová, V. Špička, B. Velický, Fast transient current response to switching events in short chains of molecular islands J. Superc. and Novel Magnetism **26**, 773 (2013).
176. Yu. B. Ivanov, J. Knoll, D. N. Voskresensky, Self-consistent approximations to non-equilibrium many-body theory, Nuc. Phys. A **657**, 413 (1999).
177. Yu. B. Ivanov, J. Knoll, D. N. Voskresensky, Resonance transport and kinetic entropy, Nuc. Phys. A **672**, 313 (2000).
178. W. Cassing, S. Juchem, Semiclassical transport of particles with dynamical spectral functions, Nuc. Phys. A **665**, 377 (2000).
179. J. Knoll, Yu. B. Ivanov, and D. N. Voskresensky, Exact Conservation Laws of the Gradient Expanded Kadanoff-baym Equations, Ann. Phys. **293**, 126 (2001).
180. S. Juchem, W. Cassing, and C. Greiner, Quantum dynamics and thermalization for out-of-equilibrium field theory, Phys. Rev. D **69**, 025006 (2004).
181. S. Juchem, W. Cassing, and C. Greiner, Non-equilibrium Quantum field dynamics and off-shell transport for ϕ^4 -theory in 2+1 dimensions, Nuc. Phys. A **743**, 92 (2004).
182. M. Galperin, A. Nitzan, M. A. Ratner, Inelastic tunneling effects on noise properties of molecular junctions, Phys. Rev. B **74**, 075326 (2006).
183. A. O. Gogolin, A. Komnik, Towards full counting statistics for the Anderson impurity model, Phys. Rev. B **73**, 195301 (2006).
184. D. A. Bagrets, Y. Utsumi, D. S. Golubev, G. Schon, Full counting statistics of interacting electrons, Fortschr. Phys. B **54**, 917 (2006).
185. F. M. Souza, A. P. Jauho, and J. C. Egues, Spin-polarized current and shot noise in the presence of spin flip in a quantum dot via nonequilibrium Green's functions, Phys. Rev. B **78**, 155303 (2008).
186. F. Haupt, T. Novotny, W. Belzig, Current noise in molecular junctions: Effects of the electron-phonon interaction, Phys. Rev. B **82**, 165441 (2010).
187. H. S. Kohler, N. H. Kwong, and H. A. Yousif, A Fortran code for solving the Kadanoff-Baym equations for a homogeneous fermion system, Comp. Phys. Commun. **123**, 123 (1999).

188. N. H. Kwong, and H. S. Kohler, Phys. Stat. Sol. B **206**, 197 (1998).
189. M. Bonitz, D. Kremp, D. C. Scott, R. Binder, W. D. Kraeft, and H. S. Kohler, Numerical analysis of non-Markovian effects in charge-carrier scattering: One time versus two-time kinetic equations, J. Phys. Cond. Matt. **8**, 6057 (1996).
190. H. S. Kohler, Memory and correlation effects in nuclear collisions, Phys. Rev. C **51**, 3232 (1995).
191. H. S. Kohler, Memory and correlation effects in the quantum theory of thermalization, Phys. Rev. E **53**, 3145 (1996).
192. H. S. Kohler, N. H. Kwong, R. Binder, D. Semkat, M. Bonitz, Numerical Solutions of the Kadanoff-Baym Equations, p. 464, in M. Bonitz (ed.), *Progress in Nonequilibrium Green's Functions*, (World Scientific, Singapore, 2000).
193. S. Fujita, Resolution of the Hierarchy of Green's Functions for Fermions, Phys. Rev. A **4**, 1114 (1971).
194. A. G. Hall, Non-equilibrium Green functions : generalized Wick's theorem and diagrammatic perturbation theory with initial correlations, J. Phys. A **8**, 214 (1975).
195. Yu. A. Kukharenko, S. G. Tikhodeev, Diagram technique in the theory of relaxation processes, Soviet. Phys. JETP **56**, 831 (1982).
196. M. Wagner, Expansions of non-equilibrium Green's functions, Phys. Rev. B **44**, 6104 (1991).
197. D. Semkat, D. Kremp, and M. Bonitz, Kadanoff-Baym equations with initial correlations, Phys. Rev. E **59**, 1557 (1999).
198. H. S. Kohler, K. Morawetz, Formation of correlations and energy conservation at short times Green's functions, Eur. Phys. J. A **64**, 291 (1999).
199. V. G. Morozov, G. Ropke, The "Mixed" Greens Function Approach to Quantum Kinetics with Initial Correlations, Ann. Phys. 278, 127 (1999).
200. C. Niu, D. L. Lin, and T. H. Lin, Equation of motion for non-equilibrium Green functions, J. Phys. Cond. Matt. **11**, 1511 (1999).
201. D. Semkat, D. Kremp, and M. Bonitz, Kadanoff-Baym equations and non-Markovian Boltzmann equation in generalized T-matrix approximation, J. Math. Phys. **41**, 7458 (2000).
202. H. S. Kohler, K. Morawetz, Correlations in many-body systems with two-time Green's functions, Phys. Rev. C **64**, 024613 (2001).
203. K. Morawetz, M. Bonitz, V. G. Morozov, G. Ropke, and D. Kremp, Short-time dynamics with initial correlations, Phys. Rev. E **63**, 020102 (2001).
204. D. Semkat, M. Bonitz, D. Kremp, M. S. Murrilo, and D. O. Gericke, Correlation induced heating and cooling of many-body systems, p. 83, in M. Bonitz, D. Semkat (ed.), *Progress in Non-equilibrium Green's Functions II* (World Scientific, Singapore, 2003).
205. D. Semkat, D. Kremp, and M. Bonitz, Relaxation of a quantum many-body system from a correlated initial state. A general and consistent approach, Contrib. Plasma Phys. **43**, 321 (2003).
206. G. Stefanucci, C. O. Almbladh, Time-dependent partition-free approach in resonant tunneling systems, Phys. Rev. B **69**, 195318 (2004).
207. D. Kremp, D. Semkat and M. Bonitz, Short-time kinetics and initial correlations in quantum kinetic theory, J. Phys.: Conference Series **11**, 1 (2005).
208. D. Kremp, D. Semkat, Th. Bornath, Short-time kinetics and initial correlations in non-ideal quantum systems, Cond. Matt. Phys. **9**, 431 (2006).
209. P. Myohanen, A. Stan, G. Stefanucci, R. van Leeuwen, A many body approach to quantum transport dynamics: Initial correlations and memory effects, Euro. Phys. Lett. **84**, 115107 (2008).

210. P. Myohanen, A. Stan, G. Stefanucci, R. van Leeuwen, Kadanoff-Baym approach to quantum transport through interacting nanoscale systems: From transient to the steady state regime, Phys. Rev. B **80**, 115107 (2009).

211. P. Lipavský, V. Špička, and B. Velický, Generalized Kadanoff-Baym Ansatz for deriving quantum transport equations, Phys. Rev. B **29**, 6933 (1986).

212. H. Chim Tso, N. J. M. Horing, Exact functional differential variant of the generalized Kadanoff-Baym Ansatz, Phys. Rev. B **44**, 1451 (1991).

213. M. Bonitz, D. Kremp, D. C. Scott, R. Binder, W. D. Kraeft, and H. S. Kohler, Numerical analysis of non-Markovian effects in charge-carrier scattering: one-time versus two-time kinetic equations, J. Phys.: Cond. Matt. **8**, 6057 (1996).

214. D. Kremp, Th. Bornath, M. Bonitz, and M. Schlanges, Quantum kinetic theory of plasmas in strong laser fields, Phys. Rev. B **60**, 4725 (1999).

215. P. Lipavský, V. Špička, and K. Morawetz, Approximative Constructions of the Double-time Correlation function from the Single-time Distribution Function, p. 63, in *Progress in Non-equilibrium Green's Functions*, M. Bonitz (ed.) (World Scientific, Singapore, 2000).

216. P. Gartner, L. Banyai, H. Haug, Two-time Electron-LO-Phonon Quantum Kinetics and the Test of the Generalized Kadanoff-Baym Relation, p. 464, in M. Bonitz (ed.), *Progress in Non-equilibrium Green's Functions* (World Scientific, Singapore, 2000).

217. K. Balzer, S. Hermanns, and M. Bonitz, The generalized Kadanoff-Baym Ansatz. Computing nonlinear response properties of finite systems, J. Phys.: Conference Series **427**, 012006 (2013).

218. B. Velický, A. Kalvová and V. Špička, Quasiparticle states of electron systems out of equilibrium, Phys. Rev. B **75**, 195125 (2007).

219. S. Nakajima, On quantum theory of transport phenomena: Steady diffusion, Prog. Theor. Phys. **20**, 948 (1958).

220. R. Zwanzig, On the Identity of Three Generalized Master Equations, Physica **30**, 1109 (1964).

221. H. Mori, Transport, Collective Motion, and Brownian Motion, Prog. Theor. Phys. **33**, 423 (1965).

222. R. Balescu, *Equilibrium and Nonequilibrium Statistical Mechanics* (Wiley, New York, 1975).

223. M. Tokuyama and H. Mori, Statistical-Mechanical Theory of Random Frequency Modulations and Generalized Brownian Motions, Prog. Theor. Phys. **20**, 411 (1976).

224. H. Grabert, *Projection Operator Techniques in Nonlinear Statistical Mechanics* (Berlin, Springer, 1982).

225. W. R. Frensley, Boundary Conditions for Open Quantum Systems Driven Far from Equilibrium, Rev. Mod. Phys. **69**, 745 (1990).

226. J. Rau, B. Muller, From Reversible Quantum Microdynamics to Irreversible Quantum Transport, Phys. Rep. **272**, 1 (1996).

227. K. Blum, *Density Matrix Theory and Applications* (Plenum Press, N.Y., 2nd edition, 1996).

228. J. W. Dufty, Ch. S. Kim, M. Bonitz, and R. Binder, Density Matrix Methods for Semiconductor Coulomb Dynamics, Int. J. Quant. Chem. **65**, 929 (1997).

229. V. May, O. Kuhn, *Charge and Energy Transfer Dynamics in Molecular Systems* (Wiley, Berlin 2000).

230. R. Zwanzig, *Nonequilibrium Statistical Mechanics* (Oxford University Press, New York, 2001).

231. H. P. Breuer, F. Petruccione, *The Theory of Open Quantum Systems* (Oxford, New York, 2002).

232. C. Jacoboni, R. Brunetti, and S. Monastra, Quantum dynamics of polaron formation with the Wigner-function approach, Phys. Rev. B **68**, 125205 (2003).

233. *Irreversible Quantum Dynamics, Lecture Notes in Physics 622*, F. Benatti, R. Floreanini (Eds.) (Springer, Berlin, 2003).

234. R. Brunetti, S. Monastra, and C. Jacoboni, Quantum dynamics of polaron formation with the Wigner-function approach, Semicon. Sci. Technol. **19**, S250 (2004).

235. I. Knezevic, D. K. Ferry, Partial-trace-free time-convolutionless equation of motion for the reduced density matrix, Phys. Rev. E **66**, 016131 (2002).

236. I. Knezevic, D. K. Ferry, Memory effects and nonequlibrium transport in open many-particle systems for the reduced density matrix, Phys. Rev. E **67**, 066122 (2003).

237. I. Knezevic, D. K. Ferry, Quantum transport and memory effects in mesoscopic structures for the reduced density matrix, Physica E **19**, 71 (2003).

238. I. Knezevic, D. K. Ferry, Open system evolution and memory dressing, Phys. Rev. A **69**, 012104 (2004).

239. I. Knezevic, D. K. Ferry, Open system non-equilibrium Green's functions and quantum transport in the transient regime, Semicon. Sci. Technol. **19**, S220 (2004).

240. M. Gell-Mann, F. Low, Bound states in quantum field theory, Phys. Rev. **848**, 350 (1951).

241. V. M. Galitskii, A. B. Migdal, ZhETF 34, 139 (1957); [Soviet Phys. JETP 7, 96 (1958)].

242. G. C. Wick, The Evaluation of the Collision Matrix, Phys. Rev. **80**, 268 (1950).

243. D. J. Thouless, Use of field theory techniques in quantum statistical physics, Phys. Rev. **107**, 1162 (1957).

244. C. Bloch, C. de Dominicis, Un development du potential de Gibbs dun systeme quantique compose dun grand nombre de particules, Nucl. Phys. **10**, 5099 (1959).

245. M. Gaudin, Une demonstration simplifiee du theoreme de Wick en mecanique statistique, Nucl. Phys. **15**, 89 (1960).

246. T. S. Evans, D. A. Steer, Wick's theorem at finite temperature, Nucl. Phys. B **474**, 89 (1996).

247. R. van Leeuwen, G. Stefanucci, Wick theorem for general initial states, Phys. Rev. B **85**, 115119 (2012).

248. G. Baym, L. P. Kadanoff, Conservation laws and correlation functions, Phys. Rev. **124**, 287 (1961).

249. G. Baym, Self-consistent approximation in many-body systems, Phys. Rev. **127**, 1391 (1962).

250. M. Revzen, T. Toyoda, Y. Takahashi, F. C. Khanna, Baym-Kadanoff criteria and the Ward-Takahashi relations in many-body theory, Phys. Rev. B **40**, 769 (1989).

251. A. Gonis, *Theoretical Materials Science* (Materials Research Society, Warrendale, 2000).

252. P. Gartner, L. Banyai, and H. Haug, Coulomb screening in the two time Keldysh-Green-function formalism, Phys. Rev. B**62**, 7116 (2000).

253. R. Ramazashvili, Ward identities for disordered metals and superconductors, Phys. Rev. B **66**, 220503(R) (2002).

254. A. I. Larkin, Y. N. Ovchinnikov, Nonlinear conductivity of superconductors in the mixed state, Zh. Eksp. Teor. Fiz. **68**, 1915 (1975), [Sov. Phys. - JETP **41**, 960 (1975)].

255. O. V. Konstantinov, V. I. Perel, Sov. Phys. - JETP **12**, 142 (1961).

256. R. Fukuda, M. Komachiya, S. Yokojima, Y. Suzuki, K. Okomura, T. Imagaki, Novel use of Legerdre transformation in field theory and many particle systems, Prog. Theor. Phys. Suppl. **121**, 1 (1995).

257. J. Koide, Quantum kinetic equation in the closed-time-path formalism, Phys. Rev. E **62**, 5953 (2000).

258. V. Špička, P. Lipavský, Quasi-particle Boltzmann Equation in Semiconductors, Phys. Rev. Lett **73**, 3439 (1994).

259. V. Špička, P. Lipavský, Quasi-particle Boltzmann Equation in Semiconductors, Phys. Rev. B **29**, 14615 (1995).

260. Th. Bornath, D. Kremp, W. D. Kraeft, and M. Schlanges, Kinetic equations for a nonideal quantum system, Phys. Rev. E **54**, 3274 (1996).

261. D. Kremp, M. Bonitz, W. D. Kraeft, and M. Schlanges, Non-Markovian Boltzmann Equation, Ann. Phys. **258**, 320 (1997).

262. V. Špička, P. Lipavský, and K. Morawetz, Quasi-particle transport equation with collision delay: I. Phenomenological approach, Phys. Rev. B **55**, 5084 (1997).

263. V. Špička, P. Lipavský, and K. Morawetz, Quasi-particle transport equation with collision delay: II. Microscopic theory, Phys. Rev. B **55**, 5095 (1997).

264. V. Špička, P. Lipavský, and K. Morawetz, Space-time versus particle-hole symmetry in quantum Enskog equations, Phys. Rev. E **64**, 046107 (2001).

265. V. Špička, P. Lipavský, and K. Morawetz, Nonlocal corrections to the Boltzmann equation for dense Fermi Systems, Phys. Lett. A **240**, 160 (1998).

266. P. Lipavský, V. Špička, and K. Morawetz, Noninstaneous collisions and two concepts of quasi-particles, Phys. Rev. E **59**, R1291 (1999).

267. P. Lipavský, K. Morawetz, and V. Špička Kinetic equation for strongly interacting dense Fermi systems, Ann. Phys. Fr. **26**, 1 (2001).

268. I. B. Levinson, Translational invariance in uniform fields and the equation for the density matrix in the Wigner representation, Zh. Eksp. Teor. Fiz.**57**, 660 (1969), [Sov. Phys. JETP **30**, 362 (1970)].

Chapter 4

A Geometric Approach to Dislocation
Densities in Semiconductors

Knut Bakke and Fernando Moraes*

*Departamento de Física, CCEN, Universidade Federal da Paraíba,
Caixa Postal 5008, 58059-900, João Pessoa, PB, Brazil*

Dislocation densities threading semiconductor crystals are a problem for device developers. Among the issues presented by the defect density is the appearance of the so called shallow levels. In this work we introduce a geometric model to explain the origin of the observed shallow levels. We show that a uniform distribution of screw dislocations acts as an effective uniform magnetic field which yields electronic bound states even in the presence of a repulsive Coulomb-like potential. This introduces energy levels within the band gap, increasing the carrier concentration in the region threaded by the dislocation density and adding additional recombination paths other than the near band-edge recombination. Our results suggest that one might use a magnetic field to destroy the dislocation density bound states and therefore minimize its effects on the charge carriers.

1. Introduction

Dislocations in semiconductors introduce states which trap electrical charge, reducing the number of available carriers. The study of the influence of dislocations on charge carrier mobility in semiconductors was made by Dexter and Seitz,[1] and Shockley[2] who suggested that dangling bonds in the dislocation core may act as traps (deep levels). Besides the dangling bond traps, electrons and holes may be trapped in more extended states (shallow levels) due to the elastic deformation field of a dislocation or of a density of dislocations. Recently, a great deal of works have investigated the effects of dislocations in the properties of semiconductors.[3,4]

In the elastic continuum, the strain and stress introduced by a topological defect distribution are related to a torsion field.[5-7] The equivalence between the continuum theory of defects in elastic solids and three-dimensional gravity with torsion was shown by Katanaev and Volovich,[7] whose deformation introduced by the defect is described geometrically by a metric which corresponds to a particular solution of the 3D Einstein-Cartan equation.[5,7,8] Interesting studies of the influence of topological defects on quantum systems using the Katanaev-Volovich approach have been made in quantum scattering,[9] Landau levels[10] and in holonomic quantum computation.[11]

---

*moraes@fisica.ufpb.br

In this short communication, we introduce a geometric model to describe the effect of a density of screw dislocations on a spin-half charged particle. We show that the torsion associated with a defect distribution acts as a magnetic field on the particles, giving rise to bound Landau-like levels even in the presence of a repulsive electric field. Therefore, we study the behavior of a spin-half charged quantum particle under the influence of a Coulomb-like potential in an elastic medium containing a uniform distribution of screw dislocations. We show that the uniform distribution of screw dislocations plays the role of an effective uniform magnetic field, and yields bound states for a spin-half quantum particle even when the Coulomb-like potential is repulsive.

2. Geometric model

Let us start by considering a charged particle moving in the presence of a torsion field. From the point of view of torsion, there is no difference between electrons and holes in a semiconductor device. Therefore, the torsion associated with a distribution of defects does not affect the charge, but it couples with the spin of the charged particle. We intend to build a system as close as possible to the problem of carrier motion in the presence of a distribution of screw dislocations, then, we consider a system consisted of a spin-1/2 charged particle moving in the presence of a uniform distribution of screw dislocations in the presence of a Coulomb-like potential. The Coulomb-like potential is given by:

$$V\left(\rho\right) = \frac{f}{\rho} = \pm\frac{|f|}{\rho}, \tag{1}$$

where f is a constant. Note that the plus (minus) sign in (1) means that the Coulomb-like potential is repulsive (attractive). We shall see that, bound states can be achieved for either sign of the Coulomb-like potential (1) due to the influence of the uniform distribution of screw dislocations.

Based on the Katanaev-Volovich approach,[7] the geometry corresponding to a uniform distribution of parallel screw dislocations is described by the line element[12] (in cylindrical coordinates):

$$ds^2 = d\rho^2 + \rho^2 d\varphi^2 + \left(dz + \Omega\rho^2 d\varphi\right)^2, \tag{2}$$

where $\Omega = b_z \frac{A}{2}$, with A being the area density of dislocations and b_z the Burgers vector. The z-axis was chosen to lie parallel to the screw dislocations.

By following the discussion about spinors and torsion made by Shapiro,[13] the torsion tensor can be represented in terms of three irreducible components, where only one of these components couples to spinors. This component is called the axial vector and is defined by $S^\mu = \epsilon^{\mu\alpha\beta\nu}T_{\alpha\beta\nu}$, where $T_{\alpha\beta\nu}$ is the torsion tensor. Thereby, the Schrödinger-Pauli equation in the presence of curvature and torsion is given by[14]

$$i\frac{\partial\psi}{\partial t} = \frac{1}{2m}\left[\vec{p} + \vec{\Xi}\right]^2 \psi + \frac{1}{8}\vec{\sigma}\cdot\vec{S}\,\psi + V\left(\rho\right)\psi, \tag{3}$$

where the vector $\vec{\Xi}$ is defined in such a way that its components are given in the local reference frame by $\Xi_k = \frac{1}{2} \sigma^3 e^{\varphi}_{\ k}(x) - \frac{1}{8} S^0 \sigma_k$. As discussed by Shapiro,[13] the components of the vector $\vec{\sigma}$ can be considered as internal degrees of freedom, that is, $\frac{1}{2}\vec{\sigma}$ corresponds to the spin of the particle. In this way, the coupling between the 4-vector S^μ and spinors gives rise to the term $\frac{1}{8}\vec{\sigma} \cdot \vec{S}$ in Eq. (3), which is called the spin-torsion coupling and is analogous to the Zeeman spin-magnetic field coupling which introduces a splitting of each of the states into a pair, one for spin up particles and the other for spin down particles. Also, we have a term coupling the linear momentum to the spin, $-\frac{1}{4}S_0\vec{\sigma} \cdot \vec{p}$, analogous to helicity.

By using the formulation of the spinor theory in curved spacetime,[15] we can show that the geometry described by the line element (2) yields only one non-null component of the axial 4-vector S^μ, which is $S^0 = -4\Omega$.[16,17] In this way, by considering the spin being aligned with the symmetry axis of the screw dislocations (z-axis), the Schrödinger-Pauli equation (3) becomes

$$
i\frac{\partial\psi}{\partial t} = -\frac{1}{2m}\left[\frac{\partial^2}{\partial\rho^2} + \frac{1}{\rho}\frac{\partial}{\partial\rho} + \frac{1}{\rho^2}\frac{\partial^2}{\partial\varphi^2} - 2\Omega\frac{\partial^2}{\partial z\partial\varphi} + (1+\Omega^2\rho^2)\frac{\partial^2}{\partial z^2}\right]\psi
$$

$$
+\frac{1}{2m}\frac{i\sigma^3}{\rho^2}\frac{\partial\psi}{\partial\varphi} - \frac{i\sigma^3}{2m}\Omega\frac{\partial\psi}{\partial z} + \frac{1}{8m\rho^2}\psi + \frac{\Omega^2}{8m}\psi + \frac{f}{\rho}\psi. \tag{4}
$$

We can see in Eq. (4) that ψ is an eigenfunction of σ^3, whose eigenvalues are $s = \pm 1$ and the Hamiltonian corresponding to Eq. (4) commutes with the operators[18] $\hat{J}_z = -i\partial_\varphi$ and $\hat{p}_z = -i\partial_z$, thus, we can write the solution of equation (4) in terms of the eigenfunctions of the operators $\hat{J}_z$ and $\hat{p}_z$, that is, $\psi_s = e^{-i\mathcal{E}t}\, e^{i(l+\frac{1}{2})\varphi}\, e^{ikz}\, R_s(\rho)$, where $l = 0, \pm 1, \pm 2, \ldots$ and k is a constant which corresponds to the momentum in the z-direction. After some calculations, we have that the radial wave function which is regular at the origin is given by

$$
R_s(\zeta) = e^{-\frac{\zeta^2}{2}}\zeta^{|\gamma_s|}H_s(\zeta), \tag{5}
$$

where $\zeta = \sqrt{\Omega k}\,\rho$. In particular, we take $k > 0$ since, as we can see in Eq. (5), for $k < 0$ the minus sign of the exponent of the Gaussian function becomes positive and we no longer have bound states.[17] This asymmetry is due to the choice of the Burgers vector orientation. Furthermore, the function $H_s(\zeta)$ written in Eq. (5) is the solution of the following second order differential equation:[17]

$$
\frac{d^2H_s}{d\zeta^2} + \left[\frac{\alpha}{\zeta} + 2\zeta\right]\frac{dH_s}{d\zeta} + \left[g - \frac{f'}{\zeta\sqrt{\Omega k}}\right]H_s = 0, \tag{6}
$$

which is known as the Heun biconfluent equation.[17,19,20] Besides, we have defined in Eqs. (5) and (6) the parameters $\gamma_s = l + \frac{1}{2}(1-s)$, $\alpha = 2|\gamma_s| + 1$, $g = \frac{\beta_s}{\Omega k} - 2|\gamma_s| - 2$, $\beta_s = 2m\left[\mathcal{E} - \frac{1}{2m}\left(k + s\frac{\Omega}{2}\right)^2 - \frac{\Omega k}{m}\gamma_s\right]$ and $f' = 2mf$. Therefore, the function H_s is the Heun biconfluent function:

$$H_s(\zeta) = H\left[2\,|\gamma_s|\,,\,0,\,\frac{\beta_s}{\Omega k},\,\frac{2f'}{\sqrt{\Omega k}},\,\zeta\right].\tag{7}$$

In order to look for bound states, for both signs of the Coulomb-like potential given in (1), we use the Frobenius method[21,22] to write the solution of Eq. (6) as a power series expansion around the origin, that is, $H(\zeta) = \sum_{m=0}^{\infty} a_m\,\zeta^m$. Substituting this series into (6), we obtain the recurrence relation:[17]

$$a_{m+2} = \frac{f'}{\sqrt{\Omega k}}\frac{a_{m+1}}{(m+2)\,(m+\alpha+1)} - \frac{(g-2m)}{(m+2)\,(m+\alpha+1)}\,a_m.\tag{8}$$

Observe that the recurrence relation (8) is valid for both signs of the Coulomb-like potential, that is, we can specify the signs of the Coulomb-like potential by making $f \to \pm|f|$ in (8). Hence, in order to obtain finite solutions everywhere which represent bound state solutions, we need that the power series expansion or the Heun biconfluent series become a polynomial of degree n. We can see in Eq. (8) that the power series expansion becomes a polynomial of degree n if we impose the conditions:

$$g = 2n \quad \text{and} \quad a_{n+1} = 0,\tag{9}$$

where $n = 1, 2, 3, \ldots$. From the condition $g = 2n$, we obtain the expression for the energy levels of the bound states, which is given by

$$\mathcal{E}_{n,l,s} = \omega_{n,l,s}\left[n + |\gamma_s| + \gamma_s + 1\right] + \frac{1}{2m}\left[k + s\frac{\Omega}{2}\right]^2,\tag{10}$$

where the angular frequency is given by $\omega_{n,l,s} = \frac{\Omega k}{m}$.

On the other hand, the condition $a_{n+1} = 0$ allows us to obtain a expression involving the angular frequency and the quantum numbers $\{n, l, s\}$. Observe that we have considered k being a positive constant, then, we can choose any value of k in such a way that the condition $a_{n+1} = 0$ and write $k = k_{n,l,s}$. We also note that by writing k in terms of the quantum numbers $\{n, l, s\}$, that is, $k = k_{n,l,s}$, it does not mean that k is quantized. Writing $k = k_{n,l,s}$ means that the choice of the values of $k > 0$ depends on the quantum numbers $\{n, l, s\}$ in order to satisfy the condition $a_{n+1} = 0$. Therefore, we can write the angular frequency as $\omega_{n,l,s} = \frac{\Omega k_{n,l,s}}{m}$. For instance, the value of the angular frequency for $n = 1$ is given by:[17]

$$\omega_{1,l,s} = \frac{\Omega k_{1,l,s}}{m} = \frac{f'^2}{2m\,(2\,|\gamma_s|+1)}.\tag{11}$$

Hence, bound state solutions of the Schrödinger equation (4) can be achieved for both attractive and repulsive Coulomb-like potential (1) in the presence of a uniform distribution of screw dislocations.[17] The asymmetry introduced by the direction of the Burgers vector permits the appearance only of bound states related to the motion in the positive z-direction. These bound states stem from the coupling between the angular variable φ and the linear variable z introduced by the screw dislocation density. If there is no motion along the z-axis, there is no coupling and hence, no bound state.

3. Summary

In this short contribution, we have shown that a uniform distribution of screw dislocations plays the role of an effective uniform magnetic field, and yields bound states for a spin-$1/2$ charged particle even when the Coulomb-like potential is repulsive. We have seen that a density of screw dislocations may leave charge carriers free to propagate along the direction of the Burgers vector (but not in the contrary direction) and can bind charge carriers in Landau-like levels in the transversal plane. It is interesting to observe that the application of a uniform magnetic field may selectively cancel this effect since the magnetic field couples to electric charge but the screw dislocation density does not. It means that the torsion field (due to the dislocation density) puts both electron and holes to move in cyclotron orbits in the same sense. The magnetic field does the same with electrons and holes moving in reversed senses. Depending on the direction of the applied field either electrons or holes are freed from their bound states. It is worth mentioning that proposals of using screw dislocations as active one-dimensional conductive channels have been proposed[23-25] for applications in semiconductor devices. Also, there exists a possibility of having one-dimensional topologically protected conducting modes associated with screw dislocations in topological insulators.[26]

Acknowledgments

The authors are grateful to CNPq, CNPq-MICINN bilateral, INCTFCx, CAPES and CAPES/NANOBIOTEC for financial support and to Prof. Eugenio B. de Mello for enlightening discussions on Heun functions.

References

1. D. L. Dexter and F. Seitz, Effects of dislocations on mobilities in semiconductors, *Phys. Rev.* **86**, 964–965, (1952).
2. W. Shockley, Dislocations and edge states in the diamond crystal structure, *Phys. Rev.* **91**, 228–228 (1953).
3. T. Figielski, Dislocations as electrically active centres in semiconductorshalf a century from the discovery, *J. Phys.: Condens. Matter* **14**, 12665–12672 (2002); R. Jaszek, Carrier scattering by dislocations in semiconductors, *J. Mater. Sci.: Mater. Electron.* **12**, 1–9 (2001).
4. M. Kittler and M. Reiche, *Structure and Properties of Dislocations in Silicon* in *Crystalline Silicon - Properties and Uses*, S. Basu (Ed.), (InTech, 2011).
5. F. Moraes, Condensed Matter Physics as a laboratory for gravitation and cosmology, *Braz. J. Phys* **30**, 304–308 (2000).
6. H. Kleinert, *Gauge Fields in Condensed Matter, Vol. 2* (World Scientific, Singapore, 1989).

7. M. O. Katanaev and I. V. Volovich, Theory of defects in solids and 3-dimensional gravity, *Ann. Phys. (NY)* **216**, 1–28 (1992).

8. K. C. Valanis and V. P. Panoskaltsis, Material metric, connectivity and dislocations in continua, *Acta Mechanica* **175**, 77–103 (2005).

9. C. Furtado, V. B. Bezerra and F. Moraes, Quantum scattering by a magnetic flux screw dislocation, *Phys. Lett. A* **289**, 160–166 (2001).

10. C. Furtado and F. Moraes, Landau levels in the presence of a screw dislocation, *Europhys. Lett.* **45**, 279–282 (1999).

11. K. Bakke and C. Furtado, One-qubit quantum gates associated with topological defects in solids, *Quantum Inf. Process.* **12**, 119–128 (2013).

12. A. L. Silva Netto and C. Furtado, Elastic Landau levels, *J. Phys.: Cond. Mat.* **20**, 125209 (2008).

13. I. L. Shapiro, Physical aspects of the space-time torsion, *Phys. Rep.* **357**, 113–213 (2002).

14. K. Bakke *et al.*, Gravitational geometric phase in the presence of torsion, *Eur. Phys. J. C* **60**, 501–507 (2009).

15. S. Weinberg, *Gravitation and Cosmology: Principles and Applications of the General Theory of Relativity* (IE-Wiley, New York, 1972).

16. K. Bakke, Discrete Energy Spectrum for a Spin-1/2 Quantum Particle Under the Influence of a Constant Force Field due to the Presence of Topological Defects, *Braz. J. Phys.* **41**, 167–170 (2011).

17. K. Bakke and F. Moraes, Threading dislocation densities in semiconductor crystals: A geometric approach, *Phys. Lett. A* **376**, 2838–2841 (2012).

18. P. Schlüter, K.-H. Wietschorke and W. Greiner, The Dirac-equation in orthogonal coordinate systems .1. The local representation, *J. Phys. A* **16**, 1999–2016 (1983).

19. A. Ronveaux, *Heun's differential equations* (Oxford University Press, Oxford, 1995).

20. E. R. Figueiredo Medeiros and E. R. Bezerra de Mello, Relativistic quantum dynamics of a charged particle in cosmic string spacetime in the presence of magnetic field and scalar potential, *Eur. Phys. J. C* **72**, 2051 (2012).

21. G. B. Arfken and H. J. Weber, *Mathematical Methods for Phisicists, sixth edition* (Elsevier Academic Press, New York, 2005).

22. C. Furtado *et al.*, Landau-levels in the presence of disclinations, *Phys. Lett. A* **195**, 90–94 (1994).

23. S. Mil'shtein, Application of dislocation-induced electric potentials in Si and Ge, *J. Physique Coll.* **40**, C6 207–211 (1979).

24. M. Kittler *et al.*, Regular dislocation networks in silicon as a tool for nanostructure devices used in optics, biology, and electronics, *SMALL* **3**, 964–973 (2007).

25. M. Reich *et al.*, Dislocation-Based Si-Nanodevices, *Jpn. J. Appl. Phys.* **49**, 04DJ02 (2010).

26. Y. Ran, Y. Zhang, and A. Vishwanath, One-dimensional topologically protected modes in topological insulators with lattice dislocations, *Nature Phys.* **5**, 298–303 (2009).

Chapter 5

Quantum Transport: A Unified Approach via a Multivariate Hypergeometric Generating Function

Ailton F. Macedo-Junior*

Departamento de Física, Universidade Federal Rural de Pernambuco Recife, 52171-900 Pernambuco, Brazil

Antonio M. S. Macêdo[†]

Departamento de Física, Universidade Federal de Pernambuco Recife, 50670-901 Pernambuco, Brazil

We introduce a characteristic function method to describe charge counting statistics in phase coherent systems that directly connects the three most successful approaches to quantum transport: random-matrix theory, the nonlinear σ-model and the trajectory-based semiclassical method. The central idea is the construction of a generating function based on a multivariate hypergeometric function, which can be naturally represented in terms of quantities that are well defined in each approach. We illustrate the power of our scheme by obtaining exact analytical results for the first four cumulants of charge counting statistics in a chaotic quantum dot coupled ideally to electron reservoirs via perfectly conducting leads with arbitrary number of open scattering channels.

1. Introduction

Quantum transport in nanoscale devices based on semiconductor heterostructures,[1] such as quantum dots, quantum wires and more recently quantum chains[2–4] and quantum networks,[5–7] has had great impact in both academic and technological research. Powerful theoretical approaches have been developed, motivated in part by the great experimental flexibility in varying control parameters, such as barrier transparencies and external fields. An exciting perspective is the possibility of building artificial quantum simulators of complex many-body systems from such devices.[8,9]

In Fig. 1 we show schematically a typical two-probe setup used in quantum transport experiments to detect electrical current. We may interpret the measured quantum current as a stochastic process in which single-particle excitations are

---

*ailton@df.ufrpe.br
[†] amsmacedo@df.ufpe.br

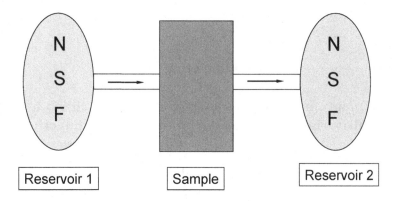

Fig. 1. A pictorial representation of a typical two-probe quantum transport setup. The letters have the following meaning: N=normal metal, S=superconductor, F=ferromagnet.

transferred through the device according to quantum mechanical laws. A quantitative description of this process can be made by using the concept of charge counting statistic (CCS),[10] a fermionic version of the photon counting statistics developed by Glauber in 1963.[11] The CCS is described in terms of the probability, $P_n(T_{\mathrm{ob}})$, that n units of charge are transferred during an observation time T_{ob}. The intrinsic random nature of the quantum transmission process and the Pauli principle play important roles in determining the value of $P_n(T_{\mathrm{ob}})$. The CCS of a phase coherent conductor is usually characterized by specifying a characteristic or cumulant generating function, $\Phi(\lambda)$, defined via the Fourier series

$$e^{\Phi(\lambda)} \equiv \sum_n P_n(T_{\mathrm{ob}})e^{in\lambda}. \tag{1}$$

The cumulant generating function of a quantum device can be conveniently expressed in terms of its transmission eigenvalues $\{\tau_i\}$, i.e. the eigenvalues of $tt^\dagger$, where t is the transmission matrix of the device. More recently, Klich and Levitov[12] found a interesting connection between the CCS of a quantum point contact and the entanglement entropy, which may be used to detect entanglement via the fluctuations of electrical current in transport measurements. Another interesting result is the reconstruction of the full entanglement spectrum, i.e. the full set of eigenvalues of the reduced density matrix, using high-order CCS cumulants.[13]

With the consolidation of CCS as an important theoretical tool, much effort has been made to relate $P_n(T_{\mathrm{ob}})$ and/or $\Phi(\lambda)$ to various quantum transport models. We briefly describe two of such efforts, which have in common their independent attempt to achieve some kind of unification. In Ref. 14, fluctuation theorems were used to build an approach to describe non-equilibrium properties, which include CCS, of open quantum systems. Using the concept of two-point projective measurements, it was shown that several known approaches to quantum transport can be recovered by taking particular cases of measurement schemes. Another general way to obtain the CCS characteristic function of a quantum device was put forward in Ref. 15. In this work, the microscopic description of the whole system, including the detectors, is treated in a quantum many-body language. The dynamics of the system is obtained via the projective technique of non-equilibrium statistical mechanics, combined with an extended in time measurement scheme. The main result obtained from this approach is a general formula expressing $\Phi(\lambda)$ in terms of the many-body Green's function of the device. There are interesting relationships between the results of Refs. 14 and 15 which may deserve further investigations.

If the quantum device's dynamics is chaotic in the classical limit, then an important universal transport regime exists. In this case, a random scattering matrix approach[16] applies, which predicts that the transmission eigenvalues of the system are strongly correlated random variables. This means that the generating function $\Phi(\lambda)$ can no longer be considered a complete characterization of the stochastic process associated with charge transfer, because of its own random fluctuations. In the semiclassical regime, characterized by a large number of open transmission channels, we may neglect such fluctuations and calculate the average generating function $S(\lambda) = \langle\Phi(\lambda)\rangle$ using various procedures, such as quantum circuit theory (QCT),[1] the trajectory-based semiclassical diagrammatic expansion[17–19] and random scattering matrix diagrammatic perturbation theory.[20] Quantum circuit theory can be derived from the saddle-point equations of non-linear σ-models.[21] Its main advantage is the fact that $S(\lambda)$ can be related directly to the system's control parameters, without having to obtain first the joint distribution of transmission eigenvalues. In Fig. 2 we give a schematic view of the relationships between different approaches to quantum transport.

This paper is organized as follows. In Sec. 2 we give a brief introduction to the three most popular approaches to quantum transport by applying them to electron

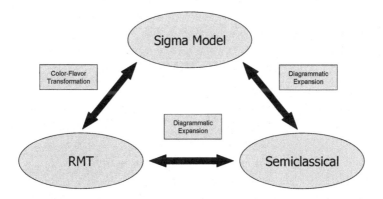

Fig. 2.   Different approaches to quantum transport and their mutual relationships: The nonlinear σ-model, random matrix theory (RMT) and the trajectory-based semiclassical approach.

transport through a ballistic chaotic cavity. In Sec. 3 we present a generating function technique based on a multivariate generalization of the hypergeometric function and apply it to calculate statistical properties of several transport observables. In Sec. 4, a method based on the direct use of Jack polynomials is presented together with some applications. Conclusions are presented in Sec. 5.

2. Three approaches to quantum transport

We shall discuss the three most common approaches to quantum transport using as a model system a ballistic chaotic cavity connected via ideal contacts to two perfectly conducting leads supporting N_1 and N_2 open scattering channels. In the scattering formalism,[1,22] the cavity is described by its unitary scattering matrix, which has the following block structure

$$S = \begin{pmatrix} r & t' \\ t & r' \end{pmatrix}, \tag{2}$$

where $r(r')$ and $t(t')$ denote reflection and transmission matrices respectively. The unitarity of S, due to flux conservation, implies that the hermitian matrices $t^\dagger t$ and $t'^\dagger t'$ have the same set of $N = \min(N_1, N_2)$ non-zero eigenvalues $(\tau_1, \ldots, \tau_N) \in (0,1)^N$, known as transmission eigenvalues.

The cumulant generating function of the charge counting statistics, $\Phi(\lambda)$, for a voltage-biased conductor can be conveniently written in terms of the transmission eigenvalues. In particular, at zero temperature, the cumulants are given by the Levitov-Lesovik formula[10,23]

$$q_k = \sum_{j=1}^{N} \left(\tau(1-\tau)\frac{d}{d\tau} \right)^{k-1} \tau \Bigg|_{\tau=\tau_j}, \quad k = 1, 2, \ldots. \tag{3}$$

Below we write explicitly the first four cumulants

$$q_1 = g = \sum_{j=1}^{N} \tau_j \tag{4}$$

$$q_2 = p = \sum_{j=1}^{N} \tau_j(1 - \tau_j) \tag{5}$$

$$q_3 = \kappa = \sum_{j=1}^{N} \tau_j(1 - \tau_j)(1 - 2\tau_j) \tag{6}$$

$$q_4 = \zeta = \sum_{j=1}^{N} \tau_j(1 - \tau_j)(1 - 6\tau_j + 6\tau_j^2). \tag{7}$$

The first two cumulants are simply the dimensionless conductance and shot-noise power, respectively. The third and forth cumulants have been also received much attention in the recent literature.[13,24]

The underlying chaotic dynamics causes the elements of the scattering matrix to fluctuate strongly as a function of energy, which in turn implies that the transmission eigenvalues are well described by correlated random variables. There are many ways in which to extract the statistical characteristics of these fluctuations, and the selection of a particular one is an important feature of the various approaches to quantum transport. We give below a brief description of three of such formalisms: random scattering matrix theory, the nonlinear σ-model and the trajectory-based semiclassical technique.

In the random matrix approach, averages are performed over an ensemble of S matrices, constructed by maximizing the information entropy subjected to the constraints of flux conservation, analyticity and certain discrete symmetries, such as time-reversal (TR) invariance and/or spin-rotation (SR) invariance.[22] For ideal contacts, the resulting S-matrix distribution is uniform over the unitary group, i.e. the differential probability density is simply proportional to the Haar measure $d\mu(S)$ of the corresponding group. Such random matrix scattering ensembles are known as Dyson's circular ensembles. According to presence or absence of TR and SR symmetries there are three circular ensembles characterized by the Dyson index β: circular ortogonal ensemble (COE, $\beta = 1$), circular unitary ensemble (CUE, $\beta = 2$) and Circular Symplectic Ensemble (CSE, $\beta = 4$). We have $\beta = 1$ for systems with TR and SR symmetry, $\beta = 2$ for systems in the absence of TR symmetry, and $\beta = 4$ for systems with broken SR symmetry but in the presence of TR symmetry. The characteristics of the circular ensembles are summarized in Table 1.

Table 1. Dyson's classification of circular ensembles. TRS = time reversal symmetric, SRS = spin reversal symmetric, COE = circular orthogonal ensemble, CUE = circular unitary ensemble, CSE = circular symplectic ensemble.

β	TRS	SRS	S-matrix	Ensemble
1	Yes	Yes	Unitary symmetric	COE
2	No	Yes(No)	Unitary	CUE
4	Yes	No	Unitary self-dual	CSE

Using a polar decomposition for the S-matrix, it is possible to derive from the Haar measure the joint distribution of transmission eigenvalues[16]

$$P(\{\tau\}) = \frac{1}{S_N(\mu, 0, \beta)} |\Delta(\{\tau\})|^\beta \prod_{i}^{N} \tau_i^\mu, \qquad (8)$$

where $\Delta(\{\tau\}) = \prod_{i<j}(\tau_i - \tau_j)$ is a Vandermonde determinant and $\mu = \beta(|N_1 - N_2|+1)/2 - 1$. The joint distribution (8) can be identified with the Jacobi ensembles of random matrix theory (RMT) and the normalization constant can be obtained

from the corresponding Selberg integral,[25] yielding

$$S_N(\mu, 0, \beta) = \prod_{i=1}^{N} \frac{\Gamma(1 + i\beta/2)\Gamma(\mu + 1 + (i-1)\beta/2)\Gamma(1 + (i-1)\beta/2)}{\Gamma(1 + \beta/2)\Gamma(\mu + 2 + (N + i - 2)\beta/2)}. \tag{9}$$

The knowledge of the joint distribution allows us, in principle, to obtain the statistical characteristics of transport observables, such as conductance, shot-noise power and higher order cumulants of the charge counting statistics. The average of an arbitrary function of transmission eigenvalues, for instance, can be written as

$$\langle f \rangle = \int_{[0,1]^N} f(\{\tau\}) P(\{\tau\}) d^N \tau. \tag{10}$$

To be more specific, consider the following function defined by an average over transmission eigenvalues

$$\Psi(x) = \left\langle \prod_{i=1}^{N} (1 - x\tau_i) \right\rangle. \tag{11}$$

Using Eqs. (8) and (10), we recognize that Eq. (11) is, for $\beta = 2$, simply a multidimensional representation of the standard hypergeometric function

$$\Psi(x) = F(-N_1, -N_2; -N_T; x), \tag{12}$$

where $N_T = N_1 + N_2$. We may thus obtain a closed expression for the average dimensionless conductance via the formula

$$\langle g \rangle = \langle \text{Tr}\left(t^\dagger t\right) \rangle = \sum_{i=1}^{N} \langle \tau_i \rangle = -\Psi'(0) = \frac{N_1 N_2}{N_T}. \tag{13}$$

There is a two-dimensional representation of the hypergeometric function that can be used to make contact with the non-linear σ-model approach. It reads

$$\Psi\left(\sin^2\left(\frac{\phi}{2}\right)\right) = \frac{N_T + 1}{2^{N_T}} \int \frac{d\mu(\hat{n})}{4\pi} (1 + \hat{n}_\phi \cdot \hat{n})^{N_1} (1 + \hat{n} \cdot \hat{n}_0)^{N_2}, \tag{14}$$

where $\hat{n}$ is a variable unit vector, $d\mu(\hat{n})$ is the solid angle element and $\hat{n}_\phi = \sin\phi \hat{e}_2 + \cos\phi \hat{e}_3$, $\hat{n}_0 = \hat{e}_3$ are fixed unit vectors. Introducing the matrices

$$Q = \frac{1}{1 + |z|^2} \begin{pmatrix} 1 - |z|^2 & -2z \\ -2z^* & |z|^2 - 1 \end{pmatrix}; \quad Q_\phi = \begin{pmatrix} 0 & e^{-i\phi} \\ e^{i\phi} & 0 \end{pmatrix} \tag{15}$$

and $Q_0 \equiv Q_{\phi=0}$, which have the properties $\text{Tr}(Q) = 0 = \text{Tr}(Q_\phi)$ and $Q^2 = 1 = Q_\phi^2$, we may rewrite $\Psi(x)$ as

$$\Psi\left(\sin^2\left(\frac{\phi}{2}\right)\right) = \int d\mu(Q)\left[\det(1_2 + Q_\phi Q)\right]^{N_1}\left[\det(1_2 + QQ_0)\right]^{N_2} \equiv \int d\mu(Q)e^{-S(Q)}$$

(16)

where

$$d\mu(Q) = \frac{2(N_T + 1)}{4^{N_T}\pi}\frac{d(\mathrm{Re}z)d(\mathrm{Im}z)}{(1 + |z|^2)^2}.$$

(17)

We remark that the matrix function $S(Q)$ coincides with the action of the zero-dimensional non-linear σ-model, which can be obtained from a microscopic Hamiltonian description using either the Keldysh or the supersymmetric approach.[21] We can make the above connection between random-matrix theory and the non-linear σ-model more explicit by introducing the matrix

$$\Lambda_\phi = \begin{pmatrix} e^{i\phi}1_{N_1} & 0 \\ 0 & 1_{N_2} \end{pmatrix}$$

(18)

and by writing $\Psi(x)$ as

$$\Psi\left(\sin^2\left(\frac{\phi}{2}\right)\right) = \frac{1}{2^{N_T}}\int d\mu(S)\det\left(1_{N_T} + \Lambda_\phi^\dagger S\Lambda_\phi S^\dagger\right).$$

(19)

The exact map

$$\int \frac{d\mu(S)}{2^{N_T}}\det\left(1_{N_T} + \Lambda_\phi^\dagger S\Lambda_\phi S^\dagger\right) = \int d\mu(Q)\left[\det(1_2 + Q_\phi Q)\right]^{N_1}\left[\det(1_2 + QQ_0)\right]^{N_2}$$

is one of the many identities that can be obtained from a technique known as the color-flavor transformation.[26]

The trajectory-based semiclassical approach starts from the following representation of an S-matrix element

$$S_{ba}(E) = \sum_{\gamma(a\to b)} A_\gamma(E)e^{iS_\gamma(E)/\hbar},$$

(20)

where $\gamma(a \to b)$ is a classical scattering trajectory connecting channel a to channel b, $A_\gamma(E)$ is its stability amplitude and $S_\gamma(E)$ its classical action. The strong fluctuations of $S_{ba}(E)$ as a function of energy, caused by the chaotic dynamics, can be smoothed by performing an average over an appropriate energy window near the Fermi energy. For instance, the average dimensionless conductance can be written as

$$\left\langle \mathrm{Tr}\left(t^\dagger t\right)\right\rangle_{sc} = \frac{1}{T_H}\sum_{a=1}^{N_1}\sum_{b=1}^{N_2}\sum_{\gamma(a\to b)}\sum_{\lambda(a\to b)}\left\langle A_\gamma A_\lambda^* e^{i(S_\gamma - S_\lambda)/\hbar}\right\rangle$$

(21)

where $T_H = \Omega/(2\pi\hbar)$ is the Heisenberg time with Ω denoting the volume of the energy shell. This representation has been used to formulate a trajectory-based diagrammatic perturbation theory. The leading term of this semiclassical expansion is obtained by selecting the diagonal contribution, i.e. the one with $\gamma(a \to b) = \lambda(a \to b)$. Using the Richter and Sieber sum rule,[17] which is a consequence of the ergodicity of the classical dynamics, we get

$$\left\langle \mathrm{Tr}\left(t^\dagger t\right)\right\rangle_{sc} \simeq \frac{N_1 N_2}{T_H} \int_0^\infty dt\, e^{-N_T t/T_H} = \frac{N_1 N_2}{N_T}, \tag{22}$$

which coincides with the exact result shown in Eq. (13). Using recent combinatorial results,[18] it is in principle possible to calculate with semiclassical trajectories all coefficients of the expansion of $\Psi(x)$ in powers of x. The calculations are certainly lengthy, but the result is expected to coincide with those of random-matrix theory and the nonlinear σ-model. We may thus formally write

$$\Psi(\sin^2(\frac{\phi}{2})) = \frac{1}{2^{N_T}}\left\langle \det\left(1_{N_T} + \Lambda_\phi^\dagger S\Lambda_\phi S^\dagger\right)\right\rangle_{sc} = F(-N_1, -N_2; -N_T; \sin^2(\frac{\phi}{2})). \tag{23}$$

In the following subsection we shall explicitly verify for our model system the equivalence of random matrix theory, the non-linear σ-model and the trajectory-based semiclassical approach in the calculation of the leading term in the asymptotic expansion, for $N_1, N_2 \gg 1$, of averages of the form

$$\langle A\rangle = \left\langle \mathrm{Tr}\left(a(t^\dagger t)\right)\right\rangle = \sum_{n=1}^N \langle a(\tau_n)\rangle, \tag{24}$$

where $a(\tau)$ is an arbitrary continuous function.

2.1. Asymptotic equivalence

We start our analysis with the observation that if we introduce the average density of transmission eigenvalues

$$\rho(\tau) = \left\langle \mathrm{Tr}\delta(\tau - tt^\dagger)\right\rangle, \tag{25}$$

then Eq. (24) can be written as

$$\langle A\rangle = \int_0^1 d\tau \rho(\tau) a(\tau). \tag{26}$$

Our task is thus reduced to the evaluation of the dominant contribution in the asymptotic expansion of $\rho(\tau)$, for $N_1, N_2 \gg 1$, in all three approaches and to show that they yield identical expressions.

(i) *Random Matrix Theory*

The semiclassical evaluation of $\rho(\tau)$ in random matrix theory is based on a asymptotic, i.e. large N_1 and N_2, diagrammatic perturbation theory.[20] Define the functions

$$f_j(z) = \text{Tr}\,(F_j(z)); \quad j = 1, 2, \tag{27}$$

where

$$F_1(z) = \left\langle \Lambda_1 \left(z - S^\dagger \Lambda_2 S \Lambda_1\right)^{-1} \right\rangle, \tag{28}$$

$$F_2(z) = \left\langle \Lambda_2 \left(z - S^\dagger \Lambda_1 S \Lambda_2\right)^{-1} \right\rangle \tag{29}$$

and

$$\Lambda_1 = \begin{pmatrix} 1_{N_1} & 0 \\ 0 & 0 \end{pmatrix}; \quad \Lambda_2 = \begin{pmatrix} 0 & 0 \\ 0 & 1_{N_2} \end{pmatrix}. \tag{30}$$

The average density is obtained from

$$\rho(\tau) = -\frac{1}{\pi} \text{Im}\,\left(f_j(\tau + i0^+)\right); \quad j = 1, 2. \tag{31}$$

The selection of planar diagrams in the perturbative series[20] gives the following set of equations

$$z(z - 1)f_1^2 - (z - 1)(N_1 - N_2)f_1 - N_1 N_2 = 0, \tag{32}$$
$$z(z - 1)f_2^2 - (z - 1)(N_2 - N_1)f_2 - N_1 N_2 = 0, \tag{33}$$

from which $\rho(\tau)$ can be readily calculated. For the purpose of our equivalence proof, let us define the auxiliary function

$$h(z) = \left\langle \text{Tr}\,\left(\frac{t^\dagger t}{1 - z t^\dagger t}\right) \right\rangle, \tag{34}$$

which is related to $f_j(z)$ through

$$z h(z) = \frac{1}{z} f_j(\frac{1}{z}) - N_j; \quad j = 1, 2. \tag{35}$$

Using Eqs. (32) and (35) we obtain

$$z(z - 1)h^2(z) + (z - 1)N_T h(z) + N_1 N_2 = 0. \tag{36}$$

Equation (36) is the final result of the RMT calculation and will be used below to prove the asymptotic equivalence of the three approaches.

(i) *Nonlinear σ-Model*

Our starting point is the nonlinear σ-model representation of the function $\Psi(x)$

$$\Psi\left(\sin^2\left(\frac{\phi}{2}\right)\right) = \frac{N_T + 1}{2^{N_T}} \int \frac{d\Omega}{4\pi} e^{S_\phi(\hat{n})}, \tag{37}$$

where

$$S_\phi(\hat{n}) = N_1 \ln(1 + \hat{n}_\phi \cdot \hat{n}) + N_2 \ln(1 + \hat{n} \cdot \hat{n}_0). \tag{38}$$

For large values of N_1 and N_2 we may use the following approximation

$$\Psi(x) = \left\langle \exp\left(\mathrm{Tr}\ln(1 - xtt^\dagger)\right)\right\rangle \simeq e^{S(x)}, \tag{39}$$

where

$$S(x) = \left\langle \mathrm{Tr}\ln(1 - xtt^\dagger)\right\rangle. \tag{40}$$

Note that the function $h(z)$ defined in Eq. (34) can be obtained from the relation

$$h(z) = -\frac{d\Phi(z)}{dz}. \tag{41}$$

For later convenience, we introduce a "pseudo-current" function $I(\phi)$, which is related to $h(z)$ via

$$I(\phi) = \sin\phi \, h(\sin^2(\frac{\phi}{2})). \tag{42}$$

We shall now evaluate $\Psi(x)$ using the saddle-point approximation. The saddle-point can be located from the path of steepest descend, which we parametrize as a geodesic on the surface of a unit sphere

$$\hat{n}_\theta = \frac{\sin(\phi - \theta)}{\sin\phi}\hat{n}_0 + \frac{\sin\theta}{\sin\phi}\hat{n}_\phi, \tag{43}$$

then

$$S_\phi(\hat{n}_\theta) = N_1 \ln(1 + \cos(\phi - \theta)) + N_2 \ln(1 + \cos\theta). \tag{44}$$

The saddle-point equation is thus given by

$$\frac{\partial}{\partial\theta}S_\phi(\hat{n}_\theta) = 0, \tag{45}$$

which can be written in the form of a pseudo-current conservation law

$$I(\phi) = I_1(\phi - \theta) = I_2(\theta), \tag{46}$$

where

$$I_j(\alpha) = 2N_j \tan(\frac{\alpha}{2}). \tag{47}$$

Equation (46) can be rewritten as the following single equation for $I(\phi)$

$$I^2(\phi) + 2N_T \cot(\frac{\phi}{2})I(\phi) - 4N_1 N_2 = 0, \tag{48}$$

which after using Eq. (42), yields (36) in agreement with RMT.

(i) *Trajectory-based Semiclassical Approach*

From the definition of $h(z)$ we may introduce coefficients M_n, denoted transport moments, from the following series expansion

$$h(z) = \left\langle \text{Tr} \left(\frac{t^\dagger t}{1 - z t^\dagger t} \right) \right\rangle = \sum_{n=0}^{\infty} z^n M_{n+1}. \tag{49}$$

The leading semiclassical contributions to M_n can be obtained from a set of diagrammatic rules[19] that can be used to compute auxiliary functions $\phi_j(z)$, $j = 1, 2$, which are related to $h(z)$ via the equations

$$zh(z) = \frac{\alpha N_j \phi_j(z)}{1 - \alpha \phi_j(z)}; \quad j = 1, 2; \quad \alpha = \sqrt{z}. \tag{50}$$

Implementing the diagrammatic rules of Ref. 19 one obtains the following set of equations

$$\alpha N_2 \phi_1^2 + ((N_1 - N_2)\alpha^2 - N_T)\phi_1 + \alpha N_2 = 0$$
$$\alpha N_1 \phi_2^2 + ((N_2 - N_1)\alpha^2 - N_T)\phi_2 + \alpha N_1 = 0,$$

which implies that $h(z)$ satisfies Eq. (36).

We have thus established, for the simple system treated here, the asymptotic equivalence of the three main methods used in quantum transport. We close this section by exhibiting a direct relation between the auxiliary functions $f_j(z)$, defined in RMT, and $\phi_j(z)$

$$z f_j(z)\left(1 - \alpha \phi_j(\frac{1}{z})\right) = N_j; \quad j = 1, 2. \tag{51}$$

We believe that relations such as (51) can be very useful to build connections between RMT and semiclassical methods in more complex systems, such as networks of quantum dots.

In the next section, we introduce a multivariable generalization of the generating function $\Psi(x)$ and explore its connection to the multivariate hypergeometric function to obtain exact results for several transport observables.

3. Generating function technique

A remarkable property of the charge counting cumulants, see Eq. (3), is their invariance under the exchange of the transmission eigenvalues, i.e.,

$$q_k(\ldots, \tau_i, \ldots, \tau_j, \ldots) = q_k(\ldots, \tau_j, \ldots, \tau_i, \ldots),$$

which implies that the cumulants of CCS are symmetric polynomials in the variables $\tau_1, \ldots, \tau_N$.

Symmetric functions are ubiquitous in mathematics and mathematical-physics. They are important in elementary algebra, in representation theory, combinatorics[27] and multivariate statistics.[28] According to the fundamental theorem of symmetric functions,[27] we can write any symmetric polynomial in the variables $\tau_1, \ldots, \tau_N$ as a polynomial in the elementary symmetric polynomials, defined by

$$e_n = \sum_{i_1 < i_2 < \cdots < i_n} \tau_{i_1} \tau_{i_2} \cdots \tau_{i_n}. \tag{52}$$

The polynomials e_n contain the sums of all products of n distinct variables τ_i extracted from $\{\tau_1, \tau_2, \ldots, \tau_N\}$, and by definition $e_0 = 1$. They can be easily obtained from the generating function[27]

$$\prod_{j=1}^{N}(1 - x\tau_j) = \sum_{n=0}^{N}(-1)^n e_n x^n. \tag{53}$$

We give in Table 2 an example of symmetric polynomials e_n in four variables $\{\tau_1, \tau_2, \tau_3, \tau_4\}$.

Table 2. Elementary symmetric polynomials in variables $\{\tau_1, \tau_2, \tau_3, \tau_4\}$.

$e_0 = 1$
$e_1 = \tau_1 + \tau_2 + \tau_3 + \tau_3$
$e_2 = \tau_1\tau_2 + \tau_1\tau_3 + \tau_1\tau_4 + \tau_2\tau_3 + \tau_2\tau_4 + \tau_3\tau_4$
$e_3 = \tau_1\tau_2\tau_3 + \tau_1\tau_2\tau_4 + \tau_1\tau_3\tau_4 + \tau_2\tau_3\tau_3$
$e_4 = \tau_1\tau_2\tau_3\tau_4$

The elementary symmetric polynomials (52) are the basic building block of symmetric polynomials. In particular, we can rewrite the cumulants of CCS, Eqs. (4)-(7) as follows

$$g = e_1, \tag{54}$$

$$p = e_1 - e_1^2 + 2\,e_2, \tag{55}$$

$$\kappa = e_1 - 3e_1^2 + 2e_1^3 + 6e_2 - 6e_1e_2 + 6e_3, \tag{56}$$

$$\zeta = e_1 - 7e_1^2 + 12e_1^3 - 6e_1^4 + 14e_2 - 36e_1e_2$$
$$+ 24e_1^2e_2 - 12e_2^2 + 36e_3 - 24e_1e_3 + 24e_4. \tag{57}$$

Similar decompositions can be made to any power of g, p, κ and ζ.

We proceed by introducing the following multivariable generalization of the generating function defined in Eq. (11)

$$\Psi(\vec{x}) = \left\langle \prod_{i=1}^{M} \det(1 - x_i tt^\dagger) \right\rangle \tag{58}$$

$$= \left\langle \prod_{i=1}^{M} \prod_{j=1}^{N} (1 - x_i \tau_j) \right\rangle, \tag{59}$$

where $\langle \cdots \rangle$ indicates an average performed according to the prescriptions given by the chosen quantum transport approach. From (58) and (53) it is clear that derivatives of $\Psi(\vec{x})$ give the average of products of elementary symmetric functions in the transmission eigenvalues via the formula

$$\langle e_{n_1} \dots e_{n_M} \rangle = \frac{(-1)^{n_1 + \dots + n_M}}{n_1! \dots n_M!} \frac{\partial^{n_1 + \dots + n_M}}{\partial x_1^{n_1} \dots \partial x_M^{n_M}} \Psi(\vec{x}) \bigg|_{\vec{x}=0}. \tag{60}$$

In the random matrix approach, we represent $\Psi(\vec{x})$ as a multidimensional integral with a Jacobi weight

$$\Psi(\vec{x}) = \frac{1}{S_N(\mu, 0, \beta)} \int_{[0,1]^N} d\tau^N |\Delta(\{\tau\})|^\beta \prod_{i=1}^{M} \prod_{j=1}^{N} \tau_i^\mu (1 - x_i \tau_j). \tag{61}$$

This expression is directly related to an integral representation of the multivariate hypergeometric function recently introduced by Kaneko.[29] We can thus simply write

$$\Psi(\vec{x}) = \mathcal{F}^{(\beta/2)} \left(-N_1, -N_2; 1 - \frac{2}{\beta} - N_T; X \right), \tag{62}$$

where $\mathcal{F}^{(\alpha)}(a, b; c; X)$ denotes the multivariate hypergeometric function. We use X as a shorthand for $x_1, x_2, \dots, x_N$. We stress that Eq. (62) is an exact nonperturbative result that does not depend on the particular choice of quantum transport approach (RMT, nonlinear σ-model and trajectory-based semiclassical), albeit they provide representations of $\Psi(\vec{x})$ which may have different domains of validity. Therefore, the use of $\Psi(\vec{x})$ in the present problem can be interpreted as a single unified nonperturbative approach.

In order to be a useful tool, we must be able to evaluate the derivatives of $\Psi(\vec{x})$. This can be achieved by using a representation of the hypergeometric function in terms of a series of Jack polynomials[29]

$$\mathcal{F}^{(\alpha)}(a,b;c;X) = \sum_\lambda \frac{[a]_\lambda^{(\alpha)} [b]_\lambda^{(\alpha)}}{|\lambda|! \, [c]_\lambda^{(\alpha)}} C_\lambda^{(\alpha)}(X), \tag{63}$$

We now describe the meaning of each term in Eq. (63). First, note that the sum is over partitions. A partition is a sequence of nonnegative integers $\lambda = (\lambda_1, \lambda_2, \lambda_3, \ldots)$ in decreasing order $\lambda_1 \geq \lambda_2 \geq \cdots \geq 0$. Partitions that differ only by a string of zeros at the end are considered equals. The number of nonzero terms in λ is the length of λ, denoted by $\ell(\lambda)$, and the sum of all parts $|\lambda| = \sum_i \lambda_i$ is the weight of λ. A partition is represented graphically by a Young tableaux $\{(i,j) \in \mathbb{Z}^2 / 1 \leq i \leq \ell(\lambda), 1 \leq j \leq \lambda_i\}$. It is also convenient to define the conjugate of a partition λ, denoted by λ', as the transpose of the Young tableaux of λ. An example of a partition, its transpose and the corresponding Young tableaux is shown in Fig. 3.

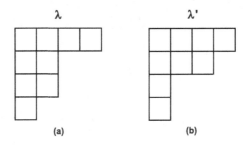

Fig. 3.   Young tableaux for the partition $\lambda = (4,2,2,1)$ and its conjugate $\lambda' = (4,3,1,1)$.

Table 3.   Jack polynomials associated to $|\lambda| = 3$.

Partition λ	Jack polynomial $C_\lambda^{(\alpha)}$
(3)	$C_{(3)}^{(\alpha)} = m_{(3)} + \dfrac{3}{1+2\alpha} m_{(2,1)} + \dfrac{6}{(1+2\alpha)(1+\alpha)} m_{(1,1,1)}$
(2,1)	$C_{(2,1)}^{(\alpha)} = \dfrac{6\alpha}{1+2\alpha} m_{(2,1)} + \dfrac{36\alpha}{(1+2\alpha)(2+\alpha)} m_{(1,1,1)}$
(1,1,1)	$C_{(1,1,1)}^{(\alpha)} = \dfrac{6\alpha^2}{(1+\alpha)(2+\alpha)} m_{(1,1,1)}$

A partition λ appears in the definition of the generalized Pochhammer symbol

$$[a]_\lambda^{(\alpha)} = \prod_{i=1}^{\ell(\lambda)} \left(a - \frac{1}{\alpha}(i-1) \right)_{\lambda_i} = \prod_{i=1}^{\ell(\lambda)} \frac{\Gamma(a + \lambda_i - (i-1)/\alpha)}{\Gamma(a - (i-1)/\alpha)}, \tag{64}$$

and as an index of the Jack polynomials $C_\lambda^{(\alpha)}(X)$. These multivariate polynomials constitute another basis of symmetric polynomials and interpolates between several types of symmetric polynomials by varying the parameter α.[30]

Jack polynomials naturally appear in the wave function of the trigonometric Calogero-Sutherland model[31] and are also used in multivariate generalizations of the classical orthogonal polynomials.[32] They are usually defined in terms of monomial symmetric functions m_λ as[30]

$$C_\lambda^{(\alpha)}(X) = \sum_{\mu \le \lambda} v_{\lambda,\mu}(\alpha) m_\mu. \tag{65}$$

The sum over partitions μ such that $\mu < \lambda$ means partitions of order $|\mu| = |\lambda|$ in which $\sum_{j=1}^p \mu_j < \sum_{j=1}^p \lambda_j$, $\forall p < \max(\ell(\mu), \ell(\lambda))$. The coefficients $v_{\lambda,\mu}(\alpha)$ are rational functions of the parameter α. As an example, we show in Table 3 the Jack polynomials associated with partitions of weight $|\lambda| = 3$.

The monomial symmetric function is defined by

$$m_\lambda = \sum_p x_{p(1)}^{\lambda_1} x_{p(2)}^{\lambda_2} \cdots x_{p(n)}^{\lambda_n}, \tag{66}$$

which is the sum of all distinct monomials in x_i with exponents $\lambda_1, \lambda_2, \ldots, \lambda_n$. We show in Table 4 the monomial symmetric functions indexed by the partitions of weight $|\lambda| = 3$ in four variables.

Table 4. Monomial symmetric functions associated to partitions of weight $|\lambda| = 3$ and four variables.

Partition λ	Monomial symmetric function m_λ
(3)	$x_1^3 + x_2^3 + x_3^3 + x_4^3$
(2,1)	$x_1^2 x_2 + x_1^2 x_3 + x_1^2 x_4 + x_2^2 x_1 + x_2^2 x_3 + x_2^2 x_4 + x_3^2 x_1 + x_3^2 x_2 + x_3^2 x_4$ $+ x_4^2 x_1 + x_4^2 x_2 + x_4^2 x_3$
(1,1,1)	$x_1 x_2 x_3 + x_1 x_2 x_4 + x_2 x_3 x_4$

We are now in a position to calculate the average of a product of elementary symmetric functions, Eq. (60). By inserting the definition of Jack polynomials, Eq. (65), in the series expansion of the generating function, Eq. (63), we can write

$$\langle e_{n_1} e_{n_2} \cdots e_{n_M} \rangle \propto \sum_\lambda \sum_{\mu \le \lambda} \gamma_\lambda^{(\alpha)} v_{\mu,\lambda}(\alpha) \left. \frac{\partial^n}{\partial x_1^{n_1} \partial x_1^{n_1} \cdots \partial x_M^{n_M}} m_\lambda \right|_{\vec{x}=0}, \tag{67}$$

where $n = n_1 + n_2 + \ldots + n_M$ and $\gamma_\lambda^{(\alpha)} = \frac{[a]_\lambda^{(\alpha)} [b]_\lambda^{(\alpha)}}{|\lambda|! \, [c]_\lambda^{(\alpha)}}$. Since we make $\vec{x} = 0$ at the end
of the calculation, we only need to expand $\Psi(\vec{x})$ with a sufficient number of terms
compatible with the order n of the derivatives we need to evaluate. We developed a
program in Maple to construct such a generating function and evaluate the derivatives. Our program uses the package MOPS[33] to generate the Jack polynomials. In
the next section and in Appendix A we apply this procedure to evaluate the average
values of the first four charge counting cumulants.

3.1. Conductance

The dimensionless conductance, $g = \sum_i \tau_i$, is the first and the simplest CCS cumulant. Since $g = e_1$, we can obtain all of its moments from the formula

$$\langle g^n \rangle = \langle e_1^n \rangle = (-1)^n \left. \frac{\partial^n \Psi(\vec{x})}{\partial x_1 \partial x_2 \ldots \partial x_n} \right|_{\vec{x}=0}. \tag{68}$$

Note that the appropriate generating function has n free variables and from the
series representation (63) we need to evaluate an n-order derivative of the form

$$\left. \frac{\partial^n}{\partial x_1 \partial x_2 \ldots \partial x_n} C_\lambda^{(\alpha)}(x_1, x_2 \ldots, x_n) \right|_{\vec{x}=0}, \tag{69}$$

which is nonzero only for partitions with weight $|\lambda| = n$. Furthermore, only the
monomials $m_{(1^n)} = x_1 x_2 \ldots x_n$ in the expansion (65) contributes to the sum. Since
the coefficient of the monomials $m_{(1^n)}$ in Jack polynomial expansion is known[30] to
be

$$v_{\lambda,1^n}(\alpha) = \frac{\alpha^{|\lambda|} |\lambda|! \, n!}{j_\lambda(\alpha)} = \frac{\alpha^n (n!)^2}{j_\lambda(\alpha)},$$

we can write the moments of conductance as

$$\langle g^n \rangle = (-1)^n n! \left(\frac{\beta}{2} \right)^n \sum_{|\lambda|=n} \frac{[-N_1]_\lambda^{(\beta/2)} [-N_2]_\lambda^{(\beta/2)}}{j_\lambda(\beta/2)[1 - 2/\beta - N_T]_\lambda^{(\beta/2)}}. \tag{70}$$

This exact expression can be used to obtain all cumulants of the conductance distribution. In particular, we write explicitly the average and variance of g

$$\langle g \rangle = \frac{N_1 N_2}{N_T - 1 + 2/\beta}, \tag{71}$$

$$\frac{\text{var}(g)}{\langle g \rangle^2} = \frac{2 \, (N_1 - 1 + 2/\beta) \, (N_2 - 1 + 2/\beta)}{\beta N_1 N_2 \, (N_T - 2 + 2/\beta) \, (N_T - 1 + 4/\beta)}. \tag{72}$$

A method based on Selberg's integral was put forward in Ref. 34 to obtain the
first four cumulants of the conductance distribution for all β values. Equation (70),
on the other hand, provides all moments and consequently all cumulants, it has
the advantage to be compact and it can be easily implemented in computer algebra
systems. In Appendix A, we apply the generating function method to calculate
statistical properties of higher order CCS cumulants.

4. Direct use of Jack polynomials

There is an alternative approach which is particularly useful to obtain certain types of closed expressions for transport observables. It explores the fact that any symmetric polynomial can also be written in terms of Jack polynomials and the result presented in corollary 2 of Ref. 29 which gives the average of a Jack polynomial over the Jacobi weight

$$\left\langle C_\lambda^{(\alpha)}(\{\tau\})\right\rangle = \frac{\alpha^{2|\lambda|}|\lambda|!}{j_\lambda(\alpha)} \frac{[N_1/\alpha]_\lambda^{(\alpha)}[N_2/\alpha]_\lambda^{(\alpha)}}{[1+(N_T-1)/\alpha]_\lambda^{(\alpha)}}, \tag{73}$$

where $\alpha = 2/\beta$ and

$$j_\lambda(\alpha) = \prod_{(i,j)\in\lambda} (\lambda_j' - i + \alpha(\lambda_i - j + 1))(\lambda_j' - i + 1 + \alpha(\lambda_i - j)).$$

We give below some examples of this procedure.

4.1. *Moments of conductance*

The Jack polynomials of type C^a have the following normalization

$$(\tau_1 + \ldots + \tau_N)^n = \sum_{|\lambda|=n} C_\lambda^{(\alpha)}(\{\tau\}), \tag{74}$$

where the sum is over all partitions of weight n. Since the left hand side of Eq. (74) is the n-th power of the conductance, g^n, we obtain the following closed form expression for the n-th moment of the conductance

$$\langle g^n \rangle = n! \left(\frac{2}{\beta}\right)^{2n} \sum_{|\lambda|=n} \frac{[N_1\beta/2]_\lambda^{(2/\beta)}[N_2\beta/2]_\lambda^{(2/\beta)}}{j_\lambda(2/\beta)[1+\beta(N_T-1)/2]_\lambda^{(2/\beta)}}. \tag{75}$$

Moments of conductance were obtained by Novaes[35] by using similar arguments but restricted to $\beta = 2$, where the Jack's polynomials reduce to the Schur's functions. In the next section we show how this approach can be used to obtain the transport moments, M_n, used in trajectory-based semiclassical approaches.

4.2. *Transport moments*

As pointed out in Sec. 2, the trajectory-based semiclassical methods apply well to the calculation of transport moments, defined as $M_n = \langle \mathcal{T}_m \rangle$, where

$$\mathcal{T}_n = \sum_{i=1}^{N} \tau_i^n. \tag{76}$$

They appear in the series expansion of the auxiliary function $h(z)$, see Eq. (49), used to calculate the average density of transmission eigenvalues. Here we show

---

[a]Besides the "C" normalization of Jack polynomials used in this work, we commonly find in literature the "P" and "J" normalization. See Ref. 33 for conversions between these normalizations.

how to obtain M_n from the average of Jack polynomials. We start by representing $\mathcal{T}_n$ in terms of Jack polynomials[36]

$$\mathcal{T}_n = \frac{1}{\alpha^{n-1}(n-1)!} \sum_{|\lambda|=n} b_\lambda(\alpha) C_\lambda^{(\alpha)}, \qquad (77)$$

where

$$b_\lambda(\alpha) = \prod_{(i,j)\in\lambda}' \left(\alpha(j-1)-(i-1)\right),$$

with the prime indicating $(i,j) \neq (1,1)$. In particular $b_{(1)}(\alpha) = 1$. Therefore, after using (73) to perform the average, we obtain

$$M_n = n \left(\frac{2}{\beta}\right)^{n+1} \sum_{|\lambda|=n} \frac{b_\lambda(2/\beta)}{j_\lambda(2/\beta)} \frac{[N_1\beta/2]_\lambda^{(2/\beta)}[N_2\beta/2]_\lambda^{(2/\beta)}}{[1+\beta(N_T-1)/2]_\lambda^{(2/\beta)}}. \qquad (78)$$

As a consistency check, note that Eqs. (75) and (78) coincide in the particular case $n = 1$. We conclude by remarking that (78), being an exact nonperturbative expression, we can be very useful in improving and extending trajectory-based semiclassical methods.

5. Conclusions

We have introduced a powerful generating function method to calculate analytically the statistical properties of transport observables in a chaotic quantum dot coupled ideally to electron reservoirs via perfectly conducting leads with arbitrary numbers of open scattering channels. The technique uses multivariate hypergeometric functions and has the potential to unify the methods of the three most common approaches to quantum transport: random matrix theory, the nonlinear σ-model and the trajectory-based semiclassical approach.

Acknowledgments

This work was partially supported by CNPq and FACEPE (Brazilian Agencies).

Appendix A. Higher order cumulants

A.1. Shot-noise power

From its definition, the shot-noise power can be written as

$$p = e_1 - e_1^2 + 2e_2, \qquad (A.1)$$

which allows us to express its average as

$$\langle p \rangle = \left(\frac{\partial^2 \Psi}{\partial x_1^2} - \frac{\partial^2 \Psi}{\partial x_1 \partial x_2} - \frac{\partial \Psi}{\partial x_1} \right)_{\bar{x}=0}, \tag{A.2}$$

which implies that the generating function needs only two free variables. Below we give explicitly the first terms of its expansion in the basis of Jack polynomials

$$\Psi(x_1, x_2) = \mathcal{F}^{(\alpha)}(a, b; c; x_1, x_2)$$

$$= 1 + \frac{[a]_{(1)}^{(\alpha)} [b]_{(1)}^{(\alpha)}}{[c]_{(1)}^{(\alpha)}} C_{(1)}^{(\alpha)} + \frac{[a]_{(2)}^{(\alpha)} [b]_{(2)}^{(\alpha)}}{[c]_{(2)}^{(\alpha)}} \frac{C_{(2)}^{(\alpha)}}{2} + \frac{[a]_{(1,1)}^{(\alpha)} [b]_{(1,1)}^{(\alpha)}}{[c]_{(1,1)}^{(\alpha)}} \frac{C_{(1,1)}^{(\alpha)}}{2} + \dots$$

$$= 1 + \frac{ab}{c}(x_1 + x_2) + \frac{a(a+1)b(b+1)}{c(c+1)} \frac{(x_1 + x_2)^2 + \alpha(x_1^2 + x_2^2)}{2(1+\alpha)}$$

$$+ \frac{a(a - 1/\alpha)b(b - 1/\alpha)}{c(c - 1/\alpha)} \frac{\alpha x_1 x_2}{1 + \alpha} + \dots.$$

Taking the derivatives and setting the parameters $\alpha = \beta/2$, $a = -N_1$, $b = -N_2$ and $c = 1 - 2/\beta - (N_1 + N_2)$, we arrive at

$$\langle p \rangle = \frac{N_1 N_2 (N_1 - 1 + 2/\beta)(N_2 - 1 + 2/\beta)}{(N_T - 2 + 2/\beta)(N_T - 1 + 2/\beta)(N_T - 1 + 4/\beta)}, \tag{A.3}$$

which agrees with results available in literature.[37] We do not have closed expressions for moments or cumulants of the shot-noise power distribution. Nevertheless, we can apply the generating function technique to obtain particular moments of interest. For example, to evaluate the variance we have to express p^2 in terms of elementary symmetric functions

$$p^2 = e_1^2 + e_1^4 + 4e_2^2 - 2e_1^3 - 4e_1^2 e_2 + 4e_1 e_2. \tag{A.4}$$

In this case, the second moment of obtained from the derivatives

$$\langle p^2 \rangle = \left(\frac{\partial^2 \Psi}{\partial x_1 \partial x_2} + \frac{\partial^4 \Psi}{\partial x_1 \partial x_2 \partial x_3 \partial x_4} + \frac{\partial^4 \Psi}{\partial x_1^2 \partial x_2^2} + 2 \frac{\partial^3 \Psi}{\partial x_1 \partial x_2 \partial x_3} - 2 \frac{\partial^4 \Psi}{\partial x_1 \partial x_2 \partial x_3^2} \right)_{\bar{x}=0}$$

$$- 2 \left(\frac{\partial^3 \Psi}{\partial x_1 \partial x_2^2} \right)_{\bar{x}=0}.$$

The result obtained from the above equation with the appropriate generating function together with Eq. (A.3) give an exact analytical expression for the variance $\mathrm{var}(p) = \langle p^2 \rangle - \langle p \rangle^2$. Since it is very lengthly, we show here only results in the particular case of symmetric contacts, $N_1 = N_2 = N$, for $\beta = 1, 2$ and 4

$$\text{var}(p)_{(\beta=1)} = \frac{N(N+1)\left(8N^5+60N^4+142N^3+91N^2-49N-36\right)}{4(7+2N)(5+2N)(-1+2N)(1+2N)^2(3+2N)^2}, \tag{A.5}$$

$$\text{var}(p)_{(\beta=2)} = \frac{N^2\left(3-9N^2+4N^4\right)}{8(1+2N)^2(-1+2N)^2(-9+4N^2)}, \tag{A.6}$$

$$\text{var}(p)_{(\beta=4)} = \frac{N\left(85N+84N^2+1616N^4-932N^3+256N^6-1088N^5-18\right)}{4(-7+4N)(1+4N)(-5+4N)(-1+4N)^2(-3+4N)^2}. \tag{A.7}$$

For completeness, we also give expressions for the extreme quantum limit $(N_1 = N_2 = 1)$

$$\text{var}(p) = \frac{4\beta\left(\beta^3+2\beta^2-8\beta+32\right)}{(2+\beta)^2(4+\beta)^2(6+\beta)(\beta+8)},$$

and the semiclassical limit $(N_1, N_2 \gg 1)$

$$\text{var}(p) = \frac{2N_1{}^2N_2{}^2}{\beta(N_1+N_2)^8}\left(N_2{}^4+8N_2{}^2N_1{}^2-4N_2N_1{}^3+N_1{}^4-4N_1N_2{}^3\right),$$

which simplifies to the well known value $\text{var}(p) = 1/64\beta$ for $N_1 = N_2 = N$.

A.2. Third cumulant

Proceeding in the same way, from the definition of the third cumulant, we get

$$\kappa = e_1 + 6e_2 + 6e_3 - 3e_1^2 - 6e_1e_2 + 2e_1^3, \tag{A.8}$$

which means that

$$\langle\kappa\rangle = \left(-\frac{\partial\Psi}{\partial x_1}+3\frac{\partial^2\Psi}{\partial x_1^2}-\frac{\partial^3\Psi}{\partial x_1^3}-3\frac{\partial^2\Psi}{\partial x_1\partial x_2}+3\frac{\partial^3\Psi}{\partial x_1\partial x_2^2}-2\frac{\partial^3\Psi}{\partial x_1\partial x_2\partial x_3}\right)_{\bar{x}=0}. \tag{A.9}$$

By constructing the appropriate generating function, we obtain a closed expression which we write in the compact form

$$\frac{\langle\kappa\rangle}{\langle p\rangle} = -\frac{(N_1-N_2)^2-(1-2/\beta)^2}{(N_1+N_2-1+6/\beta)(N_1+N_2-3+2/\beta)}, \tag{A.10}$$

in agreement with Ref. 38. As in the previous section we also give expressions for some limits of interest. First, in the extreme quantum limit $N_1 = N_2 = 1$, we have

$$\langle\kappa\rangle = -\frac{2\beta(\beta-2)}{(2+\beta)(\beta+4)(6+\beta)}.$$

In the semiclassical limit, $N_1, N_2 \gg 1$, we have the expansion

$$
\langle \kappa \rangle = -\frac{(N_1 - N_2)^2 \, N_1{}^2 N_2{}^2}{(N_1 + N_2)^5}
$$
$$
+ \frac{\left(N_1{}^2 - 6\,N_1 N_2 + N_2{}^2\right)(N_1 - N_2)^2 \,(\beta - 2)\, N_1 N_2}{\beta \,(N_1 + N_2)^6}
$$
$$
+ \frac{N_1 N_2 \left(f(N_1^4 + N_2^4) + g(N_1^3 N_2 + N_1 N_2^3) + h N_1^2 N_2^2\right)}{\beta^2 (N_1 + N_2)^7}
$$
$$
+ \dots
$$

with coefficients given by $f = 7(\beta-2)^2$, $g = -156+146\beta-39\beta^2$ and $h = 4(68-63\beta+ 17\beta^2)$. Note that the first two terms of this expansion vanishes for $N_1 = N_2 = N$. In this case, the first non-vanishing term is

$$
\langle \kappa \rangle = \frac{(\beta - 2)^2}{32\beta^2 N}. \tag{A.11}
$$

A.3. Fourth cumulant

Finally, writing the fourth cumulant in terms of elementary symmetric functions, we obtain

$$
\zeta = e_1 + 14e_2 + 36e_3 + 24e_4 - 7e_1^2 + 12e_1^3 - 6e_1^4
$$
$$
-12e_2^2 - 36e_1e_2 - 24e_1e_3 + 24e_1^2e_2. \tag{A.12}
$$

Therefore, the ensemble average is obtained from the generating function using

$$
\langle \zeta \rangle = \left(-\frac{\partial \psi}{\partial x_1} + 7\frac{\partial^2 \psi}{\partial x_1^2} - 6\frac{\partial^3 \psi}{\partial x_1^3} + \frac{\partial^4 \psi}{\partial x_1^4} - 7\frac{\partial^2 \psi}{\partial x_1 \partial x_2} - 12\frac{\partial^3 \psi}{\partial x_1 \partial x_2 \partial x_3} \right.
$$
$$
- 6\frac{\partial^4 \psi}{\partial x_1 \partial x_2 \partial x_3 \partial x_4} - 3\frac{\partial^4 \psi}{\partial x_1^2 \partial x_2^2} + 18\frac{\partial^3 \psi}{\partial x_1 \partial x_2^2} - 4\frac{\partial^4 \psi}{\partial x_1 \partial x_2^3}
$$
$$
\left. + 12\frac{\partial^4 \psi}{\partial x_1 \partial x_2 \partial x_3^2} \right)_{\vec{x}=0}. \tag{A.13}
$$

Performing the derivatives we get an exact but very lengthy expression. In particular, for symmetric contacts $N_1 = N_2 = N$ we obtain

$$
\langle \zeta \rangle_{\beta=1} = -\frac{N\,(N+1)\left(2\,N^4 + 13\,N^3 + 21\,N^2 + 2\,N - 2\right)}{4\,(3 + 2\,N)\,(5 + 2\,N)\,(7 + 2\,N)\,(-1 + 4\,N^2)},
$$
$$
\langle \zeta \rangle_{\beta=2} = -\frac{N^3\left(N^2 - 3\right)}{4\,(-1 + 4\,N^2)\,(-9 + 4\,N^2)},
$$
$$
\langle \zeta \rangle_{\beta=4} = -\frac{N\,(2\,N - 1)\left(16\,N^4 - 52\,N^3 + 42\,N^2 - 2\,N - 1\right)}{2\,(-3 + 4\,N)\,(-5 + 4\,N)\,(-7 + 4\,N)\,(-1 + 16\,N^2)}.
$$

The simplest particular case is the extreme quantum limit, $N_1 = N_2 = 1$, which gives

$$\langle \zeta \rangle = \frac{2\,(\beta - 10)\,\beta^2}{(2 + \beta)\,(4 + \beta)\,(6 + \beta)\,(\beta + 8)}. \tag{A.14}$$

In the semiclassical limit $N_1, N_2 \gg 1$ we have the expansion

$$
\langle \zeta \rangle = \frac{\left(N_2{}^4 + 12\,N_2{}^2 N_1{}^2 - 8\,N_2 N_1{}^3 + N_1{}^4 - 8\,N_1 N_2{}^3\right) N_1{}^2 N_2{}^2}{(N_1 + N_2)^7}
$$
$$
- \frac{(\beta - 2)\,N_1 N_2\,(N_1 - N_2)^2\,\left(N_1{}^4 - 20\,N_2 N_1{}^3 + 54\,N_2{}^2 N_1{}^2 - 20\,N_1 N_2{}^3 + N_2{}^4\right)}{\beta\,(N_1 + N_2)^8}
$$
$$
- \frac{N_1 N_2 \left(f(N_1^6 + N_2^6) + g(N_1^5 N_2 + N_1 N_2^5) + \Psi(N_1^4 N_2^2 + N_1^2 N_2^4) + h N_2^3 N_1^3\right)}{\beta^2 (N_1 + N_2)^9}
$$
$$
+ \dots,
$$

with coefficients given by $f = 15(\beta - 2)^2$, $g = -191\beta^2 + 726\beta - 764$, $h = -6\left(181\beta^2 - 666\beta + 724\right)$ and $\Psi = 721\beta^2 - 2676\beta + 2884$. Note that the subdominant term, usually called the weak localization correction, vanishes for $\beta = 2$ and/or for symmetric contacts $N_1 = N_2 = N$. In the latter case, the expansion simplifies to

$$\langle \zeta \rangle = -\frac{N}{64} - \frac{\beta^2 - 6\beta + 4}{128\beta^2 N} + \dots \tag{A.15}$$

References

1. For a review, see Yu. V. Nazarov, and Ya. M. Blanter, *Quantum Transport: Introduction to Nanoscience* (Cambridge University Press, Cambridge, 2009).
2. S. Oberholzer, E. V. Sukhorukov, C. Strunk, C. Schönenberger, T. Heinzel, and M. Holland, Shot Noise by Quantum Scattering in Chaotic Cavities, *Phys. Rev. Lett.* **86**, 2114 (2001).
3. S. Oberholzer, E. V. Sukhorukov, C. Strunk, and C. Schönenberger, Shot noise of series quantum point contacts intercalating chaotic cavities, *Phys. Rev. B.* **66**, 233304 (2002).
4. W. Song, A. K. M. Newaz, J. K. Son, and E. E. Mendez, Drastic Reduction of Shot Noise in Semiconductor Superlattices, *Phys. Rev. Lett.* **96**, 126803 (2006).
5. K. K. Gomes, W. Mar, W. Ko, F. Guinea, and H. C. Manoharan, Designer Dirac fermions and topological phases in molecular graphene, *Nature (London).* **483**, 306 (2012).
6. A. Singha, M. Gibertini, B. Karmakar, S. Yuan, M. Polini, G. Vignale, M. I. Katsnelson, A. Pinczuk, L. N. Pfeiffer, K. W. West, and V. Pellegrini, Two-Dimensional Mott-Hubbard Electrons in an Artificial Honeycomb Lattice, *Science* **332**, 1176 (2011).
7. G. De Simoni, A. Singha, M. Gibertini, B. Karmakar, M. Polini, V. Piazza, L. N. Pfeiffer, K. W. West, F. Beltram and V. Pellegrini, Delocalized-localized transition in a semiconductor two-dimensional honeycomb lattice, *Appl. Phys. Lett.* **97**, 132113 (2010).
8. Iulia Buluta and Franco Nori, Quantum Simulators, *Science* **326**, 108 (2009).

9. T. Byrnes, N. Y. Kim, K. Kusudo, and Y. Yamamoto, Quantum simulation of Fermi-Hubbard models in semiconductor quantum-dot arrays, *Phys. Rev. B* 0 **78**, 075320 (2008).

10. L. S. Levitov, H. Lee, and G. B. Lesovik, Electron counting statistics and coherent states of electric current, *J. Math. Phys.* **37**, 4845 (1996).

11. R. J. Glauber, Coherent and Incoherent States of the Radiation Field, *Phys. Rev.* **131**, 2766 (1963).

12. I. Klich and L. S. Levitov, Quantum Noise as an Entanglement Meter, *Phys. Rev. Lett.* **102**, 100502 (2009).

13. H. F. Song, S. Rachel, C. Flindt, I. Klich, N. Laflorencie, and K. Le Hur, Bipartite fluctuations as a probe of many-body entanglement, *Phys. Rev. B* **85**, 035409 (2012).

14. M. Esposito, U. Harbola, and S. Mukamel, Nonequilibrium fluctuations, fluctuation theorems, and counting statistics in quantum systems, *Rev. Mod. Phys.* **81**, 1665 (2009).

15. A. M. S. Macêdo, Transport theory of interacting mesoscopic systems: A memory-function approach to charge-counting statistics, *Phys. Rev. B* **69**, 155309 (2004).

16. C. W. J. Beenakker, Random-matrix theory of quantum transport, *Rev. Mod. Phys.* **69**, 731 (1997).

17. K. Richter and M. Sieber, Semiclassical Theory of Chaotic Quantum Transport, *Phys. Rev. Lett.* **89**, 206801 (2002).

18. G. Berkolaiko and J. Kuipers, Universality in chaotic quantum transport: The concordance between random-matrix and semiclassical theories, *Phys. Rev. E* **85**, 045201 (2012).

19. J. Kuipers and K. Richter, Transport moments and Andreev billiards with tunnel barriers, *J. Phys. A: Math. Theor.* **46**, 055101 (2013).

20. P. W. Brouwer and C. W. J. Beenakker, Diagrammatic method of integration over the unitary group, with applications to quantum transport in mesoscopic systems, *J. Math. Phys.* **37**, 4904 (1996).

21. G. C. Duarte-Filho, A. F. Macedo-Junior, and A. M. S. Macêdo, Circuit theory and full counting statistics of charge transfer through mesoscopic systems: A random-matrix approach, *Phys. Rev. B* **76**, 075342 (2007).

22. P. A. Mello and N. Kumar, *Quantum Transport in Mesoscopic Systems: Complexity and Statistical Fluctuations* (Oxford University Press, New York, 2004).

23. H. Lee, L. S. Levitov, and A. Yu. Yakovets , Universal statistics of transport in disordered conductors, *Phys. Rev. B.* **51**, 4079 (1995).

24. S. Gustavsson, R. Leturcq, M. Studer, I. Shorubalko, T. Ihn, K. Ensslin, D.C. Driscoll, and A. C. Gossard, Electron counting in quantum dots, *Surf. Sci. Rep.* **64**, 191 (2009).

25. M. L. Mehta, *Random Matrices* (Elsevier, Amsterdan, 2004), 3rd edition.

26. M. R. Zirnbauer, Supersymmetry for systems with unitary disorder: circular ensembles, *J. Phys. A: Math. Gen.* **29**, 7113 (1996).

27. I. G. Macdonald, *Symmetric Functions and Hall Polynomials* (Oxford University Press, Oxford, 1998).

28. R. J. Muirhead, *Aspects of Multivariate Statistical Theory* (John Wiley & Sons, New York, 1982).

29. J. Kaneko, Selberg integrals and hypergeometric functions associated with jack polynomials, *SIAM J. Math. Anal.* **24**, 1086 (1993).

30. R. P. Stanley, Some combinatorial properties of Jack symmetric functions, *Adv. Math.* **77**, 76 (1989).

31. Z. N. C. Ha, Exact Dynamical Correlation Functions of Calogero-Sutherland Model and One-Dimensional Fractional Statistics, *Phys. Rev. Lett.* **73**, 1574 (1994).

32. T. H. Baker and P. J. Forrester, The Calogero-Sutherland Model and Generalized Classical Polynomials, *Commun. Math. Phys.* **188**, 175 (1997).

33. I. Dumitriu, A. Edelman and G. Shuman, MOPS: Multivariate orthogonal polynomials (symbolically), *Journal of Symbolic Computation.* **42**, 587 (2007). Package Downloadable from: http://www.math.washington.edu/~dumitriu/mopspage.html

34. D. V. Savin, H.-J. Sommers and W. Wieczorek, Nonlinear statistics of quantum transport in chaotic cavities, *Phys. Rev. B* **77**, 125332 (2008).

35. M. Novaes, Statistics of quantum transport in chaotic cavities with broken time-reversal symmetry, *Phys. Rev. B* **78**, 035337 (2008).

36. P. J. Hanlon, R. P. Stanley and J. R. Stembridge, Some combinatorial aspects of the spectra of normally distributed random matrices, *Contemporary Mathematics* **138**, 151 (1992).

37. D. V. Savin and H.-J. Sommers, Shot noise in chaotic cavities with an arbitrary number of open channels, *Phys. Rev. B* **73**, 081307 (2006).

38. M. Novaes, Full counting statistics of chaotic cavities with many open channels, *Phys. Rev. B* **75**, 073304 (2007).

Chapter 6

Time-Dependent Coupled Harmonic Oscillators: Classical and Quantum Solutions

Diego Ximenes Macedo*‡ and Ilde Guedes†

*Departamento de Ensino, Instituto Federal de Educação,
Ciência e Tecnologia do Ceará, Campus Crateús,
63700-000, Crateús, CE, Brasil

†Departamento de Física, Universidade Federal do Ceará,
Campus do PICI, Caixa Postal 6030, 60455-760,
Fortaleza, CE, Brasil

In this work we present the classical and quantum solutions for an arbitrary system of time-dependent coupled harmonic oscillators, where the masses (m), frequencies (ω) and coupling parameter (k) are functions of time. To obtain the classical solutions we use a coordinate and momentum transformations along with a canonical transformation to write the original Hamiltonian as the sum of two Hamiltonians of uncoupled harmonic oscillators with modified time-dependent frequencies and unitary masses. To obtain the exact quantum solutions we use a unitary transformation and the Lewis and Riesenfeld invariant method. The exact wave functions are obtained by solving the respective Milne-Pinney equation for each system. We obtain the solutions for the system with $m_1 = m_2 = m_0 e^{\gamma t}$, $\omega_1 = \omega_{01} e^{-\gamma t/2}$, $\omega_2 = \omega_{02} e^{-\gamma t/2}$ and $k = k_0$.

1. Introduction

For quantum and classical collective processes, it has become very important to find the exact solution for a system of coupled harmonic oscillators. For instance, in 1988 Yeon et al.[1] used the Feynman path integral method to obtain the propagators for time-independent coupled harmonic oscillators with and without a time-dependent driven force $F(t)$. The propagators for coupled harmonic oscillators with $F(t) = 0$ were therefore used to obtain explicitly the energy expectation value, which is given by the sum of two energy expectation values corresponding to the quantum states of two oscillators. However, they have not evaluated the wave functions of both systems. In 1999, Zhukov and Zhukova[2] also used the Feynman path integral method to obtain an exact propagator for the coupled harmonic oscillators with exponentially changing masses.

---

‡diegoximenes@fisica.ufc.br

The model of coupled harmonic oscillator has been widely used to study the quantum effects in mesoscopic coupled electric circuits. These quantum effects arise owing to the miniaturization of electric devices towards atomic scale, where the transport length reaches the inelastic coherence length of the charge carrier. In 2001, Wang *et al.*[3] investigated the quantum fluctuations of charge and current of a mesoscopic inductance (L) coupled circuit in a displaced squeezed Fock state. The system consists of two LC(C, capacitance) circuits coupled by an inductor and can be modeled by a time-independent Hamiltonian, which is the sum of the Hamiltonians of two independent harmonic oscillators with different frequencies. They observed that the quantum fluctuations of charge and current in each component circuit depend on the device of two circuits and squeezing parameters, while the fluctuations do not depend on the displacement parameters.

In 2001, Zhang *et al.*[4] calculated the quantum uncertainties of a mesoscopic capacitance coupled circuit. The system consists of two RLC circuits coupled by a capacitor. Due to the resistance (R), the Hamiltonian describing the system is now time-dependent. They used the Lewis and Riesenfeld (LR) quantum invariant theory[5,6] and a unitary transformation[7,8] to obtain the eigenfunction of the quantum invariant operator which is given by the Fock states. The wave function of the system differs only by a time-dependent phase factor from the eigenfunction of quantum invariant operator. They observed that the uncertainty relations between charges and currents do not satisfy the minimum uncertainty relation even though the resistances are zero.

In 2002, Zhang *et al.*[9] studied the quantum squeezing effect of a mesoscopic capacitance-inductance-resistance coupled circuit. Again, due to the resistance (R) the Hamiltonian describing the system is also time-dependent. They used a canonical and a unitary transformation to write the Hamiltonian of the mesoscopic capacitance-inductance-resistance coupled circuit to that of mutually independent two-dimensional simple harmonic oscillator. For the ground state, they verified that the quantum variance of the charge (current) decreases (increases) exponentially as time goes by. They also observed that even if the resistances are zero, the uncertainty relations do not satisfy minimum uncertainties relations.

The model of time-dependent coupled oscillators has also been used to investigate the dynamics of charged particle motion in the presence of time-varying magnetic fields. This system has been used in different areas of physics, as condensed matter and plasma. Recently, Menouar *et al.*[10,11] have reported the "exact" wave functions of two different systems describing the motion of a charged particle in a variable magnetic field. In Ref. 10 they considered a system taking into account the static xy coupling term, while in Ref. 11, the system also includes the dynamical $p_x p_y$ coupling term.

In both papers, they considered a time-dependent Hamiltonian, by supposing that the parameters such as the effective mass of the charged particle vary explicitly with time in the presence of variable magnetic field. They used several unitary transformations to write the time-dependent coupled Hamiltonian as the sum of two uncoupled time-dependent Hamiltonians, corresponding to harmonic oscillators with different time-dependent frequencies and unitary masses. To obtain the quantum states, they used the LR dynamical method.[5,6] As it is well known the exact wave function of the system is the same as the eigenstate of the invariant operator, except for some time-dependent phase factor. For each oscillator, the wave function is expressed in terms of a c-number quantity (ρ) which is solution of the Milne-Pinney (MP) equation.[12-14] According to the authors, the complete knowledge for the behavior of the system within the scope of quantum mechanics will be taken only when the solutions of the MP equations are solved.

This paper is outlined as follows. In Section 2 we use a coordinate and momentum transformations along with a canonical transformation to write the Hamiltonian of the coupled harmonic oscillators as the sum of two Hamiltonians of uncoupled simple harmonic oscillators with modified time-dependent frequencies and unitary masses. We find the solution of the equations of motion for $m_1 = m_2 = m_0 e^{\gamma t}$, $\omega_1 = \omega_{01} e^{-\gamma t/2}$, $\omega_2 = \omega_{02} e^{-\gamma t/2}$ and $k = k_0$, where m_0, $\omega_{0i}(i = 1, 2)$, k_0 and γ are positive constants. In Section 3 we use a unitary transformation and the LR method to obtain the invariant and the general wave function of the time-dependent coupled harmonic oscillators. The exact wave functions are obtained by solving the respective MP equation. Therefore we obtain the wave function and the probability distribution for the system stated above. We end with a summary in Sec. 4.

2. Classical solutions

To describe the system of two statically coupled time-dependent harmonic oscillators, we consider the following Hamiltonian

$$H(x_1, x_2, t) = \frac{p_{1x}^2}{2m_1(t)} + \frac{p_{2x}^2}{2m_2(t)} + \frac{m_1(t)\omega_1^2(t)x_1^2}{2} + \frac{m_2(t)\omega_2^2(t)x_2^2}{2} + \frac{k(t)(x_2 - x_1)^2}{2},$$

$$(1)$$

where $m_i(i = 1, 2)$ $\omega_i(i = 1, 2)$, and $k(t)$ are the time-dependent mass, frequency, and the coupling parameter, respectively.

To decouple Eq. (1) and write it as the sum of two Hamiltonians representing time-dependent uncoupled oscillators, we proceed as follows. First, consider the coordinate and momentum transformations

$$q_1 = \left(\frac{m_1(t)}{m_2(t)}\right)^{1/4} x_1,$$

$$(2)$$

$$q_2 = \left(\frac{m_2(t)}{m_1(t)}\right)^{1/4} x_2,$$

$$(3)$$

$$p_1 = \left(\frac{m_2(t)}{m_1(t)}\right)^{1/4} p_{1x}, \tag{4}$$

$$p_2 = \left(\frac{m_1(t)}{m_2(t)}\right)^{1/4} p_{2x}. \tag{5}$$

By substituting Eqs. (2)–(5) into Eq. (1), we have

$$H = \frac{p_1^2 + p_2^2}{2m} + \frac{b_1(t)q_1^2 + b_2(t)q_2^2 + b_3(t)q_1q_2}{2}, \tag{6}$$

where

$$m(t) = (m_1(t)m_2(t))^{1/2}, \tag{7}$$

$$b_1(t) = \left(\frac{m_2(t)}{m_1(t)}\right)^{1/2} (m_1(t)\omega_1(t)^2 + k(t)), \tag{8}$$

$$b_2(t) = \left(\frac{m_1(t)}{m_2(t)}\right)^{1/2} (m_2(t)\omega_2(t)^2 + k(t)), \tag{9}$$

$$b_3(t) = -2k(t). \tag{10}$$

To eliminate the term q_1q_2 in Eq. (6), let us consider a time-dependent canonical transformation whose generating function is

$$\Phi(q_1, q_2, P_1, P_2, t) = \sqrt{m}q_1(P_1 \cos\theta + P_2 \sin\theta) + \sqrt{m}q_2(-P_1 \sin\theta + P_2 \cos\theta)$$

$$- \frac{\dot{m}}{2}(q_1^2 + q_2^2). \tag{11}$$

From Eq. (11) we obtain the new canonical variables

$$p_1 = \frac{\partial \Phi}{\partial q_1} = \sqrt{m}(P_1 \cos\theta + P_2 \sin\theta) - \frac{\dot{m}}{2}q_1, \tag{12}$$

$$p_2 = \frac{\partial \Phi}{\partial q_2} = \sqrt{m}(-P_1 \sin\theta + P_2 \cos\theta) - \frac{\dot{m}}{2}q_2, \tag{13}$$

$$Q_1 = \frac{\partial \Phi}{\partial P_1} = \sqrt{m}(q_1 \cos\theta - q_2 \sin\theta), \tag{14}$$

$$Q_2 = \frac{\partial \Phi}{\partial P_2} = \sqrt{m}(q_1 \sin\theta + q_2 \cos\theta), \tag{15}$$

and the new Hamiltonian, $H_N = \frac{\partial \Phi}{\partial P_2} + H(t)$, reads

$$H_N = \frac{P_1^2}{2} + \frac{\Omega_1^2(t)Q_1^2}{2} + \frac{P_2^2}{2} + \frac{\Omega_2^2(t)Q_2^2}{2}$$
$$+ Q_1 Q_2 \left(\frac{\sin\theta}{2m}(b_1 - b_2) + \frac{b_3}{2m}\cos\theta \right) + \frac{\dot{\theta}}{2}(Q_1 P_2 - Q_2 P_1), \qquad (16)$$

where the dots represents the derivatives and

$$\Omega_1^2(t) = \frac{\dot{m}^2}{4m^2} - \frac{\ddot{m}}{2m} + \frac{b_1}{m}\cos^2\theta + \frac{b_2}{m}\sin^2\theta - \frac{b_3}{2m}\sin 2\theta, \qquad (17)$$

$$\Omega_2^2(t) = \frac{\dot{m}^2}{4m^2} - \frac{\ddot{m}}{2m} + \frac{b_1}{m}\sin^2\theta + \frac{b_2}{m}\cos^2\theta + \frac{b_3}{2m}\sin 2\theta. \qquad (18)$$

We observe from Eq. (16) that the separation of variables is complete for $\dot{\theta} = 0$ and

$$\tan 2\theta = \frac{b_3}{b_2 - b_1} = \text{constant}. \qquad (19)$$

Equation (19) imposes a constraint on the possible values of m_i, ω_i and $k(t)$ to be considered. Equation (16) finally reads

$$H_N = \frac{P_1^2}{2} + \frac{\Omega_1^2(t)Q_1^2}{2} + \frac{P_2^2}{2} + \frac{\Omega_2^2(t)Q_2^2}{2}. \qquad (20)$$

Equation (20) represents the sum of two uncoupled Hamiltonians of simple harmonic oscillators with time-dependent modified frequency ($\Omega_t(t)$) and unitary mass. From the Hamilton equations, one finds

$$\frac{d^2 Q_1}{dt^2} + \Omega_1^2(t)Q_1 = 0, \qquad (21)$$

and

$$\frac{d^2 Q_2}{dt^2} + \Omega_2^2(t)Q_2 = 0. \qquad (22)$$

By using Eqs. (2), (3), (14), and (15), the expressions for $x_1(t)$ and $x_2(t)$ read

$$x_1(t) = \frac{1}{\sqrt{m_1(t)}}[Q_1 \cos\theta + Q_2 \sin\theta], \qquad (23)$$

and

$$x_2(t) = \frac{1}{\sqrt{m_2(t)}}[-Q_1 \sin\theta + Q_2 \cos\theta]. \qquad (24)$$

Due to the constraint imposed by Eq. (19), we will consider the following coupled harmonic oscillators, namely: $m_1 = m_2 = m_0 e^{\gamma t}$, $\omega_1 = \omega_{01} e^{-\gamma t/2}$, $\omega_2 = \omega_{02} e^{-\gamma t/2}$ and $k = k_0$. For this case we get from Eq. (19)

$$\theta = \frac{1}{2} \arctan \left[\frac{2k_0}{m_0(\omega_{01}^2 - \omega_{02}^2)} \right]. \tag{25}$$

From eqs. (17) and (18) we have

$$\Omega_1^2(t) = -\frac{\gamma^2}{4} + \omega_{01}^2 \cos^2 \theta + \omega_{02}^2 \sin^2 \theta + \frac{k_0}{m_0}(1 + \sin 2\theta), \tag{26}$$

$$\Omega_2^2(t) = -\frac{\gamma^2}{4} + \omega_{01}^2 \sin^2 \theta + \omega_{02}^2 \cos^2 \theta + \frac{k_0}{m_0}(1 - \sin 2\theta), \tag{27}$$

which are both constants. In this case, the solution of the Eqs. (21) and (22) is straightforward, and we obtain

$$x_1(t) = \frac{e^{-\gamma t/2}}{\sqrt{m_0}} \left[A_1 \cos \theta \sin(\Omega_1 t + \phi_1) + A_2 \sin \theta \sin(\Omega_2 t + \phi_2) \right], \tag{28}$$

$$x_2(t) = \frac{e^{-\gamma t/2}}{\sqrt{m_0}} \left[-A_1 \sin \theta \sin(\Omega_1 t + \phi_1) + A_2 \cos \theta \sin(\Omega_2 t + \phi_2) \right], \tag{29}$$

where, $A_i (i = 1, 2)$ and $\phi_i (i = 1, 2)$ are constants to be determined from the initial conditions.

The solutions correspond to the superposition of the motion of two simple harmonic oscillators with different frequencies, phases and exponentially decaying amplitudes. This decay is owing to the fact that the mass for both oscillators increases exponentially with time. Such systems are called the Caldirola-Kanai oscillators and were introduced in literature to a prototype to described damped systems.[15, 16]

3. Quantum solutions

The problem we want to solve is

$$i\hbar \frac{\partial \psi(x_1, x_2, t)}{\partial t} = H(x_1, x_2, t)\psi(x_1, x_2, t), \tag{30}$$

where $H(x_1, x_2, t)$ is given by Eq. (1). In Section 2 we showed that Eq. (1) can be transformed under a coordinate and canonical transformation into Eq. (20), which represents the sum of two uncoupled Hamiltonians of harmonic oscillators with time- dependent frequency $(\Omega_i(t))$ and unitary mass.

There exist several methods to obtain the solution of the Schrödinger equation (SE) for a harmonic oscillator with time-dependent frequency and unitary mass. Here we use the LR method[5, 6] method described below.

Consider the SE

$$i\hbar\frac{\partial\psi(q,t)}{\partial t} = H(t)\psi(q,t), \tag{31}$$

where $H(t)$ is given by

$$H(t) = \frac{p^2}{2} + \frac{1}{2}\Omega(t)q^2. \tag{32}$$

Lewis and Riesenfeld (LR)[5, 6] considered that there exists an invariant operator $I(t)$ which is hermitian $I = I^\dagger$, and satisfies the Heisenberg equation

$$\frac{dI}{dt} = \frac{\partial I}{\partial t} + \frac{1}{i\hbar}[I, H] = 0. \tag{33}$$

By applying Eq. (33) in the state $\psi(q,t)$ and using Eq. (31), LR find

$$i\hbar\frac{\partial}{\partial t}(I\psi(q,t)) = H(t)(I\psi(q,t)), \tag{34}$$

which implies that the action of the invariant operator on a Schrödinger wave function produces another solution of the Schrödinger equation. As pointed out by LR this result is valid for any invariant, even if the latter involves the operation of time differentiation. If the invariant does not involve time differentiation, then one is able to derive a simple and explicit rule for choosing the phases of the eigenstates of $I(t)$ such that these states eff themselves satisfy the Schrödinger equation. In Ref.,[6] LR assumed that $I(t)$ does not involve time differentiation.

Let $\phi_n(q,t)$ be the eingenfunctions of $I(t)$, which are assumed to form a complete orthonormal set ($\langle\phi_{n'}(q,t)|\phi_n(q,t)\rangle = \delta_{n'n}$) with time independent discrete eigenvalues, λ_n, i.e.,

$$I(t)\phi_n(q,t) = \lambda_n\phi_n(q,t). \tag{35}$$

According to LR[6] the solution $\psi_n(q,t)$ of the SE is related to the eigenfunction ($\phi_n(q,t)$) of I, by

$$\psi_n(q,t) = e^{i\alpha_n(t)}\phi_n(q,t), \tag{36}$$

where the phase function $\alpha_n(t)$ satisfies the equation

$$\hbar\frac{d\alpha_n(t)}{dt} = \langle\phi_n(q,t)|\left[i\hbar\frac{\partial}{\partial t} - H(t)\right]|\phi_n(q,t)\rangle. \tag{37}$$

Next, we have to find out $I(t)$ for $H(t)$ given by Eq. (31). Since $H(t)$ has a quadratic form, they considered $I(t) = A(t)q^2 + B(t)p^2 + C(t)(qp+pq)$, where $A(t)$,

$B(t)$ and $C(t)$ are functions of time. By using Eq. (33) and recalling that $I = I^\dagger$, LR found out that I for Eq. (32) reads

$$I = \frac{1}{2}\left[\left(\frac{q}{\rho}\right)^2 + (\rho p - \dot{\rho}q)^2\right],$$ (38)

where $q(t)$ satisfies the equation of motion

$$\ddot{q} + \Omega^2(t)q = 0,$$ (39)

while $\rho(t)$, which has to be a real function such that $I = I^\dagger$ holds, satisfies the Milne-Pinney (MP)[12-14] equation

$$\ddot{\rho} + \Omega^2(t)\rho = \frac{1}{\rho^3},$$ (40)

Under the unitary transformation[7,8]

$$\phi'_n(q, t) = U\phi_n(q, t),$$ (41)

where

$$U = \exp\left[-\left(i\frac{\dot{\rho}}{2\hbar\rho}\right)q^2\right],$$ (42)

becomes the eigenvalue equation for I'

$$I'\varphi_n(\xi) = \left[-\left(\frac{\hbar^2}{2}\right)\frac{\partial^2}{\partial\xi^2} + \left(\frac{\xi^2}{2}\right)\right]\varphi_n(\xi) = \lambda_n\varphi_n(\xi),$$ (43)

where $I' = UIU^\dagger$, $\xi = \frac{q}{\rho}$ and $\frac{\varphi_n(q,t)}{\rho^{1/2}} = \phi'_n(q,t)$. The factor $\rho^{1/2}$ warrants the normalization condition

$$\int \phi'^*_n(q,t)\phi'_n(q,t)dq = \int \varphi^\star_n(\xi)\varphi_n(\xi)d\xi = 1.$$ (44)

The solution of Eq. (42) corresponds to that of the time-independent harmonic oscillator with $\lambda_n = \left(n + \frac{1}{2}\right)\hbar$. Therefore, $\phi_n(q, t)$ is given by

$$\phi_n(q, t) = \left[\frac{1}{\pi^{1/2}\hbar^{1/2}n!2^n\rho}\right]^{1/2}\exp\left[\frac{i}{2\hbar}\left(\frac{\dot{\rho}}{\rho} + \frac{i}{\rho^2}\right)q^2\right] \times H_n\left[\frac{q}{\sqrt{\hbar}\rho}\right],$$ (45)

where H_n is the usual Hermite polynomial of order n.

Applying U to the right-hand side of Eq. (37) and after some algebra, we obtain

$$\alpha_n(t) = -\left(n + \frac{1}{2}\right) \int_{t_0}^{t} \frac{1}{\rho^2} dt', \tag{46}$$

and, using Eqs. (36) and (42) the exact solution of the SE reads

$$\psi_n(q, t) = e^{i\alpha_n(t)} \left[\frac{1}{\pi^{1/2} \hbar^{1/2} n! 2^n \rho}\right]^{1/2} \exp\left[\frac{i}{2\hbar} \left(\frac{\dot{\rho}}{\rho} + \frac{i}{\rho^2}\right) q^2\right] H_n\left[\frac{q}{\sqrt{\hbar}\rho}\right]. \tag{47}$$

Since $H_N = \frac{P_1^2}{2} + \frac{\Omega_1^2(t)Q_1^2}{2} + \frac{P_2^2}{2} + \frac{\Omega_2^2(t)Q_2^2}{2}$, an invariant for H_N is $I = I_1 + I_2$, where

$$I_i = \frac{1}{2} \left[\left(\frac{Q_i}{\rho_i}\right)^2 + (\rho_i P_i - \dot{\rho}_i Q_i)^2,\right] \quad (i = 1 \text{ and } 2), \tag{48}$$

and ρ_i satisfies the equation

$$\ddot{\rho}_i + \Omega_i^2(t)\rho_i = \frac{1}{\rho_i^3}, \quad (i = 1 \text{ and } 2). \tag{49}$$

The solution of the SE for H_N is $\psi_{n_1 n_2}(Q_1, Q_2, t) = \psi_{n_1}(Q_1, t)\psi_{n_2}(Q_2, t)$, i.e.,

$$\psi_{n_1 n_2}(Q_1, Q_2, t) = e^{i\alpha(n_1, n_2, t)} \sqrt{\frac{1}{\pi \hbar n_1! n_2! 2^{n_1 + n_2} \rho_1 \rho_2}} \tag{50}$$

$$\times \exp\left[\frac{i}{2\hbar} \left(\frac{\dot{\rho}_1}{\rho_1} + \frac{i}{\rho_1^2}\right) Q_1^2 + \frac{i}{2\hbar} \left(\frac{\dot{\rho}_2}{\rho_2} + \frac{i}{\rho_2^2}\right) Q_2^2\right] H_{n_1}\left[\frac{Q_1}{\sqrt{\hbar}\rho_1}\right] H_{n_2}\left[\frac{Q_2}{\sqrt{\hbar}\rho_2}\right],$$

where

$$\alpha(n_1, n_2, t) = -\left(n_1 + \frac{1}{2}\right) \int_0^t \frac{1}{\rho_1^2} dt' - \left(n_2 + \frac{1}{2}\right) \int_0^t \frac{1}{\rho_2^2} dt' \tag{51}$$

In terms of the original variables we have

$$\psi_{n_1 n_2}(x_1, x_2, t) = e^{i\alpha(n_1, n_2, t)} \sqrt{\frac{\sqrt{m_1 m_2}}{\pi \hbar n_1! n_2! 2^{n_1 + n_2} \rho_1 \rho_2}}$$

$$\times \exp\left[\frac{i}{2\hbar} \left(\frac{\dot{\rho}_1}{\rho_1} + \frac{i}{\rho_1^2}\right) (m_1 \cos^2 \theta x_1^2 + m_2 \sin^2 \theta x_2^2 - \sqrt{m_1 m_2} \sin(2\theta) x_1 x_2)\right]$$

$$\times \exp\left[\frac{i}{2\hbar} \left(\frac{\dot{\rho}_2}{\rho_2} + \frac{i}{\rho_2^2}\right) (m_1 \sin^2 \theta x_1^2 + m_2 \cos^2 \theta x_2^2 + \sqrt{m_1 m_2} \sin(2\theta) x_1 x_2)\right]$$

$$\times H_{n_1}\left[\frac{\sqrt{m_1} \cos \theta x_1 - \sqrt{m_2} \sin \theta x_2}{\sqrt{\hbar}\rho_1}\right] H_{n_2}\left[\frac{\sqrt{m_1} \sin \theta x_1 + \sqrt{m_2} \cos \theta x_2}{\sqrt{\hbar}\rho_2}\right]. \tag{52}$$

The factor $\sqrt{m_1 m_2}$ is introduced into Eq. (52) so that the normalization condition holds. It should be observed that for $k(t) = 0$, $\psi_{n_1 n_2}(x_1, x_2, t) = \psi_{n_1}(x_1, t)\psi_{n_2}(x_2, t)$.

The solution of the MP equation for the system: $m_1 = m_2 = m_0 e^{\gamma t}$, $\omega_1 = \omega_{01} e^{-\gamma t/2}$, $\omega_2 = \omega_{02} e^{-\gamma t/2}$ and $k = k_0$ is

$$\rho_j = \frac{1}{\sqrt{\Omega_j}} (j = 1, 2), \tag{53}$$

where Ω_1 and Ω_2 are given by Eqs. (26) and (27), respectively. Therefore, the wave function reads

$$\psi_{n_1 n_2}(x_1, x_2, t) = e^{i\alpha(n_1, n_2, t)} \sqrt{\frac{m_0 \exp{(\gamma t)}\sqrt{\Omega_1 \Omega_2}}{\pi \hbar n_1! n_2! 2^{n_1 + n_2} \rho_1 \rho_2}}$$

$$\times \exp\left[\frac{-\Omega_1}{2\hbar} m_0 \exp{(\gamma t)}(\cos^2 \theta x_1^2 + \sin^2 \theta x_2^2 - \sin(2\theta)x_1 x_2)\right]$$

$$\times \exp\left[\frac{-\Omega_2}{2\hbar} m_0 \exp{(\gamma t)}(\sin^2 \theta x_1^2 + \cos^2 \theta x_2^2 + \sin(2\theta)x_1 x_2)\right]$$

$$\times H_{n_1}\left[\sqrt{\frac{m_0 \Omega_1}{\hbar}} \exp{(\frac{\gamma t}{2})}(\cos \theta x_1 - \sin \theta x_2)\right]$$

$$\times H_{n_2}\left[\sqrt{\frac{m_0 \Omega_2}{\hbar}} \exp{(\frac{\gamma t}{2})}(\sin \theta x_1 + \cos \theta x_2)\right]. \tag{54}$$

The wave function of the system in the state $n_1 = n_2 = 0$ can be written as

$$\psi_{00}(x_1, x_2, t) = \exp\left[-\frac{i}{2}\left(\frac{1}{\rho_1^2} + \frac{1}{\rho_2^2}\right)\right] \sqrt{\frac{m_0 \exp{(\gamma t)}\sqrt{\Omega_1 \Omega_2}}{\pi \hbar}}$$

$$\times \exp\left[\frac{-\Omega_1}{2\hbar} m_0 \exp{(\gamma t)}(\cos^2 \theta x_1^2 + \sin^2 \theta x_2^2 - \sin(2\theta)x_1 x_2)\right]$$

$$\times \exp\left[\frac{-\Omega_2}{2\hbar} m_0 \exp{(\gamma t)}(\sin^2 \theta x_1^2 + \cos^2 \theta x_2^2 + \sin(2\theta)x_1 x_2)\right]. \tag{55}$$

The probability distribution (P_i) for the position of one oscillator is obtained by integrating the probability density over the other oscillator coordinate. The result for P_1 and P_2 are, respectively, given by

$$P_1(x_1, t) = \int_{-\infty}^{\infty} |\psi_{00}|^2 dx_2 = \sqrt{\frac{m_0 \exp{(\gamma t)}}{\pi \hbar(\rho_1^2 \cos^2 \theta + \rho_2^2 \sin^2 \theta)}} \exp\left[\frac{-m_0 \exp{(\gamma t)}x_1^2}{\hbar(\rho_1^2 \cos^2 \theta + \rho_2^2 \sin^2 \theta)}\right], \tag{56}$$

$$P_2(x_2, t) = \int_{-\infty}^{\infty} |\psi_{00}|^2 dx_1 = \sqrt{\frac{m_0 \exp(\gamma t)}{\pi \hbar (\rho_1^2 \sin^2 \theta + \rho_2^2 \cos^2 \theta)}} \exp\left[\frac{-m_0 \exp(\gamma t) x_2^2}{\hbar (\rho_1^2 \sin^2 \theta + \rho_2^2 \cos^2 \theta)}\right].$$

(57)

These are Gaussian probability distributions with standard deviation

$$\sigma_1^{(0)} = \sqrt{\frac{\hbar (\rho_1^2 \cos^2 \theta + \rho_2^2 \sin^2 \theta)}{2 m_0 \exp(\gamma t)}},$$

(58)

$$\sigma_2^{(0)} = \sqrt{\frac{\hbar (\rho_1^2 \sin^2 \theta + \rho_2^2 \cos^2 \theta)}{2 m_0 \exp(\gamma t)}}.$$

(59)

For the time-independent single harmonic oscillator the standard deviation reads

$$\sigma^{(0)} = \sqrt{\frac{\hbar}{2m\omega}}.$$

(60)

By comparing Eqs. (58), (59) and (60) we observe that the standard deviation (position uncertainty) decreases with increasing time, reflecting the fact that the particle becomes more localized as time goes on. Also, the coupling decreases the position uncertainty. Conversely, as shown in Ref. 17 the net effect of the coupling is to increase the momentum uncertainty in such way that the uncertainty product remains constant.

4. Concluding remarks

In this paper, we used a coordinate and momentum transformations along with a canonical transformation to write the Hamiltonian of the time-dependent coupled harmonic oscillators as the sum of two Hamiltonians of time-dependent uncoupled simple harmonic oscillators with modified time-dependent frequencies and unitary masses. To do so, we impose a constraint on the possible values of m_i , ω_i and $k(t)$ to be considered. Therefore we obtained the classical and quantum solutions of the system $m_1 = m_2 = m_0 e^{\gamma t}$, $\omega_1 = \omega_{01} e^{-\gamma t/2}$, $\omega_2 = \omega_{02} e^{-\gamma t/2}$ and $k = k_0$. By solving the equations of motion, we observed that the classical solutions correspond to the superposition of the motion of two simple harmonic oscillators with different frequencies, phases and exponentially decaying amplitudes. This decay is owing to the fact that the mass for both oscillators increases exponentially with time.

The quantum solutions were obtained by using the LR invariant method. In terms of the original variables (x_1, x_2), the exact wave function is given by Eq. (52). In the limit $k(t) = 0$, $\psi_{n_1 n_2}(x_1, x_2, t)$ corresponds exactly to $\psi_{n_1 n_2}(x_1, x_2, t) = \psi_{n_1}(x_1, t) \psi_{n_2}(x_2, t)$. The wave function is expressed in terms of ρ_1 and ρ_2, which are solutions of the respective MP equation, namely $\rho_j = \frac{1}{\sqrt{\Omega_j}}$ $(j = 1$ and $2)$

where Ω_1 and Ω_2 are given by Eqs. (26) and (27), respectively. We also calculated the probability distribution and the standard deviation in the state $n_1 = n_2 = 0$. While the coupling reduces the standard deviation in position, it increases the standard deviation in momentum. However, although the system depends on time, its uncertainty product does not.

Acknowledgments

This work was supported by national council of scientific and technological development (CNPq). Founded by the brazilian ministry of science and technology (MCT).

References

1. K. M. Yeon, C. I. Um, W. H. Kahng and T. F. George, Propagators for driven coupled harmonic oscillators, *Phys. Rev. A* **38**, 6224 (1988).
2. A. V. Zhukov and P. N. Zhokova, Quantum theory for coupled harmonic oscillators with exponentially changing masses, *J. Phys. A* **32**, 1779 (1999).
3. J. S. Wang, J. Feng and M. S. Zhang, Quantum fluctuations of a non-dissipative mesoscopic inductance coupling circuit in a displaced squeezed Fock state, *Phys. Lett. A* **281**, 341 (2001).
4. S. Zhang, J. R. Choi, C. I. Um and K. H. Yeon, Quantum uncertainties of mesoscopic capacitance coupled circuit, *Phys. Lett. A* **289**, 257 (2001).
5. H. R. Lewis, Class of exact invariants for classical and quantum time-dependent harmonic oscillators, *J. Math. Phys.*, **9**, 1976 (1968).
6. H. R. Lewis and W. B. Riesenfeld, An exact quantum theory of the time-dependent harmonic oscillator and of a charged particle in a time-dependent electromagnetic field, *J. Math. Phys.*, **10**, 1458 (1969).
7. J. R. Ray and J. L. Reid, Noether's theorem, time-dependent invariants and nonlinear equations of motion, *J. Math. Phys.*, **20**, 2054 (1979).
8. J. R. Ray and J. L. Reid, Ermakov systems, nonlinear superposition, and solutions of nonlinear equations of motion, *J. Math. Phys.*, **21**, 1583 (1980).
9. S. Zhang, J. R. Choi, C; I. Um and K. H. Yeon, Quantum squeezing effect of mesoscopic capacitance-inductance-resistance coupled circuit, *Phys. Lett. A* **294**, 319 (2002).
10. S. Menouar, M. Maamache and J. R. Choi, The time-dependent coupled oscillator model for the motion of a charged particle in the presence of a time-varying magnetic field, *Physica Scripta* **82**, 065004 (2010).
11. S. Menouar, M. Maamache and J. R. Choi, An alternative approach to exact wave functions for time-dependent coupled oscillator model of charged particle invariable magnetic field, *Ann. Phys.* **325**, 1708 (2010).
12. W. E. Milne, The numerical determination of characteristic numbers, *Phys. Rev.* **35**, 863 (1930).
13. E. Pinney, The nonlinear differential equation $y'' + p(x)y + cy^{-3} = 0$, *Proc. of the American Math. Soc.* **1**, 681 (1950).
14. J. F. Cariñena and Javier de Lucas, A nonlinear superposition rule for solutions of the Milne-Pinney equation, *Phys. lett. A* **372**, 5385 (2008).
15. P. Caldirola, Forze non conservative nella meccanica quantistica, *Nuovo Cimento* **18**, 393 (1941).

16. E. Kanai, On the quantization of the dissipative systems, *Prog. of Theor. Phys.* **3**, 440 (1948).

17. D. X. Macedo and I. Guedes, Time-dependent coupled harmonic oscillators, *J. Math. Phys.* **53**, 052101 (2012).

Chapter 7

Event-Based Simulation of Quantum Physics Experiments

Kristel Michielsen[*,†,§] and Hans De Raedt[‡,¶]

*Institute for Advanced Simulation, Jülich Supercomputing Centre,
Forschungszentrum Jülich, D-52425 Jülich, Germany
and
†RWTH Aachen University, D-52056 Aachen, Germany
‡Department of Applied Physics, Zernike Institute for Advanced Materials,
University of Groningen, Nijenborgh 4,
NL-9747 AG Groningen, The Netherlands

We review an event-based simulation approach which reproduces the statistical distributions of wave theory not by requiring the knowledge of the solution of the wave equation of the whole system but by generating detection events one-by-one according to an unknown distribution. We illustrate its applicability to various single photon and single neutron interferometry experiments and to two Bell-test experiments, a single-photon Einstein-Podolsky-Rosen experiment employing post-selection for photon pair identification and a single-neutron Bell test interferometry experiment with nearly 100% detection efficiency.

1. Introduction

The statistical properties of a vast number of laboratory experiments with individual entities such as electrons, atoms, molecules, photons, ... can be extremely well described by quantum theory. The mathematical framework of quantum theory allows for a straightforward calculation of numbers which can be compared with experimental data as long as these numbers refer to statistical averages of measured quantities, such as for example an interference pattern, the specific heat and magnetic susceptibility.

However, as soon as an experiment records individual clicks of a detector which contribute to the statistical average of a quantity then a fundamental problem appears. Quantum theory provides a recipe to compute the frequencies for observing

---

§k.michielsen@fz-juelich.de
¶h.a.de.raedt@rug.nl

events but it does not account for the observation of the individual events themselves, a manifestation of the quantum measurement problem.[1,2] Examples of such experiments are single-particle interference experiments in which the interference pattern is built up by successive discrete detection events and Bell-test experiments in which two-particle correlations are computed as averages of pairs of individual detection events recorded at two different detectors and seen to take values which correspond to those of the singlet state in the quantum theoretical description.

An intriguing question to be answered is why individual entities which do not interact with each other can exhibit the collective behavior that gives rise to the observed interference pattern and why two particles, which only interacted in the past, after individual local manipulation and detection can show correlations corresponding to those of the singlet state. Since quantum theory postulates that it is fundamentally impossible to go beyond the description in terms of probability distributions, an answer in terms of a cause-and-effect description of the observed phenomena cannot be given within the framework of quantum theory.

We provide an answer by constructing an event-based simulation model that reproduces the statistical distributions of quantum (and Maxwell's) theory without solving a wave equation but by modeling physical phenomena as a chronological sequence of events whereby events can be actions of an experimenter, particle emissions by a source, signal generations by a detector, interactions of a particle with a material and so on.[3–5] The underlying assumption of the event-based simulation approach is that current scientific knowledge derives from the discrete events which are observed in laboratory experiments and from relations between those events. Hence, the event-based simulation approach concerns what we can say about these experiments but not what "really" happens in Nature. This underlying assumption strongly differs from the premise that the observed discrete events are signatures of an underlying objective reality which is mathematical in nature.

The general idea of the event-based simulation method is that simple rules define discrete-event processes which may lead to the behavior that is observed in experiments. The basic strategy in designing these rules is to carefully examine the experimental procedure and to devise rules such that they produce the same kind of data as those recorded in experiment, while avoiding the trap of simulating thought experiments that are difficult to realize in the laboratory. Evidently, mainly because of insufficient knowledge, the rules are not unique. Hence, the simplest rules can be used until a new experiment indicates otherwise. On the one hand one may consider the method being entirely classical since it only uses concepts of the macroscopic world, but on the other hand one could consider the method being nonclassical because some of the rules are not those of classical Newtonian dynamics.

Obviously, using trial and error to find discrete-event rules that reproduce experimental results is unlikely to be successful. Instead, we started our search for useful rules by asking ourselves the question "by what kind of discrete-event rule should a beam splitter operate in order to mimic the build-up, event-by-event, of the interference pattern observed in the single-photon Mach-Zehnder experiments performed by Grangier *et al.*[6]?" The simplest rule (discussed below) that performs this task seems to be rather generic in the sense that it can be used to construct discrete-event processes that reproduce the results of many interference experiments. Of course, for some experiments, the simple rule is "too simple" and more sophisticated, backwards compatible variants are required. However, the guiding principle for designing the latter is the same as for the simple rule.

The event-based approach has successfully been used for discrete-event simulations of the single beam splitter and Mach-Zehnder interferometer experiments of Grangier *et al.*[6] (see Refs. 3,7,8), Wheeler's delayed choice experiment of Jacques *et al.*[9] (see Refs. 3,10,11), the quantum eraser experiment of Schwindt *et al.*[12] (see Refs. 3,13,14), two-beam single-photon interference experiments and the single-photon interference experiment with a Fresnel biprism of Jacques *et al.*[15] (see Refs. 3,4,16), quantum cryptography protocols (see Ref. 17), the Hanbury Brown-Twiss experiment of Agafonov *et al.*[18] (see Refs. 3,19,20), universal quantum computation (see Refs. 21,22), Einstein-Podolsky-Rosen-Bohm (EPRB)-type of experiments of Aspect *et al.*[23,24] and Weihs *et al.*[25] (see Refs. 3,4,26–31), the propagation of electromagnetic plane waves through homogeneous thin films and stratified media (see Refs. 3,32), and neutron interferometry experiments (see Refs. 4,5).

In this paper, we review the applicability of the event-based simulation method to various single-photon and single-neutron interferometry experiments and to Bell-test experiments. The paper is organized as follows. Section 2 is devoted to the single-particle two-slit experiment, one of the most fundamental experiments in quantum physics. We first discuss Feynman's thought experiment, demonstrating single-electron interference, and briefly review its laboratory realizations. We then describe the two-beam experiment with single-photons, a variant of Young's double slit experiment. It is seen that for these single-particle interference experiments quantum theory gives a recipe to compute the observed interference pattern after many detection events are registered, but quantum theory does not account for the one-by-one build-up process of the pattern in terms of the individual detection events. Hence, as formulated in section 3, the challenge is to come up with a set of rules which allow to produce detection events with frequencies which agree with a given distribution (in this particular case a two-slit interference pattern) without these rules referring, in any way, to the distribution itself. The event-based simulation method solves this challenging problem by modeling various physical phenomena as a chronological sequence of different events, such as actions of the experimenter, particles emitted by a source, signals generated by a detector and so on. In section 4 we explain the basis of the event-based simulation method by specifying rules which allow to reproduce the results of the quantum theoretical description of the idealized Stern-Gerlach experiment and of a single-photon experiment with a linearly birefringent crystal demonstrating Malus' law, without making any use of quantum theoretical concepts. In this section, we also discuss the efficiency of two types of single-particle detectors used in the event-based simulation method. In section 5 we show that a similar set of rules can be used to simulate single-particle interference. We demonstrate this on the basis of the single-photon two-beam experiment thereby also exactly simulating Feynman's thought experiment, the Mach-Zehnder interferometer experiment, Wheeler's delayed choice experiment and a single-neutron interferometry experiment with a Mach-Zehnder type of interferometer. We explain why the event-based simulation method can produce interference without solving a wave problem. Section 6 is devoted to the event-based simulation of EPRB-type of experiments with correlated photon pairs and with neutrons with correlated spatial and spin degrees of freedom. Since both experiments are Bell-test experiments testing whether or not a Bell-CHSH (Clauser-Horne-Shimony-Holt) inequality can be violated, we also elaborate on the conclusions that can be drawn from such a violation. For both experiments we explain why the event-based model, a classical causal model, can produce the results of quantum theory. A discussion is given in section 7.

2. Two-slit and two-beam experiments

One of the most fundamental experiments in quantum physics is the single-particle double-slit experiment. Feynman stated that the phenomenon of electron diffraction by a double-slit structure is "impossible, *absolutely* impossible, to explain in any classical way, and has in it the heart of quantum mechanics. In reality it contains the only mystery."[33] While Young's original double-slit experiment helped establish the wave theory of light,[34] variants of the experiment over the years with electrons (see below), single photons (see below), neutrons,[35,36] atoms[37,38] and molecules[39-41] helped the development of ideas on concepts such as wave-particle duality in quantum theory.[2]

Two prevailing variants of the double-slit experiments can be recognized, one consists of a source S and a screen with two apertures and another one consists of a source S and a biprism. The first one is a real two-slit experiment in which the two slits can be regarded as two virtual sources S_1 and S_2, the latter one is a two-beam experiment which can also be replaced by a system with two virtual sources S_1 and S_2.[42] In contrast to the two-slit experiment in which diffraction or scattering and interference phenomena play a role, the phenomenon of diffraction or scattering is absent in the two-beam experiment, except for the diffraction or scattering at the sources themselves.

A brief note on the difference in usage of the words diffraction, scattering and interference is here in place. Feynman mentioned in his lecture notes that "no-one has ever been able to define the difference between interference and diffraction satisfactorily. It is just a question of usage, and there is no specific, important physical difference between them."[43] In classical optics, diffraction is the effect of a wave bending as it passes through an opening or goes around an object. The amount of bending depends on the relative dimensions of the object or opening compared to the wavelength of the wave. Interference is the superposition of two or more waves resulting in a new wave pattern. Therefore a double-slit, as well as a single-slit structure illuminated by (classical) light yields an interference (or diffraction) pattern due to diffraction *and* interference. In principle, diffraction and interference are phenomena observed only with waves. However, an interference pattern identical in form to that of classical optics can be observed by collecting many detector spots or clicks which are the result of electrons, photons, neutrons, atoms or molecules travelling one-by-one through a double-slit structure. In these experiments the so-called interference pattern is the statistical distribution of the detection events (spots at or clicks of the detector). Hence in these particle-like experiments, only the correlations between detection events reveal interference. Misleadingly this interference pattern is often called a diffraction pattern in analogy with classical optics where both the phenomena of diffraction and interference are responsible for the resulting pattern. In the particle-like experiment it would be better to replace the word diffraction by scattering because scattering refers to the spreading of a beam of particles (or a beam of rays) over a range of directions as a result of collisions with other particles or objects. In what follows we use the term interference pattern for the statistical distribution of detection events.

2.1. *Two-slit experiment with electrons*

In 1964 Feynman described a thought experiment consisting of an electron gun emitting individual electrons in the direction of a thin metal plate with two slits in it behind which is placed a movable detector.[33] Feynman made the following observations:

- Sharp identical "clicks" which are distributed erratically, are heard from the detector.
- The probability $P_1(x)$ or $P_2(x)$ of arrival, through one slit with the other slit closed, at position x is a symmetric curve with its maximum located at the centre position of the open slit.
- The probability $P_{12}(x)$ of arrival through both slits looks like the intensity of water waves which propagated through two holes thereby forming a so-called "interference pattern" and looks completely different from the curve $P_1(x) + P_2(x)$, a curve that would be obtained by repeating the experiment with bullets.

which lead him to the conclusions:

- Electrons arrive at the detector in identical "lumps", like particles.
- The probability of arrival of these lumps is distributed like the distribution of intensity of a wave propagating through both holes.

- It is in this sense that an electron behaves "sometimes like a particle and sometimes like a wave".

Note that Feynman made his reasoning with probabilities $P_1(x)$, $P_2(x)$, $P_{12}(x)$, which he said to be proportional to the average rate of clicks $N_1(x)$, $N_2(x)$, $N_{12}(x)$. However, one cannot simply add $P_1(x)$ and $P_2(x)$ and compare the result with $P_{12}(x)$ because these are probabilities for different conditions (different "contexts"), namely only slit 1 open, only slit 2 open and both slits 1 and 2 open, respectively.[2] Hence, no conclusions can be drawn from making the comparison between $P_{12}(x)$ and $P_1(x) + P_2(x)$.

Although Feynman wrote "you should not try to set up this experiment" because "the apparatus would have to be made on an impossibly small scale to show the effects we are interested in", advances in (nano)technology made possible various laboratory implementations of his fundamental thought experiment. The first electron interference pattern obtained with an electron-biprism, the analog of a Fresnel biprism in optics, was reported in 1955.[44,45] In 1961 Jönsson performed the first electron interference experiment with multiple (up to five) slits in the micrometer range.[46] However, these were not single-electron interference experiments since there was not just one electron in the apparatus at any one time. The first real single-electron interference experiments that were conducted were electron-biprism experiments (for a review see Refs. 47,48) in which single electrons either pass to the left or to the right of a conducting wire (there are no real slits in this type of experiments).[49–51] In these experiments the interference pattern is built up from many independent detection events. Electron-electron interaction plays no role in the interference process since the electrons pass the wire one-by-one. More recently, single-electron interference experiments have been demonstrated with one-, two-, three and four slits fabricated by focused ion beam milling.[52–54] However, in these experiments only the final recorded electron intensity is shown. In a follow-up single-electron two-slit experiment a fast-readout pixel detector was used which allows the measurement of the distribution of the electron arrival times and the observation of the build-up of the interference pattern by individual detection events.[55] Hence, this experiment comes very close to Feynman's thought experiment except that the two electron distributions for one slit open and the other one closed are not measured. Note that one of these distributions was measured in Ref. 52 by a non-reversible process of closing one slit and without using the fast-readout pixel detector. Very recently, it has been reported that a full realization of Feynman's thought experiment has been performed.[56] In this experiment a movable mask is placed behind the double-slit structure to open/close the slits. Unfortunately, the mask is positioned behind the slits and not in front of them, so that all electrons always encounter a double-slit structure and are filtered afterwards by the mask. Hence, one could say that anno 2014 Feynman's thought experiment has yet to be performed.

2.2. *Two-beam experiment with photons*

Another interesting variant of Young's double slit experiment involves a very dim light source so that on average only one photon is emitted by the source at any time. Inspired by Thomson's idea that light consists of indivisible units that are more widely separated when the intensity of light is reduced,[57] in 1909 Taylor conducted an experiment with a light source varying in strength and illuminating a needle thereby demonstrating that the diffraction pattern observed with a feeble light source (exposure time of three months) was as sharp as the one obtained with an intense source and a shorter exposure time.[58] In 1985, a double-slit experiment was performed with a low-pressure mercury lamp and neutral density filters to realize a very low-light level.[59] It was shown that at the start of the measurement bright dots appeared at random positions on the detection screen and that after a couple of minutes an interference pattern appeared. Demonstration versions of double-slit experiments illuminated by strongly attenuated lasers are reported in Refs. 60,61 and in figure 1 of Ref. 62. However, attenuated laser sources are imperfect single-photon sources. Light from these sources attenuated to the single-photon level never antibunches, which means that the anticorrelation parameter $\alpha \geq 1$. For a real single-photon source $0 < \alpha < 1$. In 2005, a variation of Young's experiment was performed with a Fresnel biprism and a single-photon source based on the pulsed, optically excited photoluminescence of a single N-V colour centre in a diamond nanocrystal.[15] In this two-beam experiment there is always only one photon between the source and the detection plane. Is was observed that the interference pattern gradually builds up starting from a couple of dots spread over the screen for small exposure times. A time-resolved two-beam experiment has been reported in Refs. 63,64. Recently, a temporally and spatially resolved two-beam experiment has been performed with entangled photons, providing insight in the dynamics of the build-up process of the interference pattern.[65]

2.3. *The experimental observations and their quantum theoretical description*

The common observation in these single-particle interference experiments, where "single particle" can be read as electron, photon, neutron, atom or molecule, is that individual detection events gradually build up an interference pattern and that the final interference pattern can be described by wave theory. In trying to give a pictorial (cause-and-effect) view of what is going on in these experiments, it is commonly assumed that there is a one-to-one correspondence between an emission event, "the departure of a single particle from the source" and a detection event, "the arrival of the single particle at the detector". This assumption might be wrong. The only conclusion that can be drawn from the experiments is that there is some relation between the emission and detection events.

In view of the quantum measurement problem,[1,2,66] a cause-and-effect description of the observed phenomena is unlikely to be found in the framework of quantum theory. Quantum theory provides a recipe to compute the frequencies for observing events and thus to compute the final interference pattern which is observed after the experiment is finished. However, it does not account for the observation of the individual detection events building up the interference pattern. In fact quantum theory postulates that it is fundamentally impossible to go beyond the description in terms of probability distributions. Of course, one could simply use pseudo-random numbers to generate events according to the probability distribution that is obtained by solving the time-dependent Schrödinger equation. However, that is not the problem one has to solve as it assumes that the probability distribution of the quantum mechanical problem is known, which is exactly the knowledge that one has to generate without making reference to quantum theory. If we would like to produce, event-by-event, the interference pattern from Maxwell's theory and do not want to generate events according to the known intensity function we would face a similar problem.

3. Theoretical challenge and paradigm shift

In general, the challenge is the following. Given a probability distribution of observing events, construct an algorithm which runs on a digital computer and produces events with frequencies which agree with the given distribution without the algorithm referring, in any way, to the probability distribution itself. Traditionally, the behavior of systems is described in terms of mathematics, making use of differential or integral equations, probability theory and so on. Although that this traditional modeling approach has been proven to be very successful it does not seem capable of tackling this challenge. This challenge requires something as disruptive as a paradigm shift. In scientific fields different from (quantum) optics or quantum mechanics in general, a paradigm shift has been realized in terms of a discrete-event approach to describe the often very complex collective behavior of systems with a set of very simple rules. Examples of this approach are the lattice Boltzmann model to describe the flow of (complex) fluids and the cellular automata of Wolfram.[67]

We have developed a discrete-event simulation method to solve the above mentioned challenging problem by modeling physical phenomena as a chronological sequence of events whereby events can be actions of the experimenter, particles emitted by a source, signals generated by a detector, particles impinging on material, and so on. The basic idea of the simulation method is to try to invent an algorithm which uses the same kind of events (data) as in experiment and reproduces the statistical results of quantum or wave theory without making use of this theory. An overview of the method and its applications can be found in Refs. 3–5. The method provides an "explanation" and "understanding" of what is going on in terms of elementary events, logic and arithmetic. Note that a cause-and-effect simulation on a digital computer is a "controlled experiment" on a macroscopic device which is logically equivalent to a mechanical device. Hence, an event-by-event simulation that reproduces results of quantum theory shows that there exists a macroscopic, mechanical model that mimics the underlying physical phenomena. This is completely in agreement with Bohr's answer "There is no quantum world. There is only an abstract quantum mechanical description. It is wrong to think that the task of physics is to find out how nature is. Physics concerns what we can say about nature." to the question whether the algorithm of quantum mechanics could be considered as somehow mirroring an underlying quantum world.[68] Although widely circulated, these sentences are reported by Petersen[68] and there is doubt that Bohr actually used this wording.[69]

4. Event-by-event simulation method

4.1. *Stern-Gerlach experiment*

We explain the basics of the event-by-event simulation method using the observations made in the Stern-Gerlach experiment.[70] The experiment shows that a beam of silver atoms directed through an inhomogeneous magnetic field splits into two components. The conclusion drawn by Gerlach and Stern is that, independent of any theory, it can be stated, as a pure result of the experiment, and as far as the exactitude of their experiments allows them to say so, that silver atoms in a magnetic field have only two discrete values of the component of the magnetic moment in the direction of the field strength; both have the same absolute value with each half of the atoms having a positive and a negative sign respectively.[71]

In quantum theory, the stationary state of the two-state system, which is the representation of the statistical experiment, is described by the density matrix $\rho = (1 + \mathbf{S} \cdot \sigma)/2$, where $\sigma = (\sigma^x, \sigma^y, \sigma^z)$ denotes the Pauli vector and $\mathbf{S}$ denotes the average direction of magnetic moments. The average measured magnetic moment in the direction $\mathbf{a}$ is given by $\mathbf{S} \cdot \mathbf{a} = \mathrm{Tr}\rho\sigma \cdot \mathbf{a}$.

The fundamental question is how to go from the averages to the events observed in the experiment. Application of Born's rule gives the probability to observe an atom in the beam (anti-)parallel to the direction $\mathbf{a}$

$$P(w|\mathbf{S} \cdot \mathbf{a}) = \frac{1 + w\mathbf{S} \cdot \mathbf{a}}{2}, \tag{1}$$

where $w = +1$ ($w = -1$) refers to the beam parallel (anti-parallel) to $\mathbf{a}$.

Given the probability in Eq. (1) the question is how to generate a sequence of "true" random numbers $w_1, w_2, \ldots, w_N$, each taking values ± 1, such that $\sum_{n=1}^{N} w_n/N \approx \mathbf{S} \cdot \mathbf{a}$. Probability theory postulates that such a procedure exists but is silent about how the procedure should look like. In practice one could use a probabilistic processor, a device which responds to and processes input in a probabilistic way, employing pseudo-random number generators to generate a uniformly distributed pseudo-random number $0 < r_n < 1$ to produce $w_n = +1$ if

$r_n < (1 + \mathbf{S} \cdot \mathbf{a})/2$ and $w_n = -1$ otherwise. Repeating this procedure N times gives $\sum_{n=1}^{N} w_n/N \approx \mathbf{S} \cdot \mathbf{a}$. However, the form of $P(w|\mathbf{S} \cdot \mathbf{a}) = (1 + w\mathbf{S} \cdot \mathbf{a})/2$ with $w = \pm 1$ is postulated and the procedure is deterministic thereby only giving the illusion of randomness to everyone who does not know the details of the algorithm and the initial state of the pseudo-random generator. Hence, we accomplished nothing and the question is whether we can do better than by using this probabilistic processor.

Let us consider a deterministic processor, a deterministic learning machine (DLM),[8,72] that receives input in the form of identical numbers

$$0 \leq u_n \equiv u = (1 + \mathbf{S} \cdot \mathbf{a})/2 \leq 1, \tag{2}$$

for $n = 1, \ldots, N$. The processor has an internal state represented by a variable $0 \leq v_n \leq 1$ which adapts to the received input u in a manner such that the difference with the input is minimal, namely

$$v_n = \gamma v_{n-1} + (1 - \gamma)\Delta_n, \tag{3}$$

where $\Delta_n = \Theta(|\gamma v_{n-1} + (1 - \gamma) - u| - |\gamma v_{n-1} - u|)$ with $\Theta(.)$ denoting the unit step function taking only the value 0 or 1 and $0 \leq \gamma < 1$ is a learning parameter controlling both the speed and accuracy with which the processor learns the input value u. The initial value v_0 of the internal state is chosen at random. The output numbers generated by the processor are

$$w_n = 2\Delta_n - 1 = \pm 1. \tag{4}$$

In general the behavior of the deterministic processor defined by Eq. (3) is difficult to analyze without a computer. However, the operation of the processor can be easily translated in simple computer code:

```
u1 = gamma * y
u2 = u1 + 1 - gamma
if(abs(v - u1) < abs(v - u2))then
w = -1
u = u1
else
w = +1
u = u2
endif                                                    (5)
```

Also without computer this code allows getting a quick notion on how the internal state of the processor adapts to the input. Taking as an example $u = 5/8$, $\gamma = 0.5$ and $v_n = 4/8$ gives $v_{n+1} = 6/8$, $v_{n+2} = 7/8$, $v_{n+3} = 7/16$, ... From this step-by-step analysis it can be seen how v_n comes closer to u, goes further away from it to come closer again in a next step and how v_n keeps oscillating around u in the stationary regime. A detailed mathematical analysis of the dynamics of the processor defined by the rule Eq. (3) is given in Ref. 73. For $\gamma \to 1^-$ we find that $\sum_{n=1}^{N} w_n/N \approx 2u - 1 = \mathbf{S} \cdot \mathbf{a}$.

In conclusion, we designed an event-by-event process which can reproduce the results of the quantum theoretical description of the idealized Stern-Gerlach experiment without making use of any quantum theoretical concepts. The strategy employed by the processor is to minimize the distance between two numbers thereby "learning" the input number. Hence, at least one of the results of quantum theory seems to emerge from an event-based process, a dramatic change in the paradigm of the quantum science community.

4.2. *Malus' law*

The important question is whether this event-based approach can also be applied to other experiments which up to now are exclusively described in terms of wave or quantum theory. To scrutinize this question we consider a basic optics experiment with a linearly birefringent crystal, such as calcite acting as a polarizer. A beam of linearly polarized monochromatic light impinging on a calcite crystal along a direction not parallel to the optical axis of the crystal is split into two beams travelling in different directions and having orthogonal polarizations. The two beams are referred to as the ordinary and extraordinary beam, respectively.[42] The intensity of the beams is given by Malus' law, which has experimentally been established in 1810,

$$I_o = I \sin^2(\psi - \phi), \quad I_e = I \cos^2(\psi - \phi), \tag{6}$$

where I, I_o and I_e are the intensities of the incident, ordinary and extraordinary beam, respectively, ψ is the polarization of the incident light and ϕ specifies the orientation of the crystal.[42] Observations in single-photon experiments show that Malus' law is also obeyed at the single-photon level.

In the quantum theoretical description of these single-photon experiments in which the photons are detected one-by-one in either the ordinary beam (represented by a detection event $w = 0$) or in the extraordinary beam (represented by a detection event $w = 1$) it is postulated that the polarizer sends a photon to the extraordinary direction with probability $\cos^2(\psi - \phi)$ and to the ordinary direction with probability $\sin^2(\psi - \phi)$. Hence, quantum theory postulates that $\lim_{N \to \infty} \sum_{n=1}^{N} w_n/N \to \cos^2(\psi - \phi)$.

Following a procedure similar to that of the Stern-Gerlach experiment it is obvious that we can construct a simple probabilistic processor employing pseudo-random numbers to generate a uniform random number $0 < r_n < 1$ and send out a $w_n = 0$ ($w_n = 1$) event if $\cos^2(\psi - \phi) \leq r_n$ ($\cos^2(\psi - \phi) > r_n$) so that after repeating this procedure N times we indeed have $\lim_{N \to \infty} \sum_{n=1}^{N} w_n/N \to \cos^2(\psi - \phi)$. However, again, by doing this we accomplished nothing because Malus' law has been postulated from the start in the form $P(w|\psi - \phi) = w \cos^2(\psi - \phi) + (1 - w) \sin^2(\psi - \phi)$ with $w = 0, 1$. Moreover, this probabilistic processor has a relatively poor performance[73] and therefore in what follows we design and analyze a much more efficient DLM that generates events according to Malus' law.

The DLM mimicking the operation of a polarizer has one input channel, two output channels and one internal vector with two real entries. The DLM receives as input, a sequence of angles ψ_n for $n = 1, \ldots, N$ and knows about the orientation of the polarizer through the angle ϕ. Using rotational invariance, we represent these input messages by unit vectors

$$\mathbf{u}_n = (u_{0,n}, u_{1,n}) = (\cos(\psi_n - \phi), \sin(\psi_n - \phi)). \tag{7}$$

Instead of the random number generator that is part of the probabilistic processor, the DLM has an internal degree of freedom represented by the unit vector $\mathbf{v}_n = (v_{0,n}, v_{1,n})$. The direction of the initial internal vector $\mathbf{v}_0$ is chosen at random. As the DLM receives input data, it updates its internal state. The update rules are defined by

$$v_{0,n} = \pm\sqrt{1 + \gamma^2(v_{0,n-1}^2 - 1)}, \quad v_{1,n} = \gamma v_{1,n-1}, \tag{8}$$

corresponding to the output event $w_n = 0$ and

$$v_{0,n} = \gamma v_{0,n-1}, \quad v_{1,n} = \pm\sqrt{1 + \gamma^2(v_{1,n-1}^2 - 1)}, \tag{9}$$

corresponding to the output event $w_n = 1$. The parameter $0 < \gamma < 1$ controls the learning process of the DLM. The $\pm$-sign takes care of the fact that the DLM has to decide between two quadrants. The DLM selects one of the four possible outcomes for $\mathbf{v}_n = (v_{0,n}, v_{1,n})$ by minimizing the cost function defined by

$$C = -\mathbf{v}_n \cdot \mathbf{u}_n = -(v_{0,n}u_{0,n} + v_{1,n}u_{1,n}). \tag{10}$$

Obviously, the cost C is small (close to -1), if the vectors $\mathbf{u}_n$ and $\mathbf{v}_n$ are close to each other. In conclusion, the DLM generates output events $w_n = 0, 1$ by minimizing the distance between the input vector and its internal vector by means of a simple, deterministic decision process.

In general, the behavior of the DLM defined by the rules Eqs. (8)–(10) is difficult to analyze without using a computer. However, for a fixed input vector $\mathbf{u}_n = (u_0, u_1)$ for $n = 1, \ldots, N$, the DLM will minimize the cost Eq. (10) by rotating its internal vector $\mathbf{v}_n$ towards $\mathbf{u}_n$ but $\mathbf{v}_n$ will not converge to the input vector $\mathbf{u}_n$ and will keep oscillating about $\mathbf{u}_n$. This is the stationary state of the machine. An example of a simulation is given in Fig. 1. Once the DLM has reached the stationary state the number of $w_n = 0$ output events divided by the total number of output events is $\cos^2(\psi_n - \phi)$ and thus in agreement with Malus' law if we interpret the $w_n = 0$ output events as corresponding to the extraordinary beam. Note that the details of the approach to the stationary state depend on the initial value of the internal vector $\mathbf{v}_0$, but the properties of the stationary state do not. A detailed stationary-state analysis is given in Ref. 72.

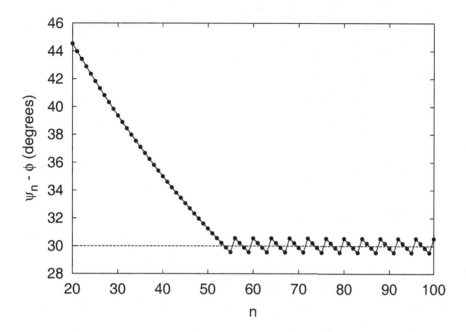

Fig. 1. The angle $\psi_n - \phi$ representing the internal vector $\mathbf{v}_n$ of the DLM defined by Eqs. (8) and (10) as a function of the number of events n. The input events are vectors $\mathbf{u}_n = (\cos 30°, \sin 30°)$. The direction of the initial internal vector $\mathbf{v}_0$ is chosen at random. In this simulation $\gamma = 0.99$. For $n > 60$ the ratio of the number of 0 events to 1 events is $1/3$, which is $(\sin 30° / \cos 30°)^2$. Data for $1 \leq n < 20$ lie on the decaying line but have been omitted to show the oscillating behavior more clearly. Lines are guides to the eye.

4.3. *Single particle detection*

In the event-based simulation of the Stern-Gerlach experiment and of the experiment demonstrating Malus' law the two-valued output events w_n $(n = 1, \ldots, N)$ can be processed by two detectors placed behind the DLM modeling the Stern-Gerlach magnet and the calcite crystal, respectively. It can be easily seen that in these two experiments the only operation the detectors have to perform is to simply count every incoming output event w_n. However, real single-particle detectors are often more complex devices with diverse properties. In our event-based simulation approach we model the main characteristics of these devices by rules as simple as possible to obtain similar results as those observed in a laboratory experiment. So far, we have designed two types of detectors, simple particle counters and adaptive threshold devices.[3] The adaptive threshold detector can be employed in the simulation of all single-photon experiments we have considered so far[3] but is absolutely essential in the simulation of for example the two-beam single photon experiment (see Sect. 5.1).

The efficiency, defined as the ratio of detected to emitted particles, of our model detectors is measured in an experiment with one single-particle point source placed far away from the detector. If the detector is a simple particle counter then the efficiency is 100%, if it is an adaptive threshold detector then the efficiency is nearly 100%. Since no absorption effects, dead times, dark counts, timing jitter or other effects causing particle miscounts are simulated, these model detectors are highly idealized versions of real single-photon detectors.

Evidently, the efficiency of a detector plays an important role in the overall detection efficiency in an experiment, but it is not the only determining factor. Also the experimental configuration, as well in the laboratory experiment as in the event-based simulation approach, in which the detector is used plays an important role. Although the adaptive threshold detectors are ideal and have a detection efficiency of nearly 100%, the overall detection efficiency can be much less than 100% depending on the experimental configuration. For example, using adaptive threshold detectors in a Mach-Zehnder interferometry experiment leads to an overall detection efficiency of nearly 100% (see Sect. 5.2.1), while using the same detectors in a single-photon two-beam experiment (see Sect. 5.1.1) leads to an overall detection efficiency of about 15%.[3,16] For the simple particle counters the configuration has no influence on the overall detection efficiency. Apart from the configuration, also the data processing procedure which is applied after the data has been collected may have an influence on the final detection efficiency. An example is the postselection procedure with a time-coincidence window which is employed to group photons, detected in two different stations, into pairs.[25] Even if in the event-based simulation approach simple particle counters with a 100% detection efficiency are used and thus all emitted photons are accounted for during the data collection process, the final detection efficiency is less than 100% because some detection events are omitted in the post-selection data procedure using a time-coincidence window.

In conclusion, even if ideal detectors with a detection efficiency of 100% would be commercially available, then the overall detection efficiency in a single-particle experiment could still be much less than 100% depending on (i) the experimental configuration in which the detectors are employed and (ii) the data analysis procedure that is used after all data has been collected.

5. Single particle interference

The particle-like behavior of photons has been shown in an experiment composed of a single 50/50 beam splitter (BS), of which only one input port is used, and a source emitting single photons and pairs of photons.[6] The wave mechanical character of the collection of photons has been demonstrated in single-particle interference experiments such as the single-photon two-beam experiment[15] (see Sect. 5.1), an experiment which shows, with minimal equipment, interference in its purest form (without diffraction), and the single-photon Mach-Zehnder interferometer (MZI) experiment[6] (see Sect. 5.2).

The three experiments have in common that, if one analyzes the data after collecting N detection events, long after the experiment has finished, the averages of the detection events agree with the results obtained from wave theory, that is with the classical theory of electrodynamics (Maxwell theory). In the first experiment one obtains a constant intensity of 0.5 at both detectors placed at the output ports of the BS, in the other two experiments one obtains an interference pattern. However, since the source is not emitting waves but so-called single photons[6,15] the question arises how to interpret the output which seems to show particle or wave character depending on the circumstances of the experiment. This question is not limited to photons. Already in 1924, de Broglie introduced the idea that also matter can exhibit wave-like properties.[74]

To resolve the apparent behavioral contradiction, quantum theory introduces the concept of particle-wave duality.[1] As a result, these single-particle experiments are often considered to be quantum experiments. However, the pictorial description using concepts from quantum theory, when applied to individual detection events (not to the averages) leads to conclusions that defy common sense: The photon (electron, neutron, atom, molecules, ...) seems to change its representation from a particle to a wave while traveling from the source to the detector in the single-photon interference experiments.

In 1978, Wheeler proposed a gedanken experiment,[75] a variation on Young's double slit experiment, in which the decision to observe wave or particle behavior is postponed until the photon has passed the slits. An experimental realization of Wheeler's delayed choice experiment with single-photons traveling in an open or closed configuration of an MZI has been reported in Refs. 9,76. The outcome, that is the average result of many detection events, is in agreement with wave theory (Maxwell or quantum theory). However, the pictorial description using concepts of quantum theory to explain the experimental facts[9] is even more strange than in the above mentioned experiments: The decision to observe particle or wave behavior influences the behavior of the photon in the past and changes the representation of the photon from a particle to a wave.

A more sensical description of the observation of individual detection events **and** of an interference pattern after many single detection events have been collected in single-particle interference experiments, can be given in terms of the event-based simulation approach. This finding is not in contradiction with Feynman's statement that electron (single particle) diffraction by a double-slit structure is "impossible, *absolutely* impossible, to explain in any classical way, and has in it the heart of quantum mechanics".[33] Reading "any classical way" as "any classical Hamiltonian mechanics way", Feynman's statement is difficult to dispute. However, taking a broader view by allowing for dynamical systems that are outside the realm of clas-

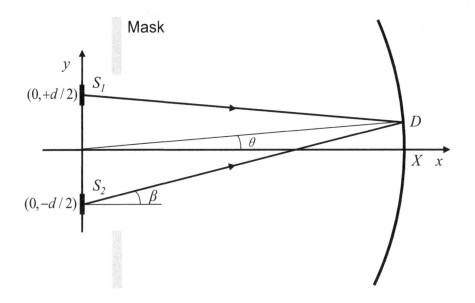

Fig. 2. Schematic diagram of a two-beam experiment with single-particle sources S_1 and S_2 of width a, separated by a center-to-center distance d. In a first experiment, which can be seen as a variant of Young's double slit experiment, N single particles leave the sources S_1 and S_2 one-by-one, at positions y drawn randomly from a uniform distribution over the interval $[-d/2 - a/2, -d/2 + a/2] \cup [+d/2 - a/2, +d/2 + a/2]$ and travel in the direction given by the angle β, a uniform pseudo-random number between $-\pi/2$ and $\pi/2$. In a second experiment, a movable mask is placed behind the sources which can block either S_1 or S_2. The sources S_1 and S_2 alternately emit M particles one-by-one, until a total of N particles has been emitted ($M \leq N/2$ and $kM = N$ with k an integer number). In both experiments, particles are emitted one-by-one either from S_1 or from S_2 and at any time there is only one particle traveling from source to detector. The particles are recorded by detectors D positioned on a semi-circle with radius X and center $(0, 0)$. The angular position of a detector is denoted by θ.

sical Hamiltonian dynamics, it becomes possible to model the gradual appearance of interference patterns through the event-by-event simulation method.

5.1. *Two-beam experiment*

We consider the experiment sketched in Fig. 2. Single particles coming from two coherent beams gradually build up an interference pattern when the particles arrive one-by-one at a detector screen. This two-beam experiment can be viewed as a simplification of Young's double-slit experiment in which the slits are regarded as the virtual sources S_1 and S_2 (see Ref. 42) and can be used to perform Feynman's thought experiment in which both slits are open or one is open and the other one closed. In the event-based model of this experiment particles are created one at a time at one of the sources and are detected by one of the detectors forming the screen. We assume that all these detectors are identical and cannot communicate among each other. We also do not allow for direct communication between the particles. This implies that this event-by-event model is locally causal by construction. Then, if it is indeed true that individual particles build up the interference pattern one-by-one, just looking at Fig. 2 leads to the logically unescapable conclusion that the interference pattern can only be due to the internal operation of the detector.[77] Detectors which simply count the incoming particles are not sufficient to explain the appearance of an interference pattern and apart from the detectors there is nothing else that can cause the interference pattern to appear. Making use of the statistical property of quantum theory one could assume that if a detector is replaced by another one as soon as it has detected one particle, one obtains similar interference patterns if the detection events of all these different detectors are combined or if only one detector detects all the particles. However, since there is no experimental evidence confirming this assumption and since our event-based approach is based on laboratory experimental setups and observations we do not consider this being a realistic option. Thus, logic dictates that a minimal event-based model for the two-beam experiment requires an algorithm for the detector that does a little more than just counting particles.

5.1.1. *Event-based model*

In what follows we specify the event-by-event model for the single-photon two-beam experiment (see Fig. 2) in sufficient detail such that the reader who is interested can reproduce the simulation results (a Mathematica implementation of a slightly more sophisticated algorithm[16] can be downloaded from the Wolfram Demonstration Project web site[78]).

- *Source and particles:* In the first experiment described in Fig. 2, N photons leave the sources one-by-one, at positions y drawn randomly from a uniform distribution over the interval $[-d/2 - a/2, -d/2 + a/2] \cup [+d/2 - a/2, +d/2 + a/2]$. In the second experiment the sources alternately emit M photons one-by-one until a total of N photons has been emitted. Here, $M \leq N/2$ and $kM = N$, where k denotes an integer number. The photons are regarded as messengers, traveling in the direction specified by the angle β, being a uniform pseudo-random number between $-\pi/2$ and $\pi/2$. Each messenger carries a message

$$\mathbf{u}(t) = (\cos(2\pi f t), \sin(2\pi f t)), \tag{11}$$

represented by a harmonic oscillator which vibrates with frequency f (representing the "color" of the light). The internal oscillator operates as a clock to encode the time of flight t, which is set to zero when a messenger is created, thereby modeling the coherence of the two single-particle beams.

This pictorial model of a "photon" was used by Feynman to explain quantum electrodynamics.[79] The event-based approach goes one step further in that it specifies in detail, in terms of a mechanical procedure, how the "amplitudes" which appear in the quantum formalism get added together. In Feynman's path integral formulation of light propagation, which is essentially quantum mechanical, the amplitude was obtained by summing over all possible paths.[79]

The time of flight of the particles depends on the source-detector distance. Here, we discuss as an example, the experimental setup with a semi-circular detection screen (see Fig. 2) but in principle any other geometry for the detection screen can be considered. The messenger leaving the source at $(0, y)$ under an angle β will hit the detector screen of radius X at a position determined by the angle θ given by $\sin \theta = (y \cos^2 \beta + \sin \beta \sqrt{X^2 - y^2 \cos^2 \beta})/X$, where $|y/X| < 1$. The time of flight is then given by $t = \sqrt{X^2 - 2yX \sin \theta + y^2}/c$, where c is the velocity of the messenger. The messages $\mathbf{u}(t)$ together with the explicit expression for the time of flight are the only input to the event-based algorithm.

- *Detector:* Here we describe the model for one of the many identical detectors building up the detection screen. Microscopically, the detection of a particle involves very intricate dynamical processes.[66] In its simplest form, a light detector consists of a material that can be ionized by light. This signal is then amplified, usually electronically, or in the case of a photographic plate by chemical processes. In Maxwell's theory, the interaction between the incident electric field $\mathbf{E}$ and the material takes the form $\mathbf{P} \cdot \mathbf{E}$, where $\mathbf{P}$ is the polarization vector of the material.[42] Assuming a linear response, $\mathbf{P}(\omega) = \chi(\omega)\mathbf{E}(\omega)$ for a monochromatic wave with frequency ω, it is clear that in the time domain, this relation expresses the fact that the material retains some memory about the incident field, $\chi(\omega)$ representing the memory kernel that is characteristic for the material used.

In line with the idea that an event-based approach should use the simplest rules possible, we reason as follows. In the event-based model, the nth message $\mathbf{u}_n = (\cos 2\pi f t_n, \sin 2\pi f t_n)$ is taken to represent the elementary unit of electric field $\mathbf{E}(t)$. Likewise, the electric polarization $\mathbf{P}(t)$ of the material is represented by the vector $\mathbf{v}_n = (v_{0,n}, v_{1,n})$. Upon receipt of the nth message this vector is updated according to the rule

$$\mathbf{v}_n = \gamma \mathbf{v}_{n-1} + (1 - \gamma)\mathbf{u}_n, \tag{12}$$

where $0 < \gamma < 1$ and $n > 0$. Obviously, if $\gamma > 0$, a message processor that operates according to the update rule Eq. (12) has memory, as required by Maxwell's theory. It is not difficult to prove that as $\gamma \to 1^-$, the internal vector $\mathbf{v}_n$ converges to the average of the time-series $\{\mathbf{u}_1, \mathbf{u}_2, \ldots\}$.[3,16] By reducing γ, the number of messages needed to adapt decreases but also the accuracy of the DLM decreases. In the limit that $\gamma = 0$, the DLM learns nothing, it simply echoes the last message that it received.[7,8] The parameter γ controls the precision with which the DLM defined by Eq. (12) learns the average of the sequence of messages $\mathbf{u}_1, \mathbf{u}_2, \ldots$ and also controls the pace at which new messages affect the internal state $\mathbf{v}$ of the machine.[7] Moreover, in the continuum limit (meaning many events per unit of time), the rule given in Eq. (12) translates into the constitutive equation of the Debye model of a dielectric,[16,80] a model used in many applications of Maxwell's theory.[81]

After updating the vector $\mathbf{v}_n$, the DLM uses the information stored in $\mathbf{v}_n$ to decide whether or not to generate a click. As a highly simplified model for the bistable character of the real photodetector or photographic plate, we let the machine generate a binary output signal w_n according to

$$w_n = \Theta(\mathbf{v}_k^2 - r_n), \tag{13}$$

where $\Theta(.)$ is the unit step function and $0 \leq r_n < 1$ is a uniform pseudo-random number. Note that the use of pseudo-random numbers is convenient but not essential.[3] Since in experiment it cannot be known whether a photon has gone undetected, we discard the information about the $w_n = 0$ detection events and define the total detector count as $N' = \sum_{j=1}^{n'} w_j$, where n' is the number of messages received. N' is the number of clicks (one's) generated by the processor. The efficiency of the detector model is determined by simulating an experiment that measures the detector efficiency, which for a single-photon detector is defined as the overall probability of registering a count if a photon arrives at the detector.[82] In such an experiment a point source emitting single particles is placed far away from a single detector. As all particles that reach the detector have the same time of flight (to a good approximation), all the particles that arrive at the detector will carry the same message which is encoding the time of flight. As a result $\mathbf{v}_n$ (see Eq. (12)) rapidly converges to the vector corresponding to this message, so that the detector clicks every time a photon arrives. Thus, the detection efficiency, as defined for real detectors,[82] for our detector model is very close to 100%. Hence, the model is a highly simplified and idealized version of a single-photon detector. However, although the detection efficiency of the detector itself may be very close to 100%, the overall detection efficiency, which is the ratio of detected to emitted photons in the simulation of an experiment, can be much less than one. This ratio depends on the experimental setup.

- *Simulation procedure:* Each of the detectors of the circular screen has a predefined spatial window within which it accepts messages. As a messenger hits a detector, this detector updates its internal state $\mathbf{v}$, (the internal states of all other detectors do not change) using the message $\mathbf{u}_n$ and then generates the event w_n. In the case $w_n = 1$ ($w_n = 0$), the total count of the particular detector that was hit by the nth messenger is (not) incremented by one and the messenger itself is destroyed. Only after the messenger has been destroyed, the source is allowed to send a new messenger. This rule ensures that the whole simulation complies with Einstein's criterion of local causality. This process of creating and destroying messengers is repeated many times, building up the interference pattern event by event. Note that the number of emitted photons N is larger than the sum of the number of clicks generated by all the detectors forming the detection screen although no photons are lost during their travel from source to detector.

5.1.2. *Simulation results*

In Fig. 3(a), we present simulation results for the first experiment for a representative case for which the analytical solution from wave theory is known. Namely, in the Fraunhofer regime ($d \ll X$), the analytical expression for the light intensity at the detector on a circular screen with radius X is given by[42]

$$I(\theta) = A \sin^2 \left(\frac{qa \sin \theta}{2} \right) \cos^2 \left(\frac{qd \sin \theta}{2} \right) / \left(\frac{qa \sin \theta}{2} \right)^2, \qquad (14)$$

where A is a constant, $q = 2\pi f/c$ denotes the wavenumber with f and c being the frequency and velocity of the light, respectively, and θ denotes the angular position of the detector D on the circular screen, see Fig. 2. Note that Eq. (14) is only used for comparison with the simulation data and is by no means input to the model. From Fig. 3(a) it is clear that the event-based model reproduces the results of wave theory and this without taking recourse of the solution of a wave equation.

As the detection efficiency of the event-based detector model is very close to 100%, the interference patterns generated by the event-based model cannot be attributed to inefficient detectors. It is therefore of interest to take a look at the ratio of detected to emitted photons, the overall detection efficiency, and compare the detection counts, observed in the event-by-event simulation of the two-beam interference experiment, with those observed in a real experiment with single photons.[15] In the simulation that yields the results of Fig. 3(a), each of the 181 detectors making up the detection area is hit on average by 55×10^3 photons and the total number of clicks generated by the detectors is 0.16×10^7. Hence, the ratio of the total number of detected to emitted photons is of the order of 0.16, two orders of magnitude larger than the ratio 0.5×10^{-3} observed in single-photon interference experiments.[15]

In Fig. 3(b), we show simulation results for the experiment in which first only source S_1 emits $N = 5 \times 10^6$ photons (downward triangles) while S_2 is blocked by the mask. Then in a new experiment (all detectors are reset) S_2 emits $N = 5 \times 10^6$ photons while S_1 is blocked (upward triangles). The sum of the two resulting detection curves is given by the curve with open squares. It is clear that this curve is completely different from the curve depicted in Fig. 3(a), as is also described in Feynman's thought experiment (see Sect. 2.1). Also in Fig. 3(b) we present the simulation results for the experiment in which first the source S_1 emits a group of $M = 5 \times 10^6$ particles one-by-one and then the source S_2 emits $M = 5 \times 10^6$ particles one-by-one (no resetting of the detectors). The resulting detection curve is drawn with closed circles. For small values of θ there is a difference between the curves with open squares and closed circles. This difference is due to the memory effect which is present in the detector model. Obviously this difference depends on

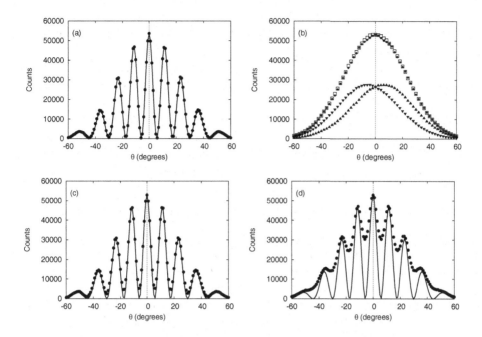

Fig. 3. Detector counts (markers) as a function of θ as obtained from the event-based simulation of the two-beam interference experiments described in Fig. 2. Simulation parameters: $N = 10^7$ so that on average, each of the 181 detectors, positioned on the semi-circular screen with an angular spacing of $1°$ in the interval $[-90°, 90°]$, receives about 55×10^3 particles, $\gamma = 0.999$, $a = c/f$, $d = 5c/f$, $X = 75c/f$, where c denotes the velocity and f the frequency of the particles ($c/f = 670$ nm in our simulations). (a): first experiment in which sources S_1 and S_2 in random order emit in total N particles one-by-one. This experiment resembles Young's (and Feynman's) two-slit experiment. (b): first experiment in which only source S_1 or S_2 emits $N = 5 \times 10^6$ particles one-by-one (downward and upward triangles, respectively). The open squares are the sum of the detector counts of the two experiments with one source emitting and the other one blocked. This experiment resembles Feynman's two-slit experiment with first slit S_2 blocked and then slit S_1 blocked. The closed circles are the result of the second experiment in which first S_1 and then S_2 emit a group of $M = 5 \times 10^6$ particles one-by-one. (c): second experiment with $M = 10^6$. (d): second experiment with $M = 25 \times 10^5$. The solid line in (a), (c) and (d) is a least-square fit of the simulation data of (a) to the prediction of wave theory, Eq. (14), with only one fitting parameter.

γ and the detector model that is used. For more complicated detector models than the one given by Eq. (12) this small difference disappears (results not shown).

Figs. 3(c),(d) depict simulation results of the experiment in which sources S_1 and S_2 alternately emit M particles one-by-one with $M = 10^6$ and $M = 25 \times 10^5$, respectively. It is seen that except for very large values of M ($M \gtrsim 10^6$), the interference pattern is the same as the one shown in Fig. 3(a). Nevertheless, for these large values of M interference can still be observed. This is a result of the memory effects built in the detector model. However, for any value of M, a simple quantum theoretical calculation would predict no interference pattern but an intensity pattern which is the sum of two single slit patterns, as the particles pass through one or the other slit, and never through both. Hence, for this type of experiment the predictions of quantum theory and of the event-based model differ.

Although we are not aware of any experiment that precisely tests the above described scenario, one experimental study in which only one slit was available to each photon[83] produced intriguing results. In that study, an opaque barrier, all the way from the laser source to the obstacle between the two slits, was used to make sure that photons had one or the other slit available to them. The interference pattern observed was nevertheless essentially unchanged despite the presence of the barrier. We are, however, not aware of any follow-up work on that study.

5.1.3. *Why is interference produced without solving a wave problem?*

As mentioned earlier, using simple particle counters as detectors would not result in an interference pattern. Essential to produce an interference pattern is to account for the information about the differences in the times of flight (or phase differences) of the particles which encode the distance the particles travelled from one of the two sources to one of the detectors constituting the circular detection screen. Simple particle counters do nothing with the information which is encoded in the messages carried by the particles and produce a click for each incoming particle. Since, in the single-photon two-beam experiment the detectors are the only apparatuses available that can process these phase differences (there are no other apparatuses present except for the source) we necessarily need to employ an algorithm for the detector that exploits this information in order to produce the clicks that gradually build up the interference pattern. A collection of about two hundred independent adaptive threshold detectors defined by Eq. (12) and Eq. (13) and each with a detection efficiency of nearly 100% is capable of doing this. As pointed out earlier, the reason why, in this particular experiment, this is possible is that not every particle that impinges on the detector yields a click.

5.2. *Mach Zehnder interferometer experiment*

5.2.1. *Event-based model*

The DLM network that simulates a single-photon MZI experiment (see Fig. 4 (left)) consists of a source, two identical BSs two phase shifters and two detectors. The network of processing units is a one-to-one image of the experimental setup.[6] Note that the two mirrors in the MZI simply bend the paths of the photons by $\pi/2$ without introducing a phase change or loss of particles and therefore they do not need to be considered in the event-based simulation network. In what follows we specify the processing units in sufficient detail such that the reader who is interested can reproduce the simulation results. We require that the processing units for

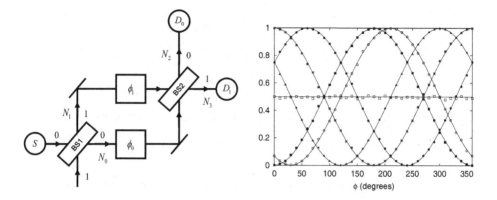

Fig. 4. Left: Schematic diagram of a Mach-Zehnder interferometer (MZI) with a single-photon source S. The MZI consists of two beam splitters, BS1 and BS2, two phase shifters ϕ_0 and ϕ_1 and two mirrors. N_0 (N_2) and N_1 (N_3) count the number of events in the output channel 0 of BS1 (BS2) and in the output channel 1 of BS1 (BS2), respectively. Dividing N_i for $i = 0, \ldots, 3$ by the total count N yields the relative frequency of finding a photon in the corresponding arm of the interferometer. Since photon detectors operate by absorbing photons, in a real laboratory experiment only N_2 and N_3 can be measured by detectors D_0 and D_1, respectively. Right: Simulation results for the normalized detector counts (markers) as a function of $\phi = \phi_0 - \phi_1$. Input channel 0 receives $(\cos \psi_0, \sin \psi_0)$ with probability one. One uniform random number in the range $[0, 360]$ is used to choose the angle ψ_0. Input channel 1 receives no events. The parameter $\gamma = 0.98$. Each data point represents 10000 events ($N = N_0 + N_1 = N_2 + N_3 = 10000$). Initially the rotation angle $\phi_0 = 0$ and after each set of 10000 events, ϕ_0 is increased by 10°. Open squares: N_0/N; solid squares: N_2/N for $\phi_1 = 0$; open circles: N_2/N for $\phi_1 = 30°$; solid circles: N_2/N for $\phi_1 = 240°$; asterisks: N_3/N for $\phi_1 = 0$; solid triangles: N_3/N for $\phi_1 = 300°$. Lines represent the results of quantum theory.[86]

identical optical components should be reusable within the same and within different experiments. Demonstration programs, including source codes, are available for download.[84,85]

- *Source and particles:* In a pictorial description of the experiment depicted in Fig. 4 (left) the photons, leaving the source S one-by-one, can be regarded as particles playing the role of messengers. Each messenger carries a message

$$\mathbf{u}_{k,n} = (\cos(2\pi f t_{k,n}), \sin(2\pi f t_{k,n})), \tag{15}$$

where f denotes the frequency of the light source and $t_{k,n}$ the time that particles need to travel a given path. The subscript $n > 0$ numbers the consecutive messengers and k labels the channel of the BS at which the messenger arrives (see below). Note that in this experiment no explicit information about distances and frequencies is required since we can always work with relative phases.

When a messenger is created its internal clock time is set to zero ($t_{k,n} = 0$) and since the source is connected to the $k = 0$ input channel of the first BS the messenger gets the label $k = 0$ (see Fig. 4 (left)).

- *Beam splitter* (BS): A BS is an optical component that partially transmits and partially reflects an incident light beam. Dielectric plate BSs are often used as 50/50 BSs. From classical electrodynamics we know that if an electric field is applied to a dielectric material the material becomes polarized.[42] Assuming a linear response, the polarization vector of the material is given by $\mathbf{P}(\omega) = \chi(\omega)\mathbf{E}(\omega)$ for a monochromatic wave with frequency ω. In the time domain, this relation expresses the fact that the material retains some memory about the incident field, $\chi(\omega)$ representing the memory kernel that is characteristic for the material used. We use this kind of memory effect in our algorithm to model the BS.

A BS has two input and two output channels labeled by 0 and 1 (see Fig 4 (left)). Note that in case of the MZI experiment, for beam splitter BS1 only entrance port $k = 0$ is used. In the event-based model, the BS has two internal registers $\mathbf{R}_{k,n} = (R_{0,k,n}, R_{1,k,n})$ (one for each input channel) and an internal vector $\mathbf{v}_n = (v_{0,n}, v_{1,n})$ with the additional constraints that $v_{i,n} \geq 0$ for $i = 0, 1$ and that $v_{0,n} + v_{1,n} = 1$. As we only have two input channels, the latter constraint can be used to recover $v_{1,n}$ from the value of $v_{0,n}$. We prefer to work with internal vectors that have as many elements as there are input channels. These three two-dimensional vectors $\mathbf{v}_n$, $\mathbf{R}_{0,n}$ and $\mathbf{R}_{1,n}$ are labeled by the message number n because their content is updated every time the BS receives a message. Before the simulation starts we set $\mathbf{v}_0 = (v_{0,0}, v_{1,0}) = (r, 1 - r)$, where r is a uniform pseudo-random number. In a similar way we use pseudo-random numbers to set $\mathbf{R}_{0,0}$ and $\mathbf{R}_{1,0}$.

When the nth messenger carrying the message $\mathbf{u}_{k,n}$ arrives at entrance port $k = 0$ or $k = 1$ of the BS, the BS first stores the message in the corresponding register $\mathbf{R}_{k,n}$ and updates its internal vector according to the rule

$$\mathbf{v}_n = \gamma \mathbf{v}_{n-1} + (1 - \gamma)\mathbf{q}_n, \tag{16}$$

where $0 < \gamma < 1$ is a parameter that controls the learning process and $\mathbf{q}_n = (1, 0)$ ($\mathbf{q}_n = (0, 1)$) if the nth event occurred on channel $k = 0$ ($k = 1$). By construction $v_{i,n} \geq 0$ for $i = 0, 1$ and $v_{0,n} + v_{1,n} = 1$. Hence the update rule Eq. (16) preserves the constraints on the internal vector. Obviously, these constraints are necessary if we want to interpret the $v_{k,n}$ as (an estimate of) the frequency for the occurrence of an event of type k. Note that the BS stores information about the last message only. The information carried by earlier messages is overwritten by updating the internal registers. From Eq. (16), one could say that the internal vector $\mathbf{v}$ (corresponding to the material polarization $\mathbf{P}$) is the response of the BS to the incoming messages (photons) represented by the vectors $\mathbf{q}$ (corresponding to the elementary unit of electric field $\mathbf{E}$). Therefore, the BS "learns" so to speak from the information carried by the photons. The characteristics of the learning process depend on the parameter γ (corresponding to the response function χ). Next, in case of a 50/50 BS, the BS uses the six numbers stored in $\mathbf{R}_{0,n}$, $\mathbf{R}_{1,n}$

and $\mathbf{v}_n$ to calculate four numbers $g_{0,n} = (R_{0,0,n}\sqrt{v_{0,n}} - R_{1,1,n}\sqrt{v_{1,n}})/\sqrt{2}$, $g_{1,n} = (R_{0,1,n}\sqrt{v_{1,n}} + R_{1,0,n}\sqrt{v_{0,n}})/\sqrt{2}$, $g_{2,n} = (R_{0,1,n}\sqrt{v_{1,n}} - R_{1,0,n}\sqrt{v_{0,n}})/\sqrt{2}$, and $g_{3,n} = (R_{0,0,n}\sqrt{v_{0,n}} + R_{1,1,n}\sqrt{v_{1,n}})/\sqrt{2}$. These four real-valued numbers can be considered to represent the real and imaginary part of two complex numbers $g_{0,n} + ig_{1,n}$ and $g_{2,n} + ig_{3,n}$ which are obtained by the following matrix-vector multiplication

$$
\begin{pmatrix} g_{0,n} + ig_{1,n} \\ g_{2,n} + ig_{3,n} \end{pmatrix} = \frac{1}{\sqrt{2}} \begin{pmatrix} \sqrt{v_{0,n}}(R_{0,0,n} + iR_{1,0,n}) + i\sqrt{v_{1,n}}(R_{0,1,n} + iR_{1,1,n}) \\ i\sqrt{v_{0,n}}(R_{0,0,n} + iR_{1,0,n}) + \sqrt{v_{1,n}}(R_{0,1,n} + iR_{1,1,n}) \end{pmatrix}
$$
$$
= \frac{1}{\sqrt{2}} \begin{pmatrix} 1 & i \\ i & 1 \end{pmatrix} \begin{pmatrix} \sqrt{v_{0,n}} & 0 \\ 0 & \sqrt{v_{1,n}} \end{pmatrix} \begin{pmatrix} R_{0,0,n} + iR_{1,0,n} \\ R_{0,1,n} + iR_{1,1,n} \end{pmatrix}, \tag{17}
$$

Identifying a_0 with $\sqrt{v_{0,n}}(R_{0,0,n} + iR_{1,0,n})$ and a_1 with $\sqrt{v_{1,n}}(R_{0,1,n} + iR_{1,1,n})$ it is clear that the computation of the four numbers $g_{i,n}$ for $i = 0, \ldots, 3$ plays the role of the matrix-vector multiplication in the quantum theoretical description of a beam-splitter

$$
\begin{pmatrix} b_0 \\ b_1 \end{pmatrix} = \frac{1}{\sqrt{2}} \begin{pmatrix} a_0 + ia_1 \\ a_1 + ia_0 \end{pmatrix} = \frac{1}{\sqrt{2}} \begin{pmatrix} 1 & i \\ i & 1 \end{pmatrix} \begin{pmatrix} a_0 \\ a_1 \end{pmatrix}, \tag{18}
$$

where (a_0, a_1) and (b_0, b_1) denote the input and output amplitudes, respectively. Note however that the DLM for the BS computes the four numbers $g_{i,n}$ for $i = 0, \ldots, 3$ for each incoming event thereby always updating $\mathbf{v}_n$ and $\mathbf{R}_{0,n}$ or $\mathbf{R}_{1,n}$. Hence, a_0 and a_1, and thus also b_0 and b_1, are constructed event-by-event and only under certain conditions ($\gamma \to 1^-$, sufficiently large number of input events N, stationary sequence of input events) they correspond to their quantum theoretical counterparts $a_0 = \sqrt{p_0}e^{i\psi_0}$, $a_1 = \sqrt{p_1}e^{i\psi_1}$ with $p_1 = 1 - p_0$ ($0 \le p_0, p_1 \le 1$) and $b_0 = a_0 + ia_1$, $b_1 = a_1 + ia_0$ (see Eq. (18)). In a final step the BS uses $g_{i,n}$ for $i = 0, \ldots, 3$ to create an output event. Therefore it generates a uniform random number r_n between zero and one. If $g_{0,n}^2 + g_{1,n}^2 > r_n$, the BS sends a message

$$
\mathbf{w}_{0,n} = (g_{0,n}, g_{1,n})/\sqrt{g_{0,n}^2 + g_{1,n}^2}, \tag{19}
$$

through output channel 0. Otherwise it sends a message

$$
\mathbf{w}_{1,n} = (g_{2,n}, g_{3,n})/\sqrt{g_{2,n}^2 + g_{3,n}^2}, \tag{20}
$$

through output channel 1.

- *Phase shifters:* These devices perform a plane rotation on the vectors (messages) carried by the particles. As a result the phase of the particles is changed by ϕ_0 or ϕ_1 depending on the route followed.
- *Detector:* Detector $D_0(D_1)$ registers the output events at channel 0 (1). The detectors are ideal particle counters, meaning that they produce a click for each incoming particle. Hence, we assume that the detectors have 100% detection efficiency. Note that also adaptive threshold detectors can be used (see Sect. 5.1.1) equally well.[3]

- *Simulation procedure:* When a messenger is created we wait until its message has been processed by one of the detectors before creating the next messenger. This ensures that there can be no direct communication between the messengers and that our simulation model (trivially) satisfies Einsteins criterion of local causality. We assume that no messengers are lost. Since the detectors are ideal particle counters the number of clicks generated by the detectors is equal to the number of messengers created by the source. For fixed $\phi = \phi_0 - \phi_1$, a simulation run of N events generates the data set $\Gamma(\phi) = \{w_n | n = 1, \ldots, N\}$. Here $w_n = 0, 1$ indicates which detector fired (D_0 or D_1). Given the data set $\Gamma(\phi)$, we can easily compute the number of 0 (1) output events N_2 (N_3).

5.2.2. Simulation results

In Fig. 4 (right), we present a few simulation results for the MZI and compare them to the quantum theoretical result. According to quantum theory, the amplitudes (b_0, b_1) in the output modes 0 and 1 of the MZI are given by[87]

$$\begin{pmatrix} b_0 \\ b_1 \end{pmatrix} = \frac{1}{2} \begin{pmatrix} 1 & i \\ i & 1 \end{pmatrix} \begin{pmatrix} e^{i\phi_0} & 0 \\ 0 & e^{i\phi_1} \end{pmatrix} \begin{pmatrix} 1 & i \\ i & 1 \end{pmatrix} \begin{pmatrix} a_0 \\ a_1 \end{pmatrix}, \tag{21}$$

where a_0 and a_1 denote the input amplitudes. For the particular choice $a_0 = 1$ and $a_1 = 0$, in which case there are no particles entering BS1 via channel 1, it follows from Eq. (21) that

$$|b_0|^2 = \sin^2(\frac{\phi_0 - \phi_1}{2}), \quad |b_1|^2 = \cos^2(\frac{\phi_0 - \phi_1}{2}). \tag{22}$$

For the results presented in Fig. 4 (right) we assume that input channel 0 receives $(\cos \psi_0, \sin \psi_0)$ with probability one and that input channel 1 receives no events. This corresponds to $(a_0, a_1) = (\cos \psi_0 + i \sin \psi_0, 0)$. We use a uniform random number to determine ψ_0. Note that this random number is used to generate all input events. The data points are the simulation results for the normalized intensity N_i/N for $i = 0, 2, 3$ as a function of $\phi = \phi_0 - \phi_1$. Note that in an experimental setting it is impossible to simultaneously measure $(N_0/N, N_1/N)$ and $(N_2/N, N_3/N)$ because photon detectors operate by absorbing photons. In the event-based simulation there is no such problem. From Fig. 4 (right) it is clear that the event-based processing by the DLM network reproduces the probability distribution of quantum theory, see Eq. (22) with $|b_0|^2$ ($|b_1|^2$) corresponding to N_2/N (N_3/N).

5.2.3. Why is interference produced without solving a wave problem?

We consider BS2 of the MZI depicted in Fig. 4 (left), the beam splitter at which, in a wave picture, the two beams join to produce interference. The DLM simulating a BS requires two pieces of information to send out particles such that their distribution matches the wave-mechanical description of the BS. First, it needs an estimate of the ratio of particle currents in the input channels 0 and 1 (paths 0 and 1 of the MZI), respectively. Second, it needs to have information about the time of flight (phase difference) along the two different paths of the MZI. The first piece of information is provided for by the internal vector $\mathbf{v} = (v_0, v_1)$. Through the update rule Eq. (16), for a stationary sequence of input events, v_0 and v_1 converge to the average of the number of events on input channels 0 and 1, respectively. Thus, the ratio of the particles (corresponding to the intensities of the waves) in the two input beams are encoded in the vector $\mathbf{v}$. Note that this information is accurate only if the sequence of input events is stationary. After one particle arrived at port 0 and another one arrived at port 1, the second piece of information is available in the registers $\mathbf{R}_0$ and $\mathbf{R}_1$. This information plays the role of the phase of the waves in the two input beams. Hence, all the information (intensity and phase) is available to compute the probability for sending out particles. This is done by calculating the numbers g_i for $i = 0, \ldots, 3$ which, in the stationary state, are identical to the wave amplitudes obtained from the wave theory of a beam splitter.[42]

5.3. *Wheeler's delayed choice experiment*

In a recent experimental realization of Wheeler's delayed-choice experiment by Jacques *et al.*[76] linearly polarized single photons are sent through a polarizing beam splitter (PBS) that together with a second, movable, variable output PBS with adjustable reflectivity $\mathcal{R}$ forms an interferometer (see Fig. 5). In the first realization[9] two 50/50 BSs were used.

Tilting the PBS of the variable output BS induces a time-delay in one of the arms of the MZI, symbolically represented by the variable phase $\phi_1(x)$ in Fig. 5, and thus varies the phase shift $\phi(x) = \phi_0 - \phi_1(x)$ between the two arms of the MZI. A voltage applied to an electro-optic modulator (EOM) controls the reflectivity $\mathcal{R}$ of the variable beam splitter BS_{output}. If no voltage is applied to the EOM then $\mathcal{R} = 0$. Otherwise, $\mathcal{R} \neq 0$ (see Eq. (2) in Ref. 76) and the EOM acts as a wave plate which rotates the polarization of the incoming photon by an angle depending on the value of $\mathcal{R}$. The voltage applied to the EOM is controlled by a set of pseudo-random numbers generated by the random number generator RNG. The key point in this experiment is that the decision to apply a voltage to the EOM is made after the photon has passed BS_{input}.

For $0 \leq \mathcal{R} \leq 0.5$ measured values of the interference visibility[88] V and the path distinguishability[76] D, a parameter that quantifies the which-path information (WPI), were found to fulfill the complementary relation $V^2 + D^2 \leq 1$.[76] For $(V = 0, D = 1)$ and $(V = 1, D = 0)$, obtained for $\mathcal{R} = 0$ and $\mathcal{R} = 0.5$, respectively, full and no WPI was found, associated with particle like and wavelike behavior, respectively. For $0 \leq \mathcal{R} \leq 0.5$ partial WPI was obtained while keeping interference with limited visibility.[76]

Although the detection events (detector "clicks") are the only experimental facts and logically speaking one cannot say anything about what happens with the photons traveling through the setup, Jacques *et al.*[9,76] gave the following

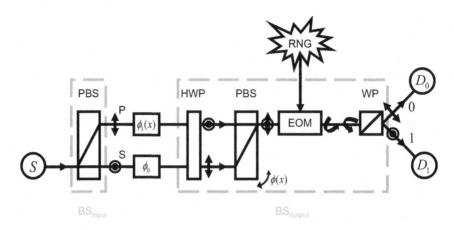

Fig. 5. Schematic diagram of the experimental setup of Wheeler's delayed-choice experiment with single photons.[9,76] S: single-photon source; PBS: polarizing beam splitter; HWP: half-wave plate; EOM: electro-optic modulator; RNG: random number generator; WP: Wollaston prism ($=$ PBS); D_0 and D_1: detectors; P, S: polarization state of the photons; $\phi(x) = \phi_0 - \phi_1(x)$: phase shift between paths 0 and 1. The diagram is that of a Mach-Zehnder interferometer composed of a 50/50 input beam splitter (BS_{input}) and a variable output beam splitter (BS_{output}) with adjustable reflectivity $\mathcal{R}$.

pictorial description: Linearly polarized single photons are sent through a 50/50 PBS (BS_{input}), spatially separating photons with S polarization (path 0) and P polarization (path 1) with equal frequencies. After the photon has passed BS_{input}, but before the photon enters the variable BS_{output} the decision to apply a voltage to the EOM is made. The PBS of BS_{output} merges the paths of the orthogonally polarized photons travelling paths 0 and 1 of the MZI, but afterwards the photons can still be unambiguously identified by their polarizations. If no voltage is applied to the EOM then $\mathcal{R} = 0$ and the EOM does nothing to the photons. Because the polarization eigenstates of the Wollaston prism correspond to the P and S polarization of the photons travelling path 0 and 1 of the MZI, each detection event registered by one of the two detectors D_0 or D_1 is associated with a specific path (path 0 or 1, respectively). Both detectors register an equal amount of detection events, independent of the phase shift $\phi(x)$ in the MZI. This experimental setting clearly gives full WPI about the photon within the interferometer (particle behavior), characterized by $D = 1$. In this case no interference effects are observed and thus $V = 0$. When a voltage is applied to the EOM, then $\mathcal{R} \neq 0$ and the EOM rotates the polarization of the incoming photon by an angle depending on $\mathcal{R}$. The Wollaston prism partially recombines the polarization of the photons that have travelled along different optical paths with phase difference $\phi(x)$ and interference appears ($V \neq 0$), a result expected for a wave. The WPI is partially washed out, up to be totally erased when $\mathcal{R} = 0.5$. Hence, the decision to apply a voltage to the EOM after the photon left BS_{input} but before it passes BS_{output}, influences the behavior of the photon in the past and changes the representation of the photon from a particle to a wave.[9]

5.3.1. *Event-based model*

We construct a model for the messengers representing the linearly polarized photons and for the processing units representing the optical components in the experimental setup (see Fig. 5) thereby fulfilling the requirements that the processing units for identical optical components should be reusable within the same and within different experiments and that the network of processing units is a one-to-one image of the experimental setup. Although, in contrast to the experiments we have considered so far, in this experiment it is necessary to include the polarization in the model for the messengers representing the photons. These more general messengers can also be used in a simulation of the experiments discussed previously. In the event-based simulation of these experiments the polarization component of the message is simply not used in the DLMs modeling the optical components of their experimental setup. In what follows we describe the elements of the model in more detail.

- *Source and particles:* The polarization can be included in the model for the messengers representing the photons by adding to the message a second harmonic oscillator which also vibrates with frequency f. There are many different but equivalent ways to define the message. As in Maxwell's and quantum theory, it is convenient (though) not essential to work with complex valued vectors, that is with messages represented by two-dimensional unit vectors

$$\mathbf{u} = (e^{i\psi^{(1)}} \sin\xi, e^{i\psi^{(2)}} \cos\xi), \tag{23}$$

where $\psi^{(i)} = 2\pi f t + \delta_i$, for $i = 1, 2$. The angle ξ determines the relative magnitude of the two components and $\delta = \delta_1 - \delta_2 = \psi^{(1)} - \psi^{(2)}$, denotes the phase difference between the two components. Both ξ and δ determine the polarization of the photon. Hence, the photon can be considered to have a polarization vector $\mathbf{P} = (\cos\delta \sin 2\xi, \sin\delta \sin 2\xi, \cos 2\xi)$. The third degree of freedom in Eq. (23) is used to account for the time of flight of the photon. Within the present model, it is thus postulated that the state of the photon is fully determined by the angles $\psi^{(1)}$, $\psi^{(2)}$ and ξ and by rules (to be specified), by which these angles change as the photon travels through the network.

A messenger with message $\mathbf{u}$ at time t and position $\mathbf{r}$ that travels with velocity $v = c/n$, where c denotes the velocity of light and n is the index of refraction of the material, along the direction $\mathbf{q}$ during a time interval $t' - t$, changes its message according to $\psi^{(i)} \to \psi^{(i)} + \phi$ for $i = 1, 2$, where $\phi = 2\pi f(t' - t)$. This suggests that we may view the two-component vectors $\mathbf{u}$ as the coordinates of two local oscillators, carried along by the messengers and that the messenger encodes its time of flight in these two oscillators.

It is evident that the representation used here maps one-to-one to the plane-wave description of a classical electromagnetic field,[42] except that we assign these properties to each individual photon, not to a wave. As there is no communication/interaction between the messengers there can be no wave equation (partial differential equation) that enforces a relation between the messages carried by different messages.

When the source creates a messenger, its message needs to be initialized. This means that the three angles $\psi^{(1)}$, $\psi^{(2)}$ and ξ need to be specified. The specification depends on the type of light source that has to be simulated. For a coherent light source, the three angles are different but the same for all the messengers being created. Hence, three random numbers are used to specify $\psi^{(1)}$, $\psi^{(2)}$ and ξ for all messengers.

In this section we will demonstrate explicitly that in the event-based model (in general, not only for this experiment) photons always have full WPI even if interference is observed by giving the messengers one extra label, the path label having the value 0 or 1. The information contained in this label is not accessible in the experiment.[76] We only use it to track the photons in the network of processing units. The path label is set in the input BS and remains unchanged until detection. Therefore we do not consider this label in the description of the processing units but take it into account when we detect the photons.

- *Polarizing beam splitter* (PBS): A PBS is used to redirect photons depending on their polarization. For simplicity, we assume that the coordinate system used to define the incoming messages coincides with the coordinate system defined by two orthogonal directions of polarization of the PBS.

In general, a PBS has two input and two output channels labeled by 0 and 1, just like an ordinary BS (see Sect. 5.2.1). Note that in case of Wheeler's delayed choice experiment, the first PBS has only one input channel labeled by $k = 0$ and therefore the second PBS has only one output channel labeled by $k = 0$. In the event-based model, the PBS has a similar structure as the BS. Therefore, in what follows we only mention the main ingredients to construct the processing unit for the PBS. For more details we refer to Sect. 5.2.1.

The PBS has two internal registers $\mathbf{R}_{k,n} = (R_{0,k,n}, R_{1,k,n})$ with $R_{i,k,n}$ for $i = 0, 1$ representing a complex number, and an internal vector $\mathbf{v}_n = (v_{0,n}, v_{1,n})$, where $v_{i,n} \geq 0$ for $i = 0, 1$, $v_{0,n} + v_{1,n} = 1$ and n denotes the message number. Before the simulation starts uniform pseudo-random numbers are used to set $\mathbf{v}_0$, $\mathbf{R}_{0,0}$ and $\mathbf{R}_{1,0}$.

When the nth messenger carrying the message $\mathbf{u}_{k,n}$ arrives at entrance port $k = 0$ or $k = 1$ of the PBS, the PBS first copies the message in the corresponding register $\mathbf{R}_{k,n}$ and updates its internal vector according to

$$\mathbf{v}_n = \gamma \mathbf{v}_{n-1} + (1 - \gamma)\mathbf{q}_n, \qquad (24)$$

where $0 < \gamma < 1$ and $\mathbf{q}_n = (1, 0)$ ($\mathbf{q}_n = (0, 1)$) represents the arrival of the nth messenger on channel $k = 0$ ($k = 1$). Note that the DLM has storage for exactly ten real-valued numbers.

Next the PBS uses the information stored in $\mathbf{R}_{0,n}$, $\mathbf{R}_{1,n}$ and $\mathbf{v}_n$ to calculate four complex numbers

$$
\begin{pmatrix} h_{0,n} \\ h_{1,n} \\ h_{2,n} \\ h_{3,n} \end{pmatrix} = \begin{pmatrix} 1 & 0 & 0 & 0 \\ 0 & 1 & 0 & 0 \\ 0 & 0 & 0 & i \\ 0 & 0 & i & 0 \end{pmatrix} \begin{pmatrix} \sqrt{v_{0,n}} & 0 & 0 & 0 \\ 0 & \sqrt{v_{1,n}} & 0 & 0 \\ 0 & 0 & \sqrt{v_{0,n}} & 0 \\ 0 & 0 & 0 & \sqrt{v_{1,n}} \end{pmatrix} \begin{pmatrix} R_{0,0,n} \\ R_{0,1,n} \\ R_{1,0,n} \\ R_{1,1,n} \end{pmatrix}
$$

$$
= \begin{pmatrix} \sqrt{v_{0,n}} R_{0,0,n} \\ \sqrt{v_{1,n}} R_{0,1,n} \\ i\sqrt{v_{1,n}} R_{1,1,n} \\ i\sqrt{v_{0,n}} R_{1,0,n} \end{pmatrix}, \tag{25}
$$

and generates a uniform random number r_n between zero and one. If $|h_{0,n}|^2 + |h_{2,n}|^2 > r_n$, the PBS sends a message

$$
\mathbf{w}_{0,n} = (h_{0,n}, h_{2,n})/\sqrt{|h_{0,n}|^2 + |h_{2,n}|^2}, \tag{26}
$$

through output channel 1. Otherwise it sends a message

$$
\mathbf{w}_{1,n} = (h_{1,n}, h_{3,n})/\sqrt{|h_{1,n}|^2 + |h_{3,n}|^2}, \tag{27}
$$

through output channel 0.

- *Half-wave plate* (HWP): A HWP not only changes the polarization of the light but also its phase. In optics, a HWP is often used as a retarder. In the event-based model, the retardation of the wave corresponds to a change in the time of flight (and thus the phase) of the messenger. In contrast to the BS and PBS, a HWP may be simulated without DLM. The device has only one input and one output port (see Fig. 5). A HWP transforms the nth input message $\mathbf{u}_n$ into an output message

$$
\mathbf{w}_n = -i(u_{0,n}\cos 2\theta + u_{1,n}\sin 2\theta, u_{0,n}\sin 2\theta - u_{1,n}\cos 2\theta), \tag{28}
$$

where θ denotes the angle of the optical axis with respect to the laboratory frame. Hence, in order to change S polarization into P polarization, or vice versa, a HWP is used with its optical axis oriented at $\pi/4$. This changes the phase of the photon by $-\pi/2$.

- *Electro-optic modulator* (EOM): An EOM rotates the polarization of the photon by an angle depending on the voltage applied to the modulator. In the laboratory experiment, the EOM is operated such that when a voltage is applied it acts as a HWP that rotates the input polarizations by $\pi/4$. We use a pseudo-random number to mimic the experimental procedure to control the EOM, but any other (systematic) sequence to control the EOM can be used as well.

- *Wollaston prism* (WP): The WP is a PBS with one input channel and two output channels and is simulated as the PBS described earlier.

- *Detector:* Detector $D_0(D_1)$ counts the output events at channel 0 (1) of the Wollaston prism. The detectors are ideal particle counters, meaning that they produce a click for each incoming particle. Hence, we assume that the detectors have 100% detection efficiency. Note that in this experimental configuration adaptive threshold detectors (see Sect. 5.1.1) can be used equally well because their detection efficiency is 100%.[3]
- *Simulation procedure:* When a messenger is created we wait until its message has been processed by one of the detectors before creating the next messenger (Einstein's criterion of local causality). During a simulation run of N events the data set $\Gamma(\phi(x)) = \{w_n, d_n, r_n | n = 1, \ldots, N; \phi(x) = \phi_0 - \phi_1(x)\}$ is generated, where $w_n = 0, 1$ indicates which detector fired (D_0 or D_1), $d_n = 0, 1$ indicates through which arm of the MZI the messenger (photon) came that generated the detection event (note that d_n is only measured in the simulation, not in the experiment), and r_n is a pseudo-random number that is chosen after the nth message has passed the first PBS, determining which voltage is applied to the EOM. Note that in one run of N events a choice is made between no voltage (open MZI configuration) or a particular voltage (closed MZI configuration) corresponding to a certain reflectivity $\mathcal{R}$ of the output BS (see Eq. (2) in Ref. 76). These choices are made such that on average the MZI configuration is as many times open as it is closed. The angle $\phi(x)$ denotes the phase shift between the two interferometer arms. This phase shift is varied by applying a plane rotation on the phase of the particles entering channel 0 of the second PBS. This corresponds to tilting the second PBS in the laboratory experiment.[76] For each $\phi(x)$ and MZI configuration the number of 0 (1) output events N_0 (N_1) is calculated.

5.3.2. *Simulation results*

We first demonstrate that our model yields full WPI of the photons. Fig. 6(a) shows the number of detection events at D_0 as a function of ϕ (($\phi \equiv \phi(x)$ for a given fixed position of the PBS in BS$_{\text{output}}$) for $\mathcal{R} = 0.5$. The events generated by photons following paths 0 and 1 of the MZI are counted separately. It is clear that the number of photons that followed paths 0 (squares) and 1 (triangles) is equal and that the total intensity in output channel 0 (open circles) shows a sinusoidal function of ϕ. Hence, although the photons have full WPI for all ϕ they can build an interference pattern by arriving one-by-one at a detector. Next, we calculate for $\mathcal{R} = 0.05$ and $\mathcal{R} = 0$ and for each phase shift ϕ and configuration (open or closed) of the MZI the number of events registered by the two detectors behind the output BS, just like in the experiment. Figures 6(b),(c) depict the normalized detection counts at D_0 (open circles) and D_1 (closed circles). The simulation data quantitatively agree with the averages calculated from quantum theory and qualitatively agree with experiment (see Fig. 3 in Ref. 76). Calculation of D as described in Ref. 76 gives the results for D^2 and V^2 shown in Fig. 6(d). Comparison with Fig. 4 in Ref. 76 shows excellent qualitative agreement.

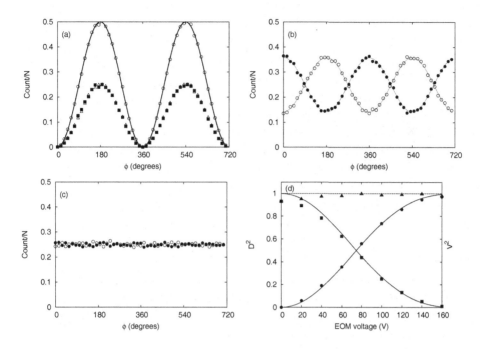

Fig. 6. Event-by-event simulation results of the normalized detector counts for different values of $\mathcal{R}$ ((a)-(c)) and of V^2, D^2 and $V^2 + D^2$ as a function of the EOM voltage (d). (a) Markers give the results for the normalized intensity N_0/N as a function of the phase shift ϕ, N_0 denoting the number of events registered at detector D_0. Squares (triangles, hardly visible because they overlap with the squares) represent the detection events generated by photons which followed path 0 (1). Open circles represent the total number of detection events. (b)-(c) Open (closed) circles give the results for the normalized intensities N_0/N (N_1/N) as a function of the phase shift ϕ, N_0 (N_1) denoting the number of events registered at detector D_0 (D_1), for (b) $\mathcal{R} = 0.05$ ($V \approx 0.45$) and (c) $\mathcal{R} = 0$ ($V = 0$). For each value of ϕ, the number of input events $N = 10000$. The number of detection events per data point is approximately the same as in experiment. Dashed lines represent the results of quantum theory. (d) Squares, circles and triangles present the simulation results for V^2, D^2 and $V^2 + D^2$, respectively. Lines represent the theoretical expectations obtained from Eqs. (2), (3) and (7) in Ref. 76 with $\beta = 24°$ and $V_\pi = 217$V.

5.4. *Single neutron interferometry*

Now that we have demonstrated the event-based simulation approach for the event-by-event realization of an interference pattern in various single-photon interference experiments, we consider in this section one of the basic experiments in neutron interferometry, namely a Mach-Zehnder type of interferometer. In neutron optics there exist various realizations of the Mach-Zehnder type of interferometer, but we only consider a triple Laue diffraction type silicon perfect single crystal interferometer.[36,89,90]

Figure 7 (left) shows the experimental configuration. The three crystal plates, named the splitter, mirror and analyzer plate, are assumed to be identical, which

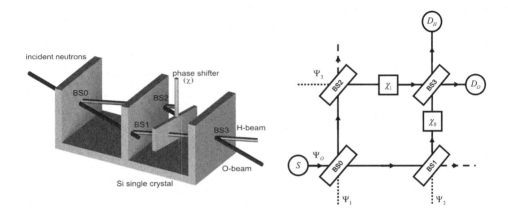

Fig. 7.   Left: Schematic picture of the silicon-perfect-crystal neutron interferometer.[89] BS0, ...,
BS3: beam splitters; phase shifter χ: aluminum foil; neutrons that are transmitted by BS1 or
BS2 leave the interferometer and do not contribute to the interference signal. Detectors count
the number of neutrons in the O- and H-beam. Right: Event-based network of the interferometer
shown on the left. S: single neutron source; BS0, ... , BS3: beam splitters; χ_0, χ_1: phase
shifters; D_O, D_H: detectors counting all neutrons that leave the interferometer via the O- and
H-beam, respectively. In the experiment and in the event-based simulation, neutrons enter the
interferometer via the path labeled by Ψ_0 only. The wave amplitudes labeled by Ψ_1, Ψ_2, and Ψ_3
(dotted lines) are used in the quantum theoretical treatment only (see text). Particles leaving the
interferometer via the dashed lines are not counted.

means that they have the same transmission and reflection properties.[36] The three
crystal plates have to be parallel to high accuracy[89] and the whole device needs to
be protected from vibrations in order to observe interference.[91] A monoenergetic
neutron beam is split by the splitter plate (BS0). Neutrons refracted by beam splitters BS1 and BS2 (mirror plate) are directed to the analyzer plate (BS3), also acting
as a BS, thereby first passing through a rotatable-plate phase shifter (e.g. aluminum
foil[36]). Absorption of neutrons by the aluminum foil is assumed to be negligible.[36]
Minute rotations of the foil about an axis perpendicular to the base plane of the
interferometer induce large variations in the phase difference $\chi = \chi_0 - \chi_1$.[36,92] Finally, the neutrons are detected by one of the two detectors placed in the so-called
H-beam or O-beam. In contrast to single-photon detectors, neutron detectors can
have a very high, almost 100%, efficiency.[36] Neutrons which are not refracted by
BS1 and BS2 leave the interferometer and are not counted. The intensities in the
O- and H-beam, obtained by counting individual neutrons for a certain amount of
time, exhibit sinusoidal variations as a function of the phase shift χ, a characteristic
of interference.[36]

   The experiment could be interpreted in different ways. In the quantum-
corpuscular view a wave packet is associated with each individual neutron. At
BS0 the wave packet splits in two components, one directed towards BS1 and one

towards BS2. At BS1 and BS2 these two components each split in two. Two of the in total four components leave the interferometer and the other two components are redirected towards each other at BS3 where they recombine. At BS3 the recombined wave packet splits again in two components. Only one of these two components triggers a detector. It is a mystery how four components of a wave packet can conspire to do such things. Assuming that only a neutron, not merely a part of it can trigger the nuclear reaction that causes the detector to "click", on elementary logical grounds, the argument that was just given rules out a wave-packet picture for the individual neutron (invoking the wave function collapse only adds to the mystery). In the statistical interpretation of quantum mechanics there is no such conflict of interpretation.[2,66] As long as we consider descriptions of the statistics of the experiment with many neutrons, we may think of one single "probability" wave propagating through the interferometer and as the statistical interpretation of quantum theory is silent about single events, there is no conflict with logic either.[36]

In what follows we demonstrate that as in the case of the single-photon interference experiments, it is possible to construct a logically consistent, cause-and-effect description in terms of discrete-event, particle like processes which produce results that agree with those of neutron interferometry experiments (individual detection events **and** an interference pattern after many single detection events have been collected) and the quantum theory thereof (interference pattern only).

5.4.1. *Event-based model*

We construct a model for the messengers representing the neutrons and for the processing units representing the various components in the experimental setup (see Fig. 7 (right)).

- *Source and particles:* In analogy to the event-based model of a polarized photon (see Sect. 5.3.1), a neutron is regarded as a messenger carrying a message represented by the two-dimensional unit vector

$$\mathbf{u} = (e^{i\psi^{(1)}} \cos(\theta/2), e^{i\psi^{(2)}} \sin(\theta/2)), \tag{29}$$

where $\psi^{(i)} = \nu t + \delta_i$, for $i = 1, 2$. Here, t specifies the time of flight of the neutron and ν is an angular frequency which is characteristic for a neutron that moves with a fixed velocity v. A monochromatic beam of incident neutrons is assumed to consist of neutrons that all have the same value of ν, that is: they have the same velocity.[36] Both θ and $\delta = \delta_1 - \delta_2 = \psi^{(1)} - \psi^{(2)}$ determine the magnetic moment of the neutron, if the neutron is viewed as a tiny classical magnet spinning around the direction $\mathbf{m} = (\cos \delta \sin \theta, \sin \delta \sin \theta, \cos \theta)$, relative to a fixed frame of reference defined by a magnetic field. The third degree of freedom in Eq. (29) is used to account for the time of flight of the neutron. Within the present model, the state of the neutron is fully determined by the angles $\psi^{(1)}$, $\psi^{(2)}$ and θ and by rules (to be specified), by which these angles change as the neutron travels through the network.

A messenger with message $\mathbf{u}$ at time t and position $\mathbf{r}$ that travels with velocity v, along the direction $\mathbf{q}$ during a time interval $t' - t$, changes its message according to $\psi^{(i)} \to \psi^{(i)} + \phi$ for $i = 1, 2$, where $\phi = \nu(t' - t)$.

In the presence of a magnetic field $\mathbf{B} = (B_x, B_y, B_z)$, the magnetic moment rotates about the direction of $\mathbf{B}$ according to the classical equation of motion. Hence, in a magnetic field the message $\mathbf{u}$ is changed into the message $\mathbf{w} = e^{ig\mu_N T \sigma \cdot \mathbf{B}} \mathbf{u}$, where g denotes the neutron g-factor, μ_N the nuclear magneton, T the time during which the neutron experiences the magnetic field, and σ denotes the Pauli vector (here we use the isomorphism between the algebra of Pauli matrices and rotations in three-dimensional space).

When the source creates a messenger, its message needs to be initialized. This means that the three angles $\psi^{(1)}$, $\psi^{(2)}$ and θ need to be specified. The specification depends on the type of source that has to be simulated. For a fully coherent spin-polarized beam of neutrons, the three angles are different but the same for all the messengers being created. Hence, three random numbers are used to specify $\psi^{(1)}$, $\psi^{(2)}$ and θ for all messengers.

- *Beam splitters* BS0, ... , BS3: Exploiting the similarity between the magnetic moment of the neutron and the polarization of a photon, we use a similar model for the BS as the one used in Sect. 5.3.1 for polarized photons. The only difference is that we assume that neutrons with spin up and spin down have the same reflection and transmission properties, while photons with horizontal and vertical polarization have different reflection and transmission properties.[42] Hence, what needs to be changed with respect to Sect. 5.3.1 are the complex numbers $h_{0,n}, \ldots, h_{3,n}$. For the neutrons we have

$$
\begin{pmatrix} h_{0,n} \\ h_{1,n} \\ h_{2,n} \\ h_{3,n} \end{pmatrix} = \begin{pmatrix} \sqrt{\mathcal{T}} & i\sqrt{\mathcal{R}} & 0 & 0 \\ i\sqrt{\mathcal{R}} & \sqrt{\mathcal{T}} & 0 & 0 \\ 0 & 0 & \sqrt{\mathcal{T}} & i\sqrt{\mathcal{R}} \\ 0 & 0 & i\sqrt{\mathcal{R}} & \sqrt{\mathcal{T}} \end{pmatrix}
$$
$$
\times \begin{pmatrix} \sqrt{v_{0,n}} & 0 & 0 & 0 \\ 0 & \sqrt{v_{1,n}} & 0 & 0 \\ 0 & 0 & \sqrt{v_{0,n}} & 0 \\ 0 & 0 & 0 & \sqrt{v_{1,n}} \end{pmatrix} \begin{pmatrix} R_{0,0,n} \\ R_{0,1,n} \\ R_{1,0,n} \\ R_{1,1,n} \end{pmatrix}
$$
$$
= \begin{pmatrix} \sqrt{v_{0,n}}\sqrt{\mathcal{T}}R_{0,0,n} + i\sqrt{v_{1,n}}\sqrt{\mathcal{R}}R_{0,1,n} \\ i\sqrt{v_{0,n}}\sqrt{\mathcal{R}}R_{0,0,n} + \sqrt{v_{1,n}}\sqrt{\mathcal{T}}R_{0,1,n} \\ \sqrt{v_{0,n}}\sqrt{\mathcal{T}}R_{1,0,n} + i\sqrt{v_{1,n}}\sqrt{\mathcal{R}}R_{1,1,n} \\ i\sqrt{v_{0,n}}\sqrt{\mathcal{R}}R_{1,0,n} + \sqrt{v_{0,n}}\sqrt{\mathcal{T}}R_{1,1,n} \end{pmatrix}, \tag{30}
$$

where the reflectivity $\mathcal{R}$ and transmissivity $\mathcal{T} = 1 - \mathcal{R}$ are real numbers which are considered to be parameters to be determined from experiment.

- *Phase shifter* χ_0, χ_1: In the event-based model, a phase shifter is simulated without DLM. The device has only one input and one output port and transforms

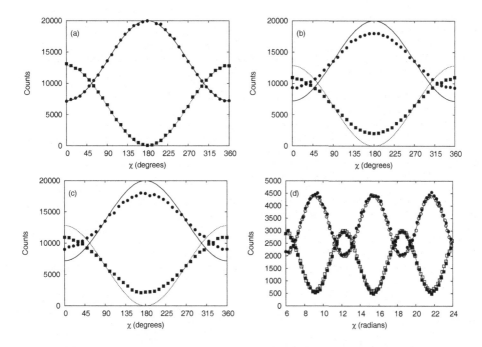

Fig. 8. (a)-(c) Event-by-event simulation results of the number of neutrons leaving the interferometer via the H-beam (circles) and O-beam (squares) as a function of the phase difference χ between the two paths inside the interferometer. For each value of χ, the number of particles generated in the simulation is $N = 100000$. The lines are the predictions of quantum theory. Solid line: p_H, see Eq. (34); dotted line: p_O, see Eq. (35). (a) Model parameters: $\mathcal{R} = 0.2$, $\gamma = 0.99$, $\delta_1 = \delta_2 = 0$. (b) Same as (a) except that $\gamma = 0.5$, reducing the accuracy and increasing the response time of the DLM. (c) Same as (a) except that to mimic the partial coherence of the incident neutron beam, the initial message carried by each particle has been modified by choosing δ_1 and δ_2 uniformly random from the interval $[-\pi/3, \pi/3]$, reducing the amplitude of the interference. (d) Comparison between the counts of neutrons per second and per square cm in the beams of a neutron interferometry experiment[91] (open symbols) and the number of neutrons per sample leaving the interferometer in an event-by-event simulation (solid symbols). Circles: counts in the H-beam; squares: counts in the O-beam. The experimental data has been extracted from Figure 2 of Ref. 91. The simulation parameters $\mathcal{R} = 0.22$ and $\gamma = 0.5$ have been adjusted by hand to obtain a good fit and the number of incident particles in the simulation is $N = 22727$ per angle χ. Lines through the data points are guides to the eye.

the nth input message $\mathbf{u}_n$ into an output message

$$\mathbf{w}_n = e^{i\chi_j}\mathbf{u}_n \quad j = 0, 1. \tag{31}$$

- *Detector:* Detectors count all incoming particles. Hence, we assume that the neutron detectors have a detection efficiency of 100%. This is an idealization of real neutron detectors which can have a detection efficiency of 99% and more.[91]

5.4.2. *Simulation results*

In Fig. 8 we present a few simulation results for the neutron MZI and compare them to the quantum theoretical result ((a)-(c)) and to experiment (d). A quantum theoretical treatment of the neutron MZI depicted in Fig. 7 is given in Ref. 93. The quantum statistics of the neutron interferometry experiment is described in terms of the state vector

$$|\Psi\rangle = (\Psi_{0\uparrow}, \Psi_{0\downarrow}, \Psi_{1\uparrow}, \Psi_{1\downarrow}\Psi_{2\uparrow}, \Psi_{2\downarrow}, \Psi_{3\uparrow}, \Psi_{3\downarrow})^T, \tag{32}$$

where the components of this vector represent the complex-valued amplitudes of the wave function. The first subscript labels the pathway and the second subscript denotes the direction of the magnetic moment relative to some **B**-field. The latter is not relevant for the neutron MZI experiment since the outcome of this experiment does not depend on the magnetic moment of the neutron. As usual, the state vector is assumed to be normalized, meaning that $\langle\Psi|\Psi\rangle = 1$. In the abstract representation of the experiment (see Fig. 7(right)) we use the notation $\Psi_j = (\Psi_{j\uparrow}, \Psi_{j\downarrow})$ for $j = 0, \ldots, 3$.

As the state vector propagates through the interferometer, it changes according to

$$|\Psi'\rangle = \begin{pmatrix} t^* & r \\ -r^* & t \end{pmatrix}_{5,7} \begin{pmatrix} t^* & r \\ -r^* & t \end{pmatrix}_{4,6} \begin{pmatrix} e^{i\phi_1} & 0 \\ 0 & e^{i\phi_1} \end{pmatrix}_{6,7} \begin{pmatrix} e^{i\phi_0} & 0 \\ 0 & e^{i\phi_0} \end{pmatrix}_{4,5}$$

$$\times \begin{pmatrix} t^* & r \\ -r^* & t \end{pmatrix}_{3,7} \begin{pmatrix} t^* & r \\ -r^* & t \end{pmatrix}_{2,6} \begin{pmatrix} t & -r^* \\ r & t^* \end{pmatrix}_{1,5} \begin{pmatrix} t & -r^* \\ r & t^* \end{pmatrix}_{0,4}$$

$$\times \begin{pmatrix} t & -r^* \\ r & t^* \end{pmatrix}_{1,3} \begin{pmatrix} t & -r^* \\ r & t^* \end{pmatrix}_{0,2} |\Psi\rangle, \tag{33}$$

where t and r denote the common transmission and reflection coefficients, respectively, and the subscripts i, j refer to the pair of elements of the eight-dimensional vector on which the matrix acts. Conservation of probability demands that $|t|^2 + |r|^2 = 1$.

In neutron interferometry experiments, particles enter the interferometer via the path corresponding to the amplitude Ψ_0 only (see Fig. 7 (right)), meaning that $|\Psi\rangle = (1, 0, 0, 0, 0, 0, 0, 0)^T$. The probabilities to observe a particle leaving the interferometer in the H- and O-beam are then given by

$$p_{\mathrm{H}} = |\Psi'_2|^2 = \mathcal{R}\left(\mathcal{T}^2 + \mathcal{R}^2 - 2\mathcal{R}\mathcal{T}\cos\chi\right), \tag{34}$$

$$p_{\mathrm{O}} = |\Psi'_3|^2 = 2\mathcal{R}^2\mathcal{T}\left(1 + \cos\chi\right), \tag{35}$$

where $\chi = \chi_0 - \chi_1$ is the relative phase shift, $\mathcal{R} = |r|^2$ and $\mathcal{T} = |t|^2 = 1 - \mathcal{R}$. Note that p_{H} and p_{O} do not depend on the imaginary part of t or r, leaving only one free model parameter (for instance $\mathcal{R}$). In the case of a 50-50 beam splitter ($\mathcal{T} = \mathcal{R} = 0.5$), Eqs. (34) and (35) reduce to the familiar expressions $p_{\mathrm{H}} = (1/2)\sin^2\chi/2$ and $p_{\mathrm{O}} = (1/2)\cos^2\chi/2$, respectively. The extra factor two is due to the fact that one half of all incoming neutrons, that is the neutrons that are transmitted by BS1 or BS2 (see Fig. 7), leave the interferometer without being counted.

The simulation results presented in Fig. 8(a) demonstrate that the event-by-event simulation reproduces the results of quantum theory if γ approaches one.[3,5,7,8] Indeed, there is excellent agreement with quantum theory. In this example, the reflectivity of the beam splitters is taken to be $\mathcal{R} = 0.2$. The parameter γ which controls the learning pace of the DLM-based processor can be used to account for imperfections of the neutron interferometer. This is illustrated in Fig. 8(b) which shows simulation results for $\gamma = 0.5$.

The quantum theoretical treatment assumes a fully coherent beam of neutrons. In the event-based approach, the case of a coherent beam may be simulated by assuming that the degree of freedom that accounts for the time of flight of the neutron takes the same initial value each time a message is created ($\delta_1 = \delta_2 = 0$). In the event-based approach, we can mimic a partially coherent beam by simply adding some noise to the message, that is when a message is created, δ_i for $i = 1, 2$ is chosen random in a specified range. In Fig. 8(c), we present simulation results for the case that δ_i is drawn randomly and uniformly from the interval $[-\pi/3, \pi/3]$, showing that reducing the coherence of the beam reduces the visibility, as expected on the basis of wave theory.[42] Comparing Fig. 8(b) and Fig. 8(c), we conclude that the same reduced visibility can be obtained by either reducing γ or by adding noise to the messages. On the basis of this interferometry experiment alone, it is difficult to exclusively attribute the cause of a reduced visibility to one of these mechanisms.

Conclusive evidence that the event-based model reproduces the results of a single-neutron interferometry experiment comes from comparing simulation data with experimental data. In Fig. 8(d), we present such a comparison using experimental data extracted from Fig. 2 of Ref. 91. It was not necessary to try to make the best fit: the parameters $\mathcal{R}$ and γ and the offset of the phase χ were varied by hand. As shown in Fig. 8(d), the event-based simulation model reproduces, quantitatively, the experimental results reported in Fig. 2 of Ref. 91.

6. Entanglement

In quantum theory entanglement is the property of a state of a two or many-body quantum system in which the states of the constituting bodies are correlated. The most prominent example is the singlet state of two spin-$\frac{1}{2}$ particles

$$|\Psi\rangle = \frac{1}{\sqrt{2}} \left(|\uparrow\downarrow\rangle - |\downarrow\uparrow\rangle \right), \tag{36}$$

which cannot be written as a product state. According to quantum theory, if the singlet state describes the correlation between the spins of the two particles and if we perform a measurement of both spins along the same direction, we observe that the particles have opposite but otherwise random values of their spins. Thus, in the quantum theoretical description, the state of the two spin-$\frac{1}{2}$ particles may be

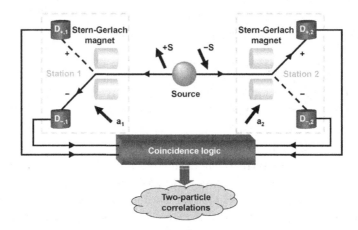

Fig. 9. Schematic diagram of the Einstein-Podolsky-Rosen-Bohm (EPRB) experiment with magnetic particles.[94] The source emits charge-neutral pairs of particles with opposite magnetic moments $+S$ and $-S$. One of the particles moves to station 1 and the other one to station 2. As the particle arrives at station $i = 1, 2$, it passes through a Stern-Gerlach magnet which deflects the particle, depending on the orientation of the magnet $\mathbf{a}_i$ and the magnetic moment of the particle. As the particle leaves the Stern-Gerlach magnet, it generates a signal in one of the two detectors $D_{\pm,i}$. Coincidence logic pairs the detection events of station 1 and station 2 so that they can be used to compute two-particle correlations.

correlated even though the particles are spatially and temporally separated and do not necessarily interact. Note however that this is a statistical interpretation which does not support the assumption that this singlet state is a property of each pair of particles and does not support the idea that changing the state of one particle has a causal effect on the state of the other.

6.1. *Einstein-Podolsky-Rosen-Bohm thought experiment*

In 1935, Einstein, Podolsky and Rosen (EPR) designed a thought experiment demonstrating the "incompleteness" of quantum theory.[95] The thought experiment involves the measurement of the position and momentum of two particles which interacted in the past but not at the time of measurement. Since this experiment is not suited for designing a laboratory experiment, Bohm proposed in 1951 a more realistic experiment which measures the intrinsic angular momentum of a correlated pair of atoms one-by-one.[94] A schematic diagram of the experiment is shown in Fig. 9. A source emits charge-neutral pairs of particles with opposite magnetic moments $+S$ and $-S$. The two particles separate spatially and propagate in free space to an observation station in which they are detected. As the particle arrives at station $i = 1, 2$, it passes through a Stern-Gerlach magnet. The magnetic moment of a particle interacts with the inhomogeneous magnetic field of the Stern-Gerlach magnet. The Stern-Gerlach magnet deflects the particle, depending on the orientation of the magnet $\mathbf{a}_i$ and the magnetic moment of the particle. The Stern-Gerlach magnet divides the beam of particles in two, spatially well-separated parts. As the particle leaves the Stern-Gerlach magnet, it generates a signal in one of the two detectors $D_{\pm,i}$. The firing of a detector corresponds to a detection event.

According to quantum theory of the Einstein-Podolsky-Rosen-Bohm (EPRB) thought experiment, the results of repeated measurements of the system of two spin-$\frac{1}{2}$ particles in the spin state $|\Psi\rangle = \alpha_0 |\uparrow\uparrow\rangle + \alpha_1 |\downarrow\uparrow\rangle + \alpha_2 |\uparrow\downarrow\rangle + \alpha_3 |\downarrow\downarrow\rangle$ with $\sum_{j=0}^{3} |\alpha_j|^2 = 1$ are given by the single-spin expectation values

$$\widehat{E}_1(\mathbf{a}_1) = \langle\Psi|\sigma_1 \cdot \mathbf{a}_1|\Psi\rangle = \langle\Psi|\sigma_1|\Psi\rangle \cdot \mathbf{a}_1,$$
$$\widehat{E}_2(\mathbf{a}_2) = \langle\Psi|\sigma_2 \cdot \mathbf{a}_2|\Psi\rangle = \langle\Psi|\sigma_2|\Psi\rangle \cdot \mathbf{a}_2, \qquad (37)$$

and the two-particle correlations $\widehat{E}(\mathbf{a}_1, \mathbf{a}_2) = \langle\Psi|\sigma_1 \cdot \mathbf{a}_1 \sigma_2 \cdot \mathbf{a}_2|\Psi\rangle = \mathbf{a}_1 \cdot \langle\Psi|\sigma_1 \cdot \sigma_2|\Psi\rangle \cdot \mathbf{a}_2$, where $\mathbf{a}_1$ and $\mathbf{a}_2$ are unit vectors specifying the directions of the analyzers, σ_i denote the Pauli vectors describing the spin of the particles $i = 1, 2$, and $\langle X\rangle = \mathrm{Tr}\rho X$ with ρ being the 4x4 density matrix describing the two spin-$\frac{1}{2}$ particle system. We have introduced the notation $\widehat{}$ to make a distinction between the quantum theoretical results and the results obtained by analysis of data sets from a laboratory experiment and from an event-based simulation (see Sect. 6.3). Quantum theory of the EPRB thought experiment assumes that $|\Psi\rangle$ does not depend on $\mathbf{a}_1$ or $\mathbf{a}_2$. Therefore, from Eq. (37) it follows immediately that $\widehat{E}_1(\mathbf{a}_1)$ does not depend on $\mathbf{a}_2$ and that $\widehat{E}_2(\mathbf{a}_2)$ does not depend on $\mathbf{a}_1$. Note that this holds for any state $|\Psi\rangle$. For later use, it is expedient to introduce the function

$$S \equiv S(\mathbf{a}_1, \mathbf{a}_2, \mathbf{a}_1', \mathbf{a}_2') = E(\mathbf{a}_1, \mathbf{a}_2) - E(\mathbf{a}_1, \mathbf{a}_2') + E(\mathbf{a}_1', \mathbf{a}_2) + E(\mathbf{a}_1', \mathbf{a}_2'), \qquad (38)$$

for which it can be shown that $|S| \leq 2\sqrt{2}$, independent of the choice of ρ.[96]

The quantum theoretical description of the EPRB experiment assumes that the state of the two spin-$\frac{1}{2}$ particles is described by the singlet state Eq. (36). For the singlet state, $\widehat{E}_1(\mathbf{a}_1) = \widehat{E}_2(\mathbf{a}_2) = 0$, $\widehat{E}(\mathbf{a}_1, \mathbf{a}_2) = -\mathbf{a}_1 \cdot \mathbf{a}_2$ and the maximum value of $|S|$ is $2\sqrt{2}$. Note that the singlet state is fully characterized by the three quantities $\widehat{E}_1(\mathbf{a}_1)$, $\widehat{E}_2(\mathbf{a}_2) = 0$, and $\widehat{E}(\mathbf{a}_1, \mathbf{a}_2)$. Hence, in any laboratory experiment, thought experiment or computer simulation of such an experiment, which has the goal to measure effects of the system being represented by a singlet state, these three quantities have to be measured and computed.

6.2. *Bell and Boole inequalities*

Quantum theory yields statistical estimates for $\widehat{E}_1$, $\widehat{E}_2$ and $\widehat{E}_{12}$ and cannot say anything about individual measurements.[1] Nevertheless, for the singlet state quantum theory predicts that, if measurement of the component $\sigma_1 \cdot \mathbf{a}_1$ with $\mathbf{a}_1$ being a unit vector, yields the value $+1$, then measurement of $\sigma_2 \cdot \mathbf{a}_1$ must yield the value -1 and vice versa. The fundamental question is how to relate the statistical results of quantum theory and the individual measurements.

6.2.1. Bell's model and inequality

Bell made the following assumptions in constructing his model and deriving his inequality:[97]

(1) $A(\mathbf{a_1}, \lambda) = \pm 1$ and $B(\mathbf{a_2}, \lambda) = \pm 1$, where A (B) denotes the result of measuring $\sigma_1 \cdot \mathbf{a_1}$ $(\sigma_2 \cdot \mathbf{a_2})$ and λ denotes a variable or a set of variables which only depend on the preparation (source) and not on the measurement (magnet settings) of the spin components. Note that this assumption already includes the hypothesis that the orientation of one magnet does not influence the measurement result obtained with the other magnet (often referred to as the locality condition). In other words, A (B) does not depend on $\mathbf{a_2}$ $(\mathbf{a_1})$.

(2) If $\rho(\lambda)$ is the probability distribution of λ $(\int \rho(\lambda)d\lambda = 1)$ then the expectation value of the product of the two components $\sigma_1 \cdot \mathbf{a_1}$ and $\sigma_2 \cdot \mathbf{a_2}$ can be written as $P(\mathbf{a_1}, \mathbf{a_2}) = \int d\lambda \rho(\lambda) A(\mathbf{a_1}, \lambda) B(\mathbf{a_2}, \lambda)$. Note that one could also introduce variables λ' and λ'' depending on the characteristics of the instruments on both sides. Averaging over these instrument dependent variables would result in new variables having values between -1 and $+1$. However, this is only the case if λ' and λ'' are completely independent. For example, if λ' and λ'' are sets of variables including the detection times, used for coincidence measurements in a laboratory experiment, then assumption 2 does not hold.[98]

(3) $A(\mathbf{a_1}, \lambda) = -B(\mathbf{a_1}, \lambda)$ so that $P(\mathbf{a_1}, \mathbf{a_2}) = -\int d\lambda \rho(\lambda) A(\mathbf{a_1}, \lambda) A(\mathbf{a_2}, \lambda)$. This assumption follows from the observation that $P(\mathbf{a_1}, \mathbf{a_2}) = \int d\lambda \rho(\lambda) A(\mathbf{a_1}, \lambda) B(\mathbf{a_2}, \lambda)$ reaches -1 at $\mathbf{a_1} = \mathbf{a_2}$ only if $A(\mathbf{a_1}, \lambda) = -B(\mathbf{a_1}, \lambda)$. Note that $P(\mathbf{a_1}, \mathbf{a_1}) = -1$ if and only if $A(\mathbf{a_1}, \lambda) = -B(\mathbf{a_1}, \lambda)$, making both these assumptions equivalent. Hence, what Bell assumed is that the results of the measurements at both sides of the source can be represented by one and the same symbol "A" that depends only on the respective magnet setting and on λ. Moreover, also the measurement outcomes of an experiment with another setting of (only one of) the magnets, can be represented by the same symbol "A".

Using the above hypotheses and considering a third unit vector $\mathbf{a_3}$ Bell derived the inequality[97]

$$|P(\mathbf{a_1}, \mathbf{a_2}) - P(\mathbf{a_1}, \mathbf{a_3})| \leq 1 + P(\mathbf{a_2}, \mathbf{a_3}), \qquad (39)$$

which is violated for certain magnet settings $\mathbf{a_1}, \mathbf{a_2}, \mathbf{a_3}$ if $P(\mathbf{a_1}, \mathbf{a_2})$ is replaced by $\widehat{E}(\mathbf{a_1}, \mathbf{a_2}) = -\mathbf{a_1} \cdot \mathbf{a_2}$, the quantum theoretical two-particle expectation value describing the averaged two-particle correlations obtained in EPRB experiments. Note that 1, 2 and 3 are sufficient conditions for the Bell inequality to be obeyed. Hence, if the Bell inequality is obeyed then one cannot say anything about the validity of the assumptions, but if it is violated then one can say that at least one of the assumptions must be false, thereby refuting Bell's model. It is worth mentioning that Bell analyzed a very restricted class of classical models, namely models which do not account for (i) the physics of the detection process and/or (ii) the use of time-coincidences to define particle pairs (see below). Although the above conclusion is the only logical conclusion that can be drawn, it is common but erroneous practice to take a violation of a Bell inequality as a "proof" of the quantum nature of the system under study. Far reaching conclusions drawn from Bell's results, such as violations of Bell-like inequalities having implications for action-on-a-distance, locality, realism , ..., have all been shown to be logical fallacies.[29,99–114]

6.2.2. *Boole inequality for the correlations of two-valued variables*

We consider two-valued variables $S(x, n) = \pm 1$ where x can be considered to represent the orientations $\mathbf{a}_1, \mathbf{a}_2, \mathbf{a}_3$ of the magnets in an EPRB experiment and $n = 1, \ldots N$ simply numbers the measurements in an experimental run. From the variables $S(x, n)$ with $x = \mathbf{a}_1, \mathbf{a}_2, \mathbf{a}_3$ we compute the averages $F_{\mathbf{a}_1, \mathbf{a}_2} = \sum_{n=1}^{N} S(\mathbf{a}_1, n) S(\mathbf{a}_2, n)/N$, $F_{\mathbf{a}_1, \mathbf{a}_3} = \sum_{n=1}^{N} S(\mathbf{a}_1, n) S(\mathbf{a}_3, n)/N$ and $F_{\mathbf{a}_2, \mathbf{a}_3} = \sum_{n=1}^{N} S(\mathbf{a}_2, n) S(\mathbf{a}_3, n)/N$. According to Boole[115] it is impossible to violate

$$|F_{\mathbf{a}_1, \mathbf{a}_2} \pm F_{\mathbf{a}_1, \mathbf{a}_3}| \leq 1 \pm F_{\mathbf{a}_2, \mathbf{a}_3}, \tag{40}$$

if there is a one-to-one correspondence between the two-valued variables $S(\mathbf{a}_1, n)$, $S(\mathbf{a}_2, n)$, $S(\mathbf{a}_3, n)$ of the mathematical description and each triple $\{X(\mathbf{a}_1, n), X(\mathbf{a}_2, n), X(\mathbf{a}_3, n)\}$ of binary data collected in the experimental run denoted by n. This one-to-one correspondence is a necessary and sufficient condition for the inequality to be obeyed. Note that inequalities Eq. (39) and Eq. (40) have the same structure. We emphasize that it is essential that the correlations $F_{\mathbf{a}_1, \mathbf{a}_2}$, $F_{\mathbf{a}_1, \mathbf{a}_3}$ and $F_{\mathbf{a}_2, \mathbf{a}_3}$ have been calculated from one data set that contains triples instead of from three sets in which the data has been collected in pairs.[113]

6.2.3. *An inequality within quantum theory*

From the algebraic identity $(1 \pm xy)^2 = (x \pm y)^2 + (1 - x^2)(1 - y^2)$ it follows that $|x \pm y| \leq 1 \pm xy$ for real numbers x and y with $|x| \leq 1$ and $|y| \leq 1$. From this inequality it immediately follows that

$$|xz \pm yz| \leq 1 \pm xy, \tag{41}$$

for real numbers x, y, z such that $|x| \leq 1$, $|y| \leq 1$ and $|z| \leq 1$.

If we now assume that the two spin-$\frac{1}{2}$ particle system is in a product state $|\Psi\rangle = |\Psi\rangle_1 |\Psi\rangle_2$ with $|\Psi\rangle_j = \alpha_{0,j} |\uparrow\rangle_j + \alpha_{1,j} |\uparrow\rangle_j$ with $|\alpha_{0,j}|^2 + |\alpha_{1,j}|^2 = 1$ for

$j = 1, 2$, then

$$\widehat{E}_1(\mathbf{a}_1) = \langle \Psi | \sigma_1 | \Psi \rangle_1 \cdot \mathbf{a}_1,$$

$$\widehat{E}_2(\mathbf{a}_2) = \langle \Psi | \sigma_2 | \Psi \rangle_2 \cdot \mathbf{a}_2,$$

$$\widehat{E}(\mathbf{a}_1, \mathbf{a}_2) = \langle \Psi | \sigma_1 | \Psi \rangle_1 \cdot \mathbf{a}_1 \langle \Psi | \sigma_2 | \Psi \rangle_2 \cdot \mathbf{a}_2 = \widehat{E}_1(\mathbf{a}_1)\widehat{E}_2(\mathbf{a}_2), \qquad (42)$$

and the correlation $\widehat{E}(\mathbf{a}_1, \mathbf{a}_2) - \widehat{E}_1(\mathbf{a}_1)\widehat{E}_2(\mathbf{a}_2) = 0$. Using Eq. (41) and unit vectors $\mathbf{a}_1$, $\mathbf{a}_2$, $\mathbf{a}_3$ we obtain a Bell-type inequality

$$|\widehat{E}(\mathbf{a}_1, \mathbf{a}_2) - \widehat{E}(\mathbf{a}_1, \mathbf{a}_3)| \leq 1 + \widehat{E}(\mathbf{a}_2, \mathbf{a}_3), \qquad (43)$$

and similarly the Bell-CHSH inequality[116]

$$|S| = |\widehat{E}(\mathbf{a}_1, \mathbf{a}_2) - \widehat{E}(\mathbf{a}_1, \mathbf{a}_2') + \widehat{E}(\mathbf{a}_1', \mathbf{a}_2) + \widehat{E}(\mathbf{a}_1', \mathbf{a}_2')| \leq 2, \qquad (44)$$

for unit vectors $\mathbf{a}_1$, $\mathbf{a}_1'$, $\mathbf{a}_2$, and $\mathbf{a}_2'$.

Hence, if the state of the two spin-$\frac{1}{2}$ particle system is a product state, then the Bell and Bell-CHSH inequality hold. On the other hand, if the Bell or Bell-CHSH inequality is violated then the two-particle quantum system is not in a product state. Note that these logical statements are made entirely within the framework of quantum theory.

6.2.4. Bell inequality tests

In a typical ideal EPRB experiment three runs are performed in which N detection events are collected on both sides (referred to by 1 and 2) of the source. The outcomes of the detection events take the values $+1$ or -1 and are represented by the symbol X. This results in the three data sets

$$\Gamma_{\mathbf{a},\mathbf{b}} = \{X(\mathbf{a}, n, 1), X(\mathbf{b}, n, 2) | n = 1, \ldots, N\},$$

$$\widetilde{\Gamma}_{\mathbf{a},\mathbf{c}} = \{\widetilde{X}(\mathbf{a}, \widetilde{n}, 1), \widetilde{X}(\mathbf{c}, \widetilde{n}, 2) | \widetilde{n} = 1, \ldots, N\},$$

$$\widehat{\Gamma}_{\mathbf{b},\mathbf{c}} = \{\widehat{X}(\mathbf{b}, \widehat{n}, 1), \widehat{X}(\mathbf{c}, \widehat{n}, 2) | \widehat{n} = 1, \ldots, N\}. \qquad (45)$$

Note that in real experiments the measurement outcomes are also labeled by the time of measurement but for simplicity we omit this label here. Using these data sets for testing the validity of Bell's inequality Eq. (39) and of the structurally equivalent Boole inequality Eq. (40), requires making the following assumptions:

(1) The   same   symbol
    X can be used for all the data collected in the three runs. This results in the data set $\Upsilon = \{X(\mathbf{a}, n, 1), X(\mathbf{a}, \widetilde{n}, 1), X(\mathbf{b}, n, 2), X(\mathbf{b}, \widetilde{n}, 1), X(\mathbf{c}, \widetilde{n}, 2),$ $X(\mathbf{c}, \widehat{n}, 2) | n, \widetilde{n}, \widehat{n} = 1, \ldots, N\}$.

(2) The data can be rearranged such that $X(\mathbf{a}, n, 1) = X(\mathbf{a}, \widetilde{n}, 1)$, $X(\mathbf{b}, \widehat{n}, 1) = X(\mathbf{b}, n, 1)$ and $X(\mathbf{c}, \widetilde{n}, 2) = X(\mathbf{c}, \widehat{n}, 2) = X(\mathbf{c}, n, 2)$. This results in the data set $\Upsilon' = \{X(\mathbf{a}, n, 1), X(\mathbf{b}, n, 2), X(\mathbf{b}, n, 1), X(\mathbf{c}, n, 2) | n = 1, \ldots, N\}$, a data set containing quadruples, not yet triples, as used in the derivation of Bell's inequality and as required by Boole for his inequality to be obeyed. Reduction to a set of triples requires the extra assumption:

(3) $X(\mathbf{b}, n, 1) = X(\mathbf{b}, n, 2)$

Since the data in EPRB laboratory experiments are not collected as one set of triples but as three sets of pairs, in case a violation of Boole's inequality Eq. (40) is found, at least one of the assumptions 1, 2 or 3 is false. In other words, if the data sets collected in an EPRB experiment satisfy these three conditions, the one-to-one correspondence between the two-valued variables in the mathematical description and the observed two-valued experimental data is guaranteed, and hence Boole's and thus also Bell's inequality are satisfied. If the Bell inequality is violated then at least one of the sufficient conditions 1, 2 or 3 to derive the Bell inequality is false, but then also at least one of the assumptions listed above is false.

6.2.5. *Summary*

One could ask the question how to translate the inequality Eq. (43) together with its accompanying assumptions, derived within the context of quantum theory, into an experimental test. The answer is one simply cannot. It is not legitimate to replace the quantum theoretical expectations that appear in Eq. (43) by certain empirical data, simply because Eq. (43) has been derived within the mathematical framework of quantum theory, not for sets of data collected, grouped and characterized by experimenters. Since the collected data have values $+1$ or -1 they can be tested against the Boole inequalities only and the conclusions that follow from their violation (see Sect. 6.2.4) have no bearing on the quantum theoretical model, without making additional assumptions which are not self-evident.

In conclusion, an inequality cannot be blindly applied to any set of experimental data, a model or theory. The inequality should be derived in the proper context and conditions and conclusions belonging to the respective derivations cannot simply be mixed.

6.3. *EPRB experiment with single photons*

In this experiment, the polarization of each photon plays the role of the spin-$\frac{1}{2}$ degree-of-freedom in Bohm's version[94] of the EPR thought experiment.[95] Using the fact that the two-dimensional vector space with basis vectors $\{|H\rangle, |V\rangle\}$, where H and V denote the horizontal and vertical polarization of the photon, respectively, is isomorphic to the vector space with basis vectors $\{|\uparrow\rangle, |\downarrow\rangle\}$ of spin-$\frac{1}{2}$ particles, we may use the quantum theory of the latter to describe the EPRB experiments with photons. The expressions for the single-photon expectation values and the two-photon correlations are similar to those of the genuine spin-$\frac{1}{2}$ particle problem except for the restriction of $\mathbf{a}_1$ and $\mathbf{a}_2$ to lie in planes orthogonal to the direction of propagation of the photons and that the polarization is defined modulo π, not modulo 2π as in the case of the spin-$\frac{1}{2}$ particles. The latter results in a multiplication of the angles by a factor of two. For simplicity it is often assumed that $\mathbf{a}_i =$

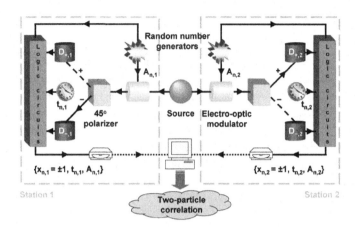

Fig. 10. Schematic diagram of the EPRB experiment with single photons.[25,117] The source emits pairs of photons. The photon pair splits and one of the photons moves to station 1 and the other one to station 2. As the photon arrives at station $i = 1, 2$ it first passes through an electro-optic modulator (EOM) which rotates the polarization of the photon by an angle θ_i depending on the voltage applied to the EOM. This voltage is controlled by a binary variable A_i, which is chosen at random. As the photon leaves the EOM, a polarizing beam splitter directs it to one of the two detectors $D_{\pm,i}$. The detector produces a signal $x_{n,i} = \pm 1$ where the subscript n labels the nth detection event. Each station has its own clock which assigns a time-tag $t_{n,i}$ to each detection signal. A data set $\{x_{n,i}, t_{n,i}, A_{n,i} | n = 1, \ldots, N_i\}$ is stored on a hard disk for each station. Long after the experiment is finished both data sets can be analyzed and among other things, two-particle correlations can be computed.

$(\cos a_i, \sin a_i, 0)$ for $i = 1, 2$. For the singlet state we then have $\widehat{E}_1(\mathbf{a}_1) = \widehat{E}_2(\mathbf{a}_2) = 0$ and $\widehat{E}(\mathbf{a}_1, \mathbf{a}_2) = -\cos 2(a_1 - a_2)$.

We take the EPRB experiment with single photons, carried out by Weihs *et al.*,[25,117] as a concrete example. We first describe the data collection and analysis procedure of the experiment and present results demonstrating that the conclusion that the experimental results can be described by quantum theory is premature. Next we illustrate how to construct an event-based model of an idealized version of this EPRB experiment which reproduces the predictions of quantum theory for the single and two-particle averages for a quantum system of two spin-$\frac{1}{2}$ particles in the singlet state and a product state,[3,31] without making reference to concepts of quantum theory.

- *Data collection:* Figure 10 shows a schematic diagram of the EPRB experiment with single photons carried out by Weihs *et al.*[25,117] The source emits pairs of photons. The photon pair splits and each photon travels in free space to an observation station, labeled by $i = 1$ or $i = 2$, in which it is manipulated and detected. The two stations are assumed to be identical and are separated spatially and temporally. Hence, the observation at station 1 (2) cannot have a causal effect on the data registered at station 2 (1).[25] As the photon arrives at station $i = 1, 2$ it first passes through an electro-optic modulator (EOM) which rotates the polarization of the photon by an angle θ_i depending on the voltage applied to the EOM.[25,117] This voltage is controlled by a binary variable A_i, which is chosen at random.[25,117] Optionally, a bias voltage is added to the randomly varying voltage.[25,117] The relation between the voltage applied to the EOM and the resulting rotation of the polarization is determined experimentally, hence there is some uncertainty in relating the applied voltage to the rotation angle.[25,117] As the photon leaves the EOM, a polarizing beam splitter directs it to one of the two detectors. The detector produces a signal $x_{n,i} = \pm 1$ where the subscript n labels the nth detection event. Each station has its own clock which assigns a time-tag $t_{n,i}$ to each signal generated by one of the two detectors.[25,117] Effectively, this procedure discretizes time in intervals, the width of which is determined by the time-tag resolution τ. In the experiment, the time-tag generators are synchronized before each run.[25,117]

The firing of a detector is regarded as an event. At the nth event at station i, the dichotomic variable $A_{n,i}$, controlling the rotation angle $\theta_{n,i}$, the dichotomic variable $x_{n,i}$ designating which detector fires, and the time tag $t_{n,i}$ of the detection event are written to a file on a hard disk, allowing the data to be analyzed long after the experiment has terminated.[25,117] The set of data collected at station i may be written as

$$\Upsilon_i = \{x_{n,i}, t_{n,i}, \theta_{n,i} | n = 1, \ldots, N_i\}, \tag{46}$$

where we allow for the possibility that the number of detected events N_i at stations $i = 1, 2$ need not (and in practice is not) to be the same and we have used the rotation angle $\theta_{n,i}$ instead of the corresponding experimentally relevant dichotomic variable $A_{n,i}$ to facilitate the comparison with the quantum theoretical description.

- *Data analysis procedure:* A laboratory EPRB experiment requires some criterion to decide which detection events are to be considered as stemming from a single or two-particle system. In EPRB experiments with photons, this decision is taken on the basis of coincidence in time.[25,118] Here we adopt the procedure employed by Weihs *et al.*[25,117] Coincidences are identified by comparing the time differences $t_{n,1} - t_{m,2}$ with a window W,[25,117,118] where $n = 1, \ldots, N_1$ and $m = 1, \ldots, N_2$. By definition, for each pair of rotation angles a_1 and a_2, the number of coincidences between detectors $D_{x,1}$ ($x = \pm 1$) at station 1 and detectors $D_{y,2}$ ($y = \pm 1$) at station 2 is given by

$$\begin{aligned} C_{xy} &= C_{xy}(a_1, a_2) \\ &= \sum_{n=1}^{N_1} \sum_{m=1}^{N_2} \delta_{x,x_{n,1}} \delta_{y,x_{m,2}} \delta_{a_1,\theta_{n,1}} \delta_{a_2,\theta_{m,2}} \Theta(W - |t_{n,1} - t_{m,2}|), \end{aligned} \tag{47}$$

where $\Theta(t)$ denotes the unit step function. In Eq. (47) the sum over all events has to be carried out such that each event ($=$ one detected photon) contributes only once. Clearly, this constraint introduces some ambiguity in the counting procedure as there is a priori, no clear-cut criterion to decide which events at stations $i = 1$ and $i = 2$ should be paired. One obvious criterion might be to choose the pairs such that C_{xy} is maximum, but such a criterion renders the data analysis procedure (not the data production) acausal. It is trivial though to analyze the data generated by the experiment of Weihs *et al.* such that conclusions do not suffer from this artifact.[80] In general, the values for the coincidences $C_{xy}(a_1, a_2)$ depend on the time-tag resolution τ and the window W used to identify the coincidences.

The single-particle averages and correlation between the coincidence counts are defined by

$$E_1(a_1, a_2) = \frac{\sum_{x,y=\pm 1} x C_{xy}}{\sum_{x,y=\pm 1} C_{xy}} = \frac{C_{++} - C_{--} + C_{+-} - C_{-+}}{C_{++} + C_{--} + C_{+-} + C_{-+}}$$

$$E_2(a_1, a_2) = \frac{\sum_{x,y=\pm 1} y C_{xy}}{\sum_{x,y=\pm 1} C_{xy}} = \frac{C_{++} - C_{--} - C_{+-} + C_{-+}}{C_{++} + C_{--} + C_{+-} + C_{-+}}$$

$$E(a_1, a_2) = \frac{\sum_{x,y=\pm 1} xy C_{xy}}{\sum_{x,y=\pm 1} C_{xy}} = \frac{C_{++} + C_{--} - C_{+-} - C_{-+}}{C_{++} + C_{--} + C_{+-} + C_{-+}}, \tag{48}$$

where the denominator $N_c = N_c(a_1, a_2) = C_{++} + C_{--} + C_{+-} + C_{-+}$ in Eq. (48) is the sum of all coincidences.

In practice, coincidences are determined by a four-step procedure:[117]

(1) Compute a histogram of time-tag differences $t_{n,1} - t_{m,2}$ of pairs of detection events.

(2) Determine the time difference Δ_G for which this histogram shows a maximum.

(3) Add Δ_G to the time-tag data $t_{n,1}$, thereby moving the position of the maximum of the histogram to zero.

(4) Determine the coincidences using the new time-tag differences, each photon contributing to the coincidence count at most once.

The global offset, denoted by Δ_G, may be attributed to the loss of synchronization of the clocks used in the stations 1 and 2.[117]

Local-realistic treatments of the EPRB experiment assume that the correlation, as measured in the experiment, is given by[119]

$$C_{xy}^{(\infty)}(a_1, a_2) = \sum_{n=1}^{N} \delta_{x,x_{n,1}} \delta_{y,x_{n,2}} \delta_{a_1,\theta_{n,1}} \delta_{a_2,\theta_{m,2}}, \tag{49}$$

which is obtained from Eq. (47) (in which each photon contributes only once) by assuming that $N = N_1 = N_2$, pairs are defined by $n = m$ and by taking the limit $W \rightarrow \infty$. However, the working hypothesis that the value of W should not matter because the time window only serves to identify pairs may not apply

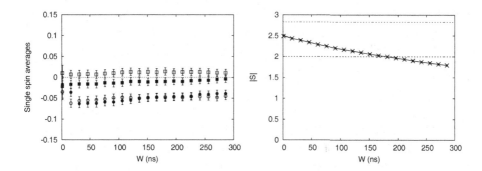

Fig. 11. Analysis of the data set **newlongtime2**. Left: Selected single-particle averages as a function of W for $\Delta_G = 0$ and $a_1 = 0$, $a_1' = \pi/4$, $a_2 = \pi/8$ and $a_2' = 3\pi/8$. Open squares: $E_1(a_1, a_2)$; open circles: $E_1(a_1, a_2')$; solid squares: $E_2(a_1, a_2)$; solid circles: $E_2(a_1', a_2)$. The error bars correspond to 2.5 standard deviations. Right: $|S| = |E(a_1, a_2) - E(a_1, a_2') + E(a_1', a_2) + E(a_1', a_2')|$ as a function of the time window W. The dashed lines represent the maximum value for a quantum system of two S = 1/2 particles in a separable (product) state ($|S| = 2$) and in a singlet state ($|S| = \sqrt{2}/2$), respectively. Crosses: $\Delta_G = 0$; solid circles connected by the solid line: $\Delta_G = 0.5$ns.

to real experiments. The analysis of the data of the experiment of Weihs *et al.* shows that the average time between pairs of photons is of the order of 30μs or more, much larger than the typical values (of the order of a few nanoseconds) of the time-window W used in the experiments.[117] In other words, in practice, the identification of photon pairs does not require the use of W's of the order of a few nanoseconds.

- *Data analysis results:* Here, we present only a very limited set of results of our analysis of the experimental data of Weihs *et al.*. This data has already been analyzed in Refs. 29,31,80,120–126.

In order to test whether the experimental results are compatible with the predictions of quantum theory for a system of two spin-$\frac{1}{2}$ particles we first check whether $E_1(a_1, a_2)$ is independent of a_2 and $E_2(a_1, a_2)$ is independent of a_1 because quantum theory predicts that this is the case independent of the state of the two-particle system (see Eq. (37)). Since we are dealing with real data we need a criterion to decide whether the data complies with this quantum theoretical prediction. We consider the data $E_1(a_1, a_2)$ ($E_2(a_1, a_2)$) to be in conflict with the quantum theoretical prediction if the data show a dependency on a_1 (a_2) that exceeds five times the upper bound $1/\sqrt{N_C(a_1, a_2)}$ to the standard deviation σ_{N_c}.

We analyze a selection of single-particle expectations as a function of W for the dataset **newlongtime2** (see Fig. 11(left)). For small W, the total number of coincidences is too small to yield statistically meaningful results. For $W > 20$ns it is clear that the curves for $E_1(a_1 = 0, a_2 = \pi/8)$ and $E_1(a_1, a_2' = 3\pi/8)$ (open symbols), and for $E_2(a_1 = 0, a_2 = \pi/8)$ and $E_2(a_1' = \pi/4 = 0, a_2 = \pi/8)$

(closed symbols) are not independent of the settings a_2 and a_1, respectively. The change of these single-spin averages observed in station 1 (station 2) when the settings are changed in station 2 (station 1), systematically exceeds five standard deviations, clearly violating our criterion for the data to be compatible with the prediction of quantum theory of the EPRB model. According to standard practice of hypothesis testing, the likelihood that this data set can be described by the quantum theory of the EPRB experiment should be considered as extremely small. An analysis of in total 23 data sets produced by the experiment of Weihs *et al.* shows that none of these data sets satisfies our hypothesis test for being compatible with the predictions of quantum theory of the EPRB model. Based on the observation of dependency of $E_1(a_1, a_2)$ on a_2 and $E_2(a_1, a_2)$ on a_1 one could conclude that the data exhibits a spurious kind of "non-locality" which cannot be described by the quantum theory of the EPRB experiment. In trying to find an explanation for this "non-locality" we demonstrated elsewhere[80,127] that including a model for the detection efficiencies of the detectors cannot resolve the conflict between the experimental data of Weihs *et al.* and the quantum theoretical description of the EPRB experiment.

Although the results for the single particle expectations demonstrate that the experimental data cannot be described by a quantum theoretical model of two spin-$\frac{1}{2}$ particles (independent of the state which the two photons are in), in what follows we nevertheless investigate the function S (see Eq. (38)) as a function of the time window W. Our motivation to do this is two-fold. First, the goal of the experiment of Weihs *et al.* was to demonstrate a violation of the Bell-CHSH inequality. We show that the amount of violation depends on W, a parameter absent in the data collection procedure but chosen in the data analysis procedure. Second, in Sect. 6.3.1 we demonstrate that the Bell-CHSH inequality can also be violated in an event-based model, a classical dynamical system outside the realm of classical Hamiltonian dynamics, of the type of EPRB experiment performed by Weihs *et al.*.

Figure 11(right) shows results of the function S as a function of W for the dataset **newlongtime2**. For $W < 150$ ns, the Bell-CHSH inequality $|S| \leq 2$ is clearly violated. For $W > 200$ ns, much less than the average time ($> 30\mu$s) between two coincidences, the inequality $|S| \leq 2$ is satisfied, demonstrating that the "nature" of the emitted pairs is not an intrinsic property of the pairs themselves but also depends on the choice of W made by the experimenter. For $W > 20$ ns, there is no significant statistical evidence that the "noise" on the data depends on W but if the only goal is to maximize $|S|$, it is expedient to consider $W < 20$ ns.

In other words, depending on the value of W, chosen by the experimenter when analyzing the data, the inequality $|S| \leq 2$ may or may not be violated. Hence, also the conclusion about the state of the system depends on the value of W. Analysis of the data of the experiment by Weihs *et al.* shows that W can be as large as 150 ns for the Bell-CHSH inequality to be violated and in the time-stamping EPRB experiment of Agüero *et al.*[124] $|S| \leq 2$ is clearly violated for $W < 9\mu s$. Hence, the use of a time-coincidence window does not create a "loophole". Nevertheless, very often it is mentioned that these single-photon Bell test experiments suffer from the fair sampling loophole, being the result of the usage of a time window W to filter out coincident photons or being the result of the usage of inefficient detectors.[122] The detection loophole was first closed in an experiment with two entangled trapped ions[128] and later in a single-neutron interferometry experiment[129] and in an experiment with two entangled qubits.[130] However, the latter three experiments are not Bell test experiments performed according to the CHSH protocol[116] because the two degrees of freedom are not manipulated and measured independently.

The narrow time window W in the experiment by Weihs *et al.* mainly acts as a filter that selects pairs of which the individual photons differ in their time tags by the order of nanoseconds. The possibility that such a filtering mechanism can lead to correlations that are often thought to be a characteristic of quantum systems only was, to our knowledge, first pointed out by P. Pearle[131] and later by A. Fine,[101] opening the route to a description in terms of locally causal, classical models. A concrete model of this kind was proposed by S. Pascazio who showed that his model approximately reproduces the correlation of the singlet state[132] with an accuracy that seems beyond what is experimentally achievable to date. Larson and Gill showed that Bell-like inequalities need to be modified in the case that the coincidences are determined by a time-window filter.[98] We found models that exactly reproduce the results of quantum theory for the singlet and uncorrelated state.[3,26,28,31] Here, we closely follow Refs. 4,28,31.

6.3.1. *Event-based simulation*

A minimal, discrete-event simulation model of the EPRB experiment by Weihs *et al.* (see Fig. 10) requires a specification of the information carried by the particles, of the algorithm that simulates the source and the observation stations, and of the procedure to analyze the data. Since in the description of the experiment the orientation of the polarization vectors and the orientations of the optical axis of the polarizers $\mathbf{a}_i = (\cos a_i, \sin a_i, 0)$ for $i = 1, 2$ is limited to the xy-plane we omit the z-component in the simulation.

- *Source and particles:* Each time, the source emits two particles which carry a vector $\mathbf{u}_{n,i} = (\cos(\xi_n + (i-1)\pi/2), \sin(\xi_n + (i-1)\pi/2))$, representing the polarization of the photons. This polarization is completely characterized by the angle ξ_n and the direction $i = 1, 2$ to which the particle moves. A uniform pseudo-random number generator is used to pick the angle $0 \leq \xi_n < 2\pi$. Clearly, the source emits two particles with a mutually orthogonal, hence correlated but otherwise random polarization. Note that for the simulation of this experiment it is not necessary that the particles carry information about the phase $2\pi f t_{i,n}$, although it would be possible. In this case the time of flight $t_{i,n}$ is determined by the time-tag model (see below).

- *Electro-optic modulator* (EOM): The EOM in station $i = 1, 2$ rotates the polarization of the incoming particle by an angle θ_i, that is its polarization angle becomes $\xi'_{n,i} \equiv \text{EOM}_i(\xi_n + (i-1)\pi/2, \theta_i) = \xi_n + (i-1)\pi/2 - \theta_i$ symbolically. Mimicking the experiment of Weihs *et al.* in which θ_1 can take the values a_1, a'_1 and θ_2 can take the values a_2, a'_2, we generate two binary uniform pseudo-random numbers $A_i = 0, 1$ and use them to choose the value of the angles θ_i, that is $\theta_1 = a_1(1 - A_1) + a'_1 A_1$ and $\theta_2 = a_2(1 - A_2) + a'_2 A_2$.

- *Polarizing beam splitter:* The simulation model for a polarizing beam splitter is defined by the rule

$$x_{n,i} = \begin{cases} +1 \text{ if } r_n \le \cos^2 \xi'_{n,i} \\ -1 \text{ if } r_n > \cos^2 \xi'_{n,i} \end{cases}, \tag{50}$$

where $0 < r_n < 1$ are uniform pseudo-random numbers. It is easy to see that for fixed $\xi'_{n,i} = \xi'_i$, this rule generates events such that

$$\lim_{N \to \infty} \frac{1}{N} \sum_{n=1}^{N} x_{n,i} = \cos^2 \theta_{n,i}, \tag{51}$$

with probability one, showing that the distribution of events complies with Malus law. Note that this model for the PBS does not make use of a DLM and is therefore much more simple than the event-based model of the PBS described in Sect. 5.3.1. This simplified mathematical model suffices to simulate the EPRB experiment but cannot be used to simulate other optics experiments (for instance Wheeler's delayed choice experiment). However, the PBS described in Sect. 5.3.1 can be used to simulate the EPRB experiment.[3]

- *Time-tag model:* As is well-known, as light passes through an EOM (which is essentially a tuneable wave plate), it experiences a retardation depending on its initial polarization and the rotation by the EOM. However, to our knowledge, time delays caused by retardation properties of waveplates, being components of various optical apparatuses, have not yet been explicitly measured for single photons. Therefore, in the case of single-particle experiments, we hypothesize that for each particle this delay is represented by the time tag[28,31]

$$t_{n,i} = \lambda(\xi'_{n,i})r'_n, \tag{52}$$

that is, the time tag is distributed uniformly ($0 < r'_n < 1$ is a uniform pseudo-random number) over the interval $[0, \lambda(\xi'_{n,i})]$. For $\lambda(\xi'_{n,i}) = T_0 \sin^4 2\xi'_{n,i}$ this time-tag model, in combination with the model of the polarizing beam splitter, rigorously reproduces the results of quantum theory of the EPRB experiments in the limit $W \to 0$.[28,31] We therefore adopt the expression $\lambda(\xi'_{n,i}) = T_0 \sin^4 2\xi'_{n,i}$ leaving only T_0 as an adjustable parameter.

- *Detector:* The detectors are ideal particle counters, meaning that they produce a click for each incoming particle. Hence, we assume that the detectors have

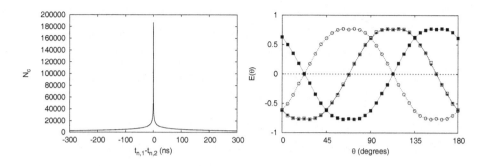

Fig. 12.   Simulation results using the time-tag model Eq. (52) with $T_0 = 1000$ ns. The total number of pairs generated by the source is 3×10^5, roughly the same as in experiment.[25]   (a) Coincidence count N_c as a function of the time-tag difference $t_{n,1} - t_{n,2}$. (b) Two-particle correlations as a function of θ for $W = 50$ ns. Open squares: $E(\theta) = E(a_1 = \theta, a_2 = \pi/8)$; open circles: $E(\theta) = E(a'_1 = \theta + \pi/4, a_2 = \pi/8)$; solid squares: $E(\theta) = E(a_1 = \theta, a'_2 = 3\pi/8)$; solid circles: $E(\theta) = E(a'_1 = \theta + \pi/4, a'_2 = 3\pi/8)$.

100% detection efficiency. Note that adaptive threshold detectors can be used (see Sect. 5.1.1) equally well.[3]

- *Simulation procedure:* The simulation algorithm generates the data sets Υ_i, similar to the ones obtained in the experiment (see Eq. (46)). In the simulation, it is easy to generate the events such that $N_1 = N_2$. We analyze these data sets in exactly the same manner as the experimental data are analyzed, implying that we include the post-selection procedure to select photon pairs by a time-coincidence window W. The latter is crucial for our simulation method to give results that are very similar to those observed in a laboratory experiment. Although in the simulation the ratio of detected to emitted photons is equal to one, the final detection efficiency is reduced due to the time-coincidence post-selection procedure.

6.3.2. *Simulation results*

In Fig. 12(a) we present simulation results for the distribution of time-tag differences, as obtained by using time-tag model Eq. (52). The distribution is sharply peaked and displays long tails, in qualitative agreement with experiment.[117] The single-particle averages $E_1(a_1, a_2)$ and $E_2(a_1, a_2)$ (results not shown) are zero up to the usual statistical fluctuations and do not show any statistically relevant dependence on a_2 or a_1, respectively, in concert with a rigorous probabilistic treatment of this simulation model.[31]

Some typical simulation results for the two-particle correlations are depicted in Fig. 12(b) for $W = 50$ ns. For this value of the time-window W, the minimum and maximum value of the two-particle correlations is not -1 and $+1$, respectively, as would be expected from the quantum theoretical description. Moreover, the two-particle correlations look more like flattened cosine functions. For $W = 50$ ns

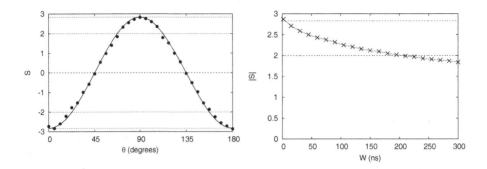

Fig. 13. Simulation results for the function $S = E(a_1, a_2) - E(a_1, a'_2) + E(a'_1, a_2) + E(a'_1, a'_2)$ using the time-tag model Eq. (52) with $T_0 = 1000$ ns. The total number of pairs generated by the source is 3×10^5, roughly the same as in experiment.[25] (a) S as a function of θ with $a_1 = \theta$, $a'_1 = \pi/4 + \theta$, $a_2 = \pi/8$ and $a'_2 = 3\pi/8$ for a time window $W = 2$ ns. The line connecting the solid circles is the result $-2\sqrt{2}\cos\theta$ predicted by quantum theory. (b) $|S|$ as a function of W for $a_1 = 0$, $a'_1 = \pi/4$, $a_2 = \pi/8$ and $a'_2 = 3\pi/8$. The line connecting the crosses is a guide to the eye only. The dashed horizontal lines indicate the maximum value for a quantum system of two spin-$\frac{1}{2}$ particles in a product state ($|S| = 2$) and in a singlet state ($|S| = 2\sqrt{2}$).

we find $|S| = 2.62$ which compares very well with the values between 2 and 2.57 extracted from different sets of experimental data of Weihs et al.. However, for $W = 2$ ns (results not shown), the results for the two-particle correlations fit very well to the prediction of quantum theory for the EPRB experiment. From these data we extract $|S| = 2.82$.

Figure 13(a) depicts $S(\theta)$ for $W = 2$ ns and shows that the event-based model reproduces the result predicted by quantum theory for the singlet state (solid line), namely $S = -2\sqrt{2}\cos\theta$. Note that the comparison between the simulation results and quantum theory becomes perfect if more pairs are generated by the source (10^6 pairs is sufficient for most purposes).

From Fig. 13(b), it follows that a violation of the Bell-CHSH inequality $|S| \leq 2$ depends on the choice of W, a parameter which is absent in the quantum theoretical description of the EPRB thought experiment. There are two limiting cases for which S become independent of W. If $W \to \infty$, it is impossible to let a digital computer violate the inequality $|S| \leq 2$ without abandoning the rules of Boolean logic or arithmetic.[113] For relatively small W ($W < 150$ns), the inequality $|S| \leq 2$ may be violated. When $W \to 0$ the discrete-event models which generate the same type of data as real EPRB experiments, reproduce exactly the single- and two-spin averages of the singlet state and therefore also violate the inequality $|S| \leq 2$. Obviously, as the discrete-event model does not rely on any concept of quantum theory, a violation of the inequality $|S| \leq 2$ does not say anything about the "quantumness" of the system under observation.[111,113,133] Similarly, a violation of this inequality cannot say anything about locality and realism.[111–113,133] Clearly, the event-based model is contextual, literally meaning "being dependent of the (experimental) measurement arrangement". The fact that the event-based model reproduces, for instance, the correlations of the singlet state without violating Einstein's local causality criterion suggests that the data $\{x_{n,1}, x_{n,2}\}$ generated by the event-based model cannot be represented by a single Kolmogorov probability space. This complies with the idea that contextual, non-Kolmogorov models can lead to violations of Bell's inequality without appealing to nonlocality or nonobjectivism.[112,134,135]

In conclusion, event-based simulation models provide a cause-and-effect description of real EPRB experiments at a level of detail which is not covered by quantum theory, such as the effect of the choice of the time-window. Some of these simulation models exactly reproduce the results of quantum theory of the EPRB experiment, indicating that there is no fundamental obstacle for an EPRB experiment to produce data that can be described by quantum theory. However, as we have shown, it is highly unlikely that quantum theory describes the data of the EPRB experiment by Weihs *et al.* This suggests that in the real experiment, there may be processes at work which have not been identified yet.

6.3.3. *Why is Bell's inequality violated?*

In Ref. 31, we have presented a probabilistic description of our simulation model that (i) rigorously proves that for up to first order in W it exactly reproduces the single particle averages and the two-particle correlations of quantum theory for the system under consideration; (ii) illustrates how the presence of the time-window W introduces correlations that cannot be described by the original Bell-like "hidden-variable" models.[119] Here, we repeat the discussion presented in Ref. 4.

The time-coincidence post-selection procedure with the time-window W filters out the "coincident" photons based on the time-tags $t_{n,i}$ thereby reducing the final detection efficiency to less than 100%, although in the simulation a measurement always returns a +1 or −1 for both photons in a pair (100% detection efficiency of the detectors). Hence, even in case of a perfect detection process the data set that is finally retained consists only of a subset of the entire ensemble of correlated photons that was emitted by the source, exactly as in the laboratory experiments.

We briefly elaborate on point (ii) (see Ref. 31 for a more extensive discussion). Let us assume that there exists a probability $P(x_1, x_2, t_1, t_2|\theta_1, \theta_2)$ to observe the data $\{x_i, t_i\}$ conditional on the settings θ_i at stations i for $i = 1, 2$. The probability $P(x_1, x_2, t_1, t_2|\theta_1, \theta_2)$ can always be expressed as an integral over the mutually exclusive events ξ_1, ξ_2, representing the polarization of the photons

$$P(x_1, x_2, t_1, t_2|\theta_1, \theta_2) = \frac{1}{4\pi^2} \int_0^{2\pi} \int_0^{2\pi} P(x_1, x_2, t_1, t_2|\theta_1, \theta_2, \xi_1, \xi_2)$$
$$\times P(\xi_1, \xi_2|\theta_1, \theta_2) d\xi_1 d\xi_2. \tag{53}$$

We now assume that in the probabilistic version of our simulation model, for each event, (i) the values of $\{x_1, x_2, t_1, t_2\}$ are independent of each other, (ii) the values of $\{x_1, t_1\}$ ($\{x_2, t_2\}$) are independent of θ_2 and ξ_2 (θ_1 and ξ_1), (iii) ξ_1 and ξ_2 are independent of θ_1 or θ_2. With these assumptions Eq. (53) becomes

$$
\begin{aligned}
P(x_1, x_2, t_1, t_2|\theta_1, \theta_2) &\overset{(i)}{=} \frac{1}{4\pi^2} \int_0^{2\pi} \int_0^{2\pi} P(x_1, t_1|\theta_1, \theta_2, \xi_1, \xi_2) \\
&\quad \times P(x_2, t_2|\theta_1, \theta_2, \xi_1, \xi_2) P(\xi_1, \xi_2|\theta_1, \theta_2) d\xi_1 d\xi_2 \\
&\overset{(ii)}{=} \frac{1}{4\pi^2} \int_0^{2\pi} \int_0^{2\pi} P(x_1, t_1|\theta_1, \xi_1) P(x_2, t_2|\theta_2, \xi_2) \\
&\quad \times P(\xi_1, \xi_2|\theta_1, \theta_2) d\xi_1 d\xi_2 \\
&\overset{(i)}{=} \frac{1}{4\pi^2} \int_0^{2\pi} \int_0^{2\pi} P(x_1|\theta_1, \xi_1) P(t_1|\theta_1, \xi_1) P(x_2|\theta_2, \xi_2) \\
&\quad \times P(t_2|\theta_2, \xi_2) P(\xi_1, \xi_2|\theta_1, \theta_2) d\xi_1 d\xi_2 \\
&\overset{(iii)}{=} \frac{1}{4\pi^2} \int_0^{2\pi} \int_0^{2\pi} P(x_1|\theta_1, \xi_1) P(t_1|\theta_1, \xi_1) P(x_2|\theta_2, \xi_2) \\
&\quad \times P(t_2|\theta_2, \xi_2) P(\xi_1, \xi_2) d\xi_1 d\xi_2,
\end{aligned} \tag{54}
$$

which is the probabilistic description of our simulation model. According to our simulation model, the probability distributions that describe the polarizers are given by $P(x_i|\theta_i, \xi_i) = [1 + x_i \cos 2(\theta_i - \xi_i)]/2$ and those for the time-delays t_i that are distributed randomly over the interval $[0, \lambda(\xi_i + (i-1)\pi/2 - \theta_i)]$ are given by $P(t_i|\theta_i, \xi_i) = \Theta(t_i)\Theta(\lambda(\xi_i + (i-1)\pi/2 - \theta_i) - t_i)/\lambda(\xi_i + (i-1)\pi/2 - \theta_i)$, where $\Theta(.)$ denotes the unit step function. In the experiment[25] and therefore also in our simulation model, the events are selected using a time window W that the experimenters try to make as small as possible.[117] Accounting for the time window, that is multiplying Eq. (54) by a step function and integrating over all t_1 and t_2, the expression for the probability for observing the event (x_1, x_2) reads

$$
P(x_1, x_2|\theta_1, \theta_2) = \int_0^{2\pi} \int_0^{2\pi} P(x_1|\theta_1, \xi_1) P(x_2|\theta_2, \xi_2) \rho(\xi_1, \xi_2|\theta_1, \theta_2) d\xi_1 d\xi_2, \tag{55}
$$

where the probability density $\rho(\xi_1, \xi_2|\theta_1, \theta_2)$ is given by

$$
\rho(\xi_1, \xi_2|\theta_1, \theta_2) = \\
\frac{\int_{-\infty}^{+\infty} \int_{-\infty}^{+\infty} P(t_1|\theta_1, \xi_1) P(t_2|\theta_2, \xi_2) \Theta(W - |t_1 - t_2|) P(\xi_1, \xi_2) dt_1 dt_2}{\int_0^{2\pi} \int_0^{2\pi} \int_{-\infty}^{+\infty} \int_{-\infty}^{+\infty} P(t_1|\theta_1, \xi_1) P(t_2|\theta_2, \xi_2) \Theta(W - |t_1 - t_2|) P(\xi_1, \xi_2) d\xi_1 d\xi_2 dt_1 dt_2}. \tag{56}
$$

The simple fact that $\rho(\xi_1, \xi_2|\theta_1, \theta_2) \neq \rho(\xi_1, \xi_2)$ brings the derivation of the original Bell (CHSH) inequality to a halt. Indeed, in these derivations it is assumed that the probability distribution for ξ_1 and ξ_2 does not depend on the settings θ_1 or θ_2.[2,119]

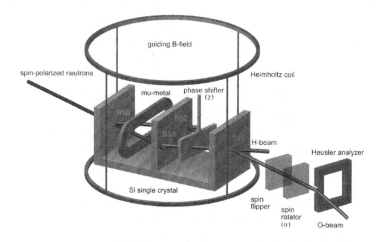

Fig. 14.   Top: Schematic picture of the single-neutron interferometry experiment to test a Bell inequality violation (see also Fig. 1 in Ref. 129). BS0, ..., BS3: beam splitters; phase shifter χ: aluminum foil; neutrons that are transmitted by BS1 or BS2 leave the interferometer and do not contribute to the interference signal. Detectors count the number of neutrons in the O- and H-beam.

By making explicit use of the time-tag model (see Eq. (52)) it can be shown that[31] (i) if we ignore the time-tag information ($W > T_0$), the two-particle probability takes the form of the hidden variable models considered by Bell,[119] and we cannot reproduce the results of quantum theory,[119] (ii) if we focus on the case $W \rightarrow 0$ the single-particle averages are zero and the two-particle average $E(\theta_1, \theta_2) = -\cos 2(\theta_1 - \theta_2)$.

Although our simulation model and its probabilistic version Eq. (54) involve local processes only, the filtering of the detection events by means of the time-coincidence window W can produce correlations which violate Bell-type inequalities.[98,101,132] Moreover, for $W \rightarrow 0$ our classical, local and causal simulation model can produce single-particle and two-particle averages that correspond with those of a singlet state in quantum theory.

6.4. *Bell-test experiment with single neutrons*

The single-neutron interferometry experiment of Hasegawa *et al.*[129] demonstrates that the correlation between the spatial and spin degree of freedom of neutrons violates a Bell-CHSH inequality. In this section we construct an event-based model that reproduces this correlation by using detectors that count every neutron and without using any post-selection procedure. We show that the event-based model reproduces the exact results of quantum theory if $\gamma \rightarrow 1^-$ and that by changing γ it can also reproduce the numerical values of the correlations, as measured in experiments.[129,136] Note that this Bell-test experiment involves two degrees of free-

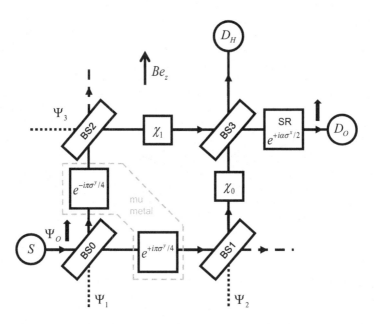

Fig. 15. Event-based network of the experimental setup shown in Fig. 14. S: single neutron source; BS0, ... , BS3: beam splitters; $e^{+i\pi\sigma^y/4}$, $e^{-i\pi\sigma^y/4}$: spin rotators modeling the action of a mu-metal; χ_0, χ_1: phase shifters; SR $e^{i\alpha\sigma^x/2}$: spin rotator; D_O, D_H: detectors counting all neutrons that leave the interferometer via the O- and H-beam, respectively. In the experiment and in the event-based simulation, neutrons with spin up (magnetic moment aligned parallel with respect to the guiding magnetic field **B**) enter the interferometer via the path labeled by Ψ_0 only. The wave amplitudes labeled by Ψ_1, Ψ_2, and Ψ_3 (dotted lines) are used in the quantum theoretical treatment only (see text). Particles leaving the interferometer via the dashed lines are not counted.

dom of one particle, while the EPRB thought experiment[94] and EPRB experiments with single photons[25,117,120,124] involve two degrees of freedom of two particles. Hence, the Bell-test experiment with single neutrons is not performed according to the CHSH protocol[116] because the two degrees of freedom of one particle are not manipulated and measured independently.

    Figure 14 shows a schematic picture of the single-neutron interferometry experiment. Incident neutrons pass through a magnetic-prism polarizer (not shown) which produces two spatially separated beams of neutrons with their magnetic moments aligned parallel (spin up), respectively anti-parallel (spin down) with respect to the magnetic axis of the polarizer which is parallel to the guiding field **B**. The spin-up neutrons impinge on a silicon-perfect-crystal interferometer.[36] On leaving the first beam splitter BS0, neutrons are transmitted or refracted. A mu-metal spin-turner changes the orientation of the magnetic moment of the neutron from parallel to perpendicular to the guiding field **B**. Hence, the magnetic moment of the neutrons following path H (O) is rotated by $\pi/2$ ($-\pi/2$) about the y axis. Before the two paths join at the entrance plane of beam splitter BS3, a difference between the time of flights along the two paths can be manipulated by a phase shifter. The neutrons which experience two refraction events when passing through the interferometer form the O-beam and are analyzed by sending them through a spin rotator and a Heusler spin analyzer. If necessary, to induce an extra spin rotation of π, a spin flipper is placed between the interferometer and the spin rotator. The neutrons that are selected by the Heusler spin analyzer are counted with a neutron detector (not shown) that has a very high efficiency ($\approx 99\%$). Note that neutrons which are not refracted by the mirror plate leave the interferometer without being detected.

The single-neutron interferometry experiment yields the count rate $N(\alpha, \chi)$ for the spin-rotation angle α and the difference χ of the phase shifts of the two different paths in the interferometer.[129] The correlation $E(\alpha, \chi)$ is defined by[129]

$$
E(\alpha, \chi) = \frac{N(\alpha, \chi) + N(\alpha + \pi, \chi + \pi) - N(\alpha + \pi, \chi) - N(\alpha, \chi + \pi)}{N(\alpha, \chi) + N(\alpha + \pi, \chi + \pi) + N(\alpha + \pi, \chi) + N(\alpha, \chi + \pi)}. \tag{57}
$$

6.4.1. *Event-based model*

A minimal, discrete event simulation model of the single-neutron interferometry experiment requires a specification of the information carried by the particles, of the algorithm that simulates the source and the interferometer components (see Fig. 15), and of the procedure to analyze the data. Various ingredients of the simulation model have been described in Sect. 5.4.1. In the following, we specify the action of the remaining components, namely the magnetic-prism polarizer (not shown), the mu-metal spin-turner, the spin-rotator and spin analyzer.

- *Magnetic-prism polarizer:* This component takes as input a neutron with an unknown magnetic moment and produces a neutron with a magnetic moment that is either parallel (spin up) or antiparallel (spin down) with respect to the z-axis (which by definition is parallel to the guiding field **B**). In the experiment, only a neutron with spin up is injected into the interferometer. Therefore, as a matter of simplification, we assume that the source S only creates messengers with spin up. Hence, we assume that $\theta = 0$ in Eq. (29).

- *Mu-metal spin turner:* This component rotates the magnetic moment of a neutron that follows the H-beam (O-beam) by $\pi/2$ ($-\pi/2$) about the y axis. The processor that accomplishes this takes as input the direction of the magnetic moment, represented by the message **u** and performs the rotation $\mathbf{u} \rightarrow e^{\pm i \pi \sigma^y / 4} \mathbf{u}$. We emphasize that we use Pauli matrices as a convenient tool to express rotations in three-dimensional space, not because in quantum theory the magnetic moment of the neutron is represented by spin-$\frac{1}{2}$ operators.

- *Spin-rotator and spin-flipper:* The spin-rotator rotates the magnetic moment of a neutron by an angle α about the x axis. The spin flipper is a spin rotator with $\alpha = \pi$.

- *Spin analyzer:* This component selects neutrons with spin up, after which they are counted by a detector. The model of this component projects the magnetic moment of the particle on the z axis and sends the particle to the detector if the projected value exceeds a pseudo-random number r.

6.4.2. *Simulation results*

In Fig. 16(left) we present simulation results for the correlation $E(\alpha, \chi)$, assuming that the experimental conditions are very close to ideal and compare them to the quantum theoretical result.

The quantum theoretical description of the experiment[129] requires a four-state system for the path and another two-state system to account for the spin-$\frac{1}{2}$ degree-of-freedom. Thus, the statistics of the experimental data is described by the state vector Eq. (32). In the experiment,[129] the neutrons that enter the interferometer all have spin up, relative to the direction of the guiding field $\mathbf{B}$ (see Fig. 14). Thus, the state describing the incident neutrons is $|\Psi\rangle = (1,0,0,0,0,0,0,0)^T$, omitting irrelevant phase factors. As the state vector propagates through the interferometer and the spin rotator (see Fig. 15), it changes according to

$$
\begin{aligned}
|\Psi'\rangle = {}& \begin{pmatrix} \cos(\alpha/2) & i\sin(\alpha/2) \\ i\sin(\alpha/2) & \cos(\alpha/2) \end{pmatrix}_{6,7} \begin{pmatrix} t^* & r \\ -r^* & t \end{pmatrix}_{5,7} \begin{pmatrix} t^* & r \\ -r^* & t \end{pmatrix}_{4,6} \\
& \times \begin{pmatrix} e^{i\phi_1} & 0 \\ 0 & e^{i\phi_1} \end{pmatrix}_{6,7} \begin{pmatrix} e^{i\phi_0} & 0 \\ 0 & e^{i\phi_0} \end{pmatrix}_{4,5} \begin{pmatrix} t^* & r \\ -r^* & t \end{pmatrix}_{3,7} \begin{pmatrix} t^* & r \\ -r^* & t \end{pmatrix}_{2,6} \\
& \times \begin{pmatrix} t & -r^* \\ r & t^* \end{pmatrix}_{1,5} \begin{pmatrix} t & -r^* \\ r & t^* \end{pmatrix}_{0,4} \begin{pmatrix} 1/\sqrt{2} & 1/\sqrt{2} \\ -1/\sqrt{2} & 1/\sqrt{2} \end{pmatrix}_{2,3} \begin{pmatrix} 1/\sqrt{2} & -1/\sqrt{2} \\ 1/\sqrt{2} & 1/\sqrt{2} \end{pmatrix}_{0,1} \\
& \times \begin{pmatrix} t & -r^* \\ r & t^* \end{pmatrix}_{1,3} \begin{pmatrix} t & -r^* \\ r & t^* \end{pmatrix}_{0,2} |\Psi\rangle, \quad\quad (58)
\end{aligned}
$$

where the subscripts i,j refer to the pair of elements of the eight-dimensional vector on which the matrix acts. Reading backwards, the first pair of matrices in Eq. (58) represents beam splitter BS0, the second pair the mu-metal (a spin rotation about the y-axis by $\pi/4$ and $-\pi/4$, respectively), the third and fourth pair beam splitters BS1 and BS2, respectively, the fifth pair the phase shifters, the sixth pair beam splitter BS3, and the last matrix represents the spin rotator SR.

From Eq. (58), it follows that the probability to detect a neutron with spin up in the O-beam is given by

$$
p_O(\alpha, \chi) = |\Psi'_{3,\uparrow}|^2 = \mathcal{T}\mathcal{R}^2 \left[1 + \cos(\alpha + \chi)\right], \quad\quad (59)
$$

where $\chi = \chi_0 - \chi_1$. From Eq. (59) it follows that the correlation $E_O(\alpha, \chi)$ is given by[129]

$$
\begin{aligned}
E_O(\alpha, \chi) \equiv {}& \frac{p_O(\alpha, \chi) + p_O(\alpha + \pi, \chi + \pi) - p_O(\alpha + \pi, \chi) - p_O(\alpha, \chi + \pi)}{p_O(\alpha, \chi) + p_O(\alpha + \pi, \chi + \pi) + p_O(\alpha + \pi, \chi) + p_O(\alpha, \chi + \pi)} \\
= {}& \cos(\alpha + \chi), \quad\quad (60)
\end{aligned}
$$

independent of the reflectivity $\mathcal{R} = |r|^2 = 1 - \mathcal{T}$ of the beam splitters (which have been assumed to be identical). The fact that $E_O(\alpha, \chi) = \cos(\alpha + \chi)$ implies that the state of the neutron cannot be written as a product of the state of the spin and the phase. In other words, in quantum language, the spin- and phase-degree-of-freedom are entangled.[129,137] Repeating the calculation for the probability of detecting a neutron in the H-beam shows that $E_H(\alpha, \chi) = 0$, independent of the direction of the spin. If the mu-metal would rotate the spin about the x-axis instead of about the y-axis, then we would find $E_O(\alpha, \chi) = \cos \alpha \cos \chi$, a typical expression for a quantum system in a product state.

As shown by the markers in Fig. 16 (left), disregarding the small statistical fluctuations, there is close-to-perfect agreement between the event-based simulation data for nearly ideal experimental conditions ($\gamma = 0.99$ and $\mathcal{R} = 0.2$) and quantum theory. However, the laboratory experiment suffers from unavoidable imperfections, leading to a reduction and distortion of the interference fringes.[129] In the event-based approach it is trivial to incorporate mechanisms for different sources of imperfections by modifying or adding update rules. However, to reproduce the available data it is sufficient to use the parameter γ to control the deviation from the quantum theoretical result. For instance, for $\gamma = 0.55$, $\mathcal{R} = 0.2$ the simulation results for $E(\alpha, \chi)$ are shown in Fig. 16 (right).

In order to quantify the difference between the simulation results, the experimental results and quantum theory it is customary to form the Bell-CHSH function[116,119]

$$S = S(\alpha, \chi, \alpha', \chi')$$
$$= E_O(\alpha, \chi) + E_O(\alpha, \chi') - E_O(\alpha', \chi) + E_O(\alpha', \chi'), \qquad (61)$$

for some set of experimental settings α, χ, α', and χ'. If the quantum system can be described by a product state, then $|S| \leq 2$. If $\alpha = 0$, $\chi = \pi/4$, $\alpha' = \pi/2$, and $\chi' = \pi/4$, then $S \equiv S_{max} = 2\sqrt{2}$, the maximum value allowed by quantum theory.[96]

For $\gamma = 0.55$, $\mathcal{R} = 0.2$ the simulation results yield $S_{max} = 2.05$, in excellent agreement with the value 2.052 ± 0.010 obtained in experiment.[129] For $\gamma = 0.67$, $\mathcal{R} = 0.2$ the simulation yields $S_{max} = 2.30$, in excellent agreement with the value 2.291 ± 0.008 obtained in a similar, more recent experiment.[138]

In conclusion, since experiment shows that $|S| > 2$, according to quantum theory it is impossible to interpret the experimental result in terms of a quantum system in the product state.[2] The system must be described by an entangled state. However, the event-based simulation which makes use of classical, Einstein-local and causal event-by-event processes can reproduce all features of this entangled state.

6.4.3. *Why are results from quantum theory produced?*

From Ref. 3 we know that the event-based model for the beam splitter produces results corresponding to those of classical wave or quantum theory when applied in

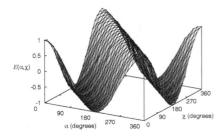 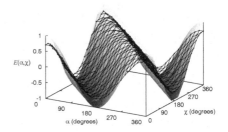

Fig. 16. Left: correlation $E(\alpha, \chi)$ between spin and path degree of freedom as obtained from an event-based simulation of the experiment depicted in Fig. 14. Solid surface: $E(\alpha, \chi) = \cos(\alpha + \chi)$ predicted by quantum theory; circles: simulation data. The lines connecting the markers are guides to the eye only. Model parameters: reflection percentage of BS0, ..., BS3 is 20% and $\gamma = 0.99$. For each pair (α, χ), four times 10000 particles were used to determine the four counts $N(\alpha, \chi)$, $N(\alpha + \pi, \chi + \pi)$, $N(\alpha, \chi + \pi)$ and $N(\alpha + \pi, \chi + \pi)$. Right: same as figure on the left but $\gamma = 0.55$.

interferometry experiments. Important for this outcome is that the phase difference χ between the two paths in the interferometer is constant for a relatively large number of incoming particles. If, for each incoming neutron, we pick the angle χ randomly from the same set of predetermined values to produce Fig. 16, an event-based simulation with $\gamma = 0.99$ yields (within the usual statistical fluctuations) the correlation $E(\alpha, \chi) \approx [\cos(\alpha + \chi)]/2$, which does not lead to a violation of the Bell-CHSH inequality (results not shown). Thus, if the neutron interferometry experiment could be repeated with random choices for the phase shifter χ for each incident neutron, and the experimental results would show a significant violation of the Bell-CHSH inequality, then the event-based model that we have presented here would be ruled out.

7. Discussion

We have presented an event-based simulation method which allows for a mystery-free, particle-only description of interference and entanglement phenomena observed in various single-photon experiments and single-neutron interferometry experiments. The statistical distributions which are observed in these single-particle experiments and which are usually thought to be of quantum mechanical origin, are shown to emerge from a time series of discrete events generated by causal adaptive systems, which in principle could be build using macroscopic mechanical parts.

As shown in the examples, in the stationary state (after processing many events), the event-based model reproduces the statistical distributions of quantum theory. This might raise questions about the efficiency of the method. Although the event-based simulation method can be used to simulate a universal quantum computer,[21,22] the so-called "quantum speed-up" cannot be obtained. This by itself is no surprise because the quantum speed-up is the result of a mathematical construct in which each unitary operation on the state of the quantum computer is counted as one operation and in which preparation and read-out of the quantum computer are excluded. Whether or not this mathematical construct is realized in Nature is an open question.

We hope that our simulation results will stimulate the design of new dedicated single-photon and neutron interferometry experiments which help extending and refining our event-based approach.

Acknowledgments

We would like to thank K. De Raedt, K. Keimpema, F. Jin, S. Miyashita, S. Yuan, and S. Zhao for many thoughtful comments and contributions to the work on which this review is based.

References

1. D. Home, *Conceptual Foundations of Quantum Physics* (Plenum Press, New York, 1997).
2. L. E. Ballentine, *Quantum Mechanics: A Modern Development* (World Scientific, Singapore, 2003).
3. K. Michielsen, F. Jin, and H. De Raedt, Event-based corpuscular model for quantum optics experiments, *J. Comp. Theor. Nanosci.* **8**, 1052–1080 (2011).
4. H. De Raedt and K. Michielsen, Event-by-event simulation of quantum phenomena, *Ann. Phys. (Berlin)* **524**, 393–410 (2012).
5. H. De Raedt, F. Jin, and K. Michielsen, Event-based simulation of neutron interferometry experiments, *Quantum Matter* **1**, 1–21 (2012).
6. P. Grangier, G. Roger, and A. Aspect, Experimental evidence for a photon anticorrelation effect on a beam splitter: A new light on single-photon interferences, *Europhys. Lett.* **1**, 173–179 (1986).
7. H. De Raedt, K. De Raedt, and K. Michielsen, Event-based simulation of single-photon beam splitters and Mach-Zehnder interferometers, *Europhys. Lett.* **69**, 861–867 (2005).
8. K. De Raedt, H. De Raedt, and K. Michielsen, Deterministic event-based simulation of quantum interference, *Comp. Phys. Comm.* **171**, 19–39 (2005).
9. V. Jacques, E. Wu, F. Grosshans, F. Treussart, P. Grangier, A. Aspect, and J.-F. Roch, Experimental realization of Wheeler's delayed-choice gedanken experiment, *Science* **315**, 966–968 (2007).
10. S. Zhao, S. Yuan, H. De Raedt, and K. Michielsen, Computer simulation of Wheeler's delayed choice experiment with photons, *Europhys. Lett.* **82**, 40004 (2008).
11. K. Michielsen, S. Yuan, S. Zhao, F. Jin, and H. De Raedt, Coexistence of full which-path information and interference in Wheeler's delayed choice experiment with photons, *Physica E* **42**, 348–353 (2010).
12. P. D. D. Schwindt, P. G. Kwiat, and B.-G. Englert, Quantitative wave-particle duality and nonerasing quantum erasure, *Phys. Rev. A* **60**, 4285–4290 (1999).

13. F. Jin, S. Zhao, S. Yuan, H. De Raedt, and K. Michielsen, Event-by-event simulation of a quantum eraser experiment, *J. Comp. Theor. Nanosci.* **7**, 1771–1782 (2010).

14. H. De Raedt, M. Delina, F. Jin, and K. Michielsen, Corpuscular event-by-event simulation of quantum optics experiments: application to a quantum-controlled delayed-choice experiment, *Phys. Scr.* **T151**, 014004 (2012).

15. V. Jacques, E. Wu, T. Toury, F. Treussart, A. Aspect, P. Grangier, and J.-F. Roch, Single-photon wavefront-splitting interference – An illustration of the light quantum in action, *Eur. Phys. J. D* **35**, 561–565 (2005).

16. F. Jin, S. Yuan, H. De Raedt, K. Michielsen, and S. Miyashita, Particle-only model of two-beam interference and double-slit experiments with single photons, *J. Phys. Soc. Jpn.* **79**, 074401 (2010).

17. S. Zhao and H. De Raedt, Event-by-event simulation of quantum cryptography protocols, *J. Comp. Theor. Nanosci.* **5**, 490–504 (2008).

18. I. N. Agafonov, M. V. Chekhova, T. S. Iskhakov, and A. N. Penin, High-visibility multiphoton interference of Hanbury Brown-Twiss type for classical light, *Phys. Rev. A* **77**, 053801 (2008).

19. F. Jin, H. De Raedt, and K. Michielsen, Event-by-event simulation of the Hanbury Brown-Twiss experiment with coherent light, *Commun. Comput. Phys.* **7**, 813–830 (2010).

20. K. Michielsen, F. Jin, M. Delina, and H. De Raedt, Event-by-event simulation of nonclassical effects in two-photon interference experiments, *Phys. Scr.* **T151**, 014005 (2012).

21. H. De Raedt, K. De Raedt, and K. Michielsen, New method to simulate quantum interference using deterministic processes and application to event-based simulation of quantum computation, *J. Phys. Soc. Jpn. Suppl.* **76**, 16–25 (2005).

22. K. Michielsen, K. De Raedt, and H. De Raedt, Simulation of quantum computation: a deterministic event-based approach, *J. Comput. Theor. Nanosci.* **2**, 227–239 (2005).

23. A. Aspect, P. Grangier, and G. Roger, Experimental realization of Einstein-Podolsky-Rosen-Bohm gedankenexperiment: A new violation of Bell's inequalities, *Phys. Rev. Lett.* **49**, 91–94 (1982).

24. A. Aspect, J. Dalibard, and G. Roger, Experimental test of Bell's inequalities using time-varying analyzers, *Phys. Rev. Lett.* **49**, 1804–1807 (1982).

25. G. Weihs, T. Jennewein, C. Simon, H. Weinfurther, and A. Zeilinger, Violation of Bell's inequality under strict Einstein locality conditions, *Phys. Rev. Lett.* **81**, 5039–5043 (1998).

26. K. De Raedt, K. Keimpema, H. De Raedt, K. Michielsen, and S. Miyashita, A local realist model for correlations of the singlet state, *Eur. Phys. J. B* **53**, 139–142 (2006).

27. H. De Raedt, K. De Raedt, K. Michielsen, K. Keimpema, and S. Miyashita, Event-based computer simulation model of Aspect-type experiments strictly satisfying Einstein's locality conditions, *J. Phys. Soc. Jpn.* **76**, 104005 (2007).

28. K. De Raedt, H. De Raedt, and K. Michielsen, A computer program to simulate Einstein-Podolsky-Rosen-Bohm experiments with photons, *Comp. Phys. Comm.* **176**, 642–651 (2007).

29. H. De Raedt, K. De Raedt, K. Michielsen, K. Keimpema, and S. Miyashita, Event-by-event simulation of quantum phenomena: Application to Einstein-Podolosky-Rosen-Bohm experiments, *J. Comp. Theor. Nanosci.* **4**, 957–991 (2007).

30. H. De Raedt, K. Michielsen, S. Miyashita, and K. Keimpema, Reply to Comment on "A local realist model for correlations of the singlet state", *Eur. Phys. J. B* **58**, 55–59 (2007).

31. S. Zhao, H. De Raedt, and K. Michielsen, Event-by-event simulation model of Einstein-Podolsky-Rosen-Bohm experiments, *Found. Phys.* **38**, 322–347 (2008).
32. B. Trieu, K. Michielsen, and H. De Raedt, Event-based simulation of light propagation in lossless dielectric media, *Comp. Phys. Comm.* **182**, 726–734 (2011).
33. R. P. Feynman, R. B. Leighton, and M. Sands, *The Feynman Lectures on Physics, Vol. 3* (Addison-Wesley, Reading MA, 1965).
34. T. Young, On the theory of light and colors, *Phil. Trans. R. Soc. Lond.* **92**, 12 (1802).
35. A. Zeilinger, R. Gähler, C. G. Shull, W. Treimer, and W. Mampe, Single and double slit diffraction of neutrons, *Rev. Mod. Phys.* **60**, 1067–1073 (1988).
36. H. Rauch and S. A. Werner, *Neutron Interferometry: Lessons in Experimental Quantum Mechanics* (Clarendon, London, 2000).
37. D. W. Keith, C. R. Ekstrom, Q. A. Turchette, and D. E. Pritchard, An interferometer for atoms, *Phys. Rev. Lett.* **66**, 2693–2696 (1991).
38. O. Carnal and J. Mlynek, Young's double-slit experiment with atoms: a simple atom interferometer, *Phys. Rev. Lett.* **66**, 2689–2692 (1991).
39. M. Arndt, O. Nairz, J. Vos-Andreae, C. Keller, G. van der Zouw, and A. Zeilinger, Wave-particle duality of C60 molecules, *Nature* **401**, 680–682 (1999).
40. B. Brezger, L. Hackermüller, S. Uttenthaler, J. Petschinka, M. Arndt, and A. Zeilinger, Matter-wave interferometer for large molecules, *Phys. Rev. Lett.* **88**, 100404 (2002).
41. T. Juffmann, A. Milic, M. Müllneritsch, P. Asenbaum, A. Tsukernik, J. Tüxen, M. Mayor, O. Cheshnovsky, and M. Arndt, Real-time single molecule imaging of quantum interference, *Nature Nanotechnology* **7**, 297–300 (2012).
42. M. Born and E. Wolf, *Principles of Optics* (Pergamon, Oxford, 1964).
43. R. Feynman, *Lectures in Physics, Vol. 1* (Addison Wesley Publishing Company Reading, Mass, 1963).
44. G. Möllenstedt and H. Düker, Fresnelscher Interferenzversuch mit einem Biprisma für Elektronenwellen, *Naturwissenschaften* **42**, 41 (1955).
45. G. Möllenstedt and H. Düker, Beobachtungen und Messungen an Biprisma-Interferenzen mit Elektronenwellen, *Z. Phys.* **145**, 377–397 (1956).
46. C. Jönsson, Elektroneninterferenzen an mehreren künstlich hergestellten Feinspalten, *Z. Phys.* **161**, 454–474 (1961).
47. F. Hasselbach, Progress in electron- and ion-interferometry, *Rep. Prog. Phys.* **73**, 016101 (2010).
48. R. Rosa, The Merli-Missiroli-Pozzi two-slit electron-interference experiment, *Phys. Perspect.* **14**, 178–195 (2012).
49. O. Donati, G. P. Missiroli, and G. Pozzi, An experiment on electron interference, *Am. J. Phys.* **41**, 639–644 (1973).
50. P. G. Merli, G. F. Missiroli, and G. Pozzi, On the statistical aspect of electron interference phenomena, *Am. J. Phys.* **44**, 306–307 (1976).
51. A. Tonomura, J. Endo, T. Matsuda, T. Kawasaki, and H. Ezawa, Demonstration of single-electron buildup of an interference pattern, *Am. J. Phys.* **57**, 117–120 (1989).
52. S. Frabboni, G. C. Gazzadi, and G. Pozzi, Nanofabrication and the realization of Feynman's two-slit experiment, *Appl. Phys. Lett.* **93**, 073108 (2008).
53. S. Frabboni, C. Frigeri, G. C. Gazzadi, and G. Pozzi, Four slits interference and diffraction experiments, *Ultramicroscopy* **110**, 483–487 (2010).
54. S. Frabboni, C. Frigeri, G. C. Gazzadi, and G. Pozzi, Two and three slit electron interference and diffraction experiments, *Am. J. Phys.* **79**, 615–618 (2011).
55. S. Frabboni, A. Gabrielli, G. C. Gazzadi, F. Giorgi, G. Matteucci, G. Pozzi, N. S. Cesari, M. Villa, and A. Zoccoli, The Young-Feynman two-slits experiment with

single electrons: Build-up of the interference pattern and arrival-time distribution using a fast-readout pixel detector, *Ultramicroscopy* **116**, 73–76 (2012).

56. R. Bach, D. Pope, S. Liu, and H. Batelaan, Controlled double-slit electron diffraction, *New J. Phys.* **15**, 033018 (2013).

57. J. J. Thomson, On the ionization of gases by ultra- violet light and on the evidence as to the structure of light afforded by its electrical effects., *Proc. Camb. Phil. Soc.* **14**, 417–424 (1908).

58. G. I. Taylor, Interference fringes with feeble light, *Proc. Cambridge Phil. Soc.* **15**, 114–115 (1909).

59. Y. Tsuchiya, E. Inuzuka, T. Kurono, and M. Hosoda, Photon-counting imaging and its application, *Adv. Electron. Electron Phys.* **64A**, 21–31 (1985).

60. S. Parker, A single-photon double slit interference experiment, *Am. J. Phys.* **39**, 420–424 (1971).

61. A. Weis and R. Wynands, Three demonstration experiments on the wave and particle nature of light, *PhyDid* **1/2**, S67–S73 (2003).

62. T. L. Dimitrova and A. Weis, The wave-particle duality of light: A demonstration experiment, *Am. J. Phys.* **76**, 137–142 (2008).

63. I. G. Saveliev, M. Sanz, and N. Garcia, Time-resolved Young's interference and decoherence, *J. Opt. B: Quantum Semiclass. Opt.* **4**, S477–S481 (2002).

64. N. Garcia, I. G. Saveliev, and M. Sharonov, Time-resolved diffraction and interference: Young's interference with photons of different energy as revealed by time resolution, *Phil. Trans. R. Soc. Lond. A* **360**, 1039–1059 (2002).

65. P. Kolenderski, C. Scarcella, K. Johnsen, D. Hamel, C. Holloway, L. Shalm, S. Tisa, A. Tosi, K. Resch, and T. Jennewein, Time-resolved double-slit experiment with entangled photons. arXiv:1304.4943.

66. A. E. Allahverdyan, R. Balian, and T. M. Nieuwenhuizen, Understanding quantum measurement from the solution of dynamical models, *Phys. Rep.* **525**, 1–166 (2013).

67. S. Wolfram, *A New Kind of Science* (Wolfram Media Inc., 2002).

68. A. Petersen, The philosophy of Niels Bohr, *Bull. Atomic Scientists* **19**, 8–14 (1963).

69. A. Plotnitsky, The Art and Science of Experimentation in Quantum Physics, *AIP Conf. Proc.* **1232**, 128–142 (2010).

70. W. Gerlach and O. Stern, Der experimentelle Nachweis der Richtungsquantelung im Magnetfeld, *Z. Phys.* **9**, 349–352 (1922).

71. W. Gerlach and O. Stern, Uber die Richtungsquantelung im Magnetfeld, *Ann. Phys.* **74**, 673–699 (1924).

72. H. De Raedt, K. De Raedt, K. Michielsen, and S. Miyashita, Efficient data processing and quantum phenomena: Single-particle systems, *Comp. Phys. Comm.* **174**, 803–817 (2006).

73. H. De Raedt and K. Michielsen, Computational Methods for Simulating Quantum Computers. In eds. M. Rieth and W. Schommers, *Handbook of Theoretical and Computational Nanotechnology*, pp. 2–48. American Scientific Publishers, Los Angeles (2006).

74. L. de Broglie, Recherches sur la théorie des quanta, *Annales de Physique* **3**, 22 (1925).

75. J. A. Wheeler (1983), in: Mathematical foundations of quantum theory, Proc. New Orleans Conf. on The mathematical foundations of quantum theory, ed. A.R. Marlow (Academic, New York, 1978) [reprinted in Quantum theory and measurements, eds. J. A. Wheeler and W. H. Zurek (Princeton Univ. Press, Princeton, NJ, 1983) pp. 182–213].

76. V. Jacques, E. Wu, F. Grosshans, F. Treussart, P. Grangier, A. Aspect, and J.-F. Roch, Delayed-choice test of quantum complementarity with interfering single photons, *Phys. Rev. Lett.* **100**, 220402 (2008).

77. R. L. Pfleegor and L. Mandel, Interference of independent photon beams, *Phys. Rev.* **159**, 1084–1088 (1967).

78. http://demonstrations.wolfram.com/ EventByEventSimulationOfDoubleSlitExperimentsWithSinglePhoto/.

79. R. P. Feynman, *QED - The Strange Theory of Light and Matter* (Princeton University Press, 1985).

80. H. De Raedt, K. Michielsen, and F. Jin, Einstein-Podolsky-Rosen-Bohm laboratory experiments: Data analysis and simulation, *AIP Conf. Proc.* **1424**, 55–66 (2012).

81. A. Taflove and S. Hagness, *Computational Electrodynamics: The Finite-Difference Time-Domain Method* (Artech House, Boston, 2005).

82. R. H. Hadfield, Single-photon detectors for optical quantum information applications, *Nature Photonics* **3**, 696–705 (2009).

83. D. L. Alkon, "Either-Or"Two-Slit interference: Stable coherent propagation of individual photons through separate slits, *Biophys. J.* **80**, 2056–2061 (2001).

84. http://demonstrations.wolfram.com/ EventByEventSimulationOfTheMachZehnderInterferometer/.

85. Sample Fortran and Java programs and interactive programs that perform event-based simulations of a beam splitter, one Mach-Zehnder interferometer, and two chained Mach-Zehnder interferometers can be found at http://www.compphys.net/.

86. We make a distinction between quantum theory and quantum physics. We use the term *quantum theory* when we refer to the mathematical formalism, i.e., the postulates of quantum mechanics (with or without the wave function collapse postulate)[2] and the rules (algorithms) to compute the wave function. The term *quantum physics* is used for microscopic, experimentally observable phenomena that do not find an explanation within the mathematical framework of classical mechanics.

87. G. Baym, *Lectures on Quantum Mechanics* (W.A. Benjamin, Reading MA, 1974).

88. T. Hellmuth, H. Walther, A. Zajonc, and W. Schleich, Delayed-choice experiments in quantum interference, *Phys. Rev. A* **72**, 2532–2541 (1987).

89. H. Rauch, W. Treimer, and U. Bonse, Test of a single crystal neutron interferometer, *Phys. Lett. A* **47**, 369–371 (1974).

90. Y. Hasegawa and H. Rauch, Quantum phenomena explored with neutrons, *New J. Phys.* **13**, 115010 (2011).

91. G. Kroupa, G. Bruckner, O. Bolik, M. Zawisky, M. Hainbuchner, G. Badurek, R. J. Buchelt, A. Schricker, and H. Rauch, Basic features of the upgraded S18 neutron interferometer set-up at ILL, *Nucl. Instrum. Methods Phys. Res. A* **440**, 604–608 (2000).

92. H. Lemmel and A. G. Wagh, Phase shifts and wave-packet displacements in neutron interferometry and a nondispersive, nondefocusing phase shifter, *Phys. Rev. A* **82**, 033626 (2010).

93. H. Rauch and M. Suda, Intensitätsberechnung für ein Neutronen-Interferometer, *Phys. Stat. Sol. A* **25** (2), 495–505 (1974).

94. D. Bohm, *Quantum Theory* (Prentice-Hall, New York, 1951).

95. A. Einstein, A. Podolsky, and N. Rosen, Can quantum-mechanical description of physical reality be considered complete?, *Phys. Rev.* **47**, 777–780 (1935).

96. B. S. Cirel'son, Quantum generalizations of Bell's inequality, *Lett. Math. Phys.* **4**, 93–100 (1980).

97. J. S. Bell, On the Einstein-Podolsky-Rosen paradox, *Physics* **1**, 195–200 (1964).

98. J.-Å. Larsson and R. D. Gill, Bell's inequality and the coincidence-time loophole, *Europhys. Lett.* **67**, 707–713 (2004).
99. L. de la Peña, A. M. Cetto, and T. A. Brody, On hidden-variable theories and Bell's inequality, *Lett. Nuovo Cim.* **5**, 177–181 (1972).
100. A. Fine, On the completeness of quantum theory, *Synthese* **29**, 257–289 (1974).
101. A. Fine, Some local models for correlation experiments, *Synthese* **50**, 279–294 (1982).
102. T. Brody, *The Philosphy Behind Physics* (Springer, Berlin, 1993).
103. M. Kupczyński, On some tests of completeness of quantum mechanics, *Phys. Lett. A* **116**, 417–419 (1986).
104. E. T. Jaynes, Clearing up mysteries - The original goal, in ed. J. Skilling, *Maximum Entropy and Bayesian Methods*, vol. 36, pp. 1–27, Dordrecht (1989), Kluwer Academic Publishers.
105. L. Sica, Bell's inequalities I: An explanation for their experimental violation, *Opt. Comm.* **170**, 55–60 (1999).
106. K. Hess and W. Philipp, Bell's theorem and the problem of decidability between the views of Einstein and Bohr, *Proc. Natl. Acad. Sci. USA* **98**, 14228–14233 (2001).
107. K. Hess and W. Philipp, Bell's theorem: Critique of proofs with and without inequalities, *AIP Conf. Proc.* **750**, 150–157 (2005).
108. A. F. Kracklauer, Bell's inequalities and EPR-B experiments: Are they disjoint?, *AIP Conf. Proc.* **750**, 219–227 (2005).
109. E. Santos, Bell's theorem and the experiments: Increasing empirical support to local realism?, *Phil. Mod. Phys.* **36**, 544–565 (2005).
110. T. M. Nieuwenhuizen, Where Bell went wrong, *AIP Conf. Proc.* **1101**, 127–133 (2009).
111. K. Hess, K. Michielsen, and H. De Raedt, Possible experience: from Boole to Bell, *Europhys. Lett.* **87**, 60007 (2009).
112. T. M. Nieuwenhuizen, Is the contextuality loophole fatal for the derivation of Bell inequalities?, *Found. Phys.* **41**, 580–591 (2011).
113. H. De Raedt, K. Hess, and K. Michielsen, Extended Boole-Bell inequalities applicable to quantum theory, *J. Comp. Theor. Nanosci.* **8**, 1011–1039 (2011).
114. K. Hess, H. De Raedt, and K. Michielsen, Hidden assumptions in the derivation of the theorem of Bell, *Phys. Scr.* **T151**, 014002 (2012).
115. G. Boole, On the theory of probabilities, *Phil. Trans. R. Soc. Lond.* **152**, 225–252 (1862).
116. J. F. Clauser, M. A. Horne, A. Shimony, and R. A. Holt, Proposed experiment to test local hidden-variable theories, *Phys. Rev. Lett.* **23**, 880–884 (1969).
117. G. Weihs, *Ein Experiment zum Test der Bellschen Ungleichung unter Einsteinscher Lokalität*, PhD thesis, University of Vienna (2000). http://www.uibk.ac.at/exphys/photonik/people/gwdiss.pdf.
118. J. F. Clauser and M. A. Horne, Experimental consequences of objective local theories, *Phys. Rev. D* **10**, 526–535 (1974).
119. J. S. Bell, *Speakable and Unspeakable in Quantum Mechanics* (Cambridge University Press, Cambridge, 1993).
120. A. Hnilo, A. Peuriot, and G. Santiago, Local realistic models tested by the EPRB experiment with variable analyzers, *Found. Phys. Lett.* **15**, 359–371 (2002).
121. A. A. Hnilo, M. D. Kovalsky, and G. Santiago, Low dimension dynamics in the EPRB experiment with variable analyzers, *Found. Phys.* **37**, 80–102 (2007).
122. G. Adenier and A. Y. Khrennikov, Is the fair sampling assumption supported by EPR experiments, *J. Phys. B: At. Mol. Opt. Phys.* **40**, 131–141 (2007).
123. J. H. Bigelow, A close look at the EPR data of Weihs et al. arXiv: 0906.5093v1.

124. M. B. Agüero, A. A. Hnilo, M. G. Kovalsksy, and M. A. Larotonda, Time stamping in EPRB experiments: application on the test of non-ergodic theories, *Eur. Phys. J. D* **55**, 705–709 (2009).

125. H. De Raedt, S. Zhao, S. Yuan, F. Jin, K. Michielsen, and S. Miyashita, Event-by-event simulation of quantum phenomena, *Physica E* **42**, 298–302 (2010), Proceedings of the international conference Frontiers of Quantum and Mesoscopic Thermodynamics FQMT '08.

126. J. H. Bigelow, Explaining counts from EPRB experiments: Are they consistent with quantum theory? arXiv: 1112.3399.

127. H. De Raedt, F. Jin, and K. Michielsen, Data analysis of Einstein-Podolsky-Rosen-Bohm laboratory experiments, *Proc. SPIE* **8832**, 88321N1–11 (2013).

128. M. A. Rowe, D. Kielpinski, V. Meyer, C. A. Sackett, W. M. Itano, C. Monroe, and D. J. Wineland, Experimental violation of a Bell's inequality with efficient detection, *Nature* **401**, 791–794 (2001).

129. Y. Hasegawa, R. Loidl, G. Badurek, M. Baron, and H. Rauch, Violation of a Bell-like inequality in single-neutron interferometry, *Nature* **425**, 45–48 (2003).

130. M. Ansmann, H. Wang, R. C. Bialczak, M. Hofheinz, E. Lucero, M. Neeley, A. D. O'Connell, D. Sank, M. Weides, J. Wenner, A. N. Cleland, and J. M. Martinis, Violation of Bell's inequality in Josephson phase qubits, *Nature* **461**, 504–506 (2009).

131. P. M. Pearle, Hidden-variable example based upon data rejection, *Phys. Rev. D* **2**, 1418–1425 (1970).

132. S. Pascazio, Time and Bell-type inequalities, *Phys. Lett. A* **118**, 47–53 (1986).

133. K. Hess, K. Michielsen, and H. De Raedt, Reply to Comment by A. J. Leggett and Anupam Garg, *Europhys. Lett.* **91**, 40002 (2010).

134. A. Y. Khrennikov, *Contextual Approach to Quantum Formalism* (Springer, Berlin, 2009).

135. A. Y. Khrennikov, On the role of probabilistic models in quantum physics: Bell's inequality and probabilistic incompatibility, *J. Comp. Theor. Nanosci.* **8**, 1006–1010 (2011).

136. C. Bartsch and J. Gemmer, Dynamical typicality of quantum expectation values, *Phys. Rev. Lett.* **102**, 110403 (2009).

137. S. Basu, S. Bandyopadhyay, G. Kar, and D. Home, Bell's inequality for a single spin-1/2 particle and quantum contextuality, *Phys. Lett. A* **279**, 281–286 (2001).

138. H. Bartosik, J. Klepp, C. Schmitzer, S. Sponar, A. Cabello, H. Rauch, and Y. Hasegawa, Experimental test of quantum contextuality in neutron interferometry, *Phys. Rev. Lett.* **103**, 040403 (2009).

Chapter 8

Lectures on Dynamical Models for Quantum Measurements

Theo M. Nieuwenhuizen[*,§], Martí Perarnau-Llobet[†,¶] and Roger Balian[‡]

*Institute for Theoretical Physics, University of Amsterdam,
Science Park 904, Postbus 94485,
1090 GL Amsterdam, The Netherlands*

†*ICFO – The Institute of Photonic Sciences, Mediterranean Technology Park,
Av. Carl Friedrich Gauss, 3, 08860 Castelldefels (Barcelona), Spain*
‡*Institute de Physique Théorique, CEA Saclay,
91191 Gif-sur-Yvette cedex, France*

In textbooks, ideal quantum measurements are described in terms of the tested system only by the collapse postulate and Born's rule. This level of description offers a rather flexible position for the interpretation of quantum mechanics. Here we analyse an ideal measurement as a process of interaction between the tested system S and an apparatus A, so as to derive the properties postulated in textbooks. We thus consider within standard quantum mechanics the measurement of a quantum spin component $\hat{s}_z$ by an apparatus A, being a magnet coupled to a bath. We first consider the evolution of the density operator of S+A describing a large set of runs of the measurement process. The approach describes the disappearance of the off-diagonal terms ("truncation") of the density matrix as a physical effect due to A, while the registration of the outcome has classical features due to the large size of the pointer variable, the magnetisation. A quantum ambiguity implies that the density matrix at the final time can be decomposed on many bases, not only the one of the measurement. This quantum oddity prevents to connect individual outcomes to measurements, a difficulty known as the "measurement problem". It is shown that it is circumvented by the apparatus as well, since the evolution in a small time interval erases all decompositions, except the one on the measurement basis. Once one can derive the outcome of individual events from quantum theory, the so-called "collapse of the wave function" or the "reduction of the state" appears as the result of a selection of runs among the original large set. Hence nothing more than standard quantum mechanics is needed to explain features of measurements. The employed statistical formulation is advocated for the teaching of quantum theory.

§t.m.nieuwenhuizen@uva.nl
¶marti.perarnau.llobet@gmail.com

1. Introduction

Quantum mechanics is our most fundamental theory at the microscopic level, and its successes are innumerable, see, e.g., Ref. 1. However, although one century has passed since its beginnings, its interpretation is still subject to discussions. What is the status of wave functions facing reality? Are they just a tool for making predictions,[2] or do they describe individual objects? How should we understand strange features such as Bell's inequalities? To answer such questions, we have to elucidate the only point of contact between theory and reality, to wit, measurements. Thus, a proper understanding of quantum measurements may provide useful lessons for a sensible interpretation of quantum theory, lessons not learnable from a "black box" approach where only the measurement postulates are employed.

A measurement should be analyzed as a dynamical process in which the tested quantum system S interacts with another quantum system, the apparatus A. This apparatus reaches at the end of the process one among several possible configurations. They are characterized by the indication of a pointer, that is, by the value of a *pointer variable* of A which we can observe or register, and which provides us with information about the initial state of S. This transfer of information from S to A, allowed by the coupling between S and A, thus involves a perturbation of A. Moreover, in quantum mechanics, the interaction process also modifies S in general; this is understandable since the apparatus is much larger than the system.[a]

For conceptual purposes, it is traditional to consider *ideal measurements*, although these can rarely be performed in actual experiments. Ideal measurements are those which produce the weakest possible modification of S. In textbooks, ideal quantum measurements are usually treated, without caring much about the apparatus, by postulating two properties about the fate of the tested system. *Born's rule* provides the probability of finding the eigenvalue s_i of the observable $\hat{s}$ which is being measured. The resulting final state of S is expressed by von Neumann's *collapse*; it is obtained by projecting the initial state over the eigenspace of $\hat{s}$ associated with s_i. Clearly, there is a gap with the practice of reading off the pointer variable of a macroscopic apparatus in a laboratory. Moreover, it is not satisfactory to complement the principles of quantum mechanics with such "postulates". In a laboratory, the apparatus itself is a quantum system coupled to S, and a measurement is a dynamical process involving S+A, so that one deals with two coupled quantum systems and therefore hopes to be able, without introducing new postulates, to describe the evolution of the coupled system S+A and its outcome by just solving its quantum equations of motion. The above properties of ideal measurements will then appear not as postulates but as mere consequences of quantum theory applied to the system S+A. Dynamical models for measurements have therefore been studied, with various benefits. The literature on this subject has been reviewed in Ref. 3.

---

[a]When we speak about "the system", we always mean: an ensemble of identically prepared systems, and for "the measurement" an ensemble of measurements performed on the ensemble of systems. As in classical thermodynamics, the ensemble can be real or Gedanken.

In particular, a rich enough but still tractable model has been introduced a decade ago, the Curie–Weiss model for the measurement of the z–component of a spin $\frac{1}{2}$ by an apparatus that itself consists of a piece of matter containing many such spins coupled to a thermal phonon bath.[4]

In most of such models, the apparatus is a *macroscopic object* having several stable states, each of which is characterised by some value of the pointer variable. Its initial state is metastable; by itself it would go after a very long time to one among these stable states. In the presence of a coupling with S, such a transition is triggered by the measurement process, in such a way that the eigenvalues s_i of $\hat{s}$ and the indications A_i of the pointer become fully correlated and can be read off – the two purposes of the measurement.

In an ideal measurement, the tested observable $\hat{s}$ commutes with the Hamiltonian, implying that in the diagonal basis $\{|i\rangle\}$ of $\hat{s}$ the various sectors of the density matrix remain decoupled during the whole measurement. They will thus evolve independently, driven by different aspects of the physics. The off-diagonal blocks of the density matrix of S+A are the ones for which S is described by $|i\rangle\langle j|$ with $j \neq i$; they are sometimes called "Schrödinger cat terms". In the considered models, they evolve due to a dephasing mechanism known from NMR (MRI) physics and/or due to a decoherence mechanism produced by a coupling of the pointer with a thermal bath. As a consequence, the effects of these off-diagonal blocks disappear in incoherent sums of phase factors, so for all practical purposes they can be considered as tending to zero (see, e.g., Ref. 5), because their contributions to the state of S are suppressed. As for each of the diagonal blocks $|i\rangle\langle i|$, its evolution describes the phase transition of the apparatus from its initial metastable state to its stable state correlated with the measured eigenvalues s_i; this process, which involves a decrease of free energy, requires a dumping of energy in a bath. The time scales of the two processes are different: the truncation happens rather fast; it involves no energy transfer and has resemblance to the $\mathcal{T}_2$ time of NMR physics, while the registration does involve energy transfer to the bath, with resemblance to the $\mathcal{T}_1$ process on its longer time scale.

If we regard the bath, introduced in most models, as being part of the apparatus, we can treat S+A as an *isolated system*. If we were dealing with pure states, its dynamics would be governed by the Schrödinger equation. However, the apparatus being macroscopic, we have to resort to *quantum statistical mechanics*.[6] We therefore rely on a formulation of quantum mechanics, recalled below in Section 1.1, which encompasses ordinary quantum mechanics but is also adapted to describe macroscopic systems, for instance in solid state physics, in the same way as classical statistical mechanics is adapted to describe large classical systems. The state of S+A is therefore not a pure state, but a statistical mixture. Wave vectors for A are thus replaced by density operators, describing mixed states. As S+A is an isolated system, the evolution of its state (i.e., its density operator) is governed by the Liouville–von Neumann equation, which replaces the Schrödinger equation. We then run into the *irreversibility paradox*. Both above-mentioned evolutions, diagonal and off-diagonal, are obviously irreversible, whereas the Liouville–von Neumann evolution is *unitary and therefore reversible*. Then, how can this equation give rise to an increase of entropy for S+A? As usual, we will solve below this paradox more or less implicitly, by relying on standard methods of statistical mechanics. In particular, acknowledging that our interest lies only in properties that can be observed on practical timescales, we are allowed to discard correlations between a macroscopic number of degrees of freedom; we are also allowed to forget about recurrences that would occur after a very large recurrence time.

1.1. *Outline*

The present course focuses on the *Curie–Weiss model*, presented in Section 2 below and already studied together with its extensions in Ref. 3. But the latter article is too detailed for a pedagogical access. We will therefore restrict ourselves to a simplified presentation. By accounting for the dynamics of the process for S+A in the framework of quantum statistical mechanics, we wish to explain for this model, within the most standard quantum theory, all the features currently attributed to ideal measurements.

Such features arise due to the physical interaction between S and A, and they are independent of the different interpretations of quantum mechanics. The state of the compound system S+A is therefore represented by a time-dependent density operator $\hat{\mathcal{D}}$ which evolves according to the Liouville–von Neumann equation. At the initial time, it is the product of the state $\hat{r}(0)$ of S that we wish to test, by the metastable state $\hat{\mathcal{R}}(0)$ of A prepared beforehand and ready to evolve towards a stable state[b]. While $\hat{\mathcal{D}}(t)$ encompasses our whole information about S+A, it is a mathematical object, the interpretation of which will only emerge at the end of the measurement process, since we can reach insight about the reality of S only through observation of the outcomes (see Section 6 below).

It is important to realize that pure states, or wave functions, are not proper descriptions of macroscopic systems.[c] Quantum mechanics deals with our *information* about systems, which can be coded only in density operators representing statistical mixtures.[d] Although it is our most precise theory, it does not deal with properties of individual systems, and thus has a status comparable to statistical classical mechanics. In a measurement, the apparatus is macroscopic and measurement theories cannot rely on pure states. In the statistical formulation employed

---

[b]The initial metastable state realises a "ready" state of the pointer, "ready" to give an indication when a measurement is performed. Metastability occurs typically in apparatuses, for example in photo multipliers and in our retina. Through its phase transition towards a stable state, it allows a macroscopic registration of a microscopic quantum signal.

[c]One of the present authors has termed "the right of every system to have its own wave function" the "fallacy of democracy in Hilbert space".

[d]Hence the "collapse postulate": after the measurement we can update our information about the system.

in the present paper, this is regarded as unphysical, because only a few degrees of freedom for the ensemble of systems can be controlled in practice, so that only ensembles of small systems can be in a pure state. Nevertheless, one encounters many pure-state discussions of measurement in the literature, in particular when it is postulated that the apparatus is initially in a pure state. Likewise, it is absurd to assume that a cat, also when termed "Schrödinger cat", can be described by a pure state, being 'in a quantum superposition of alive and dead".[e]

In this *statistical formulation* of quantum mechanics, advocated in Ref. 3, a density operator, or "state" $\hat{\mathcal{D}}$ presents an analogy with a standard probability distribution, but it has a specifically quantum feature: It is represented by a matrix rather than by a measure over ordinary random variables. The random physical quantities $\hat{O}$, or observables, are also represented by matrices, and quantities like Tr $\hat{\mathcal{D}}\hat{O}$ will come out as expectation values in experiments. Thus, as the ordinary probability theory and the classical statistical mechanics, quantum theory in its statistical formulation does not deal with individual events (see hereto e.g., Ref. 7), but with *statistical ensembles* of events. The state $\hat{\mathcal{D}}(t)$ of S+A which evolves during the measurement process describes only a generic situation. If we wish to think of a single measurement, we should regard it as a sample among a large set of runs, all prepared under the same conditions. A problem then arises because, contrary to ordinary probability theory, quantum mechanics is irreducibly probabilistic due to the non-commutative nature of the observables. After having determined the density operator of the ensemble at the final time but without other information, we *cannot make statements about individual measurements*. In particular, this knowledge is not sufficient to explain the observation that each run of a measurement yields a unique answer, the so-called *measurement problem*, which has remained unsolved till recently.

Anyhow, a first task is necessary, solving the above-mentioned equations of motion, so as to show that standard quantum statistical mechanics is sufficient to provide the outcome $\hat{\mathcal{D}}(t_{\mathrm{f}})$ expected for ideal measurements. These equations are written in Section 4, and their solution is worked out in Sections 4 and 5 for the off-diagonal and diagonal blocks of $\hat{\mathcal{D}}$, respectively.

At this stage, we shall have determined the state $\hat{\mathcal{D}}(t_{\mathrm{f}})$ of S+A which accounts for the *whole set of runs* of the measurement, and which involves the expected correlations between tested eigenvalues of $\hat{s}$ and the indications A_i of the apparatus. We will exhibit in Section 6 the difficulty that prevents us from inferring properties of individual runs from this mixed state. To overcome this difficulty without going beyond quantum theory, we will consider *subensembles of runs*, which can still be studied within standard quantum theory. If we are able to select a subensemble characterised by outcomes corresponding to a given value of the pointer, we expect to be able to *update our knowledge*, and hence to describe the selected population of

---

[e]Indeed, one can never have so much information that the Gedanken ensemble of cats may be described by a pure state.

compound systems S+A by a *new density operator*. The possibility of performing such a selection is a subtle question, which we tackle through considerations about *dynamical stability*. We will thus give an idea of a solution of the long standing measurement problem.

The solution of the model thus relies on several steps. First, the density matrix of S+A associated with the full ensemble of runs is truncated, to wit, it loses its off-diagonal blocks (Section 4). Then, its diagonal blocks relax to equilibrium, thus allowing registration into the apparatus of the information included in the diagonal elements of the initial density matrix of S (Section 5). Next we show that a special type of relaxation yields the needed result for the density operator of any subensemble of runs of the measurement (Sections 6.2 and 6.3). Finally, the structure of these density operators affords a natural interpretation of the process for individual runs in spite of quantum difficulties (Section 6.4).

The Curie–Weiss model is sufficiently simple so as to allow interesting generalisations. In Section 7, we present a model which involves two apparatuses that attempt to *measure two non-commuting observables*, namely the components of the spin on two different directions. We shall see that, although this measurement is not ideal and although it seems to involve two incompatible observables, performing a large number of runs can provide statistical information on both.

2. A Curie–Weiss model for quantum measurements

In this section we give a detailed description of the Curie–Weiss model for a quantum measurement, which was introduced a decade ago.[4] We take for S, the system to be measured, the simplest quantum system, namely a spin $\frac{1}{2}$. The observable to be measured is its third Pauli matrix $\hat{s}_z = \mathrm{diag}(1, -1)$, with eigenvalues s_i equal to ± 1. For an ideal measurement we assume that $\hat{s}_z$ commutes with the Hamiltonian of S + A. This ensures that the statistics of the measured observable are preserved in time, a necessary condition to satisfy Born rule.

We take as apparatus A = M + B, a model that simulates a *magnetic dot*: The magnetic degrees of freedom M consist of $N \gg 1$ spins with Pauli operators $\hat{\sigma}_a^{(n)}$ ($n = 1, 2, \cdots, N$; $a = x, y, z$), which read for each n

$$\hat{\sigma}_x = \begin{pmatrix} 0 & 1 \\ 1 & 0 \end{pmatrix}, \quad \hat{\sigma}_y = \begin{pmatrix} 0 & -i \\ i & 0 \end{pmatrix}, \quad \hat{\sigma}_z = \begin{pmatrix} 1 & 0 \\ 0 & -1 \end{pmatrix}, \quad \hat{\sigma}_0 = \begin{pmatrix} 1 & 0 \\ 0 & 1 \end{pmatrix}, \quad (2.1)$$

where $\hat{\sigma}_0$ is the corresponding identity matrix; $\hat{\boldsymbol{\sigma}} = (\hat{\sigma}_x, \hat{\sigma}_y, \hat{\sigma}_z)$ denotes the vector spin operator. The non-magnetic degrees of freedom such as phonons behave as a thermal bath B (Fig. 1). As pointer variable we take the order parameter, which is the magnetization in the z-direction (within normalization), as represented by the quantum observable

$$\hat{m} = \frac{1}{N} \sum_{n=1}^{N} \hat{\sigma}_z^{(n)}. \quad (2.2)$$

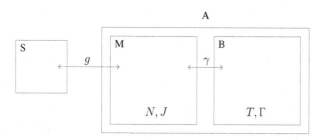

Fig. 1. The first version of the Curie-Weiss measurement model and its parameters. The system S is a spin-$\frac{1}{2}$ $\hat{s}$. The apparatus A includes a magnet M and a bath B. The magnet, which acts as a pointer, consists of N spins-$\frac{1}{2}$ coupled to one another through an Ising interaction J. The phonon bath B is characterized by its temperature T and a Debye cutoff Γ. It interacts with M through a spin-boson coupling γ. The process is triggered by the interaction g between the measured observable $\hat{s}_z$ and the pointer variable, the magnetization per spin, $\hat{m}$, of the pointer. To consider the measurement problem, certain weak terms will be added later within the apparatus.

We let N remain finite, which will allow us to keep control of the equations of motion. It should however be sufficiently large so as to ensure the existence of thermal equilibrium states with well defined magnetization (i.e., fluctuations of the order of $1/\sqrt{N}$). At the end of the measurement, the value of the magnetization (either positive or negative) is linked to the two possible outcomes of the measurement, $s_i = \pm 1$.

2.1. *The Hamiltonian*

We consider the tested system S and the apparatus A as two quantum systems, that are coupled at qtime $t = 0$ and decoupled at time t_f. The full Hamiltonian can be decomposed into terms associated with the system, with the apparatus and with their coupling:

$$\hat{H} = \hat{H}_S + \hat{H}_{SA} + \hat{H}_A. \tag{2.3}$$

Textbooks treat measurements as instantaneous, which is an idealization. If they are at least very fast, the tested system will hardly undergo dynamics by its own, so the tested quantity $\hat{s}$ is practically constant. For an ideal measurement the observable $\hat{s}$ should not proceed at all, so it should commute with $\hat{H}$. The simplest self-Hamiltonian that ensures this property (no evolution of S without coupling to A), is a constant field $-b_z \hat{s}_z$, which is for our aims equivalent to the trivial case $\hat{H}_S = 0$, so we consider the latter.

We take as coupling between the tested system and the apparatus,

$$\hat{H}_{SA} = -g\hat{s}_z \sum_{n=1}^{N} \hat{\sigma}_z^{(n)} = -Ng\hat{s}_z\hat{m}. \tag{2.4}$$

It has the usual form of a spin-spin coupling in the z-direction, and the constant $g > 0$ characterizes its strength. As wished, it commutes with $\hat{s}_z$.

$$\mathcal{\hat{D}}\left\{\hat{D}\left\{\begin{array}{ll}\text{S} & \hat{r} \\ \text{M} & \hat{R}_\text{M} \\ \text{B} & \hat{R}_\text{B}\end{array}\right\}\mathcal{\hat{R}}\right.$$

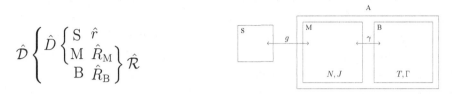

Fig. 2. Notations for the density operators of the system S + A and the subsystems M and B of A. The full density matrix $\hat{\mathcal{D}}$ is parametrized by its submatrices $\hat{\mathcal{R}}_{ij}$ (with $i, j = \pm 1$ or $\uparrow, \downarrow$), the density matrix $\hat{D}$ of S + M by its submatrices $\hat{R}_{ij}$. The marginal density operator of S is denoted as $\hat{r}$ and the one of A as $\hat{\mathcal{R}}$. The marginal density operator of M itself is denoted as $\hat{R}_\text{M}$ and the one of B as $\hat{R}_\text{B}$.

The apparatus A consists, as indicated above, of a magnet M and a phonon bath B (Fig. 2), and its Hamiltonian can be decomposed into

$$\hat{H}_\text{A} = \hat{H}_\text{M} + \hat{H}_\text{B} + \hat{H}_\text{MB}. \tag{2.5}$$

The magnetic part is chosen as

$$\hat{H}_\text{M} = -\frac{J}{4}\hat{m}^4, \tag{2.6}$$

where the magnetization operator $\hat{m}$ was defined by (2.2). It couples all spins $\hat{\sigma}^{(n)}$ symmetrically and anistropically, with the same coupling constant J. This Hamiltonian is used to describe superexchange interactions in metamagnets.

As we will show in subsequent sections, the Hamiltonian (2.6) of M, when coupled to a thermal bath at sufficiently low temperature T, leads to three locally thermal states for M: a metastable (paramagnetic) state $\hat{\mathcal{R}}(0)$ and two stable (ferromagnetic) states, $\hat{\mathcal{R}}_{\Uparrow}$ and $\hat{\mathcal{R}}_{\Downarrow}$. A first order transition can then occur from $\hat{\mathcal{R}}(0)$ to one of the more stable ferromagnetic states (for a more realistic set up including first and second order transition we refer the reader to Ref. 3). An advantage of a first-order transition is the local stability of the paramagnetic state, even below the transition temperature, which ensures a large lifetime. It is only by the measurement, i.e., by coupling to the tested spin, that a fast transition to one of the stable states is triggered. This is well suited for a measurement process, which requires the lifetime of the initial state of the apparatus to be larger than the overall measurement time.

The Hamiltonian of the phonon bath, $H_\text{M} + H_\text{MB}$, is described in full detail in the Appendix A. The bath plays a crucial role in the Curie-Weiss model, as it induces thermalization in the states of M. Nevertheless, the degrees of freedom of the bath will be traced out as we are not interested in their specific evolution (recall that the magnetization is the pointer variable). This induces a non-unitary evolution into the subspace of S+M arising from the unitary evolution of the whole closed system.

If we assume a very large bath weakly coupled to M, then all the relevant information is compressed in the spectrum of the bath, which we choose to be quasi-Ohmic:[8–11]

$$\tilde{K}(\omega) = \frac{\hbar^2}{4} \frac{\omega e^{-|\omega|/\Gamma}}{e^{\beta\hbar\omega} - 1}. \tag{2.7}$$

where $\beta = 1/k_B T$ is the inverse temperature of the bath, the dimensionless parameter γ is the strength of the interaction; and Γ is the Debye cutoff, which characterizes the largest frequencies of the bath, and is assumed to be larger than all other frequencies entering our problem.

The spin-boson coupling (A.1) between M and B will be sufficient for our purpose up to Section 6. This interaction, of the so-called Glauber type, does not commute with $\hat{H}_M$, a property needed for registration, since M has to release energy when relaxing from its initial metastable paramagnetic state (having $\langle \hat{m} \rangle = 0$) to one of its final stable ferromagnetic states at the temperature T (having $\langle \hat{m} \rangle = \pm m_F$). However, the complete solution of the measurement problem presented in Section 6 will require more complicated interactions. We will therefore later add a small but random coupling between the spins of M, and in Subsection 6.3 a more realistic small coupling between M and B, of the Suzuki type (that is to say, having terms $\hat{\sigma}_x^{(n)}\hat{\sigma}_x^{(n')} + \hat{\sigma}_y^{(n)}\hat{\sigma}_y^{(n')} = \frac{1}{2}(\hat{\sigma}_+^{(n)}\hat{\sigma}_-^{(n')} + \hat{\sigma}_-^{(n)}\hat{\sigma}_+^{(n')})$, where $\hat{\sigma}_\pm^{(n)} = \hat{\sigma}_x^{(n)} \pm i\hat{\sigma}_y^{(n)}$), which produces flip-flops of the spins of M, without changing the values of magnetisation and the energy that M would have with only the terms of (2.6).

2.2. Structure of the states

2.2.1. Notations

Our complete system consists of S+A, that is, S+M+B. The full state $\hat{D}$ of the system evolves according to the Liouville–von Neumann equation

$$i\hbar \frac{d\hat{D}}{dt} = [\hat{H}, \hat{D}] \equiv \hat{H}\hat{D} - \hat{D}\hat{H}, \tag{2.8}$$

which we have to solve. It will be convenient to define through partial traces, at any instant t, the following marginal density operators: $\hat{r}$ for the tested system S, $\hat{\mathcal{R}}$ for the apparatus A, $\hat{R}_M$ for the magnet M, $\hat{R}_B$ for the bath, and $\hat{D}$ for S + M after elimination of the bath (as depicted schematically in Fig. 2), according to

$$\hat{r} = \text{tr}_A \hat{D}, \qquad \hat{\mathcal{R}} = \text{tr}_S \hat{D},$$
$$\hat{R}_M = \text{tr}_B \hat{\mathcal{R}} = \text{tr}_{S,B} \hat{D}, \qquad \hat{R}_B = \text{tr}_{S,M} \hat{D}, \qquad \hat{D} = \text{tr}_B \hat{D}. \tag{2.9}$$

The expectation value of any observable $\hat{A}$ pertaining, for instance, to the subsystem S + M of S + A (including products of spin operators $\hat{s}_a$ and $\hat{\sigma}_a^{(n)}$) can equivalently be evaluated as $\langle \hat{A} \rangle = \text{tr}_{S+A} \hat{D}\hat{A}$ or as $\langle \hat{A} \rangle = \text{tr}_{S+M} \hat{D}\hat{A}$.

As indicated above, the apparatus A is a large system, treated by methods of statistical mechanics, while we need to follow in detail the microscopic degrees of freedom of the system S and their correlations with A. To this aim, we shall analyze the full state $\hat{D}$ of the system into several sectors, characterized by the eigenvalues of

$\hat{s}_z$. Namely, in the two-dimensional eigenbasis of $\hat{s}_z$ for S, $|{\uparrow}\rangle$, $|{\downarrow}\rangle$, with eigenvalues $s_i = +1$ for $i = {\uparrow}$ and $s_i = -1$ for $i = {\downarrow}$, $\hat{\mathcal{D}}$ can be decomposed into the four blocks

$$\hat{\mathcal{D}} = \begin{pmatrix} \hat{\mathcal{R}}_{\uparrow\uparrow} & \hat{\mathcal{R}}_{\uparrow\downarrow} \\ \hat{\mathcal{R}}_{\downarrow\uparrow} & \hat{\mathcal{R}}_{\downarrow\downarrow} \end{pmatrix}, \tag{2.10}$$

where each $\hat{\mathcal{R}}_{ij}$ is an operator in the space of the apparatus. We shall also use the partial traces (see again Fig. 3.2)

$$\hat{R}_{ij} = \mathrm{tr}_{\mathrm{B}}\hat{\mathcal{R}}_{ij}, \qquad \hat{D} = \mathrm{tr}_{\mathrm{B}}\hat{\mathcal{D}} = \begin{pmatrix} \hat{R}_{\uparrow\uparrow} & \hat{R}_{\uparrow\downarrow} \\ \hat{R}_{\downarrow\uparrow} & \hat{R}_{\downarrow\downarrow} \end{pmatrix} \tag{2.11}$$

over the bath; each $\hat{R}_{ij}$ is an operator in the 2^N-dimensional space of the magnet. Indeed, we are not interested in the evolution of the bath variables, and we shall eliminate B by relying on the weakness of its coupling (A.1) with M, expressed by the dimensionless variable $\gamma \ll 1$. The operators $\hat{R}_{ij}$ code our full statistical information about S and M. We shall use the notation $\hat{R}_{ij}$ whenever we refer to S + M and $\hat{R}_{\mathrm{M}}$ when referring to M alone. Tracing also over M, we are, according to (2.9), left with

$$\hat{r} = \begin{pmatrix} r_{\uparrow\uparrow} & r_{\uparrow\downarrow} \\ r_{\downarrow\uparrow} & r_{\downarrow\downarrow} \end{pmatrix} = r_{\uparrow\uparrow}|{\uparrow}\rangle\langle{\uparrow}| + r_{\uparrow\downarrow}|{\uparrow}\rangle\langle{\downarrow}| + r_{\downarrow\uparrow}|{\downarrow}\rangle\langle{\uparrow}| + r_{\downarrow\downarrow}|{\downarrow}\rangle\langle{\downarrow}|. \tag{2.12}$$

The magnet M is thus described by $\hat{R}_{\mathrm{M}} = \hat{R}_{\uparrow\uparrow} + \hat{R}_{\downarrow\downarrow}$, the system S alone by the matrix elements of $\hat{r}$, viz. $r_{ij} = \mathrm{tr}_{\mathrm{M}}\hat{R}_{ij}$. The correlations of $\hat{s}_z$, $\hat{s}_x$ or $\hat{s}_y$ with any function of the observables $\hat{\sigma}_a^{(n)}$ ($a = x, y, z$, $n = 1$, $\ldots N$) are represented by $\hat{R}_{\uparrow\uparrow} - \hat{R}_{\downarrow\downarrow}$, $\hat{R}_{\uparrow\downarrow} + \hat{R}_{\downarrow\uparrow}$, $i\hat{R}_{\uparrow\downarrow} - i\hat{R}_{\downarrow\uparrow}$, respectively. The operators $\hat{R}_{\uparrow\uparrow}$ and $\hat{R}_{\downarrow\downarrow}$ are hermitean positive, but not normalized, whereas $\hat{R}_{\downarrow\uparrow} = \hat{R}_{\uparrow\downarrow}^\dagger$. Notice that we now have from (2.9) – (2.11)

$$r_{ij} = \mathrm{tr}_{\mathrm{A}}\hat{\mathcal{R}}_{ij} = \mathrm{tr}_{\mathrm{M}}\hat{R}_{ij}, \qquad \hat{\mathcal{R}} = \hat{\mathcal{R}}_{\uparrow\uparrow} + \hat{\mathcal{R}}_{\downarrow\downarrow},$$
$$\hat{R}_{\mathrm{M}} = \hat{R}_{\uparrow\uparrow} + \hat{R}_{\downarrow\downarrow}, \qquad \hat{R}_{\mathrm{B}} = \mathrm{tr}_{\mathrm{M}}(\hat{\mathcal{R}}_{\uparrow\uparrow} + \hat{\mathcal{R}}_{\downarrow\downarrow}). \tag{2.13}$$

All these elements are functions of the time t which elapses from the beginning of the measurement at $t = 0$ when $\hat{H}_{\mathrm{SA}}$ is switched on to the final value t_{f} that we will evaluate in Section 7.

To introduce further notation, we mention that the combined system S + A = S + M + B should for all practical purposes end up in[f]

$$\hat{\mathcal{D}}(t_{\mathrm{f}}) = \begin{pmatrix} p_{\uparrow}\hat{\mathcal{R}}_{\Uparrow} & 0 \\ 0 & p_{\downarrow}\hat{\mathcal{R}}_{\Downarrow} \end{pmatrix} = p_{\uparrow}|{\uparrow}\rangle\langle{\uparrow}| \otimes \hat{\mathcal{R}}_{\Uparrow} + p_{\downarrow}|{\downarrow}\rangle\langle{\downarrow}| \otimes \hat{\mathcal{R}}_{\Downarrow} = \sum_{i={\uparrow},{\downarrow}} p_i \hat{\mathcal{D}}_i, \tag{2.14}$$

---

[f]The terms $|{\downarrow}\rangle\langle{\uparrow}|\mathcal{R}_{\uparrow\downarrow}(t)$ and $|{\downarrow}\rangle\langle{\uparrow}|\mathcal{R}_{\downarrow\uparrow}(t)$ are not strictly zero, in fact the trace of their product $\mathcal{R}_{\uparrow\downarrow}^\dagger(t)\mathcal{R}_{\uparrow\downarrow}(t)$ is even conserved in time. But when taking traces to obtain physical observables, the wildly oscillating phase factors which they carry prevent any meaningful contribution. There is clearly a discrepancy between *vanishing mathematically* and *being irrelevant physically*.

where $\hat{\mathcal{R}}_\Uparrow$ ($\hat{\mathcal{R}}_\Downarrow$) is density matrix of the thermodynamically stable state of the magnet and bath, after the measurement, in which the magnetization is up, taking the value $m_\Uparrow(g)$ (down, taking the value $m_\Downarrow(g)$); these events should occur with probabilities $p_\uparrow$ and $p_\downarrow$, respectively[g]. When, at the end of the measurement, the coupling g is turned off ($g \to 0$), the macroscopic magnet will relax to the nearby state having $m_\Uparrow(0) \approx m_\Uparrow(g)$ (viz. $m_\Downarrow(0) \approx m_\Downarrow(g)$). The Born rule then predicts that $p_\uparrow = \mathrm{tr_S}\hat{r}(0)\Pi_\uparrow = r_{\uparrow\uparrow}(0)$ and $p_\downarrow = r_{\downarrow\downarrow}(0)$.

Since no physically relevant off-diagonal terms occur in (2.14), a point that we wish to explain, and since we expect B to remain nearly in its initial equilibrium state, we may trace out the bath, as is standard in classical and quantum thermal physics, without losing significant information. It will therefore be sufficient for our purpose to show that the final state is[h]

$$\hat{D}(t_\mathrm{f}) = \begin{pmatrix} p_\uparrow\hat{R}_{\mathrm{M}\Uparrow} & 0 \\ 0 & p_\downarrow\hat{R}_{\mathrm{M}\Downarrow} \end{pmatrix} = p_\uparrow |{\uparrow}\rangle\langle{\uparrow}| \otimes \hat{R}_{\mathrm{M}\Uparrow} + p_\downarrow |{\downarrow}\rangle\langle{\downarrow}| \otimes \hat{R}_{\mathrm{M}\Downarrow}, \tag{2.15}$$

now referring to the magnet M and tested spin S alone.

Returning to Eq. (2.13), we note that from any density operator $\hat{R}$ of the magnet we can derive the *probabilities* $P_\mathrm{M}^\mathrm{dis}(m)$ *for* $\hat{m}$ *to take the eigenvalues* m, where "dis" denotes their discreteness. These $N + 1$ eigenvalues,

$$m = -1, \qquad -1 + \frac{2}{N}, \qquad \dots, \qquad 1 - \frac{2}{N}, \qquad 1, \tag{2.16}$$

have equal spacings $\delta m = 2/N$ and multiplicities

$$G(m) = \frac{N!}{[\frac{1}{2}N(1+m)]!\,[\frac{1}{2}N(1-m)]!} = e^{S(m)} \tag{2.17}$$

The entropy reads for large N

$$S(m) = N\left(-\frac{1+m}{2}\ln\frac{1+m}{2} - \frac{1-m}{2}\ln\frac{1-m}{2}\right) + \log\sqrt{\frac{2}{\pi N(1-m^2)}} \tag{2.18}$$

Denoting by $\delta_{\hat{m},m}$ the projection operator on the subspace m of $\hat{m}$, the dimension of which is $G(m)$, we have

$$P_\mathrm{M}^\mathrm{dis}(m,t) = \mathrm{tr_M}\hat{R}_\mathrm{M}(t)\delta_{\hat{m},m}. \tag{2.19}$$

where the superscript "dis" denotes that m is viewed as a *discrete* variable, over which sums can be carried out. In the limit $N \gg 1$, where m becomes basically a *continuous* variable, we shall later work with the functions $P_\mathrm{M}(m,t)$, defined as

$$P_\mathrm{M}(m,t) = \frac{N}{2}P_\mathrm{M}^\mathrm{dis}(m,t), \qquad \int_{-1}^{1} dm\, P_\mathrm{M}(m,t) = \sum_m P_\mathrm{M}^\mathrm{dis}(m,t) = 1, \tag{2.20}$$

that have a finite and smooth limit for $N \to \infty$. A similar relation will hold between $P_{\uparrow\uparrow}^\mathrm{dis}(m,t)$ and $P_{\uparrow\uparrow}(m,t)$, to be encountered further on.

---

[g]Notice that in the final state we denote properties of the tested system by $\uparrow, \downarrow$ and of the apparatus by $\Uparrow, \Downarrow$. In sums like (2.14) we will also use $i = \uparrow, \downarrow$, or sometimes $i = \pm 1$.

[h]Being the trace of (2.14) over the bath, its off-diagonal terms vanish, see footnote f.

2.2.2. Initial state

In order to describe an unbiased measurement, S and A are statistically independent in the initial state , which is expressed by $\hat{\mathcal{D}}(0) = \hat{r}(0) \otimes \hat{\mathcal{R}}(0)$. The 2×2 density matrix $\hat{r}(0)$ of S is arbitrary; by the measurement we wish to gain information about it. It has the form (2.12) with elements $r_{\uparrow\uparrow}(0)$, $r_{\uparrow\downarrow}(0)$, $r_{\downarrow\uparrow}(0)$ and $r_{\downarrow\downarrow}(0)$ satisfying the positivity and hermiticity conditions

$$r_{\uparrow\uparrow}(0) + r_{\downarrow\downarrow}(0) = 1, \qquad r_{\uparrow\downarrow}(0) = r_{\downarrow\uparrow}^*(0),$$
$$r_{\uparrow\uparrow}(0) \, r_{\downarrow\downarrow}(0) \geq r_{\uparrow\downarrow}(0) \, r_{\downarrow\uparrow}(0). \tag{2.21}$$

At the initial time, the bath is set into equilibrium at the temperature[i] $T = 1/\beta$. The corresponding density operator is,

$$\hat{R}_{\mathrm{B}}(0) = \frac{1}{Z_{\mathrm{B}}} e^{-\beta \hat{H}_{\mathrm{B}}}, \tag{2.22}$$

where $\hat{H}_{\mathrm{B}}$ is given in Appendix A and Z_{B} is the partition function. The connection between the initial state of the bath and its spectrum (2.7) is described in Appendix B.

According to the discussion in Section 2.1.1, the initial density operator $\hat{\mathcal{R}}(0)$ of the apparatus describes the magnetic dot in a metastable paramagnetic state and a bath. As justified below, we take for it the factorized form

$$\hat{\mathcal{R}}(0) = \hat{R}_{\mathrm{M}}(0) \otimes \hat{R}_{\mathrm{B}}(0), \tag{2.23}$$

where the bath is in the Gibbsian equilibrium state (B.1), at the temperature $T = 1/\beta$ *lower* than the transition temperature of M, while the magnet with Hamiltonian (2.5) is in a paramagnetic equilibrium state at a temperature $T_0 = 1/\beta_0$ *higher* than its transition temperature:

$$\hat{R}_{\mathrm{M}}(0) = \frac{1}{Z_{\mathrm{M}}} e^{-\beta_0 \hat{H}_{\mathrm{M}}}. \tag{2.24}$$

How can the apparatus be actually initialized in the non-equilibrium state (2.23) at the time $t = 0$? This *initialization* takes place during the time interval $-\tau_{\mathrm{init}} < t < 0$. The apparatus is first set at earlier times into equilibrium at the temperature T_0. Due to the smallness of γ, its density operator is then factorized and proportional to $\exp[-\beta_0(\hat{H}_{\mathrm{M}} + \hat{H}_{\mathrm{B}})]$. At the time $-\tau_{\mathrm{init}}$ the phonon bath is suddenly cooled down to T. We shall evaluate in Section 5 the *relaxation time* of M towards its equilibrium ferromagnetic states under the effect of B at the temperature T. Due to the weakness of the coupling γ, this time this time is long and *dominates the duration of the experiment.* We can safely assume τ_{init} to be much shorter than this relaxation time so that M remains unaffected by the cooling. On the other hand, the quasi continuous nature of the spectrum of B can allow the

---

[i]We use units where Boltzmann's constant k_B is equal to one; otherwise, T and $\beta = 1/T$ should be replaced throughout by $k_B T$ and $1/k_B T$, respectively.

phonon-phonon interactions (which we have disregarded when writing (A.2)) to establish the equilibrium of B at the temperature T within a time shorter than τ_{init}. It is thus realistic to imagine an initial state of the form (2.23).

An alternative method of initialization consists in applying to the magnetic dot a *strong radiofrequency field*, which acts on M but not on B. The bath can thus be thermalized at the required temperature, lower than the transition temperature of M, while the populations of spins of M oriented in either direction are equalized. The magnet is then in a paramagnetic state, as if it were thermalized at an *infinite* temperature T_0 in spite of the presence of a cold bath. In that case we have the initial state (see Eq. (2.1))

$$\hat{R}_{\text{M}}(0) = \frac{1}{2^N} \prod_{n=1}^{N} \hat{\sigma}_0^{(n)}. \tag{2.25}$$

The initial density operator (2.24) of M being simply a function of the operator $\hat{m}$, we can characterize it as in (2.19) by the probabilities $P_{\text{M}}^{\text{dis}}(m, 0)$ for $\hat{m}$ to take the values (2.16). Those probabilities are the normalized product of the degeneracy (2.17) and the Boltzmann factor,

$$P_{\text{M}}^{\text{dis}}(m, 0) = \frac{1}{Z_0} G(m) \exp\left[\frac{NJ}{4T_0} m^4\right], \quad Z_0 = \sum_m G(m) \exp\left[\frac{NJ}{4T_0} m^4\right]. \tag{2.26}$$

For sufficiently large N, the distribution $P_{\text{M}}(m, 0) = \frac{1}{2} N P_{\text{M}}^{\text{dis}}(m, 0)$ is peaked around $m = 0$, with the Gaussian shape

$$P_{\text{M}}(m, 0) \simeq \frac{1}{\sqrt{2\pi} \Delta m} e^{-m^2/2\Delta m^2}. \tag{2.27}$$

This peak, which has a narrow width of the form

$$\Delta m = \sqrt{\langle m^2 \rangle} = \frac{1}{\sqrt{N}}, \tag{2.28}$$

involves a large number, of order $\sqrt{N}$, of eigenvalues (2.16), so that the spectrum can be treated as a continuum (except in Section 6.3).

2.2.3. *Ferromagnetic equilibrium states of the magnet*

The measurement will drive M from its initial metastable state to one of its stable ferromagnetic states. The final state (2.14) of S + A after measurement will thus involve the two ferromagnetic equilibrium states $\hat{R}_i$, $i = \Uparrow$ or $\Downarrow$. As above these states $\hat{R}_i$ of the apparatus factorize, in the weak coupling limit ($\gamma \ll 1$), into the product of (B.1) for the bath and a ferromagnetic equilibrium state $\hat{R}_{\text{M}i}$ for the magnet M. The point of this section is to study the properties of such equilibrium states, whose temperature $T = 1/\beta$ is induced by the bath.

Let us thus consider the equilibrium state of M, which depends on β and on its Hamiltonian

$$\hat{H}_{\text{M}} = -Nh\hat{m} - NJ\frac{\hat{m}^4}{4}, \tag{2.29}$$

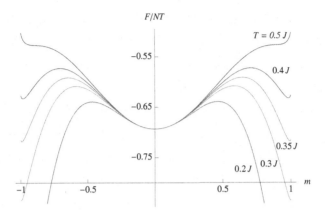

Fig. 3.   The free energy F in units of NT, evaluated from Eq. (2.31) with $h = 0$, as function of the magnetization m at various temperatures. There is always a local paramagnetic minimum at $m = 0$. A first-order transition occurs at $T_c = 0.363J_4$, below which the ferromagnetic states associated with the minima at $\pm m_F$ near ± 1 become the most stable.

where we introduced an external field h acting on the spins of the apparatus for latter convenience.[j] As in (2.19) we characterize the canonical equilibrium density operator of the magnet $\hat{R}_M = (1/Z_M)\exp[-\beta\hat{H}_M]$, which depends only on the operator $\hat{m}$, by the probability distribution

$$P_M(m) = \frac{\sqrt{N}}{Z_M\sqrt{8\pi}}e^{-\beta F(m)}, \qquad (2.30)$$

where m takes the discrete values m_i given by (2.16); the exponent of (2.30) introduces the *free energy function*

$$F(m) = -NJ\frac{m^4}{4} - Nhm + NT\left(\frac{1+m}{2}\ln\frac{1+m}{2} + \frac{1-m}{2}\ln\frac{1-m}{2}\right), \quad (2.31)$$

which arises from the Hamiltonian (2.29) and from the multiplicity $G(m)$ given by (2.17). It is displayed in Fig. 3. The distribution (2.30) displays narrow peaks at the minima of $F(m)$, and the *equilibrium free energy* $-T\ln Z_M$ is equal for large N to the absolute minimum of (2.31). The function $F(m)$ reaches its extrema at values of m given by the self-consistent equation

$$m = \tanh\left[\beta\left(h + Jm^3\right)\right]. \qquad (2.32)$$

In the vicinity of a minimum of $F(m)$ at $m = m_i$, the probability $P_M(m)$ presents around each m_i a nearly Gaussian peak, given within normalization by

$$P_{Mi}(m) \propto \exp\left\{-\frac{N}{2}\left[\frac{1}{1 - m_i^2} - 3\beta Jm_i^2\right](m - m_i)^2\right\}. \qquad (2.33)$$

[j]In Section 5 we shall identify h with $+g$ in the sector $\hat{R}_{\uparrow\uparrow}$ of $\hat{D}$, or with $-g$ in its sector $\hat{R}_{\downarrow\downarrow}$, where g is the coupling between S and A.

This peak has a width of order $1/\sqrt{N}$ and a weak asymmetry. The possible values of m are dense within the peak, with equal spacing $\delta m = 2/N$. With each such peak $P_{\mathrm{M}i}(m)$ is associated through (2.19), (2.20), a density operator $\hat{R}_i$ of the magnet M which may describe a locally stable equilibrium. Depending on the values of J and on the temperature, there may exist one, two or three such locally stable states. We note the corresponding average magnetizations m_i, for arbitrary h, as m_{P} for a paramagnetic state and as $m_{\Uparrow}$ and $m_{\Downarrow}$ for the ferromagnetic states, with $m_{\Uparrow} > 0$, $m_{\Downarrow} < 0$. We also denote as $\pm m_{\mathrm{F}}$ the ferromagnetic magnetizations for $h = 0$. When h tends to 0 (as happens at the end of the measurement where we set $g \to 0$), m_{P} tends to 0, $m_{\Uparrow}$ to $+m_{\mathrm{F}}$ and $m_{\Downarrow}$ to $-m_{\mathrm{F}}$, namely

$$m_{\Uparrow}(h > 0) > 0, \qquad m_{\Downarrow}(h > 0) < 0, \qquad m_{\Uparrow}(-h) = -m_{\Downarrow}(h),$$
$$m_{\mathrm{F}} = m_{\Uparrow}(h \to +0) = -m_{\Downarrow}(h \to +0). \tag{2.34}$$

For $h = 0$, the system M is invariant under change of sign of m. This invariance is spontaneously broken below some temperature. The two additional ferromagnetic peaks $P_{\mathrm{M}\Uparrow}(m)$ and $P_{\mathrm{M}\Downarrow}(m)$ appear around $m_{\Uparrow} = m_{\mathrm{F}} = 0.889$ and $m_{\Downarrow} = -m_{\mathrm{F}}$ when the temperature T goes below $0.496J$. As T decreases, m_{F} given by $m_{\mathrm{F}} = \tanh \beta J m_{\mathrm{F}}^3$ increases and the value of the minimum $F(m_{\mathrm{F}})$ decreases; the weight (2.30) is transferred from $P_{\mathrm{M}0}(m)$ to $P_{\mathrm{M}\Uparrow}(m)$ and $P_{\mathrm{M}\Downarrow}(m)$. A first-order transition occurs when $F(m_{\mathrm{F}}) = F(0)$, for $T_{\mathrm{c}} = 0.363J$ and $m_{\mathrm{F}} = 0.9906$, from the paramagnetic to the two ferromagnetic states, although the paramagnetic state remains locally stable. The spontaneous magnetization m_{F} is always very close to 1, behaving as $1 - m_{\mathrm{F}} \sim 2\exp(-2J/T)$.

Strictly speaking, the canonical equilibrium state of M below the transition temperature, characterized by (2.30), has for $h = 0$ and finite N the form

$$\hat{R}_{\mathrm{Meq}} = \frac{1}{2}(\hat{R}_{\mathrm{M}\Uparrow} + \hat{R}_{\mathrm{M}\Downarrow}). \tag{2.35}$$

However this state is not necessarily the one reached at the end of a relaxation process governed by the bath B, when a field h, even weak, is present: this field acts as a source which breaks the invariance. The determination of the state $\hat{R}_{\mathrm{M}}(t_{\mathrm{f}})$ reached at the end of a relaxation process involving the thermal bath B and a weak field h *requires a dynamical study* which will be worked out in Section 5. This is related to the ergodicity breaking: if a weak field is applied, then switched off, the full canonical state (2.35) is still recovered, but only after an unrealistically long time (for $N \gg 1$). For finite times the equilibrium state of the magnet is to be found by restricting the full canonical state (2.35) to its component having a magnetization with the definite sign determined by the weak external field. This is the essence of the spontaneous symmetry breaking. However, for our situation this well-known recipe should be supported by dynamical considerations, since we have to show that the thermodynamically expected states will be reached dynamically.

In our model of measurement, the situation is similar, though slightly more complicated. The system-apparatus coupling (2.4) plays the rôle of an operator-

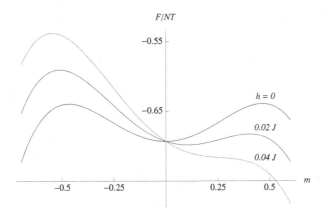

Fig. 4.   The effect of a positive field h on $F(m)$ for $q = 4$ at temperature $T = 0.2J$. As h increases the paramagnetic minimum m_P shifts towards positive m. At the critical field $h_c = 0.0357J$ this local minimum disappears, and the curve has an inflexion point with vanishing slope at $m = m_c = 0.268$. For larger fields, like in the displayed case $g = 0.04J$, the locally stable paramagnetic state disappears, and there remain only the two ferromagnetic states, the most stable one with positive magnetization $m_\Uparrow \simeq 1$ and the metastable one with negative magnetization $m_\Downarrow \simeq -1$.

valued source, with eigenvalues behaving as a field $h = g$ or $h = -g$. We shall determine in Section 6 towards which state M is driven under the conjugate action of the bath B and of the system S, depending on the parameters of the model.

As a preliminary step, let us examine here the effect on the free energy (2.31) of a small positive field h. Consider first the minima of $F(m)$.[12,13] The two ferromagnetic minima $m_\Uparrow$ and $m_\Downarrow$ given by (2.32) are slightly shifted away from m_F and $-m_F$, and $F(m_\Uparrow) - F(m_F)$ behaves as $-Nhm_F$. Hence, as soon as $\exp\{-\beta [F(m_\Uparrow) - F(m_\Downarrow)]\} \sim \exp(2\beta Nhm_F) \gg 1$, only the single peak $P_{M\Uparrow}(m)$ around $m_\Uparrow \simeq m_F$ contributes to (2.30), so that the canonical equilibrium state of M has the form $\hat{R}_{Meq} = \hat{R}_{M\Uparrow}$. The shape of $F(m)$ will also be relevant for the dynamics. If h is sufficiently small, $F(m)$ retains its paramagnetic minimum, the position of which is shifted as $m_P \sim h/T$; the paramagnetic state $\hat{R}_M(0)$ remains locally stable. It may decay towards a stable ferromagnetic state only through mechanisms of thermal activation or quantum tunneling, processes with very large characteristic times, of exponential order in N. In such cases A is not a good measuring apparatus. However, there is a threshold h_c above which this paramagnetic minimum of $F(m)$, which then lies at $m = m_c$, disappears. The value of h_c is found by eliminating $m = m_c$ between the equations $d^2F/dm^2 = 0$ and $dF/dm = 0$. We find $2m_c^2 = 1 - \sqrt{1 - 4T/3J}$, $h_c = \frac{1}{2}T \ln[(1 + m_c)/(1 - m_c)] - Jm_c^3$. At the transition temperature $T_c = 0.363J$, we have $m_c = 0.375$ and $h_c = 0.0904J$; for $T = 0.2J$, we obtain $m_c = 0.268$ and $h_c = 0.036J$; for $T \ll J$, m_c behaves as $\sqrt{T/3J}$ and h_c as $\sqrt{4T^3/27J}$. Provided $h > h_c$, $F(m)$ has now a negative slope in the whole interval $0 < m < m_F$. We can thus expect, in our measurement problem, that the registration will take place in a reasonable delay for a first order transition if the coupling g is larger than h_c.[k]

We have stressed already that the apparatus A should lie initially in a metastable state,[12,13] and finally in either one of several possible stable states (see Section 2 for other models of this type). This suggests to take for A, a quantum system that may undergo a phase transition with *broken invariance*. The initial state $\hat{\mathcal{R}}(0)$ of A is the metastable phase with unbroken invariance. The states $\hat{\mathcal{R}}_i$ represent the stable phases with broken invariance, in each of which registration can be permanent. The symmetry between the outcomes prevents any bias.

The initial state $\hat{\mathcal{R}}(0)$ of A is the metastable paramagnetic state. We expect the final state (2.15) of S + A to involve for A the two stable ferromagnetic states $\hat{\mathcal{R}}_i$, $i = \uparrow$ or $\downarrow$, that we denote as $\hat{\mathcal{R}}_\Uparrow$ or $\hat{\mathcal{R}}_\Downarrow$, respectively.[l] The equilibrium temperature T will be imposed to M by the phonon bath[8,9] through weak coupling between the magnetic and non-magnetic degrees of freedom. Within small fluctuations, the order parameter (2.2) vanishes in $\hat{\mathcal{R}}(0)$ and takes two opposite values in the states $\hat{\mathcal{R}}_\Uparrow$ and $\hat{\mathcal{R}}_\Downarrow$, $A_i \equiv \langle \hat{m} \rangle_i$ equal to $+m_F$ for $i = \uparrow$ and to $-m_F$ for $i = \downarrow$.[m] As in real magnetic registration devices, information will be stored by A in the form of the sign of the magnetization.

3. Dynamical equations

In this section we present the basic steps that lead us to solvable evolution equations. The Hamiltonian $\hat{H}_0$ in the space S + M gives rise to two Hamiltonians $\hat{H}_\uparrow$ and $\hat{H}_\downarrow$ in the space M, which according to (2.4) and (2.6) are simply two functions of the observable $\hat{m}$, given by

$$\hat{H}_i = H_i(\hat{m}) = -gN s_i \hat{m} - N\frac{J}{4}\hat{m}^4, \qquad (i = \uparrow, \downarrow) \tag{3.1}$$

with $s_i = +1$ (or -1) for $i = \uparrow$ (or $\downarrow$). These Hamiltonians $\hat{H}_i$, which describe interacting spins $\hat{\sigma}^{(n)}$ in an external field $g s_i$, occur in (2.8) both directly and through the operators

$$\hat{\sigma}_a^{(n)}(u, i) = e^{-i\hat{H}_i u/\hbar} \hat{\sigma}_a^{(n)} e^{i\hat{H}_i u/\hbar}. \tag{3.2}$$

The equation (2.8) for $\hat{D}(t)$ which governs the joint dynamics of S + M thus reduces to the four differential equations in the Hilbert space of M (we recall that

---

[k]The set of conditions on parameters of A for being a good apparatus is reminiscent of the requirements that realistic apparatuses have to fulfil.

[l]Here and in the following, single arrows $\uparrow, \downarrow$ will denote the spin S, while double arrows $\Uparrow, \Downarrow$ denote the magnet M.

[m]Note that the values $A_i = \pm m_F$, which we wish to come out associated with the eigenvalues $s_i = \pm 1$, are determined from equilibrium statistical mechanics; they are not the eigenvalues of $\hat{A} \equiv \hat{m}$, which range from -1 to $+1$ with spacing $2/N$, but thermodynamic expectation values around which small fluctuations of order $1/\sqrt{N}$ occur. For low T they would be close to ± 1.

$i, j = \uparrow, \downarrow$ or ± 1):

$$\frac{d\hat{R}_{ij}(t)}{dt} - \frac{\hat{H}_i \hat{R}_{ij}(t) - \hat{R}_{ij}(t)\hat{H}_j}{i\hbar} = \tag{3.3}$$

$$\frac{\gamma}{\hbar^2} \int_0^t du \sum_{n,a} \left\{ K(u) \left[\hat{\sigma}_a^{(n)}(u, i) \hat{R}_{ij}(t), \hat{\sigma}_a^{(n)} \right] + K(-u) \left[\hat{\sigma}_a^{(n)}, \hat{R}_{ij}(t)\hat{\sigma}_a^{(n)}(u, j) \right] \right\}.$$

The action of the bath is compressed in $K(u)$, which is related to its spectrum (defined in (2.7)) through a Fourier transform:

$$K(t) = \frac{1}{2\pi} \int_{-\infty}^{+\infty} d\omega \, e^{i\omega t} \tilde{K}(\omega), \qquad \tilde{K}(\omega) = \int_{-\infty}^{+\infty} dt \, e^{-i\omega t} K(t). \tag{3.4}$$

To obtain the left hand side from the Liouville-von Neumann equation (2.8) is a straightforward exercise, but the right hand side, giving the action of the bath to lowest order in γ, involves several subtle steps explained in Ref. 1, that we reproduce here in Appendix C.

3.1. *The Born rule*

Taking the trace of (3.3) in the $2^N \times 2^N$ dimensional Hilbert space of the magnet and over the bath, and using that the trace over the commutators vanishes, one obtains

$$i\hbar \frac{d\hat{r}_{ij}(t)}{dt} = \text{tr} \, (\hat{H}_i - \hat{H}_j)\hat{R}_{ij}(t) = -gN(s_i - s_j)\text{tr} \, \hat{m}\hat{R}_{ij}(t). \tag{3.5}$$

Thus for $i = j$ one gets the conservation $r_{\uparrow\uparrow}(t) = r_{\uparrow\uparrow}(0)$ and $r_{\downarrow\downarrow}(t) = r_{\downarrow\downarrow}(0)$. This is the *Born rule* stating that the probabilities for outcomes is given by the state at the beginning of the measurement. It is exactly obeyed, so in this aspect the Curie-Weiss model describes an ideal measurement. Various other features desired for ideal measurements will be satisfied in good approximation under suitable conditions on the system parameters.

An equivalent but simpler way to derive the Born rule is to notice that $i\hbar d\hat{s}_z/dt = [\hat{s}_z, \hat{H}] = 0$, so that $\hat{s}_z$ is conserved, and with it the diagonal part $\frac{1}{2}(1 + \langle \hat{s}_z \rangle \hat{s}_z)$ of the density matrix $\hat{r}(t)$.

The off-diagonal terms $\hat{r}_{ij}$ with $i \neq j$, that is to say, $r_{\uparrow\downarrow}(t)$ and $r_{\downarrow\uparrow}(t)$ or, equivalently, $\langle s_x \rangle$ and $\langle s_y \rangle$, do evolve and actually go to zero, as discussed next. In popular terms this is called "disappearance of Schrödinger cat terms". Eq. (3.5) shows that the principle culprit is the coupling g between tested spin and magnet, not the ferromagnetic interaction or the bath. Hence this step is a dephasing, not a decoherence.

4. Decay of off-diagonal terms

Focusing on the Curie-Weiss model, we present here a derivation of the processes which first lead to truncation of the off-diagonal elements of the density operator and which prevent recurrences from occurring. We show in Section 6 and Appendix D of Ref. 3 that the interactions with strength $\sim J$ between the spins $\hat{\sigma}^{(n)}$ of M play little role here, so that we neglect them. We further assume that M lies initially in the most disordered state (2.25), that we write out, using the notation (2.1), as

$$\hat{R}_{\mathrm{M}}(0) = \frac{1}{2^N}\hat{\sigma}_0^{(1)} \otimes \hat{\sigma}_0^{(2)} \otimes \cdots \otimes \hat{\sigma}_0^{(N)}. \tag{4.1}$$

Then, since the Hamiltonian $\hat{H}_{\mathrm{SA}}+\hat{H}_{\mathrm{B}}+\hat{H}_{\mathrm{MB}}$ is a sum of independent contributions associated with each spin $\hat{\sigma}^{(n)}$, it can be shown from the Liouville-von Neumann equation (2.8) that, due to neglect of the coupling J, the spins of M behave independently at all times, and that the off-diagonal block $\hat{R}_{\uparrow\downarrow}(t)$ of $\hat{D}(t)$ has the form

$$\hat{R}_{\uparrow\downarrow}(t) = r_{\uparrow\downarrow}(0)\,\hat{\rho}^{(1)}(t) \otimes \hat{\rho}^{(2)}(t) \otimes \cdots \otimes \hat{\rho}^{(N)}(t), \tag{4.2}$$

where $\hat{\rho}^{(n)}(t)$ is a 2×2 matrix in the Hilbert space of the spin $\hat{\sigma}^{(n)}$. This matrix will depend on $\hat{\sigma}_z^{(n)}$ but not on $\hat{\sigma}_x^{(n)}$ and $\hat{\sigma}_y^{(n)}$, and it will neither be hermitean nor normalized, except for $t = 0$ where it equals $\frac{1}{2}\hat{\sigma}_0^{(n)}$.

4.0.1. *Dephasing*

The first step in the dynamics of the off-diagonal terms happens at times where the bath is still inactive, the only active term in the Hamiltonian being the coupling to the tested spin. Here spin n processes as

$$\frac{\mathrm{d}\hat{\rho}^{(n)}(t)}{\mathrm{d}t} = \frac{2ig}{\hbar}\hat{\rho}^{(n)}\hat{\sigma}_z^{(n)} \tag{4.3}$$

with solution $\hat{\rho}^{(n)}(t) = \frac{1}{2}\exp(2igt\hat{\sigma}_z^{(n)}/\hbar) = \frac{1}{2}\mathrm{diag}[\exp(2igt/\hbar), \exp(-2igt/\hbar)]$. One can easily deduce the related $P_{\uparrow\downarrow}(m)$ defined by (4.2) and (2.19). Using that result or directly from (4.2) it is simple to show that

$$r_{\uparrow\downarrow}(t) = r_{\uparrow\downarrow}(0)\left(\cos\frac{2gt}{\hbar}\right)^N \tag{4.4}$$

For large N this expression decays quickly in time,

$$r_{\uparrow\downarrow}(t) = r_{\uparrow\downarrow}(0)\,e^{-(t/\tau_{\mathrm{trunc}})^2}, \tag{4.5}$$

or equivalently

$$\langle\hat{s}_a(t)\rangle = \langle\hat{s}_a(0)\rangle\,e^{-(t/\tau_{\mathrm{trunc}})^2}, \qquad (a = x,y), \tag{4.6}$$

where we introduced the truncation time

$$\tau_{\mathrm{trunc}} \equiv \frac{\hbar}{\sqrt{2}\,Ng\Delta m} = \frac{\hbar}{\sqrt{2N}\,\delta_0 g}. \tag{4.7}$$

Although $P_{\uparrow\downarrow}(m,t)$ is merely an oscillating function of t for each value of m, the summation over m has given rise to extinction. This property arises from the dephasing that exists between the oscillations for different values of m. There are undesired recurrences, however, when $2gt/\hbar = n\pi$, $n = 1, 2, \cdots$, which can be suppressed by a spread in the coupling g (see below) or by the action of the bath.

4.0.2. Decoherence

It is generally believed that Schrödinger cat terms (here: $\hat{r}_{\uparrow\downarrow}$ and $\hat{r}_{\downarrow\uparrow}$) disappear due to a coupling to a bath (environment). However, we stress that the basis in which the off-diagonal blocks of the density matrix of S+M disappear is not selected by the interaction with the environment (here with the bath B), but by the coupling between S and M. Moreover, for the present model, we have seen in the previous section that the main phenomenon which lets the off-diagonal blocks decay rapidly is dephasing. Here we look at the subsequent role of decoherence, while still neglecting J. We leave open the possibility for the coupling g_n to be random, whence the coupling between S and A reads $\hat{H}_{SA} = -\hat{s}_z \sum_{n=1}^{N} g_n \hat{\sigma}_z^{(n)}$ instead of (2.4). Each factor $\hat{\rho}^{(n)}(t)$, initially equal to $\frac{1}{2}\hat{\sigma}_0^{(n)}$, evolves according to the same equation, since in absence of J, the Hamiltonian is a sum of single apparatus-spin terms. It can be found by inserting the product structure (4.2) into (3.3), or by taking the latter for $N = 1$. Let us denote $2g_n/\hbar = \Omega_n$. In the limit $J \to 0$ one can show that

$$\hat{\sigma}_x^{(n)}(u, i) = \cos\Omega_n u\, \hat{\sigma}_x^{(n)} - s_i \sin\Omega_n u\, \hat{\sigma}_y^{(n)},$$
$$\hat{\sigma}_y^{(n)}(u, i) = \cos\Omega_n u\, \hat{\sigma}_y^{(n)} + s_i \sin\Omega_n u\, \hat{\sigma}_x^{(n)}, \qquad (4.8)$$

while of course $\hat{\sigma}_z^{(n)}(u, i) = \hat{\sigma}_z^{(n)}$ is conserved. Each $\hat{\rho}^{(n)}$ is only a function of Ω_n and t, viz. $\hat{\rho}^{(n)}(t) = \hat{\rho}(\Omega_n, t)$, having the diagonal form $\hat{\rho}(t) = \frac{1}{2}[\rho_0(t)\hat{\sigma}_0 + i\rho_3(t)\hat{\sigma}_3]$.

The effect of the bath is relevant only at times $t \gg \tau_T = \hbar/2\pi T$, where $\hat{\rho}(\Omega, t)$ evolves according to

$$\frac{d\hat{\rho}(t)}{dt} = i\Omega\hat{\rho}\hat{\sigma}_z + \frac{2i\gamma\hat{\sigma}_z}{\hbar^2} \int_0^t du[K(u) + K(-u)](\rho_0 \sin\Omega u - \rho_3 \cos\Omega u). \quad (4.9)$$

This encodes the scalar equations

$$\dot{\rho}_0 = -\Omega\rho_3, \qquad \dot{\rho}_3 = \Omega\rho_0 + \mu\rho_0 - 2\lambda\rho_3 \qquad (4.10)$$

where

$$\lambda = \frac{2\gamma}{\hbar^2} \int_0^t du[K(u) + K(-u)] \cos\Omega u,$$

$$\mu = \frac{4\gamma}{\hbar^2} \int_0^t du[K(u) + K(-u)] \sin\Omega u. \qquad (4.11)$$

For times larger than $\tau_T = 2\pi\hbar/T$ the integrals may be taken to infinity, so that λ and μ become constants. The Ansatz $\rho_0 = A\exp xt$, $\rho_3 = C\exp xt$ then yields

$$x_\pm = -\lambda \pm i\Omega', \qquad \Omega' = \sqrt{\Omega^2 + \Omega\mu - \lambda^2}. \qquad (4.12)$$

and, taking into account the initial conditions, the solution reads

$$\rho_0 = \frac{\Omega' \cos \Omega' t + \lambda \sin \Omega' t}{\Omega'} e^{-\lambda t}, \qquad \rho_3 = \frac{\Omega + \mu}{\Omega'} \sin \Omega' t \, e^{-\lambda t}. \qquad (4.13)$$

For small γ they imply

$$\hat{\rho}(t) = \frac{1}{2} e^{-\lambda t + i\Omega \hat{\sigma}_z t}. \qquad (4.14)$$

For $t \gg \tau_T$ the coefficient λ is equal to

$$\lambda \equiv \lambda(\infty) = \frac{\gamma}{\hbar^2}[\tilde{K}(\Omega) + \tilde{K}(-\Omega)] = \frac{\gamma \Omega}{4} \coth \frac{1}{2}\beta\hbar\Omega = \frac{\gamma g_n}{2\hbar} \coth \frac{g_n}{T}, \qquad (4.15)$$

where we could neglect the cutoff Γ. The coefficient μ, only occurring as a small frequency shift in (4.12), is less simple. After a few straightforward steps one has

$$\mu(t) = \frac{\gamma}{2\pi} \int_0^\infty d\omega \, \omega \, e^{-\omega/\Gamma} \coth \frac{\beta\hbar\omega}{2} \left(\frac{1 - \cos(\omega - \Omega)t}{\omega - \Omega} - \frac{1 - \cos(\omega + \Omega)t}{\omega + \Omega} \right). \qquad (4.16)$$

Its $t \to \infty$ limit is obtained by dropping the cosines. Inserting $\coth = 1 + (\coth -1)$ and splitting the integral, one gets from the first part $(\gamma\Omega/\pi)(\log \Gamma/\Omega - \gamma_E)$ with Euler's constant $\gamma_E = 0.577215$, while one may put $1/\Gamma \to 0$ in the second part. In fact, a further splitting $\coth -1 = (\tanh -1) + (\coth - \tanh)$ may be done to separate a possible logarithm in $\beta\hbar\Omega$, while one may perform a contour integration in the last part.

By inserting (4.14) into (4.2) and tracing out the pointer variables, one finds the transverse polarization of S as

$$\frac{1}{2}\langle \hat{s}_x(t) - i\hat{s}_y(t) \rangle \equiv \text{tr}_{S,A} \hat{\mathcal{D}}(t) \frac{1}{2}(\hat{s}_x - i\hat{s}_y) = r_{\uparrow\downarrow}(t) \equiv r_{\uparrow\downarrow}(0) \, \text{Evol}(t), \qquad (4.17)$$

where the temporal evolution is coded in

$$\text{Evol}(t) \equiv \left(\prod_{n=1}^N \cos \frac{2g_n t}{\hbar} \right) \exp \left(-\sum_{n=1}^N \frac{\gamma g_n}{2\hbar} \coth \frac{g_n}{T} t \right). \qquad (4.18)$$

To see what this describes, one can first take $g_n = g$, $\gamma = 0$ and plot the factor $|\text{Evol}(t)|$ from $t = 0$ to $5\tau_{\text{recur}}$, where $\tau_{\text{recur}} = \pi\hbar/2g$ is the time after which $|r_{\uparrow\downarrow}(t)|$ has recurred to its initial value $|r_{\uparrow\downarrow}(0)|$. By increasing N, e.g., $N = 1, 2, 10, 100$, one convince himself that the decay near $t = 0$ becomes close to a Gaussian decay, over the characteristic time $\tau_{\text{trunc}} = \hbar/\sqrt{2N}g$. One may demonstrate this analytically by setting $\cos 2g_n t/\hbar \approx \exp(-2g_n^2 t^2/\hbar^2)$ for small t. This time characterizes dephasing, that is, disappearance of the off-diagonal blocks of the density matrix while still phase coherent; we called it "*truncation time*" rather than "*decoherence time*" to distinguish it from usual decoherence, which is induced by a thermal environment and coded in the second factor of $\text{Evol}(t)$.

In order that the model describes a faithful quantum measurement, it is mandatory that $|\text{Evol}| \ll 1$ at $t = \tau_{\text{recur}}$. To this aim, keeping $\gamma = 0$, one can in the first factor of Evol decompose $g_n = g + \delta g_n$, where δg_n is a small Gaussian random variable with $\langle \delta g_n \rangle = 0$ and $\langle \delta g_n^2 \rangle \equiv \delta g^2 \ll g^2$, and average over the δg_n. The Gaussian decay (4.5) will thereby be recovered, which already prevents recurrences. One may also take e.g. $N = 10$ or 100, and plot the function to show this decay and to estimate the size of Evol at later times.

Next by taking $\gamma > 0$ the effect of the bath in (4.18) can be analyzed. For values γ such that $\gamma N \gg 1$ the bath will lead to a suppression called *decoherence*, as is exemplified by the dependence on the bath temperature T. It is ongoing, not once-and-for-all.[3] Several further aspects can be easily considered now: Take all g_n equal and plot the function Evol(t); take a small spread in them and compare the results; make the small-g_n approximation $g_n \coth g_n/T \approx T$, and compare again.

At least one of the two effects (spread in the couplings or suppression by the bath) should be strong enough to prevent recurrences, that is, to make $|r_{\uparrow\downarrow}(t)| \ll |r_{\uparrow\downarrow}(0)|$ at any time $t \gg \tau_{\text{trunc}}$, including the recurrence times.[n] In the dynamical process for which each spin $\hat{\sigma}^{(n)}$ of M independently rotates and is damped by the bath, the truncation, which destroys the expectation values $\langle \hat{s}_a \rangle$ and all correlations $\langle \hat{s}_a \hat{m}^k(t) \rangle$ ($a = x$ or y, $k \geq 1$), arises from the precession of the tested spin $\hat{s}$ around the z-axis; this is caused by the conjugate effect of the many spins $\hat{\sigma}^{(n)}$ of M, while the suppression of recurrences is either due to dephasing if the g_n are non-identical, or due to damping by the bath.

Finally, one may go back to the *time-dependent* expressions (4.11) for λ and μ and deduce how the initial growth at small t can, for large N, already induce the decoherence.[3]

5. Dynamics of the registration process

The purpose of a measurement is the registration of the outcome, which can then be read off. For the description of the registration process we need to study $P_{ii}(m,t)$ defined in terms of $\hat{R}_{ii}(t)$ in (2.19). The equations for $P_{ij}(m,t)$ follow from (3.3) and are derived in Appendix B of Ref. 3.

The integrals over u produce the functions $\tilde{K}_{t>}(\omega)$ and $\tilde{K}_{t>}(\omega)$

$$\tilde{K}_{t>}(\omega) = \int_0^t du e^{-i\omega u} K(u) = \frac{1}{2\pi i} \int_{-\infty}^{+\infty} d\omega' \tilde{K}(\omega') \frac{e^{i(\omega'-\omega)t} - 1}{\omega' - \omega}, \quad (5.1)$$

and

$$\tilde{K}_{t<}(\omega) = \int_0^t du e^{i\omega u} K(-u) = \int_{-t}^0 du e^{-i\omega u} K(u) = \left[\tilde{K}_{t>}(\omega) \right]^*, \quad (5.2)$$

where ω takes, depending on the considered term, the values $\Omega_{\uparrow}^+$, $\Omega_{\uparrow}^-$, $\Omega_{\downarrow}^+$, $\Omega_{\downarrow}^-$, given by

$$\hbar\Omega_i^{\pm}(m) = H_i(m \pm \delta m) - H_i(m), \qquad (i = \uparrow, \downarrow), \quad (5.3)$$

---

[n]The condition *strong enough* poses constraints on the parameters for the apparatus to function properly. In contrast, the interaction of the billions of solar neutrinos that pass our body every second is *weak enough* to prevent the destruction of life.

in terms of the Hamiltonians (C.8) and of the level spacing $\delta m = 2/N$. They satisfy the relations $\Omega_i^{\pm}(m \mp \delta m) = -\Omega_i^{\mp}(m)$. The quantities (5.3) are interpreted as excitation energies of the magnet M arising from the flip of one of its spins in the presence of the tested spin S (with value s_i); the sign $+$ $(-)$ refers to a down-up (up-down) spin flip. Their explicit values are:

$$\hbar\Omega_i^{\pm}(m) = \mp 2gs_i + 2J(\mp m^3 - \frac{3m^2}{N} \mp \frac{4m}{N^2} - \frac{2}{N^3}), \tag{5.4}$$

with $s_{\uparrow} = 1$, $s_{\downarrow} = -1$.

The operators $\hat{\sigma}_x^{(n)}$ and $\hat{\sigma}_y^{(n)}$ which enter (3.3) are shown in Appendix B to produce a flip of the spin $\hat{\sigma}^{(n)}$, that is, a shift of the operator $\hat{m}$ into $\hat{m} \pm \delta m$. We introduce the notations

$$\Delta_{\pm} f(m) = f(m_{\pm}) - f(m), \qquad m_{\pm} = m \pm \delta m, \qquad \delta m = \frac{2}{N}. \tag{5.5}$$

The resulting dynamical equations for $P_{ij}(m,t)$ take different forms for the diagonal and for the off-diagonal components. On the one hand, the first *diagonal block* of $\hat{D}$ is parameterized by the *joint probabilities* $P_{\uparrow\uparrow}(m,t)$ to find S in $|\uparrow\rangle$ and $\hat{m}$ equal to m at the time t. In the Markov regime $t \sim J/\gamma$ these probabilities evolve according to

$$\frac{dP_{\uparrow\uparrow}(m,t)}{dt} = \frac{\gamma N}{\hbar^2} \left\{ \Delta_+ \left[(1+m) \tilde{K} \left(\Omega_{\uparrow}^{-}(m) \right) P_{\uparrow\uparrow}(m,t) \right] \right. \tag{5.6}$$
$$\left. + \Delta_- \left[(1-m) \tilde{K} \left(\Omega_{\uparrow}^{+}(m) \right) P_{\uparrow\uparrow}(m,t) \right] \right\},$$

with initial condition $P_{\uparrow\uparrow}(m,0) = r_{\uparrow\uparrow}(0) P_M(m,0)$ given by (2.27) and boundary condition $P_{\uparrow\uparrow}(-\delta m) = P_{\uparrow\uparrow}(1 + \delta m) = 0$; likewise for $P_{\downarrow\downarrow}(m)$, which involves the frequencies $\Omega_{\downarrow}^{\mp}(m)$. The factor $\tilde{K}$ is introduced in Eq. (2.7). On times $t \ll T/\gamma$, Eq. (5.6) should actually involve the more complicated form $\tilde{K}_t(\omega)$, given by

$$\tilde{K}_t(\omega) \equiv \tilde{K}_{t>}(\omega) + \tilde{K}_{t<}(\omega) = \int_{-t}^{+t} du\, e^{-i\omega u} K(u) = \int_{-\infty}^{\infty} \frac{d\omega'}{\pi} \frac{\sin(\omega' - \omega)t}{\omega' - \omega} \tilde{K}(\omega'). \tag{5.7}$$

This expression is real too and tends to $\tilde{K}(\omega)$ at times t larger than the range $\hbar/2\pi T$ of $K(t)$[8,9,o] as may be anticipated from the relation $\sin[(\omega'-\omega)t]/(\omega'-\omega) \to \pi\delta(\omega'-\omega)$ for $t \to \infty$. Fortunately, the dynamics of the relaxation process which moves the magnet from its initial paramagnetic phase to one of the stable ferromagnetic phases takes place on times $t \sim T/\gamma$, after which $\tilde{K}_t(\omega)$ has relaxed to the simpler expression $\tilde{K}(\omega)$, so this evolution is to a very good approximation given by (5.6). This makes it possible to solve the difference equations (5.6) numerically for $N = 10$, 100, 1000 or larger. (One should keep in mind that $P = 0$ for $m = 1 + 2/N$ or $-1 - 2/N$.) Figure 5 presents the result at different times for $N = 1000$.

---

oStudents with numerical skills may check this by programming the integral; those with analytical skills may replace the cutoff factor $\exp(-|\omega|/\Gamma)$ of $\tilde{K}(\omega)$ in (2.7) by the quasi-Lorentzian $4\tilde{\Gamma}^4/(\omega^4 + 4\tilde{\Gamma}^4)$ and do a contour integral in the upper half plane. See also Appendix D of Ref. 3.

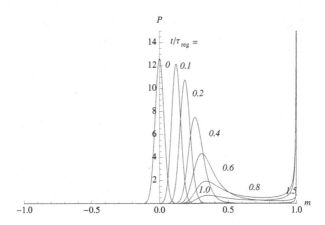

Fig. 5.   The registration process for quartic Ising interactions. The probability density $P(m,t) = P_{\uparrow\uparrow}(m,t)/r_{\uparrow\uparrow}(0)$ as function of m is represented at different times up to $t = 1.5\,\tau_{\mathrm{reg}}$. The parameters are chosen as $N = 1000$, $T = 0.2J$ and $g = 0.045J$ as in Fig 7.4. The time scale is here the registration time $\tau_{\mathrm{reg}} = 38\tau_J = 38\hbar/\gamma J$, which is large due to the existence of a bottleneck around $m_c = 0.268$. The coupling g exceeds the critical value $h_c = 0.0357J$ needed for proper registration, but since $(g - h_c)/h_c$ is small, the drift velocity has a low positive minimum at 0.270 near m_c (Fig. 7.2). Around this minimum, reached at the time $\frac{1}{2}\tau_{\mathrm{reg}}$, the peak shifts slowly and widens much. Then, the motion fastens and the peak narrows rapidly, coming close to ferromagnetism around the time τ_{reg}, after which equilibrium is exponentially reached.

One may also proceed analytically. It takes a few steps (see Ref. 3) to approximate (5.6) for large N by the Fokker-Planck equation

$$\frac{\partial P_{\uparrow\uparrow}}{\partial t} \approx \frac{\partial}{\partial m}\left[-v\left(m,t\right)P_{\uparrow\uparrow}\right] + \frac{1}{N}\frac{\partial^2}{\partial m^2}\left[w\left(m,t\right)P_{\uparrow\uparrow}\right], \qquad (5.8)$$

where

$$v\left(m,t\right) = \frac{2\gamma}{\hbar^2}\left[(1-m)\,\tilde{K}_t\left(-2\omega_{\uparrow}\right) - (1+m)\,\tilde{K}_t\left(2\omega_{\uparrow}\right)\right], \qquad (5.9)$$

$$w\left(m,t\right) = \frac{2\gamma}{\hbar^2}\left[(1-m)\,\tilde{K}_t\left(-2\omega_{\uparrow}\right) + (1+m)\,\tilde{K}_t\left(2\omega_{\uparrow}\right)\right]. \qquad (5.10)$$

One would be inclined to leave out the diffusion term of order $1/N$. Indeed, if we keep aside the shape and the width of the probability distribution, which has a narrow peak for large N, the center $\mu(t)$ of this peak moves according to the mean-field equation

$$\frac{\mathrm{d}\mu(t)}{\mathrm{d}t} = v[\mu(t)], \qquad (5.11)$$

where $v(m)$ is the local drift velocity of the flow of m,

$$v(m) = \frac{\gamma}{\hbar}(g + Jm^{q-1})\left(1 - m\coth\frac{g + Jm^{q-1}}{T}\right). \qquad (5.12)$$

This result can be derived by multiplying (5.8) by m and integrating over it, while the narrowness of $P(m)$ around its peak at μ allows to replace m by μ inside v.

If the coupling g is large enough, the resulting dynamics will correctly describe the transition of the magnetization from the initial paramagnetic value $m = 0$ to the final ferromagnetic value $m = m_F$. As a task, one can determine the minimum value of the coupling g below which the registration cannot take place. Approaching this threshold from above, one observes the slowing down of the process around the crossing of the bottleneck.

Focussing on $\mu(t) = \langle m(t) \rangle$ overlooks the broadening and subsequent narrowing of the profile at intermediate times, which is relevant for finite values of N. This can be studied by numerically solving the time evolution of $P(m, t)$, i.e., the whole registration process, at finite N, taking in the rate equations Eq. (5.6) e.g. $N = 10$, 100 and 1000. For the times of interest, $t \sim 1/\gamma$, one is allowed to employ the simplified form of the rates that arise from setting $\tilde{K}_t(\omega) \to \tilde{K}(\omega)$ and employing (5.4). The relevant rate coefficients are

$$\frac{\gamma N}{\hbar^2} \tilde{K}(\omega) = \frac{N\hbar\omega}{8J\,\tau_J} \left[\coth\left(\frac{1}{2}\beta\hbar\omega \right) - 1 \right] \exp\left(-\frac{|\omega|}{\Gamma} \right), \tag{5.13}$$

where the timescale $\tau_J = \hbar/\gamma J$ can be taken as a unit of time. The variable ω in $\tilde{K}(\omega)$ takes the values $\Omega_i^\pm$, with $i = j = \uparrow$ or $\downarrow$, which are explicitly given by (5.4) in terms of the discrete variable m. It can be verified that, for $\Gamma \gg J/\hbar$, the omission of the Debye cut-off in (5.13) does not significantly affect the dynamics.

6. The quantum measurement problem and the elements of its solution

In the measurement postulates of textbooks it is taken for granted that individual measurements yield individual outcomes. However, on a theoretical level this is a non-trivial feature to be explained, know as the "measurement problem".

6.1. *Why the task is not achieved: the quantum ambiguity*

We have shown in Sections 4 and 5 that, for suitable values of the parameters entering the Hamiltonian, S+A ends up for the Curie–Weiss model in an equilibrium state represented by the density operator

$$\hat{\mathcal{D}}(t_f) = \sum_i p_i \hat{r}_i \otimes \hat{\mathcal{R}}_i. \tag{6.1}$$

The index i takes two values associated with up or down spins; the weights p_i are equal to the diagonal elements $r_{\uparrow\uparrow}(0)$ or $r_{\downarrow\downarrow}(0)$ of the initial state of S; the states $\hat{r}_i$ of S are the projection operators $|\uparrow\rangle\langle\uparrow|$ or $|\downarrow\rangle\langle\downarrow|$ on the eigenspaces associated with the values +1 or -1 of s_z; the states $\hat{\mathcal{R}}_\Uparrow$ or $\hat{\mathcal{R}}_\Downarrow$ are the ferromagnetic equilibrium states of A. This state (6.1) exhibits the required one-to-one correspondence between the eigenvalue of $\hat{s}$ and the indication of the pointer.

It is essential to remember, as we stressed in the introduction, that the density operator (6.1) is a formal object, which encompasses the statistical properties of the outcomes of a *large ensemble* $\mathcal{E}$ of runs issued from the initial state $\hat{\mathcal{D}}(0) = \hat{r}(0) \otimes \hat{\mathcal{R}}(0)$, but which has no direct interpretation. In order to understand the various features of a measurement, we need not only to describe globally this ensemble, but to account for properties of *individual runs*. For instance, we need to explain why each individual run provides a well-defined answer, up or down, and why the coefficient $p_\uparrow$ which enters the expression (6.1) can be interpreted as Born's probability, that is, as the relative number of individual runs having provided the result up within the large ensemble $\mathcal{E}$ described by (6.1). This question is known as the "quantum measurement problem"[P]: Can we make theoretical statements about individual quantum measurements, in spite of the irreducibly probabilistic nature of quantum mechanics which deals only with ensembles of runs?

In fact, what we have derived dynamically within the statistical formulation of quantum mechanics is only the global expression (6.1), whereas we would like to know whether its two parts have *separately* a physical meaning. At first sight, this question looks innocuous. It is tempting to assume that the ensemble $\mathcal{E}$ described by (6.1) is the union of two subensembles, with relative sizes $p_\uparrow$ and $p_\downarrow$, described by the states $\hat{\mathcal{D}}_\uparrow = \hat{r}_\uparrow \otimes \hat{\mathcal{R}}_\Uparrow$ and $\hat{\mathcal{D}}_\downarrow = \hat{r}_\downarrow \otimes \hat{\mathcal{R}}_\Downarrow$, respectively. All the runs in the first subensemble would then be characterised by a value up of the pointer, and correlatively by a spin S in the collapsed state $\uparrow\rangle$. However this intuitive statement is fallacious due to a specific *quantum ambiguity*, as we now show.

As an illustration, consider first a large set of coins, thrown at random. It is correct to state that this set can be split into two subsets, with coins on the heads and tails sides, respectively. Going from random bits to random q-bits, consider now a large set of non-polarised spins. By analogy, we might believe in the existence of two subsets of spins, pointing in the $s_z = +1$ and $s_z = -1$ directions, respectively. However, we are not allowed to make such an intuitive statement. Indeed, we might as well have believed in the existence of two subsets, pointing in the $s_x = +1$ and $s_x = -1$ directions, respectively. Then there would exist individual spins pointing simultaneously in two orthogonal directions, which is absurd. Whereas we can ascertain, for the ordinary probability distribution of an ensemble of coins, that observing an individual coin will provide a well-defined result, head or tails, our uncertainty remains complete as regard individual spins characterised by the quantum distribution of their ensemble. Due to such an ambiguity, which arises from the matrix nature of quantum states, we cannot give a meaning, in terms of subensembles, to the separate terms of a decomposition of a mixed density operator. This forbids us to make any statement about individual systems in the absence of further information.

The same ambiguity prevails for the measurement model that we are considering. Although the decomposition (6.1) of $\hat{\mathcal{D}}(t_f)$ as a sum of two terms is suggestive, and although a naive interpretation of each term seems to provide the expected

---

[P]In the literature there exist various definitions of the measurement problem. We follow Laloë in Ref. 14.

result, an *infinity of other decompositions exist*, which are mathematically allowed, and of which *none has a priori a physical meaning*. Our sole determination of this expression is not sufficient to provide an interpretation of each term of the decomposition (6.1), and hence to justify, as we wish, the so-called postulates of ideal measurements. At this stage, the measurement problem remains open. We have to rely on further arguments for its solution, while our only hope can lie in properties of the apparatus.

6.2. *The strategy*

Starting from some time t_{split} at which the final state (6.1) has already been reached, we consider *all possible decompositions* into two terms,

$$\hat{\mathcal{D}} = k\hat{\mathcal{D}}_{\text{sub}} + (1-k)\hat{\mathcal{D}}_{C\text{sub}} \qquad (6.2)$$

of the density operator found above, where $0 < k < 1$ and where $\hat{\mathcal{D}}_{\text{sub}}$ and where $\hat{\mathcal{D}}_{C\text{sub}}$ have the mathematical properties of density operators (hermiticity, normalisation and non-negativity). The above quantum ambiguity does not entitle us to ascribe separately a physical meaning to each of the two terms of (6.2) – and in particular not to each associated with the two terms of (6.1). We are not allowed to regard $\hat{\mathcal{D}}_{\text{sub}}$ as a density operator of some real subensemble of $\mathcal{E}$. However, if, conversely, the full ensemble $\mathcal{E}$ of real runs of the measurement described by $\hat{\mathcal{D}}$ is split into a real subensemble $\mathcal{E}_{\text{sub}}$ of runs and its complement $\mathcal{E}_{C\text{sub}}$, each of these must be described by *genuine density operators* $\hat{\mathcal{D}}_{\text{sub}}$ and $\hat{\mathcal{D}}_{C\text{sub}}$ that satisfy (6.2) at the time t_{split} and that are later on governed by the Hamiltonian $\hat{H}$.[q]

Although we cannot identify whether an operator $\hat{\mathcal{D}}_{\text{sub}}$ issued from a decomposition (6.2) describes the state of S+A for some physical subensemble $\mathcal{E}_{\text{sub}}$, or whether it is only an element of a mathematical identity, we will take it as an initial condition at the time t_{split} and solve the equations of motion for $\hat{\mathcal{D}}_{\text{sub}}(t)$ at subsequent times. This step can again be treated, at least formally, as a process of quantum statistical mechanics; its ideas and outcome are presented in Section 6.3.

It turns out that, for a suitable choice of the Hamiltonian of the apparatus, *any operator* $\hat{\mathcal{D}}_{\text{sub}}(t)$ issued from a decomposition (6.2) of (6.1) tends, over a short time, to

$$\hat{\mathcal{D}}_{\text{sub}}(t) \mapsto \hat{\mathcal{D}}_{\text{sub}}(t_{\text{f}}) = \sum_i q_i \hat{r}_i \otimes \hat{\mathcal{R}}_i, \qquad (6.3)$$

which has the same form as (6.1) except for the values of the weights $q_\uparrow \geq 0$ and $q_\downarrow = 1 - q_\uparrow \geq 0$. The relaxation time is sufficiently short so that this form is attained at the time t_{f} determined in Section 5. The operators (6.3) are the only *dynamically stable* ones.

---

[q]Here it is essential to realize that subensembles are also ensembles, thus satisfying the same evolution though with different initial conditions, while the linearity of the Liouville-von Neumann equation allows the split up (6.2) of $\hat{\mathcal{D}}$ in separate terms.

We have stressed that the operators $\hat{\mathcal{D}}_{\text{sub}}(t)$ have not necessarily a physical meaning, but that *their class encompasses any physical density operator* describing some subset $\mathcal{E}_{\text{sub}}$ of runs. Since all candidates for such physical density operators reach the form (6.3) at the time t_{f}, we are ascertained that *the state of* S+A associated with *any real subensemble* $\mathcal{E}_{\text{sub}}$ of runs *relaxes* as shown in Section 6.3 and *ends up in the form* (6.3).

The collection of all subensembles $\mathcal{E}_{\text{sub}}$ of $\mathcal{E}$ possesses the following *hierarchic structure*. When two disjoint subensembles $\mathcal{E}_{\text{sub}}^{(1)}$ and $\mathcal{E}_{\text{sub}}^{(2)}$ merge into a new subensemble $\mathcal{E}_{\text{sub}}$, the corresponding numbers of runs $\mathcal{N}$ and weights q_i ($i =\uparrow$ or $\downarrow$) satisfy the standard addition rule

$$\mathcal{N} q_i = \mathcal{N}^{(1)} q_i^{(1)} + \mathcal{N}^{(2)} q_i^{(2)}, \qquad \mathcal{N} = \mathcal{N}^{(1)} + \mathcal{N}^{(2)}. \tag{6.4}$$

Thus, one can prove within the framework of quantum statistical dynamics, not only that the state of S+A describing the full set $\mathcal{E}$ of runs is expressed by (6.1), but also that the states describing all of its physical subsets $\mathcal{E}_{\text{sub}}$ have the form (6.3), where the weights q_i are related to one another by the hierarchic structure (6.4). In the minimalist formulation of quantum mechanics which deals only with statistical ensembles, this is the most detailed result that can be obtained about the ideal Curie–Weiss measurement process. An extrapolation is necessary to draw conclusions about individual systems, as will be discussed in Section 6.4.

6.3. *Subensemble relaxation*

We consider here the evolution for $t \geq t_{\text{split}}$ of an operator $\hat{\mathcal{D}}_{\text{sub}}(t)$, defined at an initial time t_{split} through some mathematical decomposition (6.2) of the density operator (6.1), already reached at the time t_{split} for the full ensemble of runs. We have seen in Section 5 that, during the last stage of the registration, S and A can be decoupled. Indeed, each of the two terms of the final state (6.1) of S+A is factorised, so that (6.1) describes a thermodynamic equilibrium in which S and A are correlated only through the equality between the signs of s_z and of the magnetisation of A. After decoupling of S and A, the evolution of $\hat{\mathcal{D}}_{\text{sub}}(t)$ is governed by the Hamiltonian $\hat{H}_{\text{A}}$ *of the apparatus alone*, and the above correlation will be preserved within $\hat{\mathcal{D}}_{\text{sub}}(t)$ at all times $t \geq t_{\text{split}}$.

We first show that the initial condition $\hat{\mathcal{D}}_{\text{sub}}(t_{\text{split}})$, although undetermined, must satisfy constraints imposed by the form of the equations (6.1) and (6.2) from which it is issued. As the apparatus is macroscopic, we can represent $\hat{\mathcal{R}}_{\Uparrow}$ (or $\hat{\mathcal{R}}_{\Downarrow}$) as a *microcanonical equilibrium state* characterised by the order parameter $+m_{\text{F}}$ (or $-m_{\text{F}}$). In the Hilbert space of A, we denote as $|i, \eta\rangle$ (with $i =\Uparrow$ or $\Downarrow$) a basis for the microstates that underlie each microcanonical state $\hat{\mathcal{R}}_i$, where the energy and magnetization are taken as constants. We then have

$$\hat{\mathcal{R}}_i = \frac{1}{G} \sum_{\eta} |i, \eta\rangle \langle i, \eta|, \tag{6.5}$$

where G is the large number of values taken by the index η.[r] Denoting by $|i, \eta\rangle$ (with $i = \uparrow$ or $i = \downarrow$) the two states $s_z = +1$ or $s_z = -1$ of S, we see that the density matrix (6.1) is diagonal in the Hilbert subspace of S+A spanned by the correlated kets $|i\rangle|i, \eta\rangle$ and that it has no element in the complementary subspace. The *non negativity of the two terms* of (6.2) then implies that the Ref. 3 latter property must also be satisfied by the operator $\hat{\mathcal{D}}_{\mathrm{sub}}(t_{\mathrm{split}})$, which has therefore the form

$$\hat{\mathcal{D}}_{\mathrm{sub}}(t_{\mathrm{split}}) = \sum_{i,i',\eta,\eta'} |i\rangle|i, \eta\rangle K(i, \eta; i', \eta')\langle i'|\langle i', \eta'| \tag{6.6}$$

The matrix K is Hermitean, non negative and has unit trace.

Let us now turn to the Hamiltonian $\hat{H}_{\mathrm{A}}$ that governs the subsequent evolution of $\hat{\mathcal{D}}_{\mathrm{sub}}(t)$. We assume here that it contains small terms which produce *transitions among the microstates* $|\Uparrow, \eta\rangle$, which have nearly the same energy and nearly the same magnetisation – likewise among the microstates $|\Downarrow, \eta\rangle$. Although these terms are small, they are very efficient because they practically conserve the energy. Their occurrence does not affect the derivations of Sections 4 and 5, and conversely the present "*quantum collisional process*" is governed solely by the rapid transitions between the kets $|i\rangle|i, \eta\rangle$ having the same i but different η.[s]

Such a dynamics keeps the form of (6.6) unchanged but modifies the matrix K. For a large apparatus, it produces an irreversible process which generalises the *microcanonical relaxation* to an intricate situation involving two different microcanonical states. It has been worked out in Ref. 3 (Section 12); the result is the following. Over a short delay, all the matrix elements with $i \neq i'$ (that is, the combinations $\uparrow\downarrow$ and $\downarrow\uparrow$) of $K(i, \eta; i', \eta', t)$ tend to 0. Over the same delay, its elements $\uparrow\uparrow$ with $\eta \neq \eta'$ also tend to 0, while the diagonal elements $\uparrow\uparrow$ with $\eta = \eta'$ all tend to one another, their sum remaining constant – likewise for its elements $\downarrow\downarrow$. Hence, using (6.5), we find that $\hat{\mathcal{D}}_{\mathrm{sub}}(t)$ rapidly tends to

$$\hat{\mathcal{D}}_{\mathrm{sub}}(t) \mapsto \sum_i q_i \hat{r}_i \otimes \hat{\mathcal{R}}_i, \tag{6.7}$$

where $\hat{r}_i = |i\rangle\langle i|$ and $q_i = \sum_\eta K(i, \eta; i, \eta)$. This relaxation holds for any mathematically allowed decomposition (6.2) of (6.1), and in particular for any physical decomposition associated with the splitting of the ensemble of runs of the measurement into subensembles.

6.4. *Emergence of classicality*

It remains to solve the quantum measurement problem, that is, to understand how we can make statements about individual runs of the process, although quantum

---

[r]In some models one may now disregard the bath, so that η denotes states of the magnet M (see the random matrix model of Section 11.2.3 of Ref. 3); in general models it denotes states of M+B
[s]Again, a good apparatus must satisfy the proper requirements for this aspect of the dynamics. Ref. 3 discusses that it is realistic to assume that apparatuses satisfy them in practice.

theory, in its minimalist statistical formulation, deals only with ensembles. We have already succeeded to determine, for ideal Curie–Weiss measurements treated within this theoretical framework, the expressions (6.3) and (6.4) which embody the strongest possible results about the final states of S+A for *arbitrary subensembles* of runs.

In order to extrapolate this result to the *individual runs* which constitute these subensembles, we note that the common form (6.3) of the states $\hat{\mathcal{D}}_{\text{sub}}$ and the hierarchic structure (6.4) of the weights are exactly the same as in ordinary probability theory. On the one hand, the difficulties arising from the quantum ambiguity have been overcome owing to a dynamical property, the subensemble relaxation, which produced the stable final states (6.3). On the other hand, the relation (6.4) satisfied by the weights q_i is one of the axioms that define classical probabilities as *frequencies of occurrence* of individual events.[15] It is therefore natural to interpret each coefficient q_i associated with a given subset of runs as the proportion of runs of this subset that have yielded the result i. In particular, for the full ensemble $\mathcal{E}$, we recover *Born's rule*: We had found above $p_\uparrow$ only as a weight that occurred in the decomposition (6.1) of $\hat{\mathcal{D}}(t_\text{f})$; we can now interpret it as a classical probability, defined as the relative frequency of occurrence of $+m_\text{F}$ in all the individual runs of $\mathcal{E}$.

We are then led to *interpret* $\hat{r}_\uparrow \otimes \hat{\mathcal{R}}_\Uparrow$ as the density operator associated with the subset for which $q_\uparrow = 1$, $q_\downarrow = 0$ – it is here where *interpretation* enters our approach. With now having a homogeneous (pure) subensemble at hand, we can associate this density operator *with any individual run* of this subset. Thus, contrary to the first stages of the measurement process, the truncation and the registration, the so-called "*collapse*" is not a physical process. It appears merely as a subsequent *updating of the density operator* which results from the selection of a subensemble, made possible by the effectively vanishing of the off-diagonal terms of the density operator of the full system.

Here again, the apparatus plays a major rôle. It is only the observation in a given run of its indication $+m_\text{F}$ which allows us to predict, owing to the correlations between S and A, that this run constitutes a preparation of S in the state $|\uparrow\rangle$. The emergence in a measurement process of classical concepts, uniqueness of the outcome for an individual event, classical probabilities, classical correlations between S and A, relies on the *macroscopic size of the apparatus*.

7. An attempt to simultaneously measure non-commuting variables

Textbooks in quantum mechanics (artificially) describe measurements as an instantaneous process, which rules out the possibility of even *trying* to simultaneously measure two non-commuting observables. Nevertheless, in the Curie-Weiss model, the measurement is described as a physical interaction between the measured system and the apparatus. An interesting scenario appears then if one lets the measured system interact with two such apparatuses *simultaneously*, each of which is attempting to measure a different spin component.[3]

At this point one may argue that this process is meant to fail. Indeed, even if both apparatuses would yield results for their respective measurements, it is clear that a quantum state can not have two well definite values for two non-commuting observables (the two different spin components). However, the point of this section is precisely to find out in which sense this process differs from an ideal measurement, and to give a good interpretation of the obtained results.

In order to set the problem in technical terms, let us consider a general spin state

$$\hat{\rho}(0) = \frac{1}{2} \left\{ \mathbb{I} + \langle \hat{\mathbf{s}}(0) \rangle \cdot \hat{\mathbf{s}} \right\}. \tag{7.1}$$

It will simultaneously interact with two apparatuses A and A$'$, which attempt to measure $\hat{s}_z$ and $\hat{s}_x$, respectively. By reading the pointers of A and A$'$, we aim to achieve some information about *both* $\langle \hat{s}_z(0) \rangle$ and $\langle \hat{s}_x(0) \rangle$ in every run of the experiment. As in any measurement in quantum mechanics, many runs of the experiment will be needed to know $\langle \hat{s}_z(0) \rangle$ and $\langle \hat{s}_x(0) \rangle$ with good precision.

Finally, notice that if we were to measure $\hat{s}_z$ and $\hat{s}_x$ sequentially, then the second measurement would be completely uninformative. For instance, starting form the general state (7.1), after measuring $\hat{s}_z$ the state is $\frac{1}{2} \{ \mathbb{I} + \langle \hat{s}_z(0) \rangle \hat{s}_z \}$, which has no memory about $\langle \hat{s}_x(0) \rangle$.

7.1. *The Hamiltonian*

We extend the Curie–Weiss model by adding a new apparatus A$'$ made up of a magnet M$'$ and a bath B$'$, with parameters J', g', N'.... The total Hamiltonian is then given by $\hat{H}_T = \hat{H}_{SA} + \hat{H}_{SA'} + \hat{H}_A + \hat{H}_{A'}$, with $\hat{H}_{SA} = -Ng\hat{m}\hat{s}_z$ and $\hat{H}_{SA'} = -N'g'\hat{m}'\hat{s}_x$; so each component of the spin is interacting with a different apparatus. The internal Hamiltonians H_A, $H_{A'}$ can be found from (2.5). Although the apparatuses are not necessarily identical, we assume them to be similar, i.e., N, J, g, γ are of the same order of N', J', g', γ' respectively.

It will turn out to be very useful to define a direction **u** where the interacting Hamiltonian is diagonal, that is:

$$H_{SAA'} = \hat{H}_{SA} + \hat{H}_{SA'} = \frac{\hbar}{2} w(\hat{m}, \hat{m}') \hat{s}_{\mathbf{u}}(\hat{m}, \hat{m}') \tag{7.2}$$

with $\hat{s}_{\mathbf{u}}(m, m') = \mathbf{u}(m, m') \cdot \hat{\mathbf{s}}$, and

$$\mathbf{u}(m, m') = \frac{2Ngm}{\hbar w} \hat{\mathbf{z}} + \frac{2N'g'm'}{\hbar w} \hat{\mathbf{x}} \tag{7.3}$$

$$w(m, m') = \frac{2}{\hbar} \sqrt{(Ngm)^2 + (N'g'm')^2}. \tag{7.4}$$

Therefore, effectively the spin acts on both apparatuses as a global field w in the direction $\mathbf{u}$. Finally, let us define a direction $\mathbf{v}$ perpendicular to $\mathbf{u}$ and y,

$$\hat{s}_{\mathbf{v}} = \mathbf{v}(m, m') \cdot \hat{\mathbf{s}} = u_z \hat{s}_x - u_x \hat{s}_z. \tag{7.5}$$

7.2. The state

The joint state of S + M + M' will be denoted by $\hat{D}(\hat{m}, \hat{m}', t)$, and it can be characterized as:

$$\hat{D}(\hat{m}, \hat{m}', t) = \frac{1}{2G(\hat{m})G(\hat{m}')} \left[P(\hat{m}, \hat{m}', t) + \mathbf{C}(\hat{m}, \hat{m}', t) \cdot \hat{\mathbf{s}} \right]. \tag{7.6}$$

In order to interpret this description, consider

$$\mathrm{tr}\left\{ \delta_{\hat{m},m} \delta_{\hat{m}',m'} \hat{D} \right\} = P(m, m', t), \qquad \mathrm{tr}\left\{ \delta_{\hat{m},m} \delta_{\hat{m}',m'} \hat{s}_i \hat{D} \right\} = C_i(m, m', t). \tag{7.7}$$

where $\delta_{\hat{m},m}$ is a projector on the subspace with magnetization m. Therefore, $P(m, m', t)$ is the joint probability distribution of the magnetization of the apparatuses and $C_i(m, m', t)$, with $i = x, y, z$ or $i = x, u, v$; brings information about the correlations between $\hat{s}_i$ and the apparatuses.

Initially, the system and the apparatuses are uncorrelated, thus being in a product state $\hat{r}(0) \otimes \hat{R}_\mathrm{M}(0) \otimes \hat{R}_{\mathrm{M}'}(0)$ with $\hat{R}_\mathrm{M}(0)$ given in (2.25). $P_\mathrm{M}(m)$, given in (2.27), is the probability distribution associated to $\hat{R}_\mathrm{M}(0)$, and the initial state of the correlators is $C_i(0) = \langle \hat{s}_i(0) \rangle P_\mathrm{M}(m) P_\mathrm{M}(m')$.

7.3. Disappearance of the off-diagonal terms

In the Curie-Weiss model, truncation, or the disappearance of the off-diagonal terms, was shown to be a dephasing effect due to the interacting Hamiltonian. Let us thus focus only on the action of $H_{\mathrm{SAA}'}$, as defined in (7.2), and disregard the other terms of the total Hamiltonian. Obviously then the $\mathbf{u}$-component of the spin is preserved in time, analogously to $\hat{s}_z$ for the one apparatus case. On the other hand, by inserting the Ansatz (7.6) into the Liouville-von Neumann equation of motion we find

$$i \frac{\partial \mathbf{C} \cdot \hat{\mathbf{s}}}{\partial t} = -\frac{1}{2} \left[w \hat{s}_{\mathbf{u}}, \mathbf{C} \cdot \hat{\mathbf{s}} \right] \tag{7.8}$$

where we also projected onto subspaces with given magnetizations. Using the commuting properties of the Pauli matrices these equations can be readily solved, yielding:

$$P(t) = P(0)$$
$$C_u(t) = C_u(0)$$
$$C_y(t) = C_y(0) \cos(wt)$$
$$C_v(t) = C_v(0) \sin(wt) \tag{7.9}$$

which shows how the correlators C_y and C_v rapidly rotate because of the external field w. This situation should be compared with the precessing of the spins in the magnet for the case of one apparatus, see (4.3), which lead to the decay of the off-diagonal terms (4.6). The same mechanism is responsible now for the fast decay of $\langle \hat{s}_y \rangle$ and $\langle \hat{s}_v \rangle$. Furthermore, the bath-induced decoherence at later times will only increase this effect, yielding the actual suppression of the correlators C_y and C_v.[16,17]

Therefore, truncation will now occur in the **u** direction. Notice however that **u** is a function of m and m', which in turn will evolve in time as the registration takes place. Therefore, the preferred basis is not fixed, but it keeps changing during the measurement; and the collapse basis will depend on each particular run of the process (i.e, on the final values of m and m'). This is a signature of the non-ideality of the considered measurement.

7.4. *Registration*

During the registration the magnets are expected to reach ferromagnetic states due to the combined effect of the spin system and the baths. This takes place in a longer time scale than the truncation, and it can be described by solving the equations of motion for $P(m, m', t)$ and $C_i(m, m', t)$ including the terms arising from the baths. The corresponding equations become notably complex, particularly because $P(m, m', t)$ becomes coupled to all C_i; and we refer the reader to Refs. 16,17 for a detailed analysis of the dynamics. Here instead we will focus our attention on the final state. Since it is an equilibrium state, much can be said about its characteristics by studying the free energy function.

Notice from (7.2) that the action of the spin on the magnets can be seen as an external field w, thus the joint free energy function for the both magnets can be written as

$$\mathcal{F}(m, m') = \frac{\hbar}{2} w + F(m) + F(m') \tag{7.10}$$

where $F(m)$ is the free energy of one apparatus in absence of interactions, as given in (2.31) with $h = 0$. In order to find the local stable points where the states of the magnets are expected to evolve to, the student can find the local minima of (7.10). Initially setting $w = 0$ and $T < 0.496 J$, one can find a local minima around $(m, m') = (0, 0)$; four local minima at $(0, \pm m_F)$ and $(\pm m_F, 0)$; and four global minimima at $(\pm m_F, \pm m_F)$. The paramagnetic state is the initial state for the magnets, which is metastable. On the other hand, if the final state is centered at $(0, \pm m_F)$ or $(\pm m_F, 0)$, then only one of the magnets has achieved registration; whereas if if is in one the global minima at $(\pm m_F, \pm m_F)$, then *both* of them have. Finally, one can find the minimum coupling necessary to allow for a rapid transition between the paramagnetic and the ferromagnetic states, i.e., the minimum g, g' so that the free energy barriers disappear.

7.5. The final state and its interpretation

We are interested in the final probability distribution of $P(m, m', t_f)$, from which we can extract information about the measured observables $\hat{s}_x$ and $\hat{s}_z$. Our study of the free energy function shows that the most stable points are found in $(m, m') = (\pm m_F, \pm m_F)$, and for a sufficiently large coupling we expect the final magnetization of the magnets to evolve towards such points. These four points are associated with the four possible outcomes of the measurement: $(s_z = \pm\frac{\hbar}{2}, s_x = \pm\frac{\hbar}{2})$. The final state thus has the form

$$P(m, m', t_f) = \sum_{\epsilon=\pm 1} \sum_{\epsilon'=\pm 1} \mathcal{P}_{\epsilon\epsilon'} \delta_{m,\epsilon m_F} \delta_{m',\epsilon' m'_F} \qquad (7.11)$$

where $\delta_{m,x}$ represents a narrow (normalized) peak at $m = x$. $\mathcal{P}_{\epsilon\epsilon'}$, which are the weights of each peak, represent the probabilities of getting one of the 4 possible outcomes.

Let us discuss the dependence of the weights $\mathcal{P}_{\epsilon\epsilon'}$ on the initial conditions of S. It has been argued that the correlators C_v and C_y disappear due to a dephasing effect together with a decoherence effect at later times. Therefore only $C_u(m, m', 0)$ contributes, which is a linear combination of $\langle \hat{s}_x(0) \rangle$ and $\langle \hat{s}_z(0) \rangle$. Since the equations of motion are linear, the final result for $P(m, m', t)$ (and C_u) will still be a linear combination of $\langle \hat{s}_x(0) \rangle$ and $\langle \hat{s}_z(0) \rangle$. On the other hand, if $\langle \hat{s}_x(0) \rangle = \langle \hat{s}_z(0) \rangle = 0$, then we have $\mathcal{P}_{\epsilon\epsilon'} = 1/4$ due to the symmetry $m \leftrightarrow -m$ and $m' \leftrightarrow -m'$. Putting everything together, we can write the general form:

$$\mathcal{P}_{\epsilon\epsilon'} = \frac{1}{4}[1 + \epsilon\lambda\langle \hat{s}_z(0) \rangle + \epsilon'\lambda'\langle \hat{s}_x(0) \rangle] \qquad (7.12)$$

where $\epsilon, \epsilon' = \pm 1$; and λ, λ' are the proportionality factors. We term such factors the *efficiency* factors.

Consider now a particular case where the tested spin is initiall pointing at $+z$, i.e., $\langle s_x(0) \rangle = 0$ and $\langle s_z(0) \rangle = 1$. Then, the probability that A, the apparatus measuring $\hat{s}_z$, ends up pointing at $+m$ is $\mathcal{P}_{++} + \mathcal{P}_{+-} = \frac{1}{2}(1 + \lambda)$; whereas there is a probability $\mathcal{P}_{-+} + \mathcal{P}_{--} = \frac{1}{2}(1 - \lambda)$ to end up at $-m$, thus yielding a *wrong* indication. Indeed, according to Born rule if $\langle s_z(0) \rangle = 1$ then a device measuring $\hat{s}_z$ will always yield the same outcome, whereas in the current case there is a probability $\frac{1}{2}(1 - \lambda)$ of failure. Finally, notice that it must hold $\lambda \in [0, 1]$ and similarly it can be shown $\lambda' \in [0, 1]$.

Since $\mathcal{P}_{\epsilon\epsilon'}$ must be non-negative for any initial state of S, and because (7.12) has the form $\frac{1}{4}(1 + \mathbf{a} \cdot \hat{\mathbf{s}})$ with $|\mathbf{a}| \leq 1$, we reach the condition:

$$\lambda^2 + \lambda'^2 \leq 1. \qquad (7.13)$$

Therefore, we can already say that both measurements can not be ideal. In the case of two identical apparatuses, such a condition yields: $\lambda \leq \frac{1}{\sqrt{2}}$. For example, if $\lambda = \lambda' = 1/\sqrt{2}$, starting with a spin pointing in the z direction, $\langle \hat{s}_z(0) \rangle = 1$, there is a probability of $(1 - 1/\sqrt{2})/2 \approx 0.15$ to read the result $-\hbar/2$ in the apparatus measuring $\hat{s}_z$. Nevertheless, how much information can we extract from the results of the apparatuses?

Notice that relation (7.12) can be inverted:

$$\langle \hat{s}_z(0) \rangle = \frac{1}{\lambda}(\mathcal{P}_{++} + \mathcal{P}_{+-} - \mathcal{P}_{-+} - \mathcal{P}_{--})$$

$$\langle \hat{s}_x(0) \rangle = \frac{1}{\lambda'}(\mathcal{P}_{++} - \mathcal{P}_{+-} + \mathcal{P}_{-+} - \mathcal{P}_{--})$$

as long as λ and λ' do not vanish.[t] Therefore, by counting the different results $\{++, +-, -+, --\}$ of the experiment we can obtain the weights P_{ij} $(i, j = \pm)$ and thus $\langle \hat{s}_x(0) \rangle$ *and* $\langle \hat{s}_z(0) \rangle$ with arbitrary precision. The fact that we need many runs of the experiment to determine the measured observables $\hat{s}_z$ and $\hat{s}_z$ is a feature of any measurement in quantum mechanics. In conclusion, although the process is not ideal (the apparatuses can yield false indications), it is completely informative.

8. General conclusions

The very interpretation, conceptually essential, of quantum mechanics requires an understanding of quantum measurements, experiments which give us access to the microscopic reality through macroscopic observations. In a theoretical approach, measurements should be treated as dynamical processes for which the tested system and the apparatus are coupled. Since the apparatus is macroscopic, and since the elucidation of the problems related to measurements requires an analysis of time scales, we must resort to non equilibrium quantum statistical mechanics.

This programme has been achieved above for two models, the Curie–Weiss model for ideal measurements (Sections 2–6), and a modified model which exhibits the possibility of drawing information about two non commuting observables of S through a large set of runs of non ideal measurements (Section 7). It turns out that the questions to be solved pertain to the physics of the apparatus rather than to the physics of the system itself, whether we consider the diagonal or the off-diagonal contributions to a density matrix. It is the specific properties of the apparatus and of its coupling with the system which ensure that an experiment can be regarded as a measurement providing faithful information about this system.

Our theoretical analysis relies solely on standard quantum statistical mechanics. Through such an approach we can acknowledge the emergence of qualitatively new phenomena when passing from a microscopic to a macroscopic scale. For instance, in classical statistical mechanics, the irreversibility observed at our scale emerges from the microscopic equations of motion that are reversible. This looks paradoxical, but can be explained by the possibility of neglecting correlations between a large number of microscopic constituents, which have no physical relevance, and

---

[t]In Ref. 17 it is shown how λ, λ' do not vanish and can take values close to $1/\pi$.

Table 1. The steps of ideal quantum measurements.

Descriptive level	full ensemble	full ensemble	subensembles	individual systems
Process	truncation	registration	relaxation	reduction
Mechanism(s)	dephasing decoherence	phase transition energy dump	decoherence	selection
Approach	Q stat mech	Q stat mech	Q stat mech	interpretation

For the full ensemble the initial state is the one to be measured, for the subensembles the initial conditions are unknown but constrained by positivity. The so-called "reduction of the state" or "collapse of the wave function" is the result of selection of measurement outcomes.

by the inaccessibly large value of recurrence times. The irreversibility of quantum measurement processes has the same origin.

Moreover, the same type of approximations, legitimate owing to the macroscopic size of the apparatus and to the properties of its Hamiltonian, allows us to understand another kind of emergence. The quantum formalism, which governs objects at the microscopic scale, presents abstract, counterintuitive features foreign to our daily experience. In its minimalist formulation, quantum theory deals with statistical ensembles, wave functions or density operators are not reducible to ordinary probability distributions; quantities like "quantum correlations" cannot be regarded as ordinary probabilistic correlations since they violate Bell's inequalities.[u] The quantum theoretical analysis of measurement processes allows us to grasp the emergence of a classical description of their outcome and of classical concepts, in apparent contradiction with the underlying quantum concepts (Section 6). In particular the possibility of assigning ordinary probabilities to individual events through observation of the apparatus provides a solution to the so called measurement problem.

We thus conclude that our analysis of ideal quantum measurements involves three steps: study of the dynamics of the full ensemble of runs (including truncation and registration), study of the final evolution of arbitrary subensembles, and inference towards individual systems. See Table 1.

We advocate the statistical formulation for the teaching of quantum theory, since it works for our discussion of ideal measurements where an interpretation of the "quantum probabilities" emerges. The concept of state is simple to grasp by being in spirit close to classical statistical physics. States described by wave functions should be regarded only as special cases, since pure and mixed states both describe ensembles. Non intuitive features of quantum mechanics remain concentrated in the non commutation of the observables representing the physical quantities.

---

[u]See Ref. 18 for the opinion that Bell inequality violation implies only that quantum mechanics works, without any statement about presence or absence of local realism.

Acknowledgments

The authors thank the students at the Advanced School on Quantum Foundations and Open Quantum Systems in Joaõ Pessoa, Brazil, 2012, for their enthusiastic participation and critical remarks, which helped to deepen this presentation.

Appendices

A. The phonon bath

The interaction between the magnet and the bath, which drives the apparatus to equilibrium, is taken as a standard spin-boson Hamiltonian[8–10]

$$\hat{H}_{\text{MB}} = \sqrt{\gamma} \sum_{n=1}^{N} \left(\hat{\sigma}_x^{(n)} \hat{B}_x^{(n)} + \hat{\sigma}_y^{(n)} \hat{B}_y^{(n)} + \hat{\sigma}_z^{(n)} \hat{B}_z^{(n)} \right) \equiv \sqrt{\gamma} \sum_{n=1}^{N} \sum_{a=x,y,z} \hat{\sigma}_a^{(n)} \hat{B}_a^{(n)}, \tag{A.1}$$

which couples each component $a = x$, y, z of each spin $\hat{\sigma}^{(n)}$ with some hermitean linear combination $\hat{B}_a^{(n)}$ of phonon operators. The dimensionless constant $\gamma \ll 1$ characterizes the strength of the thermal coupling between M and B, which is weak.

For simplicity, we require that the bath acts independently for each spin degree of freedom n, a. (The so-called independent baths approximation.) This can be achieved (*i*) by introducing Debye phonon modes labelled by the pair of indices k, l, with eigenfrequencies ω_k depending only on k, so that the bath Hamiltonian is

$$\hat{H}_{\text{B}} = \sum_{k,l} \hbar \omega_k \hat{b}_{k,l}^{\dagger} \hat{b}_{k,l}, \tag{A.2}$$

and (*ii*) by assuming that the coefficients C in

$$\hat{B}_a^{(n)} = \sum_{k,l} \left[C\left(n,a;k,l\right) \hat{b}_{k,l} + C^*\left(n,a;k,l\right) \hat{b}_{k,l}^{\dagger} \right] \tag{A.3}$$

are such that

$$\sum_l C\left(n,a;k,l\right) C^*\left(m,b;k,l\right) = \delta_{n,m} \delta_{a,b}\, c\left(\omega_k\right). \tag{A.4}$$

This requires the number of values of the index l to be at least equal to $3N$. For instance, we may associate with each component a of each spin $\hat{\sigma}^{(n)}$ a different set of phonon modes, labelled by k, n, a, identifying l as (n, a), and thus define $\hat{H}_{\text{B}}$ and $\hat{B}_a^{(n)}$ as

$$\hat{H}_{\text{B}} = \sum_{n=1}^{N} \sum_{a=x,y,z} \sum_{k} \hbar \omega_k \hat{b}_{k,a}^{\dagger(n)} \hat{b}_{k,a}^{(n)}, \tag{A.5}$$

$$\hat{B}_a^{(n)} = \sum_{k} \sqrt{c\left(\omega_k\right)} \left(\hat{b}_{k,a}^{(n)} + \hat{b}_{k,a}^{\dagger(n)} \right). \tag{A.6}$$

We shall see in Section B that the various choices of the phonon set, of the spectrum (A.2) and of the operators (A.3) coupled to the spins are equivalent, in the sense that the joint dynamics of S + M will depend only on the spectrum ω_k and on the coefficients $c(\omega_k)$.

B. Equilibrium state of the bath

At the initial time, the bath is set into equilibrium at the temperature[v] $T = 1/\beta$. The density operator of the bath,

$$\hat{R}_{\rm B}(0) = \frac{1}{Z_{\rm B}} e^{-\beta \hat{H}_{\rm B}}, \tag{B.1}$$

when $\hat{H}_{\rm B}$ is given by (A.2), describes the set of phonons at equilibrium in independent modes.

As usual, the bath will be involved in our problem only through its *autocorrelation function* in the equilibrium state (B.1), defined in the Heisenberg picture by

$$\mathrm{tr}_{\rm B}\left[\hat{R}_{\rm B}(0)\,\hat{B}_a^{(n)}(t)\,\hat{B}_b^{(p)}(t')\right] = \delta_{n,p}\delta_{a,b}\,K(t - t'), \tag{B.2}$$

$$\hat{B}_a^{(n)}(t) \equiv \hat{U}_{\rm B}^\dagger(t)\,\hat{B}_a^{(n)}\hat{U}_{\rm B}(t), \tag{B.3}$$

$$\hat{U}_{\rm B}(t) = e^{-i\hat{H}_{\rm B}t/\hbar}, \tag{B.4}$$

in terms of the evolution operator $\hat{U}_{\rm B}(t)$ of B alone. The bath operators (A.3) have been defined in such a way that the equilibrium expectation value of $B_a^{(n)}(t)$ vanishes for all $a = x, y, z$.[8–10] Moreover, the condition (A.4) ensures that the equilibrium correlations between different operators $\hat{B}_a^{(n)}(t)$ and $\hat{B}_b^{(p)}(t')$ vanish, unless $a = b$ and $n = p$, and that the autocorrelations for $n = p$, $a = b$ are all the same, thus defining a unique function $K(t)$ in (B.2). We introduce the Fourier transform and its inverse,

$$\tilde{K}(\omega) = \int_{-\infty}^{+\infty} {\rm d}t\, e^{-i\omega t} K(t), \qquad K(t) = \frac{1}{2\pi} \int_{-\infty}^{+\infty} {\rm d}\omega\, e^{i\omega t} \tilde{K}(\omega) \tag{B.5}$$

and choose for $\tilde{K}(\omega)$ the simplest expression having the required properties, namely the quasi-Ohmic form[8–11]

$$\tilde{K}(\omega) = \frac{\hbar^2}{4} \frac{\omega e^{-|\omega|/\Gamma}}{e^{\beta\hbar\omega} - 1}. \tag{B.6}$$

The temperature dependence accounts for the quantum bosonic nature of the phonons.[8–10] The Debye cutoff Γ characterizes the largest frequencies of the bath, and is assumed to be larger than all other frequencies entering our problem. The normalization is fixed so as to let the constant γ entering (A.1) be dimensionless. Since $\tilde{K}(\omega)$ is real, it holds that $K(-t) = K^*(t)$.

---

[v]We use units where Boltzmann's constant is equal to one; otherwise, T and $\beta = 1/T$ should be replaced throughout by k_BT and $1/k_BT$, respectively.

C. Elimination of the bath

Taking $\hat{H}_0 = \hat{H}_S + \hat{H}_{SA} + \hat{H}_M$ and $\hat{H}_B$ as the unperturbed Hamiltonians of S + M and of B, respectively, and denoting by $\hat{U}_0 = \exp(-i\hat{H}_0/\hbar)$ and $\hat{U}_B = \exp(-i\hat{H}_B/\hbar)$ the corresponding evolution operators, we consider the full evolution operator associated with $\hat{H} = \hat{H}_0 + \hat{H}_B + \hat{H}_{MB}$ in the interaction representation. In general we can expand it to first order in $\sqrt{\gamma}$ as

$$\hat{U}_0^\dagger (t) \hat{U}_B^\dagger (t) e^{-i\hat{H}t/\hbar} \approx \hat{I} - i\hbar^{-1} \int_0^t dt' \hat{H}_{MB} (t') + \mathcal{O}(\gamma), \tag{C.1}$$

where the coupling in the interaction picture is

$$\hat{H}_{MB} (t) = \hat{U}_0^\dagger (t) \hat{U}_B^\dagger (t) H_{MB} \hat{U}_B (t) \hat{U}_0 (t) = \sqrt{\gamma} \sum_{n,a} \hat{U}_0^\dagger (t) \hat{\sigma}_a^{(n)} \hat{U}_0 (t) \hat{B}_a^{(n)} (t), \tag{C.2}$$

with $\hat{B}_a^{(n)} (t)$ defined by (B.3).

We wish to take the trace over B of the exact equation of motion eq. (C.3)

$$i\hbar \frac{d\hat{\mathcal{D}}}{dt} = \left[\hat{H}, \hat{\mathcal{D}} \right], \tag{C.3}$$

for $\hat{\mathcal{D}}(t)$, so as to generate an equation of motion for the density operator $\hat{D}(t)$ of S + M. In the right-hand side the term $\mathrm{tr}_B \left[\hat{H}_B, \hat{\mathcal{D}} \right]$ vanishes and we are left with

$$i\hbar \frac{d\hat{D}}{dt} = \left[\hat{H}_0, \hat{D} \right] + \mathrm{tr}_B \left[\hat{H}_{MB}, \hat{\mathcal{D}} \right]. \tag{C.4}$$

The last term involves the coupling $\hat{H}_{MB}$ both directly and through the correlations between S + M and B which are created in $\mathcal{D}(t)$ from the time 0 to the time t. In order to write (C.4) more explicitly, we first exhibit these correlations. To this aim, we expand $\mathcal{D}(t)$ in powers of $\sqrt{\gamma}$ by means of the expansion (C.1) of its evolution operator. This provides, using $\hat{U}_0(t) = \exp[-i\hat{H}_0 t/\hbar]$,

$$\hat{U}_0^\dagger (t) \hat{U}_B^\dagger (t) \hat{\mathcal{D}} (t) \hat{U}_B (t) \hat{U}_0 (t) \approx \hat{\mathcal{D}}(0) - i\hbar^{-1} \left[\int_0^t dt' \hat{H}_{MB} (t'), \hat{D}(0) \hat{R}_B (0) \right] + \mathcal{O}(\gamma). \tag{C.5}$$

Insertion of the expansion (C.5) into (C.4) will allow us to work out the trace over B. Through the factor $\hat{R}_B (0)$, this trace has the form of an equilibrium expectation value. As usual, the elimination of the bath variables will produce memory effects as obvious from (C.5). We wish these memory effects to bear only on the bath, so as to have a short characteristic time. However the initial state which enters (C.5) involves not only $\hat{R}_B (0)$ but also $\hat{D}(0)$, so that a mere insertion of (C.5) into (C.4) would let $\hat{D}(t)$ keep an undesirable memory of $\hat{D}(0)$. We solve this difficulty by re-expressing perturbatively $\hat{D}(0)$ in terms of $\hat{D}(t)$. To this aim we note that the trace of (C.5) over B provides

$$\hat{U}_0^\dagger (t) \hat{D} (t) \hat{U}_0 (t) = \hat{D}(0) + \mathcal{O}(\gamma). \tag{C.6}$$

We have used the facts that the expectation value over $\hat{R}_B(0)$ of an odd number of operators $\hat{B}_a^{(n)}$ vanishes, and that each $\hat{B}_a^{(n)}$ is accompanied in $\hat{H}_{MA}$ by a factor $\sqrt{\gamma}$. Hence the right-hand side of (C.6) as well as that of (C.4) are power series in γ rather than in $\sqrt{\gamma}$.

We can now rewrite the right-hand side of (C.5) in terms of $\hat{D}(t)$ instead of $\hat{D}(0)$ by means of inserting (C.6), then insert the resulting expansion of $\hat{\mathcal{D}}(t)$ in powers of $\sqrt{\gamma}$ into (C.4). Noting that the first term in (C.5) does not contribute to the trace over B, we find

$$\frac{d\hat{D}}{dt} - \frac{1}{i\hbar}\left[\hat{H}_0, \hat{D}\right] = -\frac{1}{\hbar^2}\mathrm{tr}_B \int_0^t dt' \tag{C.7}$$

$$\times \left[\hat{H}_{MB}(0), \hat{U}_B(t)\hat{U}_0(t)\left[\hat{H}_{MB}(t'), \hat{U}_0^\dagger(t)\hat{D}(t)\hat{U}_0(t)\hat{R}_B(0)\right]\hat{U}_0^\dagger(t)\hat{U}_B^\dagger(t)\right] + \mathcal{O}(\gamma^2),$$

where $\hat{H}_{MB}(0)$ is just equal to $\hat{H}_{MB}$, see eq. (C.4). Although the effect of the bath is of order γ, the derivation has required only the first-order term, in $\sqrt{\gamma}$, of the expansion (C.5) of $\mathcal{D}(t)$.

The bath operators $\hat{B}_a^{(n)}$ appear through $\hat{H}_{MB}$ and $\hat{H}_{MB}(t')$, and the evaluation of the trace thus involves only the equilibrium autocorrelation function (B.2). Using the expressions (A.1) and (C.2) for $\hat{H}_{MB}$ and $\hat{H}_{MB}(t')$, denoting the memory time $t - t'$ as u, and introducing the operators $\hat{\sigma}_a^{(n)}(u)$ defined by (C.9), we finally find the differential equation (2.8) for $\hat{D}(t)$.

Notice that by using (C.6) we have written an equation which self consistently couples the time derivative of $\hat{D}(t)$ to $\hat{D}(t)$ at the same time, at lowest order in γ. The method is akin to the derivation of the renormalization group equation.

In our model, the Hamiltonian commutes with the measured observable $\hat{s}_z$, hence with the projection operators $\hat{\Pi}_i$ onto the states $|\uparrow\rangle$ and $|\downarrow\rangle$ of S. The equations for the operators $\hat{\Pi}_i\hat{D}\hat{\Pi}_j$ are therefore decoupled. We can replace the equation (2.8) for $\hat{D}$ in the Hilbert space of S + M by a set of four equations for the operators $\hat{R}_{ij}$ defined by (2.11) in the Hilbert space of M. We shall later see (Section 8.2) that this simplification underlies the ideality of the measurement process.

The Hamiltonian $\hat{H}_0$ in the space S + M gives rise to two Hamiltonians $\hat{H}_\uparrow$ and $\hat{H}_\downarrow$ in the space M, which according to (2.4) and (2.6) are simply two functions of the observable $\hat{m}$, given by

$$\hat{H}_i = H_i(\hat{m}) = -gNs_i\hat{m} - N\frac{J}{4}\hat{m}^4, \qquad (i = \uparrow, \downarrow) \tag{C.8}$$

with $s_i = +1$ (or -1) for $i = \uparrow$ (or $\downarrow$). These Hamiltonians $\hat{H}_i$, which describe interacting spins $\hat{\sigma}^{(n)}$ in an external field gs_i, occur in (2.8) both directly and through the operators

$$\hat{\sigma}_a^{(n)}(u, i) = e^{-i\hat{H}_i u/\hbar}\hat{\sigma}_a^{(n)}e^{i\hat{H}_i u/\hbar}, \tag{C.9}$$

obtained by projection of (C.10)

$$\hat{\sigma}_a^{(n)}(u) \equiv \hat{U}_0(t)\,\hat{U}_0^\dagger(t')\,\hat{\sigma}_a^{(n)}\hat{U}_0(t')\,\hat{U}_0^\dagger(t) = \hat{U}_0(u)\,\hat{\sigma}_a^{(n)}\hat{U}_0^\dagger(u). \qquad (C.10)$$

with $\hat{\Pi}_i = |i\rangle\langle i|$ and reduction to the Hilbert space of M, with $i = \uparrow, \downarrow$.

The equation (2.8) for $\hat{\mathcal{D}}(t)$ which governs the joint dynamics of S + M thus reduces to the four differential equations (3.3) in the Hilbert space of M.

References

1. V. Spicka, Th. M. Nieuwenhuizen and P. D. Keefe, Physics at the FQMT'11 conference, *Physica Scripta* T151, 014001 (2012).
2. W. M. de Muynck, *Foundations of quantum mechanics, an empiricist approach* (Kluwer Academic Publishers, Dordrecht, 2002).
3. A. E. Allahverdyan, R. Balian and Th. M. Nieuwenhuizen, Understanding quantum measurement from the solution of dynamical models, *Phys. Rep.* **525**, 1 (2013).
4. A. E. Allahverdyan, R. Balian and Th. M. Nieuwenhuizen, Curie-Weiss model of the quantum measurement process, *Europhys. Lett.* **61**, 453 (2003).
5. N. G. van Kampen, Physica A, **153**, 97 (1988).
6. R. Balian, *From microphysics to macrophysics: Methods and applications of statistical physics, I, II* (Springer, Berlin, 2007).
7. K. Michielsen, Fengping Jin, M. Delina and H. De Raedt, Event-by-event simulation of nonclassical effects in two-photon interference experiments, Physica Scripta T151, 014005 (2012).
8. H.-P. Breuer and F. Petruccione, *The Theory of Open Quantum Systems* (Oxford University Press, Oxford, 2002).
9. U. Weiss, *Quantum Dissipative Systems* (World Scientific, Singapore, 1993)
10. C. W. Gardiner, *Quantum Noise* (Springer-Verlag, Berlin, 1991).
11. A. O. Caldeira and A. J. Leggett, Quantum Tunnelling in a Dissipative System, *Ann. Phys.* **149**, 374 (1983).
12. D. A. Lavis and G. M. Bell, *Statistical Mechanics of Lattice Systems* (Springer-Verlag, Berlin, 1999), Vol. 1.
13. L. D. Landau and E. M. Lifshitz, *Statistical Physics* (Pergamon Press, Oxford, 1978), Vol. 1.
14. F. Laloë, *Do we really understand quantum mechanics?* (Cambridge University Press, Cambdrige UK, 2012).
15. R. von Mises, *Probability, Statistics and Truth* (Macmillan, London, 1957).
16. A. E. Allahverdyan, R. Balian and Th. M. Nieuwenhuizen, Simultaneous measurement of non-commuting observables, *Physica E* **42**, 339 (2010).
17. M. Perarnau-Llobet, *An attempt to simultaneously measure two non-commuting variables*. Master Thesis, University of Amsterdam, August 2012.
18. Th. M. Nieuwenhuizen, Is the Contextuality Loophole Fatal for the Derivation of Bell Inequalities?, *Found. Phys.* **41**, 580 (2010).

Chapter 9

Towards a Wave Resolution of the Wave-Particle Duality

Andrei Yu. Khrennikov*

*International Center for Mathematical Modelling in
Physics and Cognitive Sciences,
Linnaeus University, S-35195, Växjö, Sweden*

We developed a purely field model of microphenomena — prequantum classical statistical field theory (PCSFT). This model reproduces important probabilistic predictions of QM including correlations for entangled systems. Hence, the wave-particle duality can be resolved in favor of a purely wave model. In PCSFT "particles" are just clicks of detectors.

1. Introduction

The present wave of interest in quantum foundations is caused by the tremendous development of quantum information science and its applications to quantum computing and quantum communication. Nowadays this interest even increases, because *it became clear that some of the difficulties encountered in realizations of quantum information processing (especially creation of quantum computers and design of new quantum algorithms) are not simply technicalities, but instead have roots at the very fundamental level.* To solve such difficult problems, quantum theory has to be reconsidered. In particular, some prejudices must be discarded; first of all the prejudice on completeness of QM. (The latter was strongly criticized by A. Einstein during all his life.)

This paper is dedicated to Einstein's vision[1,2] of physics and specifically his hope for what quantum theory could and, in his view, should be. In particular, two of Einstein's dreams about the future of quantum theory are realized in this paper: a reduction of quantum randomness to classical ensemble randomness and the total elimination of particles from quantum mechanics (QM): the creation of a classical field model[3-21] of quantum phenomena. Thus, contrary to a number of the so-called no-go arguments and theorems advanced throughout the history of quantum theory[22] (such as those of von Neumann, Kochen-Specker, and Bell), quantum probabilities and correlations can be described in a classical manner.

---

*Andrei.Khrennikov@lnu.se

There is, however, a crucial proviso. While we argue that QM can be interpreted as a form of classical statistical mechanics (CSM), this classical statistical theory is not that of particles, but of fields. This means that the mathematical formalism of QM must be translated into the mathematical formalism of CSM on the infinite-dimensional phase space. The infinite dimension of the phase space of this translation is a price of classicality. From the mathematical viewpoint this price is very high, because in this case the theories of measure, dynamical systems, and distributions are essentially more complicated than in the case of the finite-dimensional phase space found in a CSM of particles. However, at the model level (similar to quantum information theory) one can proceed with the finite-dimensional phase space by approximating physical prequantum fields by vectors with finite number of coordinates.

On the other hand, from the physical and philosophical viewpoints, considering QM as a CSM of fields can resolve the basic interpretational problems of QM. For example and in particular, quantum correlations of entangled systems can be reduced to correlations of classical random fields. From this perspective, quantum entanglement is not mysterious at all, since quantum correlations are no longer different from the classical ones.[a]

The main difficulty is that the classical situation is very tricky by itself. All quantum correlations contain the irreducible contribution of a background field (vacuum fluctuations). Roughly speaking quantum systems are classical random signals that are measured against a sufficiently strong random background. The data of QM is a result of our ignorance concerning the contribution of this random background. Thus quantum probabilities and correlations are not simply classical quantities. They are obtained from classical quantities by means of a renormalization procedure: a subtraction of the contribution of the background field. Accordingly, the reduction of quantum randomness to classical is not totally straightforward. Nevertheless, it is possible. For example, the otherwise mysterious nature of the Heisenberg uncertainty principle can be resolved in the following way. Quantum dispersions are not simply classical dispersions, but the results of the subtraction of the dispersion of vacuum fluctuations.[19] There is nothing mysterious in the fact that renormalized quantities satisfy this type of inequality. This is not so unusual from the viewpoint of the classical probability theory.

Already Max Planck emphasized the role of a random background field, the concept that was widely used in stochastic electrodynamics. (In his letter to Einstein

---

[a]In fact, the situation is more complicated. Averages and correlations provided by the classical field theory are related to continuous signals, but the quantum ones are based on statistics of discrete clicks of detectors. However, it is possible to discretize continuous signals with the aid of the *threshold type detectors* and transform probabilistic quantities for continuous signals into statistics of discrete clicks which coincide with quantum ones. The main part of this paper is devoted to representation of quantum averages and correlations with the aid of continuous random signals. Already this step is nontrivial both from physical and mathematical viewpoints. The corresponding discretization model was presented in papers Refs. 16, 21. We call this model *the threshold detection model* (TSD).

he pointed out that spontaneous emission can be easily explained by taking into account the background field.) This field also plays an essential role in our model, CSM of classical fields, also termed as *prequantum classical statistical field theory* (PCSFT).[3-21] This model is purely that of the field type. A classical random field is associated with each type of quantum particles. There are, for example, the electronic, neutronic, and protonic fields. The photonic field is simply the classical electromagnetic field. We speculate that all these classical fields can be unified through a single prequantum field. However, this is still the open problem. This "grand-unification" can include the gravitational field. Thus, instead of quantization of gravity, this paper advertises "dequantization" of QM as the first step towards unification of QM with gravity.

There is also a deep-going analogy between the present approach and the *classical theory of random signals*. Quantum measurements can be described as measurements for classical random signals with a noisy background. This analogy between QM and classical signal theory had been explored from the reverse perspective, for example, in using quantum information theory in the theory of classical Gaussian random signals (the model of Ohya-Watanabe[23]).

I hope that the paper will stimulate research that aims to demystify QM and to create a purely field model of quantum reality, and even to go beyond QM and find classical wave phenomena behind the basic laws of QM, such as Born's rule for probabilities. In PCSFT, prequantum random fields fluctuate on a time scale that is essentially finer than the time scale of quantum measurements. The fundamental question is whether this time scale is physically approachable remains open. In particular, if the prequantum time scale were the Planck scale, the PCSFT-level would be inapproachable. There would be no hope to monitor prequantum waves and show experimentally how the quantum statistics of clicks of detectors is produced through interaction of such waves with threshold-type (macroscopic) detectors. In such a situation PCSFT can be considered as a metaphysical theory, as e.g., string theory. However, if the prequantum time scale is essentially coarser than the Planck scale, one might dream of finding experimental confirmation of derivability of quantum laws (which are fundamentally probabilistic) from behaviour of prequantum (random) waves. Either way, however, PCSFT provides an adequate theoretical model of reduction of QM to CSM of fields.

1.1. *Einstein's dream of the pure field model*

Main message: *Particles are illusions and fields are reality.*

It is not well known that Einsteinian "beyond quantum world" was not imagined as a micro-copy of our macroscopic world populated by particles. Einstein's greatest dream was to *eliminate particles totally from coming fundamental theory*. New theory should be a purely field model of reality: no particles, but only fields (or may be just one field). He considered particles as the relict of the old mechanistic model of reality. It may be not so well known, but for Einstein QM was not a theory too novel (so that he even could not understand it, as some people claim), but, in contrast, it was too old fashioned to be considered a new fundamental theory.

In book[1] he discussed a lot Bohr's principle of complementarity, so to say, wave-particle duality. He was not happy with the quantum jargon mixing waves and particles. Einstein was sure that the wave-particle duality will be finally resolved in favor of a purely wave model:

"But the division into matter and field is, after the recognition of the equivalence of mass and energy, something artificial and not clearly defined. Could we not reject the concept of matter and build a pure field physics? What impresses our senses as matter is really a great concentration of energy into a comparatively small space. We could regard matter as the regions in space where the field is extremely strong. In this way a new philosophical background could be created. Its final aim would be the explanation of all events in nature by structure laws valid always and everywhere. A thrown stone is, from this point of view, a changing field, where the states of greatest field intensity travel through space with the velocity of the stone. There would be no place, in our new physics, for both field and matter, field being the only reality. This new view is suggested by the great achievements of field physics, by our success in expressing the laws of electricity, magnetism, gravitation in the form of structure laws, and finally by the equivalence of mass and energy. Our ultimate problem would be to modify our field laws in such a way that they would not break down for regions in which the energy is enormously concentrated. But we have no so far succeeded in fulfilling this program convincingly and consistently. The decision, as to whether it is possible to carry it out, belongs to the future. At present we must still assume in all our actual theoretical constructions two realities: field and matter.", see the book of Einstein and Infeld,[1] p. 242–243. Then they discussed QM and a possibility to interpret the wave function, the probability wave, as a physical field:

"For one elementary particle, electron or photon, we have probability waves in a three-dimensional continuum, characterizing the statistical behavior of the system if the experiments are often repeated. But what about the case of not one but two interacting particles, for instance, two electrons, electron and photon, or electron and nucleus? We cannot treat them separately and describe each of them through a probability wave in three dimensions, just because of their mutual interaction. Indeed, it is not very difficult to guess how to describe in quantum physics a system composed of two interacting particles. We have to descend one floor, to return for a moment to classical physics. The position of two material points in space, at any moment, is characterized by six numbers, three for each of the points. All possible positions of the two material points form a six-dimensional continuum and not a three-dimensional one as in the case of one point. If we now again ascend one floor, to quantum physics, we shall have probability waves in a six-dimensional continuum and not in a three-dimensional continuum as in the case of one particle.

Similarly, for three, four, and more particles the probability waves will be functions in a continuum of nine, twelve, and more dimensions. This shows clearly that the probability waves are more abstract than the electromagnetic and gravitational field existing and spreading in our three-dimensional space," see book,[1] p. 290–291.

This discussion is very important for our further studies of a possibility of creation of purely wave picture of physical reality. It was directly emphasized that one of the main problem is the impossibility to realize quantum "waves of probability" for composite quantum systems on physical space. Finally, Einstein and Infeld concluded,[1] p. 293:

"But there is also no doubt that quantum physics must still be based on the two concepts: matter and field. It is, in this sense, a dualistic theory and does not bring our old problem of reducing everything to the field concept even one step nearer realization."

As we have seen, Einstein's picture of electron or neutron is very simple: these are fields densely concentrated in small areas of space. These are classical fields (not quantum!). Hence, Einstein was sure that the classical space-time picture of reality could be combined with QM. (We state again that classical has the meaning classical field theory and not at all classical mechanics of particles.)

1.2. *Anti-photon*

Main message: *A photon is a pulse of the classical electromagnetic field. Discreteness is an illusion produced by detectors.*

Now we discuss the notion of photon. It is well known that Albert Einstein invented the notion of the *quantum of light* which was later called photon.[b]

Bohr was not happy with the invention of light quanta. In particular, the *Bohr-Kramers-Slater theory* [25] was an attempt to describe the interaction of matter and electromagnetic radiation without using the notion of photon. We also mention the strong opposition to the notion of photon from two fathers of QM: Alfred Lande[26] (in particular, this name is associated with Lande *g*-factor and the first explanation for the anomalous Zeeman effect) and Willis E. Lamb[24] (e.g., Lamb shift). Their views on electromagnetism differ crucially from the view of Albert Einstein (at least, young Einstein, see Section 1.1.5 below for the evolution of the Einstein views). The latter wrote (in 1910),[27] p. 207: "What we understand by the theory of "light quanta" may be formulated in the following fashion: a radiation of frequency ν can be emitted or absorbed only in a well defined quantum of magnitude $h\nu$. The theoreticians have not yet even come to an agreement in regard to the following

---

[b]Originally the concept of photon was invented by physical chemist G. N. Lewis who really considered photons as light particles that transmit radiation from one atom to another. Wave-like properties of photon were attributed to guiding ghost field. See Lamb's "Anti-photon",[24] p. 201–211, for more details. We underscore the difference, the photon-terminology, unlike the quantum of light terminology, is not so innocent as one may think. By calling the quantum of light the photon people emphasized the role of a particle picture of light.

question: Can the light quanta be accounted for entirely by a characteristic of the emitting or absorbing substance, or should the electromagnetic radiation itself be assigned, besides a wave structure, such that the energy of the radiation itself is already divided in definite quanta? I believe that I have proven that this latter view should be adopted."

Both Lande and Lamb rejected the existence of discrete quanta of electromagnetic field. They were sure that one can proceed in the so-called *semiclassical approach*, describing the interaction of classical electromagnetic field with quantum matter. We cite Lamb,[24] p. 211: "It is high time to give up the use of the word "photon", and of a bad concept which will shortly be a century old. Radiation does not consist of particles..." For adherents of the semi-classical approach quantization of the electromagnetic field is done by detectors; it is not present in electromagnetic field propagating in the vacuum.[c] We also recall that Max Planck opposed Einstein's idea of quantum of light from the beginning, and remained a champion of the unquantized Maxwell field throughout his life. In 1907 in a letter to Einstein, he said: "I am not seeking the meaning of the quantum of action (light quantum) in the vacuum but rather in places where emission and absorption occur, and I assume that what happens in the vacuum is rigorously described by Maxwell's equations."[29]

We also mention *stochastic electrodynamics* (SED) – a variant of classical electrodynamics which postulates the existence of a classical Lorentz-invariant radiation field (zero point field, Marshall and Brafford, Boyer, de la Pena and Cetto, Coli; see, e.g., the works[30–34] and the paper[35]). The presence of this field plays the crucial role in SED's description of quantum effects. We recall that already in 1911 Planck introduced the hypothesis of the *zero point electromagnetic field* in an effort to avoid Einstein's ideas about discontinuity in the emission and absorption processes.

It is important for our further considerations, that neither the semiclassical approach, nor SED, resolve the wave-particle dualism. Neither was it resolved by Bohmian mechanics, the modern version of De Broglie's double solution approach. Bohmian mechanics reduces the quantum randomness to classical ensemble randomness, and particles are the basic objects of this theory.

1.3. *Schrödinger's wave mechanics*

Main message: *Wave mechanics is an alternative to the Laplacian deterministic model of particles' motion*

.By now, the reader might be wondering that Schrödinger's name has not yet been mentioned in the discussion of classical and quantum wave mechanics. I refrained from it till this chapter to have enough place to consider not only Schrödinger's wave mechanics, but also his philosophic doctrine, that played an important role in my own theory.

---

[c]The semiclassical approach can describe a number of quantum effects, e.g., the photoelectric effect (G. Wentzel and G. Beck, 1926; see W. E. Lamb and M. O. Scully[28] for more detailed calculations).

At the beginning Schrödinger considered the squared wave function of the electron (multiplied by its electric charge $e < 0$) as the density of its charge:

$$p(t, x) = -e|\psi(t, x)|^2.$$

The solution $\psi(t, x)$ of Schrödinger's equation for, e.g., hydrogen atom describes oscillations of such electronic cloud, which induce electromagnetic radiation with frequencies and intensities matching the experiment.[d] This picture of a quantum particle as a field, in this case electronic field, coincides with the picture from Einstein's field dream. Unfortunately, Schrödinger was not able to proceed in this way. He understood, as well as Einstein, that already for two electrons the wave function cannot be interpreted as a field on a physical space: $\psi(t, x, y)(x = (x_1, x_2, x_3), y = (y_1, y_2, y_3))$, is defined on $\mathbf{R}^6$. Although formally Schrödinger gave up and accepted Born's interpretation of the wave function, he did not like the Copenhagen interpretation, as Einstein did neither, especially, Bohr's thesis on completeness of QM. Schrödinger dreamed to go beyond QM, and to refind purely wave resolution of wave-particle duality. Nor, however, did Schrödinger accept Einstein's ensemble interpretation of the wave function. No wonder why! Schrödinger would rather have a wave associated with an individual quantum system, and the wave function was the best candidate for such wave. Therefore, he rejected Einstein's idea to associate the wave function with an ensemble of quantum systems, see their correspondence.[2]

As I mentioned, Einstein did not want the comeback to the Laplacian determinism, and Schrödinger did neither: Schrödinger's views on scientific description of physical reality were based on a well elaborated approach, the so-called *Bild-conception tradition*, see D'Agostino,[36] p. 351, for details:

Schrödinger called "the classical ideal of uninterrupted continuous description", at both observables' and theoretical levels, an "old way", meaning, of course, that this ideal is no longer attainable. He acknowledged that this problem was at the center of the scientific debate in the Nineteenth and Twentieth centuries as well:

"Very similar declarations...(were) made again and again by competent physicists a long time ago, all through the Nineteenth Century and the early days of our century...they were aware that the desire for having a clear picture necessarily led one to encumber it with unwarranted details",[37] p. 24.

I would like now to cite a rather long passage from D'Agostino,[36] pp. 351–352, presenting philosophic views of Schrödinger on two levels of description of reality: observational (empiricist) and theoretical.

"The competent physicists are almost certainly Hertz, Boltzmann and their followers. One can thus argue that Schrödinger's two-level conception above is, at bottom and despite its "amazing" appearance, part of the tradition of the nineteenth-

---

[d]Unfortunately, I was not able to find in Schrödinger's papers any explanation of the impossibility to divide this cloud into a few smaller clouds, i.e., no attempt to explain the fundamental discreteness of the electric charge.

century Bild-conception of physics, formulated by Hertz in his 1894 Prinzipien der Mechanik, and also discussed by Boltzmann, Einstein et altri. He partially inherited this tradition from his teacher Exner and he deepened this conception through his intense study of Boltzmann's work. One of the main features of the above tradition is its strong anti-inductionism. If theory is not observation-depended - in the sense that it is not constructed on (or starting from) observations - it consequently possesses a sort of distinction as regards observations. This distinction may be pushed to various degrees of independence. Hertz implied that a term-to-term correspondence between concepts and observables was not needed when he introduced hidden quantities among the theory's visible ones. In his often quoted dictum, Boltzmann asserted that only one half of our experience is ever experienced. Basically, Schrödinger was thus orthodox in his assertions that theory and observations are not necessarily related in a term-to-term correspondence and a certain degree of independence exists between them. However, when he added the further qualification that a repugnancy might exist between them, he stretched this independence to its extreme consequences, introducing a quasidichotomy between a pure theory and an observational language.

This extreme position was not acceptable to the majority of his contemporaries and to Einstein in particular. Causal gaps, even if limited to the observables level, could not be accepted by Einstein and other scientists. In fact, Einstein's completeness implied a *bi-univocal* correspondence between concepts and observables. It followed from Einstein's premises that, if Schrödinger's wave function did not correspond to a complete description of the system, the reason was to be sought in its statistical (in Einstein's sense!) features: i.e. Schrödinger's wave function refers to an ensemble not to an individual system. Differently, Schrödinger thought that incompleteness in description was generated by an illegitimate (due to indistinguishability) individualization of classical or quasi-classical particles in microphysics. On the other hand, Schrödinger could not accept Heisenberg's and Bohr's Copenhagenism, because, for him, their position represented a concession to an old conception of the theory-observations relation, implying that causality-gaps and discontinuities on the observation-level would forbid the construction of a complete theory (a complete model). One can thus argue that Schrödinger considered the fundamental defect of the Copenhagen view its missing the distinction between the two levels of language, the descriptive and the purely theoretical level. From the QM impossibility of a continuous descriptive language on the observable level, the Copenhagenists would have rushed to conclude the uselessness of a continuous purely theoretical language."

In this paper I present a theoretical (causal and continuous) model of physical reality, *prequantum classical statistical field theory* – PCSFT.[e] Since my starting

---

[e]In principle, causality is approachable in the PCFT-framework, but the situation is quite complicated, because of the presence of vacuum fluctuations; we shall come back to this problem in Section 2.7

point was not the observations, my model does not rely completely on the descriptive language of QM, which fact is in total accordance with views of Boltzmann, Hertz, Exner, and Schrödinger on relation between theoretical and observational models. The correspondence between concepts of PCSFT and QM is not straightforward, see paper[21] for coupling of PCSFT and QM through a measurement theory for PCSFT.

Let us return to the views of Schrödinger on QM and physical reality. I cite from Lockwood,[38] pp. 385–386:

"Two possibilities then present themselves. One possibility (a) is that individual physical systems do, after all, possess determinate states in essentially the classical sense. That is to say, the classical dynamical variables do have well-defined values at every moment, arbitrary precise simultaneous knowledge of which is, however, in principle unattainable. Consequently, we have to fall back on statistical statements. The assertions of quantum mechanics should accordingly be understood to refer, as in statistical mechanics, to the distribution of values of these variables within an ideal ensemble of similarly prepared systems. Schrödinger assumed this to be Einstein's position. The other possibility (b) is that the quantum-mechanical description, as embodied in the ψ-function, is a complete specification of an objectively "fuzzy" state. On this conception, quantum mechanics does offer a model of reality; but the model it presents us with is of an objectively "blurred" reality. The difference between these two interpretations, Schrödinger regards as analogous to the difference between an out-of-focus photograph of something with perfectly sharp outlines, and a properly focused photograph of something lacking sharp outlines, such as a patch of fog. Having set up these alternatives, Schrödinger then, disconcertingly, proceeds to argue that neither is tenable."

In fact, the viewpoint to QM generated by PCSFT in combination with the corresponding threshold detection model (TSD) does not match neither with (a), the classical statistical mechanical viewpoint to QM, nor with (b), the completeness viewpoint to QM. The aforementioned two level description of reality based on the combination of theoretical and observational models is sufficiently close to the one given by PCSFT/TSD. However, opposite to Schrödinger and other adherents of the Bild concept, I think that all basic features of observational model have to be derivable from the theoretical model.

1.4. *Bohr-Kramers-Slater theory*

In paper[25] Bohr, Kramers, and Slater (BKS) tried to treat the interaction of matter and electromagnetic radiation without photons. By their model atoms produce a virtual field (induced by virtual oscillators) which induces the emission and absorbsion processes. This virtual field contains contributions of all atoms and hence each transition in a single atom is determined by processes in all atoms nearby. The BKS-theory can be coupled with PCSFT. In the latter any "quantum particle" is represented by a classical random field. In particular, any atom is nothing else than an atomic field. A group of atoms induces a collective atomic field. Therefore we might try to interpret the virtual BKS-field as the real atomic field of PCSFT. Any transition in atom (by the QM-terminology from one level to another) is a completely causal process of evolution of this field. Fields of various types of "quantum particles" can interact with each other or better to say there is a single fundamental prequantum field which have various configurations: photonic (electromagnetic), electronic, atomic,.... In 'PCSFT we stress a role of the background field, vacuum fluctuations. This field is present even in the absence of "quantum particles". The presence of the background field may solve one of the main problems of the BKS-theory, namely, possible violation of the law of conservation of energy on the individual level: the impossibility to account for conservation of energy in a process of de-excitation of an atom followed by excitation of a neighboring one. In PCSFT the energy of the fundamental prequantum field is not changed in this process.

The BKS-theory was an attempt to unify wave and particle pictures on the basis of the classical field theory. This was an attempt of *causal continuous description* of quantum jumps in the processes of absorbtion and emission. We remark that Bohr elaborated his principle of complementarity only because he was not able to construct a satisfactory causal field-type model. Later he advertised completeness of QM.[39] Roughly speaking he tried to stop studies similar to his own in 1924th. (The Freudian background of such behavior is evident.)

1.5. *On the evolution of Einstein's views:*
From classical electrodynamics to photon and back

Einstein has views, as presented respectively in Sections 1.1 and 1.2, appear to be in conflict. On the one hand, as discussed in Section 1.1, he was the discoverer of the particle of light, the photon, as it eventually became known, and thus advocated the particle-like model of the behavior of light in certain circumstances, a view confirmed by the Compton scattering experiment in 1923, shortly before the discovery of quantum mechanics. On the other hand, as discussed in Section 1.2, he championed the classical-like field model as the best, if not the only model, for fundamental physics, which, given the continuous character of the classical field theory, is difficult to reconcile with the concept of photon. This discrepancy leads one to suspect that these two positions reflect the views of two different "Einsteins," especially given that they correspond to two different periods of Einstein's work, the first, roughly between 1905–1920, and second, from roughly 1920 to his death in 1955. His book with Infeld, discussed in 1.2, was originally published in 1938 and, thus, it might be added, was written shortly after the EPR argument, which solidified Einstein's critical assessment of quantum mechanics. The evolution of Einstein's views is instructive and one might sketch this evolution roughly as follows.[f]

---

[f]The account of this evolution sketched here is courtesy of Arkady Plotnitsky [private communication] (also reading of his works[40−42] played an important role in establishing my views on Einstein's position (in fact, positions)). See also Pais[43] for a discussion of the development of

It may be argued that Einstein's primary model for doing fundamental physics had always been Maxwell's electrodynamics as a field theory, which grounds special relativity, introduced in 1905, the same year he introduced the idea of photon. It also appears, however, that his thinking at the time was more flexible as concerns what type of physical theory one should or should not use. His approach was determined more by the nature of the experimental phenomena with which he was concerned, or in his own later words, his attitude was more "opportunistic",[45] p. 684, rather than guided by a given set of philosophical preferences, as in his later works. In this respect, the term "opportunistic" may no longer easily apply to his later thinking, or at least his opportunism was conditioned by his philosophical inclinations toward a classical-like field-theoretical approach to fundamental physics. Einstein appears to have introduced the concept of photon under the pressure of experimental evidence, such as that reflected in Planck's law or the law of photoeffect (for which Einstein was actually awarded his Nobel Prize). He went further than Planck by proposing that the photon was a real particle (rather than a mathematical convenience), the idea that took a while, until 1920s and much additional experimental evidence, most especially, again, Compton's scattering experiments, to accept. Intriguingly, not only Planck but also Bohr was among the skeptics, and Bohr only accepted the idea in view of these experiments. Planck, who, as discussed earlier, strongly resisted Einstein's introduction of the concept of the photon, had never reconciled himself to the idea. Thus, it appears that until roughly 1920, Einstein did not have a strongly held philosophical position of the type he developed later on, first, following his work on general relativity (a classical-like field theory) and, secondly and most especially, in the wake of quantum mechanics. It is worth noting in this connection that he initially resisted Minkowsky's concept of spacetime as insufficiently physical, but eventually came to appreciate its significance, again, especially in view of its effectiveness in general relativity. It is true, however, that theoretical physics at the time, including quantum theory (the "old" quantum theory), was still more classically oriented, as against quantum mechanics in the Heisenbergian approach. In addition, given that, in some circumstances, light would still exhibit wave behavior, Einstein also believed at the time (until even 1916 or so), that a kind of new synthesis of the particle-like and the wave-like theory of radiation would be necessary. However, this hope had not materialized in any form that he found acceptable, and he was especially dissatisfied with Heisenberg's approach, developed into the matrix mechanics by Born and Jordan, or related schemes, such as Dirac's one. The success of general relativity as a classical-like field theory was significantly responsible for strengthening Einstein's field-theoretical predilections, and shaped his program of the unified field theory (with a unification of gravity and electromagnetism as the first task), which he pursued for the rest of his life. The problems

---

Einstein's views on fundamental physics, from his earlier work to his work on general relativity and beyond; and for Einstein's earlier views, see Don Howard and John Stachel[44] and also work.[27] For Einstein's later views, see especially both of his contributions to the Schilpp volume.[45]

of quantum mechanics and his debate on the subject with Bohr continued to pre-occupy him as well, as reflected in particular in his persistent thinking concerning the EPR experiment, on which he commented virtually until his death. His view of fundamental physics following his work on relativity was also more mathematically oriented than the earlier one. In particular, he came to believe that it is a free mathematical conceptual construction, such as those of Riemann's geometry and tensor calculus in the case of general relativity and indeed of a similar classical-like field-theoretical type, that should and, he even argued, will allow us to come closest to capturing, in a realist manner, the ultimate character of physical reality. He expressly juxtaposed this approach to that of the Copenhagen-Göttingen approach in quantum mechanics,[45] pp. 83–85. In sum, Einstein had come to be convinced that a strictly field-like theory unifying the fundamental forces of nature should be pursued. He saw this kind of theory as the best and even, to him, the only truly acceptable program for the ultimate theory of nature, while he believed quantum mechanics to be a provisional theory, eventually to be replaced by a field theory of the type he envisioned.

It may be remarked that the idea of particle poses difficulties for this view, especially the particle nature of radiation, initially represented in the idea of photon. This is why Einstein preferred and saw as more promising (than matrix mechanics) Schrödinger's wave mechanics, or why earlier he liked de Broglie's approach (which he used in his work on the Bose-Einstein theory). It is true that the latter does retain the concept of particle and, as such, represents an attempt at a synthesis of the wave and the particle pictures, which, as noted above, Einstein contemplated initially. Later on, however, he did not like Bohmian mechanics, which pursued a similar line of thinking, although his negative attitude appears to have been determined by a complex set of factors. Eventually it became apparent that Schrödinger's formalism could not quite be brought under the umbrella of Einstein's unified field-theoretical program, a la Maxwell, although in his later years (in 1940s–1950s) Schrödinger return to his initial ideas concerning wave mechanics. Quantum electrodynamics and then other quantum field theories appeared even more difficult to reconcile with this approach. Even general relativity posed certain significant problems for Einstein's vision, such as singularities, eventually leading to ideas such as black holes, although the full measure of these difficulties became apparent only later on, after Einstein's death.

There is thus quite a bit of irony to this history. While Einstein was fundamen-tally responsible for several theoretical ideas that eventually led others to quantum mechanics, he had developed grave doubts about quantum mechanics as a "useful point of departure for future development",[45] p. 83. Since, however, our fundamen-tal physics remained incomplete at the time, Einstein thought that his vision might ultimately be justified. It might yet be, since our fundamental physics still remains incomplete, and in particular, is defined by a manifest conflict between relativity and quantum mechanics or higher-level quantum theories. It would be curious to contemplate whether Einstein would have liked something like the string and brane theories, or any other currently advanced programs for fundamental physics and cosmology.

2. Prequantum classical statistical field theory

Now I turn to my model, PCSFT, which is based on the unification of two Einstein's dreams: to reduce quantum randomness to classical randomness and to create a purely wave model of physical reality. I emphasize from the very beginning that the majority of PCSFT-structures are already present in QM, but in PCSFT they obtain a new (classical signal) interpretation. Therefore the introduction in PCSFT presented in this section can be considered as a short dictionary that establishes a correspondence between terms of QM and PCSFT. However, PCSFT not only reproduces QM, but provides a possibility to go beyond it. Therefore, advanced structures of PCSFT do not have counterparts in QM.

2.1. *Classical fields as hidden variables*

Main message: *Quantum randomness is reducible to randomness of classical fields.*

Classical fields are selected as the hidden variables. (PCSFT is not a deterministic-type model with hidden variables. By fixing a classical "prequantum" field we cannot determine the values of observables. These values can be predicted with probabilities which are determined by the prequantum field.) Mathematically, they are functions $\phi : \mathbf{R}^3 \to \mathbf{C}$ (or, more generally, $\to \mathbf{C}^k$) which are square-integrable, i.e., elements of the L_2-space. The latter condition is standard in the classical signal theory.

In particular, for electromagnetic field, this is just the condition of the finiteness of energy

$$\int_{\mathbf{R}^3} (E^2(x) + B^2(x))dx = \int_{\mathbf{R}^3} |\phi(x)|^2 < \infty, \tag{1}$$

where

$$\phi(x) = E(x) + iB(x) \tag{2}$$

is the *Riemann-Silbertstein vector* (the complex representation of the electromagnetic field).

Thus, the state space of our prequantum model is $H = L_2(\mathbf{R}^3)$. Formally, the same space is used in QM, but we couple it with the classical signal theory. For example, the quantum wave function satisfies the normalization condition

$$\int_{\mathbf{R}^3} |\phi(x)|^2 = 1, \tag{3}$$

but any vector ϕ in H can be selected as a PCSFT-state. These prequantum waves evolve in accordance with Schrödinger's equation; formally, the only difference is that the initial condition ϕ_0 is not normalized by 1, see Section 2.4, equation (13). Thus, these PCSFT-waves are closely related to Schrödinger's quantum waves. However, opposite to Schrödinger and to the orthodox Copenhagen interpretation, the wave function of the QM-formalism is not a state of a quantum system. In the complete accordance with Einstein's dream of reducibility of quantum randomness, wave function is associated with an ensemble. The ensemble, however, not of quantum systems, but the ensemble of classical fields, or, more precisely, a *classical random field*, random signal. It is appropriate to say that, although our model supports Einstein's views on the origin of quantum randomness, it also matches von Neumann's views[46] on individual quantum randomness. By using ergodicity we can switch from ensemble description to individual signal description and vice versa. We state again that such a possibility of peaceful combination of Einstein's and von Neumann's views on quantum randomness is a consequence of the rejection of the corpuscular model in the complete accordance with the views of "late Einstein." (It seems that at first he wanted to reduce quantum randomness to randomness of ensembles of particles.)

A random field (at a fixed instant of time) is a function $\phi(x, \omega)$, where ω is a random parameter. Thus for each ω_0, we obtain the classical field, $x \mapsto \phi(x, \omega_0)$. Another picture of the random field is the H-valued random variable, each fixed ω_0 determines a vector $\phi(\omega_0) \in H$. A random field is given by a probability distribution on H. For simplicity, we can consider a finite-dimensional Hilbert space instead of $L_2(\mathbf{R}^3)$ (as people often do in quantum information theory). In this case, PCSFT considers H-valued random vectors, where $H = \mathbf{C}^n$. However, we strongly emphasize the role of the physical state space $H = L_2(\mathbf{R}^3)$.

This is the ensemble model of the random field. In the rigorous mathematical framework it is based on the *Kolmogorov probability space* $(\Omega, \mathcal{F}, \mathbf{P})$, where Ω is a set and $\mathcal{F}$ is the σ-algebra of its subsets, $\mathbf{P}$ is a probability measure on $\mathcal{F}$. It is always possible to choose $\Omega = H$ and $\mathcal{F}$ as the σ-algebra of Borel subsets of H, and probability is a measure on the Hilbert space H. We remark that such measures are used in classical signal theory as probability distributions of random signals.

In the classical signal theory one can move from the ensemble description of randomness to the time series description – under the ergodicity hypothesis. Random signals are widely used e.g. in radio-physics; these are electromagnetic fields depending of a random parameter; by using the Riemann-Silberstein representation stationary radio-signals can be represented in the complex form: $\phi(x, \omega) = E(x, \omega) + iB(x, \omega)$.

2.2. *Covariance operator interpretation of wave function*

Main message: *The wave function is not a field of probabilities or a physical field. It encodes correlations between degrees of freedom of a prequantum random field.*

For simplicity, in this introductory section we consider the case of a single, i.e., noncomposite, system, e.g., the electron. We proceed in nonrelativistic framework, since the present PCSFT is a nonrelativistic theory and we neglect for a moment (again for simplicity) fluctuations of vacuum which will play an important role in our further consideration.

In our model the wave function Ψ of the QM-formalism encodes a class of pre-quantum random fields having the same covariance operator (determined by Ψ and determining a unique Gaussian random field.) We state again that we consider the case of a noncomposite quantum system; for composite systems, e.g., for a pair of photons or electrons, the correspondence between the wave function of QM and the covariance operator of PCSFT is more complicated.[16]

In this situation the covariance operator (normalized by dispersion) is given by the orthogonal projector on the vector Ψ :

$$D_\Psi = \Psi \otimes \Psi, \tag{4}$$

i.e., $D_\Psi u = \langle u, \Psi \rangle \Psi$, $u \in H$. Thus,

$$D_\Psi = |\Psi\rangle\langle\Psi|$$

in Dirac's notation, i.e.,

$$D_\Psi u = \langle u|\Psi \rangle \, |\Psi\rangle.$$

We also suppose that all *prequantum fields have zero average*

$$E\langle y, \phi \rangle = 0, y \in H, \tag{5}$$

where E denotes the classical *mathematical expectation* (average, mean value). By applying a linear functional y to the random vector ϕ we obtain the scalar random variable. In the L_2-case we get a family of scalar random variables:

$$\omega \mapsto \xi_y(\omega) \equiv \int_{\mathbf{R}^3} \phi(x,\omega)\overline{y(x)}dx, y \in L_2.$$

We recall that the covariance operator D of a random field (with zero average) $\phi \equiv \phi(x,\omega)$ is defined by its bilinear form

$$\langle Du, v \rangle = E\langle u, \phi \rangle\langle \phi, v \rangle, u, v \in H. \tag{6}$$

Under the additional assumption that the prequantum random fields are *Gaussian*, the covariance operator uniquely determines the field. Although this assumption seems to be quite natural both from the mathematical and physical viewpoints, we should be very careful. In the case of a single system we try to proceed as far as possible without this assumption. However, the PCSFT-description of composite systems is based on Gaussian random fields.[16] Let $H = \mathbf{C}^n$ and $\phi(\omega) = (\phi_1(\omega), ..., \phi_n(\omega))$, then zero average condition (5) is reduced to

$$E\phi_i \equiv \int_\Omega \phi_k(\omega)dP(\omega) = 0, k = 1, ..., n;$$

the covariance matrix $D = (d_{kl})$, where

$$d_{kl} = E\phi_k\bar{\phi}_l \equiv \int_\Omega \phi_k(\omega)\overline{\phi_l(\omega)}dP(\omega).$$

We also recall that the dispersion of the random variable ϕ is given by

$$\sigma_\phi^2 = E\|\phi(\omega) - E\phi(\omega)\|^2 = \sum_{k=1}^n E|\phi_k(\omega) - E\phi_k(\omega)|^2.$$

In the case of zero average we simply have

$$\sigma_\phi^2 = E\|\phi(\omega)\|^2 = \sum_{k=1}^n E|\phi_k(\omega)|^2.$$

Here it is always possible to select Ω (the set of random parameters) as $\mathbf{C}^n$. Then, the above integrals will be over $\mathbf{C}^n$. In particular, by selecting Gaussian, complex-valued, random fields we obtain Gaussian integrals over $\mathbf{C}^n$.

The case $H = L_2$ is more complicated from the measure-theoretic viewpoint, since this space is infinite-dimensional. In the case of noncomposite systems (i.e., a single photon or electron) it is also possible to select $\Omega = H$, i.e., to integrate with respect to all fields of the L_2-class. For composite systems, the situation is more complicated. Here we cannot proceed without taking into account the background field, that is of the white noise type. And the well-known fact is that the probability distribution of white noise cannot be concentrated on L_2, one has to select Ω as a space of distributions, i.e., to integrate with respect to singular fields.

We also remark that the random field $\phi(x, \omega)$ corresponding to a pure quantum state is not L_2-normalized. Its L_2-norm

$$\|\phi\|^2(\omega) \equiv \int_{\mathbf{R}^3} |\phi(x, \omega)|^2 dx \tag{7}$$

fluctuates depending on the random parameter ω. We call the quantity

$$\pi_2(\phi) \equiv \|\phi\|^2$$

the *power of the prequantum field (signal)* ϕ.

2.3. Quantum observables from quadratic forms of the prequantum field

Main message: *In spite of all no-go theorems (e.g., the Kochen-Specker theorem), a natural functional representation of quantum observables exists.*

In PCSFT quantum observables are represented by corresponding quadratic forms of the prequantum field. A self-adjoint operator $\widehat{A}$ is considered as the symbolic representation of the PCSFT-variable

$$\phi \mapsto f_A(\phi) = \langle \widehat{A}\phi, \phi \rangle. \tag{8}$$

This is a map from the L_2-space of classical prequantum fields into real numbers, a quadratic form.

We remark that f_A can be considered as a function on the phase space of classical fields: $f_A \equiv f_A(q, p)$, where $\phi(x) = q(x) + ip(x)$, i.e., it is possible to move from the complex representation to the phase space representation and vice versa. A crucial point is that the *prequantum phase space* is infinite-dimensional (and the "post-quantum phase space", i.e., the phase space of ordinary classical mechanics is finite-dimensional).

The average of this quadratic form with respect to the random field determined by the wave function Ψ coincides with the corresponding quantum average:

$$\langle f_A \rangle = \langle \widehat{A}\Psi, \Psi \rangle \tag{9}$$

or

$$\langle f_A \rangle = \langle \Psi | \widehat{A} | \Psi \rangle$$

in Dirac's notation. Here

$$\langle f_A \rangle = E f_A(\phi) = \int_H f_A(\phi) d\mu_\Psi(\phi)$$

is the classical average and μ_Ψ is the probability distribution of the prequantum random field $\phi \equiv \phi_\Psi$ determined by the pure quantum state Ψ. In the real physical case H is infinite-dimensional; the classical average is given by the integral over all possible classical fields; probabilistic weights of the fields are determined, in general, non-uniquely, by the Ψ. Thus, the quantum formula for the average of an observable was demystified:

$$\langle \widehat{A} \rangle_\Psi \equiv \langle \widehat{A}\Psi, \Psi \rangle = \int_H f_A(\phi) d\mu_\Psi(\phi) \tag{10}$$

It can be obtained via the classical average procedure.

2.4. *Quantum and prequantum interpretations of Schrödinger's equation*

Main message: *Schrödinger's equation with random initial conditions describes dynamics of the physical random field.*

Before going to the PCSFT-dynamics, we consider the Schrödinger equation in the standard QM-formalism:

$$ih\frac{\partial \Psi}{\partial t}(t, x) = \widehat{\mathcal{H}}\Psi(t, x), \tag{11}$$

$$\Psi(t_0, x) = \Psi_0(x), \tag{12}$$

where $\widehat{\mathcal{H}}$ is Hamiltonian, the energy observable. We recall that Schrödinger tried to interpret $\Psi(t, x)$ as a classical field (e.g., the electron field; the distribution of electron charge in space). However, he gave up and, finally, accepted the conventional interpretation, the probabilistic one, due to Max Born.

We recall that a time dependent random field $\phi(t, x, \omega)$ is called a *stochastic process* (with the state space $H = L_2$). Dynamics of the prequantum random field is described by the simplest stochastic process which is given by *deterministic dynamics with random initial conditions.*

In PCSFT the Schrödinger equation, but with the random initial condition, describes dynamics of the prequantum random field, i.e., the prequantum stochastic process can be obtained from the mathematical equation which is used in QM for dynamics of the wave function:

$$ih\frac{\partial \phi}{\partial t}(t, x, \omega) = \widehat{\mathcal{H}}\phi(t, x, \omega), \tag{13}$$

$$\phi(t_0, x, \omega) = \phi_0(x, \omega), \tag{14}$$

where the initial random field $\phi_0(x, \omega)$ is determined by the quantum pure state Ψ_0. Standard QM gives the covariance operator of this random field.

Roughly speaking, we combined Schrödinger's and Born's interpretations: the Ψ-function of QM is not a physical field, but for each t it determines a random physical field, i.e., the H-valued stochastic process $\phi(t, x, \omega)$.

PCSFT dynamics matches standard QM-dynamics by taking into account the PCSFT-interpretation of the wave-function, see (4). Denote by $\rho(t)$ the covariance operator of the random field $\phi(t) \equiv \phi(t, x, \omega)$, the solution of (13), (14). Then

$$\rho(t) \equiv \rho_{\Psi(t)} = \Psi(t) \otimes \Psi(t),$$

where $\Psi(t)$ is a solution of (11), (12).

Such simple description can be used only for a single system and in the absence of fluctuations of vacuum. In the general case of a composite system, e.g., a biphoton system, in the presence of vacuum fluctuations Schrödinger dynamics of the Ψ-function encodes only dynamics of the covariance operator of the prequantum stochastic process. The situation is essentially more complicated[16] than in the case of a single system.

2.5. Towards prequantum determinism?

Main message: *The background field is everywhere.*

From the PCSFT-viewpoint, the source of quantum randomness is the randomness of initial conditions (if one neglects vacuum fluctuations), i.e., impossibility to prepare a non-random initial prequantum field $\phi_0(x)$.

We expect that in future very stable and precise preparation procedures will be created. The output of such a procedure will be a deterministic field $\phi(x)$, i.e., random fluctuations will be eliminated.

However, this dream of creating supersensitive "subquantum" technologies which would recover determinism on the microlevel may never come true. In such a case PCSFT will play the role of classical statistical mechanics of prequantum fields. Unfortunately, there are a few signs that it really might happen. First of all, it might be that the scale of prequantum fluctuations is very fine, e.g., the *Planck scale*. In this case it would be really impossible to prepare a deterministic prequantum field. And there is another reason. The PCSFT-model presented up to now has been elaborated for noncomposite quantum systems, e.g., a single electron. The extension of PCSFT to composite systems, e.g., a pair of entangled photons or electrons, is based on a more complicated model of prequantum randomness.[16] We should complete the present model by considering fluctuations of the background field (zero point field, vacuum fluctuations), in the same way as in SED. In reality, these are always present. Therefore Einstein's dream of determinism cannot be peacefully combined with the presence of the background field. If this field is irreducible (as a fundamental feature of space), then deterministic prequantum fields will never be created. However, if this background field is simply noise[g] which can be eliminated, then we can dream of the creation of deterministic prequantum fields. Furthermore, a possibility to prepare such fields does not imply deterministic reduction of QM. As was already pointed out, the inter-relation between prequantum fields and quantum observables given by TSD (measurement theory of classical waves with threshold detectors) is really tricky.[16]

2.6. *Random fields corresponding to mixed states*

Main message: *A density matrix is the normalized covariance operator of a prequatum random field.*

We now consider the general quantum state given by a density operator ρ. (We still work with noncomposite quantum systems.) According to PCSFT, ρ determines the covariance operator of the corresponding prequantum field (under normalization by its dispersion)

$$D_\rho = \rho. \tag{15}$$

Dynamics of the corresponding prequantum field $\phi(t, x, \omega)$ is also described by the Schrödinger equation, see (13), (14), with the random initial condition $\phi_0(x, \omega)$. The initial random field has the probability distribution μ_{ρ_0} having zero mean value and the covariance operator

$$D(t_0) = \rho_0.$$

Under the assumption that all prequantum random fields are Gaussian, the initial probability distribution is determined in the unique way. In the general (non-Gaussian) case we lose the solid ground. The $\phi_0(x, \omega)$ can be selected in various ways, i.e., it can be any distribution having the covariance $D(t_0)$. We could not

---

[g]Hence the completely empty physical space can be really prepared, "distilled from noise".

exclude such à possibility. It would simply mean that macroscopic preparation procedures are not able to control even the probability distribution (only its covariance operator).

Denote by $\rho(t)$ the covariance operator of the random field $\phi(t) \equiv \phi(t, x, \omega)$ given by (13), (14) with ϕ_0 having the covariance operator $\rho(t_0) = \rho_0$. Then $\rho(t)$ satisfies the von Neumann equation. However, $\rho(t)$ has the classical probability interpretation as the covariance operator $D(t)$. In the Gaussian case $D(t)$ determines completely the prequantum probability distribution.

2.7. Background field

Main message: *QM is a formalism of measurement with threshold type detectors (filtering vacuum fluctuations).*

In the general PCSFT-framework the randomness of the initial conditions has to be completed by taking into account fluctuations of vacuum (to obtain a consistent PCSFT which works both for one particle system and biparticle system). In our model the background field (vacuum fluctuations) is of the white noise type. It is a Gaussian random field with zero average and the covariance operator

$$D_{\text{background}} = \varepsilon I, \ \varepsilon > 0.$$

It is a stationary field, so its distribution does not change with time.

Consider (by using the QM-language) a quantum system in the mixed state ρ_0. It determines the prequantum random field $\phi_0 \equiv \phi_0(x, \omega)$ with the covariance operator

$$\tilde{D}(t_0) = \rho_0 + \epsilon I.$$

The value of $\varepsilon > 0$ is not determined by PCSFT, but it could not be too small for a purely mathematical reason.[16] (Hence QM is a theory of filtration of strong noise.) Now consider the solution $\phi(t)$ of the Schrödinger equation (13), (14) with the initial condition ϕ_0. Its covariance operator can be easily found:

$$\tilde{D}(t) = D(t) + \varepsilon I,$$

where $D(t)$ is the covariance operator of the process in the absence of the background field, $D(t) = \rho(t)$. Here $\rho(t)$ satisfies the QM-equation for evolution of the density operator, i.e, the von Neumann equation. Thus on the level of dynamics of the covariance operator the contribution of the background field is very simple: an additive shift. However, on the level of the field dynamics the presence of vacuum fluctuations changes the field behavior crucially.

Consider the prequantum random field $\phi_0(x, \omega)$ corresponding to a pure quantum state Ψ_0. Now (in the presence of the background field) the prequantum random field $\phi_0(x, \omega)$ is not concentrated on a one-dimensional subspace[h]

$$H_{\Psi_0} = \{\phi = c\Psi_0 : c \in \mathbf{C}\};$$

the vacuum fluctuations smash it over H.

In the canonical QM the background field of the white noise type is neglected; in fact, it is eliminated "by hand" in the process of detector calibration. And it is the right strategy for a formalism describing measurements on the random background. However, in an ontic model, i.e., a model of reality as it is, this background field should be taken into account. Neglecting it induces a rather mystical picture of quantum randomness.

We shall see that in the PCSFT-formalism the background field plays the fundamental role in the derivation of Heisenberg's uncertainty relation. Roughly speaking, Heisenberg's uncertainty is a consequence of vacuum fluctuations.[16]

2.8. Coupling between Schrödinger and Hamilton equations

Main message: *The Schrödinger equation is a complex form of the Hamilton equation for a special class of quadratic Hamilton functions on an infinite-dimensional phase space.*

We remark that the Schrödinger equation can be written as the system of Hamilton equations on the (infinite-dimensional) phase space $Q \times P$, where $Q = P$ is the real Hilbert space and $H = Q \oplus iP$ is the corresponding complex Hilbert space. The prequantum field $\phi(x) = q(x) + ip(x)$, where $q(x)$ and $p(x)$ are real-valued fields (or more generally, they take values in $\mathbf{R}^m$). Consider the Hamilton function

$$\mathcal{H}(q, p) = \frac{1}{2}\langle \widehat{\mathcal{H}}\phi, \phi \rangle, \tag{16}$$

or, in Dirac's notation,

$$\mathcal{H}(\phi) = \frac{1}{2}\langle \phi | \widehat{H} | \phi \rangle.$$

In PCSFT $\mathcal{H}(q, p)$ is the energy of the prequantum field $\phi(x) = q(x) + ip(x)$. The Schrödinger equation (13) can be written as the system

$$\dot{q} = \frac{\partial \mathcal{H}}{\partial p}, \ \dot{p} = -\frac{\partial \mathcal{H}}{\partial q}, \tag{17}$$

see Strochi.[47] From the PCSFT viewpoint, there is no reason (at least mathematical) to use only quadratic Hamiltonian functions. By considering non-quadratic Hamilton functions we obtain Hamilton systems connected with the nonlinear Schrödinger equation. PCSFT naturally induces a nonlinear extension of QM.[5]

---

[h]In the absence of vacuum fluctuations the covariance operator of the random field $\phi_\Psi(x, \omega)$ corresponding to a pure state Ψ is given by the orthogonal projector on Ψ, see (4); the corresponding Gaussian measure is concentrated on a one-dimensional subspace generated by Ψ. Of course, the latter is valid only for Gaussian prequantum fields.

2.9. Nonquadratic functionals of the prequantum field and violation of Born's rule

Main message: *Nonlinear, of order higher than two, contribution of the prequantum field induces violation of Born's rule.*

In principle, there is no reason to restrict PCSFT-variables to quadratic functionals of the prequantum fields, see (9). Let us consider an arbitrary smooth functional $f(\phi), \phi \in H$, which maps the field $\phi \equiv 0$ into zero, $f(0) = 0$. Let us also consider a random field $\phi = \phi(x, \omega)$ corresponding to a quantum density operator ρ. We can find the classical average

$$\langle f \rangle_\mu = \int_H f(\phi) d\mu(\phi), \tag{18}$$

where μ is the probability distribution of the random field. We showed[3] that this classical average can be approximated by the quantum average

$$\langle \widehat{A} \rangle_\rho = \text{Tr}\rho\widehat{A}, \tag{19}$$

where

$$\widehat{A} = f''(0)/2 \tag{20}$$

is the second derivative of the field functional $f(\phi)$ at the point $\phi = 0$ (divided by the factor 2 which arises from the Taylor expansion). If a Hilbert state space is finite-dimensional, then this is the usual second derivative. Its matrix (*Hessian*) is symmetric. If a Hilbert space is infinite-dimensional (of the L_2-type), then the derivatives are so-called variations. In the rigorous mathematical framework they are Frechet derivatives, that are used, e.g., in optimization theory. In the latter case the second (variation) derivative is given by a self-adjoint operator. This is the PCSFT-origin of the representation of quantum observables by self-adjoint operators.

Quantum observables are represented by self-adjoint operators, since they correspond to Hessians of smooth functionals of the prequantum field.

Thus the QM-formalism gives approximations of classical averages with respect to the prequantum random fields by approximating field-functionals $f(\phi)$ by the quadratic terms of their Taylor expansions.

If the functional $f(\phi)$ is linear, $f(\phi) = \langle \phi, y \rangle, y \in H$, then its QM-image, the second derivative, is equal to zero. Linear field effects are too weak and they are completely ignored by QM. However, such functionals and their correlations are well described by PCSFT. Observation of such effects can be the first step beyond QM.

2.10. Wave comeback – A solution too cheap?

Main message: *Physical space exists! Hence waves propagating in this space are basic entities of nature.*

It is well known that Einstein was not happy with Bohmian mechanics. He considered this solution of the problem of completion of QM as cheap. Recently Anton Zeilinger mentioned (in his lecture at the Växjö conference-2010, "Advances in Quantum theory") that QM may be not the last theory of micro processes and in future a new fundamental theory may be elaborated. And looking me in the face, he added that those who nowadays criticize QM and dream of a prequantum theory will be terrified by this coming new theory, by its complexity and extraordinarity. They will recall the old QM-formalism, i.e., the present one, with great pleasure, since it was so close to classical mechanics.

PCSFT is a comeback to classical field theory; roughly speaking, in the spirit of early Schrödinger and late Einstein: the Maxwell classical field theory is extended to "matter waves". Of course, this comeback is not the dream of the majority of those who nowadays are not afraid to speculate on prequantum models and criticize the Copenhagen QM. Nevertheless, I do not think that PCSFT is a cheap completion of the standard QM. I hope that, in contrast to Bohmian mechanics, Einstein might accept PCSFT as one of the possible ways beyond QM. In any event the Laplacian mechanistic determinism was totally excluded from PCSFT; reality became blurred in the sense of Schrödinger. This is reality of fields and not particles, but still reality.

3. Where is discreteness?

Prequantum variables $f_A(\phi) = \langle \widehat{A}\phi, \phi \rangle$ have continuous ranges of values. On the other hand, in QM some observables have discrete spectra. Thus, although PCSFT matches precisely probabilistic predictions of QM, it violates the *spectral postulate* of QM. How can one obtain discrete spectra?

The continuous field model supports the viewpoint that "ontic reality", i.e., reality as it is, is continuous. *Discreteness of some observable data is created by our macroscopic devices* which split a prequantum signal in a number of discrete channels. Take a polarization beam splitter (PBS). Consider first a classical signal. Suppose that PBS is oriented at an angle θ. Then the classical signal is split into two channels. We can label these channels as "polarization up", $S_\theta = +1$, and "polarization down", $S_\theta = -1$, (for θ-direction). The only problem is that the classical signal is present in both channels. Thus we cannot assign to a classical signal (even to a short pulse) a concrete value of S_θ. On the contrary, for a "quantum signal" (photon), detectors never click in both channels; we get either $S_\theta = +1$ or $S_\theta = -1$. This is a standard example of quantum discreteness.

The first comment of this common description is that the situation "no double clicks" never occurred in real experiments, this is the problem of experimental estimation of the coefficient of second order coherence $g^{(2)}(0)$. There are always *double clicks!* And they are many! They are partially discarded by using the *time window.* However, this is just a remark on the standard measurement procedure. The main point is that it is possible to produce discrete clicks even from a classical continuous signal by using threshold-type detectors. My PhD-student Guillaume Adenier performed numerical simulation for the threshold detection model of classical signals. He reproduced quantum probabilities of detection and even in a more complicated framework of classical bi-signals interacting with two PBSs oriented at angles θ_1 and θ_2 the EPR-Bohm correlations; Bell's inequality was violated.[48]

We also point to works of K. Michielsen and H. de Raedt, see, e.g., paper,[49][50] exploring impact of detectors to quantum measurements.

We remark that PCSFT differs essentially from a rather popular idea that the electromagnetic field is continuous, but matter is quantized. This viewpoint was stressed in the book of Lande[26]). PCSFT does not quantize even the matter, the latter also consists of continuous fields fluctuating on very fine space-time scales. These scales are not yet approachable. In future we expect to get a possibility to monitor these fields and not only their averaged images given by quantum particles. SED-like people do not expect this. It seems that only Albert Einstein might be happy with PCSFT.

Acknowledgments

This paper was written under partial support of the grant "Quantum Bio-Informatics" of Tokyo University of Science and the author would like to thank Masanori Ohya and Noboru Watanabe for collaboration and hospitality.

References

1. A. Einstein and L. Infeld, *Evolution of Physics: The Growth of Ideas from Early Concepts to Relativity and Quanta* (Simon and Schuster, New-York, 1961).
2. A. Fine, *The Shaky Game. Einstein Realism and the Quantum Theory* (Univ. Chicago Press, Chicago and London, 1996).
3. A. Khrennikov, A pre-quantum classical statistical model with infinite-dimensional phase space, *J. Phys. A: Math. Gen.* **38**, 9051 (2005).
4. A. Khrennikov, Prequantum classical statistical field theory: Complex representation, Hamilton-Schrödinger equation, and interpretation of stationary states, *Found. Phys. Lett.* **18**, 637 (2006).
5. A. Khrennikov, Nonlinear Schrödinger equations from prequantum classical statistical field theory, *Physics Lett.* A **357**, 171 (2006).
6. A. Khrennikov, Quantum mechanics from time scaling and random fluctuations at the "quick time scale", *Nuovo Cimento B* **121**, 505 (2006).
7. A. Khrennikov, Quantum Randomness as a Result of Random Fluctuations at the Planck Time Scale? *Int. J. Theor. Phys.* **47**, 114–124 (2008).
8. A. Khrennikov, Representation of quantum field theory as classical statistical mechanics for field functionals, *Doklady Mathematics* **74**, 758–761 (2006).

9. A. Khrennikov, Prequantum classical statistical field theory: Time scale of fluctuations, *Doklady Mathematics* **75**, 456–459 (2007).

10. A. Khrennikov, Quantum mechanics as an approximation of statistical mechanics for classical fields, *Rep. Math. Phys.* **60**, 453–484 (2007).

11. A. Khrennikov, Quantum mechanics as the quadratic Taylor approximation of classical mechanics: The finite-dimensional case, *J. Theor. Math. Phys.* **152**, 1111–1121 (2007).

12. A. Khrennikov, To quantum averages through asymptotic expansion of classical averages on infinite-dimensional space, *J. Math. Phys.* **48**, Art. No. 013512 (2007).

13. A. Khrennikov, Quantum probabilities from detection theory for classical random fields, *Fluctuations and Noise Lett.* **8**, L393–L400 (2008).

14. A. Khrennikov, Born's rule from classical random fields, *Phys. Lett.* A **372**, 6588–6592 (2008).

15. A. Khrennikov, Quantum-like model for classical random electromagnetic field, *J. Modern Optics* **55**, 2257–2267 (2008).

16. A. Khrennikov, Detection model based on representation of quantum particles by classical random fields: Born's rule and beyond, *Found. Phys.* **39**, 997–1022 (2009).

17. A. Khrennikov, Entanglement's dynamics from classical stochastic process, *EPL* **88**, 40005. 1–6 (2009).

18. A. Khrennikov, M. Ohya, and N. Watanabe, Classical signal model for quantum channels, *J. Russian Laser Research* **31**, 462 (2010).

19. A. Khrennikov, Subquantum nonlocal correlations induced by the background random field, *Physica Scripta* **84**, Article Number: 045014 (2011).

20. A. Khrennikov, A classical field theory comeback? The classical field viewpoint on triparticle entanglement, *Physica Scripta* **T143**, Article Number: 014013 (2011).

21. A. Khrennikov, Violation of Bell's inequality by correlations of classical random signals. In: FQMT'11: Frontiers of Quantum and Mesoscopic Thermodynamics (Prague, Czech Republic, 25–30 July 2011); T. M. Nieuwenhuizen, P. D, Keefe and V. Spicka (eds.) *Physica Scripta* **T151**, Article Number: 014003 (2012).

22. A. Khrennikov, *Contextual Approach to Quantum Formalism* (Springer, Berlin-Heidelberg-New York, 2009).

23. M. Ohya and N. Watanabe, *Japan J. Industrial and Appl. Math.* **3**, 197 (1986).

24. W. E. Lamb, *The Interpretation of Quantum Mechanics* (Rinton Press, Inc., Princeton, 2001).

25. N. Bohr, H. A. Kramers and J. C. Slater, *Zeitschrift für Physik* **24**, 69 (1920).

26. A. Lande, *New Foundations of Quantum Mechanics* (Cambridge Univ. Press, Cambridge, 1965).

27. A. Einstein, On the theory of light quanta and the question of localization of electromagnetic energy, *Archives des sceiences physiques et naturelles* **29**, 525–528 (1910).

28. W. E. Lamb and M. O. Scully, in *Polarization, Matter and Radiation. Jubilee volume in honour of Alfred Kasiler* (Press of University de France, Paris, 1969), p. 363.

29. N. Mukunda, *Resonance* **5**, 35 (2000).

30. L. De la Pena and A. M. Cetto, *The Quantum Dice: An Introduction to Stochastic Electrodynamics* (Kluwer, Dordrecht, 1996).

31. L. De la Pena, *Found. Phys.* **12**, 1017 (1982).

32. L. De la Pena, *J. Math. Phys.* **10**, 1620 (1969).

33. L. De la Pena, A. M. Cetto, *Phys. Rev. D* **3**, 795 (1971).

34. L. De la Pena and A. M. Cetto, in *Quantum Theory: Reconsideration of Foundations-3*, eds. G. Adenier, A. Yu. Khrennikov and Th. M. Nieuwenhuizen (American Institute of Physics, Ser. Conference Proceedings Melville, NY, 2006), **810**, p. 131.

35. Th. M. Nieuwenhuizen, in *Quantum Theory: Reconsideration of Foundations-3*, eds.

G. Adenier, A. Yu. Khrennikov and Th. M. Nieuwenhuizen (American Institute of Physics, Melville, NY, 2006), **810**, p. 198.

36. S. D' Agostino, in *E. Schrödinger: Philosophy and the Birth of Quantum Mechanics*, eds. M. Bitbol and O. Darrigol (Editions Frontieres, Gif-sur-Yvette, 1992), p. 123.

37. M. Bitbol and O. Darrigol (eds.), *E. Schrödinger: Philosophy and the Birth of Quantum Mechanics* (Editions Frontieres, Gif-sur-Yvette, 1992).

38. M. Lockwood, in *E. Schrödinger: Philosophy and the Birth of Quantum Mechanics*, eds. M. Bitbol and O. Darrigol (Editions Frontieres, Gif-sur-Yvette, 1992), p. 380.

39. N. Bohr, *The philosophical writings of Niels Bohr* (Ox Bow Press, Woodbridge, Conn., 1987).

40. A. Plotnitsky, in *Foundations of probability and physics-3*, eds. G. Adenier, A. Khrennikov (American Institute of Physics, Melville, NY, 2005), **750**, p. 388.

41. A. Plotnitsky, *Reading Bohr: Physics and Philosophy* (Springer, Dordrecht, 2006).

42. A. Plotnitsky, *Epistemology and Probability: Bohr, Heisenberg, Schödinger, and the Nature of Quantum-Theoretical Thinking* (Springer, Heidelberg-Berlin-New York, 2009).

43. A. Pais, *Subtle Is the Lord: The Science and the Life of Albert Einstein* (Oxford Univ. Press, Oxford, 1983).

44. D. Howard and J. Stachel, *Einstein: The Formative Years, 1879–190* (Birkhöuser, Boston, 1994).

45. P. A. Schilpp, *Albert Einstein: Autobiographical Notes* (Open Court Publ. Company, Chicago, 1979).

46. J. Von Neuman, *Mathematical Foundations of Quantum Mechanics* (Princeton University Press, Princeton, 1955).

47. F. Strocchi, Complex coordinates and quantum mechanics. *Rev. Mod. Phys.* **38**, 36–40 (1966).

48. G. Adenier, *Local Realist Approach and Numerical Simulation of Nonclassical Experiments in Quantum Mechanics* (Växjö Univ. Press, Växjö, 2008).

49. K. Michielsen and H. De Raedt, *Quantum Theory: Reconsiderations of Foundations-5*, ed. A. Khrennikov (American Institute of Physics, Melville, NY, 2010), **1232**, p. 27.

50. K. Michielsen, F. Jin, M. Delina and H. De Raedt, Event-by-event simulation of nonclassical effects in two-photon interference experiments In: FQMT'11: Frontiers of Quantum and Mesoscopic Thermodynamics (Prague, Czech Republic, 25–30 July 2011); T. M. Nieuwenhuizen, P. D, Keefe and V. Spicka (eds.) *Physica Scripta* **T151**, Article Number: 014005 (2012).

Chapter 10

Emergence of Quantum Mechanics from a Sub-Quantum Statistical Mechanics

Gerhard Grössing*

Austrian Institute for Nonlinear Studies
Akademiehof, Friedrichstrasse 10, 1010 Vienna, Austria

A research program within the scope of theories on "Emergent Quantum Mechanics" is presented, which has gained some momentum in recent years. Via the modeling of a quantum system as a non-equilibrium steady-state maintained by a permanent throughput of energy from the zero-point vacuum, the quantum is considered as an *emergent system*. We implement a specific "bouncer-walker" model in the context of an assumed sub-quantum statistical physics, in analogy to the results of experiments by Couder and Fort on a classical wave-particle duality. We can thus give an explanation of various quantum mechanical features and results on the basis of a "21st century classical physics", such as the appearance of Planck's constant, the Schrödinger equation, etc. An essential result is given by the proof that averaged particle trajectories' behaviors correspond to a specific type of anomalous diffusion termed "ballistic" diffusion on a sub-quantum level.

It is further demonstrated both analytically and with the aid of computer simulations that our model provides explanations for various quantum effects such as double-slit or n-slit interference. We show the averaged trajectories emerging from our model to be identical to Bohmian trajectories, albeit without the need to invoke complex wave functions or any other quantum mechanical tool. Finally, the model provides new insights into the origins of entanglement, and, in particular, into the phenomenon of a "systemic" nonlocality.

1. Introduction

After many attempts to frame wave-particle duality on a single "basic" explanatory level, like on pure formalism (Heisenberg,...), pure wave mechanics (de Broglie,...), or pure particle physics (Feynman,...), for example, it is instructive to ask the following question: is perhaps the quantum a more complex *dynamical* phenomenon? One motivation for raising this question comes from the fact that we are presently witnessing a historical change with respect to wave-particle duality in double-slit interference. With regard to the latter, the standard attitude during practically the whole of the 20th century can be characterized by Richard Feynman's famous claim that it is "... a phenomenon which is impossible, absolutely impossible to explain

*ains@chello.at

in any classical way." In the present, however, we are facing the beautiful experiments by Yves Couder's group in Paris which do show wave-particle duality in a
completely classical system, i.e. with the aid of bouncers/walkers on an oscillating
fluid. (For a first impression, see the video with Yves Couder.[1])

In fact, with our approach we have in a series of papers obtained from a "classical" dynamical model several essential elements of quantum theory.[2–8] They derive
from the assumption that a "particle" of energy $E = \hbar\omega$ is actually an oscillator of
angular frequency ω phase-locked with the zero-point oscillations of the surrounding
environment, the latter of which containing both regular and fluctuating components and being constrained by the boundary conditions of the experimental setup
via the buildup and maintenance of standing waves. The "particle" in this approach
is an off-equilibrium steady-state maintained by the throughput of zero-point energy
from its "vacuum" surroundings. In other words, in our model a quantum *emerges*
from the synchronized dynamical coupling between a bouncer and its wave-like
environment. This is in close analogy to the bouncing/walking droplets in the experiments of Couder's group,[9–12] which in many respects can serve as a classical
prototype guiding our intuition. Our research program thus pertains to the scope
of theories on "Emergent Quantum Mechanics". (For the proceedings of a first
international conference exclusively devoted to this topic, see Grössing (2012).[13]
For their original models, see in particular the papers by Adler, Elze, Ord, Grössing *et al.*, Cetto *et al.*, Hiley, de Gosson, Bacciagaluppi, Budiyono, Schuch, Faber,
Hofer, 't Hooft, Khrennikov *et al.*, Wetterich, and Burinskii.) We have recently applied our model to the case of interference at a double slit,[7] thereby obtaining the
exact quantum mechanical probability distributions on a screen behind the double
slit, the average particle trajectories (which because of the averaging are shown to
be identical to the Bohmian ones), and the involved probability density currents.

Moreover, already some decades ago, Aharonov *et al.*[14,15] investigated the problem of double-slit interference "from a single particle perspective". Their explanation is based on what they call *dynamical nonlocality*, and they therewith answer
the question of how a particle going through one slit can "know" about the state
of the other slit (i.e. being open or closed, for example). This type of "dynamical"
nonlocality may lead to (causality preserving) changes in probability distributions
and is thus distinguished from the *kinematic nonlocality* implicit in many quantum
correlations, which, however, does not cause any changes in probability distributions. Section 3 of the present review extends the applicability also of our model
to a such a "dynamical" scenario, which in our terminology rather relates to a
"systemic" nonlocality, as shall be explicated below.

This paper is organized as follows. In Section 2, some main results of our sub-
quantum approach to quantum mechanics are presented. We begin by constructing
from classical physics a model that explains the quantum mechanical dispersion
of a Gaussian wave packet. As a consequence, we obtain the total velocity field
for the current emerging from a Gaussian slit, which can easily be extended to a

two-slit (and in fact to any n-slit) system, thereby also providing an explanation of the famous interference effects. With the thus obtained velocity field, one can by integration also obtain an action function that generalizes the ordinary classical one to the case that is relevant for the reproduction of the quantum results. Effectively, this is achieved by the appearance of an additional kinetic energy term, which we associate with fluxes of "heat" in the vacuum, i.e. sub-quantum currents based on the existence of the zero-point energy field. Finally, it is shown how the Schrödinger equation can be derived in a straightforward manner from the said generalized action function, or the corresponding Lagrangian, respectively. In Section 3, then, we deal with the above-mentioned problem of interference from the single particle perspective, obtaining results which are in agreement with those presented by Aharonov *et al.*,[14,15] but differing in interpretation. Finally, in Section 4 computer simulations are presented to show some examples of how our sub-quantum approach reproduces results of ordinary quantum mechanics, but eventually may go beyond these.

2. The quantum as an emergent system: A sub-quantum approach to quantum mechanics

We transfer an insight from the Couder experiments into our modeling of quantum systems and assume that the waves are a space-filling phenomenon involving the whole experimental setup. Thus, one can imagine a partial decoupling of the physics of waves and particles in that the latter still may be "guided" through said "landscape", but the former may influence other regions of the "landscape" by providing specific phase information independently of the propagation of the particle. This is why a remote change in the experimental setup, when mediated to the particle via de- and/or re-construction of standing waves, can amount to a nonlocal effect on a particle via the thus modified guiding landscape.

In earlier work[4] we presented a model for the classical explanation of the quantum mechanical dispersion of a free Gaussian wave packet. In accordance with the classical model, we now relate it more directly to a "double solution" analogy gleaned from Couder and Fort.[12] These authors used the double solution ansatz to describe the behaviors of their "bouncer"- (or "walker"-) droplets: on an individual level, one observes particles surrounded by circular waves they emit through the phase-coupling with an oscillating bath, which provides, on a statistical level, the emergent outcome in close analogy to quantum mechanical behavior (like, e.g., diffraction or double-slit interference). Originally, the expression of a "double solution" refers to an early idea of de Broglie[16] to model quantum behavior by a two-fold process, i.e. by the movement of a hypothetical point-like "singularity solution" of the Schrödinger equation, and by the evolution of the usual wavefunction that would provide the empirically confirmed statistical predictions. Now, it is interesting to observe that one can construct various forms of classical analogies to the quantum mechanical Gaussian dispersion,[17] and one of them even can be related to the double solution idea.

To establish correlations on a statistical level between individual uncorrelated particle positions x and momenta p, respectively, one considers a solution of the free Liouville equation providing a certain phase-space distribution $f(x, p, t)$. This distribution shows the emergence of correlations between x and p from an initially uncorrelated product function of non-spreading ("classical") Gaussian position distributions as well as momentum distributions. The motivation for their introduction comes exactly from what one observes in the Couder experiments. In an idealized scenario, we assume that at each point x an unbiased emission of momentum fluctuations π_0 in all possible directions takes place, thus mimicking (in a two-dimensional scenario) the circular waves emitted from the "particle as bouncer". If we compare the typical frequency of the bouncers in the Couder experiments (i.e. roughly 10^2 Hz) with that of an electron, for example (i.e. roughly 10^{20} Hz), we see that a "continuum ansatz" is very pragmatical, particularly if we are interested in statistical averages over a long series of experimental runs.

Thus, one can construct said phase-space distribution, with σ_0 being the initial x–space standard deviation, i.e. $\sigma_0 = \sigma(t = 0)$, and $\pi_0 := mu_0$ the momentum standard deviation, such that

$$f(x, p, t) = \frac{1}{2\pi\sigma_0 m u_0} \exp\left\{-\frac{(x - pt/m)^2}{2\sigma_0^2}\right\} \exp\left\{-\frac{p^2}{2m^2 u_0^2}\right\}. \tag{1}$$

Now, the above-mentioned correlations between x and p emerge when one considers the probability density in x–space. Integration over p provides

$$P(x, t) = \int f \, dp = \frac{1}{\sqrt{2\pi}\sigma} \exp\left\{-\frac{x^2}{2\sigma^2}\right\}, \tag{2}$$

with the standard deviation at time t given by

$$\sigma^2 = \sigma_0^2 + u_0^2 t^2. \tag{3}$$

In other words, the distribution (2) with property (3) describes a spreading Gaussian which is obtained from a continuous set of classical, i.e. non-spreading, Gaussian position distributions of particles whose associated momentum fluctuations also have non-spreading Gaussian distributions. One thus obtains the exact quantum mechanical dispersion formula for a Gaussian, as we have obtained also previously from a different variant of our classical ansatz. For confirmation with respect to the latter (diffusion-based) model,[4,5] we consider with (2) the usual definition of the "osmotic" velocity field, which in this case yields $u = -D\frac{\nabla P}{P} = \frac{xD}{\sigma^2}$. One then obtains, with bars denoting averages over fluctuations and positions, i.e. with $\overline{x^2} = \int P x^2 dx = \sigma^2$,

$$\overline{u^2} = D^2 \overline{\left(\frac{\nabla P}{P}\right)^2} = D^2 \int P \left(\frac{\nabla P}{P}\right)^2 dx = \frac{D^2}{\sigma^2}, \quad \text{and thus also} \quad u_0 = \frac{D}{\sigma_0}, \tag{4}$$

so that one can rewrite Eq. (3) in the more familiar form

$$\sigma^2 = \sigma_0^2 \left(1 + \frac{D^2 t^2}{\sigma_0^4}\right). \tag{5}$$

Note also that by using the Einstein relation $D = \hbar/2m = h/4\pi m$ the norm in (1) thus becomes the invariant expression (reflecting the "exact uncertainty relation"[18])

$$\frac{1}{2\pi\sigma_0 m u_0} = \frac{1}{2\pi m D} = \frac{2}{h}. \tag{6}$$

Following from (5), in Grössing *et al.*[4,7] we obtained for "smoothed-out" trajectories (i.e. averaged over a very large number of Brownian motions) a sum over a deterministic and a fluctuations term, respectively, for the motion in the x–direction

$$x_{\text{tot}}(t) = vt + x(t) = vt + x(0)\frac{\sigma}{\sigma_0} = vt + x(0)\sqrt{1 + \frac{D^2 t^2}{\sigma_0^4}}. \tag{7}$$

Thus one classically obtains with $x(t) = x_{\text{tot}}(t) - vt$ the *average velocity field* of a Gaussian wave packet as

$$v_{\text{tot}}(t) = v(t) + \frac{dx(t)}{dt} = v(t) + x(t)\frac{u_0^2 t}{\sigma^2}. \tag{8}$$

For the particular choice of $x(t) = \sigma$, taken from (5), and with $\frac{\sigma}{t} \to u_0$ for $t \to \infty$, one obtains for large t that $v_{\text{tot}}(t) = v(t) + u_0$.

Note that Eqs. (7) and (8) are derived solely from statistical physics. Still, they are in full accordance with quantum theory, and in particular with Bohmian trajectories.[17] Note also that one can rewrite Eq. (5) such that it appears like a linear-in-time formula for Brownian motion,

$$\overline{x^2} = \overline{x^2(0)} + D_{\text{t}}\, t, \tag{9}$$

where a time dependent diffusivity

$$D_{\text{t}} = u_0^2\, t = \frac{\hbar^2}{4m^2\sigma_0^2}\, t \tag{10}$$

characterizes Eq. (9) as *ballistic diffusion*. This makes it possible to simulate the dispersion of a Gaussian wave packet on a computer by simply employing coupled map lattices for classical diffusion, with the diffusivity given by Eq. (10). (For detailed discussions, see Grössing *et al.*[4,5] and the last Section of this paper.)

Moreover, one can easily extend this scheme to more than one slit, like, for example, to explain interference effects at the double slit.[7,8] For this, we chose similar initial situations as in Holland,[17] i.e. electrons (represented by plane waves in the forward y–direction) from a source passing through "soft-edged" slits 1 and 2 in a barrier (located along the x–axis) and recorded at a screen. In our model, we therefore note two Gaussians representing the totality of the effectively "heated-up" path excitation field, one for slit 1 and one for slit 2, whose centers have the distances $+X$ and $-X$ from the plane spanned by the source and the center of the barrier along the y–axis, respectively.

With the total amplitude R of two coherent waves with (suitably normalized) amplitudes $R_i = \sqrt{P_i}$, and the local phases φ_i, $i = 1$ or 2, one has, according to classical textbook wisdom, the *averaged total intensity*

$$P_{\text{tot}} := R^2 = R_1^2 + R_2^2 + 2R_1R_2 \cos\varphi_{12} = P_1 + P_2 + 2\sqrt{P_1 P_2} \cos\varphi_{12}, \qquad (11)$$

where φ_{12} is the relative phase $\varphi_{12} = \varphi_1 - \varphi_2 = (\mathbf{k}_1 - \mathbf{k}_2) \cdot \mathbf{r}$. Note that φ_{12} enters Eq. (11) only via the cosine function, such that, e.g., even if the total wave numbers (and thus also the total momenta) $\mathbf{k}_i$ were of vastly different size, the cosine effectively makes Eq. (11) independent of said sizes, but dependent only on an angle modulo 2π. This will turn out as essential for our discussion further below.

The x–components of the centroids' motions from the two alternative slits 1 and 2, respectively, are given by the "particle" velocity components

$$v_x = \pm\frac{\hbar}{m} k_x, \qquad (12)$$

respectively, such that the relative group velocity of the Gaussians spreading into each other is given by $\Delta v_x = 2v_x$. However, in order to calculate the phase difference φ_{12} descriptive of the interference term of the intensity distribution (11), one must take into account the total momenta involved, i.e. one must also include the wave packet dispersion as described above. Thus, one obtains with the displacement $\pm x\,(t) = \mp(X + v_x t)$ in Eq. (8) the total relative velocity of the two Gaussians as

$$\Delta v_{\text{tot},x} = 2\left[v_x - (X + v_x t)\frac{u_0^2 t}{\sigma^2}\right]. \qquad (13)$$

Therefore, the total phase difference between the two possible paths 1 and 2 (i.e. through either slit) becomes

$$\varphi_{12} = \frac{1}{\hbar}(m\Delta v_{\text{tot},x}\,x) = 2mv_x\frac{x}{\hbar} - (X + v_x t)x\frac{1}{D}\frac{u_0^2 t}{\sigma^2}. \qquad (14)$$

In earlier papers[2–4] we have shown that, apart from the ordinary particle current $\mathbf{J}(\mathbf{x},t) = P(\mathbf{x},t)\mathbf{v}$, we are now dealing with two additional, yet opposing, currents $\mathbf{J}_u = P(\mathbf{x},t)\mathbf{u}$, which are on average orthogonal to $\mathbf{J}$, and which are the emergent outcome from the presence of numerous momentum fluctuations, and the corresponding velocities,

$$\mathbf{u}_\pm = \mp\frac{\hbar}{2m}\frac{\nabla P}{P}. \qquad (15)$$

We denote with $\mathbf{u}_+$ and $\mathbf{u}_-$, respectively, the two opposing tendencies of the diffusion process. Moreover, one can also take the averages over fluctuations and positions to obtain a "smoothed-out" *average velocity field*,[2–5] i.e.

$$\overline{\mathbf{u}(\mathbf{x},t)} = \int P\mathbf{u}(\mathbf{x},t)\,\mathrm{d}^n x. \qquad (16)$$

In effect, the combined presence of the velocity fields **u** and **v** can be denoted as a *path excitation field*: via diffusion, the bouncer in its interaction with already existing wave-like excitations of the environment creates an "agitated", or "heated-up", thermal landscape, which can also be pictured by interacting wave configurations all along between source and detector of an experimental setup. Recall that our prototype of a "walking bouncer", i.e. from the experiments of Couder's group, is always driven by its interactions with a superposition of waves emitted at the points it visited in the past. Couder *et al.* denote this superposition of in-phase waves the "path memory" of the bouncer.[19] This implies, however, that the bouncers at the points visited in "the present" necessarily create new wave configurations which will form the basis of a "path memory" in the future. In other words, the wave configurations of the past determine the bouncer's path in the present, whereas its bounces in the present co-determine the wave configurations, and thus the probability density distribution, at any of the possible locations it will visit in the future. Therefore, by calling the latter configurations a *path excitation field*, we point to our model's physical meaning of the probability density distribution: its time evolution is to be understood as the totality of all sub-quantum currents, which may also be described as a "heated-up" thermal field.

For illustration, let us now consider a single, classical "particle" ("bouncer") following the propagation of a set of waves of equal amplitude R_i, each representing one of i possible alternatives according to our principle of path excitation, and focus on the specific role of the velocity fields. To describe the required details, each path i be occupied by a Gaussian wave packet with a "forward" momentum $\mathbf{p}_i = \hbar \mathbf{k}_i = m \mathbf{v}_i$. Moreover, due to the stochastic process of path excitation, the latter has to be represented also by a large number N of consecutive Brownian shifts as in Eq. (15), which on average, for stretches of free particle propagation $\mathbf{v}_i$, are orthogonal to $\mathbf{v}_i$. Defining an average total velocity (with indices $i = 1$ or 2 referring to the two slits, and with $+$ and $-$ referring to the right and the left from the average direction of $\mathbf{v}_i$, respectively)

$$\overline{\mathbf{v}}_{\text{tot},i} := \overline{\mathbf{v}}_i + \overline{\mathbf{u}}_{i+} + \overline{\mathbf{u}}_{i-}, \tag{17}$$

with the bars here (and further on, if not declared otherwise) denoting averages only over the spatial directions, we consider two Gaussian probability density distributions, $P_1 = R_1^2$ and $P_2 = R_2^2$, respectively. Generally, with the total amplitude R of two coherent waves with (suitably normalized) amplitudes $R_i = \sqrt{P_i}$, and the local phases φ_i, $i = 1$ or 2, one has as usual that

$$R = R_1 \cos(\omega t + \varphi_1) + R_2 \cos(\omega t + \varphi_2), \tag{18}$$

which, when squared and averaged, provides the famous formula for the intensity of the interference pattern (11). Introducing an arbitrarily chosen unit vector $\hat{\mathbf{n}}$, one may also define $\cos(\omega t + \varphi_i(\mathbf{x})) = \hat{\mathbf{n}} \cdot \hat{\mathbf{k}}_i(\mathbf{x}, t)$, such that along with the system's evolution, the emergent outcome of the time evolution of $R(\mathbf{x}, t)$ as a characteristic of our path excitation field can analogously be assumed as

$$R\left(\mathbf{x},t\right) = \hat{\mathbf{n}} \cdot \left(R_1\hat{\mathbf{k}}_1\left(\mathbf{x},t\right) + R_2\hat{\mathbf{k}}_2\left(\mathbf{x},t\right)\right). \tag{19}$$

Thus, with regard to the total amplitude R_{tot} in the double-slit case, one obtains from (19) with the appropriate normalization that

$$R_{\text{tot}}^2 = \left(R_1\overline{\hat{\mathbf{v}}}_{\text{tot},1} + R_2\overline{\hat{\mathbf{v}}}_{\text{tot},2}\right)^2. \tag{20}$$

With $P_{\text{tot}} = R_{\text{tot}}^2$, one can calculate the time development of the path excitation field, i.e. the spreading out of the total probability density in the form of the total average current $\overline{\mathbf{J}}_{\text{tot}} = P_{\text{tot}}\overline{\mathbf{v}}_{\text{tot}}$. After a few calculational steps this provides, similarly to Grössing $et\,al.$ (2012)[7] (but now with slightly different labellings which apply more generally), with the assumed orthogonality $\overline{\mathbf{v}}_i \cdot \overline{\mathbf{u}}_i = 0$ for free particle propagation ($i = 1$ or 2) that

$$\overline{\mathbf{J}}_{\text{tot}} = P_1\overline{\mathbf{v}}_1 + P_2\overline{\mathbf{v}}_2 + \sqrt{P_1 P_2}\left(\overline{\mathbf{v}}_1 + \overline{\mathbf{v}}_2\right)\cos\varphi_{12} + \sqrt{P_1 P_2}\left(\overline{\mathbf{u}}_1 - \overline{\mathbf{u}}_2\right)\sin\varphi_{12}. \tag{21}$$

An alternative, more detailed account of Eq. (21) and its extension to n slits has just been published by our group.[20] Note that Eq. (21), upon the identification of $\overline{\mathbf{u}}_i = -\frac{\hbar}{m}\frac{\nabla R_i}{R_i}$ from Eq. (15) and with $P_i = R_i^2$, turns out to be in perfect agreement with a comparable "Bohmian" derivation.[17,21] In fact, with $\overline{\mathbf{v}}_i = \nabla S_i/m$, one can rewrite (21) as

$$\overline{\mathbf{J}}_{\text{tot}} = R_1^2\frac{\nabla S_1}{m} + R_2^2\frac{\nabla S_2}{m} \tag{22}$$
$$+ R_1 R_2\left(\frac{\nabla S_1}{m} + \frac{\nabla S_2}{m}\right)\cos\varphi_{12} + \frac{\hbar}{m}\left(R_1\nabla R_2 - R_2\nabla R_1\right)\sin\varphi_{12}.$$

The formula for the averaged particle trajectories, then, simply results from

$$\overline{\mathbf{v}}_{\text{tot}} = \frac{\overline{\mathbf{J}}_{\text{tot}}}{P_{\text{tot}}}. \tag{23}$$

Although we have obtained the usual quantum mechanical results, we have so far not used the quantum mechanical formalism in any way. However, upon employment of the Madelung transformation for each path j ($j = 1$ or 2),

$$\Psi_j = R_j e^{iS_j/\hbar}, \tag{24}$$

and thus $P_j = R_j^2 = |\Psi_j|^2 = \Psi_j^*\Psi_j$, with the definitions (15) and $\overline{v_j} := \nabla S_j/m$, $\varphi_{12} = (S_1 - S_2)/\hbar$, and recalling the usual trigonometric identities such as $\cos\varphi = \frac{1}{2}\left(e^{i\varphi} + e^{-i\varphi}\right)$, etc., one can rewrite the total average current (21) immediately as

$$\overline{\mathbf{J}}_{\text{tot}} = P_{\text{tot}}\overline{\mathbf{v}}_{\text{tot}}$$
$$= (\Psi_1 + \Psi_2)^*(\Psi_1 + \Psi_2)\frac{1}{2}\left[\frac{1}{m}\left(-i\hbar\frac{\nabla(\Psi_1 + \Psi_2)}{(\Psi_1 + \Psi_2)}\right) + \frac{1}{m}\left(i\hbar\frac{\nabla(\Psi_1 + \Psi_2)^*}{(\Psi_1 + \Psi_2)^*}\right)\right]$$
$$= -\frac{i\hbar}{2m}\left[\Psi^*\nabla\Psi - \Psi\nabla\Psi^*\right] = \frac{1}{m}\text{Re}\left\{\Psi^*(-i\hbar\nabla)\Psi\right\}, \tag{25}$$

where $P_{\text{tot}} = |\Psi_1 + \Psi_2|^2 =: |\Psi|^2$. The last two expressions of (25) are the exact well-known formulations of the quantum mechanical probability current, here obtained without any quantum mechanics, but just by a re-formulation of (21). In fact, it is a simple exercise to insert the wave functions (24) into (25) to re-obtain (21).

Moreover, having thus obtained a "bridge" to the quantum mechanical formalism, one can now show how the Schrödinger equation not only complies with our classical ansatz, but actually can be derived from it. Two different ways of such a corresponding derivation have already been published by the present author (i.e. one on more general grounds,[22] the other within a more concrete model based on nonequilibrium thermodynamics[2]). Here, however, we just start with a result from the construction of the classical Gaussian with its dispersion mimicking the quantum mechanical one, i.e. more concretely, with the resulting velocity field $v_{\text{tot}}(t)$ from Eq. (8). Integrating the latter, and with $\xi(t) := x - vt$ describing the location of a particle in a Gaussian probability density distribution $P(x,t) = \frac{1}{\sqrt{2\pi}\sigma} \exp\left\{-\frac{(x-vt)^2}{2\sigma^2}\right\}$, we find an expression for the action function as

$$
\begin{aligned}
S &= \int m v_{\text{tot}}(t)\,\mathrm{d}x - \int E\,\mathrm{d}t \\
&= mvx + \frac{mu_0^2}{2}\left[\frac{\xi(t)}{\sigma(t)}\right]^2 t - Et = mvx + \frac{mu_0^2}{2}\left[\frac{\xi(0)}{\sigma_0}\right]^2 t - Et,
\end{aligned} \qquad (26)
$$

with E being the system's total energy. Note that because it generally holds that $\frac{\xi(t)}{\sigma(t)} = \frac{\xi(0)}{\sigma_0}$, the physics contained in (26) for the free particle case is essentially determined already by the initial conditions, i.e. where in the Gaussian the particle is initially located. Note also that the new term, as opposed to "ordinary" classical physics, is given by a kinetic energy, which represents a thermal fluctuation field with the kinetic temperature $kT = mu_0^2\left[\frac{\xi(t)}{\sigma(t)}\right]^2$ of what we have termed the *path excitation field*. Now, as with the above-mentioned Gaussian one has $u(x,t) = -D\frac{\nabla P}{P} = \frac{\xi(t)D}{\sigma^2} = \frac{\xi(t)\sigma_0}{\sigma^2}u_0$, one can rewrite the new kinetic energy term in Eq. (26) as

$$
\frac{mu_0^2}{2}\left[\frac{\xi(t)}{\sigma(t)}\right]^2 t = \frac{mu^2}{2}\left[\frac{\sigma(t)}{\sigma_0}\right]^2 t =: \frac{m\tilde{u}^2}{2}t, \qquad (27)
$$

where $\tilde{u} = u\frac{\sigma}{\sigma_0}$. (This is therefore in complete accordance with the starting assumption in Grössing (2008).[2]) Moreover, upon averaging over fluctuations and space, one obtains that

$$
\int P \frac{mu_0^2}{2}\left[\frac{\xi(t)}{\sigma(t)}\right]\,\mathrm{d}x = \frac{mu_0^2}{2} \equiv \frac{\hbar\omega}{2}, \qquad (28)
$$

where the latter equation describes the identity of the averaged kinetic energy term with the zero-point energy as shown in Grössing (2009).[3]

In more general terms, i.e. independently from the particular choice of P as a Gaussian distribution density, with our expression for the momentum fluctuation

$\delta \mathbf{p} = m\mathbf{u}$ as

$$\delta \mathbf{p}(\mathbf{x}, t) = \nabla(\delta S(\mathbf{x}, t)) = -\frac{\hbar}{2}\frac{\nabla P(\mathbf{x}, t)}{P(\mathbf{x}, t)}, \qquad (29)$$

we can write our additional kinetic energy term for one particle as

$$\delta E_{\text{kin}} = \frac{1}{2m}\nabla(\delta S \cdot \nabla(\delta S)) = \frac{1}{2m}\left(\frac{\hbar}{2}\frac{\nabla P}{P}\right)^2. \qquad (30)$$

Thus, writing down a classical action integral for n particles, and including this new term for each of them, yields (with Lagrangian L and external potential V)

$$A = \int L \, \mathrm{d}^m x \, \mathrm{d}t$$
$$= \int P \left[\frac{\partial S}{\partial t} + \sum_{i=1}^{n}\frac{1}{2m_i}\nabla_i S \cdot \nabla_i S + \sum_{i=1}^{n}\frac{1}{2m_i}\left(\frac{\hbar}{2}\frac{\nabla_i P}{P}\right)^2 + V\right] \mathrm{d}^m x \, \mathrm{d}t, \quad (31)$$

where $P = P(\mathbf{x}_1, \mathbf{x}_2, \ldots, \mathbf{x}_n, t)$. Using again the Madelung transformation (24) where $R = \sqrt{P}$, one has

$$\overline{\left|\frac{\nabla_i \Psi}{\Psi}\right|^2} := \int \mathrm{d}^m x \, \mathrm{d}t \left|\frac{\nabla_i \Psi}{\Psi}\right|^2 = \overline{\left(\frac{1}{2}\frac{\nabla_i P}{P}\right)^2} + \overline{\left(\frac{\nabla_i S}{\hbar}\right)^2}, \qquad (32)$$

and one can rewrite (31) as

$$A = \int L \, \mathrm{d}t = \int \mathrm{d}^m x \, \mathrm{d}t \left[|\Psi|^2\left(\frac{\partial S}{\partial t} + V\right) + \sum_{i=1}^{n}\frac{\hbar^2}{2m_i}|\nabla_i \Psi|^2\right]. \qquad (33)$$

Thus, with the identity $|\Psi|^2\frac{\partial S}{\partial t} = -\frac{i\hbar}{2}(\Psi^*\dot{\Psi} - \dot{\Psi}^*\Psi)$, one obtains the familiar Lagrange density

$$L = -\frac{i\hbar}{2}(\Psi^*\dot{\Psi} - \dot{\Psi}^*\Psi) + \sum_{i=1}^{n}\frac{\hbar^2}{2m_i}\nabla_i \Psi \cdot \nabla_i \Psi^* + V\Psi^*\Psi, \qquad (34)$$

from which by the usual procedures one arrives at the n-particle Schrödinger equation

$$i\hbar\frac{\partial \Psi}{\partial t} = \left(-\sum_{i=1}^{n}\frac{\hbar^2}{2m_i}\nabla_i^2 + V\right)\Psi. \qquad (35)$$

Note also that from (31) one obtains upon variation in P the modified Hamilton-Jacobi equation familiar from the de Broglie-Bohm interpretation, i.e.

$$\frac{\partial S}{\partial t} + \sum_{i=1}^{n}\frac{(\nabla_i S)^2}{2m_i} + V(\mathbf{x}_1, \mathbf{x}_2, \ldots, \mathbf{x}_n, t) + U(\mathbf{x}_1, \mathbf{x}_2, \ldots, \mathbf{x}_n, t) = 0, \qquad (36)$$

where U is known as the "quantum potential"

$$U(\mathbf{x}_1, \mathbf{x}_2, \ldots, \mathbf{x}_n, t) = \sum_{i=1}^{n}\frac{\hbar^2}{4m_i}\left[\frac{1}{2}\left(\frac{\nabla_i P}{P}\right)^2 - \frac{\nabla_i^2 P}{P}\right] = -\sum_{i=1}^{n}\frac{\hbar^2}{2m_i}\frac{\nabla_i^2 R}{R}. \qquad (37)$$

Moreover, with the definitions

$$\mathbf{u}_i := \frac{\delta \mathbf{p}_i}{m_i} = -\frac{\hbar}{2m_i}\frac{\nabla_i P}{P} \quad \text{and} \quad \mathbf{k}_{\mathbf{u}i} := -\frac{1}{2}\frac{\nabla_i P}{P} = -\frac{\nabla_i R}{R}, \tag{38}$$

one can rewrite U as

$$U = \sum_{i=1}^{n}\left[\frac{m_i \mathbf{u}_i \cdot \mathbf{u}_i}{2} - \frac{\hbar}{2}(\nabla_i \cdot \mathbf{u}_i)\right] = \sum_{i=1}^{n}\left[\frac{\hbar^2}{2m_i}(\mathbf{k}_{\mathbf{u}i} \cdot \mathbf{k}_{\mathbf{u}i} - \nabla_i \cdot \mathbf{k}_{\mathbf{u}i})\right]. \tag{39}$$

However, as was already detailed in Grössing (2009),[3] for the energy balance $kT = \hbar\omega$ referring to the vacuum's acting as the particle's "thermostat", $\mathbf{u}_i$ can also be written as

$$\mathbf{u}_i = \frac{1}{2\omega_i m_i}\nabla_i Q, \tag{40}$$

which thus explicitly shows its dependence on the spatial behavior of the heat flow $Q = kT\ln P$. Insertion of (40) into (39) then provides a thermodynamic formulation of the quantum potential as

$$U = \sum_{i=1}^{n}\frac{\hbar^2}{4m_i}\left[\frac{1}{2}\left(\frac{\nabla_i Q}{\hbar\omega_i}\right)^2 - \frac{\nabla_i^2 Q}{\hbar\omega_i}\right]. \tag{41}$$

3. "Systemic" nonlocality in double slit interference

In a recent paper, Tollaksen *et al.*[23] renew the discussion on interference at a double slit "from a single particle perspective", asking the following question: If a particle goes through one slit, how does it "know" whether the second slit is open or closed? We shall here first recapitulate the arguments providing these authors' answer and later provide our own arguments and answer. Of course, the question is about the phase information and how it affects the particle. We know from quantum mechanics that phases cannot be observed on a local basis and that a common overall phase has no observational meaning. Assuming that two Gaussian wave functions, Ψ_1 and Ψ_2, describe the probability amplitudes for particles emerging from slits 1 or 2, respectively, which are separated by a distance D, the total wave function for the particle exiting the double slit may be written as

$$\Psi = e^{i\alpha_1}\Psi_1 + e^{i\alpha_2}\Psi_2, \tag{42}$$

but since a common overall phase is insignificant, one multiplies (42) with $e^{-i\alpha_1}$ and writes the total wave function as

$$\Psi_\varphi = \Psi_1 + e^{i\varphi}\Psi_2, \tag{43}$$

where $\varphi := \alpha_2 - \alpha_1$ is the physically significant *relative phase*. Tollaksen *et al.* now ask where the relative phase appears in the form of (deterministic) observables that describe interference. For the following recapitulation of their argument, it must be stressed that the authors assume the two wave functions, Ψ_1 and Ψ_2, to be initially non-overlapping. When looking at the expectation value (in the one-dimensional case, for simplicity)

$$\bar{x} = \int \Psi_\varphi^* x \Psi_\varphi \, dx = \int \left(\Psi_1^* + e^{-i\varphi} \Psi_2^* \right) x \left(\Psi_1 + e^{i\varphi} \Psi_2 \right) \, dx \tag{44}$$

$$= \int \left(\Psi_1^* x \Psi_1 + \Psi_2^* x \Psi_2 \right) \, dx + \int \Psi_1^* x e^{i\varphi} \Psi_2 \, dx + \text{c.c.},$$

one sees that it is independent of φ because of the vanishing of the last term due to the assumed non-overlapping of Ψ_1 and Ψ_2. Similarly, this also holds for all moments of x, and for all moments of p as well. In particular, one has for the expectation value of the momentum

$$\bar{p} = \text{Re} \int \Psi_\varphi^* i\hbar \frac{\partial}{\partial x} \Psi_\varphi \, dx \tag{45}$$

$$= \text{Re} \int \left(\Psi_1^* i\hbar \frac{\partial}{\partial x} \Psi_1 + \Psi_2^* i\hbar \frac{\partial}{\partial x} \Psi_2 \right) \, dx + \text{Re} \int \Psi_1^* i\hbar \frac{\partial}{\partial x} e^{i\varphi} \Psi_2 \, dx + \text{c.c.},$$

where the φ–dependent term vanishes identically, because $\frac{\partial}{\partial x} \Psi_2 = 0$ for $\Psi_2 = 0$. So, again, where does the relative phase appear? The answer of Tollaksen *et al.* is given by a "shift operator" that shifts the location of, say Ψ_1, over the distance D to its new location coinciding with that of Ψ_2. The expectation value of the shift operator is thus given by

$$\overline{e^{-i\frac{pD}{\hbar}}} = \int \left(\Psi_1^* e^{-i\frac{pD}{\hbar}} \Psi_1 + \Psi_2^* e^{-i\frac{pD}{\hbar}} \Psi_2 \right) \, dx \tag{46}$$

$$+ \int \Psi_1^* e^{-i\frac{pD}{\hbar}} \Psi_2 e^{i\varphi} \, dx + \int \Psi_2^* e^{-i\varphi} e^{-i\frac{pD}{\hbar}} \Psi_1 \, dx,$$

where all but the last term vanish identically, thus providing (with the correct normalization)

$$\overline{e^{-i\frac{pD}{\hbar}}} = e^{-i\varphi}/2 \quad \text{and} \quad \overline{e^{-i\frac{pD}{\hbar}}} + \overline{e^{i\frac{pD}{\hbar}}} = \cos\varphi. \tag{47}$$

In order to physically interpret the shift operator, the authors now shift to the Heisenberg picture, thereby providing with a Hamiltonian $H = \frac{p^2}{2m} + V(x)$ the time evolution of the operator as

$$\frac{d}{dt} e^{-i\frac{pD}{\hbar}} = \frac{i}{\hbar} \left[e^{-i\frac{pD}{\hbar}}, \frac{p^2}{2m} + V(x) \right] = \frac{-iD}{\hbar} e^{-i\frac{pD}{\hbar}} \left\{ \frac{V(x) - V(x+D)}{D} \right\}. \tag{48}$$

With its dependence on the distance D between the two slits, Eq. (48) is a description of *dynamical nonlocality*: it is thus shown how a particle can "know" about the presence of the other slit. Tollaksen *et al.* maintain that it is possible to understand this dynamical nonlocality only by employing the Heisenberg picture. However, we shall now show that such an understanding is possible also within the Schrödinger picture, and even more intuitively accessible, too. For, there is one assumption in the foregoing analysis that is not guaranteed to hold in general, i.e. the non-overlapping of the wave functions Ψ_1 and Ψ_2. On the contrary, we now shall assume that the two Gaussians representing the probability amplitudes for the particle immediately after passing one of the two slits do not have any artificial cut-off, but actually extend across the whole slit system, even with only very small (and practically often negligible) amplitudes in the regions further away from the slit proper. (We shall give arguments for this assumption further below.) In other words, we now ask: what if Ψ_1 and Ψ_2 do overlap, even if only by a very small amount? To answer this question, we consider the expectation value of the momentum and obtain from Eq. (45) that the terms involving the relative phase φ provide with $\Psi_j = R_j e^{i\varphi_j}$ and $\varphi_j = \frac{S_j}{\hbar}$

$$\overline{p} = R_1 R_2 \left(\nabla S_1 + \nabla S_2\right) \cos\varphi + \hbar \left(R_1 \nabla R_2 - R_2 \nabla R_1\right) \sin\varphi. \tag{49}$$

First of all one notes upon comparison of (49) with Eqs. (21)–(23) the exact correspondence of $\overline{p}$ with our classically obtained expression for the interference terms of the emerging current, or of the expression for $m\overline{\mathbf{v}}_{\text{tot}}$, respectively. Moreover, although the product $R_1 R_2$ is in fact negligibly small for regions where only a long tail of one Gaussian overlaps with another Gaussian (i.e. such that the non-overlapping assumption would be largely justified), nevertheless the second term in (49) can be very large despite the smallness of R_1 or R_2. It is this latter part which is responsible for the genuinely "quantum mechanical" nature of the average momentum, i.e. for its nonlocal nature. This is formally similar in the Bohmian picture, but here given a more direct physical meaning in that this last term refers to a difference in diffusive currents as explicitly formulated in the last term of Eq. (21). Because of the mixing of diffusion currents from both channels, we call this dominant term in $\mathbf{J}_{\text{tot}}$ the "entangling current",[24] $\mathbf{J}_{\text{e}}$. In fact, inserting Eq. (40) for a one-particle system into the latter expression, one obtains

$$\mathbf{J}_{\text{e}} = \sqrt{P_1 P_2} \frac{\nabla\left(Q_1 - Q_2\right)}{2\omega m} \sin\varphi_{12}. \tag{50}$$

In other words, the entangling current is essentially a measure for the gradient of the vacuum's "heat" fluxes Q_i associated with the channels i.

Finally, there are substantial arguments against the non-overlapping scenario in Tollaksen *et al.* Firstly, experiments by Rauch *et al.* have shown that in interferometry interference does not only happen when the main bulks of the Gaussians overlap, but also when a Gaussian interferes with the off-bulk plane-wave components of the other wave function as well.[25] On a theoretical side, we have repeatedly stressed that the diffusion processes employed in our model must be described by nonlocal diffusion wave fields[26,27] which thus require small but non-zero amplitudes across the whole experimental setup. Moreover, and more specifically, we have shown[4,5] that quantum propagation can be identified with sub-quantum anomalous (i.e. "ballistic") diffusion which is characterized by infinite mean displacements $\overline{x} = \infty$ despite the finite drifts $\overline{x^2} = u^2 t^2$ and $u < c$.[28] In sum, these arguments speak in favor of using small, but non-zero amplitudes from the Gaussian of the "other slit" which can interfere with the Gaussian at the particle's location.

Let us now see how we can explain dynamical nonlocality within our sub-quantum approach. For example, we can ask the following question in our context: what if we start with one slit only, and when the particle should pass it we open the second slit? Let us assume for the time being, and without restriction of generality, that the x–component of the velocity v_x is zero. Then, according to (14), upon opening the second slit (with the same σ), we obtain a term proportional to the distance $2X$ between the two slits. Thus, there will be a shift in momentum on the particle passing the first slit given by

$$\frac{\Delta p}{\hbar} = \pm\frac{1}{2}\nabla\varphi_{12} = \pm\frac{1}{2\hbar}\nabla\left(S_1 - S_2\right) = \frac{\Delta p_{\text{mod}}}{\hbar}, \tag{51}$$

where one effectively uses the "modular momentum" $p_{\text{mod}} = p \bmod \frac{h}{2X} = p - 2n\pi\frac{\hbar}{2X}$ because an added or subtracted phase difference $\varphi_{12} = 2n\pi$ does not change anything. In other words, by splitting X into a component X_n providing $\varphi_{12} = 2n\pi$ on the one hand, and the modular remainder ΔX on the other hand providing $\varphi_{12} = p_{\text{mod}}x/\hbar$, one rewrites (14) with $X := X_n + \Delta X$ (and with $v_x = 0$ for simplicity) as

$$\varphi_{12} = -(X_n + \Delta X)x\frac{1}{D}\frac{u^2 t}{\sigma_0^2} =: -2n\pi - \Delta X x\frac{\hbar}{2m}\frac{t}{\sigma^2\sigma_0^2}. \tag{52}$$

Therefore, one can further on substitute Δp by Δp_{mod}, and one obtains a momentum shift (with the sign depending on whether the right or left of the two slits is opened as the second one)

$$\Delta p_{\text{mod}} = \pm\frac{\hbar}{2}\nabla\varphi_{12} = \pm\frac{\Delta X}{2}\frac{\hbar^2}{2m}\frac{t}{\sigma^2\sigma_0^2} = \pm m\Delta X\frac{D^2 t}{\sigma^2\sigma_0^2} = \pm m\Delta X\frac{\dot\sigma}{\sigma}. \tag{53}$$

It follows that, due to the "vacuum pressure" stemming from the opened second slit, there is an *emergent* nonlocal "force" which does not derive from a potential but from the impinging of the second slit's sub-quantum diffusive momenta on the particle at the first slit. As there is no additional force on the particle, we do not speak of a "dynamical" nonlocality, but of a "systemic" one instead. For more details, see Grössing *et al.* (2013).[29]

4. Quantum mechanical results obtained from classical computer simulations

For our classical simulations of quantum phenomena we make use of the physics of ballistic diffusion as given in Eq. (9), with the time-dependent diffusivity D_t from Eq. (10). We use an explicit finite difference forward scheme,[30]

$$\frac{\partial P}{\partial t} \to \frac{1}{\Delta t}\left(P[x, t+1] - P[x, t]\right), \tag{54}$$

$$\frac{\partial^2 P}{\partial x^2} \to \frac{1}{\Delta x^2}\left(P[x+1, t] - 2P[x, t] + P[x-1, t]\right), \tag{55}$$

with 1-dimensional cells. For our case of D_t being independent of x, the complete equation after reordering reads as

$$P[x, t+1] = P[x, t] + \frac{D[t+1]\Delta t}{\Delta x^2} \{P[x+1, t] - 2P[x, t] + P[x-1, t]\} \qquad (56)$$

with space x and time t, and initial Gaussian distribution $P(x, 0)$ with standard deviation σ_0.

In the following, some exemplary figures with one spatial and one time dimension are shown to demonstrate some of the present applications of our model as well as of the novel simulation protocol that comes along with it. The first three figures exhibit double-slit scenarios, with time evolving from bottom to top, and the remaining three figures show n-slit scenarios with time running from left to right. Except for the last figure, the coloring for intensity distributions is, for increasing intensity, from white through yellow and orange, and average trajectories are displayed in red.

Turning now to Fig. 1, we begin with the upper part, i.e. a quantum mechanical interference pattern obtained from a computer simulation employing classical physics only. Shown are the probability density distributions for two wave packets emerging from the Gaussian slits with opposite velocities, superimposed by flow lines which coincide with Bohmian trajectories. Note, however, that the latter appear only as the result of the averaging, whereas the sub-quantum model on which the figure is based involves wiggly, stochastic motions. For the smoothed-out trajectories, however, a "no crossing" rule applies with respect to the central symmetry axis, just like in the Bohmian picture. The figure's lower part shows the "entangling current" of Eq. (50) for the scenario above, which is in our model the decisive part of the probability density current that is responsible for the genuinely "quantum" effects. Due to the sinusoid nature, colors are used that display positive (red) and negative (blue) values. Note that after the time of maximal overlap between the two wave-packets, the order of maxima and minima is reversed, which results from the respective differences in the exchanges of "heat" according to Eq. (50).

Figure 2 shows a similar scenario as Fig. 1, albeit with zero velocity components in the transverse (i.e. $x-$) direction. However, whereas the situation of Fig. 1 is characterized by small dispersion of the Gaussians, here the dispersion is significantly higher. The effects of the "no crossing" rule are clearly visible. Note also the "kinks" of trajectories moving from the center-oriented side of one relative maximum to cross over to join more central (relative) maxima. In our model, a detailed micro-causal account of the corresponding kinematics can be given. As can be seen from the entangling current in the lower part of the figure, its extrema coincide with the areas where the kinks appear, pointing at a rapid crossing-over due to the effects of the diffusive processes involved.

Figure 3 shows almost the same scenario as Fig. 2, the only difference being that in the left channel a phase shift of $\Delta\varphi = \frac{\pi}{2}$ has been added, essentially leading to a representation of the scalar Aharonov-Bohm effect. Note that the minima and

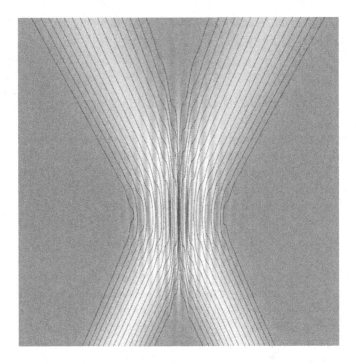

Fig. 1.   *Top:* Quantum mechanical interference pattern, with averaged trajectories, obtained from a computer simulation employing classical physics only. *Bottom:* "Entangling current" of Eq. (50) for the scenario above.

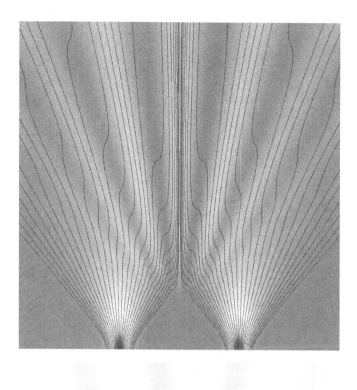

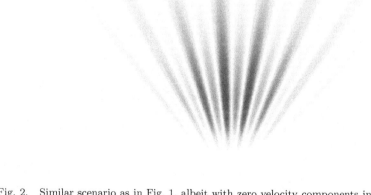

Fig. 2.   Similar scenario as in Fig. 1, albeit with zero velocity components in the transverse (i.e. $x-$) direction, and with large dispersion instead of a small one. As can be seen from the entangling current in the lower part of the figure, its extrema coincide with the areas where the trajectories' "kinks" appear, pointing at a rapid crossing-over due to the effects of the diffusive processes involved.

G. Grössing

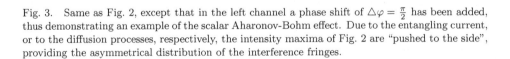

Fig. 3.   Same as Fig. 2, except that in the left channel a phase shift of $\triangle\varphi = \frac{\pi}{2}$ has been added, thus demonstrating an example of the scalar Aharonov-Bohm effect. Due to the entangling current, or to the diffusion processes, respectively, the intensity maxima of Fig. 2 are "pushed to the side", providing the asymmetrical distribution of the interference fringes.

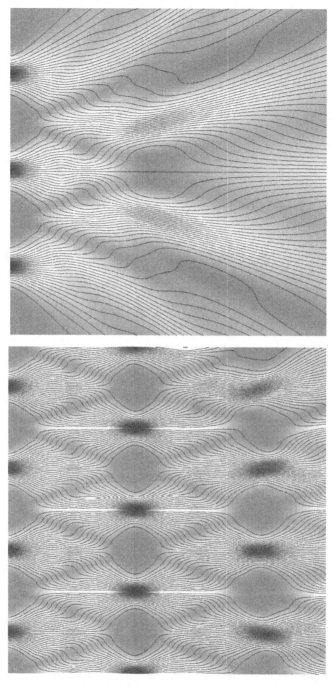

Fig. 4.   *Top:* A 3-slit system with intensity distributions and averaged trajectories. *Bottom:* Detail of a system with a large number n of slits (i.e. a so-called Talbot carpet), explicitly showing only four of them. Note that non-spreading "cells" emerge which contain the trajectories and keep them there.

Fig. 5. *Top:* Classically simulated interference pattern for a system of 9 slits with gradually diminishing initial intensity distributions. *Bottom:* Same, except that the initial intensity distributions have been chosen to have random values.

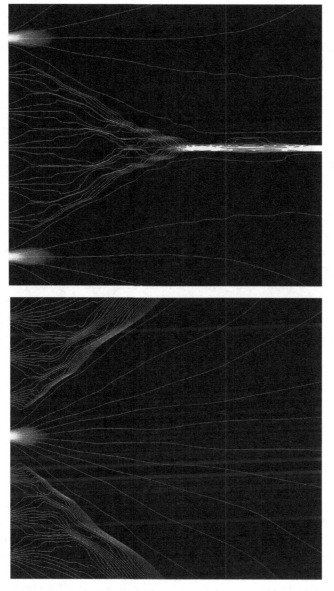

Fig. 6.   Two examples of intensity and trajectory distributions of particularly "weighted" 7-slit systems. *Top:* Quantum squeezer. *Bottom:* Quantum sweeper.

maxima of the intensity distribution are shifted accordingly. This comes along with a different behavior of the entangling current. Whereas in Fig. 2 the entangling current is asymmetrical with respect to the central symmetry line, it now is symmetrical. Effectively, this means that due to the diffusion processes providing some "vacuum pressure", the intensity maxima are "pushed to the side", providing an asymmetrical distribution of the interference fringes.

Figure 4 displays two examples which extend the double slit simulations to more than two slits. Whereas the upper picture shows a 3-slit system with intensity distributions and averaged trajectories, the lower one is a detail of a "Talbot carpet", i.e. a system with a large number n of slits, albeit exhibiting only four of them. It clearly demonstrates a result that is well-known from comparable Bohmian calculations, i.e. that for the case of large n, non-spreading "cells" emerge which contain the trajectories and keep them there.[31]

To continue with n-slit systems, Fig. 5 shows, in the upper part, a type of Talbot carpet (without trajectories) with gradually diminishing initial intensity distributions, from top to bottom. The lower figure shows the same except that the initial intensity distributions have been chosen to have random values. This is an example of a system whose complexity is considerable, despite the fact that the simulation is still very fast and simple, thus illustrating also an advantage over purely analytical tools. Eventually, this type of simulations might even apply at levels of complexity which would exceed any traditional analytical methods of quantum mechanics.

Finally, Fig. 6 shows two examples of intensity and trajectory distributions of particularly "weighted" 7-slit systems. In the upper figure, the high relative intensities at the extreme locations force the trajectories of the remaining slits to become "squeezed" into a narrow canal. The lower figure instead shows the opposite effect of a centrally positioned relatively high intensity "sweeping" away the other slits' trajectories. The two examples again illustrate the effect of the vacuum's "pressure" in our sub-quantum model.

5. Summary

An introduction to our sub-quantum approach to quantum mechanics was presented. We began by constructing from classical physics a model that explains the quantum mechanical dispersion of a Gaussian wave packet. As a consequence, we obtained the total velocity field for the current emerging from a Gaussian slit, which can easily be extended to a two-slit (and in fact to any n-slit) system, thereby also providing an explanation of the famous interference effects. With the thus obtained velocity field, one can by integration also obtain an action function that generalizes the ordinary classical one to the case that is relevant for the reproduction of the quantum results. Effectively, this is achieved by the appearance of an additional kinetic energy term, which we associate with fluxes of "heat" in the vacuum, i.e. sub-quantum currents based on the existence of the zero-point energy field. Finally, it was shown how the Schrödinger equation can be derived in a straightforward manner from the said generalized action function, or the corresponding Lagrangian, respectively. We also dealt with the problem of "interference from the single particle perspective", obtaining results which are in agreement with those presented by Aharonov *et al.*,[14,15] but differing in interpretation. Finally, computer simulations were presented to show some examples of how our sub-quantum approach reproduces results of ordinary quantum mechanics, but eventually may go beyond these.

Acknowledgments

I want to thank my friends and AINS colleagues Siegfried Fussy, Johannes Mesa Pascasio, and Herbert Schwabl for their continued enthusiasm and support in trying to work out a radically new approach to quantum mechanics. Further, I thank the organizers of the summer school, and particularly Claudio Furtado, Inacio Pedrosa, Claudia Pombo, and Theo Nieuwenhuizen for inviting me to Joao Pessoa, for their perfect organization and their great hospitality. Finally, I also gratefully acknowledge partial support by the Fetzer Franklin Fund.

References

1. Yves Couder explains wave/particle duality via silicon droplets [Through the Wormhole], (2011). URL http://www.youtube.com/watch?v=W9yWv5dqSKk.
2. G. Grössing, The vacuum fluctuation theorem: Exact Schrödinger equation via nonequilibrium thermodynamics, *Phys. Lett. A* **372**, 4556–4563 (2008). doi: 10.1016/j.physleta.2008.05.007.
3. G. Grössing, On the thermodynamic origin of the quantum potential, *Physica A* **388**, 811–823 (2009). doi: 10.1016/j.physa.2008.11.033.
4. G. Grössing, S. Fussy, J. Mesa Pascasio, and H. Schwabl, Emergence and collapse of quantum mechanical superposition: Orthogonality of reversible dynamics and irreversible diffusion, *Physica A* **389**, 4473–4484 (2010). doi: 10.1016/j.physa.2010.07.017.
5. G. Grössing, S. Fussy, J. Mesa Pascasio, and H. Schwabl, Elements of sub-quantum thermodynamics: Quantum motion as ballistic diffusion, *J. Phys.: Conf. Ser.* **306**, 012046 (2011). doi: 10.1088/1742-6596/306/1/012046.
6. G. Grössing, J. Mesa Pascasio, and H. Schwabl, A classical explanation of quantization, *Found. Phys.* **41**, 1437–1453 (2011). doi: 10.1007/s10701-011-9556-1.
7. G. Grössing, S. Fussy, J. Mesa Pascasio, and H. Schwabl, An explanation of interference effects in the double slit experiment: Classical trajectories plus ballistic diffusion caused by zero-point fluctuations, *Ann. Phys.* **327**, 421–437 (2012). doi: 10.1016/j.aop.2011.11.010.
8. G. Grössing, S. Fussy, J. Mesa Pascasio, and H. Schwabl, The quantum as an emergent system, *J. Phys.: Conf. Ser.* **361**, 012008 (2012). doi: 10.1088/1742-6596/361/1/012008.
9. Y. Couder, S. Protière, E. Fort, and A. Boudaoud, Dynamical phenomena: Walking and orbiting droplets, *Nature* **437**, 208–208 (2005). doi: 10.1038/437208a.
10. Y. Couder and E. Fort, Single-particle diffraction and interference at a macroscopic scale, *Phys. Rev. Lett.* **97**, 154101 (2006). doi: 10.1103/PhysRevLett.97.154101.
11. Y. Couder, A. Boudaoud, S. Protière, and E. Fort, Walking droplets, a form of wave-particle duality at macroscopic scale?, *Europhys. News* **41**, 5 (2010). doi: 10.1051/epn/2010101.
12. Y. Couder and E. Fort, Probabilities and trajectories in a classical wave-particle duality, *J. Phys.: Conf. Ser.* **361**, 012001 (2012). doi: 10.1088/1742-6596/361/1/012001.

13. G. Grössing, Ed., *Emergent Quantum Mechanics 2011*. Number 361/1 (IOP Publishing, Bristol, 2012). http://iopscience.iop.org/1742-6596/361/1.

14. Y. Aharonov, H. Pendleton, and A. Petersen, Modular variables in quantum theory, *Int. J. Theor. Phys.* **2**, 213–230 (1969). doi: 10.1007/BF00670008.

15. Y. Aharonov, H. Pendleton, and A. Petersen, Deterministic quantum interference experiments, *Int. J. Theor. Phys.* **3**, 443–448 (1970). doi: 10.1007/BF00672451.

16. L. V. P. R. de Broglie, *Non-Linear Wave Mechanics: A Causal Interpretation* (Elsevier, Amsterdam, 1960).

17. P. R. Holland, *The Quantum Theory of Motion* (Cambridge University Press, Cambridge, 1993). ISBN 0-521-35404-8.

18. M. J. W. Hall and M. Reginatto, Schrödinger equation from an exact uncertainty principle, *J. Phys. A: Math. Gen.* **35**, 3289–3303 (2002). doi: 10.1088/0305-4470/35/14/310.

19. E. Fort, A. Eddi, A. Boudaoud, J. Moukhtar, and Y. Couder, Path-memory induced quantization of classical orbits, *PNAS* **107**, 17515–17520 (2010). doi: 10.1073/pnas.1007386107.

20. S. Fussy, J. Mesa Pascasio, H. Schwabl, and G. Grössing, Born's rule as signature of a superclassical current algebra, *Ann. Phys.* **343**, 200–214 (2014). doi: 10.1016/j.aop.2014.02.002.

21. Á. S. Sanz and S. Miret-Artés, A trajectory-based understanding of quantum interference, *J. Phys. A: Math. Gen.* **41**, 435303 (2008). doi: 10.1088/1751-8113/41/43/435303.

22. G. Grössing, From classical hamiltonian flow to quantum theory: Derivation of the Schrödinger equation, *Found. Phys. Lett.* **17**, 343–362 (2004). doi: 10.1023/B:FOPL.0000035669.03595.ce.

23. J. Tollaksen, Y. Aharonov, A. Casher, T. Kaufherr, and S. Nussinov, Quantum interference experiments, modular variables and weak measurements, *New J. Phys.* **12**, 013023 (2010). doi: 10.1088/1367-2630/12/1/013023.

24. J. Mesa Pascasio, S. Fussy, H. Schwabl, and G. Grössing, Modeling quantum mechanical double slit interference via anomalous diffusion: Independently variable slit widths, *Physica A* **392**, 2718–2727 (2013). doi: 10.1016/j.physa.2013.02.006.

25. H. Rauch, Phase space coupling in interference and EPR experiments, *Phys. Lett. A* **173**, 240–242 (1993). doi: 10.1016/0375-9601(93)90270-A.

26. A. Mandelis, *Diffusion-wave fields: Mathematical methods and Green functions* (Springer, New York, NY, 2001). ISBN 0387951490.

27. A. Mandelis, L. Nicolaides, and Y. Chen, Structure and the Reflectionless/Refractionless nature of parabolic diffusion-wave fields, *Phys. Rev. Lett.* **87**, 020801 (2001). doi: 10.1103/PhysRevLett.87.020801.

28. J. Klafter, M. F. Shlesinger, and G. Zumofen, Beyond Brownian motion, *Phys. Today* **49**, 33–39 (1996).

29. G. Grössing, S. Fussy, J. Mesa Pascasio, and H. Schwabl, 'Systemic nonlocality' from changing constraints on sub-quantum kinematics, *J. Phys.: Conf. Ser.* **442**, 012012 (2013). doi: 10.1088/1742-6596/442/1/012012.

30. H. R. Schwarz and N. Köckler, *Numerische Mathematik* (Vieweg+Teubner, 2009), 7 edition. ISBN 978-3-8348-0683-3.

31. Á. S. Sanz and S. Miret-Artés, A causal look into the quantum Talbot effect, *J. Chem. Phys.* **126**, 234106 (2007). doi: 10.1063/1.2741555.

Chapter 11

The Zero-Point Field and the Emergence of the Quantum

Luis de la Peña*, Ana Maria Cetto[†] and Andrea Valdés-Hernándes[‡]

Instituto de Física, Universidad Nacional Autónoma de México,
Apartado Postal 20-364, México, DF, Mexico

A new way of arriving at the quantum formalism is presented, based on the recognition of the reality of the random zero-point radiation field (ZPF). The quantization of both matter and radiation field is shown to emerge as a result of the permanent interaction of matter with the ZPF. Quantum mechanics is obtained both in its Schrödinger and its Heisenberg version, under certain well-defined conditions and approximations. The theory provides for an explanation of the origin of entanglement. Further, the same physical elements and hypotheses allow us to cross the doorway and go beyond quantum mechanics, to the realm of (nonrelativistic) quantum electrodynamics.

1. Introduction

In this paper we present a new way of arriving at the quantum formalism, based on the recognition of the reality of the random zero-point radiation field (ZPF). The advantage of this approach lies in that one sees into the quantum world from outside, which affords a perspective wider than the one reachable from within the quantum theory proper. Many of the usual conceptual problems that characterize present-day quantum mechanics (QM) are thus dissolved, and new physical elements are integrated that help to clarify its physical and conceptual contents. At the same time, the fresh perspective proposed invites us to cross the doorway and go beyond the strictly quantum-mechanical realm.

When referring to the problems of quantum mechanics we have in mind basically those conceptual puzzles — frequently bordering philosophy of science, to the annoyance of some physicists, although their nature is physical — that have been under scrutiny and debate since the early days of the theory, but remain basically as unsolved now as they were eighty years ago. We recall as examples: the irreducible or unexplained indeterminism characteristic of the theory; the fact that it predicts

---

*luis@fisica.unam.mx
[†]ana@fisica.unam.mx
[‡]andreavh@fisica.unam.mx

probabilities, not outcomes; that it then requires a measurement theory for its accomplishment, which means opening the door to observers and their subjectivism, and giving birth to undefined boundaries between the quantum and the classical world. Add to this the loss of realism; two opposed laws of evolution (Schrödinger´s equation and the collapse of the wave function); nonlocality and, for many, superluminal influences, and so on. There are also some strictly technical questions, such as the lack of a clear physical explanation of the mechanism that stabilizes the atom and leads to quantized states (rather than a mere description of the phenomenon), or of the mechanism that entangles two identical noninteracting particles. The list is not short. All these are real questions, of actual interest to physics and to which hundreds of books, papers, conferences and meetings have been devoted.

The paper is structured into four parts, according to the lectures of the course for which they were prepared. The first part (section 2) presents Planck's distribution for the blackbody radiation field as a consequence of the existence of the ZPF without any quantum hypothesis, and discusses the implications of this result. The second part (section 3) presents Schrödinger's equation as a consequence of the action of the same ZPF on matter, and discusses the conditions under which quantization emerges. The third part (section 4) shows that the Heisenberg formalism of quantum theory can be derived on the basis of the same principles, and exhibits the conditions needed to arrive at this result as well as some of its implications, including the emergence of entanglement. The last part (section 5) shows that also the first-order nonrelativistic radiative corrections of QED are correctly obtained, thus confirming that the theory presented goes beyond quantum mechanics.

2. Planck's distribution and the zero-point energy

Some very fundamental properties of the equilibrium radiation field can be derived from the mere consideration of the pervasive presence of its zero-point contribution. For this purpose it is convenient to start by reviewing the thermodynamics of a harmonic oscillator of frequency ω, which can be taken to represent a mode of the field of that frequency. A careful thermo-statistical analysis of an oscillator system in thermodynamic equilibrium reveals the far-reaching implications of the existence of a fluctuating zero-point (nonthermal) energy term.

For earlier literature and details on the material presented in the following sections, see Refs. 1–10.

2.1. *The zero-point energy and energy equipartition*

Let us consider the radiation field as made of independent modes of oscillation of frequency ω, in thermal equilibrium at a given temperature T. Then according to Wien's law, the mean energy of every mode is given by an expression of the form[11]

$$U = \omega f(\omega/T) = \omega f(z). \tag{2.1}$$

Since no specific details about the oscillator are needed to arrive at this result, the function f should have a universal character. This law will be at the base of our considerations below.

In the limit $T \to 0$, Eq. (2.1) gives for the mean energy

$$\mathcal{E}_0 \equiv U(0, \omega) = \omega f(\infty) = A\omega, \tag{2.2}$$

so the zero-point energy (ZPE) $\mathcal{E}_0$ — the energy of the oscillators at absolute temperature $T = 0$ — is determined by the value that f attains at infinity. In the usual thermodynamic analysis the value of the constant $f(\infty) = A$ is *arbitrarily* chosen as zero, so there is no ZPE. However, the more general (and more natural) solution corresponds to a non-null value of A. In the case of the radiation field, this represents a physically more reasonable choice than a vacuum that is completely devoid of electromagnetic phenomena. By taking A to be nonzero we attest the existence of a ZPE that fills the whole space and is proportional to ω,

$$\mathcal{E}_0 = A\omega = \tfrac{1}{2}\hbar\omega. \tag{2.3}$$

The value of A (with dimensions of action) must be universal because it determines the (universal, according to Kirchhoff) equilibrium distribution at $T = 0$; we have put it equal to $\hbar/2$ in order to establish contact with present day knowledge.

A value of A different from zero means a violation of energy equipartition among the oscillators, since the equilibrium energy becomes a function of the oscillator frequency. This holds at any temperature, at least because the ZPE is part of the equilibrium energy. Hence the ensuing physics necessarily transcends classical physics.

Let us note that the selection made for A assigns an interesting meaning to the variable z, which so far has been written just as $z = \omega/T$. Since z should be dimensionless (as follows from Eq. (2.1)), given the parameters at hand it becomes naturally expressed as the ratio of two energies, $z = \hbar\omega/k_B T$.

2.2. *General thermodynamic equilibrium distribution*

Our aim now is to find the mean equilibrium energy U of a system of oscillators as a function of the temperature. For this purpose we revisit in the following three subsections some general thermo-statistical results, which are not exclusive of the harmonic oscillator, but hold for any physical system that is in equilibrium at temperature T.

Following a standard procedure in thermodynamics, the probability that the energy attains a value between $\mathcal{E}$ and $\mathcal{E} + d\mathcal{E}$ for a fixed temperature can be written in the general form for a canonical ensemble

$$W_g(\mathcal{E})d\mathcal{E} = \frac{1}{Z_g(\beta)} g(\mathcal{E})e^{-\beta\mathcal{E}} d\mathcal{E}, \tag{2.4}$$

$$Z_g(\beta) = \int g(\mathcal{E})e^{-\beta\mathcal{E}}d\mathcal{E}, \tag{2.5}$$

where $\beta = 1/(k_BT)$ is the inverse temperature, $Z_g(\beta)$ is the partition function that normalizes $W_g(\mathcal{E})$ to unity, $\int_0^\infty W_g(\mathcal{E})d\mathcal{E} = 1$, and $g(\mathcal{E})$ is a weight function representing the intrinsic probability of the states with energy $\mathcal{E}$, known as the structure function. The mean value $\langle f(\mathcal{E})\rangle$ of any function $f(\mathcal{E})$ is of course

$$\langle f(\mathcal{E})\rangle = \int_0^\infty W_g(\mathcal{E})f(\mathcal{E})d\mathcal{E}. \tag{2.6}$$

For the particular case $f(\mathcal{E}) = \mathcal{E}$, (2.6) gives the average value $\langle\mathcal{E}\rangle = U$ by definition.

Equation (2.4) constitutes an extension of the usual Boltzmann distribution to the general case in which the states with energy $\mathcal{E}$ can have an intrinsic probability that depends on $\mathcal{E}$. The (classical) Boltzmann distribution is obtained from (2.4) in the particular case in which $g(\mathcal{E})$ does not depend on $\mathcal{E}$ and can therefore be written in the form (with due account of the dimensions)[12]

$$g_{cl}(\mathcal{E}) = \frac{1}{s\omega} = \text{const}, \tag{2.7}$$

where s is a constant with dimensions of action, so g has the dimension of (energy)$^{-1}$. In this case one gets from the above equations:

$$W_{cl}(\mathcal{E}) = W_{g_{cl}}(\mathcal{E}) = \frac{e^{-\beta\mathcal{E}}}{\int_0^\infty e^{-\beta\mathcal{E}}d\mathcal{E}}; \tag{2.8}$$

$$\langle\mathcal{E}\rangle = U = \frac{1}{\beta} = k_BT. \tag{2.9}$$

From the last equation it follows that $\mathcal{E}_0 = \langle\mathcal{E}(T=0)\rangle = 0$. This means that to allow for the a nonthermal energy of the system a form for $g(E)$ different from that given by Eq. (2.7) must be used. Finding this $g(E)$ becomes an important task in what follows.

2.3. Thermal fluctuations of the energy

Equations (2.4), (2.5) and (2.6) lead to a series of important results. With $f(\mathcal{E}) = \mathcal{E}^r$, r a positive integer, it follows that (the prime indicates derivative with respect to β)

$$\langle\mathcal{E}^r\rangle' = -\frac{Z_g'}{Z_g}\langle\mathcal{E}^r\rangle - \langle\mathcal{E}^{r+1}\rangle, \tag{2.10}$$

and further, from (2.5),

$$\langle\mathcal{E}\rangle = U = \frac{1}{Z_g}\int_0^\infty \mathcal{E}g(\mathcal{E})e^{-\beta\mathcal{E}}d\mathcal{E} = -\frac{Z_g'}{Z_g}. \tag{2.11}$$

These two expressions combined give the recurrence relation

$$\langle \mathcal{E}^{r+1} \rangle = U \langle \mathcal{E}^r \rangle - \langle \mathcal{E}^r \rangle'. \tag{2.12}$$

With $r = 1$ this gives a most important expression for the energy variance,

$$\sigma_{\mathcal{E}}^2 \equiv \langle \mathcal{E}^2 \rangle - U^2 = -\frac{dU}{d\beta}, \tag{2.13}$$

which can be rewritten as the well-known relation[11]

$$\sigma_{\mathcal{E}}^2 = -\frac{dU}{d\beta} = k_B T^2 \left(\frac{\partial U}{\partial T} \right)_\omega = k_B T^2 C_\omega \tag{2.14}$$

in terms of the specific heat (or heat capacity) C_ω. Because C_ω $(=C_V)$ is surely finite at low temperatures, the right-hand side of this expression is zero at $T = 0$, whence

$$\sigma_{\mathcal{E}}^2(T = 0) = 0, \tag{2.15}$$

indicating that the present description does not allow for the dispersion of the energy at zero temperature. Of course, this result refers to thermal fluctuations, since the description provided by the distribution W_g is of thermodynamic nature: temperature-independent fluctuations find no place in W_g. In the particular case of a system of harmonic oscillators this becomes an important shortcoming, since the ZPF, being part of the field in the cavity, shares with it the presence of unavoidable fluctuations.

2.4. *General statistical equilibrium distribution*

To go ahead it is required to derive an expression for $\sigma_{\mathcal{E}}^2$ that allows for the non-thermal fluctuations excluded by the distribution W_g, and is expressed in terms of the mean equilibrium energy $U(\beta)$. This can be achieved by paying attention to the *statistical* distribution of the energy as a function of the mean energy, in contrast to the thermodynamic description studied in section 2.2. In this form we are able to bypass the problem presented by the still unknown function $g(\mathcal{E})$. For this purpose we look for a distribution $W_s(\mathcal{E})$ that, according to a well established principle, maximizes the entropy S_s, defined (up to an arbitrary additive constant) as[12,13]

$$S_s = -k_B \int W_s(\mathcal{E}) \ln \left[c W_s(\mathcal{E}) \right] d\mathcal{E}, \tag{2.16}$$

where c is an appropriate constant with dimensions of energy. In contrast to the thermodynamic entropy S which is defined in the phase space of the particles, with $w(p, q)$ the distribution in such space,

$$S = -k_B \int w(p, q) \ln w(p, q) dp dq, \tag{2.17}$$

the *statistical entropy* S_s is normally interpreted as a measure of the disorder present in the system. Thus the demand of maximal entropy implies maximum disorder, which is considered the "natural order" under equilibrium. In the case of interest here the system consists of an immense (practically infinite) number of independent modes of the field that for each frequency interfere among themselves, so a final state of maximal disorder is to be naturally expected.

The maximum-entropy formalism is designed to determine a distribution $W_s(\mathcal{E})$ that maximizes the function S_s subject to a set of auxiliary conditions (or constraints), which in the present case take the form (the subscript s denotes averages with respect to W_s, to be distinguished from quantities calculated with W_g)

$$\int W_s(\mathcal{E})d\mathcal{E} = 1, \quad \int \mathcal{E}W_s(\mathcal{E})d\mathcal{E} = U. \tag{2.18}$$

Of course, at the moment the mean value $U(\beta)$ is still an unknown function of β, to be determined. By applying the method of the Lagrange multipliers one arrives thus at[16]

$$W_s(\mathcal{E}) = \frac{1}{U} e^{-\mathcal{E}/U}. \tag{2.19}$$

Note that the specific choice $U = \beta^{-1}$ (corresponding to $\mathcal{E}_0 = 0$) results in the usual canonical distribution Eq. (2.9).

Equation (2.19) gives the general statistical result

$$\langle \mathcal{E}^r \rangle_s = r!U^r, \tag{2.20}$$

from which it follows, in particular, that $\langle \mathcal{E}^2 \rangle_s = 2U^2$, whence

$$(\sigma_{\mathcal{E}}^2)_s = U^2. \tag{2.21}$$

We see that W_s indeed allows for nonthermal fluctuations, since at $T = 0$

$$(\sigma_{\mathcal{E}}^2)_s\big|_0 = U^2(T = 0) = \mathcal{E}_0^2. \tag{2.22}$$

To reproduce the thermodynamic condition (2.15) we must therefore subtract from $(\sigma_{\mathcal{E}}^2)_s$ the quantity $\mathcal{E}_0^2$. This gives $\sigma_{\mathcal{E}}^2 = (\sigma_{\mathcal{E}}^2)_s - \mathcal{E}_0^2$, or

$$\sigma_{\mathcal{E}}^2 = U^2 - \mathcal{E}_0^2. \tag{2.23}$$

In combination with Eq. (2.13), this result leads to

$$-\frac{dU}{d\beta} = \sigma_{\mathcal{E}}^2 = U^2 - \mathcal{E}_0^2. \tag{2.24}$$

Now we are in possession of a differential equation for the mean energy U, which allows to determine the function $U(\beta)$. An integration of the expression

$$d\beta = -\frac{dU}{\sigma_{\mathcal{E}}^2(U)} = -\frac{dU}{U^2 - \mathcal{E}_0^2} \tag{2.25}$$

subject to the appropriate condition at infinity ($U \to \infty$ as $T \to \infty$) yields

$$
\beta = \begin{cases} \frac{1}{U} & \text{for } \mathcal{E}_0 = 0, \\ \frac{1}{\mathcal{E}_0} \coth^{-1} \frac{U}{\mathcal{E}_0} & \text{for } \mathcal{E}_0 \neq 0. \end{cases} \tag{2.26}
$$

Although the case $\mathcal{E}_0 = 0$ can be treated as a limit of the case $\mathcal{E}_0 \neq 0$, it is more illustrative to treat each case separately. The functions in Eq. (2.26) can be inverted to obtain

$$
U(\beta) = \begin{cases} \frac{1}{\beta}, & \text{for } \mathcal{E}_0 = 0; \\ \mathcal{E}_0 \coth \mathcal{E}_0 \beta, & \text{for } \mathcal{E}_0 \neq 0. \end{cases} \tag{2.27}
$$

2.5. *Mean equilibrium distribution of the oscillators*

As Eq. (2.27) shows, the functional form of the mean energy depends critically on the presence of $\mathcal{E}_0$. For $\mathcal{E}_0 = 0$ the classical energy equipartition is recovered,

$$
U_{\text{cl}} = \beta^{-1}, \tag{2.28}
$$

whereas for $\mathcal{E}_0 \neq 0$, a more complicated expression for $U(\beta)$ is obtained. In particular, for a system of harmonic oscillators, with $\mathcal{E}_0 = \hbar\omega/2$, Planck's law is obtained,

$$
U_{\text{Planck}} = \tfrac{1}{2}\hbar\omega \coth \tfrac{1}{2}\hbar\omega\beta. \tag{2.29}
$$

The ZPE is of course included, as can be seen by taking the limit $T \to 0$,

$$
U_{\text{Planck}}(\beta \to \infty) = \tfrac{1}{2}\hbar\omega = \mathcal{E}_0. \tag{2.30}
$$

This establishes Planck's law as a physical result whose ultimate meaning — or cause — is the existence of a fluctuating ZPE, whereas its absence leads to the equipartition of energy.

It is important to stress that Planck's law has been obtained without introducing any explicit quantum or discontinuity requirement. Equation (2.25), which is the result of the recurrence relation (2.12) — resulting in its turn from the general distribution $W_g(\mathcal{E})$ and the quadratic dependence of the variance in U — together with Wien's law (which establishes the frequency dependence of the zero-point energy, ZPE), suffice to obtain Planck's law. The fact that the latter is the one that opens the door to the ZPE leads to the conclusion that Wien's law (with $A \neq 0$) is crucial to obtain Planck's law and its quantum consequences. We are thus compelled to say that Wien's law (with $A \neq 0$) is an extension of classical physics that enters into the quantum domain; strictly speaking, as a precursor of Planck's distribution law it should be considered to represent historically the first quantum law. It should be enphazised that Eq. (2.29) and the ensuing consequences are of general validity, regardless of the nature of the oscillators, provided they have a nonzero energy at $T = 0$. The corroboration that the law that gave rise to quantum theory stems from the existence of a fluctuating ZPE brings to the fore the crucial importance of this nonthermal energy for the understanding of QM and more generally, of quantum theory.

A comment on the fluctuations of the zero-point energy is in place. We have seen that for $\mathcal{E}_0 \neq 0$ the thermal energy dispersion is given by

$$\sigma_{\mathcal{E}}^2(U) = U^2 - \mathcal{E}_0{}^2 \quad (U = U_{\text{Planck}}), \tag{2.31}$$

whereas in the classical case ($\mathcal{E}_0 = 0$),

$$\sigma_{\mathcal{E}}^2(U) = U^2 \quad (U = U_{\text{cl}}). \tag{2.32}$$

Whilst in the latter case the thermal fluctuations of the oscillator's energy depend solely on its thermal mean energy, U_{cl}, in the former case Eq. (2.31) relates the thermal fluctuations with the *total* mean energy U_{Planck}, which includes the temperature-independent contribution. This nonthermal contribution cannot be derived from a purely thermodynamic analysis as the one afforded by the distribution W_g. Therefore a statistical treatment was necessary, to arrive at the nonthermal fluctuations given by equation (2.22).

2.6. *Planck, Einstein and the zero-point energy*

The previous discussion suggests separating the average energy U_{Planck} (which as of now will be denoted simply by U) into a thermal contribution U_T and the temperature-independent part $\mathcal{E}_0$,

$$U = U_T + \mathcal{E}_0. \tag{2.33}$$

The first term in this equation,

$$U_T = \mathcal{E}_0 \coth \mathcal{E}_0\beta - \mathcal{E}_0 = \frac{2\mathcal{E}_0}{e^{2\mathcal{E}_0\beta} - 1}, \tag{2.34}$$

is Planck's law without ZPE.[14,15] At sufficiently low temperatures it takes the form

$$U_T = 2\mathcal{E}_0 e^{-2\mathcal{E}_0\beta}. \tag{2.35}$$

This is the (approximate) distribution that was suggested by Wien at the end of the 19th century and considered for some time to be the exact law for the blackbody distribution. In terms of (the correct) U_T Eq. (2.31) reads

$$\sigma_{\mathcal{E}}^2 = U_T^2 + 2\mathcal{E}_0 U_T. \tag{2.36}$$

Equations (2.35) and (2.36) represent the germ of quantum theory, since it is precisely on their basis that Planck and Einstein advanced the notion of the quantum (for the material oscillators and for the radiation field, respectively), by putting $2\mathcal{E}_0 = \hbar\omega$. The following comments contain a discussion of their respective points of view and of the relation with our present notions based on the reality of the ZPE. A remarkable relationship will thus be disclosed.

It is important to note that no ZPE was considered by neither Planck nor Einstein in their analysis of Eq. (2.36). Instead, Planck interpreted the term $2\mathcal{E}_0 U_T$ as a result of the discontinuities in the processes of interchange of energy between matter and field (more specifically, in the emissions, from 1912 onwards[16]). As for

Einstein, he interpreted this term as a manifestation of a corpuscular structure of the radiation field, and thus pointed to it as the key to Planck's law.[17] Now, from the point of view proposed here the consideration of the ZPE gives rise to an alternative understanding of Eq. (2.36), namely an interpretation of the term $2\mathcal{E}_0 U_T$ that does not depend on the notion of quanta. The elucidation of the term U_T^2 as a result of the interference of the modes of frequency ω of the thermal field[18] suggests to interpret the term $2\mathcal{E}_0 U_T$ as due to additional interferences, now between the thermal field and a *zero-point radiation field* that is present at all temperatures including $T = 0$, and whose mean energy is just $\mathcal{E}_0$. As is by now clear, in Eq. (2.36) there is no extra term $\mathcal{E}_0^2$ due to the interference among the modes of the ZPF themselves, because the thermodynamic analysis made by Planck and Einstein had no room for the nonthermal fluctuations.

From this new perspective the notion of intrinsic discontinuities in the energy interchange or in the field itself is unnecessary to explain either Planck's law or the linear term in Eq. (2.36). By contrast, it is the existence of a fluctuating ZPF what accounts for the equilibrium spectrum. This could of course not be Planck's or Einstein's interpretation because the ZPE was still unknown at that time, even though their results were consistent with its existence.

What is important to stress is that the interpretation made here of the linear term $2\mathcal{E}_0 U_T$ implies the existence of a fluctuating zero-point radiation *field* — the vacuum (radiation) field — with a mean energy per mode of frequency ω equal to $\mathcal{E}_0 = \hbar\omega/2$. As will be remarked in section 2.8, the existence of fluctuations of this energy is required to recover several other important characteristics of the quantum description.

2.7. *Continuous versus discrete*

We have just seen how three alternative approaches provide three quite different readings of the same quantity $U_T^2 + 2\mathcal{E}_0 U_T$. In these approaches, either the (continuous) ZPF or the quantization is identified as the notion underlying the Planck distribution. Therefore the next logical step is to inquire about the relation between the ZPE and quantization. Is quantization inevitably linked to Planck's law, or is it merely the result of a point of view, of a voluntary choice?

An answer to this question is found from an analysis of the partition function obtained from (2.27). It follows by an integration of the equation $(d \ln Z_g/d\beta) = -U(\beta)$ that the partition function is

$$Z_g = \frac{C}{\sinh \mathcal{E}_0 \beta}. \tag{2.37}$$

The value of the constant C is determined by requiring the classical result $Z_g = \beta^{-1}$ to be recovered in the limit $T \to \infty$. This leads to

$$Z_g(\beta) = \frac{\mathcal{E}_0}{s\omega \sinh \mathcal{E}_0 \beta}. \tag{2.38}$$

The constant s can be determined from the entropy, which is, up to an additive constant and with $S = S_g$,

$$S_g = k_B \ln \frac{\hbar}{s} - k_B \ln(2 \sinh \mathcal{E}_0 \beta) + k_B \beta U. \tag{2.39}$$

In the zero-temperature limit this reduces to

$$S_g(T \to 0) = k_B \ln \frac{\hbar}{s} - k_B \mathcal{E}_0 \beta + k_B \mathcal{E}_0 \beta = k_B \ln \frac{\hbar}{s}. \tag{2.40}$$

Setting the origin of the entropy at $T = 0$ leads to $s = \hbar$. The partition function takes thus the form

$$Z_g(\beta) = \frac{1}{2 \sinh \mathcal{E}_0 \beta}. \tag{2.41}$$

2.7.1. *The origin of discreteness*

We now discuss the discontinuities characteristic of the quantum description, which are hidden in the continuous description given by the distribution W_g (see Refs. 19–21 for related discussions). To this end we expand Eq. (2.38) and write

$$Z_g = \frac{1}{2 \sinh \mathcal{E}_0 \beta} = \frac{e^{-\mathcal{E}_0 \beta}}{1 - e^{-2\mathcal{E}_0 \beta}} = \sum_{n=0}^{\infty} e^{-\mathcal{E}_0 \beta (2n+1)}, \tag{2.42}$$

$$Z_g = \sum_{n=0}^{\infty} e^{-\beta \mathcal{E}_n}, \tag{2.43}$$

with

$$\mathcal{E}_n \equiv (2n+1)\mathcal{E}_0 = \hbar \omega n + \tfrac{1}{2} \hbar \omega. \tag{2.44}$$

The function $g(\mathcal{E})$ can now be determined by means of (2.5),

$$Z_g(\beta) = \int_0^{\infty} g(\mathcal{E}) e^{-\beta \mathcal{E}} d\mathcal{E} = \sum_{n=0}^{\infty} e^{-\beta E_n} = \int_0^{\infty} \sum_{n=0}^{\infty} \delta(\mathcal{E} - \mathcal{E}_n) e^{-\beta \mathcal{E}} d\mathcal{E}, \tag{2.45}$$

whence

$$g(\mathcal{E}) = \sum_{n=0}^{\infty} \delta(\mathcal{E} - \mathcal{E}_n). \tag{2.46}$$

The introduction of (2.45) and (2.46) into Eq. (2.4) finally determines the probability density $W_g(\mathcal{E})$,

$$W_g(\mathcal{E}) = \frac{1}{Z_g} \sum_{n=0}^{\infty} \delta(\mathcal{E} - \mathcal{E}_n) e^{-\beta \mathcal{E}}. \tag{2.47}$$

This distribution gives for the mean value of a function $f(\mathcal{E})$

$$\langle f(\mathcal{E}) \rangle = \int_0^{\infty} W_g(\mathcal{E}) f(\mathcal{E}) d\mathcal{E} = \frac{1}{Z_g} \sum_{n=0}^{\infty} f(\mathcal{E}_n) e^{-\beta \mathcal{E}_n} = \sum_{n=0}^{\infty} w_n f(\mathcal{E}_n), \tag{2.48}$$

with the canonical weights given by

$$w_n = \frac{e^{-\beta\mathcal{E}_n}}{Z} = \frac{e^{-\beta\mathcal{E}_n}}{\sum_{n=0}^{\infty} e^{-\beta\mathcal{E}_n}}. \tag{2.49}$$

The result (2.48) shows that the mean value of a function of the *continuous* variable $\mathcal{E}$ calculated with the distribution $W_g(\mathcal{E})$, can be obtained equivalently by averaging over the set of *discrete* indices (or *states*) n, with respective weights w_n. These weights correspond to those of a canonical ensemble, which suggests identifying $\mathcal{E}_n$ with the discrete energy levels of the quantum oscillators (including the ZPE), as follows from (2.43). Equation (2.48) can be recognized as the description afforded by the density matrix for the canonical ensemble with weights w_n (see e.g. Ref. 22).

Although the two averages in Eq. (2.48) (calculated using W_g or w_n) are formally equivalent, their descriptions are essentially different, the first one referring to an average over the *continuous variable* $\mathcal{E}$, the second one to a summation over *discrete states* (levels) with energies $\mathcal{E}_n$. Because the $\mathcal{E}_n$ are completely characterized by the states, it is natural to interpret the right-hand side of (2.48) as a manifestation of the discrete nature of the energy. The mechanism leading to this discreteness, seemingly excluding all other values of the energy, is due to the highly pathological distribution $g(\mathcal{E})$, Eq. (2.46). It is important to bear in mind that nevertheless the existence of fluctuations leading to the natural linewidth[23] and other processes, effectively dilutes the discrete distribution of energies into a somewhat smoothened-out distribution acquiring a more continuous shape. Thus $g(\mathcal{E})$ should be seen as a theoretical limiting distribution, an explanation for which will be found in section 3 below.

From the present point of view, the quantization of the radiation field does not follow from some intrinsic property, but arises as a property acquired by the field through its interaction with matter in equilibrium. In other words, quantization is here exhibited as an emergent property of matter and field in interaction, an idea that is closely examined all along the present work.

2.8. A quantum statistical distribution

As was discussed in sections 2.4 and 2.5, the existence of a fluctuating ZPE requires a more general distribution than W_g, able to account for all fluctuations of the energy, including the nonthermal contribution. Such distribution was shown to be W_s, and it led to results that are equivalent to the quantum description, in which temperature-independent fluctuations appear as a characteristic trait. A closer study of this problem will help us to establish contact with one of the most frequently used distributions in quantum statistics.

We recall that the statistical distribution appropriate to include all fluctuations is given by Eq. (2.19), namely

$$W_s(\mathcal{E}) = \frac{1}{U} e^{-\mathcal{E}/U}. \tag{2.50}$$

$W_s(\mathcal{E})$ maximizes the statistical entropy S_s, whereas $W_g(\mathcal{E})$ maximizes the thermo-dynamic entropy S defined in the phase space of the particles. The crucial point that guarantees that both distributions describe the same physical system is that in both instances the energy distribution corresponds to maximal entropy.

$W_s(\mathcal{E})$ yields for the variance of the energy at all temperatures (including $T = 0$) the expression

$$(\sigma_\mathcal{E}^2)_s = U^2. \tag{2.51}$$

This result goes back to Lorentz (see Ref. 18) when applied to the fluctuations of the thermal field. It can also be obtained by demanding that it should hold as well for the field at zero temperature, as follows by considering that thermal and zero-point field are part of the same creature. It is reinforced by the observation that the fluctuations of the field arise as a result of the interferences among the immense number of independent modes of a given frequency, so that the central-limit theorem[24,25] applies, which means that the moments (and thus the variance) correspond to those of a normal distribution.

As has been remarked, the specific choice $U = \beta^{-1}$ ($\mathcal{E}_0 = 0$) results in the usual canonical distribution and leads to the classical expression (2.32). But in presence of the nonthermal energy $\mathcal{E}_0$, U is given by Planck's law and the resulting total fluctuations are indeed (with U_T given by (2.34))

$$(\sigma_\mathcal{E}^2)_s = U^2 = (U_T + \mathcal{E}_0)^2 = U_T^2 + 2\mathcal{E}_0 U_T + \mathcal{E}_0^2. \tag{2.52}$$

Therefore, the energy does not have a fixed value at $T = 0$, but is allowed to fluctuate with variance $\mathcal{E}_0^2$. As has been stressed, this term represents the nonthermal contribution to the fluctuations of the energy.

In conformity with the present discussion the total energy can be written as consisting of two *fluctuating* parts,

$$\mathcal{E} = \mathcal{E}_T + \mathcal{E}_0, \tag{2.53}$$

$\mathcal{E}_T$ and $\mathcal{E}_0$ being the thermal and nonthermal energies, respectively. The total energy fluctuations are then given by the sum of three terms,

$$(\sigma_\mathcal{E}^2)_s = \sigma_{\mathcal{E}_T}^2 + \sigma_{\mathcal{E}_0}^2 + 2\Gamma(\mathcal{E}_T, \mathcal{E}_0), \tag{2.54}$$

where $\Gamma(\mathcal{E}_T, \mathcal{E}_0)$ is the covariance of the variables indicated by its arguments,

$$\Gamma(\mathcal{E}_T, \mathcal{E}_0) \equiv \langle \mathcal{E}_T \mathcal{E}_0 \rangle - \langle \mathcal{E}_T \rangle \langle \mathcal{E}_0 \rangle. \tag{2.55}$$

Comparing Eq. (2.54) with Eq. (2.52), we arrive at $\Gamma(\mathcal{E}_T, \mathcal{E}_0) = 0$. This shows that the fluctuations of $\mathcal{E}_T$ and $\mathcal{E}_0$ are statistically independent, as is expected due to the independence of their sources.

The statistical entropy is given, as follows from (2.50), by

$$S_s = -k_B \int W_s \ln W_s d\mathcal{E} = k_B \ln U + k_B, \qquad (2.56)$$

from where it follows that $(\partial S_s / \partial U) = (k_B / U)$. A comparison with the thermodynamic entropy, which satisfies $(\partial S_g / \partial U) = (1/T)$, shows that these two entropies coincide only when $U = k_B T$, i.e., for $\mathcal{E}_0 = 0$.

Let us now investigate how the nonthermal fluctuations become manifest in the statistical properties of the ensemble of oscillators. The usual expression for the energy of the harmonic oscillator (again with $m = 1$)

$$\mathcal{E} = \tfrac{1}{2}(p^2 + \omega^2 q^2), \qquad (2.57)$$

can be used as a starting point to effect a transformation from the energy distribution given by Eq. (2.50) to a distribution $w_s(p, q)$ defined in the oscillator's phase space (p, q). To this end we introduce the pair of variables $(\mathcal{E}, \theta)$ related to the couple (p, q) by the extended canonical transformation[26]

$$p = \sqrt{2\mathcal{E}} \cos \theta, \qquad (2.58)$$

$$q = \sqrt{\frac{2\mathcal{E}}{\omega^2}} \sin \theta, \qquad (2.59)$$

so that $w_s(p, q)$ is given by[25]

$$w_s(p, q) = W_s(\mathcal{E}(p, q), \theta(p, q)) \left| \frac{\partial(\mathcal{E}, \theta)}{\partial(p, q)} \right|, \qquad (2.60)$$

the Jacobian of the transformation being

$$\frac{\partial(p, q)}{\partial(\mathcal{E}, \theta)} = \left| \frac{\partial(\mathcal{E}, \theta)}{\partial(p, q)} \right|^{-1} = \frac{1}{\omega}. \qquad (2.61)$$

Now $W(\mathcal{E})$ is a marginal probability density that can be obtained from $W_s(\mathcal{E}, \theta)$ by integrating over the variable θ, so that

$$W(\mathcal{E}) = \int_0^{2\pi} W_s(\mathcal{E}, \theta) d\theta. \qquad (2.62)$$

For a system of harmonic oscillators in equilibrium, the trajectories (in general, the surfaces) of constant energy do not depend on θ, so all values of θ are equally probable, which means that

$$W_s(\mathcal{E}, \theta) = \frac{1}{2\pi} W(\mathcal{E}). \qquad (2.63)$$

Using Eqs. (2.57) and (2.60) we thus obtain for the distribution in phase space

$$w_s(p, q) = \frac{\omega}{2\pi} W(\mathcal{E}(p, q)) = \frac{\omega}{2\pi U} \exp\left(-\frac{p^2 + \omega^2 q^2}{2U}\right). \qquad (2.64)$$

This expression, which is known in quantum theory as the Wigner function for the harmonic oscillator whenever U corresponds to Planck's law,[27] can be factorized as a product of two normal distributions,

$$w_s(p,q) = w_p(p)w_q(q) = \frac{1}{\sqrt{2\pi\sigma_p^2}}e^{-p^2/2\sigma_p^2} \cdot \frac{1}{\sqrt{2\pi\sigma_q^2}}e^{-q^2/2\sigma_q^2}, \qquad (2.65)$$

where $\sigma_p^2 = U$ and $\sigma_q^2 = U/\omega^2$. The product of these dispersions gives

$$\sigma_q^2\sigma_p^2 = \frac{U^2}{\omega^2} = \frac{\mathcal{E}_0^2}{\omega^2} + \frac{\sigma_{\mathcal{E}_T}^2}{\omega^2} \geq \frac{\mathcal{E}_0^2}{\omega^2} = \frac{\hbar^2}{4}, \qquad (2.66)$$

where Eq. (2.31) (with $\sigma_{\mathcal{E}}^2$ written appropriately as $\sigma_{\mathcal{E}_T}^2$) was used to write the second equality and the value $\mathcal{E}_0 = \hbar\omega/2$ was introduced into the last one.

Equation (2.66) points to the fluctuating energy of the ZPF as the ultimate (and irreducible) source of the Heisenberg inequalities. The magnitude of $\sigma_q^2\sigma_p^2$ is bounded from below because of the nonthermal energy fluctuations; the minimum value $\hbar^2/4$ is reached when all thermal fluctuations have been suppressed. Therefore, descriptions afforded by purely thermal distributions such as W_g cannot account for the meaning of these inequalities. This result stresses again the fact that once a ZPE has been introduced into the theory, new distributions (specifically statistical rather than thermodynamic) are needed to include its fluctuations and to obtain the corresponding quantum statistical properties. Note that the Heisenberg inequalities should be understood as referring to ensemble averages, due to the statistical nature of (2.66).

2.9. Comments on the reality of the zero-point fluctuations

The concept of a zero-point energy of the radiation field entered into scene as early as 1912, with Planck's second derivation of the blackbody spectrum.[16] Yet further to the frustrated attempt by Einstein and Stern,[28] and despite the suggestive proposal made by Nernst in 1916[29] to consider this field as responsible for atomic stability, little or no attention was paid to its existence as a real physical entity that could have a role in the newly developing QM. Interestingly, it was the crystallographers and the physical chemists who through fine spectroscopic analysis verified the existence of the ZPE — linked however to matter, not to the field.[30-32]

Today it is well accepted that the fluctuations of the electromagnetic vacuum are responsible for important observable physical phenomena. Perhaps their best known manifestations, within the atomic domain, are the Lamb shift of energy levels and their contribution to the spontaneous transitions of the excited states to the ground state, as will be seen in section 5. By far the most accepted evidence of the reality of the ZPF is the Casimir effect, that is, the force between two parallel neutral metallic plates resulting from the modification of the field by the boundaries (see e.g. Refs. 33, 34 and section 5). Thus the existence of the ZPF can be considered a reasonably well established physical fact. In the following chapters we will have occasion to study in depth the essential role played more broadly by the fluctuating ZPF in its interaction with matter at the atomic level.

3. Emergence of quantum mechanics

The above discussion led us to an important conclusion: the radiation field in equilibrium with matter acquires a discrete energy distribution, by virtue of its zero-point component. This was interpreted to mean that the quantization of the field comes about from its interaction with matter. Then, what about matter?

As our journey progresses it will become clear that also matter is so strongly influenced by its interaction with the background field that it ends up acquiring its quantum properties. Once again, quantization is revealed as a phenomenon that emerges as a result of the matter-field interaction.

The material presented here is mostly based on previous work; detailed derivations and additional references can be found in Refs. 10 and 35–38.

3.1. *Embarking on our journey towards the Schrödinger equation*

The main actor in this section is a charged particle immersed in the zero-point field (ZPF), performing a bounded motion under the action of an external conservative force. In addition, the particle may be subject to some external radiation field. What is important, however, is that the ZPF is always present. To keep the exposition as simple as possible, a one-dimensional motion is normally considered.

We start from the Abraham-Lorentz equation of motion

$$m\ddot{\mathbf{x}} = \mathbf{f}(\mathbf{x}) + m\tau\dddot{\mathbf{x}} + e\mathbf{E}(\mathbf{x}, t) + \frac{e}{c}\mathbf{v} \times \mathbf{B}(\mathbf{x}, t). \tag{3.1}$$

The term $m\tau\dddot{\mathbf{x}}$, with $\tau = 2e^2/3mc^3$, represents the radiation-reaction force on the particle. For an electron this force is normally small ($\tau \sim 10^{-23}$ s). The fields $\mathbf{E}(\mathbf{x}, t)$ and $\mathbf{B}(\mathbf{x}, t)$ must be represented by stochastic variables.

A number of simplifications and approximations will be introduced to go ahead. First we simplify the Lorentz force considering that in the nonrelativistic regime the magnetic term becomes negligible compared with the electric force, so

$$m\ddot{x} = f(x) + m\tau\dddot{x} + eE(x, t). \tag{3.2}$$

As a second step we use the long-wavelength approximation, considering that in the region of space occupied by the particle during its motion, the electric field does not vary appreciably. The x-dependence of $E(x, t)$ can then be neglected, so that one can write

$$m\ddot{x} = f(x) + m\tau\dddot{x} + eE(t). \tag{3.3}$$

$E(t)$ is a stochastic variable with zero mean value, $\overline{E(t)}^E = 0$. The spectral energy density $\rho(\omega)$ of the field corresponds to a mean energy $\hbar\omega/2$ per mode of frequency ω, hence (see e.g. Ref. 23)

$$\rho(\omega) = \rho_{\text{modes}}(\omega) \cdot \frac{1}{2}\hbar\omega = \frac{\omega^2}{\pi^2 c^3} \cdot \frac{1}{2}\hbar\omega = \frac{\hbar\omega^3}{2\pi^2 c^3}, \tag{3.4}$$

which represents a highly colored noise. This result can also be expressed in terms of the correlation of the Fourier transform,

$$\tilde{E}(\omega) = \frac{1}{\sqrt{2\pi}} \int_{-\infty}^{+\infty} E(t)e^{i\omega t}dt, \tag{3.5}$$

$$\overline{\tilde{E}(\omega)\tilde{E}^*(\omega')}^E = \frac{2\hbar\omega^3}{3c^3}\delta(\omega - \omega'). \tag{3.6}$$

Equation (3.6) means that the Fourier components of the (stationary) ZPF that pertain to different frequencies are statistically independent.

3.2. Fokker-Planck-type equation in phase space

We write

$$m\dot{x} = p, \quad \dot{p} = f(x) + m\tau\dddot{x} + eE(t), \tag{3.7}$$

and start from the continuity equation for the total system in terms of the density $R(x_\alpha, p_\alpha, t)$ of states in the whole phase space, where $\{x_\alpha, p_\alpha\}$ denotes the set of variables of both particle and field,

$$\frac{\partial R}{\partial t} + \sum_\alpha \left[\frac{\partial}{\partial x_\alpha}(\dot{x}_\alpha R) + \frac{\partial}{\partial p_\alpha}(\dot{p}_\alpha R) \right] = 0. \tag{3.8}$$

The ensuing Fokker-Planck-type equation for $Q(x, p, t)$, which is the average of R over the field variables, is a complicated integro-differential equation, or, equivalently, a differential equation of infinite order. A good part of the complication is due to the memory developed as the system evolves in time. This equation reads[35,39]

$$\frac{\partial Q}{\partial t} + \frac{1}{m}\frac{\partial}{\partial x}pQ + \frac{\partial}{\partial p}(f + m\tau\dddot{x})Q = e^2\frac{\partial}{\partial p}\hat{D}Q, \tag{3.9}$$

where the diffusion operator $\hat{D}$ is

$$\hat{D}(x, p, t)Q(x, p, t) = \hat{P}E\hat{G}\frac{\partial}{\partial p}E\sum_{k=0}^{\infty}\left[e\hat{G}\frac{\partial}{\partial p}\left(1 - \hat{P}\right)E\right]^{2k}Q. \tag{3.10}$$

$\hat{P}$ is the smoothing (projection) operator, which averages over the stochastic field variables, so that

$$Q = \hat{P}R = \overline{R}^E, \quad \delta Q = \left(1 - \hat{P}\right)R,. \quad R = Q + \delta Q.$$

$\hat{G}$ is the operator

$$\hat{G}A(x, p) = \int_0^t e^{-\hat{L}(t-t')}A(\mathring{x}, \mathring{p})dt',$$

$\hat{L}$ is the Liouvillian operator of the particle,[§]

$$\hat{L} = \frac{1}{m}\frac{\partial}{\partial x}p + \frac{\partial}{\partial p}\left(f + m\tau\dddot{x}\right), \qquad (3.11)$$

and $\mathring{x}, \mathring{p}$ under the integral denote the values that the phase-space variables must have at time t' so as to evolve towards x, p at time t.

Note that, by virtue of (3.10), the right-hand side of the Fokker-Planck-type Eq. (3.9) contains integro-differential terms of increasing order in e^2 and in the energy density of the ZPF, which makes it virtually impossible to find an exact solution for this equation.

3.3. *Some relations for average values*

From Eq. (3.9) it is a straightforward matter to derive equations for the average values of dynamical quantities. Consider the average value of a general phase function $G(x, p)$ that has no explicit time dependence. By multiplying (3.9) by G and integrating by parts it follows that

$$\frac{d}{dt}\langle G\rangle = \frac{1}{m}\left\langle p\frac{\partial G}{\partial x}\right\rangle + \left\langle f\frac{\partial G}{\partial p}\right\rangle + m\tau\left\langle \dddot{x}\frac{\partial G}{\partial p}\right\rangle - e^2\left\langle\left(\frac{\partial G}{\partial p}\right)\hat{D}(t)\right\rangle. \qquad (3.12)$$

By taking successively $G = x, x^2, p, p^2, xp$, one obtains

$$\frac{d}{dt}\langle x\rangle = \frac{1}{m}\langle p\rangle; \qquad (3.13)$$

$$\frac{d}{dt}\langle x^2\rangle = \frac{2}{m}\langle xp\rangle; \qquad (3.14)$$

$$\frac{d}{dt}\langle p\rangle = \langle f\rangle + m\tau\langle\dddot{x}\rangle - e^2\left\langle\hat{D}(t)\right\rangle; \qquad (3.15)$$

$$\frac{d}{dt}\langle p^2\rangle = 2\langle fp\rangle + 2m\tau\langle\dddot{x}p\rangle - 2e^2\left\langle p\hat{D}(t)\right\rangle; \qquad (3.16)$$

$$\frac{d}{dt}\langle xp\rangle = \frac{1}{m}\langle p^2\rangle + \langle xf\rangle + m\tau\langle x\dddot{x}\rangle - e^2\left\langle x\hat{D}(t)\right\rangle, \qquad (3.17)$$

whence for $H = (p^2/2m) + V$, the mechanical Hamiltonian function,

$$\frac{d}{dt}\langle H\rangle = \tau\langle\dddot{x}p\rangle - \frac{e^2}{m}\left\langle p\hat{D}(t)\right\rangle. \qquad (3.18)$$

Let us consider that a stationary state is eventually reached by the system. Under stationarity the time derivative of the mean value of any variable is zero.

[§]Strictly speaking Eq. (3.11) is not a Liouvillian, due to the radiation reaction term, which here is taken as an external force, given approximately by $(\tau/m)f'p$.

It follows that $\langle p \rangle = 0$ and $\langle xp \rangle = 0$, which means that x and p end up being uncorrelated. Further, Eq. (3.17) reads

$$\frac{1}{m} \langle p^2 \rangle + \langle xf \rangle + m\tau \langle x\ddot{x} \rangle = e^2 \left\langle x\hat{D}(t) \right\rangle. \tag{3.19}$$

Neglecting the terms of order e^2 this reduces to $\frac{1}{m} \langle p^2 \rangle + \langle xf \rangle = 0$, which is the virial theorem. Thus, Eq. (3.17) is a time-dependent form of the virial theorem for the present problem. Also under stationarity, (3.18) reduces to

$$m\tau \langle \ddot{x}p \rangle = e^2 \left\langle p\hat{D}(t) \right\rangle, \tag{3.20}$$

which constitutes the *energy-balance condition*. This is an important result that will be used later.

From (3.15) we see that any net diffusion giving rise to a term $-e^2 \left\langle \hat{D}(t) \right\rangle$ develops a net force acting on the particle; this is similar to the osmotic force in the case of Brownian diffusion. As follows from (3.16), this force conveys to the particle a net kinetic energy per unit of time of value $\left(-e^2/m \right) \left\langle p\hat{D}(t) \right\rangle$, in agreement with Eq. (3.18).

3.4. *Transition to the configuration space*

At this point it is convenient to reduce the present description to the configuration space of the particle, in order to make contact with the Schrödinger description of quantum mechanics. The transition to configuration space can be performed in a systematic way with the help of the characteristic function (or momentum generating function) $\widetilde{Q}$,

$$\widetilde{Q}(x, z, t) = \int Q(x, p, t)e^{ipz}dp. \tag{3.21}$$

The probability density $\rho(x, t)$ in configuration space is a marginal probability,

$$\rho(x, t) = \int Q(x, p, t)dp = \int Q(x, p, t)e^{ipz}dp \bigg|_{z=0} = \widetilde{Q}(x, 0, t), \tag{3.22}$$

and the local moments of p are given by

$$\langle p^n \rangle (x) \equiv \langle p^n \rangle_x = \frac{1}{\rho(x)} \int p^n Q dp = (-i)^n \left(\frac{1}{\widetilde{Q}} \frac{\partial^n \widetilde{Q}}{\partial z^n} \right) \bigg|_{z=0}. \tag{3.23}$$

Note that the moments $\langle p^n \rangle (x)$ represent *partially averaged* quantities, for a given value of x. The fully averaged quantities are $\langle p^n \rangle = \int \langle p^n \rangle_x \rho(x)dx$.

By taking the Fourier transform of (3.9) and assuming that all surface terms appearing along the integrations vanish at infinity, one gets

$$\frac{\partial \widetilde{Q}}{\partial t} - i\frac{1}{m} \frac{\partial^2 \widetilde{Q}}{\partial x \partial z} - izf(x)\widetilde{Q} - \frac{\tau}{m}f'z\frac{\partial \widetilde{Q}}{\partial z} = -ie^2 z(\widetilde{DQ}). \tag{3.24}$$

It is important to observe that instead of making the transition from the phase space to the configuration space, we could have made the transit to the momentum space. In such case one gets

$$\tilde{P}(k, p, t) = \int Q(x, p, t) e^{ikx} dx, \tag{3.25}$$

$$\frac{\partial \tilde{P}}{\partial t} - i \frac{p}{m} k \tilde{P} + \frac{1}{2\pi} \frac{\partial}{\partial p} \int \tilde{K}(k - k', p) \tilde{P}(k', p) dk' = e^2 \frac{\partial}{\partial p} \widetilde{(\hat{D}P)}(k, p), \tag{3.26}$$

$$\text{with} \qquad K(x, p) = f(x) + \frac{\tau}{m} f'(x) p. \tag{3.27}$$

Things become more complicated in the p-description, since instead of a differential equation one gets an integro-differential equation (one with long memory). In what follows we shall continue to analyze the reduced description in the configuration space.

A convenient procedure to get the most of Eq. (3.24) consists in expanding it into a power series around $z = 0$, and separating the coefficients of z^k ($k = 0, 1, 2, \ldots$). The first three equations thus obtained are

$$\frac{\partial \rho}{\partial t} + \frac{1}{m} \frac{\partial}{\partial x} (\langle p \rangle_x \rho) = 0; \tag{3.28}$$

$$\frac{\partial}{\partial t} (\langle p \rangle_x \rho) + \frac{1}{m} \frac{\partial}{\partial x} (\langle p^2 \rangle_x \rho) - f\rho - \frac{\tau}{m} f' \langle p \rangle_x \rho = -e^2 \widetilde{(DQ)} \Big|_{z=0}; \tag{3.29}$$

$$\frac{\partial}{\partial t} (\langle p^2 \rangle_x \rho) + \frac{1}{m} \frac{\partial}{\partial x} (\langle p^3 \rangle_x \rho) - f \langle p \rangle_x \rho - \frac{\tau}{m} f' \langle p^2 \rangle_x \rho = 2e^2 \widetilde{(DpQ)} \Big|_{z=0}, \tag{3.30}$$

and so on. The subsequent equations (corresponding to higher powers of z) are connected to the above by the same elements $\rho, \langle p \rangle_x, \langle p^2 \rangle_x, \ldots$ in addition to contributions deriving from the term $z\widetilde{(DQ)}$. The entire set of these equations constitutes thus an infinite hierarchy of coupled nonlinear equations.

The first member of the hierarchy is the continuity equation, which describes the transfer of matter in configuration space. It follows that the mean flux of particles is $j(x) = \rho(x) \langle p \rangle_x / m$, hence the local mean velocity is given by

$$v(x) = \frac{j(x)}{\rho(x)} = \frac{1}{m} \langle p \rangle_x. \tag{3.31}$$

The function $v(x)$ refers to a locally averaged quantity; it differs essentially from the instantaneous velocity of *one* (specific) particle that visits the neighborhood of x at time t, and varies randomly from one particle to another. To avoid confusion we call $v(x)$ *local mean velocity* or *flux (flow) velocity*.

The second equation describes the transfer of momentum, or equivalently, the evolution of the current density $j(x, t) = v(x, t) \rho(x, t)$. It contains, in addition to ρ and $\langle p \rangle_x$, the second moment $\langle p^2 \rangle_x$, whose value is determined by the third equation, the one describing the transfer of kinetic energy (up to a factor $1/2m$). These moments are

$$\langle p \rangle_x = \frac{1}{\rho(x)} \int pQ dp = -i \left(\frac{1}{\widetilde{Q}} \frac{\partial \widetilde{Q}}{\partial z} \right)\Bigg|_{z=0} = -i \left(\frac{\partial}{\partial z} \ln \widetilde{Q} \right)\Bigg|_{z=0}, \tag{3.32}$$

$$\langle p^2 \rangle_x = -\left(\frac{1}{\widetilde{Q}} \frac{\partial^2 \widetilde{Q}}{\partial z^2} \right)\Bigg|_{z=0} = -\left(\frac{\partial}{\partial z} \ln \widetilde{Q} \right)^2\Bigg|_{z=0} - \left(\frac{\partial^2}{\partial z^2} \ln \widetilde{Q} \right)\Bigg|_{z=0}. \tag{3.33}$$

Combining these expressions gives

$$\langle p^2 \rangle_x - \langle p \rangle_x^2 = -\left(\frac{\partial^2}{\partial z^2} \ln \widetilde{Q} \right)\Bigg|_{z=0}. \tag{3.34}$$

With the change of variables $(x, z) \rightarrow (z_+, z_-)$,

$$z_+ = x + \eta z, \quad z_- = x - \eta z, \tag{3.35}$$

where η is a parameter with dimensions of action, whose value remains to be fixed, the above expressions transform into

$$\langle p \rangle_x = -i\eta \left(\partial_+ - \partial_- \right) \ln \widetilde{Q} \Big|_{z=0},$$

$$\langle p^2 \rangle_x - \langle p \rangle_x^2 = -\eta^2 \left(\frac{\partial^2}{\partial x^2} \ln \widetilde{Q} \right)\Bigg|_{z=0} + 4\eta^2 \left(\partial_+ \partial_- \ln \widetilde{Q} \right)\Big|_{z=0}. \tag{3.36}$$

Writing $\widetilde{Q}$ in the general form

$$\widetilde{Q}(z_+, z_-, t) = q_+(z_+, t) q_-(z_-, t) \chi(z_+, z_-, t) \tag{3.37}$$

where χ represents the nonfactorizable contribution to $\widetilde{Q}(z_+, z_-)$, and taking into account (3.22), one gets

$$\langle p \rangle_x = -i\eta \frac{\partial}{\partial x} \ln \frac{q_+(x, t)}{q_-(x, t)} + g, \tag{3.38}$$

$$\text{with} \quad g = i\eta \left(\partial_- - \partial_+ \right) \ln \chi |_{z=0}, \tag{3.39}$$

and

$$\langle p^2 \rangle_x - \langle p \rangle_x^2 = -\eta^2 \frac{\partial^2}{\partial x^2} \ln \rho + \Sigma, \tag{3.40}$$

$$\text{with} \quad \Sigma = 4\eta^2 \left(\partial_+ \partial_- \ln \chi \right)|_{z=0}. \tag{3.41}$$

The first two equations of the hierarchy, generalized to three dimensions, become thus (a sum over repeated indices is understood)

$$\frac{\partial \rho}{\partial t} + \frac{\partial}{\partial x_j} (v_j \rho) = 0, \tag{3.42}$$

$$m\frac{\partial}{\partial t}(v_i\rho) + m\frac{\partial}{\partial x_j}(v_iv_j\rho) - \frac{\eta^2}{m}\frac{\partial}{\partial x_j}\left(\rho\frac{\partial^2}{\partial x_i\partial x_j}\ln\rho\right) + \frac{1}{m}\frac{\partial}{\partial x_j}\Sigma_{ij}\rho - f_i\rho$$

$$= \tau v_i\frac{\partial f_i}{\partial x_j}\rho - e^2\left.(\widetilde{\hat{D}Q})_i\right|_{z=0}.$$

These equations will be analyzed in the following sections.

Note that according to Eq. (3.21), $\widetilde{Q}^*(\mathbf{x},\mathbf{z},t) = \widetilde{Q}(\mathbf{x},-\mathbf{z},t)$, so that from (3.37) (we come back to a three-dimensional description for convenience),

$$q_+(\mathbf{z}_-,t) = q_-^*(\mathbf{z}_-,t), \quad q_-(\mathbf{z}_+,t) = q_+^*(\mathbf{z}_+,t), \quad \chi^*(\mathbf{z}_+,\mathbf{z}_-,t) = \chi(\mathbf{z}_-,\mathbf{z}_+,t).$$
$$(3.43)$$

$\widetilde{Q}$ can be rewritten therefore as

$$\widetilde{Q}(\mathbf{z}_+,\mathbf{z}_-,t) = q(\mathbf{z}_+,t)q^*(\mathbf{z}_-,t)\chi(\mathbf{z}_+,\mathbf{z}_-,t), \tag{3.44}$$

where $q(\mathbf{z}_+,t) \equiv q_+(\mathbf{z}_+,t), \quad q^*(\mathbf{z}_-,t) \equiv q_+^*(\mathbf{z}_-,t).$ (3.45)

Further, from (3.22) and (3.44) it follows that one may write

$$\rho(\mathbf{x},t) = \widetilde{Q}(\mathbf{x},0,t) = q^*(\mathbf{x},t)q(\mathbf{x},t)\chi_0(\mathbf{x},t), \tag{3.46}$$

with $\chi_0(\mathbf{x},t) = \chi|_{\mathbf{z}=0}$ a real function that can be taken as a constant without loss of generality, absorbing its possible time and space dependence into the functions $q(\mathbf{x},t), q^*(\mathbf{x},t)$. We therefore write

$$\chi_0(\mathbf{x},t) = 1, \quad \rho(\mathbf{x},t) = q^*(\mathbf{x},t)q(\mathbf{x},t). \tag{3.47}$$

3.5. *The Schrödinger equation*

The Fokker-Planck-type Eq. (3.9) — or its equivalent in $(\mathbf{x},\mathbf{z})$-space, Eq. (3.24) — contains essentially two kinds of terms: the first three originate in the Newtonian part of the Liouvillian, and the last two originate in the matter-field interaction and describe the dissipative and stochastic (diffusive) behavior of the particle, respectively. The latter gave rise to the terms on the right-hand side of Eq. (3.4). The first one, proportional to τ (hence to e^2), is due to the radiation reaction and has a dissipative effect over the motion. The second one (also proportional to e^2, plus higher-order terms) describes a permanent fluctuating action over the motion.

We assume that the systems of present interest reach — in the time-asymptotic limit, when the irreversible processes involving the radiation field have disappeared — a state of balance between the mean absorbed and radiated powers, as expressed in Eq. (3.20). This is to be expected for all bound systems, but we leave this point open to further consideration. Under such condition the two radiative contributions essentially cancel each other in the mean. The fact that both terms are proportional to e^2 means that their remaining contribution should represent radiative corrections and can therefore be neglected in a first approximation. We describe here this situation of energy balance in the *radiationless approximation*, and leave the detailed

discussion about the energy-balance condition for subsection 3.5.2. The first two
equations of the hierarchy reduce then to

$$\frac{\partial \rho}{\partial t} + \frac{\partial}{\partial x_j}\left(v_j\rho\right) = 0, \tag{3.48}$$

$$m\frac{\partial}{\partial t}(v_i\rho) + m\frac{\partial}{\partial x_j}(v_iv_j\rho) - \frac{\eta^2}{m}\frac{\partial}{\partial x_j}\left(\rho\frac{\partial^2}{\partial x_i\partial x_j}\ln\rho\right) + \frac{1}{m}\frac{\partial}{\partial x_j}\Sigma_{ij}\rho - f_i\rho = 0. \tag{3.49}$$

A series of algebraic manipulations allows us to recast them as

$$\rho\frac{\partial}{\partial x_i}\hat{M}q + \rho\left(\frac{\partial \mathbf{g}}{\partial t} - \mathbf{v}\times(\nabla\times\mathbf{g})\right)_i + \frac{1}{m}\left[\nabla\cdot(\rho\mathbf{\Sigma})\right]_i = 0 \tag{3.50}$$

and its complex conjugate, with

$$\hat{M}q \equiv \frac{1}{2m}\left(-2i\eta\nabla + \mathbf{g}\right)^2 q + Vq - 2i\eta\frac{\partial q}{\partial t}. \tag{3.51}$$

The functions g_i and Σ_{ij} are defined through Eqs. (3.39) and (3.41). Since according
to Eq. (3.47), $q^*(\mathbf{x},t)q(\mathbf{x},t) = \rho(\mathbf{x},t)$, we see that the operator $\hat{M}$ acts over a
probability amplitude. When $\hat{M}q = 0$, Eq. (3.51) becomes the Schrödinger equation
for $q(\mathbf{x},t) \equiv \psi(\mathbf{x},t)$, containing the as yet undetermined parameter η and the vector
function $\mathbf{g}(\mathbf{x},t)$,

$$\frac{1}{2m}\left(-2i\eta\nabla + \mathbf{g}\right)^2 \psi + V\psi = 2i\eta\frac{\partial \psi}{\partial t}, \tag{3.52}$$

and

$$\rho(\mathbf{x},t) = \psi(\mathbf{x},t)\psi^*(\mathbf{x},t) \tag{3.53}$$

corresponds to the Born rule. Equation (3.52) requires that the remaining terms in
Eq. (3.50) cancel each other, i.e.,

$$\left(\frac{\partial \mathbf{g}}{\partial t} - \mathbf{v}\times(\nabla\times\mathbf{g})\right) = -\frac{1}{m}\left[\nabla\cdot(\rho\mathbf{\Sigma})\right]. \tag{3.54}$$

In other words, Eq. (3.52) is the Schrödinger equation with minimal coupling (with
$\mathbf{g} = -\frac{e}{c}\mathbf{A}$), provided that: a) the parameter η has the value $\hbar/2$, and b) condition
(3.54) holds.

The determination of the parameter η is made below, where it is shown that
indeed, its value is given correctly by $\hbar/2$. As to the functions g_i and Σ_{ij}, we
note from (3.39) and (3.41) that they depend on the first and second derivatives,
respectively, of the (nonfactorizable) function $\chi(\mathbf{z}_+, \mathbf{z}_-)$, evaluated at $\mathbf{z} = 0$. The
function $\chi(\mathbf{z}_+, \mathbf{z}_-, t)$ cannot be found exactly without solving the complete hierarchy of equations in configuration space. For the time being we shall make some
reasonable assumptions that allow us to move ahead; once we have learned the
meaning of condition (3.54) we may attempt to study the general (unconditioned)
case.

3.5.1. *Factorizable case*

Let us consider the case in which the coupling function $\mathbf{g} = 0$, i.e., $\tilde{Q}(\mathbf{z}_+, \mathbf{z}_-, t)$ is factorizable to first order in $\mathbf{z}$. Then it follows that

$$\nabla \cdot (\rho \mathbf{\Sigma}) = 0. \tag{3.55}$$

Equation (3.52) reduces, with $\psi(\mathbf{x}, t) = q(\mathbf{x}, t)$, to

$$\frac{2\eta^2}{m} \nabla^2 \psi + V\psi = 2i\eta \frac{\partial \psi}{\partial t}, \tag{3.56}$$

and the local momentum is given by

$$\langle \mathbf{p} \rangle_x = m\mathbf{v}(\mathbf{x}) = -i\eta \nabla \ln \left[\psi(\mathbf{x}, t)/\psi^*(\mathbf{x}, t) \right]. \tag{3.57}$$

A general solution to Eq. (3.55) is

$$\mathbf{\Sigma} = \frac{K}{\rho}, \tag{3.58}$$

where K is a divergencefree second-rank tensor. From Eq. (3.54) we verify that the quantity $\mathbf{\Sigma}$ determining the function χ that enters into the phase-space distribution (Eq. (3.44)), is related to radiative corrections. On the contrary, the quantity $\mathbf{\Sigma}$ in Eq. (3.58) is not necessarily a small radiative correction. Thus, it can be taken to correct, when necessary, the negative value of the quantity $\langle \mathbf{p}^2 \rangle_x - \langle \mathbf{p} \rangle_x^2$, which occurs in some quantum mechanical problems. Of course, this last requirement is not part of the usual formal baggage of quantum mechanics because functions such as $\langle \mathbf{p}^2 \rangle_x$ do not belong to the Hilbert-space formalism, and (both parts of) the function $\mathbf{\Sigma}$ are unknown to that theory.

Notice from Eqs. (3.48), (3.49) that with $\mathbf{\Sigma} = 0$ we are left with a couple of nonlinear equations for $\rho(\mathbf{x}, t)$ and $\mathbf{v}(\mathbf{x}, t)$, which is uncoupled from the rest of the hierarchy and is entirely equivalent to Schrödinger's equation:

$$\frac{\partial \rho}{\partial t} + \frac{\partial}{\partial x_j} (v_j \rho) = 0, \tag{3.59}$$

$$m\frac{\partial}{\partial t}(v_i \rho) + m \frac{\partial}{\partial x_j}(v_i v_j \rho) - \frac{\eta^2}{m} \frac{\partial}{\partial x_j} \left(\rho \frac{\partial^2}{\partial x_i \partial x_j} \ln \rho \right) - f_i \rho = 0. \tag{3.60}$$

The term that contains the $\ln \rho$ is due to the momentum fluctuations (transcribed to configuration space by the reduction process). It is nonlocal in nature — even in the single-particle case — due to its dependence on the distribution of the particles in the entire space. It encapsulates, as will become clear below, the quantum behavior of matter, including quantum fluctuations and the characteristic apparent quantum non-local effects. Since the source of the momentum fluctuations is the ZPF, it follows that *the quantum fluctuations are conventional fluctuations with a causal origin.*

3.5.2. Detailed energy balance: Determining η

We now focus on the energy-balance condition (3.20)

$$m\tau \langle \ddot{x} p \rangle = e^2 \left\langle p\hat{D}(t) \right\rangle \tag{3.61}$$

and use it to determine the value of η. According to this equation, energy balance is reached when the average power dissipated by the particle along its orbital motion (the left-hand side) is compensated by the average power absorbed by the fluctuations impressed upon the particle by the random field along the mean orbit.

The statistical description of the system is now given by Eq. (3.56), so we use it to carry out the calculations, to lowest order in e^2. The mean values on both sides of Eq. (3.61) are calculated for the particle in its ground state (ψ_0) and the background field also in its ground state, with spectral energy density given by (3.4)

$$\rho_0(\omega) = \frac{\hbar \omega^3}{2\pi^2 c^3}. \tag{3.62}$$

The left-hand side of (3.61) gives

$$m\tau \langle \ddot{x} p \rangle_0 = -m\tau \sum_k \omega_{0k}^4 |x_{0k}|^2, \tag{3.63}$$

with $\omega_{0k} = (\mathcal{E}_0 - \mathcal{E}_k)/2\eta$ and $x_{0k} = \int \psi_0^* x \psi_k dx$, where $\mathcal{E}_k$ are the energy eigenvalues and ψ_k are the corresponding eigenfunctions of Eq. (3.56). To calculate the right-hand side we write to lowest order in e^2

$$e^2 \frac{\partial}{\partial p} \hat{D}(t) Q = \frac{4\pi}{3} e^2 \int d\omega \int dt' \rho_0(\omega) \cos \omega(t-t') e^{-\hat{L}(t-t')} \frac{\partial}{\partial p} Q(t'). \tag{3.64}$$

After multiplying by p/m and integrating over the phase space we obtain

$$\frac{e^2}{m} \left\langle p\hat{D} \right\rangle = \frac{\hbar\tau}{\pi} \int_0^\infty d\omega \, \omega^3 \int_0^t dt' \cos \omega(t-t') \, I(t-t') \tag{3.65}$$

with

$$I(t-t') = \int dx' \int dp' \, p \frac{\partial}{\partial p'} Q(x', p', t') = -\left\langle \frac{\partial p}{\partial p'} \right\rangle_0, \tag{3.66}$$

where $p' = p(t')$ evolves towards $p(t)$ and the subindex 0 indicates that the mean value of the propagator $\partial p / \partial p'$ is to be calculated in the ground state. Using the solutions of Eq. (3.56) we thus write

$$\left\langle \frac{\partial p}{\partial p'} \right\rangle_0 = \frac{1}{2i\eta} \langle [\hat{x}', \hat{p}] \rangle_0 = \frac{m}{2\eta} \sum_k \omega_{k0} |x_{0k}|^2 \cos \omega_{k0}(t-t'), \tag{3.67}$$

which inserted in Eq. (3.65) gives after integrating over time, in the time-asymptotic limit ($t \to \infty$), the result

$$\frac{e^2}{m} \left\langle p\hat{D} \right\rangle_0 = -\frac{\hbar m\tau}{2\eta} \sum_k \omega_{0k}^4 |x_{0k}|^2. \tag{3.68}$$

Equating this with (3.63), we obtain for η the value

$$\eta = \frac{\hbar}{2}.$$

With this result for η, Eq. (3.56) coincides exactly with Schrödinger's equation

$$i\hbar \frac{\partial \psi}{\partial t} = -\frac{\hbar^2}{2m}\nabla^2 \psi + V\psi. \tag{3.69}$$

This is, then, the door through which Planck's constant enters into the quantum description. Note that the balance condition is seen to be satisfied not only globally, but frequency by frequency, for every ω_{0k}, so that it reflects a condition of *detailed energy balance* between particle and field, which means that the spectral density of the vacuum field at equilibrium remains unaffected. In fact the ZPF with spectrum $\rho \sim \omega^3$ *is the single one* that guarantees detailed balance and leads to equilibrium with the ground state of the material subsystem. This radically departs from the classical situation, in which detailed equilibrium is reached with the Rayleigh-Jeans spectrum, proportional to ω^2, as was established by van Vleck almost a century ago.[40,41] The structure of the Rayleigh-Jeans spectrum is tightly linked to the Maxwell-Boltzmann distribution, so the above results confirm that the quantum systems obey a nonclassical statistics.

It is clear that, in general, the equality (3.61) can hold only for a selected set of mean stationary motions. This discloses the mechanism responsible for quantization: the atomic stationary states are those for which the equality holds. Such demanding energy equilibrium can hold only for certain orbital motions. Equation (3.61) explains thus *why* atoms reach and maintain their stability. Thanks to the existence of the ZPF it becomes possible to explain this stability: the electrons radiate to the field, but at the same time they absorb energy from it.

3.5.3. *Local velocities*

From the discussion in sections 3.1 and 3.2 it becomes clear that the particles follow (stochastic) trajectories, so the study of such trajectories becomes an important subject for the theory, at least because their knowledge should help to get a better inkling on the quantum behavior of matter and its motions. As a complement to the above discussion, notice that in parallel to the (local) current velocity $v = \langle p \rangle_x / m$ one may introduce a velocity u associated with the mean local value of the momentum fluctuations (cf. Eq. (3.40)),

$$u = \frac{\hbar}{2m}\frac{\partial \ln \rho}{\partial x} = \frac{\hbar}{2m}\left(\frac{1}{\psi^*}\frac{\partial \psi^*}{\partial x} + \frac{1}{\psi}\frac{\partial \psi}{\partial x}\right). \tag{3.70}$$

This stochastic velocity (it disappears if the ZPF is disconnected) has some place in usual QM, although it goes largely unnoticed. Indeed, from combining Eqs. (3.31) and (3.70) it follows that

$$-i\hbar \frac{\partial \psi}{\partial x} = \hat{p}\psi = m(v - iu)\psi. \tag{3.71}$$

We see that an application of the quantum operator $-i\hbar\partial/\partial x$ gives more than just the expected current velocity. It also carries with it information about the diffusion taking place in the momentum subspace (projected onto the configuration space), in the form of an additional effective momentum mu (affected by the imaginary unit, i). By taking into account that the mean values satisfy $\langle v'\rangle = -(2m/\hbar)\langle uv\rangle$ and $\langle u'\rangle = -(2m/\hbar)\langle u^2\rangle$ we arrive at

$$\langle\hat{p}^2\rangle/2m = \tfrac{1}{2}m\langle v^2 + u^2\rangle. \tag{3.72}$$

This result shows that the two velocities contribute symmetrically to the mean value of the kinetic energy. From this point of view they are equally important, despite the concealed presence (or patent absence) of u in usual quantum theory. Being associated with diffusion, the velocity u is the counterpart of the osmotic velocity characteristic of Brownian processes.

Consider a stationary state with $v = 0$, so its mean kinetic energy is given by $\tfrac{1}{2}m\langle u^2\rangle$. Since from (3.70) one gets

$$\tfrac{1}{2}m\langle u^2\rangle = \frac{\hbar^2}{8m}\int\rho\left(\frac{\nabla\rho}{\rho}\right)^2 d^3x = -\frac{\hbar^2}{8m}\int\rho\nabla^2\ln\rho, \tag{3.73}$$

the total mean energy of the system is given in this case by

$$\langle H\rangle = \int\rho(x)\left[-\frac{\hbar^2}{8m}\frac{\partial^2}{\partial x^2}\ln\rho + V(x)\right]dx. \tag{3.74}$$

Under the demand that this mean energy acquire an extremum value under conservation of probability, $\int\rho(x)dx = 1$, with $\rho = \varphi^2$, this variational problem has as solution the Euler-Lagrange equation[42]

$$-\frac{\hbar^2}{2m}\frac{\partial^2\varphi}{\partial x^2} + V\varphi = \mathcal{E}\varphi, \tag{3.75}$$

where $\mathcal{E}$ is a Lagrange multiplier. This result emphasizes the remarkable role played by the averaged local dispersion of the momentum, Eq. (3.73): it leads directly to the (stationary) Schrödinger equation and guarantees that the stationary probability distribution of particles corresponds to an extremum (normally a minimum) of the mean energy of the system. These extrema of the mean energy correspond to the quantized solutions.

The observation that the energy of the stationary states is a local minimum is very suggestive. Of course every particle finds eventually its own specific (stochastic) trajectory, with a certain mean energy (averaged over the trajectory) and more or less stable. The energy over the ensemble of such trajectories acquires a minimum mean value, which corresponds to the set of most robust motions and generates a certain spatial (probability) distribution of particles. A first image that comes to mind is that the orbits are 'trapped' once they are close enough to the stationarity condition. This picture will become more transparent below, with the demonstration that the stationary states correspond to situations that satisfy an ergodic condition. The trapping of the orbits is a normal occurrence, at least in the long run (i.e. the time-asymptotic limit), because the stochastic field compels the particle to explore the neighboring phase-space regions; as long as the particle is not trapped, it will continue probing the phase space. This means that the stationary orbits are qualitatively analogous to limit cycles, although here the attraction basin is formed by the lowest average energy. Similar possibilities have been suggested earlier by several authors.[43-45]

3.6. Quantum potential and nonlocality in the single-particle case

By writing the wave function in its polar form

$$\psi = \sqrt{\rho}e^{iS/\hbar}. \tag{3.76}$$

with $S(x, t)$ real and $S/\hbar$ dimensionless, the Schrödinger Eq. (3.69) gives, after some elementary simplifications (it is advantageous to use here 3-D vector notation)

$$-\frac{\partial S}{\partial t} + \frac{i\hbar}{2\rho}\frac{\partial \rho}{\partial t} = \frac{1}{2m}(\boldsymbol{\nabla}S)^2 + V +$$

$$+\frac{\hbar^2}{4m\rho^2}(\boldsymbol{\nabla}\rho)^2 - \frac{\hbar^2}{2m\rho}\nabla^2\rho - \frac{i\hbar}{2m\rho}\boldsymbol{\nabla}\rho\cdot\boldsymbol{\nabla}S - \frac{i\hbar}{2m}\nabla^2S.$$

To disentangle this expression we write the continuity equation in the form

$$\frac{\partial \rho}{\partial t} = -\boldsymbol{\nabla}\cdot(\rho\mathbf{v}) = -\frac{1}{m}\rho\nabla^2S - \frac{1}{m}\boldsymbol{\nabla}S\cdot\boldsymbol{\nabla}\rho. \tag{3.77}$$

Equation (3.6) reduces then to

$$-\frac{\partial S}{\partial t} = \frac{1}{2m}(\boldsymbol{\nabla}S)^2 + V + \frac{\hbar^2}{4m\rho^2}(\boldsymbol{\nabla}\rho)^2 - \frac{\hbar^2}{2m\rho}\nabla^2\rho, \tag{3.78}$$

$$\text{or} \qquad \frac{\partial S}{\partial t} + \frac{1}{2m}(\boldsymbol{\nabla}S)^2 + V - \frac{\hbar^2}{2m}\frac{\nabla^2\sqrt{\rho}}{\sqrt{\rho}} = 0. \tag{3.79}$$

The point with this result is that *if* one defines an effective potential acting on the particles by means of

$$V_{\text{eff}} = V + V_Q, \quad V_Q = -\frac{\hbar^2}{2m}\frac{\nabla^2\sqrt{\rho}}{\sqrt{\rho}}, \tag{3.80}$$

where V_Q is known as the *quantum potential* (see e.g. Ref. 46; some authors call it Bohm's potential), then (3.79) takes the *form* of a Hamilton-Jacobi equation for the principal function S,

$$\frac{\partial S}{\partial t} + \frac{1}{2m}(\boldsymbol{\nabla}S)^2 + V_{\text{eff}} = 0. \tag{3.81}$$

This procedure is effectively followed in certain occasions, particularly in the causal de Broglie-Bohm interpretation of QM.[46] Whereas the rest of the terms in Eq. (3.79) are local, the function $V_Q(x)$ bears information (through ρ) about the distribution of all the members of the ensemble in the *entire* configuration space. Therefore

Eq. (3.81) differs essentially from a real Hamilton-Jacobi equation, which, by definition, describes the motion of a congruency of (single) particles acted on by local potentials. (A *congruency* refers to a single-valued trajectory field.) With the addition of V_Q, (3.81) refers to an *ensemble* of similar particles; it acquires a statistical meaning described in configuration space by $\rho(x)$. By becoming nonlocal, its meaning becomes obscure. Assume for instance that each member of the ensemble is a single particle, separated by any other member by hours, or days, and that the same experiment is performed a huge number of times until the ensemble is well approximated. Under such circumstances there is clearly no opportunity for any nonlocal physical effect to take place among the members of the ensemble. This nonlocality is not ontological, but rather a semblance, an artifact of the reduced statistical description that vanishes by going back to the full phase-space statistical description, which is as local as any true statistical description can be.

According to (3.70) and (3.80), the mean value of V_Q is equal to

$$\langle V_Q \rangle = \tfrac{1}{2} m^2 \left\langle \mathbf{u}^2 \right\rangle. \tag{3.82}$$

Therefore another form of expressing the mean energy is, using Eq. (3.72),

$$\left\langle \hat{H} \right\rangle = \frac{1}{2m} \left\langle \hat{\mathbf{p}}^2 \right\rangle + \langle V \rangle = \frac{m}{2} \left\langle \mathbf{v}^2 \right\rangle + \langle V_Q + V \rangle. \tag{3.83}$$

Here the kinetic energy of diffusion $m^2 \left\langle \mathbf{u}^2 \right\rangle /2$ becomes reinterpreted as a quantum potential energy $\langle V_Q \rangle$. This is a usual translation in the literature, clearly legitimate for the mean values, although somewhat misleading for the functions V_Q and $m\mathbf{u}^2/2$ themselves.

The fact that the most extensive line of research on quantum nonlocalities during the last decades is related to the Bell inequalities, has led to the widespread conviction that nonlocality is a property exclusive of multipartite quantum systems. It should be stressed that independently of interpretation, the Schrödinger equation contains the quantum potential — even if disguised — and hence the associated quantum nonlocalities. These are present in all cases, including the single-particle one. As for multiparticle systems, extra nonlocalities arise due to the correlations among variables that pertain to different particles, particularly for entangled states. These extra nonlocalities are responsible for the violation of the Bell inequalities. The mechanism that gives rise to the entanglement between the components of a pair of noninteracting particles within the present theory is discussed in section 4.5

3.7. *Phase-space distribution; Wigner's function*

Our starting point for the derivation of the Schrödinger equation in configuration space was the phase-space Fokker-Planck-type Eq. (3.9). We have also at our disposal the momentum representation of the Schrödinger equation. Is it then possible to proceed in the opposite sense, starting from the usual quantum description provided separately in configuration or momentum space, and recover a *unique* full phase-space description? That this question cannot be answered in the positive is a well-established fact. Let us briefly look into the matter from the present perspective and disclose the reason for this difficulty.

Equation (3.21) for $\widetilde{Q}(x, z, t)$ can be inverted and combined with Eqs. (3.35) and (3.37) to obtain

$$Q(x, p, t) = \frac{1}{2\pi} \int \widetilde{Q}(x, z, t) e^{-ipz} dz =$$

$$= \frac{1}{2\pi} \int q_+(x + \eta z, t) q_-(x - \eta z, t) \chi(x + \eta z, x - \eta z, t) \, e^{-ipz} dz.$$

By construction $Q(x, p, t)$ furnishes a true (Kolmogorovian) probability density in phase space. This means that *if* the exact solutions for q_+, q_- and χ were known for all values of z, we would have a full phase-space description for the particle. However, all we can construct is an approximate form $W(x, p, t)$ obtained by taking $\chi = 1$ and allowing z to remain as a Fourier variable, which gives

$$W(x, p, t) = \frac{1}{\pi \hbar} \int \psi(x + y, t) \psi^*(x - y, t) e^{-i2py/\hbar} dy, \tag{3.84}$$

with $y = \eta z = \hbar z / 2$. This is the well-known Wigner phase-space function.[47,48] As a result, we cannot guarantee W to be a true Kolmogorovian probability. And indeed, despite its recognized value, it is not, since as is well known it can take on negative values in some regions of phase space for almost all states and systems (the exception being the Gaussian states). The right solution to this long-standing problem is of course to recognize the intrinsic limitation of W that ensues from its approximate nature, and to revert to the full distribution $Q(x, p, t)$.

As was shown above, the assumptions and approximations made to arrive at the Schrödinger equation imply an enormous simplification of the problem at hand, since only the first two equations of the infinite hierarchy are used. But this comes with a high price: the loss of a true phase-space description. We conclude that a most important problem that remains open is the investigation of the possibilities offered by the full phase-space probability (3.7). It is clear that its study opens a wide door to some new physics.

4. Heisenberg quantum mechanics

The findings presented in the previous section suggest to explore the Heisenberg formulation of quantum mechanics from a new perspective. We achieve this by introducing the principle of ergodicity of the stationary solutions of the particle-zeropoint field problem, which ultimately leads to the matrix description. The response of the particle is found to be always linear in the field components, regardless of the nonlinearities of the external force. This is the essence of *linear stochastic electrodynamics* (LSED).[36−37,49−51]

The path followed here is perhaps less intuitive than the one presented in the preceding sections, but at the same time it is more revealing and illustrative of several of the intricacies of QM. It serves to disclose different and to some extent complementary features of the physics underlying the quantization process, that remain hidden in the Schrödinger formalism. Moreover, the extension of the theory developed to the case of two particles allows for a clear physical understanding of the mechanism leading to quantum entanglement.

4.1. *Resonant solutions in the stationary regime*

As before, we start from the approximate (non relativistic, long-wavelength) equation of motion

$$m\ddot{x} = f(x) + m\tau\dddot{x} + eE(t). \tag{4.1}$$

The stationary solutions of this equation, $x^{\text{stat}}(t)$, can be decomposed into a time-independent contribution, which coincides with the time average $(\overline{(\cdot)}^t)$ defined as

$$\overline{x^{\text{stat}}(t)}^t = \lim_{T\to\infty} \frac{1}{T} \int_0^T x^{\text{stat}}(t)dt, \tag{4.2}$$

plus an oscillatory contribution that averages to zero. We thus express $x^{\text{stat}}(t)$ in the form

$$x^{\text{stat}}(t) = \overline{x^{\text{stat}}(t)}^t + \sum_{k\neq 0} \left(\tilde{x}_k a_k e^{i\omega_k t} + \text{c.c.} \right), \tag{4.3}$$

where we have excluded from the sum the term corresponding to the null frequency, $\omega_0 = 0$, and $\tilde{x}_k$ represents an amplitude (stochastic in principle) associated with the frequency ω_k. We now define $\mathring{x}(t)$ as

$$\mathring{x}(t) = \sum_k \tilde{x}_k a_k e^{i\omega_k t} + \text{c.c.} \tag{4.4}$$

With this definition $\mathring{x}(t)$ and $x^{\text{stat}}(t)$ differ only in their time-independent term, with $\overline{\mathring{x}(t)}^t = \tilde{x}_0 a_0 + \text{c.c.} = 2\overline{x^{\text{stat}}(t)}^t$.

We assume that the external force $f(x)$ can be expanded as a power series of x in the form $f(x(t)) = \sum_{n=1} c_n x^n(t)$, and that once the stationary state is reached it can be decomposed in a form analogous to (4.3)

$$f^{\text{stat}}(t) = \overline{f^{\text{stat}}(t)}^t + \sum_{k\neq 0} \left(\tilde{f}_k a_k e^{i\omega_k t} + \text{c.c.} \right). \tag{4.5}$$

Let us now define

$$\mathring{f} = \sum_k \tilde{f}_k a_k e^{i\omega_k t} + \text{c.c.}, \tag{4.6}$$

whose relation to $f^{\text{stat}}(t)$ is analogous to that previously defined between $\mathring{x}(t)$ and $x^{\text{stat}}(t)$. The quantities as $\tilde{f}_k$ or $\tilde{x}_k$ depend in general on the set of variables $\{a_j\}$ for non-linear forces. In other words, neither Eq. (4.4) nor (4.6) are *explicit* developments in the field variables a_k; yet the time dependence in both equations is fully expressed through the factors $e^{i\omega_k t}$. Introducing Eqs. (4.4) and (4.6) into (4.1) leads, after separating the terms that oscillate with frequency ω_k, to

$$- m\omega_k^2 \tilde{x}_k e^{i\omega_k t} = \tilde{f}_k e^{i\omega_k t} - im\tau\omega_k^3 \tilde{x}_k e^{i\omega_k t} + e\tilde{E}_k e^{i\omega_k t}, \tag{4.7}$$

where we used the expansion

$$E(t) = \sum_k \tilde{E}_k a_k e^{i\omega_k t} + \text{c.c.} \tag{4.8}$$

for the component of the electric field in the direction of motion. Therefore,

$$\tilde{x}_k = -\frac{e}{m}\frac{\tilde{E}_k}{\Delta_k}, \quad \text{with} \quad \Delta_k \equiv \omega_k^2 - i\tau\omega_k^3 + \frac{\tilde{f}_k}{m\tilde{x}_k}. \tag{4.9}$$

Thus the important contributions to $\mathring{x}(t)$ come from the poles of $\tilde{x}_k$, i.e., they correspond to those frequencies that satisfy

$$\omega_k^2 \approx -\frac{\tilde{f}_k}{m\tilde{x}_k}. \tag{4.10}$$

For frequencies in the atomic range, the resonances are extremely sharp due to the small value of τ (recall that $\tau \sim 10^{-23}$ s for electrons). We will denote the set of solutions of Eq. (4.10) as $\{\omega_k\}_{\text{res}}$ and refer to its elements as *resonance frequencies*.

Because of the stochastic nature of the background field, the solutions of Eq. (4.1) constitute a stochastic process $x^{\text{stat}^{(i)}}(t)$, where the index (i) signals the dependence of $x^{\text{stat}}(t)$ on the field realization (i). When the set $\{i\}$ of all the realizations of the field is considered, an ensemble of single-particle solutions is determined. Thus the statistical set can be reproduced by considering an ensemble of particles, each of which is subject to a different realization of the field. In other words, the averages over the ensemble of realizations of the field (denoted as $\overline{(\cdot)}^{(i)}$) can alternatively be determined by averaging over the ensemble of particles. Now, according to the above discussion, in the long run the particles reach stationary states that can be labeled with an index α (which in its turn, being a one-dimensional problem, is in direct correspondence with the mechanical energy $\mathcal{E}_\alpha$). To distinguish among these different stationary states accessible to the mechanical subsystem, we decompose the ensemble $\{i\}$ into subensembles, $\{i\} = \cup_\alpha \{i_\alpha\}$, such that the energy corresponding to those particles subject to the field realization $i \in \{i_\alpha\}$ is precisely $\mathcal{E}_\alpha$.

In what follows we focus our attention on a particular subensemble $\{i_\alpha\}$, and construct the appropriate expansions for the dynamical variables corresponding to such subensemble. The frequencies that play an important role in the dynamics of the particle (to be determined by the theory itself) constitute then an α-dependent subset of all the resonance frequencies $\{\omega_k\}_{\text{res}}$. We refer to these as *relevant frequencies* and denote them by $\omega_{\alpha\beta}$, where β serves to enumerate the different frequencies of the set, so that when β varies the $\omega_{\alpha\beta}$ coincide with the different (resonance) frequencies of the solutions $\mathring{x}(t)$ in a state characterized by α.

Now, the fact that the amplitudes $\tilde{x}_k$, $\tilde{f}_k$ (or more generally $\tilde{A}_k$) and the variables a_k in expansions such as (4.4) and (4.6) are associated with the frequency ω_k, leads us to introduce the same couple of indices (α and β) — with the meaning already explained for the frequencies $\omega_{\alpha\beta}$ — in such quantities, so that we write $\tilde{E}_{\alpha\beta}$, $\tilde{x}_{\alpha\beta}$, $\tilde{f}_{\alpha\beta}$ and $a_{\alpha\beta}$ instead of $\tilde{E}_k$, $\tilde{x}_k$, $\tilde{f}_k$, and a_k, respectively. In this way, the transit from a description referred to the complete ensemble ($\{i\}$) to one restricted to the subensemble $\{i_\alpha\}$ is achieved by performing the substitutions: $\mathring{A}(t) \to \mathring{A}_\alpha(t)$, $\omega_k \to \omega_{\alpha\beta}$, and $a_k \to a_{\alpha\beta}$ in expansions of the type (4.4). Thus, for the subsystem that has attained the stationary state α we write,

$$\mathring{x}_\alpha(t) = \sum_\beta \tilde{x}_{\alpha\beta} a_{\alpha\beta} e^{i\omega_{\alpha\beta}t} + \text{c.c.}, \tag{4.11}$$

and similarly for $\mathring{f}_\alpha$ and $\mathring{E}_\alpha$. The $\omega_{\alpha\beta}$ can acquire positive or negative values, so the present expansions, at variance with those of the form (4.4), are not necessarily in terms of positive and negative frequencies. The transition from $\sum_k$ to $\sum_\beta$ requires therefore a reordering of the terms in the sum.

For the subensemble characterized by the (mean) energy $\mathcal{E}_\alpha$, we are led to rewrite Eq. (4.1) as

$$m\frac{d^2\mathring{x}_\alpha}{dt^2} = \mathring{f}_\alpha + m\tau\frac{d^3\mathring{x}_\alpha}{dt^3} + e\mathring{E}_\alpha. \tag{4.12}$$

When the expansions for $\mathring{x}_\alpha$, $\mathring{f}_\alpha$ and $\mathring{E}_\alpha$ are introduced, Eq. (4.7) rewritten for the state α becomes

$$-\sum_\beta m\omega_{\alpha\beta}^2 \tilde{x}_{\alpha\beta} a_{\alpha\beta} e^{i\omega_{\alpha\beta}t} + \text{c.c.} = \sum_\beta \left(\tilde{f}_{\alpha\beta} - im\tau\omega_{\alpha\beta}^3 \tilde{x}_{\alpha\beta} + e\tilde{E}_{\alpha\beta}\right) a_{\alpha\beta} e^{i\omega_{\alpha\beta}t} + \text{c.c.} \tag{4.13}$$

Assuming that detailed balance is satisfied for each relevant frequency $\omega_{\alpha\beta}$, we are led to

$$m\frac{d^2\tilde{x}_{\alpha\beta}(t)}{dt^2} = \tilde{f}_{\alpha\beta}(t) + m\tau\frac{d^3\tilde{x}_{\alpha\beta}(t)}{dt^3} + e\tilde{E}_{\alpha\beta}(t), \tag{4.14}$$

with

$$\tilde{A}_{\alpha\beta}(t) \equiv \tilde{A}_{\alpha\beta} e^{i\omega_{\alpha\beta}t}. \tag{4.15}$$

As for the terms oscillating with the factor $e^{-i\omega_{\alpha\beta}t}$, they give rise to an equation that is the complex conjugate of (4.14). From this latter we obtain

$$\tilde{x}_{\alpha\beta} = -\frac{e}{m}\frac{\tilde{E}_{\alpha\beta}}{\Delta_{\alpha\beta}}, \quad \Delta_{\alpha\beta} = \omega_{\alpha\beta}^2 - i\tau\omega_{\alpha\beta}^3 + \frac{\tilde{f}_{\alpha\beta}}{m\tilde{x}_{\alpha\beta}}. \tag{4.16}$$

This means that the mechanical system responds resonantly to those frequencies that solve, in analogy with Eq. (4.10), the (approximate) system of equations $\omega_{\alpha\beta}^2 \approx -\frac{\tilde{f}_{\alpha\beta}}{m\tilde{x}_{\alpha\beta}}$. The relevant frequencies that satisfy such equations are precisely the resonance frequencies associated with the subensemble α, which constitute a subset of $\{\omega_k\}_{\text{res}}$. As before, quantities such as $\tilde{x}_{\alpha\beta}$, $\tilde{f}_{\alpha\beta}$, $\omega_{\alpha\beta}$, etc. depend in principle on the set of stochastic amplitudes $\{a\}$.

The fact that Eq. (4.12) decomposes into two equations, namely (4.14) and its complex conjugate, allows us to restrict the study to the solutions of (4.14) only. This latter is but the detailed (term by term) form of the equation

$$m\ddot{x}_\alpha = f_\alpha + m\tau\dddot{x}_\alpha + eE_\alpha, \tag{4.17}$$

where we have defined generically the complex quantities

$$A_\alpha(t) = \sum_\beta \tilde{A}_{\alpha\beta} a_{\alpha\beta} e^{i\omega_{\alpha\beta}t}. \tag{4.18}$$

In what follows we work with Eq. (4.17) — which is in direct correspondence with the original equation of motion — and with expansions of the form (4.18).

4.2. *The principle of ergodicity*

Being the particle subject to the permanent interchange of energy with the random field, it appears natural to consider that once in the stationary regime, the system satisfies an ergodic principle, so that the time average of a function $A^{(i)}(t)$ coincides with its ensemble average. In this and the following sections we shall explore the consequences of introducing the demand of ergodicity.

We decompose $x_\alpha(t)$ into its time-independent contribution (which coincides with $\overline{x_\alpha(t)}^t$) plus an oscillating part that averages to zero,

$$x_\alpha(t) = \tilde{x}_{\alpha\alpha} a_{\alpha\alpha} + \sum_{\beta \neq \alpha} \tilde{x}_{\alpha\beta} a_{\alpha\beta} e^{i\omega_{\alpha\beta}t}, \tag{4.19}$$

with $\omega_{\alpha\alpha} = 0$. The second term in this expression corresponds to the deviations of $x_\alpha(t)$ from its mean value, and its modulus allows us to calculate the variance $\sigma_{x_\alpha}^2$ defined as

$$\sigma_{x_\alpha}^2 = \overline{\left| x_\alpha - \overline{x_\alpha(t)}^t \right|^2}^t. \tag{4.20}$$

Direct calculation gives

$$\sigma_{x_\alpha}^2 = \sum_{\beta \neq \alpha} \left| \tilde{x}_{\alpha\beta} a_{\alpha\beta} \right|^2. \tag{4.21}$$

According to the discussion in subsection 4.1, $x_\alpha(t)$ depends on the specific realization $(i) \in \{i_\alpha\}$ of the field. Therefore, $\tilde{x}_{\alpha\beta}$, the set $\{\omega_{\alpha\beta}\}$, and clearly $a_{\alpha\beta}$ also depend on the field realization. We therefore add the superindex (i) to these stochastic quantities in order to stress their dependence on the field realization. Thus, $\sigma_{x_\alpha}^2$ is expressed in an explicit stochastic form as

$$\sigma_{x_\alpha}^{2(i)} = \sum_{\beta \neq \alpha} \left| \tilde{x}_{\alpha\beta}^{(i)} a_{\alpha\beta}^{(i)} \right|^2. \tag{4.22}$$

By construction, $\sigma_{x_\alpha}^2$ is time-independent, although it depends on the specific realization of the field (i). On the other hand, the variance obtained by averaging over the ensemble of realizations,

$$\sigma_{x_\alpha}^2(t) = \overline{\left|x_\alpha - \overline{x_\alpha(t)}^{(i)}\right|^2}^{(i)}, \tag{4.23}$$

may depend in general on t, but not on i, by definition. At this point we introduce the ergodic hypothesis. This means that both expressions for $\sigma_{x_\alpha}^2$, Eqs. (4.22) and (4.23), coincide. Under this condition the right-hand side of (4.22) must be independent of the realization; that is, the ergodic condition implies that

$$\sigma_{x_\alpha}^2 = \sum_{\beta(\neq\alpha)} \left|\tilde{x}_{\alpha\beta}^{(i)} a_{\alpha\beta}^{(i)}\right|^2 \text{ is independent of } (i). \tag{4.24}$$

Since the right-hand side of this equation is a sum of statistically independent terms, each one being a non-negative quantity, the only possible way for condition (4.24) to be satisfied in general is that each term of the sum is independent of the realization, so that

$$\left|\tilde{x}_{\alpha\beta}^{(i)}\right|^2 \left|a_{\alpha\beta}^{(i)}\right|^2 \text{ is independent of } (i) \quad (\text{for}\beta \neq \alpha), \tag{4.25}$$

whence the moduli of both $\tilde{x}_{\alpha\beta}^{(i)}$ and $a_{\alpha\beta}^{(i)}$ are independent of (i), and their polar forms become $\tilde{x}_{\alpha\beta}^{(i)} = \chi_{\alpha\beta} e^{i\xi_{\alpha\beta}^{(i)}}$, $a_{\alpha\beta}^{(i)} = r_{\alpha\beta} e^{i\zeta_{\alpha\beta}^{(i)}}$. Since $\tilde{x}_{\alpha\beta}^{(i)}$ always appears next to $a_{\alpha\beta}^{(i)}$, any random contribution contained in the phase of $\tilde{x}_{\alpha\beta}$ can be transferred to the (random) phase of $a_{\alpha\beta}^{(i)}$, which allows us to redefine

$$\tilde{x}_{\alpha\beta}^{(i)} = \tilde{x}_{\alpha\beta}, \quad a_{\alpha\beta}'^{(i)} \equiv r_{\alpha\beta} e^{i\varphi_{\alpha\beta}^{(i)}} \quad (\beta \neq \alpha). \tag{4.26}$$

The new $a_{\alpha\beta}'^{(i)}$ represents the value acquired by the stochastic field variable through its interaction with matter. As for the term with $\beta = \alpha$, we revert to the demand of ergodicity on the trajectory $x_\alpha(t)$, which requires that

$$\overline{x_\alpha}^t = \tilde{x}_{\alpha\alpha}^{(i)} a_{\alpha\alpha}^{(i)} = \tilde{x}_{\alpha\alpha} a_{\alpha\alpha} = \overline{x_\alpha}^{(i)}. \tag{4.27}$$

Hence the above equations are also valid for $\beta = \alpha$, with $\varphi_{\alpha\alpha}^{(i)} = \varphi_{\alpha\alpha} = 0$. In summary, our results are (with $a_{\alpha\beta}^{(i)}$ instead of $a_{\alpha\beta}'^{(i)}$, for simplicity)

$$a_{\alpha\beta}^{(i)} = r_{\alpha\beta} e^{i\varphi_{\alpha\beta}^{(i)}}, \quad \varphi_{\alpha\alpha} = 0 \tag{4.28}$$

$$\tilde{x}_{\alpha\beta}^{(i)} = \tilde{x}_{\alpha\beta}. \tag{4.29}$$

Now the variance of x_α takes the form

$$\sigma_{x_\alpha}^2 = \sum_{\beta\neq\alpha} \left|\tilde{x}_{\alpha\beta}^{(i)} a_{\alpha\beta}^{(i)}\right|^2 = \sum_{\beta\neq\alpha} |\tilde{x}_{\alpha\beta}|^2 r_{\alpha\beta}^2. \tag{4.30}$$

Following a similar procedure one can calculate the variance of the momentum p_α. This allows to conclude that $\omega_{\alpha\beta}^{(i)} = \omega_{\alpha\beta}$, so that all the stochasticity has been absorbed into the phases of the $a_{\alpha\beta}^{(i)}$. Neither $\tilde{x}_{\alpha\beta}$ nor $\omega_{\alpha\beta}$ depend now on (i), which means that also

$$\tilde{f}_{\alpha\beta} \text{ is independent of } (i). \tag{4.31}$$

This is a most remarkable outcome of the principle of ergodicity, whose meaning and implications are examined below.

The fact that the $\tilde{x}_{\alpha\beta}$, $\tilde{f}_{\alpha\beta}$ and $\omega_{\alpha\beta}$ are nonstochastic variables in the present approximation means that the quantities $\Delta_{\alpha\beta}$ and $\tilde{E}_{\alpha\beta}$ are also independent of (i). Hence the Fourier developments above are in fact *explicit* developments in the variables $a_{\alpha\beta}^{(i)}$, and further,

$$x_\alpha^{(i)}(t) = -\frac{e}{m} \sum_\beta \frac{\tilde{E}_{\alpha\beta}}{\Delta_{\alpha\beta}} a_{\alpha\beta}^{(i)} e^{i\omega_{\alpha\beta}t} + \text{c.c.} \tag{4.32}$$

has become a *linear* function of the stochastic components of the field, $\tilde{E}_{\alpha\beta} a_{\alpha\beta}^{(i)}$. For this reason, the theory that ensues as a result of the condition of ergodicity is termed *Linear Stochastic Electrodynamics* (LSED).

4.2.1. *Quantization as an outcome of ergodicity*

To expand the force f_α (or more generally, any dynamical variable $A_\alpha(x)$) in terms of the amplitudes $\tilde{x}_{\alpha\beta}$ that appear in the expression for x_α,

$$x_\alpha(t) = \sum_\beta \tilde{x}_{\alpha\beta} a_{\alpha\beta} e^{i\omega_{\alpha\beta}t}, \tag{4.33}$$

we must construct the expansions that correspond to the different powers of x for a given state α. For example, for the quadratic case, the most immediate option is to write $f_\alpha = \left(x^2\right)_\alpha = (x_\alpha)^2$. In this case we get

$$f_\alpha^{(i)} = \left(x_\alpha^{(i)}\right)^2 = \sum_{\beta'} \left[\sum_{\beta''} \tilde{x}_{\alpha\beta'} \tilde{x}_{\alpha\beta''} a_{\alpha\beta'}^{(i)} a_{\alpha\beta''}^{(i)} e^{i(\omega_{\alpha\beta'} + \omega_{\alpha\beta''})t} \right]. \tag{4.34}$$

On the other hand, according to Eq. (4.5), $f_\alpha^{(i)}$ has the form

$$f_\alpha^{(i)} = \left(x^2\right)_\alpha^{(i)} = \sum_{\beta'} \widetilde{x^2}_{\alpha\beta'} a_{\alpha\beta'}^{(i)} e^{i\omega_{\alpha\beta'}t}, \tag{4.35}$$

where the quantity $\tilde{f}_{\alpha\beta'} = \widetilde{x^2}_{\alpha\beta'}$ is still to be determined. A comparison between Eqs. (4.34) and (4.35) leads to

$$\widetilde{x^2}_{\alpha\beta'} a_{\alpha\beta'}^{(i)} e^{i\omega_{\alpha\beta'}t} = \sum_{\beta''} \tilde{x}_{\alpha\beta'} \tilde{x}_{\alpha\beta''} a_{\alpha\beta'}^{(i)} a_{\alpha\beta''}^{(i)} e^{i(\omega_{\alpha\beta'} + \omega_{\alpha\beta''})t}$$

$$= x_\alpha^{(i)}(t) \tilde{x}_{\alpha\beta'} a_{\alpha\beta'}^{(i)} e^{i\omega_{\alpha\beta'}t}, \tag{4.36}$$

which means that one should make the identification $\widetilde{x^2}_{\alpha\beta'} = x_\alpha^{(i)}(t)\tilde{x}_{\alpha\beta'}$. This expression is inconsistent with Eq. (4.31), which indicates that neither $\widetilde{x^2}_{\alpha\beta'}$ nor $\tilde{x}_{\alpha\beta'}$ depend on i. The result shows that the election $(x^2)_\alpha = (x_\alpha)^2$, in spite of being the most natural, is inconsistent with the implications of the ergodic principle. Consequently the problem of determining the expansion of a given power of x requires a more careful analysis. The rather lengthy procedure of such analysis can be seen in Ref. 37, where it is shown in detail that in order to guarantee consistency with the ergodic principle one must write

$$\left(x^2\right)_\alpha = \sum_\beta \widetilde{x^2}_{\alpha\beta} a_{\alpha\beta} e^{i\omega_{\alpha\beta}t} = \sum_{\beta,\beta'} \tilde{x}_{\alpha\beta'}\tilde{x}_{\beta'\beta} a_{\alpha\beta'} a_{\beta'\beta} e^{i(\omega_{\alpha\beta'}+\omega_{\beta'\beta})t}. \tag{4.37}$$

Equation (4.37) goes hand in hand with the following conditions on the stochastic variables and the relevant frequencies (no summation over repeated indices!)

$$a_{\alpha\beta'} a_{\beta'\beta} = a_{\alpha\beta}, \tag{4.38}$$

$$\omega_{\alpha\beta'} + \omega_{\beta'\beta} = \omega_{\alpha\beta}. \tag{4.39}$$

The relation (4.38) is easily generalized to any number of factors by a successive (chained) application of it,

$$\begin{aligned}
a_{\alpha\beta'} a_{\beta'\beta''} a_{\beta''\beta'''} \cdots a_{\beta(n-1)\beta} &= \left(a_{\alpha\beta'} a_{\beta'\beta''}\right) a_{\beta''\beta'''} \cdots a_{\beta(n-1)\beta} \\
&= \left[\left(a_{\alpha\beta''}\right) a_{\beta''\beta'''}\right] \cdots a_{\beta(n-1)\beta} \\
&= \left[a_{\alpha\beta'''}\right] \cdots a_{\beta(n-1)\beta} \\
&= a_{\alpha\beta}.
\end{aligned} \tag{4.40}$$

With each $a_{\beta(n)\beta(m)}$ written in polar form according to (4.28), Eq. (4.40) can be broken down into the couple of equations

$$\varphi_{\alpha\beta'}^{(i)} + \varphi_{\beta'\beta''}^{(i)} + \ldots + \varphi_{\beta(n-1)\beta}^{(i)} = \varphi_{\alpha\beta}^{(i)}, \tag{4.41}$$

$$r_{\alpha\beta'} r_{\beta'\beta''} r_{\beta''\beta'''} \cdots r_{\beta(n-1)\beta} = r_{\alpha\beta}. \tag{4.42}$$

Since the number of factors on the left-hand side of this last equation is unrestrained, its only (nontrivial) solution is

$$r_{\delta\eta} = 1, \quad \forall \delta, \eta. \tag{4.43}$$

In its turn, the general solution to Eq. (4.41) is

$$\varphi_{\alpha\beta}^{(i)} = \phi_\alpha^{(i)} - \phi_\beta^{(i)}, \tag{4.44}$$

where each of the ϕ_λ represents a random phase. Combining Eqs. (4.43) and (4.44) with (4.28), we get

$$a_{\alpha\beta} = e^{i\varphi_{\alpha\beta}} = e^{i(\phi_\alpha - \phi_\beta)}, \tag{4.45}$$

whence $a_{\beta\alpha} = a^*_{\alpha\beta}$. Similarly, Eq. (4.39) can be generalized to an arbitrary number of terms,

$$\omega_{\alpha\beta'} + \omega_{\beta'\beta''} + \ldots + \omega_{\beta(n-1)\beta} = \omega_{\alpha\beta}, \tag{4.46}$$

in analogy with (4.41), and the general solution is therefore of the form

$$\omega_{\alpha\beta} = \Omega_\alpha - \Omega_\beta. \tag{4.47}$$

The relations (4.40) and (4.46), which are fundamental for the theory, constitute the *chain rule*. The frequencies appearing in the chain rule are the relevant frequencies, which are either frequencies of resonance or linear (chained) combinations of them, according to (4.46). With regard to the final phases of the field, Eq. (4.41) tells us that those pertaining to the relevant modes become partially correlated, which confirms that not only the material part, but also the near background field is affected during the evolution of the complete system towards equilibrium.

4.2.2. *The matrix rule*

We now use the chain rule to recast (4.37) into the form

$$\sum_\beta \widetilde{x^2}_{\alpha\beta} a_{\alpha\beta} e^{i\omega_{\alpha\beta}t} = \sum_\beta \left(\sum_{\beta'} \tilde{x}_{\alpha\beta'} \tilde{x}_{\beta'\beta} \right) a_{\alpha\beta} e^{i\omega_{\alpha\beta}t}, \tag{4.48}$$

which shows that $\widetilde{x^2}_{\alpha\beta}$ is given by the sum

$$\widetilde{x^2}_{\alpha\beta} = \sum_{\beta'} \tilde{x}_{\alpha\beta'} \tilde{x}_{\beta'\beta}, \tag{4.49}$$

embodying the rule for matrix multiplication applied to the amplitudes $\tilde{x}_{\alpha\beta}$. Thus $\tilde{x}_{\alpha\beta}$ can be identified with the element $\alpha\beta$ of a square matrix $\tilde{x}$ such that

$$\widetilde{x^2}_{\alpha\beta} = \left(\tilde{x}^2 \right)_{\alpha\beta}, \tag{4.50}$$

which leads, together with Eq. (4.49), to recast Eq. (4.35) into

$$\left(x^2 \right)_\alpha = \sum_\beta \left(\sum_{\beta'} \tilde{x}_{\alpha\beta'} \tilde{x}_{\beta'\beta} \right) a_{\alpha\beta} e^{i\omega_{\alpha\beta}t} = \sum_\beta \left(\tilde{x}^2 \right)_{\alpha\beta} a_{\alpha\beta} e^{i\omega_{\alpha\beta}t}. \tag{4.51}$$

It is now easy to verify that Eqs. (4.40) and (4.46) allow us to write for an arbitrary power of x under the ergodic condition

$$\left(x^n \right)_\alpha = \sum_\beta \left(\tilde{x}^n \right)_{\alpha\beta} a_{\alpha\beta} e^{i\omega_{\alpha\beta}t}, \tag{4.52}$$

with $\left(\tilde{x}^n \right)_{\alpha\beta}$ given by the element $\alpha\beta$ of the corresponding matrix product. Thus, every dynamical variable $A(t)$ that can be expressed as a power series of x or $\dot{x}$ — or, more generally, as a power series of the form $h(x) + g(\dot{x})$ — can be expanded, whenever the particle has reached a state α, as Eq. (4.18), and has a square matrix

$\tilde{A}$ associated with it, with elements given by the amplitudes $\tilde{A}_{\alpha\beta}$. Note that the condition of Hermiticity of the matrix $\tilde{x}$, $\tilde{x}^*_{\beta\alpha} = \tilde{x}_{\alpha\beta}$, is consistent with the property $\tilde{x}(\omega_k) = \tilde{x}^*(-\omega_k)$ satisfied by the amplitudes $\tilde{x}_k = \tilde{x}(\omega_k)$ of the original expansion (4.4).

4.3. *Consequences of the ergodic principle for the dynamics*

By reducing a product of the amplitudes $a_{\alpha\beta}$ to a linear function of such variables, the chain rule has as a direct consequence the emergence of a matrix algebra for the amplitudes $\tilde{A}_{\alpha\beta}$. Moreover, application of Eq. (4.46) shows that also the fundamental oscillators of the form (4.15) satisfy a matrix algebra. In line with the customary notation, in what follows $\hat{A}$ represents the matrix whose elements are $\tilde{A}_{\alpha\beta}$, in contrast to subsection 4.2.2, where $\tilde{A}$ denoted the corresponding matrix. With this notation we can rewrite (4.14) as

$$m\frac{d^2\hat{x}(t)}{dt^2} = \hat{f}(t) + m\tau\frac{d^3\hat{x}(t)}{dt^3} + e\hat{E}(t). \qquad (4.53)$$

It is important to stress that this is much more than a new form of expressing the equation of motion. As a consequence of the ergodic condition, neither $\tilde{x}_{\alpha\beta}$ ($\hat{x}$) nor $\tilde{f}_{\alpha\beta}$ ($\hat{f}$) depend on the coefficients $a_{\alpha\beta}$, and hence all reference to the stochastic variables has vanished from (4.53). That is, the original stochastic Eq. (4.1), written in terms of c-(stochastic-)numbers, has been transformed into a non-stochastic matrix equation (q-numbers) as a result of the ergodic principle.

Equation (4.53) is precisely the equation of motion of non-relativistic QED. This is a most important outcome, pointing to the convergence of the present theory and (non-relativistic, spinless) QED. It should be clear, however, that the equivalence between the two descriptions (QED and LSED) refers to their formal features. Although they share the same equations, they are conceptually distinct, there being important differences in their physical outlook.

The fact that the oscillators $\tilde{A}_{\alpha\beta}$ satisfy a matrix algebra allows us to determine the evolution law for every dynamical variable A that can be represented in the form (4.18). Indeed, by considering the time derivative of $A_\alpha(t)$ and using Eq. (4.47), one arrives at

$$\frac{dA_\alpha(t)}{dt} = \sum_\beta i\left[\hat{\Omega}, \hat{A}\right]_{\alpha\beta} a_{\alpha\beta} e^{i\omega_{\alpha\beta}t}, \qquad (4.54)$$

where $\hat{\Omega}$ is a diagonal matrix with elements $\hat{\Omega}_{\alpha\beta} = \Omega_\alpha\delta_{\alpha\beta}$. It follows that $\tilde{\dot{A}}_{\alpha\beta} = i\left[\hat{\Omega}, \hat{A}(t)\right]_{\alpha\beta}$, or, in closed matrix notation,

$$i\frac{d\hat{A}(t)}{dt} = \left[\hat{A}(t), \hat{\Omega}\right]. \qquad (4.55)$$

This result is a direct consequence of the structure of the expansion $A_\alpha(t)$ and of the antisymmetry of the frequencies $\omega_{\alpha\beta}$. This antisymmetry is thus at the root of the description of the evolution of the dynamical operators by an algebra of commutators. Equation (4.55) shows that the matrix $\hat{\Omega}$ is central in determining the evolution of the mechanical subsystem. As this evolution is controlled by the Hamiltonian of the particle, it follows that the matrix $\hat{\Omega}$ should be related with the matrix $\hat{H}$.

4.3.1. *Radiationless regime. Contact with quantum mechanics*

Once the quantum regime has been attained — that is, once the system has reached the stationary and ergodic limit, thanks to the combined effect of radiation reaction and the ZPF — the radiative terms can be neglected in a first (radiationless) approximation, as discussed in section (3.5). Under this approximation the Hamiltonian matrix (of the particle) reduces to

$$\hat{H} = \frac{\hat{p}^2}{2m} + V(\hat{x}), \tag{4.56}$$

where $\hat{x}$ and $\hat{p}$ evolve according to

$$\hat{p} = m\frac{d\hat{x}}{dt}, \quad \hat{f} = \frac{d\hat{p}}{dt}. \tag{4.57}$$

The stationarity condition implies, through Eq. (4.55), that $\hat{H}$ commutes with the diagonal matrix $\hat{\Omega}$, and hence is also diagonal, with elements

$$\tilde{H}_{\alpha\beta} = \mathcal{E}_\alpha \delta_{\alpha\beta}. \tag{4.58}$$

Since $\hat{\Omega}$ and $\hat{H}$ are related (diagonal) matrices, we may write $\hat{\Omega}$ as a product of the form $\hat{H}\hat{G}(\hat{H})$

$$i\frac{d\hat{A}(t)}{dt} = \left[\hat{A}(t), \hat{H}\hat{G}\right]. \tag{4.59}$$

An inspection of Eqs. (4.56) and (4.57) shows that for the function $\hat{G}(\hat{H})$ to preserve the linear correspondence between $V(\hat{x})$ and $\hat{f}(\hat{x})$, it must be independent of $\hat{H}$ and hence constant; we therefore write $\hat{\Omega} = C\hat{H}$, with C independent of $\hat{H}$, i.e., of the specific problem, and therefore

$$i\frac{d\hat{A}(t)}{dt} = C\left[\hat{A}(t), H\right]. \tag{4.60}$$

Applied to $\hat{A} = \hat{x}$, this equation leads to the commutator

$$[\hat{x}, \hat{p}] = \frac{1}{C}\mathbb{I}. \tag{4.61}$$

This is an important result, as it indicates that the value of the basic commutator fixes the scale of the time evolution in the quantum regime, according to Eq. (4.60). Hence the commutator of $\hat{x}$ and $\hat{p}$ and the equation that governs the evolution of the dynamical variables (both in the radiationless approximation) are intimately related, which endows the commutator with a dynamical meaning.

Given the universality of Eq. (4.61), one may use the problem of a harmonic oscillator of frequency ω_0 to determine the value of the constant C. For this purpose, recall from section 2.8 that the minimum value of the product of the dispersions of x and p for the oscillator is

$$\left(\sigma_x^2 \sigma_p^2\right)_{\min} = \frac{\mathcal{E}_0^2}{\omega_0^2} = \frac{\hbar^2}{4}. \tag{4.62}$$

On the other hand the variances σ_x^2 and σ_p^2 satisfy the Robertson-Schrödinger inequality,

$$\sigma_x^2 \sigma_p^2 \geq \tfrac{1}{4} \left|\langle [\hat{x}, \hat{p}] \rangle\right|^2 + \left|\langle \tfrac{1}{2} \{\hat{x}, \hat{p}\} - \langle \hat{x} \rangle \langle \hat{p} \rangle \rangle\right|^2, \tag{4.63}$$

with $\{\hat{x}, \hat{p}\} = \hat{x}\hat{p} + \hat{p}\hat{x}$. From this (strictly algebraic) expression and Eq. (4.61), it follows that

$$\left(\sigma_x^2 \sigma_p^2\right)_{\min} = \tfrac{1}{4} \left|\langle [\hat{x}, \hat{p}] \rangle\right|^2 = \tfrac{1}{4C^2}, \tag{4.64}$$

whence by comparing with (4.62) one obtains $C = \hbar^{-1}$, i.e.,

$$[\hat{x}, \hat{p}] = i\hbar \mathbb{I}. \tag{4.65}$$

The present derivation of this well-known quantum relation is important because it shows that the value of the commutator $[\hat{x}, \hat{p}]$ is determined *exclusively* by the properties of the ZPF; more specifically, by the constant that fixes the mean energy of each of its modes. It follows that as a result of the action of the ZPF *every* quantum system acquires unavoidable fluctuations, determined in general by the same universal constant. It is thus confirmed that the so-called quantum fluctuations are real and causal.

4.4. *The Heisenberg equation and energy eigenvalues. Transition to the Hilbert-space description*

With $C = \hbar^{-1}$ introduced in Eq. (4.60) we get

$$i\hbar\frac{d\hat{A}}{dt} = \left[\hat{A}, \hat{H}\right], \tag{4.66}$$

which is the Heisenberg equation for the dynamical operator $\hat{A}(t)$. On the other hand, from (4.55) and (4.58) $\Omega_\alpha = \mathcal{E}_\alpha/\hbar$, except for an additive constant, whence (4.47) can be identified with Bohr's transition rule,

$$\hbar\omega_{\alpha\beta} = \mathcal{E}_\alpha - \mathcal{E}_\beta, \tag{4.67}$$

and the resonance frequencies with the quantum transition frequencies. This identification shows that the transitions occur during resonant responses of the mechanical system driven by the ZPF. This result explains how it is that the electron 'knows' in advance the energy of the state where it will land when realizing a transition, as the energy difference is determined precisely by the resonance. The resonant response of the particle to a selected set of frequencies of the background field constitutes the physical mechanism responsible for the transition of the particle to one among the collection of accessible stationary states. Which transition will effectively take place in a given case is a question of chance, since it depends on the precise conditions of the atom and the mode of the field at the moment of the transition. This resonant phenomenon, along with the fact that the quantities featuring in Eq. (4.67) become non-stochastic, point to the ergodic condition as a quantization principle.

The diagonal form of the matrix $\hat{H}$ was used in deriving the evolution equation. However, this latter continues in force after a unitary change of basis is performed that transforms $\hat{H}$ into a non-diagonal matrix $\hat{H}' = U^\dagger \hat{H} U$. In the latter case the energy that corresponds to $\hat{H}'$ becomes a fluctuating quantity, and the states cease to be stationary. Thus Eq. (4.66) gives the general law of evolution in the quantum regime; it includes stationary and time-dependent states.

To complete the present description it becomes necessary to introduce the vectors of an appropriate Hilbert space that represent the states of the quantum system. We start by noting that the matrix $\hat{A}$ — associated to any dynamical variable that can be written in the form (4.18) and whose elements are given by the elementary oscillators $\tilde{A}_{\alpha\beta}(t) = \tilde{A}_{\alpha\beta} e^{i\omega_{\alpha\beta} t}$ — can be expanded as

$$\hat{A} = \sum_{\alpha,\beta} \tilde{A}_{\alpha\beta}(t) \, |e_\alpha\rangle \langle e_\beta| = \sum_{\alpha,\beta} \tilde{A}_{\alpha\beta} e^{i\omega_{\alpha\beta} t} \, |e_\alpha\rangle \langle e_\beta| \,, \tag{4.68}$$

in terms of a basis $\{|e_\alpha\rangle \langle e_\beta|\}$ of operators constructed from the vectors of a complete orthonormal basis $\{|e_\alpha\rangle\}$.

The time dependence of $\hat{A}$ — which lies at the core of the Heisenberg representation — can be transferred from the matrix elements to the vectors of a new basis obtained by means of the unitary transformation

$$|e_\alpha\rangle \to |\alpha(t)\rangle = e^{-i(\mathcal{E}_\alpha/\hbar)t} \, |e_\alpha\rangle \,, \quad |e_\alpha\rangle = |\alpha(0)\rangle \,, \tag{4.69}$$

so that Eq. 4.68) takes the form (we used Eq. (4.67))

$$\hat{A} = \sum_{\alpha,\beta} \tilde{A}_{\alpha\beta} e^{i\omega_{\alpha\beta} t} \, |\alpha(t)\rangle \langle \beta(t)| =$$

$$= \sum_{\alpha,\beta} \tilde{A}_{\alpha\beta} \, |\alpha(0)\rangle \langle \beta(0)| = \hat{A}(0). \tag{4.70}$$

The vectors $\{|\alpha(t)\rangle\}$ are directly related to the energy values $\{\mathcal{E}_\alpha\}$, and evolve in time according to Eq. (4.69). This allows us to establish contact with the Schrödinger representation of QM, as is usually done.

The statistical meaning of the quantities appearing in this description can be recovered as follows. From Eq. (4.70) one gets the expected relation between the matrix elements of $\hat{A}$ and the corresponding state vectors,

$$\tilde{A}_{\alpha\beta} = \langle e_\alpha| \, \hat{A}(t) \, |e_\beta\rangle = \langle \alpha(t)| \, \hat{A} \, |\beta(t)\rangle \equiv \langle \alpha| \, \hat{A} \, |\beta\rangle \,. \tag{4.71}$$

This result, along with Eq. (4.66), leads to the standard quantum-mechanical formalism. For $\hat{A}$ Hermitean, Eq. (4.27) (which holds in general for every dynamical variable of the form (4.18)) allows us to write

$$\overline{A}_\alpha^t = \langle A_\alpha \rangle = \tilde{A}_{\alpha\alpha} = \langle \alpha | \hat{A} | \alpha \rangle, \tag{4.72}$$

whence $\langle \alpha | \hat{A} | \alpha \rangle$ can legitimately be called an expectation (or mean) value. Further, Eq. (4.71) applied to $A = x$ together with Eqs. (4.30) and (4.43) leads to

$$\sigma_x^2 = \sum_{\beta(\neq\alpha)} |\langle \alpha | \hat{x} | \beta \rangle|^2 = \sum_{\beta(\neq\alpha)} \langle \alpha | \hat{x} | \beta \rangle \langle \beta | \hat{x} | \alpha \rangle = \langle \alpha | \hat{x}^2 | \alpha \rangle - \langle \alpha | \hat{x} | \alpha \rangle^2. \tag{4.73}$$

Since the variance is obtained by calculating averages either over time or over the realizations of the field, the first equality confirms that the quantities $\langle \alpha | \hat{x} | \beta \rangle$ possess a statistical connotation. An equation similar to (4.73) holds for the variance σ_A^2 of any dynamical variable A, so that although no trace of stochasticity remains in the quantities (4.71), their statistical nature has not been lost. Moreover, the last equality in (4.73) indicates that the quantity as calculated within the standard quantum formalism should be interpreted just as the statistical variance. From this it follows, in particular, that the Heisenberg inequality actually involves statistical variances, so that no reference to observations or measurements is required for its interpretation: it ensues as a direct consequence of the persistent action of the ZPF once the ergodic regime has been reached, as mentioned earlier.

Finally, with the elements at hand it is a simple matter to make the transition to the Schrödinger equation in terms of wave functions of the form $\psi_\alpha(x,t) = \langle x | \alpha(t) \rangle = e^{-i\mathcal{E}_\alpha t/\hbar} \varphi_\alpha(x)$.

4.5. *Bipartite systems. Emergence of entanglement*

The theory just presented can be generalized to the case of two noninteracting particles. Here we briefly sketch some of our main results (details can be seen in Refs. 52, 53). The starting equations of motion are

$$m_1 \ddot{x}_1 = f_1(x_1) + m_1 \tau_1 \dddot{x}_1 + e_1[E_1(t) + \frac{1}{e_2} m_2 \tau_2 \dddot{x}_2], \tag{4.74}$$

$$m_2 \ddot{x}_2 = f_2(x_2) + m_2 \tau_2 \dddot{x}_2 + e_2[E_2(t) + \frac{1}{e_1} m_1 \tau_1 \dddot{x}_1]. \tag{4.75}$$

These equations are entirely similar to the one studied in the single-particle case (Eq. (4.1)), the crucial difference being the presence of the last (radiative) terms, which couple the particles. This coupling is fundamental, and shows that each particle modifies the field acting on the other one, so that the partners cease to be independent, a fact that ultimately leads to nonclassical correlations between both particles, as will be seen below.

The radiative coupling terms in Eqs. (4.74), (4.75) superpose on the respective background field $E_i(t)$ in the vicinity of the particle $i(=1,2)$, giving rise to effective external fields $E_i^{\text{eff}} = E_i(t) + (m_j \tau_j / e_j) \dddot{x}_j$. Since the particle located at x_1 reaches a stationary state when the mean power radiated by it balances the mean power absorbed from the *effective* field E_1^{eff}, it follows that the solution $x_{1\alpha}$ (assuming the particle has reached a stationary state characterized by α) bears information regarding the state α' reached by the second particle. As a result of this coupling, the solutions of Eqs. (4.74), (4.75) must now be labeled by a compound index $A = (\alpha, \alpha')$ that is in direct (though not univocal because of the possible energy degeneracy) correspondence with the total mechanical energy $\mathcal{E}_A = \mathcal{E}_\alpha + \mathcal{E}_{\alpha'}$. Moreover, the stationary solutions depend on the stochastic variables of the background field $E_1(t)$ ($\{a_{\alpha\beta}\}$) *and also* on those of $E_2(t)$ ($\{b_{\alpha'\beta'}\}$). This ultimately leads to the substitution $x_{i\alpha}(a_{\alpha\beta}) \rightarrow x_{iA} = x_{i\alpha\alpha'}(a_{\alpha\beta}, b_{\alpha'\beta'})$ for the solutions of (4.74), (4.75), and similarly for the expansions of the dynamical variables.

Now, if F (or G) represents a dynamical variable that belongs to particle 1 (or 2) in state α (or α'), the expansion for the product variable FG in the radiationless approximation can be written as

$$(FG)_A = \tilde{F}_{\alpha\alpha}\tilde{G}_{\alpha'\alpha'} + \sum_{\beta,\beta'} \tilde{F}_{\alpha\beta}\tilde{G}_{\alpha'\beta'} a_{\alpha\beta} b_{\alpha'\beta'}\Big|_{\omega_{\alpha\beta}=-\omega_{\alpha'\beta'}\neq 0} + \mathcal{O}, \qquad (4.76)$$

where $\mathcal{O}$ contains the set of all time-dependent (oscillatory terms) that average to zero. Since $\overline{F_A}^t = \tilde{F}_{\alpha\alpha}$ and $\overline{G_A}^t = \tilde{G}_{\alpha'\alpha'}$, the covariance $\Gamma_{(FG)_A} \equiv \overline{(FG)_A}^t - \overline{F_A}^t\overline{G_A}^t$ becomes

$$\Gamma_{(FG)_A} = \sum_{\beta,\beta'} \tilde{F}_{\alpha\beta}\tilde{G}_{\alpha'\beta'} a_{\alpha\beta} b_{\alpha'\beta'}\Big|_{\omega_{\alpha\beta}=-\omega_{\alpha'\beta'}\neq 0}. \qquad (4.77)$$

Given that $\omega_{\alpha\beta}$ (or $\omega_{\alpha'\beta'}$) represents a relevant frequency of particle 1 (or 2), this equation shows that the existence of nontrivial (nonzero) *common relevant frequencies* is crucial for the existence of correlation between the (corresponding dynamical variables of the) particles. The common frequencies satisfy the condition $\omega_{\alpha\beta} = -\omega_{\alpha'\beta'}$, or equivalently $\mathcal{E}_A = \mathcal{E}_B$, as follows from Eq. (4.67). Thus, when the total energy $\mathcal{E}_A$ of the actual state (α, α') has no degeneracies, the particles are uncorrelated.

4.6. *Entangled states*

Let us focus on two-particle states A whose energy is degenerate (with $\mathcal{E}_A = \mathcal{E}_K$) and consider the following expansion, describing the variable FG not in state A, but rather in a state of (degenerate) energy $\mathcal{E}_A$:

$$
\begin{aligned}
(FG)_{\mathcal{E}_A=\mathcal{E}_K} &= \tfrac{1}{2}[(FG)_A + (FG)_K] \\
&= \tfrac{1}{2}\sum_{\beta,\beta'} \left(\tilde{F}_{\alpha\beta}\tilde{G}_{\alpha'\beta'} a_{\alpha\beta} b_{\alpha'\beta'} + \tilde{F}_{\kappa\beta}\tilde{G}_{\kappa'\beta'} a_{\kappa\beta} b_{\kappa'\beta'} \right) e^{i(\omega_{\alpha\beta}+\omega_{\alpha'\beta'})t} \\
&= \tfrac{1}{2}\sum_{\beta,\beta'} \left(\tilde{F}_{\alpha\beta}\tilde{G}_{\alpha'\beta'} + a_{\kappa\alpha} a_{\kappa'\alpha'} \tilde{F}_{\kappa\beta}\tilde{G}_{\kappa'\beta'} \right) a_{\alpha\beta} b_{\alpha'\beta'}\Big|_{\omega_{\alpha\beta}=-\omega_{\alpha'\beta'}} + \mathcal{O} \quad (4.78)
\end{aligned}
$$

Accordingly, in what follows we consider the covariance

$$\Gamma_{(FG)_{\mathcal{E}_A=\mathcal{E}_K}} \equiv \overline{(FG)_{\mathcal{E}_A=\mathcal{E}_K}}^t - \overline{F_{\mathcal{E}_A=\mathcal{E}_K}}^t \overline{G_{\mathcal{E}_A=\mathcal{E}_K}}^t. \qquad (4.79)$$

As was done in the one-particle case, we can now perform a unitary transformation that transfers the evolution from the matrices $\hat{F}(t) \otimes \hat{G}(t) \in \mathcal{H}_1 \otimes \mathcal{H}_2$ (with elements $\tilde{F}_{\alpha\beta}\tilde{G}_{\alpha'\beta'}e^{i(\omega_{\alpha\beta}+\omega_{\alpha'\beta'})t}$) to vectors of the expanded Hilbert space. In the radiationless approximation, such evolving vectors are

$$|A(t)\rangle = |\alpha(t)\rangle \, |\alpha'(t)\rangle = e^{-i(\mathcal{E}_A/\hbar)t} \, |e_A\rangle \,. \tag{4.80}$$

Expanding $\hat{F}\hat{G}$ in terms of the basis $\{|A(t)\rangle \langle B(t)|\}$ we obtain the usual expression

$$\tilde{F}_{\alpha\beta}\tilde{G}_{\alpha'\beta'} = \langle A| \, \hat{F}\hat{G} \, |B\rangle \,. \tag{4.81}$$

An essential difference with respect to the one-particle problem arises, since now the energy $\mathcal{E}_A$ is not always univocally related to a single vector of the transformed basis (4.80). More specifically, an analysis of the spectral decomposition of $(FG)_A$ allows us to conclude that for nondegenerate $\mathcal{E}_A$, the factorizable vector $|\alpha\rangle \, |\alpha'\rangle$ is univocally related to $\mathcal{E}_A$, whereas for degenerate $\mathcal{E}_A$ a new type of vector, that is not an element of the basis $\{|A(t)\rangle\}$, must be introduced. Whenever there exist common relevant frequencies $\omega_{\alpha\beta} = \omega_{\beta'\alpha'}$, the vector representing the state has the structure

$$|\alpha\rangle \, |\alpha'\rangle + \lambda_{AK} \, |\beta\rangle \, |\beta'\rangle \,, \tag{4.82}$$

where λ_{AK} is a *non-stochastic* phase factor given by

$$\lambda_{DG} \equiv a_{\delta\gamma}b_{\delta'\gamma'} = e^{i\varsigma_{\delta'\gamma'}} = e^{-i\varsigma_{\delta\gamma}}, \quad \text{for} \quad \omega_{\delta\gamma} = -\omega_{\delta'\gamma'}. \tag{4.83}$$

Thus, whenever the two particles share a common resonance frequency, a *new* class of state vector arises in the transition to a Hilbert space description, which is nonfactorizable and gives rise to *entanglement*.

Since the entangled vectors are the suitable ones for describing the mechanical system in the degenerate case, they must be associated with the expansions $(FG)_{\mathcal{E}_A=\mathcal{E}_K}$ in the form

$$(FG)_{\mathcal{E}_A=\mathcal{E}_K} \to \frac{1}{\sqrt{2}}\Big(|\alpha\rangle \, |\alpha'\rangle + \lambda_{AK} \, |\kappa\rangle \, |\kappa'\rangle \Big). \tag{4.84}$$

The factor $1/\sqrt{2}$ here originates in the factor $1/2$, representing a (balanced) statistical weight, which appears in the first line of Eq. (4.78).

The need for entangled vectors ultimately implies nonzero correlations between (certain variables of) the particles, as stated after Eq. (4.77). From here it follows that interaction with a common background field, emergence of correlations, and entanglement are tightly linked notions. On the other hand, the covariance Γ^{qm}_{FG} calculated in accordance with the quantum-mechanical rule,

$$\Gamma^{qm}_{FG} = \left\langle \hat{F}\hat{G} \right\rangle - \left\langle \hat{F} \right\rangle \left\langle \hat{G} \right\rangle \equiv \langle \Psi| \, \hat{F}\hat{G} \, |\Psi\rangle - \langle \Psi| \hat{F} |\Psi\rangle \, \langle \Psi| \hat{G} |\Psi\rangle \tag{4.85}$$

agrees with the expressions obtained for $\Gamma_{(FG)_A}$ and $\Gamma_{(FG)_{\mathcal{E}_A=\mathcal{E}_K}}$ whenever $|\Psi\rangle$ is given by $|\alpha\rangle \, |\alpha'\rangle$ and by (4.84), respectively. This exhibits the need for quantum mechanics to resort to entangled states in order to properly describe these bipartite correlations. The agreement between our results and those obtained with the usual quantum methods confirms that also in the two-particle case, the present theory correctly leads to quantum mechanics once the stationary, ergodic and radiationless limit is taken and a description in terms of vectors in a Hilbert space is adopted.

Note that the correlation due to the structure of the vector (4.84) depends in a most direct way on the phase factor λ_{AK}, whose origin goes back to the field variables, since λ_{DG} can be also written as $\lambda_{DG} = \langle a(\omega_{\delta\gamma})b(\omega_{\delta'\gamma'})\rangle$. In this regard it is important to point out that it is thanks to Eq. (4.83) — which allows to write λ_{AK} as a nonstochastic phase factor — that the background field leaves its footprint in the Hilbert-space description, even in the radiationless limit and when all explicit reference to the stochastic field variables has disappeared. The entanglement phase factor λ_{AK} appears thus as a vestige of the zero-point field, reminding us of its active role as a member of the entire system.

4.6.1. *Identical particles: unavoidable entanglement*

As stated above, the existence of a common relevant frequency is sufficient for the particles to get entangled, regardless of their nature. However, when the particles are identical and are subject to the same external potential, their entanglement becomes unavoidable, since in this case *all* the relevant frequencies (or equivalently, the sets of accesible states) are common to both particles.

The double degeneracy of all the energies $\mathcal{E}_{(\alpha,\alpha')} = \mathcal{E}_{(\alpha',\alpha)}$ leads us to associate $(FG)_{\mathcal{E}_{(\alpha,\alpha')}=\mathcal{E}_{(\alpha',\alpha)}}$ with the vector (c.f. Eq. (4.84))

$$\frac{1}{\sqrt{2}}\Big(|\alpha\rangle_1 |\alpha'\rangle_2 + \lambda_A |\alpha'\rangle_1 |\alpha\rangle_2 \Big), \quad \lambda_A \equiv a_{\alpha\alpha'}b_{\alpha'\alpha}, \ (\alpha \neq \alpha'). \tag{4.86}$$

Notice the need to introduce here the indices $1,2$ in order to avoid ambiguities with regard to the particle that is in the given state α or α'.

A natural transformation in this kind of systems is the exchange of particles, I_p. According to the present approach, exchanging the particles amounts to exchange their actual accessible states (operation $I_s : \alpha \leftrightarrow \alpha'$), plus the fields in which they are immersed (operation $I_f : a \leftrightarrow b$). Direct application of $I_p = I_s I_f$ shows that the phase factor λ_A is invariant under the exchange of particles,

$$I_p\lambda_A = I_p a_{\alpha\alpha'}b_{\alpha'\alpha} = b_{\alpha'\alpha}a_{\alpha\alpha'} = \lambda_A. \tag{4.87}$$

However, when the field variables are eliminated from the description and the transition to the product Hilbert space formalism is made, the exchange of particles reduces to the substitution $\alpha \leftrightarrow \alpha'$. The exchange of particles is then represented in this reduced description by $I_p^{\text{qm}} = I_s$, where the superindex 'qm' distinguishes this operator from the transformation I_p. Moreover, consistency with the radiationless approximation requires writing $\lambda_A = a_{\alpha\alpha'}b_{\alpha'\alpha}$ in a form that does not make explicit reference to the field variables. The required form turns out to be

$$\lambda_A = \exp(i\varsigma_{\alpha'\alpha}), \tag{4.88}$$

and this is the expression over which I_p^{qm} acts. Since λ_A must maintain its invariance properties irrespective of the specific form we use to express it (whether (4.86) or (4.88)), in addition to (4.87) we must have $I_p^{\mathrm{qm}}\lambda_A = \lambda_A$, with λ_A given by (4.88). From here it follows that $\lambda_A = \lambda_A^*$, and consequently, λ_A being a phase factor, we arrive at

$$\lambda_A = \pm 1, \quad \forall \alpha \neq \alpha'. \tag{4.89}$$

This result tells us that a system of two identical, non-interacting particles subject to the same external potential, is described by maximally entangled, hence totally (anti)symmetric states.

5. Contact with QED. Radiative corrections

In this last part we extend our journey from quantum mechanics to a domain usually considered to be the exclusive province of quantum electrodynamics: the radiative transitions and corrections to the atomic energy levels.

We achieve this by drawing further consequences from the energy-balance equation, and more generally by focusing on the effects of the radiative terms that were neglected in the transition to the Schrödinger formalism carried out above. The results obtained coincide in every case with those of nonrelativistic spinless QED. This confirms that the present theory goes *beyond* quantum mechanics. Related work can be seen in Refs. 10 and 54.

5.1. *Absence of detailed energy balance*

We have obtained already an important result from the energy-balance condition for a system in a stationary state, Eq. (3.61),

$$m\tau \langle \dot{x}\,\dddot{x} \rangle = \frac{e^2}{m} \left\langle p\hat{D} \right\rangle, \tag{5.1}$$

with $\tau = 2e^2/3mc^3 = (2/3)(\hbar/mc^2)\alpha$, $\alpha = e^2/\hbar c$. Applied to the ground state it ensures the correct value for the Planck constant in the Schrödinger equation. Now instead of considering the particle in its ground state, we assume that it is in an excited state n, the background field still being in its ground state. Then both sides of the detailed-balance condition must be recalculated, in the time-asymptotic limit. For the left-hand side one obtains

$$m\tau \langle \dot{x}\,\dddot{x} \rangle_n = -m\tau \sum_k \omega_{nk}^4 \left| x_{nk} \right|^2, \quad n > 0. \tag{5.2}$$

For the right-hand side one gets

$$\frac{e^2}{m} \left\langle p\hat{D} \right\rangle_n = -m\tau \sum_k \omega_{nk}^4 \left| x_{nk} \right|^2 \mathrm{sign}\,\omega_{kn}. \tag{5.3}$$

Equation (5.3) contains now a mixture of positive and negative terms, whilst in (5.2) all contributions have the same sign. As a result, according to Eq. (3.18) there is a net loss of mean energy,

$$
\frac{d}{dt} \langle H \rangle_n = -m\tau \sum_k \omega_{nk}^4 |x_{nk}|^2 (1 - \text{sign}\omega_{kn})
$$

$$
= -2m\tau \sum_{k<n} \omega_{nk}^4 |x_{nk}|^2 , \tag{5.4}
$$

indicating that there cannot be detailed balance between the ZPF and the particle in an excited state — as was to be expected, since the ZPF is the background radiation field in its ground state. This confirms that only the ground state of the particle ($n = 0$) is stable in the sole presence of the ZPF.

Let us investigate whether there is any background field $\rho(\omega) = \rho_0(\omega)g(\omega)$ with which a mechanical system in its excited state can be in equilibrium, where $g(\omega)$ represents either an excitation of the background field or the contribution of an external field. To respond to this question we observe that the expressions for the mean power radiated by the particle, Eq. (5.2), and for the mean power provided by the field, Eq. (5.3), contain in general mixtures of terms of different frequencies (ω_{nk} for different values of k), but with different signs, so that there is no way that detailed balance is satisfied in general. Only for the particular case in which all values of ω_{nk} coincide, the possibility of detailed balance exists.

We shall explore this possibility for the case of the harmonic oscillator, in which all $|\omega_{nk}|$ that contribute are equal and coincide with the oscillator frequency ω_0. With $|x_{nn+1}|^2 = a(n+1)$, $|x_{nn-1}|^2 = an$, $|x_{nk}|^2 = 0$ for $k \neq n\pm1$, and $a = \hbar/2m\omega_0$, Eq. (5.2) gives

$$
m\tau \langle \dot{x}\,\dddot{x} \rangle_n = -m\tau\omega_0^4 a(2n + 1) = -\frac{1}{2}\hbar\tau\omega_0^3(2n + 1). \tag{5.5}
$$

From (5.3), on the other hand, one gets, with $\rho(\omega) = \rho_0(\omega)g_n(\omega)$ where $g_n(\omega)$ is a function to be determined,

$$
\frac{e^2}{m} \langle p\hat{D} \rangle_n = -m\tau g_n(\omega_0)\omega_0^4 a(n + 1 - n) = -\frac{1}{2}\hbar\tau\omega_0^3 g_n(\omega_0). \tag{5.6}
$$

A comparison of these two expressions shows that indeed, detailed balance exists between a harmonic oscillator in its excited state n and a background field with spectral energy density

$$
\rho(\omega) = \rho_0(\omega)(2n + 1). \tag{5.7}
$$

Also this result should not come as a surprise, since this field has precisely an energy per normal mode $\frac{1}{2}\hbar\omega_0(2n + 1)$ — equal to the energy of the mechanical oscillator with which it is in equilibrium.

5.2. Radiative transitions and atomic lifetimes: Einstein's A and B coefficients

Now we investigate some important implications of the absence of detailed balance. This can be done by using again Eq. (3.18) to obtain the average energy lost (or gained) by the mechanical system due to the difference between the terms on the right-hand side. According to Eqs. (5.2) and (5.3) (but now with $\rho(\omega) = \rho_0(\omega)g(\omega)$; $\rho(\omega) = \rho(|\omega|)$), this difference is given by

$$\frac{dH_n}{dt} = -m\tau \sum_k \omega_{nk}^4 |x_{nk}|^2 \left[1 - g(|\omega_{nk}|)\mathrm{sign}\omega_{kn}\right].$$

It is convenient to rewrite the right-hand side by introducing $g(\omega) = 1 + g_a(\omega)$, in order to separate the contribution coming from the external (or additional) background field

$$\rho_a(\omega) = \rho_0(\omega)g_a(\omega), \tag{5.8}$$

$$\frac{dH_n}{dt} = -m\tau \sum_k \omega_{nk}^4 |x_{nk}|^2 \left[1 - (1 + g_a(\omega_{kn}))\mathrm{sign}\omega_{kn}\right] =$$

$$= m\tau \sum_k \omega_{nk}^4 |x_{nk}|^2 \left[(g_a)_{\omega_{kn}>0} - (2 + g_a)_{\omega_{kn}<0}\right]. \tag{5.9}$$

The term within the brackets in the second line of this equation, proportional to g_a, represents the absorptions and the second one, proportional to $2 + g_a$, the emissions. It is clear from this result that there can be absorptions only when the background field is excited or there is an external component, whilst the emissions can be either 'spontaneous' (in presence of just the ZPF) or else stimulated by the additional field (represented by g_a). The coefficients appearing in the various terms determine the respective rates of energy gain and energy loss; therefore, they should be expected to be directly related with Einstein's A and B coefficients. The coefficient A is defined as the time rate for spontaneous emissions,

$$dH_n = -\hbar \sum_k \omega_{nk} A_{nk} dt, \tag{5.10}$$

and the coefficients B, which determine the rate of energy gain or loss due to transitions induced (stimulated) by the external field, are defined through

$$dH_n = \pm\hbar \sum_k |\omega_{nk}| B_{nk,kn} \rho_a(|\omega_{nk}|) dt. \tag{5.11}$$

With the aid of these definitions Eq. (5.9) can be rewritten in the even more transparent form

$$\frac{dH_n}{dt} = \sum_{k>n} \hbar |\omega_{nk}| \left[\rho_a(|\omega_{nk}|)B_{kn}\right] - \sum_{k<n} \hbar |\omega_{nk}| \left[A_{nk} + \rho_a(|\omega_{nk}|)B_{nk}\right]. \tag{5.12}$$

whence, by comparison with (5.9),

$$A_{nk} = \frac{4e^2 |\omega_{nk}|^3}{3\hbar c^3} |x_{nk}|^2, \tag{5.13}$$

in agreement with the QED prediction.[23,55] In its turn, the coefficient B is given by the first term (in the case of absorptions) or the last one (for emissions) within the square brackets in Eq. (5.9),

$$B_{nk} = B_{kn} = \frac{m\tau\omega_{nk}^4 |x_{nk}|^2 g_a(\omega_{nk})}{\hbar |\omega_{nk}| \rho_a(\omega_{nk})} = \frac{4\pi^2 e^2}{3\hbar^2} |x_{nk}|^2, \tag{5.14}$$

a result that also coincides with the respective formula of QED (or QM).[23] This confirms the key role played by *both* radiative terms — the radiation reaction and the background field (which always contains the ZPF but can include the additional component) — in determining the rates of transition between stationary states of the mechanical system. The results also demonstrate the equivalence of the present theory and nonrelativistic QED.

The expressions for the Einstein coefficients A_{nk}, B_{nk}, and B_{kn}, involve each one the single frequency $|\omega_{nk}|$, which means that the system as a whole reaches a state of *detailed equilibrium*, i.e., equilibrium of matter with the field at each separate frequency, as was already noticed. The theory has thus led us to a transition from global equilibrium to detailed equilibrium in the quantum regime. We recall that this demand was one of Einstein's hypotheses in his pioneering work where he introduced the absorption and emission coefficients.[56]

It is pertinent to ask here at what point does quantization enter in Einstein's paper so as to arrive at the Planck distribution, an inquiry that comes to surface not infrequently. A current answer to this question is that quantization is introduced by assuming discrete atomic levels. However, this is wrong, as Einstein and Ehrenfest demonstrated some time after the initial paper by redoing the calculations with a continuous distribution of atomic levels,[57] recovering the old results. The correct answer is that quantization enters through the introduction of a source that includes at the outset the possibility of the ZPF, able to generate 'spontaneous' transitions. This can be easily verified by redoing the Einsteinian calculation (see Eq. (5.16)), but omitting any of the three terms that lead to matter-field equilibrium (stimulated absorptions and emissions, and spontaneous emissions). The absence of the latter leads to absurd results (such as atomic coefficients that depend on the temperature). It is interesting to observe that the omission of the term that describes the stimulated emissions in Eq. (5.9) (after introducing appropriate populations) leads to the approximate expression for Planck's law proposed by Wien (Eq. (2.35)), which correctly approximates it at low temperatures, so it already contains some quantum principle (as corresponds to the consideration of the ZPF). All this can be easily seen in the present context by focusing on just two states n and k, with $\mathcal{E}_n - \mathcal{E}_k = \hbar\omega_{nk} > 0$ and respective populations N_n, N_k. When the system is in thermal equilibrium the relation

$$N_k/N_n = \exp(\mathcal{E}_n - \mathcal{E}_k)/k_B T \tag{5.15}$$

holds (forgetting about possible but inconsequencial degeneracies). Since according to Eq. (5.9) the number of emissions is proportional to $N_n g_a(\omega_{nk})$ and the number of absorptions is proportional to $N_k[2 + g_a(\omega_{nk})]$, from the equilibrium condition

$$N_n g_a = N_k(2 + g_a) \tag{5.16}$$

one obtains indeed Planck's law (for the thermal field)

$$g_a(\omega_{nk}) = \frac{2}{\exp\left[(\mathcal{E}_n - \mathcal{E}_k)/k_B T\right] - 1}. \tag{5.17}$$

The ratio of the A to the B coefficients at any given temperature is

$$\frac{A_{nk}}{B_{nk}} = \frac{\hbar |\omega_{nk}|^3}{\pi^2 c^3} = 2\rho_0(|\omega_{nk}|). \tag{5.18}$$

This relation and the equality $B_{nk} = B_{kn}$ were predicted by Einstein on the basis of his statistical considerations; here they follow from the theory, as is the case in QED. Notice in particular the factor 2 in Eq. (5.18). Given the definition of the coefficients, one could expect the ratio in this equation to correspond exactly to the spectral density of the ZPF, which would mean a factor of 1. The factor 2 seems to suggest that the ZPF has double the ability of the rest of the electromagnetic field to induce transitions. The correct explanation is another: inspection of Eq. (5.9) shows that one should actually write $2\rho_0 = \rho_0 + \rho_0$. One of these two equal contributions to spontaneous decay is due to the effect of the fluctuations impressed on the particle by the field; the second one is the expected contribution due to radiation reaction, that is, Larmor radiation. Not surprisingly, they turn out to be equal: it is precisely their equality what leads to the exact balance of these two contributions when the system is in its ground state, guaranteeing the stability of this state. Yet one can frequently find in the literature that all the spontaneous decay is attributed to one or the other of these two causes, more frequently to Larmor radiation. It is an important result of both the present theory and QED that the two effects contribute with equal shares. (Interesting related discussions can be seen in Refs. 23, 58–61.)

5.3. *Radiative corrections to the energy levels*

The derivation of the Schrödinger equation presented in section 2 confirms that the radiative terms (or corrections) give rise just to corrections to the solutions of the (unperturbed) Schrödinger equation. The Einstein A and B coefficients for the lifetimes of excited states, determined by the right-hand side of Eq. (5.9), pertain to this category. A further important — even if smaller — radiative correction, one that represents a major success of QED, is the shift of the atomic levels due to another residual effect of the ZPF. Indeed, the effective work realized by the fluctuating motions of the bound particle gives rise to a tiny modification of the mean kinetic energy that affects the energy levels, as is here shown by means of a direct approach.

5.4. *The Lamb shift*

Let us go back to Eq. (3.17),

$$\frac{d}{dt} \langle xp \rangle = \frac{1}{m} \langle p^2 \rangle + \langle xf \rangle + m\tau \langle x\,\dddot{x} \rangle - e^2 \langle x\hat{D} \rangle, \tag{5.19}$$

and use it to calculate the radiative energy shift. As explained in section 3.2, Eq. (5.19) is a time-dependent version of the virial theorem, with radiative corrections included. It is interesting to observe that the average values are here taken over the ensemble, instead of over time. This is but an example of the ergodic properties acquired by the quantum states, a matter discussed at length in section 4.2.

In the stationary state, the two previously neglected terms in Eq. (5.19) represent radiative corrections to the (kinetic) energy T,

$$\delta \langle T \rangle_n = -\frac{m\tau}{2} \langle x\,\dddot{x} \rangle_n + \frac{e^2}{2} \langle x\hat{D} \rangle_n. \tag{5.20}$$

For consistency, the contribution of these terms is again calculated to lowest order in $\tau \sim e^2$, which means calculating the two average values, $\langle x\,\dddot{x} \rangle_n$ and $\langle x\hat{D} \rangle_n$, in the radiationless limit. The first one gives, using the solutions of the Schrödinger equation,

$$-\frac{m\tau}{2} \langle x\,\dddot{x} \rangle_n = \frac{\tau}{2} \langle \dot{x}\,f \rangle_n = \frac{\tau}{2} \frac{d}{dt} \langle T \rangle_n = 0, \tag{5.21}$$

which means that the Larmor radiation does not contribute to the energy shift in the mean, in a stationary state. The correction to the energy is therefore due exclusively to the diffusion produced by the interaction of the particle with the background field, represented by the second term in Eq. (5.20):

$$\frac{e^2}{2} \langle x\hat{D} \rangle_n = -\frac{2e^2}{3\pi c^3} \sum_k |x_{nk}|^2 \omega_{kn} \int_0^\infty d\omega \, \frac{\omega^3}{\omega_{kn}^2 - \omega^2}. \tag{5.22}$$

The radiative correction to the mean energy is therefore given by (in three dimensions, for comparison purposes)

$$\delta \mathcal{E}_n = \frac{e^2}{2} \langle \mathbf{x}\cdot\hat{D} \rangle_n = -\frac{2e^2}{3\pi c^3} \sum_k |\mathbf{x}_{nk}|^2 \omega_{kn} \int_0^\infty d\omega \, \frac{\omega^3}{\omega_{kn}^2 - \omega^2}. \tag{5.23}$$

This result coincides with the formula derived by Power[62] for the Lamb shift on the basis of Feynman's argument.[63] According to Feynman, the presence of the atom creates a weak perturbation on the nearby field, thereby acting as a refracting medium. The effect of this perturbation is to change the frequencies of the background field from ω to $\omega/n(\omega)$, n being the refractive index. The shift of the ZPF energy due to the presence of the atom is then[23,62]

$$\Delta \mathcal{E}_n = \sum_{\mathbf{k},\lambda} \frac{1}{2} \frac{\hbar \omega_k}{n(\omega_k)} - \sum_{\mathbf{k},\lambda} \frac{1}{2} \hbar \omega_k \simeq - \sum_{\mathbf{k},\lambda} [n(\omega_k) - 1] \frac{1}{2} \hbar \omega_k, \qquad (5.24)$$

and the refractive index is given in this approximation by (Ref. 58, chapter 9)

$$n(\omega_k) \simeq 1 + \frac{4\pi}{3\hbar} \sum_m \frac{|\mathbf{d}_{mn}|^2 \omega_{mn}}{\omega_{mn}^2 - \omega^2}, \qquad (5.25)$$

where $\mathbf{d}_{mn} = e\mathbf{x}_{mn}$ is the transition dipole moment. After an integration over the solid angle $\hat{\mathbf{k}}$ and summation over the polarizations $\lambda = 1, 2$, Power obtains in the continuum limit for ω_k the formula

$$\Delta \mathcal{E}_n = -\frac{2}{3\pi c^3} \sum_m |\mathbf{d}_{mn}|^2 \omega_{mn} \int_0^\infty d\omega \, \frac{\omega^3}{\omega_{mn}^2 - \omega^2}, \qquad (5.26)$$

which coincides with the previous result, Eq. (5.23).

The Lamb shift *proper* (called also observable Lamb shift) is obtained by subtracting from the total energy shift given by (5.23), the free-particle contribution represented by this same expression in the limit of continuous electron energies (when ω_{kn} can be ignored compared with ω in the denominator),

$$\delta \mathcal{E}_{\text{fp}} = \frac{2e^2}{3\pi c^3} \sum_k |\mathbf{x}_{nk}|^2 \omega_{kn} \int_0^\infty d\omega \, \omega = \frac{e^2 \hbar}{\pi m c^3} \int_0^\infty d\omega \, \omega. \qquad (5.27)$$

This gives for the Lamb shift proper of level n

$$\delta \mathcal{E}_{\text{L}n} = \delta \mathcal{E}_n - \delta \mathcal{E}_{\text{fp}} = -\frac{2e^2}{3\pi c^3} \sum_k |\mathbf{x}_{nk}|^2 \omega_{kn}^3 \int_0^\infty d\omega \, \frac{\omega}{\omega_{kn}^2 - \omega^2}, \qquad (5.28)$$

which agrees again with the nonrelativistic QED formula. The integral diverges logarithmically, so inserting the usual (non-relativistic) regularizing cutoff $\omega_c = mc^2/\hbar$, one gets

$$\delta \mathcal{E}_{\text{L}n} = \frac{2e^2}{3\pi c^3} \sum_k |\mathbf{x}_{nk}|^2 \omega_{kn}^3 \ln \left| \frac{mc^2}{\hbar \omega_{kn}} \right|. \qquad (5.29)$$

This is the Bethe's well-known expression. Note, however, that in the present approach (as in Power's) no mass renormalization was required; we come back to this point below.

The interpretation of the Lamb shift as a change of the atomic energy levels due to the interaction with the surrounding ZPF is fully in line with the general approach of the present theory. It constitutes one more manifestation of the influence of the particle on the field, which is then fed back on the particle. An alternative way of looking at this reciprocal influence is by considering the general relation between the atomic polarizability α and the refractive index of the medium affected by it (for $n(\omega) \simeq 1$),

$$n(\omega) = 1 + 2\pi\alpha(\omega). \tag{5.30}$$

By comparing this expression with Eq. (5.25) one obtains

$$\alpha_n(\omega) = \frac{4\pi}{3\hbar} \sum_m \frac{|\mathbf{d}_{mn}|^2 \, \omega_{mn}}{\omega_{mn}^2 - \omega^2}, \tag{5.31}$$

which is the Kramers-Heisenberg formula.[58] This indicates that the Lamb shift can also be viewed as a Stark shift associated with the dipole moment $\mathbf{d}(\omega) = \alpha(\omega)\mathbf{E}$ induced by the electric component of the ZPF on the atom,

$$\delta\mathcal{E}_{Ln} = \tfrac{1}{2} \langle \mathbf{d} \cdot \mathbf{E} \rangle_n . \tag{5.32}$$

For completeness, let us look at the radiative energy corrections as is proper (mutatis mutandis) of QED and also physically very suggestive. The (minimal) coupling of the particle to the ZPF gives the Hamiltonian

$$H = H_0 - \frac{e}{mc}\mathbf{A} \cdot \mathbf{p} + \frac{e^2}{2mc^2}\mathbf{A}^2, \tag{5.33}$$

where $H_0 = (\mathbf{p}^2/2m) + V$ is the original Hamiltonian and $\mathbf{A}$ represents the vector potential associated with the ZPF. Note that the Hamiltonian is now a stochastic variable, whence the (mean) energy shift is given by the average of Eq. (5.33) over the realizations of the field (denoted by the superscript E),

$$\delta\mathcal{E}_n = -\frac{e}{mc}\overline{\mathbf{A} \cdot \mathbf{p}}^E + \frac{e^2}{2mc^2}\overline{\mathbf{A}^2}^E . \tag{5.34}$$

With the correlation of the field given by (see section 3.1)

$$\overline{E(t)E(t')}^E = (4\pi/3) \int_0^\infty \rho_0(\omega) \cos\omega(t - t')d\omega, \tag{5.35}$$

calculation of the second term is straightforward and yields

$$\delta\mathcal{E}_{fp} = \frac{e^2}{2mc^2}\overline{\mathbf{A}^2}^E = \frac{e^2\hbar}{\pi mc^3}\int_0^\infty d\omega\, \omega, \tag{5.36}$$

since $\overline{\mathbf{A}^2}^E = (2\hbar/\pi c)\int d\omega\, \omega$ for the ZPF and $\rho_0(\omega) = \hbar\omega^3/(2\pi^2 c^3)$. This result reproduces Eq. (5.27), the free-particle correction to the energy, which does not contribute to the observable Lamb shift. Thus the free-particle Lamb shift is just the contribution due to the presence of the unperturbed ZPF; the term $\overline{\mathbf{A}^2}^E$ is of universal value, independent of the system or its state, a testimony of the ubiquitous presence of the ZPF.

For the calculation of the first term one needs to consider the effect (to first order in e, and returning to the 1-D notation) of the stochastic field A on the particle momentum p. This is most easily done by rewriting the original equations of motion in terms of the (canonical) variables (we continue however to treat the term $m\tau\dddot{x}$ has an 'external' force acting on the particle.), i.e.

$$m\dot{x} = p - (e/c)A, \quad \dot{p} = f(x) + m\tau\dddot{x} \tag{5.37}$$

whence the new (stochastic) Liouvillian is

$$\hat{L} = \frac{1}{m}(p - \frac{e}{c}A)\frac{\partial}{\partial x} + \frac{\partial}{\partial p}(f + m\tau\dddot{x}). \tag{5.38}$$

With $p(t) = e^{-\hat{L}(t-t')}p(t')$ one obtains to lowest order in $\tau \sim e^2$, after averaging over the realizations of the field, taking the time-asymptotic limit and integrating over the particle phase space,

$$-\frac{e}{mc}\langle Ap\rangle_n = \left(\frac{e}{mc}\right)^2\left(\frac{2\hbar}{3\pi c}\right)\int d\omega \int dt'\omega\cos\omega(t-t')\left\langle\frac{\partial p'}{\partial x}\right\rangle_n, \tag{5.39}$$

whence (returning to 3-D notation for comparison purposes)

$$\delta\mathcal{E}_{Ln} = -\frac{e}{mc}\langle\mathbf{A}\cdot\mathbf{p}\rangle_n = -\frac{2e^2}{3\pi c^3}\sum_k|\mathbf{x}_{nk}|^2\omega_{kn}^3\int_0^\infty d\omega\,\frac{\omega}{\omega_{kn}^2 - \omega^2}. \tag{5.40}$$

This result coincides with Eq. (5.28), thus confirming the consistency of the different approaches.

It seems convenient to point out some differences between the procedures used in SED and in QED to arrive at the formula for the Lamb shift. In the QED case, second-order perturbation theory is used, with the interaction Hamiltonian given by $\hat{H}_{\text{int}} = -(e/mc)\hat{\mathbf{A}}\cdot\hat{\mathbf{p}}$. But the energy derived from this term,[23]

$$-\frac{2e^2}{3\pi c^3}\sum_k|\mathbf{x}_{nk}|^2\omega_{kn}^2\int_0^\infty d\omega\,\frac{\omega}{\omega - \omega_{nk}}, \tag{5.41}$$

still contains the (linearly divergent) free-particle contribution

$$-\frac{2e^2}{3\pi c^3}\sum_k|\mathbf{x}_{nk}|^2\omega_{kn}^2\int_0^\infty d\omega = -\frac{4e^2}{3\pi c^3}\left(\frac{1}{2m}\sum_k|\mathbf{p}_{nk}|^2\right)\int_0^\infty d\omega \tag{5.42}$$

that must be subtracted to obtain the Lamb shift proper. Because the result is proportional to the mean kinetic energy, the ensuing correction represents a mass correction (mass renormalization),

$$\delta m = \frac{4e^2}{3\pi c^3}\int_0^\infty d\omega, \tag{5.43}$$

which with the cutoff $\omega_c = mc^2/\hbar$ becomes $\delta m = (4\alpha/3\pi)m$. On the other hand, in the derivation presented here to obtain the formula for $\delta\mathcal{E}_L$, Eq. (5.28), there was no need to renormalize the mass. The result (5.43) is just the *classical* contribution to the mass predicted by the Abraham-Lorentz equation (see Ref. 36, Eq. (3.114)); in the equations of motion (2.1) this contribution has been already subtracted, so there is no more need of mass renormalization in the SED calculation. As has been seen, however, the formula (5.28) (common to both SED and renormalized QED) still has a logarithmic divergence, which calls for the introduction of the cutoff frequency ω_c as was done by Bethe, thus leading to a very satisfactory result for the Lamb shift. From the present perspective the problem of the divergence is closely linked to the unsolved problem of the (unphysical) divergence at high frequencies of the ZPF energy density given by Eq. (3.4).

5.5. *External effects on the radiative energy corrections*

By now it is clear that some basic properties of the vacuum —such as the intensity of its fluctuations or its spectral distribution — are directly reflected in the radiative corrections. This means that a change in such properties can in principle lead to an observable modification of these corrections. The background field can be altered, for instance, by raising the temperature of the system, by adding external radiation, or by introducing objects that alter the distribution of the normal modes of the field.

Such 'environmental' effects have been studied for more than 60 years, normally within the framework of quantum theory — although some calculations have been made also within the framework of SED, in particular for the harmonic oscillator, leading to comparable results (see e.g. Refs. 64–66). Here we have the possibility of applying the formulas derived in the previous sections to the general case, without restricting the calculations to the harmonic oscillator. The task is facilitated and becomes transparent by the use of the present theory because the presence (and action) of the background radiation field is clear from the beginning.

In section 5.2 we have already come across one observable effect of a change in the background field: according to Eq. (5.9) the rates of stimulated atomic transitions are directly proportional to the spectral distribution of the external (or additional) background field, be it a thermal field or otherwise. In the case of a thermal field in particular, with $g_a(\omega_{nk})$ given by (5.17), the (induced) transition rate from state n to state k becomes (using Eqs. (5.12) and (5.14))

$$\frac{dN_{nk}}{dt} = \rho_0(\omega_{nk})\gamma_a(\omega_{nk})B_{nk} =$$

$$= \frac{4e^2 |\omega_{nk}|^3 |x_{nk}|^2}{3\hbar c^3} \frac{1}{e^{\hbar|\omega_{nk}|/k_B T} - 1}. \tag{5.44}$$

This result shows, as is well known, that no eigenstate is stable at $T > 0$, because the thermal field induces both upward and downward transitions. For downward transitions ($\omega_{nk} > 0$) we can rewrite (5.44) for comparison purposes in terms of A_{nk} as given by Eq. (5.13),

$$\frac{dN_{nk}}{dt} = \frac{A_{nk}}{e^{\hbar|\omega_{nk}|/k_B T} - 1}, \tag{5.45}$$

which indicates that the effect of the thermal field on the decay rate is hardly noticeable at room temperature ($k_B T \simeq .025$ eV), since for typical atomic frequencies $[\exp(\hbar|\omega_{nk}|/k_B T) - 1]^{-1}$ ranges between $\exp(-40)$ and $\exp(-400)$.

When the geometry or the spectral distribution of the field is modified by the presence of conducting objects, the transition rates are affected accordingly. Assume, for simplicity, that the modified field is isotropic, with the density of modes of a given frequency ω_{nk} reduced by a factor $g(\omega_{nk}) < 1$. Then according to the results of section 5.2 the corresponding spontaneous and induced transition rates are reduced by this factor, since both A and ρB are proportional to the density of modes. By enclosing the atoms in a high-quality cavity that excludes the modes of this frequency one can therefore virtually inhibit the corresponding transition. For the more general anisotropic case the calculations are somewhat more complicated, without however leading to a substantial difference from a physical point of view. These cavity effects have been the subject of a large number of experimental tests since the early works of Kleppner,[67] Goy et al.[68] and others. Concerning the physical implication of these effects, they provide a clear proof that the background field, including the ZPF, acts as a mediator between the atom and the cavity walls, so as to directly influence the (spontaneous and induced) transition rates. How else could the atom register the influence of its surroundings, even before the emission? Moreover, it is clear that, rather than the energy levels of the initial and final atomic states, it is the (resonance) frequencies that play an essential role in determining the transition rates, as expressed in the formulas for A and ρB.

The changes in the energy shift produced by the addition of an (external or thermal) background field can be calculated readily from Eqs. (5.27) and (5.28). Let ρ_a be the spectral energy density of the additional field, so that $\rho = \rho_0 + \rho_a$. Then the formulas for the variations of the (first-order) radiative corrections are obtained by determining the shifts produced by the total field and subtracting the original shifts produced by the ZPF (ρ_0). The results are

$$\Delta\left(\delta\mathcal{E}_{\text{fp}}\right) = \frac{4\pi e^2}{3\hbar} \sum_k |\mathbf{x}_{nk}|^2 \omega_{kn} \int_0^\infty d\omega \, \frac{\rho_a}{\omega^2} = \frac{e^2\hbar}{\pi mc^3} \int_0^\infty d\omega \frac{\rho_a}{\rho_0}\omega, \tag{5.46}$$

$$\Delta\left(\delta\mathcal{E}_{\text{Ln}}\right) = -\frac{2e^2}{3\pi c^3} \sum_k |\mathbf{x}_{nk}|^2 \omega_{kn}^3 \int_0^\infty d\omega \, \frac{\rho_a}{\rho_0} \frac{\omega}{\omega_{kn}^2 - \omega^2}, \tag{5.47}$$

for a homogeneous field. If, for instance, the additional field represents blackbody radiation at temperature T, i.e. $\rho_a(T) = 2\rho_0/(\epsilon - 1)$ with $\epsilon = \exp(\hbar\omega/kT)$, Eq. (5.46) gives

$$\Delta_T\left(\delta\mathcal{E}_{\text{fp}}\right) = \frac{2e^2\hbar}{\pi mc^3} \int_0^\infty d\omega \, \frac{\omega}{\epsilon - 1} = \frac{2\alpha}{\pi mc^2}(kT)^2 \int_0^\infty dy \, \frac{y}{\exp y - 1}. \tag{5.48}$$

With $\int_0^\infty dy \, \dfrac{y}{\exp y - 1} = \dfrac{\pi^2}{6}$ this gives for the change of the free-particle energy

$$\Delta_T\left(\delta\mathcal{E}_{\text{fp}}\right) = \frac{\pi\alpha}{3mc^2}(kT)^2. \tag{5.49}$$

The formula for the change of the Lamb shift is given according to Eq. (5.47) by

$$\Delta\left(\delta\mathcal{E}_{\mathrm{L}n}\right) = -\frac{4e^2}{3\pi c^3}\sum_k |\mathbf{x}_{nk}|^2\,\omega_{kn}^3\int_0^\infty d\omega\;\frac{\omega}{\omega_{kn}^2-\omega^2}\left(\frac{1}{\exp(\hbar\omega/kT)-1}\right).\quad(5.50)$$

These results coincide with the QED predictions[69,70] and the corresponding thermal shifts have been experimentally observed (see e.g. Ref. 71). From the point of view of SED (or QED) their interpretation is clear: they represent additional contributions to the kinetic energy impressed on the particle by the thermal field, according to the discussion at the beginning of section 5.4.

6. What have we learned about quantum mechanics?

We have arrived along this work at an important substantiation: the quantum properties of both matter and field emerge quite naturally from the consideration of the existence of a real, pervasive, ubiquitous zero-point radiation field. From the analysis in section 2 of the problem of the thermal radiation field in equilibrium we concluded that in presence of its zero-point component the field becomes quantized. We then made the corresponding analysis for matter, and found that also it becomes quantized. The conclusion is that the ZPF is the central piece that nature uses to perform the miracle of quantization in general. This is another form of saying that the quantum phenomenon, rather than being intrinsic to matter or to the radiation field, emerges from the matter-field interaction.

We further learned that the description provided by the Schrödinger equation ensues from a Fokker-Planck-type equation in phase space. The secret of QM lies to a large extent in the fluctuations of the momentum transferred to configuration space. This striking result can be understood by observing that the fluctuating ZPF is able to compensate for the associated dissipative effects of radiation reaction, thus allowing the particle to reach a stable dynamic state.

The Schrödinger description refers to a statistical ensemble, not to an individual particle. This conclusion appears as inescapable (see Ref. 72), and marks a clear departure from the usual (Copenhagen or orthodox) interpretation of QM, in favour of the less popular ensemble (or statistical) interpretation. Further, the theory affords a physical cause for the fluctuations characteristic of QM. The so-called *quantum fluctuations* appear as real, objective fluctuations impressed on the particle by the permanent interaction of the atomic system (or whatever quantum structure is under study) with the ZPF. This puts the quantum fluctuations on a mundane perspective.

The transition from the original equation to the Schrödinger equation is an irreversible formal procedure: relevant information is lost along the way. Thus, one cannot transit back on purely logical steps. In particular, one cannot reconstruct the true probability density in phase space from the (approximate) Wigner function. This explains the large number of existing phase-space versions of QM, resulting from the attempts to discover the real one (see, e.g., Ref. 73).

The theory is based on the notion of trajectories, which belong to subensembles characterized by the local mean velocities $v(x)$ and $u(x)$. Due to its intrinsically statistical nature, it cannot be applied in general to isolated events, so individual trajectories appear as unknown.

Even though a true Fokker-Planck-type equation in phase space exists, the appropriate (radiationless) description in configuration space requires only a pair of balance equations, one being the continuity equation, the other describing the balance of the average momentum. The latter resembles the classical Hamilton-Jacobi equation, but contains a crucial term that originates in the fluctuations in momentum space and refers to an ensemble of particles, thus changing the meaning of the equation.

The solutions of the Schrödinger equation must be consistent with the demand of energy balance, Eq. (3.20). In addition to fixing the scale of the quantum phenomenon through the introduction of Planck's constant, this condition is also the guarantor of the stationarity of the ZPF itself, by being satisfied frequency by frequency.

Another cost of the simplifications made is the nonlocal nature of the quantum description. This nonlocality, which is not ontological but the result of a cryptic description, has been the source of much quantum ado, and even of avowals bordering on mysticism. An important point to stress is that the nonlocality of QM applies even to single-particle systems. Thus nonlocality and entanglement (which requires at least a couple of particles) are not the same thing.

The simplifications made have also had the effect of deleting from the final description every explicit reference to the ZPF — the ultimate cause of the quantum behavior! As a result, the reason for the stochasticity of the system becomes hidden and the fluctuations become causeless. From this moment on, quantum mechanics cannot be understood from within quantum mechanics. Precisely by exhibiting the ZPF as the source of stochasticity, our results explain the success of a number of works within stochastic electrodynamics. From among those of relevance we recall the important numerical simulations of Cole and Zou,[74,75] leading to a correct statistical prediction of the ground-state orbit for the H-atom.

Further, our results led to a highly interesting conclusion regarding the stationary states of QM, namely that they satisfy an ergodic principle. This condition is at the core of the linear response of the mechanical system to the field: whatever the external force $f(x)$, when the system is in a quantum state α, the (resonant modes) of the background field drive it as if it were composed of a set of oscillators of frequencies $\omega_{\alpha\beta}$. The ergodic principle is thus central in defining a matrix algebra for the dynamical description of the system, and by assigning a sure (nonstochastic) value to the resonance frequencies it plays the role of a quantization principle: it selects from among all the possible stationary solutions of Eq. (4.1), those that are robust with respect to the field fluctuations, which are the quantum solutions. Further, the results of section 4 show that also the (nearby) ZPF acquires specific properties, which are elegantly encapsulated (or rather concealed) in the Heisenberg matrix formalism for the particle dynamics.

The same physical principles that allowed us to recover and reread the quantum formalism in the one-particle case, also served to disclose the physical mechanism behind the entanglement of two noninteracting particles embedded in the (common) background field. It is found that whenever the particles share one resonance frequency, correlations arise between their motions, these being induced via the background field. When the description is made in terms of state vectors of an appropriate Hilbert space, the entangled states emerge naturally as the only ones that can reproduce such correlations. (In this regard, see also Ref. 76.) Moreover, when the particles are identical the ensuing states are totally (anti) symmetric. With these results LSED reveals the physical mechanism and origin — both of them foreign to the usual quantum-mechanical description — of the quantum symmetrization postulate.

It is important to realize that although the ZPF could appear at first as a sort of collection of hidden variables introduced to complete the quantum description, this is not the case. Quite the contrary: nothing is added to QM, but the latter emerges from a more general theory that contains the ZPF. The emerging description is then naturally indeterministic, since in every case the specific realization of the field is unknown. Quantum mechanics is thus exposed as a (handy, but incomplete) description of the statistical behavior in configuration (or momentum) space of the mechanical part of the particle-field system.

Finally, the radiative terms that were neglected in the transition to quantum mechanics are identified by the theory as the source of all (nonrelativistic) radiative corrections, including the elusive Einstein A coefficient, the Lamb shift, and the cavity effects on such corrections. All this can be expressed succinctly by stating that the complete, nonapproximate theory is equivalent to nonrelativistic QED.

Since, according to the present theory, present-day QM furnishes merely an approximate, time-asymptotic, partially averaged description of the physical phenomenon, there exists plenty of room for further and deeper investigations. So far nobody has explored, for instance, the consequences of using the density $Q(x, p, t)$ given by Eq. (3.7), or the behavior of the system before it reaches the state of energy balance (the quantum regime) in which the approximations apply. What would it look like? One should expect an entirely unknown behavior of matter that can neither be classical because the $\hbar$ due to the interaction with the field is already in the picture, nor quantum-mechanical because the conditions to apply such description have yet not been reached.

Undoubtedly an exploration into this realm of physics would represent a new adventure in physics, with possibly very promising outcomes, including predictions that are experimentally testable. This adds of course an important motivation to undertake deeper and further investigations into the theory.

Acknowledgments

The authors gratefully acknowledge supportive and useful comments from Theo Nieuwenhuizen. This work was carried out with financial support provided by DGAPA-UNAM through projects IN106412 and IN112714.

References

1. T. H. Boyer, Derivation of the Blackbody Radiation Spectrum without Quantum Assumptions, *Phys. Rev.* **182**, 1374 (1969).
2. T. H. Boyer, Classical statistical thermodynamics and electromagnetic zero-point radiation, *Phys. Rev.* **186**, 1304 (1969).
3. T. H. Boyer, Derivation of the Planck radiation spectrum as an interpolation formula in classical electrodynamics with classical electromagnetic zero-point radiation, *Phys. Rev. D* **27**, 2906 (1983).
4. T. H. Boyer, Derivation of the blackbody radiation spectrum from the equivalence principle in classical physics with classical electromagnetic zero-point radiation, *Phys. Rev. D* **29**, 1096 (1984).
5. T. H. Boyer, The Classical Vacuum, *Sci. American* **253**, 56 (1985).
6. T. H. Boyer, Thermodynamics of the harmonic oscillator: Wien's displacement law and the Planck spectrum, *Am. J. Phys.* **71**, 866 (2003).
7. L. de la Peña and A. M. Cetto, Planck's law as a consequence of the zeropoint radiation field, *Rev. Mex. Fís.* **48**, Suppl. 1, 1 (2002).
8. L. de la Peña, A. Valdés-Hernández, and A. M. Cetto, Statistical consequences of the zero-point energy of the harmonic oscillator, *Am. J. Phys.* **76**, 947 (2008).
9. L. de la Peña, A. Valdés-Hernández, and A. M. Cetto, Wien's Law with Zero-Point Energy Implies Planck's Law Unequivocally, in *New Trends in Statistical Physics. Festschrift in Honor of Leopoldo García-Colín's 80th Birthday*, A. Macías and L. Dagdug, eds. (World Scientific, Singapore, 2010).
10. A. M. Cetto, L. de la Peña and A. Valdés-Hernández, Quantization as an emergent phenomenon due to matter–zeropoint field interaction, *J. Phys. Conf. Series* **361**, 012013 (2012).
11. F. Mandl, *Statistical Physics* (John Wiley & Sons, New York, 1988).
12. R. K. Pathria, *Statistical Mechanics* (Butterworth-Heinemann, Oxford, 1996).
13. E. W. Montroll and M. F. Shlesinger, Maximum entropy formalism, fractals, scaling phenomena, and 1/f noise: A tale of tails, *J. Stat. Phys.* **32**, 209 (1983).
14. M. Planck, Entropie und Temperatur strahlender Wärme, *Ann. d. Phys.* **1**, 719 (1900).
15. M. Planck, Über das Gesetz der Energieverteilung im Normalspektrum, *Ann. d. Phys.* **4**, 453 (1901).
16. M. Planck, Über die Begründung des Gesetzes der schwarzen Strahlung, *Ann. d. Phys.* **37**, 642 (1912).
17. A. Einstein, Zum gegenwärtigen Stand des Strahlungsproblems *Phys. Z.* **10**, 185 (1909).
18. V. Vedral, *Modern Foundations of Quantum Optics* (Imperial College Press, London, 2005).
19. E. Santos, Comment on 'Presenting the Planck's relation E = nhv, *Am. J. Phys.* **43**, 743 (1975).
20. O. Theimer, Blackbody spectrum and the interpretation of the quantum theory, *Am. J. Phys.* **44**, 183 (1976).

21. P. T. Landsberg, Einstein and statistical thermodynamics. II. Oscillator quantisation, *Eur. J. Phys.* **2**, 208 (1981).
22. C. Cohen-Tannoudji, B. Diu, and F. Laloë, *Quantum Mechanics* (John Wiley & Sons, New York, 1977), Vol. I.
23. P. W. Milonni, *The Quantum Vacuum* (Academic Press, New York, 1994).
24. G. R. Grimmett and D. R. Stirzaker, *Probability and Random Processes* (Clarendon Press, Oxfor, 1983), chapter 4.
25. A. Papoulis, *Probability, Random Variables, and Stochastic Processes* (McGraw-Hill, Boston, 1991).
26. H. Goldstein, *Classical Mechanics* (Addison-Wesley, Reading, 1980).
27. M. Hillery, R. F. O'Connell, M. O. Scully, and E. P. Wigner, Distribution Functions in Physics: Fundamentals, *Phys. Rep.* **106**, 121 (1984).
28. A. Einstein and O. Stern, Einige Argumente für die Annahme einer molekularen Agitation beim absoluten Nullpunkt, *Ann. d. Phys.* **40**, 551 (1913).
29. W. Nernst, Uber einen Versuch, von quantentheoretischen Betrachtungen zur Annahme stetiger Energieänderungen zurückzukehren, *Verh. Deutsch. Phys. Ges.* **18**, 83 (1916).
30. R. W. James, T. Waller, and D. R. Hartree, An Investigation into the Existence of Zero-Point Energy in the Rock-Salt Lattice by an X-Ray Diffraction Method, *Proc. Roy. Soc. A* **118**, 334 (1928).
31. E. O. Wollan, Experimental Electron Distributions in Atoms of Monatomic Gases, *Phys. Rev.* **38**, 15 (1931).
32. R. S. Mulliken, Electronic Structures of Polyatomic Molecules and Valence. V. Molecules RXn, *J. Chem. Phys.* **1**, 492 (1933).
33. T. H. Boyer, A Brief Survey of Stochastic Electrodynamics, in: *Foundations of Radiation Theory and Quantum Electrodynamics*, A. O. Barut, ed. (Plenum, New York, 1980).
34. M. Bordag, G. L. Klimchitskaya, U. Mohideen and V. M. Mostepanenko, *Advances in the Casimir Effect* (Oxford U. P. Oxford, 2009).
35. L. de la Peña and A. M. Cetto, Derivation of quantum mechanics from stochastic electrodynamics, *J. Math. Phys.* **18**, 1612 (1977).
36. L. de la Peña and A. M. Cetto, *The Quantum Dice. An Introduction to Stochastic Electrodynamics* (Kluwer Academic, Dordrecht, 1996).
37. L. de la Peña, A. Valdés-Hernández, and A. M. Cetto, Quantum Mechanics as an Emergent Property of Ergodic Systems Embedded in the Zeropoint Radiation Field, *Found. Phys.* **39**, 1240 (2009).
38. L. de la Peña, A. M. Cetto, A. Valdés-Hernández, and H. França, Genesis of quantum nonlocality, *Phys. Lett. A* **375**, 1720 (2011); L. de la Peña, A. M. Cetto and A. Valdés-Hernández, Quantum behavior derived as an essentially stochastic phenomenon, *Phys. Scripta T* **151**, 014007 (2012).
39. U. Frisch, Wave propagation in random media, in *Probabilistic Methods in Applied Mathematics*, Vol. I, A. T. Bharucha-Reid, ed. (Academic Press, New York, 1968).
40. J. H. van Vleck, The absorption of radiation by multiply periodic orbits, and its relation to the correspondence principle and the Rayleigh-Jeans law: Part II. Calculation of absorption by multiply periodic orbits, *Phys. Rev.* **24**, 347 (1924).
41. J. H. van Vleck and D. L. Huber, Absorption, emission, and linebreadths: A semihistorical perspective, *Rev. Mod. Phys.* **49**, 939 (1977).
42. S. Hassani, *Mathematical Physics. A Modern Introduction to Its Foundations* (Springer, New York, 1999).

43. M. Surdin, Absorption, emission, and linebreadths: A semihistorical perspective, *Ann. Inst. Henri Poincaré* **13**, 363 (1970).

44. G. t'Hooft, Determinism Beneath Quantum Mechanics, in *Quo vadis Quantum Mechanics?* A. Elitzur, S. Dolev, and N. Kolenda, eds. (Springer, Berlin, 2005); arXiv:quant-ph/0212095v1.

45. L. de la Peña and A. M. Cetto, Zeropoint waves and quantum particles, in *Fundamental Problems in Quantum Physics*, M. Ferrero and A. van der Merwe, eds. (Kluwer Academic, Dordrecht, 1995).

46. P. R. Holland, *The Quantum Theory of Motion* (Cambridge U. Press, Cambridge, 1993).

47. E. Wigner, On the Quantum Correction for Thermodynamic Equilibrium, *Phys. Rev.* **40**, 749 (1932).

48. J. E. Moyal, Quantum mechanics as a statistical theory, *Proc. Cambridge Phil. Soc.* **45**, 99 (1949).

49. L. de la Peña and A. M. Cetto, Quantum phenomena and the zeropoint radiation field. II, *Found. Phys.* **25**, 573 (1995).

50. L. de la Peña and A. M. Cetto, Quantum Theory and Linear Stochastic Electrodynamics, *Found. Phys.* **31**, 1703 (2001); arXiv//quant-ph: 050101v2.

51. L. de la Peña and A. M. Cetto, Recent Developments in Linear Stochastic Electrodynamics in: *Quantum Theory: Reconsideration of Foundations 3*, G. Adenier, A. Yu. Khrennikov, and Th. Nieuwenhizen, eds., AIP Conference Proceedings 810 (2006).

52. A. Valdés-Hernández, *Investigación del origen del enredamiento cuántico desde la perspectiva de la Electrodinámica Estocástica Lineal*, Ph. D. Thesis (Universidad Nacional Autónoma de México, México, 2010).

53. A. Valdés-Hernández, L. de la Peña and A. M. Cetto, Bipartite Entanglement Induced by a Common Background (Zero-Point) Radiation Field, *Found. Phys.* **41**, 843 (2011); *Physica E* **42**, 308 (2010).

54. A. M. Cetto and L. de la Peña, Radiative corrections for the matter-zeropoint field system: Establishing contact with quantum electrodynamics, *Phys. Scripta T* **151**, 014009 (2012); A. M. Cetto, L. de la Peña, and A. Valdés-Hernández, Atomic radiative corrections without QED: role of the zero-point field, *Rev. Mex. Fís.* **59**, 433 (2013); arXiv/quant-ph: 1301.6200v1.

55. L. E. Ballentine, *Quantum Mechanics* (Prentice-Hall, New York, 1989).

56. A. Einstein, Zur Quantentheorie der Strahlung, *Phys. Zeitschr.* **18**, 121 (1917).

57. A. Einstein and P. Ehrenfest, Zur Quantentheorie des Strahlungsgleichgewichts, *Phys. Zeitschr.* **19**, 301 (1923).

58. A. S. Davydov, *Quantum Mechanics* (Addison-Wesley, Reading, 1965).

59. V. M. Fain, *Soviet Physics JETP* **23**, 882 (1976).

60. V. M. Fain, and Y. L. Khanin, *Quantum Electronics, Vol. 1: Basic Theory* (Pergamon, Oxford, 1969).

61. J. Dalibard, J. Dupont-Roc, and C. Cohen-Tannoudji, Vacuum fluctuations and radiation reaction: identification of their respective contributions, *J. Phys.* (Paris) **43**, 1617 (1982).

62. E. A. Power, Zero-Point Energy and the Lamb Shift, *Am. J. Phys.* **34**, 516 (1976).

63. R. P. Feynman, in *The Quantum Theory of Fields*, The 12th Solvay Conference (Interscience, New York, 1961).

64. A. M. Cetto and L. de la Peña, Environmental effects on the Lamb shift according to stochastic electrodynamics, *Phys. Rev. A* **37**, 1952 (1988).

65. A. M. Cetto and L. de la Peña, Environmental effects on radiative lifetimes and

mass renormalization, according to stochastic electrodynamics, *Phys. Rev. A* **37**, 1960 (1988).

66. A. M. Cetto and L. de la Peña, Environmental Effects on Spontaneous Emission and Lamb Shift, According to Stochastic Electrodynamics, *Phys. Scr. T* **21**, 27 (1988).
67. D. Kleppner, Inhibited Spontaneous Emission, *Phys. Rev. Lett.* **47**, 233 (1981).
68. P. Goy, J. M. Raimond, M. Gross, and S. Haroche, Observation of Cavity-Enhanced Single-Atom Spontaneous Emission, *Phys. Rev. Lett.* **50**, 1903 (1983).
69. P. L. Knight, Effects of external fields on the Lamb shift, *J. Phys. A* **5**, 417 (1972).
70. W. Zhou and H. Yu, The Lamb shift in de Sitter spacetime, *Phys. Rev. D* **82**, 124067 (2010); arXiv:1012.4055v1 [hep − th] , 2010.
71. L. Hollberg and J. L. Hall, Measurement of the shift of Rydberg energy levels induced by blackbody radiation, *Phys. Rev. Lett.* **53**, 230 (1984).
72. A. E. Allahverdian, R. Balian and Th. M. Nieuwenhuizen, Understanding quantum measurement from the solution of dynamical models, *Phys. Rep.* **525**, 1 (2013).
73. C. K. Zachos, D. B. Fairlie, and T. L. Curtright, eds. *Quantum Mechanics in Phase Space* (World Scientific, Singapore, 2005).
74. D. C. Cole and Y. Zou, Quantum Mechanical Ground State of Hydrogen Obtained from Classical Electrodynamics, *Physics Letters A* **317**, 14 (2003).
75. D. C. Cole and Y. Zou, Simulation study of aspects of the classical hydrogen atom interacting with electromagnetic radiation: Circular orbits, *Journal of Scientific Computing* **20**, 43, 379 (2004).
76. K. Michielsen, F. Jin, M. Delina and H. De Raedt, Event-by-event simulation of non-classical effects in two-photon interference experiments, *Phys. Scripta T* **121**, 014005 (2012).

Chapter 12

Kerr-Newman Electron as Spinning Soliton

Alexander Burinskii*

*Theor. Phys. Lab., NSI, Russian Academy of Sciences,
B. Tulskaya 52, Moscow, 115191 Russia*

Measurable parameters of the electron indicate that its background should be described by the Kerr-Newman (KN) solution. The spin/mass ratio of the electron is extreme large, and the black hole horizons disappear, opening a topological defect of space-time – the Kerr singular ring of Compton size, which may be interpreted as a closed fundamental string of low energy string theory. The singular and two-sheeted structure of the corresponding Kerr space has to be regularised, and we consider the old problem of regularising the source of the KN solution. As a development of the earlier Keres-Israel-Hamity-López model, we describe the model of smooth and regular source forming a gravitating and relativistically rotating soliton based on the chiral field model and the Higgs mechanism of broken symmetry. The model reveals some new remarkable properties: (1) the soliton forms a relativistically rotating bubble of Compton radius, which is filled by the oscillating Higgs field in a pseudo-vacuum state; (2) the boundary of the bubble forms a domain wall which interpolates between the internal flat background and the external exact Kerr-Newman (KN) solution; (3) the phase transition is provided by a system of chiral fields; (4) the vector potential of the external the KN solution forms a closed Wilson loop which is quantised, giving rise to a quantised spin of the soliton; (5) the soliton is bordered by a closed string, which is a part of the general complex stringy structure.

1. Introduction and summary

It is now commonly accepted that black holes (BH) have to be associated with elementary particles. The physics of black holes is based on complex analyticity, which unites them with quantum and superstring theories and particle physics.

In spite of these evident relationships, Gravity and Quantum theory are conflicting and cannot be unified. Similarly, the path from Superstring theory to particle physics represents also a still unsolved problem, and as it was recently claimed by John Schwarz[1] that "... a realistic model of elementary particles still appears to be a distant dream...".

---

*bur@ibrae.ac.ru

The principal point of the conflict between gravity and quantum theory is related with the statement on the pointlike and structureless quantum electron. This point cannot be accepted by gravity, which requires an extended soliton-like structure of the electron, as a field distribution with a regular energy-momentum tensor in configurational space. Contrary, quantum theory suggests the claim on the pointlike structure of the electron, or its statistic description by a wave function. String theory replaced the point-like quantum particles by the extended strings and membrane-like sources. However, the principal quantum particle - electron, is still considered as point-like. In particular, Frank Wilczek writes in Ref. 2: "...There's no evidence that electrons have internal structure (and a lot of evidence against it)". Similarly, the superstring theorist Leonard Susskind notes that electron radius is "...most probably not much bigger and not much smaller than the Planck length".[3] It should be mentioned that this point of view is supported by the high energy scattering, which have not found the electron structure down to 10^{-16} cm. However notice that Quantum Electrodynamic considers an effective size of a "dressed electron", which corresponds to the Compton region of vacuum polarisation. Although the space-time structure of this region is usually not discussed, some hint is coming from relativistic quantum mechanics, which indicates the Zitterbewegung of the electron, a lightlike helicoidal motion following from the Dirac equation.

These partial indications on the peculiar role of the Compton zone of the electron find unexpectedly strong support from Kerr's gravity.

It was shown by Carter that the Kerr-Newman (KN) solution, which is exact solution of the Einstein-Maxwell gravity for a charged and rotating black-hole (BH), has the gyromagnetic ration $g = 2$ as that of the Dirac electron. Therefore, the four experimentally observable parameters of the electron: spin J, mass m, charge $-e$ and magnetic moment μ *indicate unambiguously* that the gravitational background of the electron should be described by the KN solution. Extremely large spin of the electron with respect to its mass should produce an over-rotating Kerr geometry without horizon, which displays a naked topological defect of space-time in the form of the "Kerr singular ring" of the radius $a = \hbar/2mc$, which is half of the Compton wave length $\lambda_c = \hbar/mc$. This singular ring turns out to be the branch line of space into two sheets resulting in a two-fold structure of the electron background. The corresponding gravitational and electromagnetic fields of the electron are concentrated near the Kerr ring, forming a sort of a closed string, the structure of which turns out to be close to the described by the heterotic string solution of Sen.[5,6] This contradicts statements on structureless of the electron and is very far from the its Planck size suggested by superstring theory. However, it confirms the peculiar role of the Compton zone of the "dressed" electron of Quantum Electrodynamics, matches with the known limit of the localisation of the Dirac electron[7] and indicates relationships with string theory.

There appear two questions:

(A) How does the KN gravity know about one of the principal parameters of Quantum theory? and

(B) Why does Quantum theory works successfully on flat space-time, ignoring such strong defect of the background geometry?

A small and slowly varying gravitational field could be neglected, however the stringlike KN singularity forms a branch-line of the KS space-time, and such a topological defect cannot be ignored. A natural resolution of this trouble could be the assumption that there is an underlying structure, or even the theory providing the consistency of quantum theory and gravity. We conjecture that such underlying structure is to be the Kerr geometry, complex structure of which indicates close relations to a four-dimensional version of superstring theory, the "mysterious" $N = 2$ superstring theory which is consistent in four dimensions, but does not have the standard string interpretation.[8] It has been suggested in Ref. 9 that it may be considered as an alternative to the higher dimensional superstring theory.

In this paper, we consider structure of the real source of the Kerr geometry. Starting in Sec. 2 from motivations to consider the Kerr singular ring as a closed heterotic string, we discuss in Sec. 3 peculiarities of the over-rotating Kerr geometry (without horizons), singular ring, two-sheetedness and specific properties of the Kerr coordinate system. In Sec. 4 we consider development the models of source of the KN solution, and in Sec. 5 consider regularisation of the KN solution by the Higgs fields, which creates the source of the KN solution as a *gravitating soliton* in the form of a rotating superconducting disk of the Compton radius with a closed relativistic string situating at the disk perimeter. In this section we obtain some remarkable peculiarities of the spinning soliton model:

(a) oscillations of the Higgs field with the frequency $\omega = 2mc^2/\hbar$, where m is mass of the soliton,

(b) the appearance of a quantum loop of the vector potential (Aharonov-Bohm-Wilson loop), which is wrapped around the disk-like source providing quantisation of the soliton spin.

Section 6 may be the hardest for reading, since we consider there the field aspect of this model, and show that for consistency with gravity we need to extend the Higgs field model to a chiral model containing a triplet of chiral fields. This complication provides a phase transition from the external exact KN solution to a regular source, the interior of which is a pseudo-vacuum state of the Higgs field resulting in a flat metric in the disk-like core of the KN solution. Therefore we obtain the desirable flat background in the vicinity of the KN source, answering the above-mentioned question (B).

In Sec. 7 we give an answer on the fundamental question: If the KN soliton is represented as a model of electron, the Dirac equation should be found there, but where is there hidden the Dirac equation? Although, our treatment in this section is only preliminary, we believe that the principal mechanism of formation of the

Dirac equation related with the role of the Kerr two-sheetedness, Kerr congruences, and the Yukawa coupling of the Higgs field is pointed out correctly.

Finally, in the conclusion we discuss some relations of this model with QED and superstring theory.

2. Kerr singular ring as a string, first qualitative treatment

The Kerr singular ring is generated as a caustic of the Kerr congruence or the focusing line. The Kerr-Newman (KN) gravity indicates that this string should represent one of the principal elements of the extended electron structure.

The widespread opinion that the range of interaction for gravitational field is "tremendously weak and becomes compatible to other forces only at Planck scale" is inspired by the usual analysis of the spherically symmetric Schwarzschild solution, which has a characteristic radius of the interaction (determined by the position of horizon, $r_g = 2m$) proportional to the mass parameter. The Kerr geometry breaks this predicate, and moreover, it turns this dependence into an inverted one, $r_g \sim J/m$, showing that the area of expansion of the Kerr gravitational field is inversely proportional to mass and proportional to spin of the system. This unexpected effect follows from the Kerr relation for the radius of the Kerr singular ring, $a = J/m$, which shows that a strong gravitational field may occupy very large region for objects of low mass m and large angular momentum J, which is just the case corresponding to elementary particles. This paradox has very simple explanation – the gravitational field of the Kerr solution vanishes at the centrum of the solution and concentrates in a thin vicinity of the Kerr singular ring, forming a type of "gravitational waveguide", or string, for propagation of circular waves, as was suggested more than forty years ago in Ref. 10. Although this simple model of the electron is very naive, it gives an intuitive explanation to many important facts:

(i) first of all, we obtained that the Kerr ring-like represented a waveguide for propagation of circular waves, which corresponded to circular motion of a massless particle. In the modern terms of the dual string model, there appeared a type of the four-dimensional "compactification without compactification",

(ii) mass of the particles originated as energy of the massless excitations, which is similar to origin of the mass spectrum in string theory and corresponded to the old Wheeler model of the "mass without mass ",

(iii) the process of mutual transformation of the massive and massless particles, in particular, annihilation of the electron positron pair had got natural intuitive explanation,

(iv) circular motion of the photon with the wave-length λ and the energy $E = hc/\lambda$ created relativistic increase of mass $m = m_0/\sqrt{1 - v^2/c^2}$, which followed from the simple geometric relations,

(v) there appeared natural explanation of the "Zitterbewegung" of the Dirac electron,

(vi) quantum spin could be interpreted as a consequence of the Bohr quantisation of the photon waves wrapped around the Kerr ring,

(vii) the half-integer wave-lengths corresponding to half-integer spin and the spinor two-sheetedness could be related with two-sheeted structure of the Kerr space-time,

(viii) the wave-particle dualism and origin of the de Broglie waves have got natural explanation.

It was too much for such a simple model. Of course, there appeared also problems. Principal trouble of this model is the question: What could keep the photon on the orbit of the Compton radius? Estimations of the Schwarzschild gravitational field showed that it is too weak at the Compton distances. However, contrary to the Schwarzschild solution, which has a 'range' of the gravitational field proportional to the mass of the source (radius of the horizon $r_g = 2m$),[a] the new dimensional parameter of the KN geometry $a = J/m$, which grows with angular momentum J and has the reverse mass-dependence. As a result, the zone of gravitational interaction determined by parameter a increases for the large angular momentum and small masses, and therefore, it turns out to be essential for elementary particles. The reason of that, it a specific structure of the KN gravitational field, which concentrates near the Kerr singular ring, forming a closed gravitational waveguide – a type of the closed gravitational string.[10]

Kerr-Newman (KN) solution has gyromagnetic ratio $g = 2$, as that of the Dirac electron,[4] and therefore, at least the asymptotic gravitational and electromagnetic (em) field of the electron should correspond to the KN solution with great precision. Because of that, the charged Kerr-Newman (KN) solution[11] has paid attention as a classical background of electron.[4,12-21]

3. Over-rotating Kerr geometry and two-sheetedness

The spin/mass ratio of the elementary particles $a = J/m$ is extremely high. In the dimensionless units $c = G = \hbar = 1$, it is about 10^{22}, while for $a > 1$ the BH horizons disappear. It indicates that *spinning particles should correspond to over-rotating BH solutions, for which the BH horizons disappear, and there appears a naked Kerr singular ring.* The electron background acquires a source in the form of a closed string of the Compton radius $a = \hbar/(2mc)$.

The over-rotating KN has simple representation in the Kerr-Schild (KS) form of metric,[12]

$$g_{\mu\nu} = \eta_{\mu\nu} + 2Hk_\mu k_\nu, \tag{1}$$

where $\eta_{\mu\nu}$ is metric of an auxiliary Minkowski background in Cartesian coordinates $x = x^\mu = (t, x, y, z)$, $H = H(x)$ is a scalar function, and $k^\mu = (1, \vec{k})$ is a null 4-vector field. Using signature $(-+++)$ we obtain $k_\mu k^\mu = (\vec{k})^2 - 1 = 0$, and consequently $\vec{k}$

---

[a]We use the natural units $\hbar = G = c = 1$, in which $e^2 = \alpha \approx 137^{-1}$.

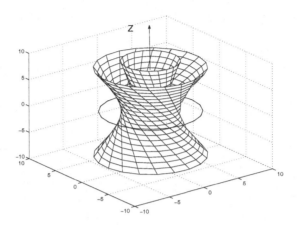

Fig. 1. Twistor null lines of the Kerr congruence are focusing on the Kerr singular ring, which forms a branch line of space in two sheets.

is a unit spacelike vector field, $(\vec{k})^2 = 1$. The use of auxiliary Minkowski background allows to avoid dependence of the metric on position of the horizon. The KS metric form is extreme simple, and one wonders how it can describe the Kerr metric which is known as very complicated. The reason is hidden in the very complicated form of the vector field $\vec{k}$, which represents a vortex of the so-called Pricipal Null Congruence (PNC), or simple Kerr congruence. The form of field $\vec{k}(x)$ is shown in Fig. 1.

Four vector $k_\mu(x)$, tangent direction to PNC, represents a family of the light-like lines, which form a skeleton of the Kerr geometry. These lines are twistors. The reader should not be frightened by word 'twistor'. Indeed, twistor theory is excessively mathematised, but physical meaning of a twistor is very simple: it is a geodesic line of a photon (lightlike or null line) passing aside of the coordinate origin or position of the observer. Twistor Kerr congruence means that each point of the Kerr space-time is polarised, and has a selected lightlike direction. All the tensor fields of the Kerr geometry turns out to be aligned with this selected lightlike direction.

The factor H in (1) is a scalar function, which for the charged Kerr-Newman (KN) solution takes the form

$$H = \frac{mr - e^2/2}{r^2 + a^2 \cos^2 \theta}, \tag{2}$$

where $r = 0$ and $\theta = 0$ are oblate spheroidal coordinates which are adapted to twisted and two-sheeted structure of the Kerr congruence. The KS formalism uses a few different coordinate systems, which allows one to adapt treatment to different aspects of the Kerr solution and simplify calculations. The oblate spheroidal coordinates represent a family of oblate ellipsoids $r = const.$ and confocal family of the hyperboloids $\theta = const.$, see Fig. 2. The Kerr oblate spheroidal coordinates r and

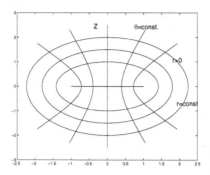

Fig. 2. Oblate coordinate system r, θ with focal points at $r = \cos\theta = 0$ covers the space-time twice, for $r > 0$ and $r < 0$.

θ and ϕ_K are related with minkowskian coordinates as follows,

$$x + iy = (r + ia)e^{i\phi_K}\sin\theta, \qquad z = r\cos\theta, \qquad \rho = t - r, \qquad (3)$$

where ρ is the additional 'light cone'or 'retarded time' coordinate, which describes propagations of the waves along the null rays of the Kerr congruence. However, it will not be essential for our treatment here.

The naked Kerr singular ring corresponds to $r = 0$ and $\theta = \pi/2$. It is a focus line of the null directions k^μ, and simultaneously, it is a branch line of the Kerr space into two sheets, $r < 0$, and $r > 0$. The oblate spheroidal coordinates are two-sheeted and adapted to two sheets of the Kerr geometry. One sees that if congruence is in-going on the sheet $r < 0$, its analytic extension is to be out-going on the sheet $r > 0$. The null directions k_μ, and metric (1) turn out to be different on the 'positive' and 'negative' sheets. We obtain that the background is spoiled at the Compton distances $a = \hbar/(2mc)$, getting a strong topological defect in the form of the Kerr singular ring and two-sheetedness. This two-sheeted space is reminiscent of the old Einstein-Rosen bridge, or the discussed later Wheeler's worm-hole, which allows one to create a 'charge without charge' preventing the space-time from formation of the singularities. Strong curvature of the background in the Compton region indicates that a *regularisation, analogous to regularisation of charge, has to be executed for the metric too.* In fact, this strong breakdown of space indicates that the Compton region has to be a zone of new physics. In accord to results of Quantum Electrodynamics this zone should be related with processes of vacuum polarisation.

4. The 'bubble' and string-like sources of KN solution

Bubble source. It is remarkable, that the Kerr geometry gives unambiguous answer on the size and shape of this zone.[15] One sees that the KS metric (1) becomes flat for the vanishing function H, and therefore, one should set $H = 0$ for regularisation of the metric in the source of KN solution. The boundary of this source is determined from (2) by the value of the Kerr ellipsoidal radial coordinate

$$r = r_e = e^2/(2m),\qquad(4)$$

inside of which, $r < r_e$, the space-time should be set as flat, $H = 0$. We obtain that the source of KN solution should represent an ellipsoidal shell with flat interior which resembles a *vacuum bubble*. The bubble has form on a highly oblate disk. One sees from the coordinate relations (3), that for $r = r_e$, radius of the disk (corresponding to $z = 0$) is $\sqrt{x^2 + y^2} = \sqrt{r_e^2 + a^2}$, while the disk thickness (corresponding to $x^2 + y^2 = 0$) is determined by $z_{disk} = r_e$, and therefore, the ratio thickness/radius turns out to be $r_e/\sqrt{r_e^2 + a^2}$. In the natural units ($\hbar = G = c = 1$) $e^2 \approx 137^{-1}$, and $r_e = e^2/(2m) \approx 137^{-1}/(2m)$. From the Kerr relation $J = ma = \hbar/2 = 1/2$ we have $a = 1/(2m)$, and therefore $r_e = a/137 \ll a$. We obtain that the ratio thickness/radius is close to $r_e/a = 137^{-1}$, and determined by the fine structure constant $\alpha = 137$. The fine structure constant acquires in Kerr's electron a geometric meaning! A very important specification was given by Hamity,[23] who noticed that the disk should be rigidly rotating reaching the velocity of the light at the edge border.

Let us look now at the electromagnetic (EM) field. The vector potential of the KN solution is given by

$$A^\mu_{KN} = \Re\frac{e}{r + ia\cos\theta}k^\mu.\qquad(5)$$

One sees that it is also proportional to the null direction k^μ, and therefore, it is also aligned with the Kerr congruence (PNC), i.e.

$$A^\mu_{KN}g_{\mu\nu}k^\nu = 0.\qquad(6)$$

This is principal property of the algebraically special solutions, that all the tensor quantities are aligned with PNC, which simplifies solutions of the field equations.

This source was suggested by López as a classical model of an extended electron in general relativity.[15] Like the other shell-like models, the López model was not able to explain the origin of Poincaré stress, and the necessary tangential stress was introduced by him phenomenologically, as a distribution over the surface of the rotating shell.

Meanwhile, the necessary tangential stress appears naturally in the field models of domain walls. The corresponding field model of the domain wall bubble was suggested in Refs. 24, 25 and developed in Ref. 18 as a gravitating soliton model, in which the Higgs field is concentrated inside the bubble in a pseudo-vacuum superconducting state and performs regularisation of the KN solution.

The Higgs field pushes the EM field from the bubble. As a result, the EM field is regularised, acquiring the cut-off parameter r_e leading to maximal value of the vector potential

$$A^\mu_{max} = \frac{e}{r_e}k^\mu = \frac{2m}{e}k^\mu,\qquad(7)$$

which is reached on the boundary of the disk corresponding to $\cos\theta = 0$.

The regularised KN space tends to flat near the source, in agreement with the requirements of Quantum theory. However, it should be noted, that performing regularisation we distorted the original KN solution. Submission of the regular EM field and metric in the Einstein-Maxwell system of the field equations will produce the additional charge, currents and matter on the surface of the bubble in the form of a density distributions described by δ-function.[15] These tensor densities take a simple diagonal form in a corotating system of coordinates, which evidences that the disk-like bubble has to be rigidly rotating, and the linear velocity at the boundary of the disk is to be close to the velocity of light. In the recent development, this model takes the form of a gravitating soliton,[18] in which the infinite thin shell of this bubble is replaced by a domain wall boundary, and the bubble is not empty, but is filled by the Higgs field in a pseudovacuum state. A special set of the chiral fields performs a phase transition from the external exact KN solution to the Higgs field inside the bubble providing the flat internal metric and pseudovacuum state of the Higgs field.

String-like source. Let's now consider stringy interpretation of the Kerr singular ring, which was initially considered as an alternative model of the source of Kerr geometry. It was suggested in Ref. 10 to consider the Kerr ring as a closed relativistic string similar to strings of the dual models. The massless equations of the relativistic strings create the massive states from energy of string excitations. This mechanism is similar to the Wheeler's idea of 'geon', gravitational-electromagnetic object with 'mass without mass'. Mass of the 'geon' is generated by energy of the electromagnetic field, in particular, by photons traveling on the circular orbits. It was suggested in Ref. 10 that the Kerr singular ring may represent a type of gravitational waveguide which keeps the electromagnetic (EM) or spinor waves in the orbital motion.[10,13,20] Twenty years later, the string-like solitonic solutions with traveling waves were considered as the fundamental string solutions to low energy string theory.[26,27] It has been shown in Ref. 28 that the field structure of the Kerr singular ring is similar to the structure of the fundamental heterotic string in the obtained by Sen solutions to low-energy string theory.[5,6] Later on, it was obtained that this string is only "a tip of the iceberg", and a wonderful stringy system related with $N = 2$ superstring theory is indeed hidden in the complex Kerr geometry.[9]

5. Regularisation of the EM field by the Higgs fields

5.1. *Basic field equations*

Regularisation of the EM field in the bubble-source model was performed by the Higgs mechanism of broken symmetry, which was also used in many particle-like models, like the 't Hooft-Polyakov monopole and the the Nielsen-Olesen field model for the string-like solution in superconductivity.[29]

The EM field in vacuum and in the Einstein-Maxwell theory is massless and long-distant. Presence of the Higgs field gives a mass to the EM field making it

short-distant. The EM field cannot deeply penetrate in the regions occupied by
the Higgs field. The depth of penetration $\delta \sim 1/m$ depends on the acquired mass
m. Because of that the Higgs field is used for description of the Meissner effect,
interaction of the EM field with ideal conductors and superconductors. There is
also an opposite influence: the strong EM field expels the Higgs field. For example, it
penetrates in a superconductor in the form of vortex filaments, as it was described by
Abrikosov solutions. Consider the Nielsen and Olesen model of the dual relativistic
string in a superconducting media.[29] The the Lagrangian is based on the complex
Higgs field $\Phi(x)$ interacting with the EM vector field A^μ,

$$\mathcal{L}_{NO} = -\frac{1}{4}F_{\mu\nu}F^{\mu\nu} + \frac{1}{2}(\mathcal{D}_\mu\Phi)(\mathcal{D}^\mu\Phi)^* + V(|\Phi|), \tag{8}$$

where $\mathcal{D}_\mu = \nabla_\mu + ieA_\mu$ are to be covariant derivatives, and $F_{\mu\nu} = \partial_\mu A_\nu - \partial_\nu A_\mu$.
We can use the similar field model. In the vicinity of the KN source we have $H \approx 0$
leading to flat space-time, which allows us to consider ∇_μ as flat derivatives and
an set $\nabla_\nu\nabla^\nu = \Box$. This Lagrangian gives rise to the system of equations for the
coupled Maxwell-Higgs system

$$\mathcal{D}_\nu^{(1)}\mathcal{D}^{(1)\nu}\Phi = \partial_{\Phi^*}V, \tag{9}$$

$$\Box A_\mu = I_\mu = e|\Phi|^2(\chi_{,\mu} + eA_\mu). \tag{10}$$

This system of the coupled equations describes an interplay of the Higgs and
Maxwell field separated by some contact boundary and their mutual penetration
through this boundary. Exact solutions of this system are known only for some particular cases of the potential V and only for flat boundary. Analysis of the solutions
requires usually numerical calculations or strong simplifications. It should also be
noticed that the most of the known solutions to the Maxwell-Higgs system describe
the localised EM field confined in a restricted region of space surrounded by the
Higgs field, i.e. a cloud of the gauge massless fields surrounded by a superconducting
media formed by. Meanwhile, *our task is to consider opposite situation, in which
the Higgs field is localised inside the bubble and surrounded by the massless EM field
extended to infinity.* At first sight, this difference seems inessential, but indeed, our
case of interest becomes much more complicated and requires introduction of the
several Higgs-like fields and the *potential V* of a special domain wall form. We will
discuss resolution of this problem in the Sec. 6, but now we consider some extra
simplifications, which will allow us to clarify basic peculiarities of the regular source
of the KN solution.

5.2. *The source without rotation*

We can further simplify the problem considering the source without rotation. By
setting $a = 0$, we obtain that the Kerr singular ring shrinks to a point and the

vector potential (5) takes the spherically symmetric form

$$A_0^\mu = \frac{e}{r} k^\mu, \tag{11}$$

where r is the usual real radial coordinate, $r = \sqrt{x^2 + y^2 + z^2}$. The null vector field k^μ turns into a spherically symmetric system of the four-vectors $k^\mu = (1, \vec{n})$, where $\vec{n} = x^\mu / r$ represents a hedgehog of the unit radial directions. The metric (1) and vector potential (11) correspond to the Reissner-Nordström solution in the Kerr-Schild form.

The bubble-source filled by Higgs field takes in this case the spherical form, and the maximal value of regularised vector potential will be again

$$A_{max}^\mu = \frac{e}{r_e} k^\mu = \frac{2m}{e} k^\mu, \tag{12}$$

which corresponds to the classical model of the electron as a charged sphere with a unique difference that the sphere is replaced by a superconducting ball, interior of which is filled by a complex Higgs field $\Phi(x) = \Phi_0 \exp\{i\chi(x)\}$, where Φ_0 is the vacuum vacuum expectation value ('vev') of the Higgs field.

Therefore, we replace the equation (9) by assumption on the sharp boundary of the bubble, setting $\Phi(x) = 0$ outside the bubble and $\Phi(x) = \Phi_0$, for $r < r_e$. The corresponding EM field is massless and has the form (5) for $r > r_e$, while penetrating inside the bubble it should satisfy the Eq.

$$\Box A_\mu = I_\mu = e|\Phi|^2 (\chi_{,\mu} + eA_\mu). \tag{13}$$

If the phase χ is a constant, this equation turns into $\Box A_\mu = m_v^2 A_\mu$, where $m_v = |e\Phi|$ is the mass acquired by vector field A_μ due to Higgs mechanism. Here we have another action of the Higgs field, which is related with compensating role of the phase χ as a space-time function $\chi(x)$. Splitting $A_\mu = (A_0, \vec{A})$ into timelike component $A_0 = \frac{e}{r}$ and the radial field $\vec{A} = \frac{e}{r}\vec{n} \equiv ed\ln r$, one can drop *formally* the radial part as a full differential, which will not products the strengths of the EM field $F_{\mu\nu}$. In accord to (13), the time-like component $A_0 = \frac{e}{r}$ could create the charge

$$\rho = I_0 = e|\Phi|^2 (\chi_{,0} + eA_0) \tag{14}$$

However, the charge and current have to be expelled from superconducting interior of the bubble to its boundary. It sets the condition $I_0 = 0$ for $r < r_e$ resulting in the the relation $\chi_{,0} + eA_0^{in} = 0$, which fixes the value of component A_0 inside the bubble as a constant determined by the frequency of oscillations of the field Higgs ω,

$$A_0^{(in)} = -\frac{1}{e}\partial_t \chi = -\frac{1}{e}\omega. \tag{15}$$

Note, that the left side of the (13), $\Box A_0^{(in)} = 0$, is satisfied by the constant value of $A_0^{(in)}$. At the boundary of the bubble r_e, the constant value $A_0^{(in)}$ has to be matched with external solution $A_0 = \frac{e}{r}$,

$$A_0^{(in)} = -\frac{\omega}{e} = \frac{e}{r_e}. \tag{16}$$

Using $r_e = e^2/(2m)$, we arrive at very important result that *the Higgs field forms a coherent vacuum state oscillating with the frequency*

$$|\omega| = 2m. \tag{17}$$

Solitonic solutions of this type were called "oscillons". Apparently, they were first obtained by Gerald Rosen.[30] Examples of the oscillon solutions are the spinning Q-balls[31,32] and the bosonic star solutions.[33]

5.3. *Inclusion of the rotation*

Let us write now the Higgs phase in the form $\chi = \omega t + n\phi + \chi_1(r)$, where we have taken into account the constant frequency ω, periodicity in ϕ and some dependence on r. The Higgs field inside the bubble can be represented as follows

$$\Phi = |\Phi| \exp\{i(\omega t + n\phi + \chi_1(r))\}, \tag{18}$$

where the azimuthal coordinate ϕ my be expressed via cartesian coordinates as follows $\phi = -i \ln[(x + iy)/\rho]$, $\rho = (x^2 + y^2)^{1/2}$.

In the rotating case the KN gauge field A_μ is twisted and given by (5), where k^μ is the tangent vector to Kerr congruence. One sees, that the basic expressions for the Kerr metric (1) and the EM field (5) are extreme simple. The complicated structure of the Kerr solution is concentrated in the form of vector field $k^\mu(x)$ which may be described in differential form in the Cartesian or the Kerr angular coordinates. The Cartesian representation is more important from theoretical point of view, since it shows relation to twistors, the Kerr theorem and superstrings.[9] For our aims here, it is enough to give k^μ in differential form expressed in the Kerr angular coordinates[12]

$$k_\mu dx^\mu = dr - dt - a\sin^2\theta d\phi_K. \tag{19}$$

The Kerr azimuthal coordinate ϕ_K has a very specific form determined by the relation (3). It is inconsistent with the standard angular coordinate ϕ of the Higgs field, and their differentials are related as follows

$$d\phi_K = d\phi + \frac{a\,dr}{r^2 + a^2}. \tag{20}$$

Figure 3 shows that the spacelike part of $k^\mu = (1, \vec{k})$, $\vec{k}$ is tangent to Kerr ring in the equatorial plane at $r = 0$. It means that the Kerr ring is lightlike, i.e. it slides along itself with the speed of light.

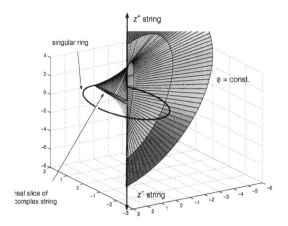

Fig. 3.　The Kerr surface $\phi = const$. The Kerr congruence is tangent to singular ring at $\theta = \pi/2$.

The KN vector potential (5) takes the form

$$A_\mu dx^\mu = \frac{er}{r^2 + a^2 \cos^2 \theta}[dr - dt - a \sin^2 \theta d\phi_K]. \tag{21}$$

Using (20), one can express it in terms of $d\phi$ and obtain

$$A_\mu dx^\mu = \frac{-er}{r^2 + a^2 \cos^2 \theta}[dt + a \sin^2 \theta d\phi] + \frac{erdr}{(r^2 + a^2)}, \tag{22}$$

which shows that radial component A_r represents a full differential and can be dropped.[b] At the external side of the bubble boundary, $r = r_e + 0 = e^2/2m$, the the potential takes the value $A_\mu(r_e)dx^\mu = \frac{-er_e}{r_e^2 + a^2 \cos^2 \theta}[dt + a \sin^2 \theta d\phi]$, and in the equatorial plane, $\cos \theta = 0$, it forms the loop wrapped along the ϕ-direction,

$$A_\mu(r_e)dx^\mu = -\frac{2m}{e}[dt + ad\phi], \tag{23}$$

with tangent component $A_\phi(r_e) = -2ma/e$. This field is penetrated inside the bubble,

$$A_\mu^{(in)} dx^\mu = -\frac{2m}{e}[dt + ad\phi]. \tag{24}$$

However, in agreement with (13), it is compensated by the phase of Higgs field, resulting in cancelling of the corresponding circular current inside the bubble,

$$I_\phi^{(in)} = 0 \Rightarrow \chi_{,\phi} = -eA_\phi^{(in)}. \tag{25}$$

It implies periodic dependence of the Higgs field on ϕ, $\Phi \sim \exp\{in\phi\}$, with integer n. There appears the closed Wilson loop of the potential along the rim of disk-like source. The integral over the loop

$$S = \oint eA_\phi(r_e)d\phi = -4\pi ma \tag{26}$$

---

[b]Inside the bubble A_r is compensated by the $\chi_1(r)$ component of the Higgs phase.

has to be cancelled by a periodic incursion of the Higgs phase $2\pi n$, and therefore, it turns out to be quantised. Using the KN relation $J = ma$, we obtain wonderful result that the quantum Wilson loop of the KN EM field $e \oint A_\phi^{(str)} d\phi = -4\pi ma$ requires **quantisation of the total angular momentum of the soliton,** $J = ma = n/2$, $n = 1, 2, 3, \ldots$

This remarkable effect is specific for the Kerr geometry and is consequence of the considered classical conditions for parameters of the bubble source $r = r_e$, resulting in $H = 0$, and as consequence, in flat geometry inside the bubble. This property was never observed in the other spinning Q-balls and the bosonic star solutions.

Extra note. By construction of the solution, the time-like and ϕ components of the obtained vector field are continuous at the boundary of the bubble r_e. However, there is discontinuity in their radial derivatives, which has to generate circular currents on the bubble boundary. Practically, the boundary is not sharp, and the vector field should have a finite depth of penetration δ_v, and derivatives should be smoothed by this skin effect. The real values of the vector potential will differ from the obtained background solution (24) in the boundary layer $r \in [r_e, r_e - \delta_r]$. Denoting this deviation as $\delta A_\mu^{(bound)} = A_\mu - A_\mu^{(in)}$ in (13) we obtain the equation

$$\Box \delta A_\mu^{(bound)} = I_\mu = e^2 |\Phi|^2 \delta A_\mu^{(bound)} \tag{27}$$

which shows that the EM field acquires via Higgs mechanism the mass $m_v = e|\Phi|$, which produce the charge and ring-like current with the depth of penetration $\delta r \sim 1/m_v$. It may be interpreted as a string-like massive and charged vector meson residing at the circular boundary of the bubble.

6. Full action: Potential, domain wall, gravity and phase transition

6.1. *Phase transition*

We have to consider now the role of the potential $V(\Phi)$ in formation of the boundary condition and in the phase transition from the external vacuum state corresponding to KN solution to some flat false-vacuum state inside the bubble-source. In this point our model differs essentially from the other solitonic models, as well as from the known oscillon and Q-ball models. The renormalisable quartic potential $V = (|\Phi|^2 - v)^2$, is generally used for the Higgs field models. In particular, is was used by Graham for the oscillon model in.[32] It is inappropriate for our model, since our Higgs field is concentrated inside the bubble and has to fall quickly off by deviation from the region of source, and in particular, it has to vanish at infinity, $\Phi_\infty = 0$. One sees that the potential $V = (|\Phi|^2 - v^2)^2$, will take at infinity non-zero vacuum expectation value (vev) $V_\infty = v^4$, which contradicts to the required asymptotic true vacuum state. In Ref. 31 the authors used the non-renormalisable potential of sixth degree in $|\Phi|$, which falls off asymptotically, but has a nonzero vev in the core region, which will prevent the formation of flat interior inside the bubble-source. The attempts to cancel the vev by oscillating Higgs field do not go in our case, since energy of these oscillations is already compensated by the time-like component of the vector potential in accord to (14).

The similar problem appears in the Vilenkin-Witten model of the superconducting cosmic string,[34] and to resolve this problem Witten used the $U(1) \times \tilde{U}(1)$ field model[34] which contains two Higgs field: one of which Φ^1 is concentrated inside the superconducting source, while another one Φ^2 takes the complementary domain extended up to infinity. These two Higgs field are charged and adjoined to two different gauge fields A^1 and A^2, so that when one of them is long-distant in some region Ω, the second one is short-distant in this region and vice-verse. This model is suitable for our case, however, in Ref. 18 we used a generalisation of this model suggested by Morris.[35] The Morris potential V depends on two charged Higgs-like complex fields Φ and Σ and one auxiliary uncharged real field Z, which are combined in the superpotential

$$W = \lambda Z(\Sigma\bar{\Sigma} - \eta^2) + (Z + \mu)\Phi\bar{\Phi}, \qquad (28)$$

where μ, η, λ are real constants. In accord to theory of the chiral superfields,[48] the potential V is determined from the superpotential W by the relation

$$V(r) = \sum_i |\partial_i W|^2, \qquad (29)$$

where $\partial_1 = \partial_\Phi$, $\partial_2 = \partial_Z$, $\partial_3 = \partial_\Sigma$, and $\Phi^{(i)}$, $i = 1, 2, 3$ forms triplet of the chiral fields[c]

$$\Phi^{(i)} = \{\Phi, Z, \Sigma\}. \qquad (30)$$

Vacuum states $V_{(vac)} = 0$ are determined by the conditions $\partial_i W = 0$. It is easy to obtain for (28) two solutions:

(I) vacuum state: $Z = -\mu$; $\Sigma = 0$; $|\Phi| = \eta\sqrt{\lambda}$, corresponds to $W_I = \lambda\mu\eta^2$,
(II) vacuum state: $Z = 0$; $\Phi = 0$; $\Sigma = \eta$, corresponding to $W_{II} = 0$.

One can identify the field Φ as the main Higgs field, and the state (I), where $|\Phi| > 0$, as the false-vacuum state of the Higgs field inside the bubble. Then the state (II), where $|\Phi| = 0$, will be identified with external vacuum state with the non-zero Higgs-like field $|\Sigma| = \eta^2 > 0$.

The coupled with gravity action reads

$$S = \int \sqrt{-g}Rd^4x(\frac{R}{16\pi G} + \mathcal{L}^{mat}), \qquad (31)$$

where the full matter Lagrangian takes the form

$$\mathcal{L}^{mat} = -\frac{1}{4}F_{\mu\nu}F^{\mu\nu} + \frac{1}{2}\sum_i (\mathcal{D}_\mu^{(i)}\Phi^{(i)})(\mathcal{D}^{(i)\mu}\Phi^{(i)})^* + V, \qquad (32)$$

which contains contribution from triplet of the chiral field $\Phi^{(i)}$.

---

[c](29) assumes that $\Phi^{(i)}$ and $\bar{\Phi}^{(i)}$ are independent fields.

The potential V (29) is positive by definition (29) and forms a domain wall interpolating between the internal false-vacuum state (I) ($V_{(int)} = 0$) and external gravi-electro-vacuum state (II) ($V_{(ext)} = 0$) corresponding to KN solutions of the Einstein-Maxwell field equations.

The covariant derivations $\mathcal{D}_\mu^{(i)} = \nabla_\mu + ieA_\mu^{(i)}$ contain in general case three different gauge fields $A^{(i)}$. However, in our case we need only one gauge field A_μ associated with the principal chiral field $\Phi^{(1)} \equiv \Phi$. Therefore, we set $F_{\mu\nu} = \partial_\mu A_\nu - \partial_\nu A_\mu$ and the corresponding covariant derivations will be $\mathcal{D}_\mu^{(1)} = \nabla_\mu + ieA_\mu$, $\mathcal{D}_\mu^{(2)} = \mathcal{D}_\mu^{(3)} = \nabla_\mu$. the field $\Phi^{(2)} \equiv Z$ is uncharged and $A^{(2)} = 0$.

It should be noted that generalisation to triplet of the chiral field may be useful for generalisation of this model to the Weinberg-Salam theory.[d]

6.2. *Problem of the exact solutions*

We have now to turn to Einstein equations

$$R_{\mu\nu} - \frac{1}{2}g_{\mu\nu} = 8\pi G T_{\mu\nu}^{(mat)} \tag{33}$$

The stress-energy tensor may be decomposed into pure EM part and contributions from the chiral fields

$$T_{\mu\nu}^{(mat)} = T_{\mu\nu}^{(em)} + \tag{34}$$
$$\delta_{i\bar{j}}(\mathcal{D}_\mu^{(i)}\Phi^i)\overline{(\mathcal{D}_\nu^{(j)}\Phi^j)} - \frac{1}{2}g_{\mu\nu}[\delta_{i\bar{j}}(\mathcal{D}_\lambda^{(i)}\Phi^i)\overline{(\mathcal{D}^{(j)\lambda}\Phi^j)} + V],$$

and we have to consider three different regions: external zone, $r \geq r_e$, transition zone $(r_e - \delta_r) \leq r_e$, and zone of flat interior $r < (r_e - \delta_r)$.

In the external zone, $r \geq r_e$, we have $V^{ext} = 0$. The unique nonzero chiral field Σ is constant, and therefore, all the derivatives $\mathcal{D}_\mu^{(i)}\Phi^{(i)}$ vanish. As a result $T_{\mu\nu}^{(tot)}$ is reduced to $T_{\mu\nu}^{(em)}$, and we obtain the Einstein-Maxwell field equations are satisfied and for the external KN electromagnetic field they result in the external KN solution.

For interior of the bubble, $r < (r_e - \delta_r)$ we have also $V^{int} = 0$, and the unique nonzero Higgs field is $\Phi(x) = |\Phi(x)|e^{i\chi(x)}$.

The Lagrangian (32) is reduced to (8) with $V(r) = V^{int} = 0$, which leads to eqs.

$$\mathcal{D}_\nu^{(1)}\mathcal{D}^{(1)\nu}\Phi = 0, \tag{35}$$

$$\nabla_\nu\nabla^\nu A_\mu = I_\mu = \frac{1}{2}e|\Phi|^2(\chi_{,\mu} + eA_\mu). \tag{36}$$

The only variable chiral field in the flat interior is the oscillating Higgs field, and we have to consider it in more details. One sees that the term

$$\mathcal{D}_t^{(1)}\Phi = (\partial_t + ieA_0^{(in)})|\Phi|\exp\{i\omega t\} = i(\omega + eA_0^{(in)})\Phi \tag{37}$$

is cancelled in agrement with (15), and therefore, (35) is satisfied.

---

[d]This idea is suggested by Theo Nieuwenhuizen.

For flat interior the second eq. (36) reduces to the system (13) and we obtain all the consequences considered in Sec. 5. Therefore, $T_{\mu\nu}^{(mat)} = 0$ and the Einstein-Maxwell equations are trivially satisfied for flat interior.

One sees that considered in Sec. 5 EM solutions together with the discussed here Einstein-Maxwell system, are consistent with the sharp boundary between the external and internal regions, and the limit $\delta_r \to 0$, may be interpreted as a *thin wall approximation*.

In this limit, external KN metric matches continuously with flat interior of the bubble and turns out to be consistent with the stress-energy tensor of the external KN solution and flat interior of the bubble. However, in the thin wall limit, there appears discontinuity in the first derivatives of the metric. Because the Einstein equations contain the second derivatives of the metric, the stress-energy tensor has a δ-function singularity at the thin wall.[38] In this case of thin wall, the analysis was usually performed on the basis of Israel's formalism of singular layers,[14,38] or in terms of generalised functions of the theory of distributions.[15] The obtained solution may be considered as a consistent with gravity thin wall approximation. However, internal structure of the wall is indeed a non-trivial and very important problem.

6.3. *Beyond the thin wall approximation*

Beyond the thin wall approximation, we have to consider as the third zone the zone of phase transition $(r_e - \delta_r) < r < r_e$.

Zone of phase transition is the practically inaccessible for analytic solutions. Up to our knowledge, analytic solutions of the *similar* problems with the Higgs field are unknown even for the simple spherical configuration. The known analytic solutions obtained for the vacuum domain walls with planar geometry have a kink-like form, and typical domain wall form of the stress-energy tensor is

$$T_\nu^\mu = \text{diag}(\rho, \ -\rho, \ -\rho, \ 0). \tag{38}$$

One expects, that in the case of planar domain wall, the solutions to the full system of the nonlinear eqs. may in principle be obtained by analytic methods or by the numerical calculations.

Information on the phase transition, produced by the domain wall of the chiral field model, is concentrated in the structure of the stress-energy tensor $T_{\mu\nu}^{(mat)}$, (35) which should be matched with phase transition in gravitational sector in accord to Einstein equations. Gravitational counterpart to this phase transition is the transfer from the flat metric inside the bubble, $r < r_e$, to the external metric of the exact KN solution. For this transfer there is a remarkable ansatz suggested by Gürses and Gürsey (GG).[36] The standard KS form of the metric (1) with the fixed Kerr congruence k^μ, is deformed only in the form of function H (2), which takes the form

$$H = f(r)/\Sigma, \quad \Sigma = (r^2 + a^2 \cos^2 \theta), \tag{39}$$

where f is arbitrary smooth function.

For the external KN metric $r > r_e$, one sets $f(r) = f_{KN} = mr - e^2/2$, while the smooth transfer to some internal metric may be provided by any its smooth extension $f(r) = f_{int}$. The GG form of metric has remarkable properties. It was shown in Refs. 24, 25 that setting for the interior $f(r) = f_{int} = \alpha r^4$, one obtains a regular version of the KN metric, in which the Kerr singularity is suppressed. The GG form of metric describes the rotating solutions with flat asymptotic as well as the rotating versions of the de Sitter and Anti-de Sitter solutions. Moreover, it allows one to match smoothly the external and internal metrics of different sorts.[24,25] On the other hand the GG form of metrics is a particular case of the Kerr-Schild solutions and inherits its remarkable properties. In particular, the electromagnetic field and its stress-energy tensor displays a partial linearisation in the Kerr-Schild and GG-space-times, and there is the exact correspondence between the rotating and non-rotating solutions which allows one to simplify analysis of the rotating solutions, by means of the analysis of its non-rotating analogs. In particular, the stress-energy tensor of the KN solution and its non-rotating analog with $a = 0$ take in the orthonormal tetrad u, l, n, m, where u is the unit timelike vector and l the radial one, the form[17,24,25,36]

$$T_{\mu\nu} = (8\pi)^{-1}[(D + 2G)g_{\mu\nu} - (D + 4G)(l_\mu l_n - u_\mu u_\nu)], \qquad (40)$$

where

$$G = \frac{f'r - f}{\Sigma^2}, \quad D = -\frac{f''}{\Sigma}, \qquad (41)$$

which may be recognised as diagonal one

$$T_\nu^\mu = \text{diag}(\rho, \ -\tau, \ -\tau, \ p), \qquad (42)$$

where the radial pressure p and tangential stress τ are given by

$$\rho = -p = \frac{1}{8\pi} 2G, \quad \tau = -\frac{1}{8\pi}(D + 2G) = -\rho - \frac{D}{8\pi}. \qquad (43)$$

This correspondence allows us to consider the phase transfer in the non-rotating case and translate the results to the rotating KN solution.

For the case of quick rotation, $a \gg r_e$, there appears a stringy contribution to the mass-energy caused by concentration of the electromagnetic field on the edge of bubble. However, analytic calculations showed (see Ref. 18) that the 'stringy' contributions from the shell and external EM field are mutually cancelled, and the total mass $m_{ADM}^{(total)} = m_{ADM}^{(int)} + m_{ADM}^{(shell)} + m_{ADM}^{(ext)}$ turns out to be equal to m. Indeed, this result could be predicted a priori, since the total ADM mass is determined *only* by the asymptotical gravitational field, i.e. only by the value of parameter m in function H, (2). Therefore, the naive stringy interpretation does not go, at least for the GG form of the metric.

The typical local structure of the stress-energy tensor for the vacuum domain wall[37,38] is

$$T_\nu^\mu = \rho \, \text{diag}(1, \ -1, \ -1, \ 0) \qquad (44)$$

shows that the surface energy density ρ is equal to tangential stress τ.

Comparison between (44) and (42) shows that the GG metric contains the domain wall contribution, however there is no full correspondence. It opens a chance to get the the correspondence and stringy effect from some generalisations of the GG form of metric. In particular, the appearance of the string tension may be related with extra dilaton or axion fields, which are typical for the metrics of domain wall models and their analogs in superstring theory.[40] One more very important generalisation of this problem may be related with transfer from the used Einstein-Maxwell-Higgs (Chiral) system of the eqs. to its remarkable natural analog in 4D Supergravity,[48] where the complex chiral fields $\phi^{(i)}$ form an extra Kähler manifold supplied by Kähler metric and Kähler potential. It is probably, that the solutions turn out to be consistent with such a generalised system.

We have to stop at this point, since neither analytic calculations for planar domain walls, nor the numerical calculations even for the Einstein-Maxwell-Higgs system were so far performed. There is a great field for activity.

7. How the Dirac equation is hidden inside solitonic source of the KN electron

As we have shown, the Kerr-Newman solution has many remarkable properties indicating its relationships with the structure of the Dirac electron. However, all these evidences cannot draw us away from the natural question how and where the Dirac equation may be residing in the solitonic source of KN geometry. In this chapter we will try to answer this question. The soliton source of the KN represents a bag confining the Higgs field. Analyzing the twosheeted structure of the KN solution, we have seen that the Kerr geometry is based on the twistorial structures of the Kerr congruence. The naked Kerr singular ring forms a branch line of space into the sheets of advanced and retarded fields, and the null vector field, $k_\mu(X)$ forms the Principal Null Congruence (PNC) $\mathcal{K}$, form of which is changed by the transfer from positive $(r > 0)$ to negative $(r < 0)$ sheet. The surface $r = 0$ represented a disklike "door" separating two different null congruences $\mathcal{K}^\pm$, creating two different metrics

$$g_{\mu\nu}^\pm = \eta_{\mu\nu} + 2Hk_\mu^\pm k_\nu^\pm \qquad (45)$$

on the same Minkowski background M^4.

It seemed that the formation of the soliton source closed this "door" and removed the problem of twosheeted space. However, this problem emerges from another side. The second sheet appeared as a sheet of advanced fields, which are related with the old Dirac problem of radiation reaction.[42] Dirac splits the expression for retarded potential A_{ret} into a half-sum and half-difference with advanced fields A_{adv} as follows

$$A_{ret} = \frac{1}{2}[A_{ret} + A_{adv}] + \frac{1}{2}[A_{ret} - A_{adv}]. \qquad (46)$$

The half-sum he takes responsible for self-interaction of the source, while the half-difference should be responsible for radiation and radiation reaction. The soliton source acquires the second sheet outside the bag as the sheet of advanced fields, which should provide self-interaction of the soliton and create a massive Dirac equation. In accord with the basic properties of the Kerr-Schild solutions, the fields A_{ret} and A_{adv} could not reside on the same physical sheet, because each of them should be aligned with the its own Kerr congruence. Considering the retarded sheet as a basic physical sheet, we fix the congruence $\mathcal{K}_{ret}$ and corresponding metric $g_{\mu\nu}^+$, which are not allowed for the advanced field A_{adv}, which is consistent with another congruence $\mathcal{K}_{adv}$, which should be positioned on a separate sheet with different metric $g_{\mu\nu'}^-$. However, the problem of their incompatibility disappears inside the bag, where the space is flat, and the both null congruences $\mathcal{K}_{ret}$ and $\mathcal{K}_{adv}$ are null not only with respect to the corresponding Kerr-Schild metrics, but also with respect to the flat Minkowski background.

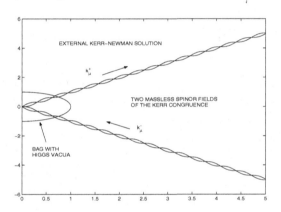

Fig. 4. Creation of the Dirac equation inside the bag by the confined Higgs field coupled to two massless spinor fields k_μ^+ and k_μ^-, generators of the Kerr principal null congruences $\mathcal{K}^+$ and $\mathcal{K}^-$.

For the sake of convenience, we replace further notations $\mathcal{K}_{ret}$ and $\mathcal{K}_{adv}$ by the notations $\mathcal{K}^\pm$, identifying $\mathcal{K}^+ = \mathcal{K}_{ret}$, $\mathcal{K}^- = \mathcal{K}_{adv}$. The external null fields $k_\mu^\pm(x)$ are extended inside the bag *on the same sheet* forming two different null congruences determined by two conjugate solutions of *the Kerr theorem* $Y^\pm(x^\mu)$.[12,43]

Now, we have to say a few words on the **Kerr theorem**. The Kerr theorem is formulated in the Minkowski background, $x^\mu = (t, x, y, z) \in M^4$, in terms of the projective twistor coordinates

$$T^A = \{Y, \ \zeta - Yv, \ u + Y\bar{\zeta}\}, \qquad A = 1, 2, 3, \tag{47}$$

where $u, v, \zeta, \bar{\zeta}$ are the null Cartesian coordinates, $\zeta = (x + iy)/\sqrt{2}$, $\bar{\zeta} = (x - iy)/\sqrt{2}$, $u = (z + t)/\sqrt{2}$, $v = (z - t)/\sqrt{2}$. The Kerr congruence is determined by solution of the equation

$$F(T^A) = 0, \tag{48}$$

where F in general case is a holomorphic function on the projective twistor space CP^3. For the Kerr solution the congruence is created by function F which is quadratic in Y and may be represented in the form

$$F(Y, x^\mu) = A(x^\mu)Y^2 + B(x^\mu)Y + C(x^\mu). \tag{49}$$

In this case the equation (48) has two explicit solutions

$$Y^\pm(x^\mu) = (-B \mp \tilde{r})/2A, \tag{50}$$

where $\tilde{r} = (B^2 - 4AC)^{1/2}$, and it was shown in Ref. 44, that these two solutions are related by antipodal correspondence

$$Y^+ = -1/\bar{Y}^-. \tag{51}$$

It should be noted, that Y is a projective spinor coordinate, $Y = \phi_1/\phi_0$, and it is equivalent to the Weyl two-component spinor $\phi_\alpha = (\phi_1, \phi_0)^T$.

In the Kerr-Schild formalism,[12] function $Y(x^\mu)$ determines the null congruence as a field of null directions $k_\mu(x^\mu)$ via differential form

$$k_\mu dx^\mu = P^{-1}(du + \bar{Y}d\zeta + Yd\bar{\zeta} - Y\bar{Y}dv), \tag{52}$$

where $P = (1 + Y\bar{Y})/\sqrt{2}$ is a normalizing factor. This form is equivalent to standard representation of the null vector via spinor ϕ and Pauli matrices $k_\mu^+ = \bar{\phi}\bar{\sigma}_\mu\phi$, and therefore, generating two conjugate null congruences, the Kerr theorem determines simultaneously two massless spinor fields of different chirality $\phi_\alpha = (\phi_1, \phi_0)^T$ and $\chi^{\dot\alpha} = (\chi^{\dot 1}, \chi^{\dot 0})^T$.

We can compare these spinor fields with the structure of the Dirac equation

$$(\gamma^\mu \hat\Pi_\mu + m)\Psi = 0, \tag{53}$$

in which $\Psi = (\phi_\alpha, \chi^{\dot\alpha})^T$ and $\hat\Pi_\mu - -i\partial_\mu - eA_\mu$, and $\hat\Pi_\mu = -i\partial_\mu - eA_\mu$, written in the Weyl basis, where it splits in two equations[45]

$$\bar\sigma^{\mu\dot\alpha\alpha}(i\partial_\mu + eA_\mu)\phi_\alpha = m\chi^{\dot\alpha}, \quad \sigma^\mu_{\alpha\dot\alpha}(i\partial_\mu + eA_\mu)\chi^{\dot\alpha} = m\phi_\alpha. \tag{54}$$

In the Standard Model these equations are called the "left handed" and the "right handed electron fields". In the massless case (54) represents two equations, connected by antipodal relation (51). The corresponding null vectors

$$k_L^\mu = \bar\phi\sigma^\mu\phi, \quad k_R^\mu = \bar\chi\bar\sigma^\mu\chi, \quad k_{\mu L}k_L^\mu = k_{\mu R}k_R^\mu = 0, \tag{55}$$

describe two principal null congruences given by antipodal solutions $Y^\pm$ of the Kerr theorem. Outside the bag these null fields should reside on different sheets of the KN solution, but penetrating inside the bag they are meeting without conflict and can reside on the same false-vacuum sheet because of flatness of the space inside the bag. They form a four-component Dirac spinor $\Psi(x)$ satisfying the massless Dirac equation

$$(\gamma^\mu \hat\Pi_\mu)\Psi = 0, \tag{56}$$

and the Higgs field Φ confined in the bag provides them by the Yukawa interaction

$$\mathcal{L}_{Yukawa}(\Phi, \Psi) = -g\bar{\Psi}\Phi\Psi. \tag{57}$$

Therefore, two antipodal twistorial congruences, obtained as two conjugate solutions of the Kerr theorem $Y^{\pm}(x)$, create two massless spinor fields ϕ_{α} and $\chi^{\dot{\alpha}}$, which are combined in a four-component massless Dirac field $\Psi = (\phi_{\alpha}, \chi^{\dot{\alpha}})^T$. The Higgs field Φ confined inside the bag connects them by the Yukawa coupling (57), which gives the mass term to the Dirac equation in full agreement with the one of principal tasks of the Higgs field in the Standard Model.

8. Outlook: Relation to superstring theory

After regularisation of the KN solution by solitonic source, the negative sheet of metric disappears and regularised metric turns out to be practically flat in concordance with quantum theory. However, the EM field is not disappeared. It is regularised and takes the form of a lightlike heterotic string on the sharp border of the bubble. The resulting solution is axially symmetric, and therefore, traveling waves are absent. Traveling waves are related with some extra EM excitations of the soliton, which may create a mass spectrum of particles or resonances. This point is not investigated so far.

Does the KN model of electron contradict to Quantum Theory? It seems "yes", if one speaks on the "bare" electron. However, in accordance with QED, vacuum polarisation creates in the Compton region a cloud of virtual particles forming a "dressed" electron. This region gives contribution to spin of the electron and is closely related with procedure of renormalisation, which determines the physical values of the electron charge and mass. The described here solitonic version of the regular source of the KN solution based on the domain wall bubble model with the confined Higgs field in a false-vacuum state represents an alternative to the known formal mathematical procedure of the renormalisation in QED. This procedure seems to be more physical, and it would be very desirable to obtain their correspondence.

Speaking about the "dressed" electron, one can say that principal contradictions between the KN soliton model and the Quantum electron in QED are absent. However, it should be mentioned that dynamics of the virtual particles in QED is chaotic, and conventionally, it could be separated from the core of "bare" electron. Contrarily, the vacuum state inside the Kerr–Newman soliton forms a *coherent oscillating state* adjoined to the electron core which has the form of a closed Kerr string. Therefore, the oscillating bubble source represents an *integral whole with the extended electron,* and its 'internal' structure cannot be separated from a "bare" particle.

By regularisation of the KN electromagnetic field, the maximal value of the vector potential is realised in the equatorial plane, on the stringy boundary of the bubble. As a result, radius of the regularised closed string is shifted to the domain wall boundary of the bubble and turns out to be slightly increased. The position of the string confirms the known old suggestion by Witten that the heterotic strings have to be formed on the boundary of a domain wall.

It should be noted that the real closed Kerr string is only the peak of the iceberg. As it was shown in the recent paper[9,44] there is very deep parallelism between the complex structure of the Kerr geometry and basic structures of superstring/M-theory. In particular, about two decades ago, the present author obtained a complex string inside of the complex 4D Kerr geometry, which together with the closed Kerr string forms the form the membrane source of the M-theory. The closed Kerr string is lightlike and similar to the Sen fundamental string solution to low-energy string theory. In the recent paper[46] Adamo and Newman reobtained these two strings analysing asymptotic form of the geodesic and shear-free congruences. Their emotionally comments are worth quoting: "...It would have been a cruel god to have layed down such a pretty scheme and not have it mean something deep."

In a recent papers,[9,44] a new remarkable fact was obtained – the appearance of a Calabi-Yau twofold (K3 surface) in the projective twistor space CP^3. Therefore, the famous K3 surface of the superstring theory represents nothing other than a twistor description of the Principal Null Congruence (PNC) of the Kerr geometry. One can suppose that this parallelism is not accidental, and that there should be an underlying superstring structure lying beyond these relationships. We suggest that it is $N = 2$ critical superstring. The $N = 2$ superstring is one of the three string theories consistent with quantum theory; number of supersymmetries $N = 0, 1, 2$ corresponds to consistent space-time dimensions $D = 26$, $D = 10$ and $D = 4$.[8] The $N = 2$ string, having complex dimension two, was very popular a few decades ago.[8] However, it has unusual signature which is in conflict with the real Minkowskian space-time. As a result, it was almost forgotten, and in particular it is not discussed in the very good modern textbook.[47] It has been shown recently in Refs. 9, 44, that the $N = 2$ string can be embedded in the complex Kerr geometry, and moreover, there is evidence that it can be considered as a complex source of the Kerr geometry.

Acknowledgments

This work is supported under the RFBR grant 13-01-00602. The author thanks Theo M. Nieuwenhuizen for permanent interest in this work and useful conversations.

References

1. J. H. Schwarz, The early history of String Theory and Supersymmetry, [arXiv:1201.0981].
2. F. Wilczek, *The Lightness Of Being* (Basic Books, 2008).
3. L. Susskind, *The Black Hole War* (Hachette Book Group US, 2008).

4. B. Carter, Global structure of the Kerr family of gravitational fields, *Phys. Rev.* **174**, 1559 (1968).
5. A. Sen, Macroscopic charged heterotic string, *Nucl.Phys.* B **388** 457 (1992).
6. A. Sen, Rotating charged black hole soltion in heterotic string theory, *Phys. Rev. Lett.* **69**, 1006 (1992).
7. J. Bjorken and S. Drell, *Relativistic Quantum Mechanics* (McGraw Hill Book, 1964).
8. M. B. Green, J. Schwarz and E. Witten, *Superstring Theory, v.I* (Cambridge Univ. Press, 1987).
9. A. Burinskii, Complex structure of the four-dimensional Kerr geometry: Stringy system, Kerr theorem, and Calabi-Yau twofold, *Adv. High Energy Phys.* 2013:509749 (2013).
10. D. D. Ivanenko and A. Ya. Burinskii, Gravitational strings in the models of elementary particles, *Izv. Vuz. Fiz.* **5**, 135 (1975).
11. E. T. Newman et al., Metric of a rotating charged mass, *J. Math. Phys.* **6**, 918 (1965).
12. G. C. Debney, R. P. Kerr and A. Schild, Solutions of the Einstein and Einstein-Maxwell equations, *J. Math. Phys.* **10**, 1842 (1969).
13. A. Ya. Burinskii, Microgeons with spins, *Soviet Phys. JETP* **39**, 193, (1974).
14. W. Israel, Source of the Kerr metric, *Phys. Rev. D* **2**, 641 (1970).
15. C. A. López, An extended model of the electron in General Relativity, *Phys. Rev. D* **30**, 313 (1984).
16. A. Burinskii, The Dirac-Kerr-Newman electron, *Grav. Cosmol.* **14**, 109 (2008).
17. I. Dymnikova, Spinning superconducting electrovacuum soliton, *Phys. Lett. B* **639**, 368 (2006).
18. A. Burinskii, Regularized Kerr-Newman solution as a gravitating soliton, *J. Phys. A: Math. Theor.* **43**, 392001 (2010).
19. Th. M. Nieuwenhuizen, The electron and the neutrino as solitons in classical Electrodynamics, in *Beyond the Quantum*, pp. 332–342, eds. Th.M. Nieuwenhuizen et al., World Scientific, Singapure, (2007).
20. A. Burinskii, Gravitational strings beyond quantum theory: Electron as a closed heterotic string, *J. Phys.: Conf. Ser.* **361**, 012032 (2012).
21. A. Burinskii, Gravity vs. Quantum theory: Is electron really pointlike? *J. Phys: Conf. Ser.* **343**, 012019 (2012).
22. H. Keres, On physical interpretation of solution of the Einstein equations, *Soviet Phys. JETP* **25**, 504 (1967).
23. V. Hamity, An interior of the Kerr metric, *Phys. Lett. A* **56**, 77 (1976).
24. A. Burinskii, Supersymmetric superconducting bag as a core of Kerr spinning particle, *Grav. Cosmol.* **8**, 261 (2002).
25. A. Burinskii, E. Elizalde, S. R. Hildebrandt and G. Magli, Regular sources of the Kerr-Schild class for rotating and nonrotating black hole solutions, *Phys. Rev. D* **65**, 064039 (2002).
26. D. Garfinkle, Black string traveling waves, *Phys. Rev. D* **46**, 4286 (1992).
27. A. Dabholkar, J. P. Gauntlett, J. A. Harvey and D. Waldram, Strings as solitons & black holes as strings, *Nucl. Phys. B* **474**, 85–121 (1996).
28. A. Burinskii, Some properties of the Kerr solution to low-energy string theory, *Phys. Rev. D* **52**, 5826 (1995).
29. H. B. Nielsen and P. Olesen, Vortex-line models for dual strings, *Nucl. Phys. B* **61**, 45 (1973).
30. G. Rosen, Particlelike solutions to nonlinear complex scalar field theories with positive definite energy densities, *J. Math. Phys.* **9**, 996 (1968).
31. M. Volkov and E. Wöhnert, Spinning Q-balls, *Phys. Rev. D* **66**, 085003 (2002).

32. N. Graham, An electroweak oscillon, *Phys. Rev. Lett.* **98**, 101801 (2007).
33. F. E. Schunck and E. W. Mielke, Rotating boson stars, in *Relativity and Scientific Computing*, edited by F. Hehl et al., pp. 138–151, Springer, Berlin (1996).
34. E. Witten, Superconducting strings, *Nucl. Phys. B* **249**, 557 (1985).
35. J. R. Morris, Supersymmetry and gauge invariance constraints in a $U(1) \times U(1)'$ Higgs superconducting cosmic string model, *Phys. Rev. D* **53**, 2078 (1996).
36. M. Gürses and F. Gürsey, Lorentz covariant treatment of the Kerr-Schild geometry. *J. Math. Phys.* **16**, 2385 (1975).
37. J. Ipser and P. Sikive, Gravitationally repulsive domain wall, *Phys. Rev. D* **30**, 712 (1984).
38. M. Cvetič, S. Griffies and H. H. Soleng, Local and global gravitational aspects of domain wall space-times, *Phys. Rev. D* **48**, 2613 (1993).
39. M. Cvetič, S. Griffies and S. Rey, Nonperturbative stability of supergravity and superstring vacua, *Nucl. Phys. B* **381**, 301 (1992).
40. M. Cvetič, F. Quevedo and S. J. Rey, Target space duality and stringy domain. walls, *Phys. Rev. Lett.* **67**, 1836 (1991).
41. M. Cvetič, S. Griffies and H. H. Soleng, Non-extreme and ultra-extreme domain walls and their global space-times, *Phys. Rev. Lett.* **71**, 670 (1993).
42. P. A. M. Dirac, Clasical theory of radiating electrons, *Proc. R. Soc. London*, Ser. A **167**, 148 (1938).
43. D. Kramer, H. Stephani, E. Herlt and M. MacCallum, *Exact Solutions of Einstein's Field Equations*, Cambridge Univ. Press, Cambridge (1980).
44. A. Burinskii, Stringlike structures in Kerr-Schild geometry: the $N = 2$ strings, twistors, and the Calabi-Yau twofold, *Theor. Math. Phys.* **177**(2), 1492–1504 (2013).
45. V. B. Berestetsky, E. M. Lifshitz and L. P. Pitaevsky, *Quantum Electrodynamics (Course Of Theoretical Physics, 4)*, Oxford, UK: Pergamon (1982).
46. T. M. Adamo and E. T. Newman, Light cones in relativity: Real, complex and virtual, with applications, *Phys. Rev. D* **83**, 044023 (2011).
47. K. Becker, M. Becker and J. Schwarz, *String Theory and M-Theory - A Modern Introduction* (Cambridge University Press, 2007).
48. J. Wess J and J. Bagger, *Supersymmetry and Supergravity* (Princeton Univ. Press, Princeton, New Jersey, 1983)

Chapter 13

Elements for the Development of a Darwinian Scheme Leading to Quantum Mechanics

Carlos Baladrón*

Departamento de Física Teórica, Atómica y Optica,
Universidad de Valladolid, 47071 Valladolid, Spain

A subquantum theory is examined in which a fundamental system has been characterized as the association of a particle with a continuous trajectory in real space and a classical probabilistic Turing machine defined on an informational space. The particle transfers information to the machine, and this steers the particle by means of self-interaction. In a certain sense, the associated Turing machine might be considered a generalization of the pilot wave function of Bohmian mechanics. The data processing capability entailed by the Turing machine makes the particle a generalized Darwinian system on which natural selection may operate. Darwinian evolution acting on the informational space should then drive the particle from random behaviour purportedly associated to an initial blank state to a possible evolutionarily stable strategy (ESS). Three regulating principles that plausibly encode an ESS are stated. The derivation of the postulates of quantum mechanics is discussed assuming that the behaviour of systems is governed by the three regulating principles. The theory also enables, within the generalized Darwinian framework, a natural characterization of entanglement through the local interaction between the Turing machines of the subsystems. Some possible future experimental and computational tests of the theory are outlined. The central aim of this scheme is to explore the possibility that generalized Darwinian natural selection might induce the emergence of quantum mechanics and its weird features from a real and local underlying description of particles supplemented with a Turing machine on an informational space, since those systems presenting quantum behaviour seem to be the most robust at a microscopic level.

1. Introduction

The classical way of thinking seems to be no longer valid when entering quantum mechanics. In some sections, classical concepts, methods, even classical logic and language appear to be of limited help to foray into the new territory to such an extent that only mathematical, abstract concepts and operations give the impression of being trustworthy, and only as long as one does not try to interpret these mathematical symbols and operations much beyond their correspondence with the final detection products in an experiment. The physical meaning of the

---

*baladron@cpd.uva.es

mathematical structure of the theory, and the explanation about what is really happening in nature, and even whether this explanation is possible or not, are still open questions to a vast number of interpretations.

The realization of the fundamental character of probabilities and complex numbers in the theory – possibly the main ingredients of quantum mechanics[1] – is a turning point in its comprehension. But perhaps the most astonishing insight comes out when – in connection with the fundamental character of probabilities – one tracks the first hints of lawlessness pointing at the very grounds of the theory. An example of this lack of law is already present in one of the first problems confronted by newcomers in quantum mechanics: the hydrogen atom. The answer to the question "in which position can the electron of a hydrogen atom in the ground state be detected?" is, namely, *anywhere* with a probability determined by the squared modulus of the wave function. The word *anywhere* suggests a certain absence of law in the quantum realm. Precisely this trait is a fundamental characteristic of the path integral or Feynman's formulation of quantum mechanics, and its central relevance in quantum mechanics was explicitly stated by Wheeler[2] and stressed by means of his famous aphorism: "law without law".

The study on the foundations of quantum mechanics presented in this work is profoundly influenced by Wheeler's ideas. First, the search of some regulating principles to describe the fundamental systems and their interactions is on the trail of the "law without law" idea. Second, the importance attached to information in the description of the physical world as it is coded in other influential Wheeler's sentence: "it from bit".

In Section 2 we briefly consider and classify some of the different approaches developed to tackle the difficulties of quantum mechanics. Then in Section 3 we explore the main sources that have inspired the theory examined in this lecture, and that was introduced in previous papers.[3–6] The theory itself is presented in Section 4. The derivation of the quantum mechanics postulates is discussed in Section 5, and the possible tests of the theory are set forth in Section 6. Finally, the conclusions are drawn in Section 7.

2. Approaches to quantum mechanics

The development of quantum mechanics has been marked on the one hand by an excellent agreement with experiments – arguably the best accord ever achieved by a theory, but on the other hand by an elusive explanation in classical rational terms of the microscopic world. For some physicists this last sentence has not a clear significance in the sense that quantum mechanics has no need of further interpretation beyond its own internal consistency.[7] For others, from the very beginning of the quantum theory, the classical edifice of science was shaken to its foundations. Schrödinger[8] summarized this state of affairs caused by the advent of the quantum theory as a threat to the two basic pillars of classical science since Ancient

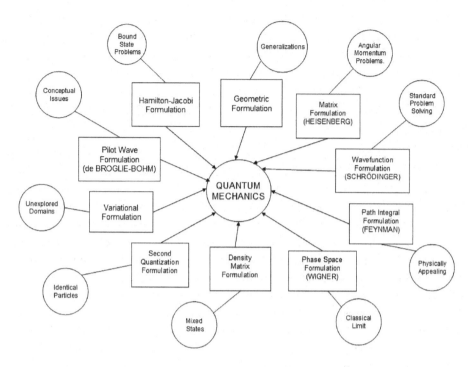

Fig. 1.   A list of several formulations of quantum mechanics with notes pointing out one specific application or feature for every formulation. The scheme is mainly based on Ref. 10.

Greece: The world around is comprehensible, and the real world is independent of the observer. The situation was plainly posed by Mückenheim[9] when asserting that after the development of quantum mechanics the scientist was obliged to sacrifice at least one of the three classical, fundamental properties in the description of nature, namely, reality, causality, and locality.

As a consequence of this unsatisfactory situation for many physicists we have witnessed from the start of quantum mechanics a proliferation of different approaches ranging from the mere mathematical reformulation of the theory to the study of possible theories underlying quantum mechanics. These different approaches to quantum mechanics can be classified in four categories by increasing order of revision depth. The border line between different classes of this catalogue is not clear-cut as it can be deduced from the fact that certain approaches, like the de Broglie-Bohm scheme, appear in different categories depending on the perspective which it is considered from.

First, the different formulations of quantum mechanics. In Fig. 1, it has been represented a collection of several formulations of quantum mechanics mainly based on the study realized by Styer *et al.*[10] This category is constituted by those approaches that mathematically recast the theory, but that yield the same numerical predictions for experiments. In principle, there is nothing specifically quantum in this category, since this kind of mathematical reconstruction is also present in classical mechanics and it is related to the simplification that some reformulations bring about certain kind of problems either technically or conceptually. But what in classical mechanics is usually just a different way of picturing a problem while maintaining the conceptual hard core of the theory, on the contrary, in quantum mechanics a pure mathematical reformulation can dramatically influence the conceptual interpretation. An example of this might be the crucial difference highlighted by Deutsch when comparing the Heisenberg picture versus the Schrödinger picture in the description of a problem. The Schrödinger formulation is the standard scheme to solve most problems due to its simpler mathematical requirements. However Deutsch stresses the fact that the Heisenberg formulation conceptually outperforms Schrödinger's to explain physically systems and processes. In particular, Heisenberg formulation allows us to track the local flow of information in any physical process[11] – criticism about this Deutsch´s claim is contended by some authors,[12] what discards nonlocality as a fundamental feature of quantum mechanics based on the argument that if you can find a formulation in which the theory is local, then it is local. Therefore, although the merit of most formulations in quantum mechanics also resides in allowing a simpler mathematical treatment of certain problems, however in some cases – as in the mentioned example – it goes much further in such a way that even formulations in quantum mechanics cannot be considered completely neutral with respect to interpretations.

The second category would be formed by interpretations of quantum mechanics. This kind would be composed of those approaches still yielding the same numerical results for any experimental set up, but in which the ontological or epistemological status of some mathematical objects in the theory can be different and certain postulates can be rephrased or modified – under the restraint of producing no observational alteration at the present experimental state-of-the-art. Copenhagen, Ensemble, de Broglie-Bohm, Many-Worlds, Time-Symmetric, Stochastic, Collapse, Transactional or Relational interpretations (see Ref. 13 and related entries therein) are just a non-exhaustive list in which might be included quite a few more. But even after trying to make a complete list of interpretations a second problem arises connected with the fact that several interpretations admit a subdivision according to the different ways in which certain concepts are treated within its own interpretation – e.g. most proponents of the de Broglie-Bohm interpretations consider the wave function as a real object, however there is a branch for which the wave function is a mere informational tool[14] that guides the real object, the particle – or even, as it is the case with the fuzzily defined Copenhagen interpretation,[10] certain assertions apparently in contradiction are ascribed to the same label.

Some interpretations seem irreconcilable, however at the same time one can interweave a network of unexpected connections between many of them. Just to mention: between de Broglie-Bohm and Many-Worlds,[15,16] between de Broglie-Bohm and Nelson's stochastic interpretation,[17] between Collapse and de Broglie-

Bohm.[18] So, in the end every new interpretation enriches somehow our knowledge of quantum mechanics, contributing to shed new light on the old problems.

A third level is constituted by those deeper schemes[19-21] whose aim is the derivation of the standard postulates of quantum mechanics from a set of simpler or more natural postulates. An improvement in the comprehension of the mysteries of quantum mechanics is expected from this perspective as well as a more profound explanation of the physical world, since it is assumed that the standard postulates are quite obscure and far from the usual structure in classical physics. In most of these new axioms studies, information plays a central role in the description of physical systems and processes.

The last and deepest stage is formed by the so-called subquantum theories, those that explore the possibility of an underlying reality to quantum mechanics. Some of these theories expect to dilute quantum weirdness by understanding the fundamental mechanism operating this hypothesized subquantum reality, and restore, at least in part, a classical rationality[8] in the description of nature. A key element characterizing this fourth and deepest category of approaches is the possibility of performing or formulating some experimental tests that allow us to discriminate between quantum mechanics and its competing theory. A scheme depicting a set of subquantum theories (see Ghirardi,[18] Nielsen,[22] Smolin,[23] Grössing *et al.*[24] and references therein) has been drawn in Fig. 2. Some of these theories have already appeared as interpretations when restricting their scope. As for interpretations of quantum mechanics, in the case of subquantum theories a network of fruitful connections may also be established between some of them.[17,25]

It would be a scientific tour the force to achieve the same degree of consensus about the meaning of quantum mechanics as classical physics raised in the past,

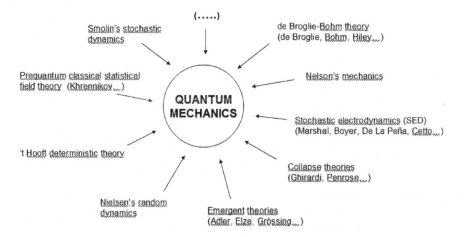

Fig. 2. A set of several subquantum theories with some of their proponents (see Ghirardi,[18] Nielsen,[22] Smolin,[23] Grössing *et al.*[24] and references therein).

and that would undoubtedly indicate a sound breakthrough in the understanding of nature. In order to reach that point every perspective plays its role.

3. Sources of the theory

There are four main elements originally involved in the development of the scheme expounded in this article. First, as was already mentioned in the introduction, the highly influential perspective of Wheeler on quantum mechanics and the central role that he attributed to information in the description of physical systems. The concept of information was formalized in a mathematical theory by Shannon.[26] Landauer[27] studied the basic relations between information and physics, and Wheeler delved into the relevance of information in the foundations of quantum mechanics, following a road that finally ushered in the development of quantum information.

A corresponding development of the concept of computation, the second main area underpinning the scheme presented in this article, was accomplished by Turing[28] – a similar work was independently realized by Church[29] – defining mathematically a logical device – the Turing machine – that provided a precise characterization of a general computing algorithm. The deep significance of computation in physics, and especially in quantum mechanics, was analyzed by Feynman[30] and Benioff[31] driving in the following decades to the field of quantum computation.

The third element in the scheme is Darwin's theory of evolution applied to these physical systems to which information is central. This seems to be a quite unusual standpoint, however Lotka[32] in 1922 already suggested the possibility that natural selection might supplement the three principles of thermodynamics with a fourth principle that would determine the way in which a thermodynamic system actually evolves.

Darwin's idea has probably been the best idea ever.[33] The spectacular success in explaining the biological world has driven to the development of the so-called universal Darwinism[34] or generalized Darwinism[35] by applying the basic structure of Darwin's theory to a huge variety of different areas ranging from economy to social science or even physics.[36,37] In this work we apply Darwin's theory of evolution by natural selection (see Ref. 38-40 for an analysis of Darwin's theory) to physical systems in which information plays a crucial role in their constitution and dynamics.

The fourth area that has a main influence on the scheme to be developed in this article analyzes the possibility that the physical laws of nature are not universal and immutable, but emergent, evolving relations.[41,42]

A sketch of the four principal areas to which our theory is related – with a list of some of the most relevant contributors to the development and the interconnections of these areas – is represented in Fig. 3.

In addition, the theory that we are going to develop also profits, as it will be seen in the following sections, from many concepts and methods introduced in previous subquantum theories – see Fig. 2 for a list of some subquantum theories.

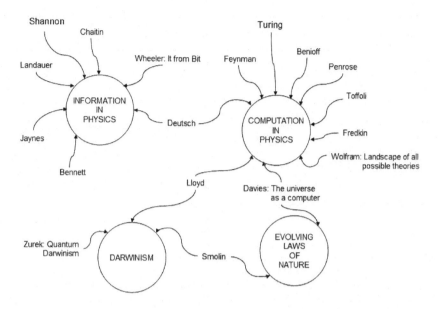

Fig. 3. Interconnections and several contributors to the four main areas related to the theory analyzed in the paper (see Davies,[41] Smolin,[42] Zenil[43] and references therein).

From this sketched fundamental scheme, the aim is to check the possibility of deducing quantum mechanics as an emergent theory, and at the same time, in the process, to better understand the intrinsic difficulties and weirdness of quantum mechanics. But, as we'll see along the development of the article, three other possible value-added results might be brought about by the implementation of this general scenario. These by-products might be: (a) a certain improved coexistence of realism, causality, and locality within quantum mechanics, (b) a particular answer to the question "why the quantum?", and (c) the generation of physical laws as a consequence of evolution under natural selection.

4. Theory

In this section a draft of the basic Darwinian subquantum theory is outlined. The constitution, state and dynamics of a fundamental particle, let's say an electron, are defined.

4.1. *Generalized Darwinian system*

A system must possess three fundamental features – variation, retention (or inheritance) and selection – in order to be describable by generalized Darwinism.[35,44] These properties are not enough to represent the detailed behaviour of the system. Specific mechanisms must be added to reflect the particular characteristics, interactions and processes of the field.

Generalized Darwinism is not based on analogies extracted from the biological realm, and then artificially fitted into a different area.[35] It rather relies on the recognition of general abstract evolutionary principles that are common to a vast variety of domains, and that despite of being central in biology, however these evolutionary principles are not inherently biological.[35,45]

The generalization of the machinery of evolution supposes the separation between the informational layer on which the development, inheritance, and selection of information focus – in which a general programme-based behaviour[35,46,47] independent of the concrete systems under scrutiny can be identified – and the material substrate on which the specific characteristics and processes of the field appear. We might thus consider the system as a dual entity formed by two interacting parts as follows: first, a material component defined on the physical space; and second, an abstract component defined on an informational space.

4.2. *Structure of a fundamental physical system*

A fundamental physical system – an electron – is defined as the association of a point-like particle in the Euclidean three-dimensional space and a methodological classical Turing machine[48] in an informational space. In short, a classical Turing machine is a computer with an unlimited amount of storage space. More in detail, it is a logical device constituted by four elements:

(1) the memory, a tape formed by a series of contiguous cells on which a binary string is written with a symbol per cell –'0' or '1'–,
(2) a read-and-write head that can move left and right along the cells of the tape,
(3) a finite list of machine states,
(4) the program, a finite list of transition rules that can also be written in binary code.

In fact, the definition of the fundamental physical system is extended by the requirement of an extra property for the Turing machine: to be probabilistic. This property implies that the transition rules are not uniquely established, but they are affected by normalized weighted factors in such a way that transitions are not deterministic, but probabilistic. This feature improves the machine computing performances for certain problems.

By means of the adopted definition, a fundamental physical system can be considered a generalized Darwinian system, since the Turing machine endows the particle with the possibility of processing data, and therefore the system, as we shall see in subsequent sections, is susceptible of variation, inheritance, and selection on the informational space.

The system has been characterized as the association of a particle and a probabilistic classical Turing machine. The aim is to reproduce quantum behaviour in real time by means of this description. This would be guaranteed by the Church-Turing principle[49] if the informational device associated to the particle were a quantum

Turing machine. However, as Timpson[50] has pointed out, the only difference between a probabilistic classical Turing machine and a quantum Turing machine is that the latter would be much more efficient in certain problems – e.g. an exponential speed-up over the known classical algorithms for factoring numbers,[50,51] but both machines can solve, in principle, the very same set of problems. Therefore, it would be necessary to add to the definition of fundamental system a procedure being able to drastically increase the intrinsic efficiency of a probabilistic classical Turing machine. This procedure is thought to be Darwinian natural selection. In fact, the main thesis of our theory is that quantum behaviour is nothing but the result of informational and computational efficiency attained by the action of generalized Dawinian evolution driven by natural selection and applied on the informational space of the particle.

We must supplement the constitution of the system with the rules that govern the particle in real space, and the mechanism of interaction between the particle and the Turing machine. It is assumed that at any time the particle has a rest mass m, a position in real space $\vec{Q}$, and a velocity $\vec{v}$, and that the interaction of the system with other particles in real space is established by the absorption and emission of energy, these processes being subject to the conservation of energy and momentum.

The particle acts on the Turing machine transferring the collected information from the absorbed photons. The information is processed according to the program P stored in the Turing machine and then, after a run of the program, the result of the computation, the output, is a command of action for the particle that takes the shape of emission of a photon of a determined energy an momentum. A parameter characterizing the time of reply of the machine t_r has to be assumed. This time – that implies an underlying discrete behaviour for the dynamics of the particle – would be proportional both to the length of the program and to the inverse of the number of instructions per second (IPS) that the Turing machine were able to execute.

This outlined mechanism of interaction closes the causal loop between material and informational spaces, since it is postulated that there isn't any exchange of energy between both spaces, but only exchange of information as raw data in one direction, and information as a command after the computation in the other direction. The informational space is thus assumed to be a methodological space. Within the material, real space the interaction between the particle and the Turing machine appears as sheer self-interaction of the particle.

The characterization of the particle as a methodological dual system suggests a singular rephrasing of Wheeler's saying "it from bit" transformed into a cyclic flow of mutual interactions "it from bit and bit from it" – see Fig. 4. In this way, the outer universal laws that dictate the behaviour of matter now, in our methodological dual description, enter the scene as possible dynamical elements that govern the behaviour of matter, but that, in turn, arise from the previous particle-to-particle

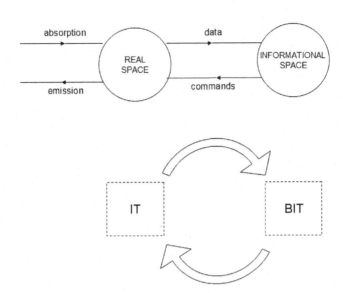

Fig. 4. Schematic description (above) of the mechanism of interaction between the material particle and the methodological Turing machine that is analyzed within the text. Symbolic characterization (below) of the cyclic flow of interactions between matter (particle) and information (Turing machine).

interactions. Thus the generalized Darwinian structure of the particle allows one to check the possibility of an evolutionary origin for the laws that rule matter, instead of being just postulated, immutable, and inaccessible, residing outside the evolving material universe. Quantum mechanics should emerge as a result of evolution in this general scenario.

4.3. State of the system

The state of the system is specified by the position that occupies the particle $\vec{Q}$, the current input data (ID) stored on the tape, and the current program P running on the Turing machine. Note that the output data – the energy and the momentum of the photon to be emitted by the particle – computed by the program P from the current input data ID allow us to deduce the velocity of the particle by simply applying the conservation principles of energy and momentum.

The description of the state of a system in our theory highly resembles that of the de Broglie-Bohm theory,[16,52,53] also called Bohmian mechanics, in which the state of a system is determined by the position of the particle $\vec{Q}$ in real space and the pilot-wave or wave function ψ of the particle defined on the configuration space that codes the active information[53] that the particle has got about the surrounding systems. In the de Broglie-Bohm theory the information in the wave function is not information about the system for observers, but information on the environment for the own system.[54] The velocity of the particle in Bohmian mechanics can be calculated from the so-called guiding equation, as we will see in next sections (see Section 5.3.3), once the pilot-wave is known. Then we could assimilate our output data (OD) – a binary string – from a program run of the Turing machine with the pilot-wave of Bohmian mechanics, and, following with the comparison, the program P in the Turing machine with the quantum mechanical equations. In fact, Bohm and Hiley[53] compare their description of the dynamics of a particle controlled by its pilot-wave with that of a ship steered by a computer. It seems that our theory can be considered as a generalization of Bohmian mechanics, since in principle it is able to accommodate Bohmian mechanics as a particular case – provided that we can demonstrate that the quantum mechanical postulates can be generated as a program in the Turing machine from generalized Darwinian evolution acting on the informational space. Therefore our system can be characterized at the time t by the triad $(\vec{Q}, \psi, P)$ where ψ is equivalent to the output binary string from the Turing machine, and the program P is equivalent to the quantum mechanical equations.

A natural generalization of the concept of genotype from genetics – the hereditary information contained in an organism (biological system) – would drive, in our theory, to the concept of program P stored in the Turing machine, and from the concept of phenotype – the particular physical properties of an organism (biological system) produced by the expression of the genotype – to that of output data OD – the wave function in the comparison with Bohmian mechanics. See Ref. 55 for more precise, technical definitions of genotype and phenotype.

In the next section we shall consider the dynamics of a fundamental system, and from this point of view we shall specify the meaning of the fundamental generalized properties of variation, inheritance and selection in the output data OD and program P of the Turing machine in the informational space of the particle.

4.4. *Dynamics of the system*

The continuous trajectory of the particle in real space is determined by the temporal sequence of absorption and emission events. The velocity of the particle after every event is established by the conservation of energy and momentum. Therefore the dynamics of the particle is determined both by the action of the surrounding systems through the absorption of photons and by self-interaction through the emission of photons.

The self-interaction of the system is specified, as previously outlined in Section 4.2, by the output OD of the Turing machine after one run of the program. The input data ID contain the information that the system has gathered about the positions of the surrounding systems from the absorption events. If the particle were a Bohmian particle we could assimilate the ID to the extra active information received by the particle during the interval Δt, OD to the wave function at time $t + \Delta t$, and the programme P would contain, for all practical purposes, both the Schrödinger equation (see Section 5.3.2) from which the wave function at time

$t + \Delta t$ is obtained and the guiding equation (see Section 5.3.3) from which the velocity of the particle is determined, and therefore the hypothetical command of self-interaction for the particle.

The sketchy mechanism of interaction is inspired by Bohmian mechanics[16,56] and stochastic electrodynamics (SED).[57] On the one hand, in the de Broglie-Bohm theory the quantum potential, responsible of the quantum behaviour of the particle, comes out from the own particle.[58] As previously mentioned, our Turing machine can be considered a generalization of those Bohmian tools as the quantum potential in the de Broglie-Bohm original second-order theory or the pilot-wave in the alternative first-order description to which the name Bohmian mechanics is usually ascribed. There is no a completely satisfactory explanation of the source and balance of energy in the original de Broglie-Bohm theory,[59] although there are subsequent, remarkable studies on this line, as that of Hall[59] in which by means of a Lagrangian formulation of the theory the conservation of energy and momentum is then assured, and also that of Grössing[56] that provides a more physical explanation of the action of the quantum potential in terms of classical diffusion waves. On the other hand, SED accounts for most quantum mechanical phenomena[57,60] by means of the interaction of the classical, stochastic electromagnetic zero point field (ZPF) with matter.

The project is to design a self-consistent mechanism in which the ZPF would provide a closure to counterbalance the possible energy loss excess in the self-interacting emission events. The connection between quantum fluctuations – the quantum version of the classical ZPF – and radiation reaction is strongly supported in quantum electrodynamics by the possibility of explaining specific phenomena – like the Lamb effect, spontaneous emission... – by different orderings of operators, corresponding every ordering to a different proportion in the weight given to radiation reaction and quantum fluctuations in the explanation of the phenomena.[61] The connection is strengthened by the impossibility of describing some phenomena (e.g. spontaneous emission) exclusively by means of either quantum fluctuations or radiation reaction.[61]

The central idea to balance the energy bookkeeping of the mechanism of interaction is to consider a variable mass as the extra source of energy for the specific quantum emission events in such a way that when a surplus of energy is required – determined by a run of the program in the Turing machine – for the emission of a photon, then this energy is taken from the mass of the particle. Subsequently, this decrease in mass has to be compensated by the absorption of energy from the ZPF. Every particle would then be an open system with a variable rest mass in a fine-grained underlying subquantum picture, but with a constant average mass in the coarse-grained quantum dynamics. The ZPF would be constituted by the photons coming from the self-interaction emissions with decrease in mass and by the emission events whose photons were not absorbed by near particles and were then lost in the far field.

This speculative mechanism has to be studied in order to first determine the feasibility of reproducing the ZPF-spectral energy density from an event-to-event emission mechanism that, in turn, has to account for the quantum dynamics of particles.

The possibility of reproducing quantum behaviour from an event-to-event scheme without solving a Schrödinger equation has been shown by De Raedt *et al.*[62] It has also been proven the compatibility of a local realistic interpretation of quantum mechanics with the results of EPR-Bell experiments[63,64] and quantum interference.[65]

Let us now consider more in depth the structure of data within the informational space. Like in de Broglie-Bohm theory, the fundamental variable is the position of the particle. In consequence, the fundamental information about outside systems is also their position. The problem posed in this way resembles the radar problem in which the point is to infer the position of target objects from raw data involving the original backscattered signals from the targets plus random noise – we shall retake this comparison in Section 5.3.1. These collected data at a certain time t_1 are the input data $ID(t_1)$ for the Turing machine. Assuming that the program P in the Turing machine simulates quantum behaviour, then, as aforementioned, the output data after a run of the program will be equal, for all practical purposes, to the wave function at time $t_1 + \Delta t$ at any position $\vec{q}$ on the space of configurations. From the calculated wave function, the system can anticipate the position of other particles and, as a consequence, produce an appearance of nonlocality. This nonlocality is manifest in the de Broglie-Bohm theory in the wave function of multiparticle systems.[16] Note that nonlocality in our theory is only apparent, an artifact of the processing data capability of the system through the Turing machine defined in the informational space. In our theory the wave function is not a real element in the description of the particle – as in some versions[14,66] of the de Broglie-Bohm theory, but only a binary string of data defined on the informational space. Matter, energy and information flows on real space are completely local within the whole theory.

At this point, after the definition of the structure and dynamics of the system, we must check the completion of the three properties that characterize a generalized Darwinian system: variation, inheritance, and selection. The program P on the Turing machine was identified as the natural generalization of the genotype in biology. Therefore, it is assumed that the program possesses the three generalized Darwinian features. First, variation on the defined fundamental system is considered to appear as a consequence of the random error rate of read-write operations during the execution of the program – another adjustable parameter of the theory that regulates the rhythm of evolution. It is a natural generalization of the random mutations rate in DNA-transcription in biology. Second, inheritance is referred to the new copy of the program after every run. Third, natural selection favours those programs that produce the fittest behaviour to get stability in the specific environment.

Bell considered necessary prior to introducing the concept of information in physics to give an adequate answer to the two following questions:[18,67] "about what information?" and "whose information?". Our theory has two unambiguous answers. The information is about positions of particles and it is for physical systems. In addition, we have seen that natural selection endows information with a context-dependent meaning that is associated with its relevance or utility for the stability of the own system within a specific environment.

4.5. *Mathematical methods*

The theory involves two basic originally non-physical subjects – information and Darwinian evolution. Both have their specific mathematical tools that will help to analyze the theory, in addition to the most traditional mathematical methods of physics.

On the one hand, evolutionary game theory[68] and genetic algorithm techniques[69] are the main tools to study evolution. Evolutionary game theory analyzes the dynamics of survival strategies in non-rational systems under the action of natural selection. The key concept of evolutionarily stable strategy (ESS) that captures the essentials of equilibrium and stability in evolutionary game theory plays[70] a central role in the simplification of the analysis.

On the other hand, it has been identified the similarity between the input data processing in the dynamics of a particle in our theory and the location of targets in a radar station. Therefore, estimation theory[71] and information geometry[72,73] are applied to study the problem, and as a consequence quantum mechanics might appear as the natural structure to solve the problem in an optimal way.

5. Results

The Darwinian scheme whose structure has been presented in the preceding sections is now going to be applied to the description of the dynamics of a fundamental particle. As aforementioned, the aim of the analysis is to observe the emergence of quantum mechanics and its intrinsic difficulties as a result of the proposed generalized Darwinian evolution, i.e. natural selection acting on physical systems should lead to quantum behaviour as an efficient method to ensure the stability of systems in their evolutionary environments.

5.1. *From the blank state to the regulating principles*

Let us assume that a fundamental system at $t = 0$ has no information at all. Then all the registers in its associated Turing machine would be zeros. Not only its input data on the memory tape are considered to be in the blank state, but also the program P on the machine is assumed to contain just the elementary instructions and operations characterizing the defining basic actions of a probabilistic classical

Turing machine, along with the command capability to interact with the material bare particle. This is defined as the blank state of the particle. Random fluctuations are expected to happen both in real space by the random emission of photons and in informational space by random read-and-write operations on the memory tape – data and instructions set. As photons impinge on the particle, information begins to be transferred to the memory tape as input data. Elementary self-interaction events, in addition to the random emissions, start to take place.

Therefore the coding error rate – read and write errors and so forth – that determines the mutation rate r_m associated to the Turing machine has to be incorporated as a new parameter to its characterization, along with the machine's time of reply t_r. The mutation rate plays an important role in the rhythm of evolution.[74] It is assumed that in the long-term, as an overall tendency, generalized Darwinian evolution increases systems' complexity as it does in biological evolution.[75] This assertion is hindered by the absence of consensus in the definition of a satisfactory magnitude to measure complexity. In our theory, Bennett's logical depth[76] is adopted as the magnitude that quantifies systems' complexity. For a binary string, Bennett's logical depth is defined as the computation time of a minimal-length program that yields the binary string. Thus the complexity of our fundamental system is equated to the logical depth of the binary sequence that codes the programme P of the Turing machine at time t.

As time increases, so does logical depth. We ought to study the evolution of systems assuming that natural selection will favour the prevail of those systems that optimize both the extraction of information from the input data and the self-interaction strategy with respect stability. The analysis of this evolution can be drastically simplified thanks to the concept of EES[70] that is central to evolutionary game theory. An ESS is a strategy that when being predominantly followed by most systems within a certain domain cannot be eradicated by any alternative strategy. An ESS can then be considered an attractor, and therefore it is supposed that the strategy of the population of systems will evolve towards the ESS over time. We expect to enunciate a set of regulating principles defining an ESS.

5.2. *Regulating principles*

The central idea steering the heuristic construction of the regulating principles is thus optimizing the stability expectations of systems in order to get an ESS. Note that natural selection is supposed to drive the evolution of the population strategy from the blank state to an ESS.

Principle 1 (Structure). *A fundamental system maximizes its complexity (Bennett logical depth).*

See Section 5.1 for a definition of Bennett's logical depth. Its maximization is done in two steps. First, by means of storing the information about the outside systems in the most efficient way in order to optimize the system's resources. And second, through maximizing the utility of the information stored in the system for the system's stability.

Principle 2 (Dynamics). *A fundamental system minimizes the information (Fisher information) sent to the environment.*

The Fisher information I contained in a function $\rho(x)$ – the probability density function of a random variable x – is defined mathematically by the following equation:[77]

$$I = \int_{-\infty}^{+\infty} \left[\left(\frac{d\rho}{dx} \right)^2 \cdot \frac{1}{\rho(x)} \right] dx \qquad (1)$$

Roughly speaking, a function represented by a flat, nearly constant-valued curve has a low Fisher information whereas a curve with a sharp maximum has high Fisher information. Fisher information is a local measure of the sharpness or smoothness of a function,[78] while Shannon information gives a global description of these properties. Therefore Fisher information is a more adequate magnitude to measure the information content of a function when the aim is to obtain a differential equation characterizing the phenomenon under study.[78]

There is a central analysis by Frank[79] that strongly supports the character of the **Principle 2** as a key element of an ESS. Frank[79] shows that Fisher information appears as the intrinsic metric to describe natural selection and the dynamics of evolution, and that natural selection drives systems to maximize the amount of Fisher information about the environment gathered by the population. Considering that information flows are zero-sum processes, then maximizing the gain of Fisher information is equivalent to minimize its loss.

Principle 3 (Interaction). *The interaction between two subsystems maximizes the complexity (Bennett's logical depth) within the whole system.*

These three regulating principles as a whole seem to constitute a reasonable candidate strategy to an ESS considering optimization in a static landscape of strategies,[68] given that the principles have been defined to optimize the information flows regarding the stability of systems. However it would be necessary to carry out a dynamical simulation in order to check the robustness of the regulating principles when allowing for mutant strategies whose payoff might depend on the strategy frequency in the population of systems.

5.3. *From the regulating principles to the postulates of quantum mechanics*

Once that an ESS has been identified, then the postulates of quantum mechanics should be deduced from the assumed regulating principles. Quantum mechanics would appear as a consequence of natural selection acting on the informational space of particles, and therefore as an efficient behaviour for stability.

Let us now consider our fundamental system characterized at time t by the triad $(\vec{Q}, \psi, P)$ where $\vec{Q}$ is the current position of the particle in real space, ψ is a binary string coding the wave function, and P is the present machine program, another binary sequence, that supposedly codes the regulating principles, since it is assumed that after enough time natural selection has driven to an ESS scenario. Our system, as mentioned above, is akin of a Bohmian particle if P has induced quantum behaviour on the particle. Therefore the most natural way to proceed is deducing the quantum mechanical postulates in the Bohmian version. In particular we are going to discuss the derivation from the regulating principles of the stochastic version of Bohmian mechanics known as Bohm-Vigier theory.[58,80] This theory is mathematically equivalent to quantum mechanics and consequently drives to the same experimental results. The postulates of the Bohm-Vigier theory are three:

(1) The system is described by its position in real space and a complex wave function defined in the configuration space (see Section 5.3.1).
(2) The wave function satisfies the Schrödinger equation (see Section 5.3.2).
(3) The velocity of the particle is determined by the stochastic guiding equation (see Section 5.3.3).

The two other usual postulates of quantum mechanics are not present in the Bohm-Vigier theory, since on the one hand, the Born rule – that establishes that the probability distribution of finding a particle in the position $\vec{q}$ at time t is the squared modulus of the wave function ψ – is asymptotically ensured by the stochastic guiding equation.[3,17,58] And on the other hand, in all versions of the de Broglie-Bohm theory, there is no collapse or measurement problem,[16] since every system has a well defined position at any moment in addition to its wave function, and any measurement, in the end, can be reduced to read the position of a pointer. Therefore a measurement is described as the process of interaction between the system of interest and a measuring device. The wave function of the process is then the superposition of all the components associated each one to every possible result of the measurement. The only physically meaningful branch of the wave function that describes the system after the measurement process is that associated with the actual value measured for the particle, since in the moment in which the measurement is registered the overlapping of the branch of the wave function actually occupied by the particle with the components associated with the other potential results of the measurement is negligible for all practical purposes. But if there is no overlapping, then those empty components are irrelevant for the future evolution of the system.[16]

5.3.1. *The wave function*

It was also previously mentioned – following the analogy originally established by Bohm and Hiley[53] for a particle in the de Broglie-Bohm theory – that the dynamics of a particle in our theory can be compared with the problem of a ship's trajectory controlled by a computer. This problem can be decomposed in three parts. First, the storage and analysis of the noisy signals backscattered – in our case, emitted photons – by the targets to be tracked – the storage and analysis of the input data, a radar-like problem. Second, the program to be run. And third, the determination of the commands to be applied to steer the ship – the output data.

In this scheme we can recognize two separate functions. On the one hand, the storage and analysis of information being studied and optimized by the theory of information retrieval[81] and estimation theory.[71] On the other hand, how to proceed, what strategy to apply, questions and elements that for a generalized Darwinian system are studied and optimized by evolutionary game theory. Darwinian evolution endows information, strategies, and the whole process with a contextual meaning. The meaning of information is referred to its relevance for the persistence of stable forms.[32]

The programme P contains the general rules determined by Darwinian evolution that control the response of the system. The wave function plays a role in both tasks – information storage and system behaviour. It codes the outside information, as well as the particular commands to apply under the specific circumstances surrounding the system.

The point now is to derive the complex Hilbert space structure for the states of a system from the **Principle 1**. In fact, it is unnecessary to go so far, since in the Bohmian version of quantum mechanics the Hilbert space structure just comes out of the collection of normalisable functions that are solutions of the linear Schrödinger equation.[82] Even exact linearity is in principle not strictly necessary as a postulate in the de Broglie-Bohm theory.[82] But we intend to deduce orthodox quantum mechanics, so that we proceed to consider the derivation of the complex Hilbert space structure from the maximization of the system's complexity (Bennett logical depth).

Let us start considering the analysis of Summhammer.[83,84] He posits the problem of an experiment and an observer. The observer obtains information about a system by means of the clicks in a collection of detectors from which the relative frequency of events can be calculated. Now by requiring maximum predictive power from these recorded data, the observer can deduce[83,84] the structure of the wave function of the system – squared root of the relative frequency, its complex character, and the linearity of the theory as optimal options among a set of considered possible solutions in order to preserve the information contained in the wave function. We now adapt the study of Summhammer, just changing the point of view. Instead of considering the direction of the information from the system to the observer, we take into account the flow of information from the environment to the system[5] – a particle plus a data processing machine. This is the Bohmian perspective in which the wave function contains active information,[53] i.e. information for the particle about the surrounding systems. Then applying to Summhammer's scheme, considering this new perspective, the **Principle 1** – maximization of the system's complexity, it is deduced[83,84,5] that the complex Hilbert space structure is a possible consequence of Darwinian natural selection acting on a particle endowed with a Turing machine, owing to the efficiency that such a structure implies in the processing of data.

Common structural properties among quantum mechanics and several disciplines in which information and data analysis play a central role have been reported and analyzed. These studies reveal that the Hilbert space structure of quantum mechanics is suitable to be applied in information retrieval,[81] network computation,[85] and signal theory.[86]

In information retrieval and network computation even the concept of entanglement or quantum correlations can be applied to the description of certain specific problems.[85] This fact suggests that the quantum statistical approach might be not only adequate, but perhaps even necessary to properly tackle data analysis questions.[85] The fitting of quantum structure to network computation is so remarkable that some weird quantum issues like nonlocality seemingly admit a more natural explanation in network computation than in physical terms.[85]

Aerts[86] has also studied the connections between optimal inference and Hilbert space structure. In particular, he has analyzed signal theory and its Hilbert space representation. Signal theory can be considered a formalization of observation through optimal statistical inference.[86] One can conclude[86] that it is the common trait of being theories of optimal observation what endows both theories, quantum mechanics and signal analysis, with their Hilbert space structure. In the present article it is also maintained this point of view, since in our theory there is even a deeper connection between quantum mechanics and signal analysis, given that the first part of the quantum particle dynamics has been assimilated to a radar-like problem from the perspective of the particle. In fact, we are going one step further asserting that the optimization in the particle of the inference procedure in the signal analysis is a consequence of natural selection acting on the informational space. Systems that optimize their estimation procedures by implementing a Hilbert space structure in the data processing have the greatest chances to persist.

Finally, we point out that in the geometric formulation of quantum mechanics[87,88] the Fubini-Study metric can be induced in the complex projective Hilbert space of quantum theory by extending with a compatible complex structure in the framework of information geometry the original unit sphere in a real Hilbert space – that can be considered the space of probability distributions – in which a statistical model has been introduced and the natural Fisher-Rao or Fisher information metric – that is proportional to the statistical distance or dissimilarity between distributions – has in turn been induced. Thus, Fisher information seems to appear in a natural way in a formulation of quantum mechanics that is quite suitable to shed new light on the differences between the classical and quantum descriptions of nature thanks to the reformulation of quantum mechanics in a geometrical language. Therefore, let us now pay attention to Fisher information and its role in the deduction of the Schrödinger equation in our theory.

5.3.2. *Schrödinger equation*

Let us consider a generalized Darwinian particle that is restrained to move on the real line, from the point of view of an observer who intends to locate its position. We are again before a classical radar-like problem as described in Section 4.4. Note that being a generalized Darwinian particle implies to have a well-defined position at any moment, although because of the process of measurement the observer registers a probability distribution function $\rho(x)$ for the position of the particle. This probability distribution has to be related to the wave function $\psi(x)$ of the particle, since in the end the state of the system in our Darwinian theory must depend on the information collected by the particle about the environment, and therefore the behaviour of the system has to reflect the gathered information. Thus, it is reasonable to assume that the probability distribution function is again the squared modulus of the wave function, $\rho(x) = |\psi(x)|^2$, as it was plausibly established by applying the **Principle 1** to determine the optimal way of storing information in the particle (see Section 5.3.1). In addition, it is an intriguing fact, as remarked by Frieden and Soffer,[78] the utilization of probability amplitudes by Fisher[89] in his statistical studies in order to identify different population classes in a more convenient way, and having no connection with quantum mechanics at all. In our theory, this fact is naturally explained as a consequence of the common trait of optimal data processing to statistical inference and quantum mechanics.

Following Frieden and Soffer,[78] we can also assume that the game of information between a system and its environment is a zero-sum game, i.e. the information lost by the particle is gained by the observer. Now, according to the ESS – therefore to generalized Darwinian natural selection, the system's dynamics should be determined by minimizing the information that the system sends to the environment, i.e. applying the **Principle 2**. But this is basically the principle of minimum Fisher information or maximum Cramer-Rao bound[77] applied to the probability distribution function for the position of the particle. Frieden[77] deduced the Schrödinger equation from this assumption, that is based on the observed convergence between estimation procedures and physical reality.[77] Let us follow his development (see Ref. 5,78 and 90 for further comments and more detailed discussions).

The procedure consists of minimizing the Fisher information of the probability distribution function $\rho(x)$ of the particle's position subject to a constraint that reflects the physical content of the problem:

$$\int_{-\infty}^{+\infty} \left[\left(\frac{d\rho}{dx} \right)^2 \cdot \frac{1}{\rho(x)} \right] dx + \lambda_c \int_{-\infty}^{+\infty} \rho(x) \left[E - V(x) \right] dx = \min. \qquad (2)$$

The first term is the Fisher information of $\rho(x)$ as defined in Eq. (1). The second term is the average kinetic energy $\langle K \rangle$ of the particle whose unknown total energy is E, and $V(x)$ is the potential energy of the particle associated to the presence of a potential. λ_c is a fixed negative parameter that determines the relative weight of each term in the minimization. The first term, Fisher information, is a non-classical ingredient in the dynamics of a particle, whereas the second one conveys the energy conservation assumption in addition to the requirement of a positive average kinetic energy $\langle K \rangle$. The lack of a physical constraint would imply a trivial minimum Fisher information equal to zero.

We can rewrite Eq. (2) in terms of a probability amplitude $\psi(x)$ – the wave function – defined by $\rho(x) = [\psi(x)]^2$. Then the expression reads

$$\int_{-\infty}^{+\infty} \left(\frac{d\psi}{dx}\right)^2 dx + \lambda \int_{-\infty}^{+\infty} [\psi(x)]^2 [E - V(x)] \, dx = \min \tag{3}$$

being $\lambda = \lambda_c/4$. The extremum problem is now reduced, as customary in variational methods in physics – e.g. see Ref. 91, to solve the Euler-Lagrange equation:

$$\frac{d}{dx}\left(\frac{\partial L}{\partial \dot{q}}\right) - \frac{dL}{dq} = 0 \tag{4}$$

in this case with $L = \dot{q}^2 + \lambda q^2 [E - V(x)]$, $\dot{q} = dq/dx$, and $q(x) = \psi(x)$, that leads, taking $\lambda = -2m/\hbar^2$, to the Schrödinger equation

$$\frac{d^2\psi}{dx^2} + \frac{2m}{\hbar^2} [E - V(x)] \, \psi(x) = 0. \tag{5}$$

Thus, the Schrödinger equation solves Eq. (3). Now, we see that the wave function being a solution of the Schrödinger equation also codes the strategy that the particle must follow in order to minimize the information sent outside.

Reginatto[92] also derives the Schrödinger equation from the principle of minimum Fisher information and a physical constraint that, this time, is the physical content assumption that a wave front can be associated with the motion of particles. This description is deeply connected with the de Broglie-Bohm theory, allowing for an analysis of the characteristic Bohmian elements in terms of Fisher information. In particular, the average of the quantum potential – the quantum potential[16] determines the quantum behaviour of particles in the Bohmian description – turns out to be proportional to the Fisher information of the probability density.[92] The study of Reginatto[92] directly drives to a time-dependent Schrödinger equation with a complex wave function.

5.3.3. *Guiding equation*

In the previous subsection we have analyzed the dynamics of the wave function, and obtained the Schrödinger equation from the regulating principles that summarize the strategy – ESS – that in the long run the generalized Darwinian evolution

has induced. The wave function was characterized in Section 4.3 as the generalized phenotype of the particle coding its properties and behaviour in the specific
environment in which the particle is located. But in our theory the ontological
element in the description of a system is the position of the particle. Therefore we
must complete the dynamical description of a fundamental system by specifying the
equation that enables us to determine the position and velocity of the particle at
any time.

According to our scheme we should derive the stochastic guiding equation of
the Bohm-Vigier theory from the **Principle 2** – the dynamical regulating principle
– in order to accomplish the deduction of the postulates of quantum mechanics in
this stochastic version of Bohmian mechanics. Goldstein[93] discusses the plausible
nomological character of the wave function – i.e. it is arguably more an element
of physical law than a physical real object – reflected, among other things, on
the similar role played by the wave function, more precisely the logarithm of the
wave function, and the classical Hamiltonian in the guiding equation of Bohmian
mechanics, the former, and in the equations of motion of classical mechanics, the
latter. As it can be deduced from our position held all along this article, we share
Goldstein's view.

Therefore, let us apply **Principle 2** to determine the trajectory of the particle.
In order to minimize the information sent outside, the particle will follow the trajectory for which the outward flow of information is a minimum. Let us consider,
following the analogy between the logarithm of the wave function and the classical
Hamiltonian, that the gradient of the logarithm of the wave function tracks this
path in average, and that the resultant current particle velocity is the addition of
$\nabla \ln \psi$ plus a Brownian-like term $\vec{\eta}$ accounting for the inclusion of a fundamental
randomness in the strategy of the system that according to evolutionary game theory[94] optimizes a sheer deterministic strategy. Then the velocity of the particle
reads as follows

$$\frac{d\vec{Q}}{dt} \propto \nabla \ln \psi + \vec{\eta} = i\nabla S + \frac{\nabla |\psi|^2}{2 |\psi|^2} + \vec{\eta} \tag{6}$$

where S is the phase of the wave function $\psi = |\psi| \exp iS$. Let us compare Eq. (6)
with the stochastic guiding equation[17]

$$\frac{d\vec{Q}}{dt} = \frac{\hbar}{m}\nabla S + \alpha\frac{\hbar}{2m}\frac{\nabla |\psi|^2}{|\psi|^2} + \sqrt{\alpha} \cdot \vec{\eta} \tag{7}$$

being $\hbar$ the reduced Planck constant, m the mass of the electron, and α a constant
whose value varies between 0 and 1, and that reproduces[17] the original de Broglie-
Bohm theory when setting $\alpha = 0$ and Nelson's stochastic mechanics for $\alpha = 1$.

There are some remarks to be considered. First, in quantum fluid dynamics[95]
the general quantum mechanical local velocity field defined from the momentum
operator is given by the expression $- (i\hbar/m) \nabla \ln \psi$ where its real part is the first
term on the right-hand side of Eq. (7) and its imaginary part is proportional

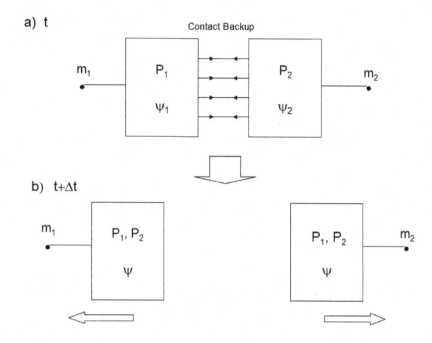

Fig. 5. Two particles (m_1 and m_2) in direct contact at time t (above) with their respective Turing machines loaded with program P_1 and wavefunction ψ_1, the first machine, and program P_2 and wavefunction ψ_2, the second one, are entangled by the complete copy of mutual data (wavefunctions and programs), according to the model of entanglement in our theory. After an interval of time Δt (below), both particles separate conveying the same data and programs in their respective Turing machines (see text for a detailed description of the process).

to the second term. This fact further grounds the analogy between the classical Hamiltonian and the logarithm of the wave function.

Second, as pointed out in Section 4.4, our theory sets forth a tentative mechanism of self-interaction – inspired by previous studies[16,56−60] – in order to explain the physical origin of the stochastic guiding equation, and preserve the balance of energy. Further work has to be done on this line.

5.4. *Entanglement*

Our theory might enable a natural explanation of quantum entanglement[96,97] – the possible non-classical correlations between spatially separated quantum systems. Let us consider two fundamental particles, e.g. two electrons, with their respective Turing machines that are in direct contact (see Fig. 5). Both particles become entangled if a complete backup or copy of data is accomplished between them. Therefore, both systems share the very same amount of data, i.e. the wave functions ψ_1 and ψ_2 and the programs P_1 and P_2 are stored on both Turing machines. The time dependent wave function of the whole system ψ can be calculated in both Turing machines by orderly applying to every part its corresponding program, and if the wave function of the whole system becomes entangled – in mathematical language, this means that the addition of all the calculated terms cannot be expressed as an ordered product of two factors, the first one formed by terms of the first particle and the second one formed by terms of the second particle–, then even in the case in which the particles separate spatially the dynamics of every particle will also depend on the current position of the other one through the entangled total wave function present in both guiding equations. Note that the behaviour of particle 1 is anticipated in the Turing machine of particle 2, and vice versa, by means of the calculation, since both particles share exactly the same program and the same data at the time in which they got entangled. Future interactions will progressively blur the calculated anticipations of every particle about the behaviour of its respective partner, owing to the incorporation of new data proceeding from the interactions with the environment and with other systems, but we are just describing the process of disentanglement.

When the pair is disentangled, the wave function of the two particles considered as a whole can be written as a product of two wave functions every one referring to its corresponding particle. The total wave function is now separable.

The current program on the Turing machine of every particle might be the result of an algorithmic version of the process of genetic recombination in which some segments of the original P_1 – the program controlling particle 1 before the copy of P_2 and ψ_2 from particle 2 – have been substituted for equivalent segments from the program P_2. This speculation points at another possible source of variation in the generalized Darwinian evolution affecting physical systems in our proposed theory. This quite interesting possibility should be further explored.

The characterization of entanglement in our theory also seems to explain in a natural way certain peculiarities of this phenomenon, like its rather strange adaptability to be described in terms of cryptographic concepts.[97] In our theory, the cryptographic communication by means of a common private key shared by two entangled parties in order to decipher coded information fits quite well, since both parties have exchanged their wave function and program by direct contact when the entanglement was established. In addition, the cryptographic communication might improve the expectations of persistence for the entangled system, and therefore it would be fostered by the regulating principles.

Finally, let us explore the possibility of entanglement being the central element that determines the way in which interaction is established between two systems. Let us assume that two entangled subsystems try to maintain or increase their entanglement. This behaviour would enhance the complexity of the system as a whole, taking into account that two entangled particles shared somehow two interconnected Turing machines admitting classical communication through interaction, and therefore its expectations of stability should also rise in comparison with those of the separated subsystems. Consequently, the subsystems will try to increase the entanglement by properly interacting. But this is just the content of **Principle 3**, that the interaction will tend to increase the complexity of the joint system, and, according to our discussion, this will be done by increasing the entanglement.

Let us try to implement the **Principle 3** in a mathematical form of extremum principle.[4] First, we consider the projective Hilbert space of a system formed by two separate subsystems, in which a natural metric has been defined. Then we take the entanglement of formation $E_F(\psi)$ as a measure of the entanglement content in a state ψ of the system. Now applying the **Principle 3** to the bipartite system we deduce that in order to maximize its complexity the system will evolve along the trajectory for which the entanglement content is a maximum, i.e. we have translated the **Principle 3** into the following variational principle:

$$\delta \left[\int_a^b E_F(\psi) dl \right] = 0 \tag{8}$$

where a and b are respectively the initial and final points of a trajectory on the projective Hilbert space of the system.

But the evolution of the system's wave function is already determined by the Schrödinger equation. Therefore from Eq. (8), in principle, we might derive the interaction Hamiltonian.

Our subquantum theory might contribute to shed light on these two especially baffling quantum features: entanglement and the origin of the Hamiltonians' shape for certain problems.[98] In our theory, both questions could be related, and, as it has been discussed, could find in the future a convenient characterization. Anyway, these difficult, long-standing problems need further investigation.

6. Tests of the theory

In this section, we are going to analyze possible tests to check the theory.

First, the variational method, Eq. (8), proposed to determine the form of interaction Hamiltonians might be applied to simple bipartite systems – e.g. simple diatomic molecules – in which the interaction Hamiltonian could be parameterized in terms of the interatomic distance, and the dependence on the interatomic distance of the electronic correlation energy – an approximate meter[99] of the electronic entanglement – could be studied.

A second group of tests may be the simulation of the behaviour of an electron within a simple scenario in which the background is represented by its potential, but instead of solving the Schrödinger equation for the electron, this is endowed with a probabilistic classical Turing machine and a program that codes the three regulating principles. It is expected that the dynamics of the electron generated from an event-by-event procedure – in a similar way to the studies of De Raedt *et al.*[62] – presents quantum features. In particular, an interesting problem, although extremely difficult, would be the simulation of a superconductor with two selected electrons endowed with their respective Turing machines and a program containing the three regulating principles. Under adequate conditions, a Cooper pair behaviour for the two electrons should be obtained from the event-by-event simulation.

Finally, the main challenge of a theory is to successfully confront the experimental results. The state-of-the-art of experimental techniques, although impressive, doesn't enable us yet to measure directly in the laboratory the fine-grain differences predicted by subquantum theories with respect to orthodox quantum mechanics. However, as Valentini[100] has proposed in the framework of Bohmian mechanics, there are some astrophysical and cosmological tests that might detect the existence of remnants of quantum non-equilibrium. On the same line of thought, we could apply this idea to our theory. According to the evolutionary scheme, and the concept of an ESS, these strategies are local minima in the landscape of strategies. Therefore there are several possible solutions depending on the frequencies and interactions of strategies in the populations systems of different isolated areas, i.e. depending on the trajectories followed in the general landscape of strategies. As a consequence, there might be certain regions in which the physical laws implemented on the Turing machines of the fundamental systems were not those of quantum mechanics, but another laws derived from a different ESS.

7. Conclusions

A fundamental particle – an electron – has been characterized as a generalized Darwinian system by endowing the particle with a probabilistic classical Turing machine that incorporates the capacity of processing information. The system transfers the information conveyed by the impinging photons to the machine, and this, in return, directs the particle by self-interaction. Darwinian natural selection now can act on the informational space and, in the long run, drive the particle from an initial informational blank state, associated with dynamical random behaviour, to an evolutionarily stable strategy.

Three plausible regulating principles, hypothetically encoding an ESS, have been stated. Then the derivation of the postulates of quantum mechanics in the stochastic Bohm-Vigier version from these regulating principles has been discussed.

Quantum behaviour seems to appear as a consequence of contextual-efficient procedures for the stability of systems, in other words, quantum behaviour seems to be a synonymous for robustness at a microscopic level. Most quantum features are then naturally explained in this general scenario: complex numbers as an efficient way of storing information, apparent nonlocality as a consequence of calculated anticipations supplied by the data processing capability, randomness as a procedure to optimize strategies, entanglement as a mutual backup of data and programs between two systems in contact as a means of gaining efficiency.

The subquantum theory outlined in the present article advocates the possibility of evolving physical laws that would govern the behaviour of matter in different ways, in different regions of space-time. The geometric formulation of quantum mechanics[87] with its connections with the geometry of information[92] displays a scenario in which quantum mechanics might appear as a particular linear case among non-linear more general theories in an overall landscape. These possibilities seem to be worth exploring on the theoretical and experimental plane.

Acknowledgment

I am deeply grateful to Claudia Pombo, Theo Nieuwenhuizen, Inácio Pedrosa, Cláudio Furtado and all the remaining organizers for their invitation to participate in the Joao Pessoa's advanced school, and for their warm support and care. I also wish to thank all the professors and students who contribute to jointly develop this scientific adventure. My deep gratitude also to Olga Baladrón for her lovely patient technical advice on electronic editing tasks, and to Marisa Zorita for our passionate discussions about biological concepts.

References

1. R. Penrose, *Shadows of the Mind* (Oxford University Press, Oxford, 1994).
2. J. A. Wheeler, On recognizing law without law, *Am. J. Phys.* **51**, 398–404 (1983).
3. C. Baladrón, In search of the adaptive foundations of quantum mechanics, *Physica E* **42**, 335–338 (2010).
4. C. Baladrón, Study on a possible Darwinian origin of quantum mechanics, *Found. Phys.* **41**, 389–395 (2011).
5. C. Baladrón, On the possibility of considering quantum mechanics as a Darwinian theory, in ed. J. P. Groffe, *Quantum Mechanics*, pp. 271–294. Nova Science, New York (2011).
6. C. Baladrón, Quantumness to survive in an evolutionary environment, *AIP Conference Proceedings* **1327**, 316–321 (2011).
7. C. A. Fuchs and A. Peres, Quantum theory needs no "interpretation", *Physics Today* **53**, 70–71 (2000).
8. E. Schrödinger, *Nature and the Greeks* (Cambridge University Press, Cambridge, 1954).
9. W. Mückenheim, A review of extended probabilities, *Phys. Rep.* **133**, 337–401 (1986).
10. D. F. Styer, M. S. Balkin, K. M. Becker, M. R. Burns, C. E. Dudley, S. T. Forth, J. S. Gaumer, M. A. Kramer, D. C. Oertel, L. H. Park, M. T. Rinkoski, C. T. Smith and T. D. Wotherspoon, Nine formulations of quantum mechanics, *Am. J. Phys.* **70**, 288–297 (2002).
11. D. Deutsch and P. Hayden, Information flow in entangled quantum systems, *Proc. R. Soc. Lond. A* **456**, 1759–1774 (2000).
12. R. E. Kastner, Quantum nonlocality: not eliminated by the Heisenberg picture, *Found. Phys.* **41**, 1137–1142 (2011).
13. J. Ismael, Quantum mechanics, in ed. Edward N. Zalta, *The Stanford Encyclopedia of Philosophy* (Fall 2009 Edition).
    http://plato.stanford.edu/archives/fall2009/entries/qm/.
14. C. Callender,Discussion: the redundancy argument against Bohm's theory, (2009).
    http://philosophyfaculty.ucsd.edu/faculty/ccallender/publications.

15. D. Deutsch, Comment on Lockwood, *British Journal for the Philosophy of Science* **47**, 222–228 (1996).
16. S. Goldstein, Bohmian mechanics, in ed. Edward N. Zalta, *The Stanford Encyclopedia of Philosophy* (Spring 2013 Edition).
    http://plato.stanford.edu/archives/spr2013/entries/qm-bohm/.
17. G. Baccigaluppi, Nelsonian mechanics revisited, *Found. Phys. Lett.* **12**, 1–16 (1999).
18. G. Ghirardi, Collapse theories, in ed. Edward N. Zalta, *The Stanford Encyclopedia of Philosophy* (Winter 2011 Edition).
    http://plato.stanford.edu/archives/win2011/entries/qm-collapse/.
19. A. Zeilinger, A foundational principle of quantum physics, *Found. Phys.* **29**, 631–643 (1999).
20. L. Hardy, Quantum theory from five reasonable axioms, arXiv: quant-ph/0101012 (2001).
21. R. Clifton, J. Bub and H. Halvorson, Characterizing quantum theory in terms of information-theoretic constraints, *Found. Phys.* **33**, 1561–1591 (2003).
22. H. B. Nielsen, Imaginary part of action, future functioning as hidden variables, *Found. Phys.* **41**, 608–635 (2011).
23. L. Smolin, Could quantum mechanics be an approximation to another theory?, arXiv:quant-ph/0609109 (2006).
24. G. Grössing, S. Fussy, J. M. Pascasio and H. Schwabl, The quantum as an emergent system, arXiv:1205.3393[quant-ph] (2012).
25. L. de la Peña, A. M. Cetto and A. Valdés-Hernández, The emerging quantum (2012).
    http://www.fisica.ufpb.br/asqf2012/.
26. C. E. Shannon, A mathematical theory of comunication, *Bell System Technical Journal* **27**, 379–423 (1948).
27. R. Landauer, Information is physical, *Phys. Today* **44**, 23–29 (1991).
28. A. M. Turing, On computable numbers, with an application to the Entscheidungsproblem, *Proc. Lond. Math. Soc.* **42**, 230–265 (1937).
29. A. Church, An unsolvable problem of elementary number theory, *American Journal of Mathematics* **58**, 345–363 (1936).
30. R. P. Feynman, Simulating physics with computers, *International Journal of Theoretical Physics* **21**, 467–488 (1982).
31. P. Benioff, Quantum mechanical Hamiltonian models of Turing machines, *J. Stat. Phys.* **29**, 515–546 (1982).
32. A. J. Lotka, Natural selection as a physical principle, *Proc. Natl. Acad. Sci.* **8**, 151–154 (1922).
33. D. Dennett, *Darwin's Dangerous Idea: Evolution and the Meanings of Life* (Simon and Schuster, New York, 1996).
34. R. Dawkins, Universal Darwinism, in ed. D. S. Bendall, *Evolution from Molecules to Man*, pp. 403–425. Cambridge University Press, Cambridge (1983).
35. H. E. Aldrich, G. M. Hodgson, D. L. Hull, T. Knudsen, J. Mokyr and V. J. Vanberg, In defence of generalized Darwinism, *J. Evol. Econ.* **18**, 577–596 (2008).
36. W. H. Zurek, Quantum Darwinism, *Nat. Phys.* **5**, 181–188 (2009).
37. L. Smolin, The status of cosmological natural selection, arXiv:hep-th/0612185 (2006).
38. J. Lennox, Darwinism, in ed. Edward N. Zalta, *The Stanford Encyclopedia of Philosophy* (Fall 2010 Edition).
    http://plato.stanford.edu/archives/fall2010/entries/darwinism/.
39. P. Sloan, Evolution, in ed. Edward N. Zalta, *The Stanford Encyclopedia of Philosophy* (Fall 2010 Edition).
    http://plato.stanford.edu/archives/fall2010/entries/evolution/.

40. R. Brandon, Natural selection, in ed. Edward N. Zalta, *The Stanford Encyclopedia of Philosophy* (Fall 2010 Edition). http://plato.stanford.edu/archives/fall2010/entries/natural-selection/.

41. P. C. W. Davies, The implication of a cosmological information bound for complexity, quantum information and the nature of physical laws (2007). http://cosmos.asu.edu/.

42. L. Smolin, Does time emerge from timeless laws, or do laws of nature emerge in time? (2011). http://www.perimeterinstitute.ca/people/lee-smolin.

43. H. Zenil, Ed., *A Computable Universe. Understanding and Exploring Nature as Computation* (World Scientific Publishing, 2012).

44. D. T. Campbell. Variation, selection and retention in sociocultural evolution, in eds. H. R. Barringer, G. I. Blanksten and R. W. Mack, *Social Change in Developing Areas: a Reinterpretation of Evolutionary Theory*, pp. 19–49. Schenkman, Cambridge, MA (1965).

45. J. S. Metcalfe, *Evolutionary Economics and Creative Destruction* (Routledge, London, 1998).

46. E. Mayr, *Toward a New Philosophy of Biology: Observations of an Evolutionist* (Harvard University Press, Cambridge, MA, 1988).

47. H. A. Simon, *Models of Man: Social and Rational* (Wiley, New York, 1957).

48. D. Barker-Plummer, Turing machines, in ed. Edward N. Zalta, *The Stanford Encyclopedia of Philosophy* (Winter 2012 Edition). http://plato.stanford.edu/archives/win2012/entries/turing-machine/.

49. D. Deutsch, Quantum theory, the Church-Turing principle and the universal quantum computer, *Proc. R. Soc. Lond. A* **400**, 97–117 (1985).

50. C. Timpson, Philosophical aspects of quantum information theory, in ed. D. Rickles, *The Ashgate Companion to the New Philosophy of Physics*. Ashgate (2008).

51. P. W. Shor, Algorithms for quantum computation: Discrete logarithms and factoring, in ed. S. Goldwasser, *Proceedings of the 35th Annual IEEE Symposium on Foundations of Computer Science*, pp. 124–134. IEEE Computer Society Press (1994).

52. P. Holland, *The Quantum Theory of Motion: An Account of the de Broglie-Bohm Causal Interpretation of Quantum Mechanics* (Cambridge University Press, Cambridge, 1993).

53. D. Bohm and B. J. Hiley, *The Undivided Universe: An Ontological Interpretation of Quantum Theory* (Routledge and Kegan Paul, London, 1993).

54. D. Gernert, Pragmatic information: historical exposition and general overview, *Mind and Matter* **4**, 141–167 (2006).

55. R. Lewontin, The genotype/phenotype distinction, in ed. Edward N. Zalta, *The Stanford Encyclopedia of Philosophy* (Summer 2011 Edition). http://plato.stanford.edu/archives/sum2011/entries/genotype-phenotype/.

56. G. Grössing, On the thermodynamic origin of the quantum potential, arXiv:0808.3539 [quant-ph] (2008).

57. L. de la Peña and A. M. Cetto, *The Quantum Dice: An Introduction to Stochastic Electrodynamics* (Kluwer Academic, Dordrecht, 1996).

58. D. Bohm, Non-locality in the stochastic interpretation of the quantum theory, *Ann. I. H. P., Sect. A* **49**, 287–296 (1988).

59. B. Hall, *Energy and Momentum Conservation in Bohm's Model for Quantum Mechanics*, PhD thesis, University of Western Sydney, Australia (2004).

60. L. de la Peña and A. M. Cetto, Contribution fron stochastic electrodynamics to the understanding of quantum mechanics, arXiv: quant-ph/0501011 (2005).

61. P. W. Milonni, *The Quantum Vacuum: An Introduction to Quantum Electrodynamics* (Academic Press, San Diego, 1994).

62. H. De Raedt, K. De Raedt, K. Michielsen, K. Keipema and S. Miyashita, Event-by-event simulation of quantum phenomena: application to Einstein-Podolsky-Rosen-Bohm experiments, *J. Comput. Theor. Nanosci.* **4**, 957–991 (2007).
63. L. Accardi, Topics in quantum probability, *Phys. Rep.* **77**, 169–192 (1981).
64. T. M. Nieuwenhuizen, Is the contextuality loophole fatal for the derivation of Bell inequalities? *Found. Phys.* **41**, 580–591 (2011).
65. A. Khrennikov and Y. Volovich, Discrete time dynamical models and their quantum-like context-dependent properties, *J. Mod. Opt.* **51**, 1113–1114 (2004).
66. D. Dürr, S. Goldstein and N. Zanghi, Bohmian mechanics and the meaning of the wave function, arXiv:quant-ph/9512031 (1995).
67. J. S. Bell, Against 'measurement', in ed. A. Miller, *Sixty-Two Years of Uncertainty*, pp. 17–30. Plenum, New York (1990).
68. M. A. Nowak and K. Sigmund, Evolutionary dynamics of biological games, *Science* **303**, 793–799 (2004).
69. J. H. Holland, *Adaptation in Natural and Artificial Systems* (University of Michigan, Ann Arbor, MI, 1975).
70. J. Maynard Smith, The theory of games and the evolution of animal conflicts, *J. Theor. Biol.* **47**, 209–221 (1974).
71. S. M. Kay, *Fundamentals of Statistical Signal Processing: Estimation Theory* (Prentice Hall, 1993).
72. C. R. Rao, Information and accuracy attainable in the estimation of statistical parameters, *Bull. Calcutta Math. Soc.* **37**, 81–91 (1945).
73. S. Amari, *Differential-Geometrical Methods in Statistics* (Springer-Verlag, Berlin, 1985).
74. A. Wagner, The role of randomness in Darwinian evolution, *Philosophy of Science* **79**, 95–119 (2012).
75. C. Adami, What is complexity? *BioEssays* **24**, 1085–1094 (2002).
76. C. H. Bennett, Logical depth and physical complexity, in ed. R. Herken, *The Universal Turing Machine. A Half-Century Survey*, pp. 227–257. Oxford University Press, Oxford (1988).
77. B. R. Frieden, Fisher information as the basis for the Schrödinger wave equation, *Am. J. Phys.* **57**, 1004–1008 (1989).
78. B. R. Frieden and B. H. Soffer, Lagrangians of physics and the game of Fisher-information transfer, *Phys. Rev. E* **52**, 2274–2286 (1995).
79. S. A. Frank, Natural selection maximizes Fisher information, *J. Evol. Biol.* **22**, 231–244 (2009).
80. D. Bohm and J. P. Vigier, Model of the causal interpretation of quantum theory in terms of a fluid with irregular fluctuations, *Phys. Rev.* **96**, 208–216 (1954).
81. C. J. van Rijsbergen, *The Geometry of Information Retrieval* (Cambridge University Press, New York, 2004).
82. M. Lienert, Pilot wave theory and quantum fields (2011). http://philsci-archive.pitt.edu/id/eprint/8710.
83. J. Summhammer, Quantum theory as efficient representation of probabilistic information, arXiv:quant-ph/0701181 (2007).
84. J. Summhammer, Maximum predictive power and the superposition principle, *Int. J. Theor. Phys.* **33**, 171–178 (1994).
85. D. Pavlovic, On quantum statistics in data analysis, arXiv:0802.1296 [cs.IR] (2008).
86. S. Aerts, An operational characterization for optimal observation of potential properties in quantum theory and signal analysis, *Int. J. Theor. Phys.* **47**, 2–14 (2008).

87. A. Ashtekar and T. A. Schilling, Geometrical formulation of quantum mechanics, arXiv:gr-qc/9706069 (1997).
88. D. C. Brody and L. P. Hughston. Geometric models for quantum statistical inference. http://theory.ic.ac.uk/brody/DCB/dcbcp2.pdf.
89. R. A. Fisher and K. Mather, The inheritance of style length in Lythrum salicaria, *Ann. Eugenics* **12**, 1–23 (1943).
90. J. M. Honig, The role of Fisher information theory in the development of fundamental laws in physical chemistry, *Journal of Chemical Education* **86**, 116–119 (2009).
91. P. M. Morse and H. Feshbach, *Methods of Theoretical Physics, Part I* (McGraw-Hill, New York, 1953).
92. M. Reginatto, Derivation of the equations of nonrelativistic quantum mechanics using the principle of minimum Fisher information, *Phys. Rev. A* **58**, 1775–1778 (1998).
93. S. Goldstein, Bohmian mechanics and quantum information, *Found. Phy.* **40**, 335–355 (2010).
94. D. Ross, Game theory, in ed. Edward N. Zalta, *The Stanford Encyclopedia of Philosophy* (Spring 2006 Edition). http://plato.stanford.edu/archives/spr2006/entries/game-theory/.
95. S. K. Ghosh and B. M. Deb, Densities, density-functionals and electron fluids, *Phys. Rep.* **92**, 1–44 (1982).
96. J. Bub, Quantum Entanglement and Information, in ed. Edward N. Zalta, *The Stanford Encyclopedia of Philosophy* (Winter 2010 Edition). http://plato.stanford.edu/archives/win2010/entries/qt-entangle/.
97. R. Horodecki, P. Horodecki, M. Horodecki and K. Horodecki, Quantum entanglement, *Rev. Mod. Phys.* **81**, 865–942 (2009).
98. E. T. Jaynes, Scattering of light by free electrons, in eds. D. Hestenes and A. Weingartshofer, *The Electron*, pp. 1–20. Kluwer, Dordrecht (1991).
99. Z. Huang, H. Wang and S. Kais, Entanglement and electron correlation in quantum chemistry calculations, *J. Mod. Opt.* **53**, 2543–2558 (2006).
100. A. Valentini, Astrophysical and cosmological tests of quantum theory, *J. Phys. A: Math. Theor.* **40**, 3285–3303 (2007).

Chapter 14

Reality, Causality, and Probability, from Quantum Mechanics to Quantum Field Theory

Arkady Plotnitsky*

Theory and Cultural Studies Program, Purdue University,
W. Lafayette, IN 47907, USA

These three lectures consider the questions of reality, causality, and probability in quantum theory, from quantum mechanics to quantum field theory. They do so in part by exploring the ideas of the key founding figures of the theory, such N. Bohr, W. Heisenberg, E. Schrödinger, or P. A. M. Dirac. However, while my discussion of these figures aims to be faithful to their thinking and writings, and while these lectures are motivated by my belief in the helpfulness of their thinking for understanding and advancing quantum theory, this project is not driven by loyalty to their ideas. In part for that reason, these lectures also present different and even conflicting ways of thinking in quantum theory, such as that of Bohr or Heisenberg vs. that of Schrödinger. The lectures, most especially the third one, also consider new physical, mathematical, and philosophical complexities brought in by quantum field theory vis-à-vis quantum mechanics. I close by briefly addressing some of the implications of the argument presented here for the current state of fundamental physics.

Prefatory remarks

These lectures are concerned with the philosophy of quantum physics. They represent, however, a different form of philosophy of physics, vis-à-vis most forms of the institutional philosophy of physics and specifically of quantum theory. This difference is reflected in my emphasis on *thinking* concerning quantum physics, both by the key figures considered in these lectures and on our own part. Physics is a product of human thought under complex material, technological, psychological, historical, and sociological conditions. Accordingly, one can pursue, as I shall do here, a philosophy of physics that attempts to understand how physicists think under these conditions, especially at the time of, and in the process of making, new discoveries. This approach is different from dealing, as is more common, especially, again, in the institutional philosophy of physics, with the logical-axiomatic structure of quantum theory or with broader epistemological or ontological questions, such as reality and causality. These questions are of course important, and they will be

---

*plotnits@purdue.edu

considered here, but considered as part of our *thinking*. Physics is thinking about nature, about what is true or probable about nature or those aspects of nature that physics considers. In all modern, post-Galilean physics, classical, relativistic, or quantum, this truth or probability is determined by means of mathematical theories connected to suitably configured numerical data obtained in experiments. How close we come to the ultimate constitution of nature in this way may depend on a given theory or on nature itself, on how far nature will allow our mathematical theories and experimental technologies to reach.

Given the aims and scope of these lectures, I will not be concerned with psychological and sociological aspects of quantum-theoretical thinking. On the other hand, history will play a significant role in my argument. History is unavoidable in physical thinking, which always builds on preceding thinking in physics, even at the time of new discoveries, however revolutionary or unexpected such discoveries may be. Every physical idea, again, no matter how original or new, has a history, some trajectories of which may be short and others long, sometimes extending to ancient thought. We create our ideas by engaging with this history, which help us to understand earlier ideas and to create our own, especially when these ideas are created by the likes of A. Einstein, N. Bohr, W. Heisenberg, E. Schrödinger, or P. A. M. Dirac, who are the main focus of these lectures. But they, too, created their ideas by engaging both more immediate history of these ideas, sometimes in each other works (as in Heisenberg's engagement with Bohr, or Dirac's with Heisenberg), and a longer history of physics and philosophy, in some respects going back as far as Aristotle and Plato, or even to the pre-Socratics.

A qualification is in order, however. When I speak of the thinking of any particular figure, such as any of those just mentioned, I do not claim to have a determinable access to this thinking. Such an access is limited even when the author is alive and could, in principle, provide one with as much information as possible concerning this thinking, or if the author had left an extensive record of this thinking, say, in letters and notes, that could supply this kind of information. Instead, I refer to thinking that one can follow and can engage in one's own thinking on the basis of certain works of a given author, and even then in a particular reading or interpretation of these works, which can be interpreted differently, and, thus, related to different ways of thinking. A proper name, such as Einstein, Bohr, Heisenberg, Schrödinger, or Dirac, is the signature underneath a given work or set of works, a signature that attests to one's role as a creator of these works, which serve as a guidance for thinking that we can pursue as a result of reading them. In the process, one can of course also gain insights into how a given author might actually have thought, but claims to that effect are, again, hard to make with certainty, although some among such claims may be probable and even highly probable. Be it as it may on this score, any theory or interpretation only becomes effective or, again, operative to begin with when it becomes part of our thinking, and a theory or interpretation advances physics, or the philosophy of physics, when it moves this thinking beyond itself. This does not mean that, helpful as they might be, the thinking and works of any particular author, no matter how great or important, or the understandings of this thinking and these works is the only path towards a better understanding of a given physical theory, here quantum theory, and, especially, advancing this understanding of the theory itself. When it comes to advancing physics one's faithfulness or loyalty to anyone's thinking becomes a secondary matter and only counts insofar as it helps this advancement. While my discussion of the figures considered here aims to be faithful to their thinking and writings as much as possible, as a matter of maintaining proper scholarly standards, and while these lectures are motivated by my belief in the helpfulness of their thinking for understanding and advancing quantum theory, this project is not driven by loyalty to their ideas. In part for that reason, these lectures also present different and even conflicting ways of thinking in quantum theory, such as that of Bohr and Heisenberg vs. that of Schrödinger. At a certain point, our thinking concerning quantum physics will inevitably have to move beyond the authors discussed here, and there is no special reason to assume that it will do so following the path established by any one of them. An entirely different trajectory, either already in place but unknown to us (some of us) or yet to be discovered, may be necessary for this task.

I thought that it would be more efficient to present my argument directly and avoid footnotes. All references are given at the end. These lectures are partly based on previously published works, in particular Ref. [1–4], which contain extended discussions of the subjects considered here and further references.

1. Bohr, Heisenberg, Schrödinger, and the epistemology of quantum mechanics

1.1. *Summary*

My point of departure and my first main subject in this lecture is the physical, mathematical, and philosophical significance of W. Heisenberg's discovery of quantum mechanics. This discovery was among the most momentous discoveries in the history of physics, comparable to Sir Isaac Newton's discovery of classical mechanics, J. C. Maxwell's discovery of his equations for electromagnetism, and A. Einstein's discoveries of special and then general relativity. However, and this fact grounds my main argument in this lecture, the relationships between the mathematical formalism invented by Heisenberg and the physical phenomena considered were entirely different in the case of Heisenberg's quantum mechanics, as he conceived of it, than in the case of other theories just mentioned or all physical theories prior to quantum mechanics. Unlike these theories, Heisenberg's new "calculus" did not describe (especially, causally) the behavior of quantum objects, but only related, in terms of probabilistic predictions, to quantum phenomena, manifested in measuring instruments impacted by quantum objects. As will explained later, the probabilistic character of quantum theory goes beyond that of classical statistical physics, in which the primitive individual constituents of the systems considered are, in contrast to primitive individual quantum systems, assumed to behave causally. Heisenberg's approach, influenced by Bohr's 1913 atomic theory and based in Bohr's correspondence principle, became then the foundation for Bohr's interpretation of quantum phenomena and quantum mechanics, in terms of what he called "complementarity." This interpretation was introduced by Bohr around 1927 and was developed by him over the next decade or so, in the course of his debate with Einstein, who found Heisenberg's and Bohr's approach to quantum mechanics and even quantum mechanics itself deeply unsatisfactory. I shall discuss the key features of Bohr's interpretation and the key questions at stake in this debate below. It is important, however, that the mathematical formalism of quantum mechanics can be developed from a different starting point and on more classical (realist and causal) lines, as was indeed done by the theory's cofounder, E. Schrödinger. This fact tells us that there may be more than one way of thinking that leads to a correct theory. Nor, it also tells us, can one be certain what theory, or what kind of theory, will be more effective in helping us to confront the new (or sometimes old) problems nature continuously poses to us. Accordingly, I shall devote the last part of this lecture to Schrödinger's thinking in his work on his wave mechanics. As is apparent in several other lectures in this summer school, this type of thinking continues to serve as inspiration for alternative approaches to quantum mechanics or to quantum phenomena themselves, resulting in theories that are different from quantum mechanics. I might add that the so-called dominance of "the Copenhagen interpretation" is largely a myth. Indeed, the very existence of "the Copenhagen interpretation" is a myth in itself, because there has never been a single such interpretation, even in Bohr's case, given that Bohr adjusted and even changed his views a few times. More important, however, is the myth of its dominance. To cite Bohr's own statement, made in 1949, after two decades of this presumed dominance: "I am afraid that I had [in advancing my views] only little success in convincing my listeners, for whom the dissent among the physicists themselves was naturally a cause of skepticism about the necessity of going so far in renouncing customary demands as regards the explanation of natural phenomena" (Ref. [5], v. 2, p. 63). In fact, if anything, this philosophical position, defining Bohr's ultimate interpretation of quantum phenomena and quantum mechanics (his previous views were somewhat less radical), is and has always been a minority view, although it is usually admitted by its opponents, beginning with Einstein, that this view is "logically possible without contradiction" (Ref. [6], p. 349).

1.2. Heisenberg: The correspondence principle, quantum variables, and probability

Heisenberg's thinking leading him to his discovery of quantum mechanics was defined by three key elements, manifested in his paper announcing his discovery (Ref. [7]). These elements are as follows:

(1) *The Mathematical Correspondence Principle.* Stemming from Bohr's correspondence principle, this principle states that one should maintain the consistency, "correspondence," between quantum mechanics and classical physics. Specifically, in cases, such as that of large quantum numbers for electrons in atoms (when electrons are far away from the nucleus), where one could use classical physics in dealing with quantum processes, the predictions of classical and quantum theory should coincide. The correspondence principle, used more heuristically by Bohr and others before quantum mechanics, was also given a more rigorous, mathematical form by Heisenberg. Thus understood, the principle requires recovering the variables of classical mechanics (the equations used are, again, formally the same) in the limited region where the old theory could be used in predicting quantum phenomena.

(2) *The Introduction of a New Type of Variables.* Heisenberg's discovery was characterized by the introduction of a new type of physical variables – matrix variables with complex coefficients, essentially operators in Hilbert spaces over *complex numbers* vs. classical physical variables, which are differential functions of *real* variables. This was a momentous mathematical move that has defined the mathematics of quantum theory from that point on and still continues to define it.

(3) *A Probabilistically Predictive Character of the Theory.* This change in *mathematical* variables was accompanied by a fundamental change in physics: the variables and equations of quantum mechanics no longer described, even by way of idealization, the properties and behavior of quantum objects themselves, in the way classical physics or relativity do for classical objects. Instead, the formalism only predicted the outcomes of events, in general probabilistically, even in the case of primitive (indecomposable) individual events, and of statistical correlations between some of these events, thus establishing the new type of relationship between mathematics and physics. Heisenberg, in this connection, spoke of "new kinematics" of quantum mechanics, vis-à-vis classical mechanics in Ref. [7]. Unlike in classical statistical physics, however, in this case (at least in the interpretation of quantum phenomena and quantum mechanics adopted here), one deals with "probability without causality."

This character of quantum mechanics became the foundation of Bohr's interpretation of the theory and of quantum phenomena themselves (by which, throughout these lectures, I mean those phenomena in considering which Planck's constant, h, must be taken into account). As Bohr said in the immediate wake of Heisenberg's discovery:

> In contrast to ordinary mechanics, the new mechanics does not deal with a space-time description of the motion of atomic particles. It operates with manifolds of quantities which replace the harmonic oscillating components of the motion and symbolize the possibilities of transitions between stationary states in conformity with the correspondence principle. These quantities satisfy certain relations

which take the place of the mechanical equations of motion and the quantization rules. (Ref. [5], v. 1, p. 48)

This suspension of "a space-time description of the motion of atomic particles" in quantum mechanics or in the corresponding interpretation of the theory (and then, as will be discussed in Lecture 3, quantum field theory, or, again, of the corresponding interpretation of it) is arguably the most controversial feature of the theory. It was famously unacceptable to Einstein and many others who followed him, such as Schrödinger, the co-founder of the theory, who, however, created his wave mechanics from a very different set of principles and by following a very different approach.

The probabilistic character of quantum mechanics is in accordance with the experimental data in question in quantum theory. For, identically prepared quantum experiments (in terms of the condition of the observable parts of the apparatuses involved), in general lead to different recordings of their outcomes, again, even in the case of primitive individual quantum objects, such as individual electrons, photons, and so forth, which fact makes predicting these outcomes unavoidably probabilistic. Certain quantum phenomena also exhibit statistical correlations not found in classical physics. In a way, quantum phenomena are more remarkable for these correlations than for the irreducible randomness of individual quantum events. Perhaps the greatest of many enigmas of quantum physics is how random individual events combine into (statistically) ordered multiplicities under certain conditions, such as those of the EPR (Einstein-Podolsky-Rosen) type experiments, considered in Bell's theorem or the Kochen-Specker theorem.

These circumstances do not exclude the possibility of causal or, in the first place, realist interpretations of quantum mechanics, or alternative causal and realist theories of quantum phenomena, which would give the underlying causality to quantum probability. Schrödinger, again, initially conceived of his wave mechanics on this model. He appears to have abandoned a belief that his equation itself can be understood in this way, and he brilliantly exposed the standard or, as it is sometimes called, orthodox (nonrealist and noncausal) view of quantum mechanics in his cat-paradox paper (Ref. [8]). The paper was written in response to A. Einstein, B. Podolsky, and N. Rosen's (EPR's) paper concerning the EPR experiment, which Schrödinger, in agreement with the authors, saw as posing a serious new challenge to quantum mechanics as a complete theory of quantum phenomena (Ref. [9]). It is less known that eventually Schrödinger returned to the possibility of interpreting his equation as a continuous wave equation describing quantum behavior of electrons in a realist and causal way, rather than, as in Bohr's and related interpretations, as a mathematical algorithm for predicting the outcomes of quantum experiments (Ref. [10]). Bohmian, hidden-variables theories (there are several versions of them) see or idealize the behavior of quantum objects in this way, but this idealization is achieved at the expense of nonlocality (in the sense of the existence of instantaneous physical connections between spatially separated events, which is in conflict with relativity), while the standard quantum mechanics may be seen as local. For the sake of full disclosure, there is some debate concerning this point in view of Bell's theorem and related findings, the validity of which has in turn been debated. It may be added that even in the case of Bohmian theories we cannot know the value of all the variables pertaining to quantum objects (the uncertainty relations still apply) and are limited to the statistical predictions of the outcomes of experiments, predictions that coincide with those of quantum mechanics. There have been a number of other attempts along these lines, and some of the lectures presented in this summer school are examples of such attempts. Apart from my discussion of Schrödinger's thinking concerning his wave mechanics, I shall not consider these attempts in these lectures. I will only be concerned with the standard versions of quantum mechanics and then quantum field theory, and the nonrealist and noncausal interpretations of these theories, especially following Bohr's and related interpretations of quantum phenomena and quantum mechanics, in "the Spirit of Copenhagen," as Heisenberg called it (Ref. [11], p. iv). The expression refers to a shared set of views on quantum physics and is preferable to "the Copenhagen interpretation," of which, as noted above, there are many, sometimes quite different, versions, even, again, in Bohr's own case. I would like to stress that I see Bohr's or any of these interpretations, as *an interpretation* of quantum phenomena, one of several possible interpretations, rather than "the truth of nature" (although Bohr sometimes appears to take a stronger position on this point). Accordingly, alternative interpretations of quantum phenomena and quantum mechanics, or alternative theories of quantum phenomena cannot be excluded and may be important for our understanding of quantum theory and for a future development of theoretical physics. Of the approaches presented in this summer school, the statistical or ensemble interpretation of quantum mechanics by A. E. Allahverdyan, R. Balian, and T. M. Nieuwenhuizen is the closest to the spirit of Copenhagen (Ref. [12]). The approach presented by A. Khrennikov here also shares certain features with Bohr's approach as concerns the role of measuring instruments in the constitution of the observed quantum phenomena, although Khrennikov's view of the ultimate reality of nature is different from that of Bohr.

With these qualifications in mind, I argue that Heisenberg's revolutionary thinking established a new way of doing theoretical physics, and, as a consequence, it redefined experimental physics as well. The practice of experimental physics no longer consists, as in classical experiments, in tracking the independent behavior of the systems considered, but in *unavoidably* creating configurations of experimental technology that reflect the fact that what happens is *unavoidably* defined by what kinds of experiments we perform, how we affect quantum objects, rather than only by their independent behavior. My emphasis on "unavoidably" reflects the fact that, while the behavior of classical physical objects is sometimes affected by experimental technology, in general we can observe classical physical objects, such as planets moving around the sun, without appreciably affecting their behavior. This does not appear to be possible in quantum experiments. That identically prepared quantum experiments lead to different outcomes, thus making our predictions unavoidably probabilistic, appears to be correlative to the irreducible role of measuring instruments in quantum experiments (e.g., Ref. [5], v. 2, p. 73).

The practice of theoretical physics no longer consists, as in classical physics or relativity, in offering an idealized mathematical description of quantum objects and their behavior. Instead it consists in developing mathematical machinery that is able to predict, in general (again, in accordance with what obtains in experiments) probabilistically, the outcomes of quantum events and of correlations between some of these events.

I shall now discuss Heisenberg's thinking and logic, just outlined, in more detail by following his argument in his paper announcing his discovery of quantum mechanics. In accordance with the main point of Bohr's correspondence principle (in its initial form), that in the regions where both classical and quantum theory could be used their predictions should coincide, Heisenberg took advantage of the following fact. In the region of large quantum numbers, corresponding to low energy levels, the behavior of electrons in the atom could be treated classically, and hence in accordance with the equations of classical mechanics, even though this behavior itself is quantum, which is an important point often stressed by both Bohr and Heisenberg. Heisenberg then assumed that *the same equations* should formally apply in the region of small quantum numbers. It follows, that this correspondence is not only matter of replacing Planck's constant, h, with zero in the classical case, but of maintaining the same relationships between the corresponding variables in quantum and classical theory, while using different types of variables for each. Dirac was among the first to make this point in his first paper on quantum mechanics: "*The correspondence between the quantum and classical theories lies not so much in the limiting agreement when h → 0 as in the fact that the mathematical operations on the two theories obey in many cases the same [formal] laws*" (Ref. [13], p. 315).

Although, especially in retrospect, Heisenberg's move may appear natural, it was entirely unexpected at the time, because it was well known that these equations failed in the region of small quantum numbers. Previous (old-quantum-theoretical) approaches to quantum phenomena, including those, such as that of Bohr or A. Sommerfeld, guided by the correspondence principle (in Bohr's initial version), were based on modifying classical equations to fit quantum data, while, however, retaining classical variables. Heisenberg, by contrast, came to realize that classical equations failed not because of the equations themselves but because of classical variables still used in the old quantum theory. Hence, his second, even more revolutionary, move was the introduction of new (matrix) variables to which he applied these equations. Heisenberg, accordingly, spoke of "new kinematics" of quantum mechanics. In the process, Heisenberg reinvented matrix algebra, the existence of which he was apparently unaware, and de facto, as became clear shortly thereafter, he introduced his variables as Hilbert-space operators, now known as "observables."

There was, however, a price to pay, which Heisenberg or Bohr was ready to do, but which many, beginning, again, with Einstein, found exorbitant and even unacceptable. Heisenberg's mechanics offered no description of the quantum-level behavior of electrons (in the way classical mechanics would for classical objects), but only predicted, in general probabilistically, the outcome of possible future events on the basis of the data obtained from the observation of earlier quantum events. I shall now trace these ideas in Heisenberg's paper introducing quantum mechanics. Heisenberg begins by stating: "[I]n quantum theory it has not been possible to associate the electron with a point in space, considered as a function of time, by means of observable quantities. However, even in quantum theory it is possible to ascribe to an electron the emission of radiation" (Ref. [7], p. 263). He then says: "In order to characterize this radiation we first need the frequencies which appear as functions of two variables. In quantum theory these functions are in the form [originally introduced by Bohr]:

$$\nu(n, n - \alpha) = \frac{W(n) - W(n - \alpha)}{h},\tag{1.1}$$

and in classical theory in the form

$$\nu(n, \alpha) = \alpha\nu(n) = \frac{\alpha}{h}\frac{dW}{dn}.\tag{1.2}$$

(Ref. [7], p. 263)

This difference leads to a difference between classical and quantum theories as concerns the combination relations for frequencies, which correspond to the Rydberg-Ritz combination rules. However, "in order to complete the description of radiation [in accordance with the Fourier representation of kinematic formulas] it is necessary to have not only frequencies but also the amplitudes" (Ref. [7], p. 263). The crucial point is that, in Heisenberg's theory and in quantum mechanics since then, these "amplitudes" are no longer amplitudes of any physical, such as orbital, motions, which makes the name "amplitude" itself an artificial, *symbolic* term. Quantum amplitudes are linked to the probabilities of transitions between stationary states: they were what we now call probability amplitudes. The corresponding probabilities are derived by a form of Born's rule for this limited case (Born's rule applies more generally). Heisenberg says:

> The amplitudes may be treated as complex vectors, each determined by six independent components, and they determine both the polarization and the phase. As the amplitudes are also functions of the two variables n and α, the corresponding part of the radiation is given by the following expressions: Quantum-theoretical:

$$\text{Re } A(n, n - \alpha)e^{i\omega(n, n - \alpha)t}.\tag{1.3}$$

Classical:

$$\operatorname{Re} A(n)e^{i\omega(n)\alpha t}. \tag{1.4}$$

(Ref. [7], p. 263)

The problem – a difficult and, "at first sight," even insurmountable problem – is now apparent: "[T]he phase contained in A would seem to be devoid of physical significance in quantum theory, since in this theory frequencies are in general not commensurable with their harmonics" (Ref. [7], pp. 263–264). Heisenberg now proceeds by inventing a new theory around the problem that appears to be insurmountable and is insurmountable within the old theory. It is a question of changing the perspective completely. Most crucial is, again, that the new theory offers the possibility of rigorous predictions of the outcomes of the experiments at the cost of abandoning the physical description of the ultimate objects considered, no longer seen as a problem but instead as a way to the solution. Heisenberg says: "However, we shall see presently that also in quantum theory the phase had a definitive significance which is *analogous* to its significance in classical theory" (Ref. [7], p. 264; emphasis added). "Analogous" could only mean here that the way it functions mathematically is analogous to the way the classical phase functions mathematically in classical theory, or analogous in accordance with the *mathematical* form of the correspondence principle, as defined above. Physically there is no analogy. As Heisenberg explains, if one considers a given quantity $x(t)$ [a coordinate as a function of time] in classical theory, this can be regarded as represented by a set of quantities of the form

$$A_\alpha(n)e^{i\omega\alpha t}, \tag{1.5}$$

which, depending upon whether the motion is periodic or not, can be combined into a sum or integral which represents $x(t)$:

$$x(n,t) = \sum_\alpha A_\alpha(n)e^{i\omega(n)\alpha t}, \tag{1.6}$$

or

$$x(n,t) = \int A_\alpha(n)e^{i\omega(n)\alpha t}d\alpha. \tag{1.7}$$

(Ref. [7], p. 264)

Heisenberg is now ready to introduce his most decisive and most extraordinary move. He first notes that "a similar combination of the corresponding quantum-theoretical quantities seems to be impossible in a unique manner and therefore not meaningful, in view of the equal weight of the variables n and $n - \alpha$ ...

"However," he says, "one might readily regard the ensemble of quantities $A(n, n - \alpha) \exp i\omega(n, n - \alpha)t$ [an infinite square matrix] as a representation of the quantity $x(t)$" (Ref. [7], p. 264). The arrangement of the data into square tables is a brilliant and, in retrospect but only in retrospect, natural way to connect the relationships (transitions) between two stationary states, and it is already a great invention. However, it does not by itself establish an *algebra* of these arrangements, for which one needs to find the rigorous rules for adding and multiplying these elements – rules without which Heisenberg cannot use his new variables in the equations of the new mechanics. To produce a *quantum-theoretical interpretation* (which, again, abandons motion and other concepts of classical physics at the quantum level) of the classical equation of motion that he considered, as applied to these new variables, Heisenberg needs to be able to construct the powers of such quantities, beginning with $x^2(t)$. The answer in classical theory is of course obvious and, for the reasons just explained, obviously unworkable in quantum theory. Now, "in quantum theory," Heisenberg proposes, "it seems that the simplest and most natural assumption would be to replace classical [Fourier] equations ... by

$$B(n, n - \beta)e^{i\omega(n,n-\beta)t} = \sum_{\alpha=-\infty}^{\infty} A(n, n - \alpha)A(n - \alpha, n - \beta)e^{i\omega(n,n-\beta)t}, \qquad (1.8)$$

or

$$= \int_{-\infty}^{\infty} A(n, n - \alpha)A(n - \alpha, n - \beta)e^{i\omega(n,n-\beta)t}d\alpha. \qquad (1.9)$$

(Ref. [7], p. 265)

This is the main postulate, the (matrix) multiplication postulate, of Heisenberg's new theory, "and in fact this type of combination is an almost necessary consequence of the frequency combination rules" (equation [1] above) (Ref. 7, p. 265; emphasis added). This combination of the particular arrangement of the data and the construction, through physics, of an algebra of multiplying his new variables is his great invention. Although it is commutative in the case of squaring a given variable, x^2, this multiplication is in general noncommutative, expressly for position and momentum variables, and Heisenberg in effect (without realizing it) uses this noncommutativity in solving his equation. The overall scheme amounts to the Hilbert-space formalism, with Heisenberg's matrices as operators, introduced by J. von Neumann shortly thereafter.

In sum, while the classical equations of motion are retained, the variables and rules of mathematically manipulating them are replaced, as are the rules on relating the resulting equations to the experiments. The latter are no longer based on describing the physical behavior (motion) of quantum objects, but only relating to the probabilistic outcome of the corresponding experiments, even in the case of individual quantum events, which is a crucial difference from classical mechanics. Planck's constant, h, enters the scheme as part of this new relation between the data in question and the mathematics of the theory. The nature of the mathematics used, that of the infinite-dimensional Hilbert spaces over complex numbers, already makes it difficult to establish such descriptive relations. On the other hand, it is possible to establish the relation to probabilities of the outcomes of the relevant experiment, via Born's or related rules, which allows us to move from complex quantities of the formalism to positive real numbers between zero and one. As stated from the outset, the implications of this epistemology are radical for our understanding of both the nature and the practice of theoretical and experimental physics alike, and the situation takes an even more radical form in quantum field theory. I have started my discussion with Bohr's commentary on Heisenberg's approach, written in the wake of Heisenberg's paper and Born and Jordan's first paper, which developed Heisenberg's argument just outlined into a full-fledged matrix mechanics, but before Schrödinger introduced his wave mechanics. It might be appropriate, before moving to Bohr's interpretation of quantum phenomena and quantum mechanics, to cite Bohr's closing remark in the same 1925 article, "Atomic Theory and Mechanics":

> It will interest mathematical circles that the mathematical instruments created by the higher algebra play an essential part in the rational formulation of the new quantum mechanics. Thus, the general proofs of the conservation theorems in Heisenberg's theory carried out by Born and Jordan are based on the use of the theory of matrices, which go back to Cayley and were developed especially by Hermite. It is to be hoped that a new era of mutual stimulation of mechanics and mathematics has commenced. To the physicists it will at first seem deplorable that in atomic problems we have apparently met with such a limitation of our usual means of visualization [intuitable representation]. This regret will, however, have to give way to thankfulness that mathematics in this field, too, presents us with the tools to prepare the way for further progress.

(Ref. [5], v. 1, p. 51)

The subsequent history has proven that Bohr was too optimistic as concerns the physicists' attitude. There has been much "thankfulness that mathematics in this field, too, presents us with the tools to prepare the way for further progress." On the other hand, the discontent with "the limitation" in question has never subsided and is still with us now. Einstein led the way. He did not find satisfactory or even acceptable either this state of affairs as concerns physics or this type of use of mathematics in physics. Schrödinger was quick to join, with many, indeed a substantial majority of, physicists and philosophers to follow. However one views the situation on that score, one has to appreciate Bohr's carefulness and precision in this statement, beginning with stressing the essential nature of the mathematics in question for quantum mechanics, especially for the rigorous proof of the conservation theorems. It is also significant that Bohr speaks of "a new era of mutual stimulation of *mechanics* and mathematics," rather than physics and mathematics, although Heisenberg's discovery redefines the relationships between them as well. At stake are *individual* quantum processes and events. The mathematical science, both descriptive and predictive, of these processes in classical physics is mechanics. It is, accordingly, classical mechanics that is now replaced by quantum mechanics, but as a theory that is only able to predict, in general in probabilistic terms, such individual events as effects of certain processes without describing these processes.

1.3. Bohr: Quantum objects, measuring instruments, and complementarity

I shall now explain Bohr's interpretation of quantum phenomena and quantum mechanics. This interpretation was based on Heisenberg's approach, just discussed, but it was developed over the course of two decades, following the development of quantum theory (including quantum electrodynamics and quantum field theory) and Bohr's debate with Einstein, which compelled Bohr to adjust and in some respects change his views. I can only present here a summary of Bohr's interpretation, and moreover, of the ultimate version of this interpretation, formulated sometimes around 1938, again, under the impact of his debate with Einstein, by this time specifically in view of their exchanges concerning the EPR experiment, introduced in 1935. I have offered a full-fledged treatment of Bohr's interpretation and his philosophy in general in Ref. [3].

Given the significance of the Bohr-Einstein debate in shaping Bohr's view, it may be appropriate to begin by briefly outlining Einstein's philosophical position. In characterizing Einstein's requirements for a proper, "complete," physical theory W. Pauli correctly notes the primary significance of the realist, rather than only causal, character of a physical theory for Einstein (Ref. [14], p. 131). Einstein would prefer causality as well, in accordance with the epistemological structure of both classical physics and relativity, but Pauli is right to stress the precedence of realism over causality for Einstein. Pauli cites one of Einstein's many statements describing his philosophical position: "There is such a thing as a real state of a physical system, which exists objectively, independently of any observation or measurement, and *can in principle be described by the methods of expression of physics*" (Ref. [14], p. 131; emphasis added; also Ref. [9], pp. 138–139). The last part of Einstein's statement is crucial (hence my emphasis), although the concept of the real state of a physical system, if conceived, as it appears to have been by Einstein, on the model of classical physics, already implies a particular, and specifically mathematical (although not only mathematical), form of expression in physics. The role of this expression is important for Einstein, because it can only be achieved by means of a free conceptual construction, rather than by means of observable facts themselves, which Einstein sees as the empiricist or positivist "philosophical prejudice," found, for example, in Ernst Mach's philosophy. "Such a misconception," he adds, "is possible because one does not easily become aware of the free choice of such concepts, which, through success and long usage, appear to be immediately connected with the empirical material" (Ref. [15], p. 47). Thus, Einstein is a realist, and not a positivist, and moreover not a naïve realist, given the role he assigns to "the free choices of ... concepts" in creating our theories.

It is, Einstein argues, this kind of expression that quantum mechanics fails to deliver, which makes it incomplete. Einstein is factually correct, insofar as quantum mechanics, at least in Bohr's and related interpretations, does not provide a description or even conception of individual quantum objects and processes. The

A. Plotnitsky

question, however, is whether nature allows us this kind of expression in the case of quantum phenomena or not. Einstein thought it should. Bohr's view was that it *might not*, which, importantly, is not the same as to say that it *never will*. This view allows Bohr to argue that quantum mechanics is complete within its proper scope: as complete as, within this scope, nature allows a theory of quantum phenomena to be, as things stand now, at least, if such a theory is local, the point equally crucial to Bohr and Einstein. Pauli sees Einstein's "requirement" as an expression of "a special ideal which is satisfied in classical particle mechanics and [classical] electrodynamics, and also in the theory of relativity, but not in the equally objective description of nature given by quantum mechanics. Einstein always emphasized anew that on this account he regards quantum mechanics as *incomplete*, and is unwilling to abandon the hope of a completion of quantum mechanics which would again restore his narrow reality requirement" (Ref. [14], pp. 131–132). By contrast, in agreement with Heisenberg, Born, and Bohr, Pauli "do[es] not share these misgivings, and regard[s] as definitive just this step, which was made in quantum mechanics, of including the observer and the conditions of experiment in a more fundamental way in the physical explanation of nature" (Ref. [14], p. 132). Pauli's invocation of "the equally objective" character of the quantum-mechanical understanding of nature is important, and I shall return to this point presently. Pauli's statement especially refers to Bohr's argument concerning the irreducible role of measuring instruments and his concept of phenomenon, which Bohr was compelled to introduce around 1949, under the impact of his, by then a decade long, debate with Einstein. According to Bohr:

> I advocated the application of the word phenomenon exclusively to refer to the observations obtained under specified circumstances, including an account of the whole experimental arrangement. In such terminology, the observational problem is free of any special intricacy since, in actual experiments, all observations are expressed by unambiguous statements referring, for instance, to the registration of the point at which an electron arrives at a photographic plate. Moreover, speaking in such a way is just suited to emphasize that the appropriate physical interpretation of the symbolic quantum-mechanical formalism amounts only to predictions, of determinate or statistical character, pertaining to individual phenomena appearing under conditions defined by classical physical concepts [describing measuring instruments]. (Ref. [5], v. 2, p. 64)

Bohr sees quantum-mechanical formalism as "symbolic" because, in this view, while the theory uses mathematical symbols analogous to those used in classical mechanics, quantum mechanics, in Bohr's interpretation, does not refer these symbols to, and hence does not describe the actual behavior of, quantum objects. Instead, the formalism only predicts, in general probabilistically, what is observed, as phenomena, in measuring instruments under the impact of quantum objects. This view and Bohr's use of the term "symbolic" in this sense emerges beginning with Heisenberg's introduction of his "new kinematics," as considered earlier. Bohr's concept of phenomena includes a rigorous specification of each arrangement, determined by the type of measurement or prediction we want to make, which specification also reflects the irreducibly individual, unique character of each phenomenon. It is also crucial that the term refers "to the observations [already] *obtained* under specified circumstances" and hence to already *registered* phenomena, rather than to what can be predicted. For one thing, such predictions are always probabilistic and, hence, what will happen can never be assured, unlike, at least ideally or in principle, in classical mechanics. We always have a free choice as concerns what kind of experiment we want to perform, in accordance with the very idea of experiment, which, as Bohr notes, defines classical physics as well (Ref. [16], p. 699). Contrary to the case of classical physics, however, implementing our decision concerning what we want to do will allow us to make only a certain type of prediction (for example, that concerning a future position measurement) and will unavoidably exclude the possibility of certain other, complementary, types of prediction (in this case, that concerning a future momentum measurement). Bohr's *concept* of complementarity reflects this kind of mutual exclusivity of certain situations of measurement or phenomena. I emphasize "concept" because eventually Bohr came to call "complementarity" his overall interpretation, based on this concept, of quantum phenomena and quantum mechanics. Complementarity is defined by (a) a mutual exclusivity of certain phenomena, entities, or conceptions; and yet

(b) the possibility of applying each one of them separately at any given point; and

(c) the necessity of using all of them at different moments for a comprehensive account of the totality of phenomena that we must consider.

Parts (b) and (c) of this definition are just as important as part (a), and to miss them, as is often done, is to miss much of the import of Bohr's concept. This definition is very general and allows for different instantiations of the concept in quantum theory and enables the application of the concept beyond physics.

In Bohr's interpretation, one only deals with complementary phenomena manifest in measuring instruments under the impact of quantum objects. One never deals with complementary properties of quantum objects themselves or their independent behavior, given that, in Bohr's view, no attribution of any such properties, single or joint, is ever possible. Indeed, no such complementary arrangements or phenomena can even be associated with a single quantum object either. One would always require two quantum objects in order to enact, in two separated experiments, two complementary arrangements, say, those associated with the position and the momentum measurement, respectively, with, in each case, the measurement itself physically pertaining strictly to the measuring instruments. Accordingly, as explained above, in this view the uncertainty relations apply to the corresponding variables physically pertaining to measuring instruments and not to quantum objects.

By the same token, experimental decisions that we make shape what will and will not happen (even if we can only expect any actual outcome with a certain degree of probability), as opposed to, in general, merely tracking, as in classical physics, what is bound to happen in any event. Thus, given that, in classical mechanics, both the position and the momentum of a given object are always determined at any given point (thus also ensuring the causal nature of the evolution of this object), which we measure at any given point does not affect the value of the other. But in quantum physics this is not the case. This is why quantum phenomena are subject to the uncertainty relations, which are independent of quantum mechanics, although they can be derived from it. Indeed, in Bohr's interpretation, the uncertainty relations mean that one cannot even define, rather than only measure, both variables simultaneously. Probability, to which the uncertainty relations are correlative, is the cost of our active, shaping role in defining physical events in quantum physics, rather than merely tracking them, as we do in classical physics.

Importantly, our freedom of choosing the experimental setup only allows us to select and control the initial setting up of a given experiment but not its outcome, which, again, can only be probabilistically estimated. This fact further reflects the "objectivity" of the situation, defined by the verifiability and, thus, the possibility of the unambiguous communication of the data involved in our experiments (both that of their setup and their outcome), and hence, the objective character of quantum mechanics in this interpretation. Bohr summarizes the argument just outlined in his 1955 "Unity of Knowledge:"

A most conspicuous characteristic of atomic physics is the novel relationship between phenomena observed under experimental conditions demanding different
elementary concepts for their description. Indeed, however contrasting such experiences might appear when attempting to picture a course of atomic processes
on classical lines, they have to be considered as complementary in the sense that
they represent equally essential knowledge about atomic systems and together
exhaust this knowledge. The notion of complementarity does in no way involve
a departure from our position as *detached observers* of nature, but must be regarded as the logical expression of our situation as regards objective description
in this field of experience. The recognition that the interaction between the measuring tools and the physical systems under investigation constitutes an integral
part of quantum phenomena has ... forced us ... to pay proper attention to the
conditions of observation. (Ref. [5], v. 2, p. 74; see also p. 73; emphasis added)

Thus, in Bohr's understanding of quantum phenomena, the observers are as de-
tached *vis-à-vis measuring instruments* as they are in classical physics *vis-à-vis classical objects*, thus ensuring the objectivity of Bohr's scheme. On the other hand, the measuring instruments used in quantum measurement can, in an act of observation or measurement, never be "detached" from quantum objects because the latter cannot be "extracted from" the *closed* observed phenomena (in Bohr's sense) containing them (Ref. [5], v. 2, p. 73). Phenomena *cannot be opened* so as to reach quantum objects by disregarding the role of measuring instruments in the way it is possible in classical physics or relativity, and thus are in conflict with Einstein's

ideal of objectivity. Hence, although quantum objects do exist independently of us and of our measuring instruments, they can never be observed independently. Nobody has ever seen, at least not thus far, a moving electron or photon as such, but only traces of this "movement" (assuming even this concept applied), traces that, in view of the uncertainty relations, do not allow us to reconstitute this movement itself in the way it is possible in classical physics or relativity. In Bohr's interpretation, quantum phenomena strictly preclude any description or even conception of quantum objects themselves and their behavior, which behavior is, nevertheless, responsible for the emergence of these phenomena.

Accordingly, the physical quantities obtained in quantum measurements can no longer be assumed to represent the corresponding properties of quantum objects, even any single such property, rather than only certain joint properties, in accordance with the uncertainty relations. In Bohr's view, an attribution even of a single property to any quantum object as such is *never possible – before, during, or after measurement*. The conditions that experimentally obtain in quantum experiments only allow us to rigorously specify measurable quantities that can, in principle, physically pertain to measuring instruments and never to quantum objects. Even when we do not want to know the momentum or energy of a given quantum object and thus need not worry about the uncertainty relations, neither the exact *position* of this object itself nor the actual time at which this "position" is established is ever available and hence in any way verifiable. These properties, assuming they could be defined (as they can be in some interpretations of quantum mechanics or in Bohmian theories), are lost in "the finite [quantum] and uncontrollable interaction" between quantum objects and measuring instruments (Ref. [16], p. 697). However, this interaction leaves a mark in measuring instruments, a mark that can be treated as a part of a permanent, objective record, which can be discussed, communicated, and so forth. The uncertainty relations remain valid, of course, but they now apply strictly to the corresponding (classical) variables of suitably prepared measuring instruments, impacted by quantum objects. We can either prepare our instruments so as to measure a change of momentum of certain parts of those instruments or so as to locate a spot impacted by a quantum object, but never do both together, which makes the uncertainty relations correlative to the complementary nature of these arrangements or phenomena.

Bohr's epistemology, as just outlined, entails an understanding of randomness or chance (or quantum correlations, which are an important part of quantum phenomena) and probability that is different from that found in classical physics, including in classical statistical physics. This understanding is defined by the suspension of the applicability of the idea of causality even to individual quantum processes and events, and thus fundamentally departs from the view adopted in classical statistical physics, where the behavior of the individual entities comprising the multiplicities considered statistically is assumed to be causal, at least ideally. Our predictions themselves only concern the effects of such processes manifest in the measuring instruments involved. The probabilistic character of our predictions concerning quantum phenomena is unavoidable, because, in Bohr's words, "one and the same experimental arrangement may yield different recordings" (Ref. [5], v. 2, p. 73). It is, again, possible to speak of "one and the same experimental arrangement," because, unlike the outcomes of experiments, we can control the measuring instruments involved, given that the observable parts of these instruments relevant for setting up our experiments can be described classically. By contrast, the state of each quantum object under investigation in each repeated experiment (say, at the time when an electron or photon is emitted from the source considered) will not, in general, be identical. Under these conditions, the probabilistic character of such predictions will also concern primitive individual quantum events. For, unlike in the case of certain classical individual events, such as a coin toss, in the case of quantum phenomena it does not appear possible – and in Bohr's interpretation, it is in principle impossible – to subdivide these phenomena into entities of different kinds, concerning which our predictions could be exact, even ideally or in principle. Any attempt to do so will require the use of an experimental setup that leads to a phenomenon or set of phenomena of the epistemologically same type (they could be different physically), concerning which we could again only make probabilistic predictions.

Accordingly, as noted earlier, rather than whether quantum phenomena entail a probabilistic theory predicting them (they manifestly do), the question is whether there is or not an underlying classical-like causal dynamics ultimately responsible for such events. If this kind of underlying causal dynamics exists, it would imply that a classical-like account of such events could in principle eventually be developed. Bohr argued, however, that quantum phenomena may disable, and in his interpretation they do, the underlying assumptions of classical statistical physics (or accounts of individual classical phenomena that we cannot sufficiently track), based on the causality of the primitive individual processes involved. In Bohr's interpretation, the lack of causality in these processes is automatic in quantum physics because quantum objects and processes are beyond any description and even conception. Causality would be a feature of such a description and hence is disallowed automatically. As Schrödinger noted: "if a classical [physical] state does not exist at any moment, it can hardly change causally" (Ref. [8], p. 154). Schrödinger made this point by way of a very different assessment of this argumentation, which he saw as "a doctrine born of distress" (Ref. [8], p. 154), an assessment defined by his very different view of physics. It was the same view that earlier guided his invention of his wave mechanics, to which I now turn.

1.4. Schrödinger: Physical reality, quantum waves, and the optical-mechanical analogy

Discovered a few months after Heisenberg's matrix mechanics, Schrödinger's wave mechanics aimed at offering, and initially appeared to be able to offer, a theory that would be realist and causal. It was expected to be able, just as classical mechanics did, both to *describe* the physical processes at a subatomic level (as wave-like processes) *and predict*, on the basis of this description, the outcomes of the corresponding experiments. Schrödinger was aware of Heisenberg's and related work on matrix mechanics, and of the successes of the theory. Apart, however, from his discontent with the mathematical cumbersomeness of the matrix version and with its epistemological implications, which his wave mechanics would avoid, his path to his wave equation for the electron and then to his more ambitious program for a wave quantum mechanics was different. He proceeded from L. de Broglie's ideas concerning matter waves and related work by Einstein. Prior to Heisenberg's introduction of quantum mechanics, and for a while even following it, the situation appeared along the lines of Schrödinger's comment in a letter to W. Wien. Schrödinger wrote:

> This is an extremely funny thing. The contrast between light-quanta and wave radiation can be traced even in atoms as: (a) single orbiting electrons; or (b) a standing vibration of the whole atomic region. In the interpretation (a) [we have]: the Hamiltonian partial differential equation; the separation of variables; and the quantization in the well-known manner. In the interpretation (b) [we have]: the wave equation; the separation of variables in the old, well-known sense that the unknown function is assumed to be, i.e., the product of a function of [radial coordinate] r, of a function of [the angle] θ, and a function of [the angle] ϕ; searching for the "normal vibrations." The frequencies emitted appear as frequency differences, i.e., as beat frequencies of the normal frequencies. Something must be hidden behind that. I hope to formulate the results soon in an organized way. (Schrödinger to Wien, 8 January 1926; cited in Ref. [17], v. 5, p. 465; translation slightly modified)

Schrödinger's equation (initially in its time-independent form) was a fulfillment of this hope. It is worth noting that this insight suggests that, while mutually exclusive, the wave and particle "pictures" may be better seen as *equivalent*, rather than *complementary* in Bohr's sense. Bohr in fact avoids the wave–particle complementarity, never invoked as such in his works, even though it often serves as the main example of complementarity (see Ref. [3], pp. 44–45). From Bohr's perspective, either visualization could only be symbolic, rather than representational, insofar as one applies it to quantum process, rather than, as Bohr eventually came to do, to the phenomena observed in measuring instruments, while assuming that the ultimate nature of subatomic processes themselves is neither wave-like nor particle-like, or again, not subject to any description or even conception. On this view, in quantum physics we only deal with *discrete* multiplicities of, sometimes correlated, phenomena, as juxtaposed to quantum objects, which are placed beyond the reach of quantum theory or even our thought itself. That, however, was not something that Schrödinger was eager to accept. He never liked the probabilistic view of quantum mechanics either, including Born's interpretation of the wave function, offered in the wake of Schrödinger's discovery of his equation. He saw this interpretation as an extension of Bohr's atomic theory and then Heisenberg's approach, and hence in essential conflict with his own, or Einstein's, desiderata for quantum theory. As he wrote to Bohr: "What is before my eyes, is only one thesis: one should, even if a hundred trials fail, not give up the hope of arriving at the goal – I do not say by means of classical pictures, but by logically consistent conceptions – of *the real structure of space-time processes*. It is extremely probable that this is possible" (Schrödinger to Bohr, 23 October 1926, cited in Ref. [17], v. 5, p. 828; emphasis added).

Schrödinger's qualification to the effect that this "real structure of space-time processes" at the quantum level may be different from that provides by "classical pictures" is worth registering. Nevertheless, given the preceding history of quantum theory, leading to Heisenberg's discovery of matrix mechanics, one could have been skeptical concerning the suitability of Schrödinger's program for dealing with quantum phenomena. Arguably most significantly, the problem of discreteness of quantum phenomena had never found an adequate resolution within this program, which indeed aimed to dispense with any discreteness at the ultimate level. For example, it was quickly noted at the time, by Heisenberg in particular, that, by virtue of viewing the charge density as a classical source of radiation, Schrödinger's wave approach was in conflict with Planck's radiation law, with which Heisenberg's approach was consistent. These difficulties ultimately compelled Schrödinger's to abandon his project of wave mechanics, although, as noted above, in the late 1940s he returned to the view that the project could be viable, as he argued in Ref. [10].

Be it as it may, Schrödinger's thinking of the ultimate nature of the physical world in terms of waves or indeed *in* waves is significant, including as an instructive attempt, successful or not, to relate continuity and discontinuity in terms of underlying continuity. The problem or set of problems that Schrödinger posed might be described as follows. Could one find a wave-type equation that would describe physical processes at the quantum level, such as the behavior of an electron in the atom, and that would enable us to predict the results of quantum-mechanical experiments? What would the (wave-like) character of the processes corresponding to this equation be, given the peculiar features of quantum physics, in particular, discreteness and indeterminism, which correspond to the outcome of experiments and which, accordingly, must be retained? Can a wave-like theory solve "the quantum riddle," as it was to be called by Einstein, who welcomed Schrödinger's theory as an important step in this direction? Would such a solution allow us to capture, at least by way of an idealized approximation the ultimate reality of nature in a wave-like picture? While, especially in retrospect, Schrödinger's program appears to have been difficult to complete, if possible at all, the possibility of these questions reveals a number of deeper aspects of modern physics, classical and quantum, and of the relationships among physics, mathematics, and philosophy, which are important to quantum mechanics and debates concerning it.

Schrödinger's philosophy compelled him to postulate the underlying causal dynamics of fundamental constituents and processes in nature, although his position, just as that of Einstein, was not a naively realist one, and he was mindful of the approximate or idealized character of all our physical theories, as he explained it in his cat-paradox paper (Ref. [8], p. 152). This dynamics then may, under certain circumstances, lead to probabilistic outcomes in predicting actual situations, similarly to the way it happens when we use probability in classical physics. This assumption is in sharp contrast with the assumption of the irreducibly noncausal (or inconceivable in general) character of primordial individual processes or events, or indeed with any probabilistic or statistical physical theory that would not assume the underlying causal physical dynamics of the type found in classical physics, including classical statistical physics. Schrödinger never gave up on this classical idea, or ideal, the *classical ideal*, as he, again, saw it in the cat-paradox paper, juxtaposing it to the *nonclassical* "doctrine" defining quantum mechanics in the spirit of Copenhagen, "the doctrine ... born of distress" (Ref. [8], p. 152). By then, in 1935, the mathematics of the doctrine would include his equation, no longer seen by him as offering a hope for an alternative to the "doctrine." This hope returned later on in Ref. [10]. Indeed, this return to his initial program might even have been initiated by EPR's paper, to which Schrödinger's cat-paradox paper responds. Be that as it may, in 1926 it appeared possible to provide the underlying causality to quantum phenomena by means of wave rather than particle motion of quantum objects, even though the observed phenomena in question would appear to correspond to particles. In other words, along with probabilistic effects and, in part correlatively, particles, too, are *effects* of the underlying "universal radiation," "the wave radiation forming the basis of the universe," as he put it (Ref. [18], p. 95; cited in Ref. [17], v. 5, p. 435]). *The wave radiation forming the basis of the universe* – no less! The phrase is as remarkable for its philosophical ambition as for its physical one. Of course, Schrödinger's "wave radiation forming the basis of the universe" was unobservable, perhaps in principle unobservable, but that need not mean that it could not exist.

Schrödinger's next step toward his equation is captured by a passage in a letter to Wien, cited above, concerning the apparent equivalence of the two descriptions, in terms of the particle and the wave motion. It is important to keep in mind that, unlike de Broglie's propagating waves, Schrödinger's original (time independent) equation was written for a standing wave, "a standing vibration of the whole atomic region." Solving this time-independent equation gives one the hydrogen spectrum in a much more immediate and mathematically easier way then matrix mechanics did, a feature, also found in his mathematical program in general, that assured the immediate success of his approach vis-à-vis the matrix one. In addition, while in Schrödinger's "picture" there are no particles at the ultimate level, in de Broglie's and then Bohm's theory a wave would be accompanying the particle in question, in this case an electron, in the manner of a pilot wave. In Schrödinger's wave mechanics, there were only waves at the level of the ultimate constitution of nature, with "particles" seen as certain singularity-like surface effects. Schrödinger sometimes stopped short of claiming that these waves actually represent the ultimate reality of nature, as opposed to providing an intuitively accessible (*anschaulich*) theory sufficiently approximating, as an idealization, this reality. While, given that all theoretical physics only deals with idealized models, it may not perhaps otherwise in general, there is always a question of what kind of model one argues for, say, a discrete one vs. a continuous one. The theory Schrödinger envisioned would also possess a powerful predictive power, even if, at least initially, a limited descriptive capacity to capture the underlying reality of the ultimate constitution of nature – limited, but not completely absent in the way it was in Bohr's or Heisenberg's approach to quantum mechanics. It also allowed one, as did Dirac's q-number scheme (but not the matrix scheme of Heisenberg, Born, and Jordan), to offer a quantum-mechanical representation of stationary states. Bohr's 1913 theory describing stationary states in terms of classical object could no longer adequately apply, and some of such states could not be so described even approximately. In Heisenberg's theory there would be no waves, but at the cost, unacceptable to Schrödinger (or Einstein), of renouncing any description of physical processes concerning electrons in space-time, eventually leading Bohr to a nonclassical epistemology of quantum theory. Waves would either be used symbolically, in conjunction with Born's probability interpretation of the wave function, or, metaphorically, in relation to certain wave-*like* effects, comprised by multiple discrete individual phenomena in certain circumstances, such as the appearance of the interference pattern, which is wave-like, in the double-slit experiment.

There are well-known and *relatively* straightforward paths to Schrödinger's equation, and Schrödinger indicated some of them in his initial article. Perhaps the most natural is to derive Schrödinger's equation via de Broglie's formulas for phase waves associated with particles, which were crucial to Schrödinger's thinking and which he used in a derivation found in one of his notebooks. This is also one of the most common ways it is done it in textbooks on quantum mechanics. De Broglie's formula for the speed of the phase wave of an electron, adjusted for the speed of the electron in the electric field of a hydrogen nucleus, is inserted into the classical relativistic wave equation for the wave function, ψ. Since de Broglie's formula conveys both the particle and the wave aspects of the behavior of quantum objects, the nature of the equation changes. Unfortunately, the resulting equation, usually known as the Klein-Gordon equation (Schrödinger appears to be the first to have written it), does not work for a relativistic electron because the predictions one makes by using it are in conflict with experimental results. One needs Dirac's equation to make correct relativistic predictions. However, if, in the procedure just described, one drops terms that are small at the nonrelativistic limit, which is easily done mathematically, one arrives at a different equation. This is, in essence, what Schrödinger appears to have done initially, as his notebooks indicate. The resulting equation, which is Schrödinger's equation, happens to offer correct predictions of experimental results in the nonrelativistic case.

In terms of theoretical justification, the situation is far more complicated. These complications arise not only and not so much because it is a nonrelativistic treatment of an object that ultimately needs to be treated relativistically, although this is important. Perhaps most significant, however, is the following problem. One derives the *right* nonrelativistic equation from a *wrong* relativistic one, at a nonrelativistic limit. Accordingly, Schrödinger's (nonrelativistic) equation becomes more of a guess, albeit a correct guess, which would need to be justified otherwise or perhaps even derived otherwise. This is what Schrödinger attempted to do and, in some measure, accomplished in his published papers, which derive his equation differently, indeed in several alternative ways. Eventually, Dirac showed Schrödinger's equation to be the nonrelativistic limit of his equation as well, which straightened out the difficulty. As will be seen in Lecture 3 below, this was a crucial part of Dirac's derivation of his equation, because by that time Schrödinger's equation was established as the correct nonrelativistic equation for the electron, regardless of derivation. In any event, while Schrödinger's equation gave correct predictions at the nonrelativistic limit, the situation was more complex in terms of his theoretical argument concerning the equation.

Schrödinger began to lay down his more ambitious program for wave mechanics in his second paper on the subject, which also offered a new derivation of his equation as well, representative of this program. Schrödinger's approach was also different mathematically, beginning with the idea of using a wave equation to describe the behavior of electron phase waves in atoms, as against de Broglie's direct geometrical treatment of the motion of electrons. Schrödinger arrived at his view of the problem of quantization as an eigenvalue problem, treated, via his equation, by variational methods, with atomic spectra derived accordingly. If one has an equation, such as Schrödinger's equation, defined by an action of an operator, in this case the energy operator H, upon a vector variable x, by transforming this variable into $H(x)$, it may be possible to find a vector X (eigenvector) such that $H(X) = EX$, where E is a number (eigennumber). In the case of the time-independent (standing-wave) Schrödinger's equation for the hydrogen atom, the eigenvalues En are possible energy levels (corresponding to the stationary states) of the electron in the hydrogen atom, which gave the equation its truth and beauty, to put it poetically. The time-dependent Schrödinger's equation has a greater generality and in principle can be written for any quantum-mechanical situation. It is a separate question, what, if anything, the equation (or for the matter the time-independent Schrödinger's equation) *describes*. We do know, however, what it predicts.

As is clear from his notebooks, Schrödinger's alternative derivation was linked and possibly motivated by the fact that some of the predictions based on his wave equation coincided with those of Bohr's and Sommerfeld's semi-classical theory, which used the standard, classical-like, Hamiltonian approach to quantum mechanics. Schrödinger explained these relationships in a way that led him to the derivation of his equation that is found in his first paper. He did so by replacing, without a real theoretical justification from the first principles (which was never achieved), the mechanical Hamilton-Jacoby equation

$$H(q, \partial S/\partial q) = E, \tag{1.10}$$

with a wave equation by substituting $S = K \ln \psi$ (K is a constant that has a dimension of action). This was a radical step, which had little, if any physical justification but which led him, via a mechanical-optical analogy, to the equation,

$$\Delta \psi + (2m/K^2)(E - V)\psi = 0, \tag{1.11}$$

and then to the right equation for the nonrelativstic hydrogen atom,

$$\Delta \psi + (2m_e/K^2)(E + e^2/r)\psi = 0. \tag{1.12}$$

Here $K^2 = h^2/4\pi^2$, h is the Planck constant and m_e the electron mass. The Hamilton-Jacoby equation considered by Schrödinger was used in the old quantum theory, specifically by Sommerfeld and P. Epstein, as well as, with proper quantum-mechanical adjustments, by Born, Jordan, and Heisenberg in matrix mechanics and by Dirac in his version of quantum mechanics. Schrödinger's key step was made via the mechanical-optical analogy and the connections between the principle of least action in mechanics and Fermat's principle in optics, and it was to give to S the wave form by $S = K \ln \psi$. In the case of the hydrogen atom, one thus also replaces the causal mechanics of the particle motion with amplitudes and then probabilities, although Schrödinger did not realize this at the time. He aimed at a (wave-like) causal picture, but ultimately arrived elsewhere.

Although Schrödinger's wave-mechanics program was enthusiastically received by some, notably by Einstein, the skepticism toward the program from the Copenhagen-Göttingen side emerged just as quickly, in part in view of Heisenberg's matrix theory and its epistemological implications. Schrödinger's equation itself was welcomed by nearly everybody, including Bohr, Born, Dirac, Pauli, and eventually by Heisenberg, who was initially skeptical even concerning the equation, rather than only Schrödinger's wave program. Thus, according to Sommerfeld, writing to Schrödinger: "The difference of the points of departure between your and Heisenberg's approaches is peculiar in the light of the same results. Heisenberg starts from the epistemological postulate not to put more into the theory than can be observed. You put in all kinds of possible frequency processes, node lines and spherical harmonics. After our epistemological knowledge has been sharpened by relativity theory, the large, unobservable ballast in your presentation also seems to be suspicious for the time being" (Sommerfeld to Schrödinger, 3 February 1926, cited in Ref. [17], v. 5, p. 502).

Schrödinger defended his approach on this last point, but, curiously, by seeing Sommerfeld's objection in terms of "possibly unnecessary assumptions," rather than "unobservable ballast." This is an important difference, arising in part from

Schrödinger's view of the situation in terms of trajectories of motions, "the electrons orbits with their loops," as represented by the "the fundamental equations of mechanics" (Schrödinger to Sommerfeld, 20 February 1926, cited in Ref. [17], v. 5, p. 502). Orbits and all classical mechanical concepts of motion were abandoned by Heisenberg in view of the just about insurmountable problems they posed. Schrödinger was aware that his own scheme by no means revolved these problems, either at the time or later when his time-dependent equation was introduced, although he did hope to avoid these problems by re-conceiving all these processes in terms of wave motion. The passage from his paper, cited above, clearly suggests both his awareness of these problems and his hopes of resolving them by appealing to "*vibration process* in the atom, which would more nearly approach reality than electronic orbits, the real existence of which is being very much questioned today" (Ref. [19], p. 371, cited in Ref. [17], v. 5, p. 533]). His theory does not appear to have ever been able to fulfill this promise. Thus, Sommerfled's point was not really countered by Schrödinger's subsequent papers, in which he promised to provide "more general foundations of the theory" (Schrödinger to Sommerfeld, 20 February 1926, cited in Ref. [17], v, 5. p. 502).

Schrödinger pressed on with his program in his second paper, which, as Mehra and Rechenberg note, "establishes the foundations and the definite outlines of what was later [in his next communication] called 'wave mechanics'" (Ref. [17], v. 5, p. 533). The program was, as I noted, to be enacted through the mechanical-optical analogy, accompanying, since Hamilton's work, the Hamilton-Jacoby framework for classical mechanics, the connections to which Bohr, too, refers on a number of occasion. Schrödinger wanted to "throw *more* light on the *general* correspondence which exists between the Hamilton-Jacoby differential equation of a mechanical problem and the 'allied' *wave equation*," that is, Schrödinger's equation (Ref. [19], p. 13; cited Ref. [17], v. 5, p. 533; emphasis on "light" added). The mathematical procedures used in the first paper are now declared "unintelligible" and "incomprehensible," from, one presumes, physical and conceptual viewpoints. It is crucial to keep in mind that, as is clear from his second paper and his other papers on wave mechanics, and from his accompanying statements and correspondence, Schrödinger was aware that his wave mechanics would have to be different from previous wave theories. Accordingly, he anticipated that some changes of the physical concepts involved were likely to be necessary, in part in order to preserve the classical-like (causal and realist) character of the theory. It is worth citing Schrödinger's notebook comments written in preparation for his second paper:

> The somewhat dark connections between the Hamiltonian differential equation $[H(q, \partial S/\partial q) = E]$ and the wave equation $[\Delta\psi + (2m/K^2)(E - V)\psi = 0]$ must be clarified. This connection is not new at all; it was, in principle, already known to Hamilton and formed the starting point of Hamilton's theory, since Hamilton's variational principle has to be considered as *Fermat's principle* for a certain wave propagation in configuration space, and the partial differential equation

of Hamilton as *Huygens' principle* for exactly this wave propagation. Equation $[\Delta\psi + (2m/K^2)(E - V)\psi = 0]$ is nothing but – or better, just a possible – wave equation for this wave process. These things are generally known, but perhaps I should recall them at this point. (Ref. [20], Microfilm No. 40, Section 6; cited as Notebook II, p. 1 in Ref. [17], v. 5, p. 543).

The situation is, again, more complicated than Schrödinger makes it sound, especially if one takes into account subtler aspects of these connections, as they apply to the quantum-mechanical, instead of the classical, case. These connections became even darker as quantum mechanics developed, in spite of its successes as a physical theory. As Schrödinger wrote to Sommerfeld:

> The ψ-vibrations are naturally not electromagnetic vibrations in the old sense. Between them some *coupling* must exist, corresponding to the coupling between the vectors of the electromagnetic field and the four-dimensional current in the Maxwell-Lorentz equations. In our case the ψ-vibrations correspond to the four-dimensional current, that is, the four-dimensional current must be replaced by something that is derived from the function ψ, say the four-dimensional gradient of ψ. But all this is my fantasy; in reality, I have not yet thought about it thoroughly. (Schrödinger to Sommerfeld, 20 February 1926, cited in Ref. [17], v. 5, p. 542).

Schrödinger does, however, close his second paper with the following speculative conclusion, which envisions a wave mechanics:

> *We know today, in fact, that our classical mechanics fails for very small dimensions of the path and for very great curvatures.* Perhaps this failure is in strict analogy with the failure of geometrical optics, i.e. "the optics of infinitely small wavelengths," that become evident as soon as the obstacles or apertures are no longer great compared with the real, finite, wavelength. Perhaps our classical mechanics is the *complete* analogy of geometrical optics and as such is wrong and not in agreement with reality; it fails whenever the radii of curvature and dimensions of the path are no longer great compared with a certain wavelength, to which, in q-space, a real meaning is attached. Then it becomes a question of searching for an undulatory mechanics, and the most obvious way is the working out of the Hamiltonian analogy on the lines of undulatory optics. (Ref. [21], pp. 496–497, cited in Ref. [17], v. 5. p. 559)

The physical analogy with optics, thus, should be carefully distinguished from the Hamiltonian one. The latter analogy only gives one the wave function in a q-space; this function then should be related to some actual physical (vibrational) process. Related considerations are found throughout the paper. Schrödinger writes, for example: "We *must* treat the matter strictly on the wave theory, i.e., we must proceed from the *wave equation* and not from fundamental equations of mechanics, in order to form a picture of the manifold of possible processes" (Ref. [21], p. 506, cited in Ref. [17], v. 5, p. 569). Schrödinger also develops a reversed argument to the effect that the true mechanical processes in nature are represented by *the wave processes* in q-space and not by the motion of *image points* in that space. This argumentation reflects the possibilities (and hopes) for wave mechanics and the potential difficulties involved, and Schrödinger's awareness of these difficulties. Schrödinger offers a number of observations concerning the nature of quantum processes as reflected in his equation and comes close to the probabilistic interpretation of it, but never quite gets there. Perhaps he could not, given his overall philosophy and agenda. For, it follows from his equation that it is difficult and perhaps impossible to speak of the path of an electron in the atom, in accordance with the ideas of classical mechanics. In this respect Schrödinger was, as he acknowledged, in agreement with the views of Bohr, Heisenberg, Born, Jordan, and Dirac. He was reluctant, however, to accept the kind of suspension, let alone prohibition (this view came later, though) of any description of the underlying quantum behavior ("reality"), which ideal of classical physics he was not about to surrender. He was more ready to give up his equation, and, as he reportedly said to Bohr at some point, he was sorry to have discovered it, and for about two decades he did gave up on it as reflecting the ultimate workings of nature. But, as I said, in the later 1940s, he appears to have changed his mind yet again and return to his original views in Ref. [10].

What is one, then, to make of Schrödinger's work on wave mechanics, beyond acknowledging, as one must, his discovery of his equation as one the great achievements of the twentieth-century physics? One cannot, I think, hope for an easy answer, given, on the one hand, the labyrinthine complexity and contortions of his project, unavoidable given the difficulties of what Schrödinger wanted to achieve, and on the other, the many interpretive decisions involved in assessing the case. Such decisions would concern Schrödinger's own thinking, the nature of his equation, wave mechanics, quantum mechanics, classical physics, relationships between physics and mathematics, the history of quantum mechanics and of classical physics, and many other things. According to Mehra and Rechenberg:

> In the beginning of the new atomic theory there stood a wave equation for the specific example of the hydrogen atom. Schrödinger essentially guessed its structure and form [in his first paper]: its derivation or – more adequately – connection with the dynamical equation of the old quantum theory of atomic structure (working with the Hamilton-Jacoby partial differential equation) had been rather artificially forced. This was soon felt by Schrödinger himself ... Did the new formulation of undulatory mechanics lead in a less arbitrary and artificial way to wave equations that described atomic systems and processes, notably the successful nonrelativistic hydrogen equation? ... Was the wonderful analogy, between Hamiltonian mechanics and higher-dimensional non-Euclidean spaces and the undulatory optics, just highbrow idealistic decoration and useless for the practical purposes of atomic theory? (Ref. [17], v. 5, pp. 571–573)

Mehra and Rechenberg are right to reject "so pessimistic a view of [Schrödinger's] achievement," suggested by the last question (Ref. [17], v. 5, p. 573). They do not, however, explore the deeper complexities of the situation or the optical-mechanical analogy of Klein and Schrödinger as reflecting these complexities. I would like to briefly comment on the subject in order to give a clearer picture of the situation.

It is worth stating the main point of this analogy in classical mechanics: *Every Hamiltonian system of classical dynamics, such as that of one or an ensemble of particles, can be considered in terms of the motion of a wave front in a suitably chosen medium, although in general in a higher-dimensional space, rather than in a three-dimensional one in which actual physical processes occur.* The approach thus defined is powerful and effective in developing the mathematical formalism of classical mechanics. It may, however, also be misleading, even in the case of classical physics. For the optical (wave or geometrical) part of the argument is a mathematical generalization of the actual physical propagation of light in the three-dimensional space. It is, one might say, a metaphor. The "space" or "medium" of propagation, or for that matter the "light" itself in question, does not physically exist or have a proper physical meaning. It is not a physical space or a physical light. Even for the mechanical systems with only three degrees of freedom, in which case the configuration space is three-dimensional, one should not think that one could interpret the physical motion of a given mechanical (particle) system in optical, wave-like, terms, even though the relevant predictions concerning both "systems" would coincide. It is a mathematical, algorithmic coincidence, which to some degree misled Schrödinger in the case of certain simple quantum systems, where one finds an analogous (but, since the physics is now quantum, not identical) coincidence, as was often noted, including by both Bohr and Heisenberg (e.g., Ref. [5], v. 1, pp. 76–77).

Schrödinger was, again, aware of the difficulties of attributing reality to the waves in the configuration or phase space, and thought of such waves "as something real *in a sense*, and the constant h universally determined their frequencies or their wave length" (Schrödinger to Wien, 22 February 1926, cited in Ref. [17], v. 5, p. 536). It was real *"in a sense,"* a sense never really established. He was also careful to caution against using his analogy as that between mechanics and physical or undulatory optics, as opposed to that between mechanics and geometrical optics (Ref. [21], p. 495; cited in Ref. [17], v. 5, p. 558). In other words, one must find a relationship and, hopefully, a classical-like *correspondence* between these waves in the phase or configuration space and some physical vibrations – correspondence, but not identification. Schrödinger continued to believe in his program, again, in a certain general sense, that is, insofar as he seemed to have thought, with Einstein and on the model of general relativity, that (classical-like) fields should be seen as primary and perhaps ultimate physical entities anyhow. As is clear from the last section of his cat-paradox paper, Schrödinger was even more suspicious of quantum field theory than of quantum mechanics (e.g., Ref. [8], p. 167). Accordingly, Schrödinger appears to have been thinking along these more cautious lines of relationships between these "waves" and actual physical vibrations corresponding to them (perhaps indirectly) than in terms of identifying both. Such identification would, as I said, not be rigorous even in the case of classical physics. In classical physics, however, one can relate this (metaphorically) "optical" machinery to actual (causal) mechanical processes, say, the motions of particles in space and time, properly described by the Hamiltonian equations for the system. In other words, this machinery relates to an idealized causal model, which both offers a good *descriptive* approximation and ensures correct *predictions* for many physical processes in nature.

It is difficult, if possible at all, to develop a descriptive physically causal model, to which the optical machinery can relate in this way, in quantum physics. One may, accordingly, have to content oneself with at most the symbolic or metaphorical "optical space," where the wave function supposedly propagates, as only related to the outcome of experiments in terms of *predictions*, which are, moreover, generally probabilistic, without describing the physical processes that lead to these outcome. In other words, this metaphorical optical space does not in any way correspond or relate to idealized (descriptive and predictive) models of the kind that we can create in classical physics and that one can then relate, as idealizations, to actual quantum objects and processes. Indeed, the type of (in the case of continuous variables) infinite-dimensional space over complex numbers used in quantum mechanics, too, can be called space only by analogy and metaphorically. This is so even in the case of discrete quantum variables, since the latter require Hilbert spaces over complex numbers, which are non-visualizable in phenomenally spatial terms, or indeed even in the case of classical physical systems with phase spaces of higher dimensions.

One can, as Born did, speak, again, metaphorically, of a wave-like propagation of probabilities, in this case by relating this "propagation" more rigorously to what physically occurs and is observed in measuring instruments, but still with much qualification. A rigorous assignment of probabilities according to Born's rule requires a very different way of looking at the situation, as Schrödinger makes clear in his cat-paradox paper. In particular, we deal with a discrete set of experimental situations or phenomena, rather than with continuous quantum processes in space and time, for each of which phenomena we have a given probability, provided by the wave function cum Born's rule.

Whether they are related to continuous or discrete variables and quantum mechanics, quantum phenomena make us confront the essentially probabilistic character of quantum predictions. These predictions are, in quantum mechanics, enabled by the particular structure of the Hilbert spaces involved, as complex Hilbert spaces (mathematically essential to quantum mechanics of both continuous and discrete variables), cum Born's or similar rules. These rules have no rigorous justification from within the formalism itself, even though the shift, defining Born's rule, from a complex quantity to its modulus, which is a real quantity (necessary to for expressing the probability of a given event), is quite natural mathematically. The justification for Born's rule is "experimental:" it works! This justification is not discountable, and classical physics, from Galileo and Newton on, is ultimately justified experimentally, too, with that crucial difference that there our predictions are defined and, in this sense, justified by the descriptive character of our mathematical models. We do of course try to develop our theories so as to provide such further justifications, to the degree we can, when, for example, we use the quantum vs. classical theory of light, or general relativity as against Newton's theory of gravity. In any event, the corresponding quantum-mechanical Hamiltonian equations (we can, accordingly, no longer speak of equations of motion) become, in Heisenberg's terms, a new kinematics and a new form of relationships between a mathematical formalism and the experimental data to which it relates. This argument applies to whatever of the mathematically equivalent formalisms one prefers: that of Heisenberg's matrix mechanics, Schrödinger's wave equation, Dirac's Hamiltonian q-numbers, von Neumann's Hilbert-space formalism, C*-algebra formalism, and so forth. Borrowed from classical physics, where it relates to representation of motion, the term "kinematic" is, again, misleading here, just as is the term "wave equation." Heisenberg, however, never intended to relate his kinematical elements to motion, as against Schrödinger's program. His "new kinematics" was truly new, in the sense of establishing the relationships between mathematics and physics, as discussed earlier. In sum, in this view the mathematics of quantum theory (in whatever form) relates to experimentally verifiable probabilities of transition from one particular physical situation to another such situation, both established by using measuring devices or equivalent macro-objects. It does not appear to, and in the present view does not, describe or even relate to any idealized models of quantum processes, which are, nevertheless, responsible for the existence of these situations and the connections between them.

Thus, while this view retains Schrödinger's mathematics, it reverses the vision that grounded and guided Schrödinger's physical program, defined by the idea of "the wave radiation forming the basis of the universe." Schrödinger's equation does not describe any physical waves (actual or idealized in terms of models), as Schrödinger initially hoped it would. Instead, quantum probabilistic predictions, enabled by Schrödinger's equation and Born's or equivalent rules for deriving probabilities from quantum amplitudes, may be physically linked to a set of discrete individual phenomena, corresponding to certain wave-*like* correlational patterns. From this perspective, a certain Hamiltonian optical-mechanical analogy or translation of a mechanical kinematic into an undulatory one could be maintained in quantum mechanics or, to begin with, in classical mechanics, and can be given a rigorous form. This can be done if one sees the "optics" involved only as predictive machinery in either case, classical or quantum. In classical mechanics, however, this machinery is also accompanied by descriptive (realist and causal) idealized models. By contrast, in quantum mechanics, such models do not appear to be available, and in the present interpretation are abandoned, making the "optics" in question strictly predictive. In this view, the quantum-mechanical Hamiltonian equations map no motion in space and time and only predict probabilities of the outcome of experiments, staged as physical situations defined by physically classical observable phenomena, manifested in measuring instruments.

Beyond explicating an important part of quantum mechanics, the preceding remarks and my overall discussion of Schrödinger's work here allows one to reassess his contribution to quantum theory and his thinking itself as thinking in terms of and indeed in waves. By offering this reassessment, this discussion also becomes a tribute to Schrödinger's work on quantum mechanics, even though, from the present perspective, some of the most significant aspects of this work emerge against the grain of his thought and his ideal of physical theory, the "classical ideal," as he, again, called it in his cat-paradox paper (Ref. [8], p. 152). The power of his ideas is apparent throughout his work on wave mechanics and in his later attempts to think through the complexities of quantum mechanics, as in his cat-paradox paper. Even if the key ideas or ideals that guided his work might not have succeeded in quantum theory, they remain important, and his analysis of the ideas of his opponents, such Heisenberg and Bohr, is valuable as well. As I stated from the outset of this lecture, however, there are also alternative perspectives, which are closer to that of Schrödinger, or Einstein, and I hope that my discussion of Schrödinger's thinking, or for that matter that of Heisenberg's and Bohr's thinking, may also be helpful to those who hold such alternative views. Certainly, these thinkers learned a great deal from each other not only, self-evidently, as far as physics is concerned, but also in developing their philosophical views, different and sometimes contrary as these views were.

2. Causality and probability in classical and quantum physics

2.1. *Summary*

This lecture explores the nature and limits of the concepts of causality in classical and quantum physics, and considers the impact of our understanding of causality on the use of probability in all physics, classical or quantum. Classical physics is customarily viewed as a causal theory, while, as discussed in my previous lecture, quantum physics, from quantum mechanics on, is often, although not uniformly, viewed as a noncausal theory. The situation, however, is more complex. First, while classical mechanics may be considered as causal in the conventional sense (explained below), the application of this concept of causality in classical mechanics is subject to important qualifications. Secondly, quantum mechanics may be considered "causal" in turn, even in the case of standard quantum mechanics (as against other theories of quantum phenomena, such as Bohmian mechanics, which are causal in the conventional sense) and even when it is interpreted as noncausal in the conventional sense, as it is in Bohr's and related interpretations. As explained in the first lecture, these interpretations renounce not only the (classical) idea of causality but also, and in the first place, the idea of reality as applicable to our understanding of nature at the quantum level. As also explained there, however, this is only an interpretation, one of several possible interpretations, rather than a theory that definitively establishes "the truth of nature" (although this claim is made sometimes). In particular, such an interpretation cannot exclude classically causal, or realist, interpretations of quantum mechanics or causal and realist alternative theories of quantum phenomena, some of which are discussed in other lectures presented in this summer school. On the other hand, if one wants to speak of "causality" in the case of (classically speaking) noncausal interpretations of quantum mechanics one needs an alternative concept of causality, and I shall discuss possible candidates for such alternative concepts and how they relate to our uses of probability in quantum physics. The possibility of such alternative concepts has significant implications for the philosophy of causality. In particular, contrary to a long-standing philosophical view, there might not be a single general concept of causality, under which all workable concepts of causality may be subsumed as special cases. We may need different concepts of causality in different areas of physics. In some cases, we may need more than one such concept, for example in those fields, such as modern cosmology, which involves both classical and quantum physics, or relativity, which is a classically causal theory.

2.2. *The classical philosophy of causality*

I begin with a profound remark of Heisenberg, made by him in his discussion of I. Kant's philosophy and Kant's argument for the ideas and the law of causality (defined below) as given a priori, rather than derived from experience. This argument, Heisenberg contends, no longer applies in "atomic [quantum] physics" (Ref. [22], pp. 89–90). He also makes a broader point, however:

> Any concepts or words which have been formed in the past through the interplay between the world and ourselves are not really sharply defined with respect to their meaning; that is to say, we do not know exactly how far they will help us in finding our way in the world. We often know that they can be applied to a wide range of inner or outer experiences, but we practically never know precisely the limits of their applicability. This is true even of the simplest and most general concepts like "existence" and "space and time." Therefore, it will never be possible by pure reason to arrive at some absolute truth [as Kant thought it might be].
> The concepts may, however, be sharply defined with regard to their connections. This is actually the fact when the concepts become a part of a system of axioms and definitions which can be expressed consistently by a mathematical scheme. Such a group of connected concepts may be applicable to a wide field of experience and will help us to find our way in this field. But the limits of the applicability will in general not be known, at least not completely. (Ref. [22], p. 92)

Helped by insights gained from quantum physics, Heisenberg's criticism of Kant has a point, and Kant, if he could have benefited from these insights, might have agreed with Heisenberg. Indeed, apart from the belief that it would be possible to arrive at some absolute truth by pure reason, Kant is not that far from Heisenberg, and Kant appears only to have argued that it *might be possible* to do so, which means that this possibility might not actually materialize. Although not embedded in "a mathematical scheme," Kant's analysis is also characterized by a search for a sharper definition and the limits of applicability of his concepts, in part through establishing the connections between them. More importantly, Kant argued that concepts, either those given *a priori*, such as space, time, and causality, or those established from experience, generally apply only in the *phenomenal* domain (of what appears to our thought), and that their application to the *noumenal* domain (of things as they actually exist in nature or mind) is limited and is never guaranteed to be adequate. The distinction between phenomena and noumena is the basis of Kant's philosophy, and his view of causality is grounded in this distinction. As indicated in Lecture 1, one of the reasons that the classical, such as the Kantian, concept of causality works in classical, but not in quantum, physics is that in classical physics the distinction between phenomenal and noumenal entities, although technically valid, could be disregarded, at least ideally or in principle. This does not appear possible to do, even ideally or in principle, in quantum physics.

Consider the following dictionary definition of causality from *The Random House Webster's Unabridged Dictionary*: "*Causality* is the relationship between an event (the *cause*) and a second event (the *effect*), where the second event is a direct consequence of the first" (Ref. [23]). This is a good definition, and it is not so easy to significantly improve on it. While most definitions in philosophical literature refine this type of definition and probe the limits within which it applies, they retain the key elements of this definition, which, it may be noted, is in turn indebted to the philosophy of causality. One can make this definition more general by extending the application of terms "cause" and "effect" to entities (individual or, since there may be more than one cause to a given effect, collective), A and B, other than events, and without requiring that A and B should be entities of the same kind. This generalization is often useful in physics. For example, the gravity of the Sun (or of other bodies in the solar system), which is cumbersome (but not impossible!) to define as a spatiotemporal event, can be seen as the cause of the motion of planets in the Solar system, and hence of any event in the course of this motion in space and time. The physical nature of gravitation may not be completely known, and it was not known at all at the time of Newton. This circumstance indicates the complexity of the relationships between causality and succession, and especially, the difficulties of using the idea of the ultimate cause in physics and elsewhere: the ultimate origins of things are not known, and generally suspended in our analyses of physical systems, even when dealing the Universe itself considered in its cosmological history.

M. Born offers the following, often cited, definition of causality: "Causality postulates that there are laws by which the occurrence of an entity B of a certain class depends on the occurrence of an entity A of another class, where the word entity means any physical object, phenomenon, situation, or event. A is called the cause, B the effect" (Ref. [24], p. 9). Although one may need a set of sequential or parallel causes for a given event, and hence the concept of the complete set of causes to properly define entity A here, Born's and most other standard definitions of causality can be easily adjusted to accommodate this requirement. Born adds two other, again, common, postulates, that of antecedence and that of contiguity: "Antecedence postulates that the cause must be prior to, or at least simultaneous with, the effect" and "Contiguity postulates that cause and effect must be in spatial contact or connected by a chain of intermediate things in contact" (Ref. [24], p. 9). By starting with "effect," Born de facto formulates the *principle* of causality, which states that if an event takes place, it has a cause of which it is an effect. This principle is crucial for both the application of and the critical analysis of causality, from Hume and Kant on.

Thus, Kant defines, similarly to the dictionary definition cited above, the relation of causality as that of, first, the cause and, secondly, the effect: "the concept of the *relation of cause and effect*, the former of which determines the latter in time, as its consequence [*Folge*], as something that could merely precede it in the imagination (or not even be perceived at all)" (Ref. [25], p. 305). By contrast, the *principle* of causality proceeds from effects to causes: "If ... we experience that something happens, then we always presuppose that something else precedes it, which it *follows* in accordance with a rule" (Ref. [25], p. 308). Kant is careful to qualify that "the logical clarity of this representation of a rule determining the series of occurrences, as that of the concept of cause, is possible if we have made use of it in experience." "But," he concludes (against Hume's empiricist view), "a consideration of [causal representation], as the condition of the synthetic unity of the appearances in time, was nevertheless the ground of experience itself, and therefore preceded it *a priori*" (Ref. [25], p. 309).

In Kant, the principle of causality also implies that under the same conditions identical events will take place and, thus, that in science identical experiments will lead to identical outcomes. As many others before and after him, Kant sees both the principle itself and this implication as, in Heisenberg's words, "the basis of all scientific work," a contention radically challenged by quantum physics (Ref. [22], pp. 89–90). Hume, too, uses, along more empiricist lines, this aspect of causality in his empirical ("regularity") theory of causality in nature, in which case he allows for a meaningful application of the concept (e.g., Ref. [26], pp. 18–21). By the same token, as Kant does, Hume appears (there is some debate concerning Hume's view on this) to allow for general causality in nature, even though the human mind can at most perceive it only partially, via the regularity of certain causal conjunctions of events. Hume and Kant also maintain the antecedent and, hence, the asymmetric relationships between cause and effect.

The Kantian framework of causality is paradigmatic of most views of causality, at least as concerns the essential features just outlined, and I shall call this type of view *classical causality*, and the models by this view the *classical models of causality*. There are several reasons to adopt this terminology, beginning with the fact that the historical period of the Enlightenment (roughly the eighteenth century), is sometimes referred to as the Classical Age, and also as the Age of Reason. (Kant was a major figure of this period, and he introduced the term "Enlightenment" itself.) The terminology also correlates with the term "classical physics," where, especially in Newtonian mechanics, or in the *idealization* defined by it, classical causality works well. I shall more properly explain the nature and role of idealization in classical physics below, merely noting here that it implies considering only those properties of natural objects that could be suitably mathematized by classical physics. Newtonian mechanics was a key development of the Classical Age, and along with Euclidean geometry, Kant's main scientific inspiration in developing his philosophy.

Technically, in accordance with Kant's scheme, the idealization of classical mechanics only applies at the level of phenomena. Classical physics, however, at least classical mechanics, can disregard this difficulty in practice and, within the limits of its idealization, treat phenomena as objects. The kinetic theory of gases, electrodynamics, and relativity, complicate this view, as both Maxwell and Einstein noted (Ref. [27]; Ref. [28]; Ref. [29], pp. 306–328), but do not change the essential role of causality in these theories. For the moment, *within these limits*, one can, for example, maintain both (a) that the configuration of the Solar system, defined by its gravity, at a given time causes any subsequent event in the motion of a given body, say, a planet, such as Mars, in this system (barring outside interferences); and (b) that, if the Sun and other bodies in the solar system were not there, the motion of a given planet would not be observed in the way it is. In other words, both the definition of cause given here and the principle of causality are permissible and are assumed to be operative within the limits of the idealization of classical mechanics. The latter, moreover, usually presupposes both antecedence ("the cause must be prior or simultaneous with the effect") and contiguity ("cause and effect must be in spatial contact or connected by a chain of intermediate things in contact"), which is why Born invokes them. These requirements were strengthened by relativity, which restricted the propagation of all physical influences by the speed of light in a vacuum. Relativity restricts causes to those occurring in the backward (past) light cone of the event that is seen as an effect of this cause, while no event can be a cause of any event outside the forward (future) light cone of that event. Sometimes the term "causality" is used in physics to designate the compliance with this requirement. The relativity requirements, or antecedence and contiguity, do not depend on the (classical) principle of causality. As will be seen, all three conditions (antecedence, continuity, and the compatibility with relativity) are satisfied in spite of the absence of (classical) causality in quantum mechanics, in the non-

causal type of interpretation adopted here. It also follows that, in accordance with
Kant's requirement for scientific practice, identically prepared experiments in classical physics lead to identical outcomes, at least, again, ideally, insofar as statistical
errors in repeated experiments can in principle be neglected, and they are usually
neglected in practice. The situation, as discussed in Lecture 1, is different in the
case of quantum phenomena, say, an emission of a photon by an electron or what –
an *effect* – we so interpret from observing a spot on a photographic plate. Although
we do know, with reasonable certainty, that such events occur, neither claim (a) nor
claim (b), specified above, could be made with certainty, as in classical mechanics,
but only with a certain degree probability. In other words, while one can speak of
quantum "effects" (one of Bohr's preferred terms), their "causes" are similar to the
way the *ultimate causes* function in philosophy or in classical physics: even if such
ultimate causes exists, their connections to effects cannot be meaningfully tracked
down. In classical physics, this difficulty is usually handled by establishing spatial
and temporal frames that limit a given case to those causes that can be meaningfully tracked down, at least, again, in principle. In quantum theory, however, this
type of framing does not appear possible, even for individual events, which, unlike
in classical mechanics, are not comprehended by law, at least not by a causal law.
In addition, it is difficult to meaningfully argue (although the assumption is made
sometimes) that such a law could in principle exist, in the way, say, one could argue
for the application of the causal laws of classical mechanics to individual molecules
in classical statistical physics. To cite W. Pauli, quantum mechanics "predicts only
the statistics of the results of an experiment, when it is repeated under a given
condition. Like an ultimate fact without any cause, the individual outcome of a
measurement is, however, in general not comprehended by laws" (Ref. [14], p. 32).

Kant was aware that the application of his concept and principle of causality has
its limits and requires qualification, including in classical physics. For one thing,
according to Kant, the principle of causality rigorously applies only at the level of
phenomena. As I noted above, both Hume and Kant appear to have believed (at
bottom, not that differently from P. S. Laplace), that the world is, or is created by
God as, governed by hidden causality. It is just that the ultimate causal architecture of nature, in its (hidden) particularities and its (manifest) complexity, is never
available to our knowledge, even if not necessarily unavailable to our thought, which,
however, could not be guaranteed to be correct in this regard (Ref. [25], p. 115).
While Hume doubted the validity of the concept of causality at the human level more
than Kant, he, as noted above, appears to have a similar view concerning nature, in
light of his empirical regularity theory (e.g., Ref. [26], pp. 18–20). There is, as indicated above, some debate on this point among scholars as concerns Hume's view,
which might be argued only to allow for "the relation of contiguity and succession
... [as] independent of, and antecedent to the operations of the [human] understanding" (Ref. [30], pp. 168–196). However, Hume's empirical regularity theory of
causality does appear to suggest the ultimate causality in nature as "independent of

our thought and reasoning" along with the antecedent nature of causality, implied by the successive nature of this relation (Ref. [30], pp. 168–69; also Ref. [25], pp. 18–20). In sum, Hume's and Kant's critique of causality applies to claims concerning causality in the human experience of and representation of the world, and not to the ontological architecture of the world itself.

Historically, this view is not surprising. Although probability theory was advanced by then, the ultimate nature of the world was usually interpreted in terms of the underlying but unknown and perhaps unknowable classical causality, as it was famously by Laplace. Indeed, until the twentieth century and quantum physics, very few, if anyone, doubted that some form of classical causality would apply at the ultimate level of the world. As I said, many still hope that this will ultimately prove to be the case. Hume and Kant, however, deserve credit for their realization of the complexities and limitations of classical causality, and for their critical analyses of causality, aimed at the lack of sensitivity to these limitations and unwarranted extrapolations of classical causality. In sum, it is not a question of abandoning classical causality, which is workable and effective within large limits, but instead that of demarcating these limits, and, beyond these limits, of possibly introducing other concepts of causality, as I shall attempt to do in Section 3. First, however, I shall consider the role and limits of classical causality in classical physics and relativity, with some comparisons with quantum physics offered by way of contrast.

2.3. *Classical physics and classical causality*

I would like to begin with Pauli's discussion of the difference between classical and quantum mechanics in his letter to Born (Ref. [31], pp. 221–223). Pauli first considers "the determination of the path of a planet" as an example: "if one is in possession of the simple *laws* for the motion of the body (for example, Newton's law of gravitation), one is able to *calculate* the path (also position *and* velocity) with as *high* an accuracy *as one likes* (and also to test the assumed law again at different times). Repeated measurements of the position with limited accuracy can therefore successfully replace *one* measurement of the position with high accuracy. The assumption of the relatively simple law of force like that of Newton (and not some irregular zig-zag motion or other on a small scale) *then appears as an idealization which is permissible in the sense of classical mechanics.*" By contrast, in quantum mechanics "the repetition of positional measurement with the same accuracy ... is of *no use at all* in predicting subsequent position measurements. For [given the uncertainty relations, $\Delta q \Delta p \simeq h$] every positional measurement to [the same] accuracy at [a given] time implies the inaccuracy [defined by the uncertainty relations] at a later time, and *destroys the possibility of using all previous positional measurements within these limits of error*! (If I am not mistaken, Bohr discussed this example with me many years ago)" (Ref. [31], p. 219; emphasis added). Bohr indeed makes this type of argument on several occasions, beginning with the so-called Como lecture (1927), introducing complementarity (Ref. [5], v. 1, p. 68).

It is especially significant that "the repetition of positional measurement with the same accuracy . . . is of *no use at all* in predicting subsequent position measurements." This circumstance reflects one of the most essential differences between classical and quantum phenomena and, as a consequence, of (the idealizations of) classical and quantum mechanics. The difference between the outcomes of identically prepared experiments and the corresponding statistical errors are, in principle, reducible in classical physics, which fact allows us, ideally, to disregard these differences and treat classical experiments as ideally repeatable, as concerns both preparations and outcomes. This, however, is not the case in quantum physics. As explained earlier, identically prepared experiments (which are possible, because we can classically control the instruments involved) in general lead to different outcomes. While we can probabilistically predict these outcomes, for example, by means of quantum mechanics, these probabilities are irreducible (we cannot improve our predictions regardless of the precision of our instruments, at least we cannot do so beyond certain point, defined by Planck's constant h). This circumstance is correlative to the uncertainty relations, $\Delta q \Delta p \simeq h$, which would, it follows, apply even to ideal instruments. By the same token, every quantum measurement renders any preceding measurement meaningless as concerns our predictions of the outcomes of any subsequent measurement.

The main reason that classical causality works in classical mechanics is that, within the idealization of the theory, the state of the object at a given point defines the states of this object at all future points, *within the range of the system's history as defined for a given case.* The state of the system is defined by its position and momentum, both of which can be, ideally, measured and predicted, and therefore considered as properly definable and determinable at the same time at any given point in the evolution of the system. Indeed, a given state is equally *defined by* the past states, and it would allow us to make definitive conclusions concerning past states and predict all future states, within the frame of a given experiment. I am reluctant to say that the present state of the system also *defines* all its past states. For, although the equations of classical mechanics allow us to know these past states, physically the relation of determination between states proceeds from the past or present to the future in most applications of this model. As I shall explain presently, while it is easier to speak of physical causes, defined by the laws of classical mechanics, in relation to which all states of the system are effects, viewing the relationships between such states themselves as those between causes and effects requires further qualifications.

These qualifications would also explain why the assumption that the cause precedes or is, at most, simultaneous with the effect is appropriate to the idealization of classical mechanics, or in relativity, where this assumption is amplified by the finite limit upon the propagation of physical influences, as explained above. Mathematically, the equations of classical mechanics or relativity are time reversible, as are the equations of quantum mechanics, although the latter requires additional qualifications. I shall bypass these qualifications, since in the present interpretation, these equations only have a predictive role and do not describe any physical processes in space and time, and as such are always future oriented: hence their time-reversible nature has no physical significance in this interpretation. This reversibility may suggest (and it does to some) that time reversal and backward-in-time causality are possible. In the present view, it seems more reasonable to exclude both from the idealization of classical mechanics or that of relativity, or that of quantum mechanics (where both ideas are sometimes entertained as well), because there is, thus far, neither experimental evidence nor, at least to the present author, other compelling reasons to consider them. I shall return to this point below.

I insist on "idealization," first, because, as noted earlier, all modern physics deals only with idealized models, even in the case of observed phenomena (since many properties of phenomena are disregarded), let alone nature itself in its ultimate constitution, although classical mechanics allows us to disregard the Kantian difference between the phenomena, as representations, and the actual physical objects considered. Secondly, the role of idealization in classical mechanics is crucial to the use of classical causality there, which is rigorously applicable only to idealized systems, while allowing for very good approximations of and predictions concerning the behavior of the actual physical systems, thus idealized, and considered within suitably demarcated spatial and temporal limits.

Some would speak of classical causality in physics and beyond as "deterministic causality," which is not out of place (e.g., Ref. [26], p. 18). I prefer to understand determinism as having to do with our capacity to make predictions concerning the behavior of a given system. The causal character of classical mechanics allows for exact, deterministic predictions concerning individual systems in many practical cases, that is, this causal nature allows us to neglect unavoidable practical deviations (in view of the limited capacity of our measuring instruments) from strictly exact predictions. On the other hand, certain sufficiently complex systems may be seen as classically causal, although we have no capacity to make, even ideally, exact predictions concerning their behavior. Consider the case of a coin toss, which may be seen as an ideally classical process (one can exclude the quantum aspects of the constitution of the coin as having no effect on the outcome of a toss). The system, however, is too sensitive to the initial conditions and outside interferences for us to be able to ever make exact predictions by means of the mechanical mathematical model we use in other cases of classical mechanics. In other words, in practice it would be very difficult or even, as things stand now, impossible to construct a good classical-mechanical model in this case. It is true that there are alternative accounts of this case, for example, those along Bayesian lines, such as by E. T. Jaynes (Ref. [32], pp. 317–320). These accounts, however, do not affect my point concerning the underlying classical causality (presupposed by Jaynes), if one assumes that quantum aspects of the constitution of the coin do not affect the situation. Similarly, the models that we use in chaos theory may give us reasonably reliable patterns of the behavior of the systems considered, not by predictive algorithms of the type more conventional or simpler cases of classical mechanics (say, those of many two-body systems) do. In principle, however, one could imagine a tracking technology that would enable us to follow the coin's trajectory and configure our mathematics accordingly so as to make exact prediction in each case, as exact as in such simpler cases. Accordingly, it is reasonable to think of classical causality as a feature of these processes, at least as concerns the ideally descriptive, even if not practically predictive, capacity of our classical models.

With due adjustments (defined, again, by the special significance of the speed of light in a vacuum, c, as independent from the speed of the motion of the source), classical causality also applies in relativity. By contrast, it is difficult to adopt a similar conception in quantum physics, and, as I said, the limitations concerning the kind of determination necessary to apply the classical models would arise even if we assume ideal measuring instruments. Unlike chaos-theoretical models, quantum mechanics, at least in the present view, provides good, albeit, generally, probabilistic predictions concerning the outcome of quantum experiments (defined by phenomena observed in measuring instruments), without providing a descriptive mathematical model, however idealized, of the behavior of quantum objects themselves.

The question of cause and effect is more complicated in classical mechanics. One could still say that, for each given measurement and the corresponding prediction, the state of the object itself in question established by this measurement at a given point is a cause of and thus determines its future state, as an effect, at any given future moment. This statement, however, is only true insofar as this causal determination is enabled and defined by the laws of motion used, say, those of Newton's laws for gravity, and the corresponding equations, which are assumed to reflect certain physical forces in nature itself. This *determination* descriptively (mathematically) idealizes the corresponding physical configuration in nature. One might say that the real physical cause for any determination, including that of the initial state that defines a given situation, is the gravitational field defined by the Sun and other corporeal objects or fields in the Solar system. From this viewpoint, a given state of any single object can only be seen as a *physical cause* of its future states insofar as the whole configuration of bodies and forces involved, which determines the law of motion and hence of causality, is considered as part of this state. On the other hand, one might see these factors as built into the state of an object as defined by its position and momentum, at each point. Then one might say that each given state is a cause of all subsequent states, as its effects, with the *law of causality* defined by the *laws of motion and forces* for this system.

In addition, the history of a classical system only goes so far in a given representation, and thus, as indicated above, involves the suspension of the ultimate cause or even many more remote causes. Newton bracketed the physical nature of and hence the causes of gravity and was (wisely) content to merely take its force into account in his law of gravity. This bracketing allows one to apply this law, including as a law of causality, as part of the overall "legislature," as it were, of causality for a given classical system, defined by Newton's law of gravity and other laws of Newton's mechanics. While this application is only possible within those limits where we need not be concerned with the physical nature of gravity itself, these limits are very broad and allow us to consider a large number of physical systems. In most applications of classical physics, the earlier history of a given system, say, that of the emergence of the Solar system, is bracketed as well, although some cases assume large spatial and temporal frames, all the way to the scale the Universe, at least up to a point. For, once one gets closer to the Big Bang, the practical use of the model become difficult, even if one remains within the classical scheme, but at least once galaxies are formed, one can have good, albeit limited, approximations and assessments, even by using Newton's theory of gravity.

Einstein's general relativity has, to some degree, resolved the problem of the nature of gravity, but only to some degree, since the ultimate nature of gravity may be and generally is assumed to be quantum. In this case, our theory of gravity may and, on the view adopted here, would require the suspension of classical causality at the ultimate level. Just as in classical physics, however, for many practical cases in which we use relativistic gravity, the *ultimate nature* of gravity is not crucial. On the other hand, the difference between general-relativistic and Newtonian laws of gravity is crucial, even in explaining the behavior of the motion of planets, such as, famously, Mercury, in the Solar system. Classical causality, however, applies, with certain qualifications, in general (or special) relativity as well.

It is, again, also usually assumed that there is no backward-in-time physical influence, even though the equations of classical physics are mathematically symmetrical with respect to time reversal. In addition, these equations or those of relativity do not provide for the concept of now, defined only from the outside by the clocks we use and our consciousness, a circumstance much pondered by Einstein throughout his life. This assumption, along with the historical framing just defined, makes classical causality related to, but not quite the same as the temporal division of past and future. Causes always precede effects: bodies, such as planets, move in a gravitational field, such as that of the solar system, because of the earlier history of this field, even though these bodies contribute to this field. Relativistic considerations, again, impose further restrictions by limiting causes to those occurring in the backward (past) light cone of the event that is seen as an effect of this cause, while no event can be a cause of any event outside the forward (future) light cone of that event.

It is argued sometimes that general relativity and quantum mechanics suggest the possibility of retroaction in time and backward-in-time causality (e.g., Ref. [26], p. 188). I do not find these arguments sufficiently compelling in either case, first of all, because there is, thus far, no experimental evidence for retroaction in time or backward-in-time causality. It is true that there are certain legitimate (hypothetical) arguments for the possibility of retroaction in time and, by implication, for the corresponding concept of causality. Among them is the existence of closed time loops in K. Gödel's solutions of the equations of general relativity ("Gödel's metric"), K. Thorne's wormhole "time-machines," the hypothetical existence of tachions (particles that only travel faster than light in a vacuum, which is not forbidden by relativity), and a few others.

It is, however, a different question *how* compelling these arguments are and to whom. The problems of the assumption remain serious on well-known logical and physical grounds, such as those having to do with possible alterations of the past incompatible with the world at the time at which a given time travel commences. Thus, in the so-called grandfather paradox, the time traveler travels into the past and kills his grandfather before the latter is married, which implies that the time traveler could not be born later on, and hence could not to travel through time and kill his grandfather. Such paradoxes and most arguments against retroaction in time do not altogether rule it out, because it is possible to design schemes circumventing them, but such schemes are artificial and, while they may be compelling to some, they are not generally accepted. In addition, there is, again, no experimental evidence supporting the idea. Its main *physical* appeal appears to be that it may "solve" certain actual or sometimes perceived problems of the current quantum theory, from quantum mechanics to quantum field theory to (as yet not developed) quantum gravity. Its main *philosophical* appeal is that some physicists entertain the idea on the grounds just stated. Because retroaction in time cannot be completely ruled out by our current theories, one could explore the corresponding notions of causality. It is also true that both relativity and quantum theory taught us that we should not trust our general (everyday) or even philosophical intuition in fundamental physics. Accordingly, one might agree with P. Dowe's *general* contention that "it will not do for the philosopher to rule out a priori what the scientist is currently contemplating as a serious hypothesis" (Ref. [26], p. 188). It does not appear to me, however, that there are sufficiently compelling physical reasons (mostly those mentioned above) to pursue the possibility of retroaction in time, which Dowe has *specifically* in mind here, while there are more compelling reasons against this possibility. In other words, *how seriously* this particular hypothesis is contemplated by scientists is not altogether clear, and I would argue that it is not very seriously entertained widely. Given that our current fundamental theories are manifestly incomplete, it is of course possible that retroaction in time or that backward-in-time causality might in one way or another be shown to be a feature of nature. For now, however, although the equations of classical physics or relativity are mathematically symmetrical with respect to time reversal, the assumption that there is no retroaction in time and no corresponding causal influence is reasonable and, within a wide range, workable within in the idealization of classical or relativistic physics.

It is worth noting that Dowe's argument for backward-in-time causality in quantum mechanics is primarily motivated by a *philosophical* discontent, which is common, with certain interpretations of the Copenhagen type and certain attempts to address the problems posed by the famous Einstein, Podolsky, and Rosen (EPR) experiment and Bell's theorem (Ref. [26], pp. 182–183). To the present author, these reasons, again, do not appear to be sufficient to resort to backward-in-time causality, given the difficulties of applying it in physics, as explained above. Dowe's gloss of the Copenhagen interpretation hardly does justice to the views of most followers of the spirit of Copenhagen, and specifically those of major figures, such as Bohr, whose position on the subject is, as I have discussed in Lecture 1 and in more detail elsewhere (Ref. [1 and 3]), nothing like the gloss offered by Dowe. Dowe's argument is shaped by a rather limited view of the history of the question of causality in physics or philosophy, a view, by and large, restricted to the Anglo-American analytic philosophical tradition, and a few earlier authors, such as Hume. Thus, remarkably, Kant is not considered. Nor are F. Nietzsche or the American pragmatists, such as C. H. Peirce and W. James, who offered important critiques of the idea of causality. None of the founding figures of quantum theory is discussed either, although, as we have seen, they all commented on the question of causality.

In sum, classical causality is workable in classical mechanics and, with certain qualifications (not fundamental in nature) elsewhere in classical physics and relativity. However, it only applies to the idealized (mathematical) models used by these theories; and, moreover, the application of these models is limited by spatial, temporal, and other frames, although one might also see these frames as parts or parameters of the model. In other words, in accordance with Heisenberg's argument with which I began, although by connecting classical causality to other concepts of classical physics we can define this concept and its limits more sharply, we still do not know, at least not completely, how far classical causality ultimately extends in physics. It does appear, however, that classical causality is likely to have a limited domain of application in physics. In particular, it may not apply at all either on very small scales, in view of quantum physics, or on very large scales, for a complex set of reasons, which, however, include the apparently quantum origins of the universe and the ultimately quantum character of gravity. In other words, a kind of causal or deterministic picture that Laplace and others envisioned for the universe is unlikely to apply. The idea of an overall causal universe, including one based on the classical concept of causality, is by no means completely abandoned, however. It was, for example, advocated by G. 't Hooft, for the reasons having to do with the EPR-Bell type experiments Ref. [33]). 't Hooft certainly does not think in terms of anything like classical mechanics in considering the behavior of individual quantum systems, even though he does want to depart from standard quantum mechanics. Similarly, while Bohmian and other causal quantum theories (such as some of those considered in other lectures in this summer school) are different from classical mechanics, they are classically causal. So are certain interpretations of standard quantum mechanics. I shall leave these interpretations aside, primarily because, unlike those interpretations that are not classically causal, such interpretations or theories tell

us little new about causality, even if one adopts backward-in-time causality, since
the latter, again, has an essentially classical architecture.

On the other hand, it appears to me that there are good reasons to ask the following question. Assuming an interpretation of quantum mechanics, such as the one
adopted here, in which quantum mechanics or quantum phenomena themselves do
not obey classical causality, and assuming both locality and the absence of retroaction in time or backward-in-time causality, is it possible to introduce a concept of
causality that is different from the classical one? The answer, I would argue is yes,
and I shall propose such a concept in the next section.

2.4. Quantum physics and quantum causality

I shall begin by stating my definition of quantum causality, which is as follows:
> Whatever happens as a quantum event and is registered in the measuring in-
> struments (thus providing us with the initial data) defines a possible set of, in
> general probabilistically, predictable outcomes of future events and irrevocably
> rules out the possibility of our predictions concerning certain other, such as and
> in particular complementary, events.

The discussion to follow is designed to explain and justify this definition. It is helpful to revisit, first, some of the main reasons for why quantum physics (independent
of any particular theory that handles quantum phenomena) makes it difficult to
maintain classical causality. Arguably, the most decisive circumstance responsible
for this situation is that, as noted earlier, identically prepared quantum experiments
(in the sense of the state of the measuring instruments involved, which we can control classically) in general lead to different outcomes. This difference is ineliminable
and automatically implies the irreducibly probabilistic nature of our quantum predictions, even when they concern individual quantum objects and events, which
cannot be subdivided further. It is crucial that we cannot improve the accuracy of
our predictions either by repeating identically prepared experiments or by improving the precision of our measuring instruments, which would allow us to speak of
ideally exact predictions in the way it is possible in classical mechanics when dealing
with individual classical objects. In quantum experiments, the probabilities of our
individual predictions will remain the same, at least beyond certain limit (defined
by Planck's constant, h), regardless of how much we improve the precision of our
measuring instruments. This can also be expressed by saying that this would remain
the case even if we had ideal instruments. The uncertainty relations, $\Delta q \Delta p \simeq h$,
which, too, would apply even if we used ideal instruments, are correlative to this
situation and indeed define this limit.

Strictly speaking, the situation is even subtler. In any single run of a given
experiment (the double-slit experiment, for example), the emission of an object,
such as an electron, is never assured. Nor, at the other end of the experiment
can a given event, marked, say, by a spot on the screen, be guaranteed to have
resulted from the collisions between the screen and an electron emitted from the

source. Statistically, however, such events can be neglected, since we do know that in a vast majority of cases traces on the screen can be correlated with, and in this respect are "caused" by, particles emitted from the source, and thus with events of emission. The outcome of each run of the experiments (in the same setup) is, again, different each time (a spot on the screen is found in a different place), while the condition of each emission is the same as concerns the state of the apparatus that makes the emission possible. To return to Pauli's formulation, there is no law that comprehends *the outcome of individual experiments*. In the present view, the emission of an electron or its interaction with a measuring device does not have a physical (mechanical) explanation, which also prevents us from being able to explain why the differences in the outcome of identically prepared experiments arise (cf., Ref. [22], pp. 89–90). In view of these circumstances, it is remarkable that there exist algorithmic procedures, such as the one provided by quantum mechanics, where we change from adding probabilities themselves to adding "probabilities amplitudes" and by applying Born's rule accordingly, that allow us to make correct probabilistic estimates for the outcomes of quantum experiments (e.g., Ref. [34], v. 3, pp. 1–11). It follows that the rules for counting probabilities in the case of quantum phenomena is different from those of classical statistical physics. M. Planck was the first to discover this fact in deriving his black body radiation law in 1900, which inaugurated quantum physics, and Einstein was the first to understand the seriousness of these difficulties, without, however, be ever willing to give up the classical ideal of causality or, to begin with, reality.

As discussed in Lecture 1, the history of the renunciation of this classical ideal begins with Heisenberg's introduction of quantum mechanics as a theory the deals with strictly the probabilities of transitions between stationary states of electrons in atoms, and suspend of the description of individual quantum processes and, by the same token, causal connections between individual quantum events. The concept of event is especially pertinent here, since this individuality, too, is only manifested at the level of observable events, or again, phenomena, and not at the level of quantum objects and processes, to which the concept of individuality may not rigorously apply any more than any other concept. In this sense, the elementary particle of a given type, such as the electron, is defined by a given set (potentially very large, but quite specific) of possible phenomena or events observable in measuring instruments associated with it. This set is the same for all possible electrons, while the correlation between any such phenomenon and any given electron can never be assured, which makes us speak of electrons as indistinguishable from one another. There are also sets of phenomena associated with multiple particles of the same type, also with large multiplicities of particles as in quantum statistics. As will be seen in the next lecture, the situation becomes especially dramatic in high-energy regimes, where we need quantum field theory. This understanding is, thus, fundamentally different from the view adopted in classical statistical physics, where the behavior of the individual entities comprising the multiplicities considered statistically is assumed to be causal, at least ideally. Finally, in the ultimate version of this approach, as developed by Bohr, our predictions themselves only concern the effects of such processes manifest in the measuring instruments involved, while the nature and behavior of quantum objects are places beyond any description or even conception. As noted earlier, the lack of causality is an automatic consequence of thus epistemology, because the assumption of causality would imply that we could conceive, however partially, of the nature of quantum objects and processes (as "causal"), which is in principle precluded by this epistemology. The fact that probabilistic predictions of quantum mechanics or higher-level quantum theories are correct is enigmatic, insofar as there does not appear to be an underlying physical justification for them, in contrast, say, to classical statistical physics. We are lucky to have them.

Bohr begun to move toward his complete "renunciation of the ideal of causality" under the impact his exchanges with Einstein, following the so-called Como lecture, "The Quantum Postulate and the Recent Development of Atomic Theory," which introduced the concept of complementarity, but which still retained, if uneasily, this ideal (Ref. [5], v. 1, pp. 52–91, also Ref. [3], pp. 41–58). Bohr appears to have arrived at his ultimate view on the subject, along with and, again, as a consequence, of his ultimate epistemology, by around 1937–1938. Bohr's exchanges with Einstein, especially those concerning the EPR experiment, played a key role in this development of Bohr's thinking (Ref. [16]). By 1937 Bohr firmly established the connections between his epistemology and "the renunciation of the ideal of causality." As he said, "the renunciation of the ideal of causality in atomic physics which has been forced on us is founded logically only on our not being any longer in a position to speak of the autonomous behavior of a physical object" (Ref. [35], p. 87). In the so-called Warsaw lecture, "The Causality Problem in Atomic Physics," given in 1938, Bohr argued as follows:

The unrestricted applicability of the causal mode of description to physical phenomena has hardly been seriously questioned until Planck's discovery of the quantum of action, which disclosed a *novel feature of atomicity* in the laws of nature supplementing in such unsuspected manner the old doctrine of the limited divisibility of matter. Before this discovery statistical methods were of course extensively used in atomic theory but merely as a practical means of dealing with the complicated mechanical problems met with in the attempt at tracing the ordinary properties of matter back to the behaviour of assemblies of immense numbers of atoms. It is true that the very formulation of the laws of thermodynamics involves an essential renunciation of the complete mechanical description of such assemblies and thereby exhibits a certain formal resemblance with typical problems of quantum theory. So far there was, however, no question of any limitation in the possibility of carrying out in principle such a complete description; on the contrary, the ordinary ideas of mechanics and electrodynamics were found to have a large field of application also proper to atomic phenomena, and above all to offer an entirely sufficient basis for the experiments leading to the isolation of the electron and the measurement of its charge and mass. Due to the essentially statistical character of the thermodynamical problems which led to the discovery of the quantum of action, it was also not to begin with realized, that the insufficiency of the laws of classical mechanics and electrodynamics in dealing with atomic problems, disclosed by this discovery, implies a shortcoming of the causality ideal itself. (Ref. [35], pp. 94–95; emphasis added)

Bohr, thus, makes a strong historical claim, which he elsewhere extends to the history of philosophy as well. This extension is not surprising, given the fundamental relationships, discussed earlier in this lecture, between classical physics (first, classical mechanics and then classical statistical physics) and the philosophy of causality, which preceded classical physics and shaped it conceptually, but was then, from Kant on, even more significantly shaped by classical physics. According to Bohr: "[E]ven in the great epoch of critical philosophy in the former century [from Hume and Kant on], there was only a question to what extent *a priori* arguments could be given for the adequacy of space-time coordination and causal connection of experience, but never a question of rational generalizations or inherent limitations of such categories of human thinking [made possible by quantum physics]" (Ref. [5], v. 2, p. 65).

There is still the question of quantum correlations, the "enigma" of correlations, defined by the circumstance that, while each individual quantum event is *always irreducibly lawless*, in certain circumstances multiple quantum events exhibit statistical correlational orders, such as that found in the EPR-type experiments. The presence of these correlations is, however, consistent with the unthinkable and, hence, noncausal character of quantum processes, and it appears to have served as a further impetus for Bohr's ultimate interpretation of quantum phenomena and quantum mechanics. I am, again, not saying that correlations would necessitate this type of interpretation. In any event, the enigmatic combination of lawlessness of each individual quantum event and statistically correlated nature of certain aggregates of such events, leaves space for probabilistic predictions, space that quantum mechanics appears to use maximally. But, in Bohr's view, it leaves little, if any, space for the description or even conception of the actual processes responsible for this situation, and, again, as a consequence, for the underlying causality behind this combination.

The relationships between different interpretations of probability itself and different interpretations of quantum mechanics are significant, and they have led to stimulating debates concerning the mathematical aspects of quantum probability. There are numerous arguments regarding what kind of probability theory – such as frequentist, Bayesian, Kolmogorovian, or contextual – is best suited to quantum theory, as discussed, for example, in (Ref. [36]). I shall, however, bypass the subject, since it does not change my main point here and in my overall discussion of "quantum causality," given that, as things stand now, quantum predictions are irreducibly probabilistic on experimental grounds, regardless of how we calculate or interpret the probabilities involved.

If, however, there is something like "quantum causality," it should be defined by quantum phenomena themselves, rather than by quantum mechanics in whatever interpretation or by any theory of these phenomena. As explained earlier (it is helpful to briefly revisit this point here), we can, ideally, *identically prepare* a given experiment in the sense of the state of our equipment in both classical and quantum physics. This identical preparation is essential for the very functioning of physics as science, since we must be able to repeat our experiments, *in this sense*, to maintain this functioning. However, unlike in classical physics, in the case of (individual) quantum experiments we can, rigorously, speak only of the identical preparations of such initial set-ups of the relevant observable parts of measuring instruments, but given different outcomes of these experiments, not of identically repeating the experiment in the sense of the behavior of quantum objects. In other words, there is never a guarantee that two quantum *objects* could ever be identically prepared under the identical conditions of the apparatus, and in general they are not. There is no experiment that would allow us to ascertain the identical initial physical states of quantum objects themselves in identically prepared quantum experiments. As a result, what we can repeat in terms of experimental outcomes are only observed probabilities and statistics of these outcomes, and these probabilities are well predicted by quantum mechanics, thereby properly confirmed experimentally as a theory of quantum phenomena.

In classical physics, we can, in principle, simultaneously ascertain both the position and the momentum of the object under investigation because we can observe this object without disturbing or interfering with it appreciably and thus apply the (idealized) realist and causal model of classical mechanics. It is the possibility, at least, again, in principle, of this definition at any point by determining *both* the position and the momentum that enables the (classically) causal character of classical mechanics as a proper theory of the individual behavior of classical objects. By contrast, in quantum physics, it is never possible to observe quantum objects independently of their interactions with measuring instruments, which thus always *interfere* with the behavior of quantum objects or disturb them. The uncertainty relations, which do not depend on quantum mechanics (although the latter is consistent with and in effect contains them or certain elements of formalism that are correlative to them) may be seen as a correlative to this impossibility. Quantum mechanics, again, reflects this situation and the difficulties of applying classical causality under these conditions.

As Bohr, who eventually preferred to speak of *interference*, came to realize, the language of "disturbance" is hardly suitable here (Ref. [5], v. 2, pp. 63–64). For, although this language is not entirely out of place, it may suggest that the independent behavior of quantum objects may be classical-like and specifically (classically) causal, and that it is only the interference of measuring instruments that introduces probability into our account of the situation. This view, often accompanied by the view that the formalism of quantum mechanics describes this classically causal behavior has not been uncommon even in standard quantum mechanics, and is found also in the work of several founding figures, including, as noted above, briefly, in Bohr (Ref. [5], v. 1, pp. 52–91; Ref. [1], pp. 191–211; Ref. [3], pp. 41–58). There is, however, no particular reason to adopt this view, especially given that (as is often acknowledged by those who hold it) that this independent causal behavior is in principle unobservable. As noted earlier, nobody has ever observed a quantum object, say, a moving photon, as such, apart from its effects on measuring instruments. There are quantum macro objects, such as Josephson's junctures, whose quantum behavior we can ascertain experimentally, as opposed to most other macro objects, which behave classically, although their ultimate constitution is quantum (or at least is generally assumed to be). However, engaging with the properly quantum behavior of such objects, that is, observing the corresponding quantum effects, requires proper measuring instruments. In other words, quantum macro objects are no more observable as quantum than are quantum micro objects. It does not appear possible, at least in Bohr's and related interpretations, to know or, again, even to conceive of what happens between quantum experiments. In Heisenberg's words: "There is no description of what happens to the system between the initial observation and the next measurement" (Ref. [22], p. 47).

It follows from the argument given in these lectures, quantum mechanics extends classical physics insofar as it is, just as classical physics, from Galileo on, and then relativity have been, the experimental-mathematical science of nature. However, quantum mechanics, at least, again, in the interpretations of the type adopted here, following Bohr, breaks with both classical physics and relativity by establishing new relationships between mathematics and physics, or mathematics and nature. The mathematics of quantum theory is able to predict correctly the experimental data in question without offering and even preventing the description of the physical processes responsible for these data. Taking advantage of and bringing together both main meanings of the word "experiment," I would argue, that, while not without some, indeed indispensable, help from nature, quantum mechanics was the first physical theory that is both, and jointly, truly experimental and truly mathematical. It is (I am indebted to G. Mauro D'Ariano on this point) truly experimental because it is not, as in classical physics, merely the independent behavior of the systems considered that we track, but what kinds of experiments we perform, how we *experiment* with nature, that defines what happens. Of course, we experiment, often with great ingenuity, in classical physics as well. There, however, our experiments do not define what happens, but essentially track what would have happened in any event. In quantum physics, for the first time, we can do something in *defining* the world by our experiments, and our experiments *cannot avoid doing so*. This last qualification is crucial because some of our classical experiments may also change the world if our interference is sufficient to significantly disturb the classical configuration involved. By the same token, quantum mechanics is truly mathematical because the mathematical formalism of the theory is not defined and hence constrained by this tracking of what would have happened anyway, but is concerned with predictions defined by our experiments. Indeed, it follows that we experiment with mathematics as well, in any event more so than in classical physics, since we invent mathematical schemes unrelated to any reality rather than refine our phenomenal perceptions or representations, which constrain us in classical physics.

Now, it is this determination, probabilistic though it is, of what can and conversely cannot happen by virtue of our experimental decisions that defines what I call "quantum causality." Or rather, since quantum events or phenomena may occur without our staging of quantum experiments, this determination is an instance of quantum causality in the case of human quantum experiments. This instance, however, provides a model of the more general definition, which I stated in the beginning of this section and which I now repeat with the preceding analysis in mind as a justification for it. *Whatever happens as a quantum event and is registered in the measuring instruments (thus providing us with the initial data) defines a possible set of, in general probabilistically, predictable outcomes of future events and irrevocably rules out the possibility of our predictions concerning certain other, such as and in particular complementary, events.*

At the same time, for the reasons explained earlier, each such event completely erases any data obtained in any preceding events as meaningful for the purposes of our predictions concerning future events from this point on. There is nothing that can help us to improve the probabilities of our predictions, neither the information previously obtained by measurements on the same object, nor a repetition of the same experiment with another quantum object of the same kind with the identical preparation, in the way it can be done in classical physics. On the one hand, no determinate connection to any past event can ever be guaranteed in the case of individual quantum events (again, no mechanical cause of any quantum event can be found) and no exact repetition that establishes classical-like regularity of events is possible. This situation clearly excludes both classical causality and, automatically, backward-in-time causality of any kind, classical or any other. On the other hand, quantum events do define future quantum events in strong, even if probabilistic, terms. In this sense the language of causality as referring to probabilistic correlations is appropriate, especially because, as thus defined, quantum causality, while irreducibly probabilistic, does refer, closer to Bayesian lines of thinking, to individual events, rather than only to statistical multiplicities of event.

The Bohr-Einstein debate concerning quantum mechanics may be considered from this perspective as well. Bohr saw quantum mechanics as a probabilistic theory of *individual* quantum processes, or more accurately, *individual* quantum phenomena or events, manifested in measuring instruments, since in his interpretation quantum mechanics does not describe quantum objects and behavior, but only probabilistically predicts the outcome of relevant experiments, in accord with quantum causality, as defined here. Einstein would prefer to see quantum mechanics as a statistical theory of multiplicities, on the model of classical statistical mechanics, under the assumption that a realist and classically causal theory, a proper mechanics, of individual quantum processes could eventually be developed. Accordingly, in Bohr's view, even though quantum mechanics did not describe quantum processes themselves, it would be a complete, as well a local, theory of these processes and observable phenomena and events they lead to, at least, as complete as nature allows our theory of quantum phenomena to be (e.g., Ref. [16]; also Ref. [1], pp. 237–278). In Einstein's view, quantum mechanics would not be a complete theory of individual quantum processes, of the type classical mechanics is, although he acknowledged, for example, in his letter to Born, that, as such a theory, quantum mechanics could be seen as local (Ref. [31], p. 205). Considered as a theory of individual quantum processes, quantum mechanics could only be seen as either incomplete or nonlocal in Einstein's view based on his analysis of the EPR-type experiment. Einstein was, as I said, correct in arguing that quantum mechanics, at least if viewed in the spirit of Copenhagen, is not a theory of individual quantum processes and events of the (realist and causal) *type* classical mechanics is. The question is, again, whether such a more complete (by Einstein's criteria) and local classical-like theory of quantum phenomena is possible, as Einstein hoped, or whether, as Bohr thought to be more likely, nature allows us only as much as quantum mechanics (within its proper scope) and higher-level quantum theories (within their scope) deliver. In this case, these theories would be complete (as well as, again, local), albeit in a sense different from that of Einstein, whose concept of completeness was modeled on classical mechanics. This question is still with us, and Einstein's view has continued to serve as an inspiration for many physicists and philosophers, Schrödinger, Bohm, J. S. Bell, and R. Penrose, among them, ever since, and it still does. Bohmian mechanics, for example, was in part inspired by this view, although the theory did not satisfy Einstein, because of its nonlocality (as strong a requirement for Einstein as for Bohr) and because it was, in his view, too close to standard quantum mechanics, in part by virtue of making exactly the same predictions as the latter. Einstein appears to have preferred to see quantum mechanics to be proven wrong one day on experimental grounds. Thus far, however, it has been amply confirmed experimentally and appears to have withstood all experimental attempts to disprove it.

As explained earlier, in order to effectively apply classical causality as part of our mathematical descriptive-predictive machinery, we impose artificial frames, most especially spatial and temporal ones, but also others, for example, by bracketing the atomic or quantum constitution of the physical objects considered. In some cases, our spatiotemporal frames may extend quite far, for example, in the history of the solar system or even in the known universe itself, nearly to its origin, some 14 billion years ago – nearly, but not quite, because the very early, pre-Big-Bang, history of the universe may be quantum. If such is the case, however, *in the present view* (assuming the same epistemology applies) our mathematical machinery will not be able to provide us with the description of this early quantum history as a quantum process. This does not mean that there is nothing that we can say about these early stages of the Universe as quantum events. Our classical observations of the traces of these early processes may be established as configurations of quantum phenomena, like the traces on the screen found in the double-slit experiments. These traces can give us information, physically classical, as concerns any actual part, each bit, for example, of this information, but organizationally quantum (just as in the case of the double-slit experiment), concerning this earlier history. This is similar to the case of the double-slit experiment, where the presence or conversely the absence of an interference pattern tells us something about the earlier conditions of the overall arrangement. A number of currently available (albeit hypothetical) theories of the early Universe, such as various versions of the "inflation" theory or the "cosmic landscape" theory, depend on an effective use this type of data. The very early history of the Universe, which is likely to have been purely quantum, might well have been erased without a trace and hence is altogether beyond our reach. The available traces might also enable us to meaningfully relate, probabilistically, different successive (classically manifest) stages of the Universe as quantum phenomena. They may even enable us to make predictions, again, probabilistic in character, concerning the future state of the Universe, by writing (which has been attempted), a Schrödinger-like equation for the state defined by these traces, once again, however, without enabling us to say anything about the quantum aspects of the process itself that will lead to this future state. We can only trace classical or (classically) relativistic aspects of this process, without, however, being able to connect or predict the corresponding events under these circumstances, that is, given the quantum nature of the processes that link these events.

It is also possible, however, that the epistemological argument offered in these lectures may only apply to quantum mechanics as a theory operative within its particular scope and limits, just as classical physics is operative within its scope and limits, or various quantum field theories are within their respective scopes and limits. Indeed, as will be discussed in the next lecture, the epistemology of quantum field theory, beginning with quantum electrodynamics, may well require still more radical departure from our classical epistemological ideas and ideals. The prospects are far less certain for the more comprehensive theories that are necessary (as our theories at present are manifestly incomplete) but yet to be developed (see, for example, Ref [37]). As we haven't heard their last word (which is to say their next word), one cannot be sure. Nature might show itself to be less mysterious at the next stage of our, it appears, interminable and interminably inconclusive encounter with it, or, just as it did in the case of quantum phenomena in the last century, it might confront us with something more mysterious than we can imagine now, even with quantum theory in hand.

3. Mathematics, physics, and philosophy in quantum field theory

3.1. *Summary*

The philosophy of quantum field theory is still far from sufficiently developed in comparison with the philosophy of quantum mechanics, even though quantum field theory itself is virtually as old as quantum mechanics is. It is true that the situation has changed somewhat more recently, as one might surmise from Ref. [38], in general, a helpful source concerning the current state of the theory. Still, one could think only of a few books devoted to the philosophy of quantum field theory, in contrast to an unending stream of books on the philosophy of quantum mechanics. This lecture will address some of the key philosophical aspects of quantum field theory, beginning with quantum electrodynamics, and the more complex (physically, mathematically, and philosophically) character of the theory vis-à-vis quantum mechanics, from the perspective, adopted throughout these lectures, of the nature of the quantum-field-theoretical vs. quantum-mechanical *way of thinking* about quantum phenomena. My main focus is Dirac's thinking that led him to his famous equation for the relativistic electron in 1928. This thinking was inspired by that of Heisenberg in his discovery of quantum mechanics, as discussed in Lecture 1, but it also displayed several highly original moves, which enabled Dirac to discover the extraordinary mathematical architecture that grounded his equation and that became the foundation for the subsequent development of quantum electrodynamics and quantum field theory. By using this type of architecture, these theories have demonstrated their remarkable capacity to handle the corresponding (high-energy) quantum phenomena throughout their history. At least in the view adopted here, quantum electrodynamics and quantum field theory are, just as is quantum mechanics, probabilistically predictive and not descriptive theories. Their predictive effectiveness is extraordinary, however. Thus, quantum electrodynamics is the best-confirmed physical theory ever. At the same time, as became clear shortly after the introduction of quantum electrodynamics and then quantum field theory, the theory contained certain troubling and, to many, Dirac was among them, unacceptable features, most especially, although not exclusively, those stemming from the presence of mathematically illegitimate infinities or divergences in the formalism of the theory. These difficulties have continued to plague the theory throughout its history, and they are still with us now. These difficulties are managed, although not, strictly speaking, resolved, by means of the so-called renormalization procedure, developed, in the case of quantum electrodynamics, by about 1950, and, for other forms of quantum field theory (which are parts of the so-called standard model of particle physics), in the 1970s. There are a number of possible reasons for these difficulties, arguably, primarily having to do with various idealizations involved in the composition of quantum field theory, some inherited from quantum mechanics or even classical physics, and others new. Some of them appear, especially if one adopts the present interpretation of quantum theory, in which measurement plays

a defining role, to arise because of the further complexities involved in our idealized treatment of measuring instruments in the (high-energy) regimes governed
by quantum field theory, vis-à-vis the (low-energy) regimes governed by quantum
mechanics. Specifically, as became apparent in the early 1930s, following Bohr and
Rosenfeld's analysis of "the measurability of electromagnetic field quantities" (Ref.
[39]), in measuring these quantities the atomic structure of measuring instruments
can have a greater impact than in the case of measuring the quantities considered
in quantum mechanics. On the other hand, quantum field theory, as it was constituted then and as it is constituted now, does not consider this structure any
more than does quantum mechanics, in which case, however, there are no infinities necessitating renormalization. I shall devote part of this lecture, Section 2, to
this situation, which is intriguing epistemologically, even if it is not uniquely or
even primarily responsible for the difficulties of quantum field theory leading to the
necessity of renormalization. Most of the lecture, however, considers the mathematical architecture of quantum field theory, for which Dirac's discovery of quantum
electrodynamics and (my main focus here) his equation for the relativistic electron
laid the foundations.

3.2. *From Heisenberg to Dirac, from quantum mechanics to quantum field theory*

In 1949, Bohr described Dirac's theory of the relativistic electron, as, on the one
hand, "a most striking illustration of the power and fertility of the general quantum-
mechanical way of description" and, on the other, as reflecting "new fundamental
features of atomicity, which ... demanded a still more radical [than in quantum me-
chanics] renunciation of explanation in terms of a pictorial representation" (Ref. [5],
v. 2, p. 63). Technically, as discussed here, even in Bohr's interpretation of low-
energy quantum phenomena, considered in quantum mechanics, quantum objects,
such as electrons, and their behavior are beyond any description we can imagine
and ultimately even conception that we can form. Accordingly, it is not clear how
one can renounce one's explanation in terms of a pictorial representation still more
radically. Instead, as I shall argue here, while quantum mechanics and Dirac's theory share (in this interpretation) this epistemology, the difference between them is
defined by a new form of multiplicity of quantum phenomena, revealed by Dirac's
discovery of antimatter. This discovery that was a consequence of his theory and
was experimentally confirmed shortly thereafter, although it took a few more years
to realize that one was in fact dealing with antimatter. The term "atomicity" in
the above quotation is used by Bohr in his special sense, essentially equivalent to
his concept of phenomena, as discussed in Lecture 1. As such, "atomicity" refers
not to the indivisible ("atomic" in the original Greek sense of the word) nature
of elemental quantum objects themselves or, as we call them, "elementary particles," such as electrons and photons, but instead to individual quantum phenomena
observed in measuring instruments, impacted by quantum objects. Phenomena in

Bohr's sense are indivisible not physically but epistemologically, because, in Bohr's interpretation, we cannot separate the behavior of quantum objects from their interactions with measuring instruments and, as a result, cannot track this behavior independently in the way it is possible in classical physics or relativity. We only observe impacts of quantum objects on measuring instruments and study the effects of such impacts, and nobody has ever observed a moving electron or photon as such. In Bohr's interpretation, the character of these effects prevents us from forming a pictorial or any other representation of the independent behavior of quantum objects in any quantum regime. The configurations of these effects, while retaining most key epistemological features of (low-energy) quantum-mechanical phenomena, reveal new complexities, especially as concerns their multiplicity, in the case of quantum phenomena in higher-energy regimes, treated by quantum field theory, beginning with quantum electrodynamics. The situation may be illustrated by comparing Heisenberg's and Dirac's discoveries of, respectively, quantum mechanics and quantum electrodynamics, and specifically Dirac's equation.

As discussed in Lecture 1, Heisenberg's thinking leading him to his discovery of quantum mechanics was defined by three key elements, manifested in his paper announcing his discovery (Ref. [7]). The same elements, I argue, defined Dirac's work on quantum electrodynamics, most especially his discovery of his relativistic equation for the electron (Ref. [40]). Dirac's derivation of the equation was significantly influenced by Heisenberg's paper. As is well known, Dirac read the paper very carefully in 1925, and it inspired his 1925–1926 work on quantum mechanics and quantum electrodynamics, from which, especially his transformation theory (independently discovered by P. Jordan), his work on his equation grew (Ref. [41]). It is true that Dirac's work on his equation was also indebted to other developments in quantum mechanics, most especially Schrödinger's equation, the transformation theory, and Pauli's spin theory (Ref. [42]). Nevertheless, Heisenberg's thinking in his 1925 paper, and specifically the three key points in question, exerted a profound influence on Dirac's work on his equation. It may be helpful to restate here the three key elements of Heisenberg's discovery, described in Lecture 1, in parallel with the role of the same type of elements in Dirac's discovery of his equation:

(1) *The Mathematical Correspondence Principle.* This principle, as used by Heisenberg, mathematically formalizes Bohr's, more heuristic, correspondence principle used by Bohr and others in the old quantum theory, and it can, I argue, be extended beyond quantum mechanics, to quantum electrodynamics and other forms of quantum field theory. Considered in this more rigorous, mathematical form, the principle requires recovering the equations and variables of the old theory, classical mechanics in the case of quantum mechanics and quantum mechanics in the case of quantum field theory, in the limit region where the old theory could be used. Rather than having become obsolete after quantum mechanics, as has been sometimes argued, the principle, in this mathematical form, has continued to play a major role in the development of

quantum theory. It still does, for example, in string and brane theories, where the corresponding limit theory is quantum field theory. As will be seen, however, the mathematical correspondence principle might have also been in part responsible for the mathematically divergent nature of quantum electrodynamics and quantum field theory, which requires renormalization to make these theories workable.

(2) *The Introduction of a New Type of Variables.* Arguably most centrally, both discoveries, that of Heisenberg and that of Dirac, were characterized by the introduction of new types of variables.

(QM) In the case of quantum mechanics, these were matrix variables with *complex* coefficients, essentially operators in Hilbert spaces over *complex* numbers (apparently, unavoidable in quantum mechanics) vs. classical physical variables, which are differential functions of *real* variables. Heisenberg formally retained the equations of classical physics.

(QED) In the case of quantum electrodynamics, these were Dirac's spinors and multicomponent wave functions, which, jointly, entail more complex operator variables, and a more complex structure of the corresponding Hilbert spaces, again, over complex numbers (equally unavoidable in quantum field theory). In contrast to Heisenberg, Dirac also introduced a new equation, Dirac's equation, formally different from Schrödinger's equation, which, in accordance with the mathematical correspondence principle, is a far nonrelativistic limit of Dirac's theory, via Pauli's theory, the immediate nonrelativistic limit of Dirac's theory. Following Dirac or, again, Heisenberg, finding new matrix variables or Hilbert-space operators became the uniquely defining mathematical element of any quantum field theory.

(3) *A Probabilistically Predictive Character of the Theory.* This change in mathematical variables was accompanied by a fundamental changes in physics: the variables and equations of quantum mechanics and quantum electrodynamics no longer described, even by way of idealization, the properties and behavior of quantum objects themselves, in the way classical physics or relativity do for classical objects. Instead, the formalism only predicts the outcomes of events, in general probabilistically (even in the case of individual events), and of statistical correlations between some of these events, thus establishing a new type of relationship between mathematics and physics.

As explained in Lecture 1, Heisenberg's revolutionary thinking established a new way of doing theoretical physics, and, as a consequence, it redefined experimental physics as well. The practice of experimental physics no longer consists, as in classical experiments, in tracking the independent behavior of the systems considered, but in *unavoidably* creating configurations of experimental technology that reflect the fact that what happens is *unavoidably* defined by what kinds of experiments we perform, how we affect quantum objects, rather than only by their independent behavior. My emphasis on "unavoidably" reflects the fact that, while the behavior of classical physical objects is sometimes affected by experimental technology, in general we can observe classical physical objects, such as planets moving around the sun, without appreciably affecting their behavior. This does not appear to be possible in quantum experiments. The practice of theoretical physics no longer consists, as in classical physics or relativity, in offering an idealized mathematical description of quantum objects and their behavior. Instead it consists in developing mathematical machinery that is able to predict, in general (again, in accordance with what obtains in experiments) probabilistically, the outcomes of quantum events and of correlations between some of these events.

The situation takes a more radical form in quantum field theory and the experimental physics in the corresponding (high) energy regimes. While retaining, at least in the present view, Heisenberg's nonrealist and noncausal epistemology of quantum mechanics, quantum field theory is characterized by, correlatively:

(1) more complex configurations of phenomena observed and hence measuring apparatuses involved, and thus more complex configurations of effects of the interactions between quantum objects and measuring instruments;

(2) a more complex nature of the mathematical formalism of theory, in part reflected in the necessity of renormalization;

(3) a more complex character of quantum-field-theoretical predictions and, hence, of the relationships between the mathematical formalism and the measuring instruments involved.

A few brief historical remarks might be in order. Quantum electrodynamics was introduced by Dirac in 1926–1927 as a theory in which both electrons and photons were treated as particles, and his relativistic equation for the free electron, discovered in 1928, followed the same approach. In deriving his equation, Dirac developed a (Hamiltonian) formalism applied to a new type of matrix variables, analogous to but more complex than those used in quantum mechanics, and based on the mathematics of the so-called spinors, which form what is known as Clifford algebras. By contrast, in 1930, Pauli and Heisenberg developed, by analogy with classical electrodynamics, a classical field theory of electromagnetic radiation, which was introduced in the nineteenth century by J. C. Maxwell and which was based on a new concept, that of field, developed by M. Faraday and Maxwell (Ref. [43]). A bit later, E. Fermi introduced a version of quantum field theory of weak nuclear forces in which the photons were treated in terms of (quantum) fields and the electrons in terms of particles (Ref. [44]). Then Heisenberg and H. Yukawa introduced the quantum field theory of strong nuclear forces, mediating quantum interactions between the constituents of nuclei (proton and neutron) and Yukawa postulated the existence of the meson, as the elementary particle corresponding to this field. In the modern version of this theory, quantum chromodynamics, the particles carrying the nuclear forces, now coming in three kinds, are gluons, which mediate the interaction between quarks. Protons, neutrons, and mesons (the particles that appeared to correspond to those predicted by Yukawa were discovered and named "mesons,"

but they eventually proved to be something other than carriers of the nuclear force)
all proved to be composite, consisting of quarks and gluons.

Eventually, it became clear that most forms of quantum field theory could be
presented mathematically equivalently in either "particle" or "field" terms, in a
(qualified) parallel with "particle" and "wave" versions in quantum mechanics. Its
name, "quantum *field* theory," notwithstanding, the theory has remained a theory
of both "particles" and "fields," allowing one to use, again, symbolically, either
"picture," or to variously combine both, depending on one's need or preference.
In this respect, the situation is analogous to the one that obtains in quantum
mechanics, where one could use either Schrödinger's "wave" equation or a more
algebraic formalism of matrix mechanics, or that of transformation theory that
combines both. From around 1930 on, the more unified and more flexible Hilbert-
space formalism, introduced by J. von Neumann, is used in quantum mechanics and
quantum field theory alike.

In the present interpretation, just as "particles" and "waves" in quantum mechanics, "particles" and "fields" in quantum field theory could, as physical terms,
only be used provisionally or by a symbolic analogy, at least in the interpretation of
both theories adopted here. Thus, while they do relate to physical entities that exist, the so-called "elementary particles" of quantum physics, cannot, at least, again,
in this interpretation, be conceptualized by means of any particle-like models currently available, certainly not that of classical mechanics. Nor, conversely, can they
be conceived on the model of classical concept of field, which was abandoned by
Pauli and Heisenberg in their pioneering paper. While Pauli and Heisenberg's approach was similar to that of Schrödinger in developing his wave mechanics, unlike
Schrödinger, Pauli and Heisenberg suspended the physical picture of wave or field
propagation, and thus of classical field from the outset. In the classical electromagnetic theory of Faraday and Maxwell, the concept of field associates with each point
of a propagating "field" a vector or a set of vectors, representing actual physical
forces, active at this point, which gives rise to a kind of geometrical picture of the
field. Nothing like that was found in Pauli and Heisenberg's theory, in this respect
extending, just as Dirac did, the approach Heisenberg used in creating quantum
mechanics, in contrast to Schrödinger's wave program, initially conceived on realist
lines. Heisenberg and Pauli used equations symbolically analogous to the (wave)
equations of classical electromagnetism, Maxwell's equations, to develop the mathematics of their theory, but again, with new quantum types of variables. These
equations, too, only predicted the probabilities of the outcomes of the experiments
in question, along the lines of Heisenberg's initial approach to quantum mechanics and, by then, Bohr's interpretation of quantum mechanics. Dirac adopted the
same view of his approach to quantum electrodynamics and in his derivation of his
equation for the electron.

Just as in the case of quantum mechanics, however, this epistemological attitude to quantum field theory met and has continued to be met with considerably

resistance. The epistemological and ontological status of "particles" and "fields" has remained the subject of the continuing and still ongoing debates, extending the debate concerning the nature of quantum phenomena and quantum objects in quantum mechanics to quantum phenomena and theories in high-energy regimes (e.g., Ref. [37 and 45]). No end of either debate appears to be in sight. Amidst these debates, quantum mechanics and quantum field theory, culminating in the so-called standard model of particle physics, remain our best theories of the ultimate constitution of nature, considered apart from gravity. We do not as yet have workable quantum theories that incorporate gravity and, thus, establish proper connections between quantum theory and general relativity. All of the theories available thus far, such as string or brane theory, that aim to achieve this goal remain highly hypothetical.

3.3. *Quantum theory and measurement*

The new complexities of quantum field theory appear to be connected to the question of measurement, which takes on new dimensions, vis-à-vis the quantum-mechanical measurement, in quantum field theory, beginning with quantum electrodynamics, or even quantum field theory in low-energy regimes. I would like to address this question by way Pauli's intriguing "conjecture" that "the observer in the present-day [1957] physics is still too completely detached, and that physics will depart still further from the classical example" (Ref. [14], p. 132). This conjecture arises in view of the departure of quantum theory from the classical (realist) ideal, that of "the detached observer," advocated by Einstein, as discussed in Lecture 1, via Pauli's article, "Phenomenon and Physical Reality," just cited (Ref. [14]). For Einstein, the observer, as theorized by quantum mechanics, is not sufficiently detached from quantum objects and their behavior. Indeed for Einstein, a proper physical theory needs to see the observer as completely detached from the objects considered, given that such a theory must account and in fact describe the independent behavior of these objects, according to Einstein, by means of freely chosen mathematical concepts, a qualification that moves Einstein's position beyond positivism or naïve realism. For Pauli, by contrast, although, quantum theory makes a "definitive ... step ... of including the observer and the conditions of experiment in a more fundamental way in the physical explanation of nature," the observer may still be too detached, even "too completely detached," for quantum theory in its present form to be fully successful (Ref. [14], p. 132). In stating this conjecture, Pauli does not refer only or perhaps even primarily to quantum field theory. There is, however, little doubt that the statement, especially Pauli's expectation "that physics will depart still further from the classical example," was made with quantum field theory in mind. In the same article, "Phenomenon and Physical Reality," Pauli sees quantum field theory as related to "some fundamental problems" still hidden in the relationships "between observer or instrument of observation and the system observed" (Ref. [14], p. 133). Pauli was involved in its developments for decades by then, and was in particular engaged in intense and sometimes heated exchanges with J. Schwinger concerning the renormalization of quantum electrodynamics (Ref. [46], pp. 345–352). The new physical and epistemological complexities, which were introduced by quantum electrodynamics and quantum field theory, appear to be connected to measurement.

Bohr addressed the subject at the early stages of the development of the theory in the 1930s, both in his influential collaboration with Rosenfeld on measurement in quantum field theory in Ref. [39]) (updated in 1950 to Ref. [47]) and other writings on quantum mechanics. Quantum mechanics is greatly helped by the fact that we can disregard the quantum constitution of measuring instruments, even though this constitution is responsible for the emergence of the (low-energy) quantum phenomena. As Bohr explains: "Although, of course, the existence of the quantum of action [h] is ultimately responsible for the properties of the materials of which the measuring instruments are built and on which the functioning of the recording devices depends, this circumstance is not relevant for the problem of the adequacy and completeness of the quantum-mechanical description in its aspects here discussed [i.e., as concerns what can be actually observed and predicted]" (Ref. [5], v. 2, p. 51). In other words, the theory disregards the atomic structure of the measuring instruments and, hence, the quantum interactions between them and quantum objects, but we are lucky to get away with it. The traces left by these interactions in the measuring instruments and thus in the classical world we perceive suffice to do the job. To be more precise, the measuring instruments are, at the time of measurement, *not detached* from quantum objects. They *interact* with quantum objects, and these interactions are quantum, even though they produce classical effects in certain observable parts of the instruments, which parts are described by means of classical physics. But by disregarding, as quantum mechanics allows us to do, these quantum interactions between quantum objects and measuring instruments, and by limiting us to the classical data thus created and observed (after the measurement has taken places and its outcome is established) makes these observable parts of the instruments and, thus, us, as human observers, detached from quantum objects. As Bohr argued already in 1930s, the situation may, however, entail new complexities in the quantum-field-theoretical situation where the quantum constitution of measuring instruments might need to be taken into account, even though quantum field theory disregarded this constitution as well, just as does quantum mechanics. This, perhaps unavoidably, is still the case now. In his 1937 "Causality and Complementarity," Bohr says:

> On closer consideration, the present formulation of quantum mechanics in spite of its great fruitfulness would yet seem to be no more than a first step in the necessary generalization of the classical mode of description, justified only by the possibility of disregarding in its domain of application the atomic structure of the measuring instruments themselves in the interpretation of the results of experiment. For a correlation of still deeper laws of nature involv-

ing not only the mutual interaction of the so-called elementary constituents of matter but also the stability of their existence, this last assumption can no longer be maintained, as we must be prepared for a more comprehensive generalization of the complementary mode of description which will demand a still more radical renunciation of the usual claims of the so-called visualization. (Ref. [35], p. 88)

As I noted earlier, in commenting on Bohr's 1949 reprisal of this point, made in expressly referring to Dirac's theory of the electron, clearly on Bohr's mind here as well, one might wonder how much more radical this renunciation might be, if all claims concerning visualization are already renounced altogether even in quantum mechanics in Bohr's interpretation. I shall return to these connections later, especially in the context of the stability or identity of elementary particles, which Dirac's discovery of antimatter no longer allows one to maintain, eventually leading to the concept of virtual particle formation. For the moment, this statement must have been brought about by Bohr's work with Rosenfeld on measurement in quantum field theory in response to L. Landau and R. Peierls's argument (Ref. [39]). The latter argument, contested by Bohr and Rosenfeld, concerned a possible inapplicability of the uncertainty relations in quantum field theory, a subject that I shall put aside. My main concern here is Bohr and Rosenfeld's analysis of measurement in quantum field theory and the disregard of "the atomic structure of the measuring instruments" there. According to Bohr and Rosenfeld:

Insofar as we can disregard all restrictions arising from the atomistic structure of the measuring instruments, it is actually possible to demonstrate a complete accord [with complementarity and the uncertainty relations]. Besides a thorough investigation of the construction and handling of the test bodies, this demonstration requires, however, consideration of certain new features of the complementary mode of description, which come to light in the discussion of the measurability question, but which were not included in the customary formulation of the indeterminacy principle in connection with non-relativistic quantum mechanics. Not only is it an essential complication of the problem of field measurements that, when comparing field averages over different space-time regions, we cannot in an unambiguous way speak about a temporal sequence of the measurement processes; but even the interpretation of individual measurement results requires a still greater caution in the case of field measurements than in the usual quantum-mechanical measurement problem. (Ref. [39], pp. 480–481)

Thus, while, contrary to Landau and Peierls, consistent with the uncertainty relation (in the form they take in quantum field theory) the measurements of quantum fields involve additional levels and complexities of idealization, related to "the atomic structure of the field sources and the measuring instruments." Ideally, one might prefer a theory to be able to take this structure into account, but quantum field theory could not do so then or perhaps, in principle, ever. It still cannot do now. Is this in fact necessary? This is one of my questions here, and my answer, as it appears, was that of Bohr and Rosenfeld, is "Perhaps!" For the moment, Bohr and Rosenfeld conclude their analysis as follows:

> We thus have arrived at the conclusion already stated at the beginning, that with respect to the measurability question the quantum theory of fields represents a consistent idealization to the extent that we can disregard all limitations due to the atomic structure of the field sources and the measuring instruments ... this result should properly be regarded as an immediate consequence of the fact that both the [quantum-electromagnetic] formalism and the viewpoints on which the possibilities of testing this formalism are to be assessed have as their common foundation the correspondence argument. Nevertheless, it would seem that the somewhat complicated character of the considerations used to demonstrate the agreement between formalism and measurability are hardly avoidable. For in the first place the physical requirements to be imposed on the measuring arrangement are conditioned by the integral form in which the assertions of the quantum-electromagnetic formalism are expressed, whereby the peculiar simplicity of the classical field theory as a purely differential theory is lost. Furthermore, as we have seen, the interpretation of the measuring results and their utilization by means of the formalism require consideration of certain features of the complementary mode of description which do not appear in the measurement problems of non-relativistic quantum mechanics. (Ref. [39], pp. 520–521)

This assessment has proven to be highly relevant the subsequent development of quantum electrodynamics and quantum field theory, including as concerns certain problems accompanying this development. These problems had begun to emerge a few years before Bohr and Rosenfeld's paper. They have primarily to do with the appearance of infinities or divergences in the theory once one attempted to use the theory, including Dirac's equation, to make calculations that would provide closer approximations matching certain experimentally observed data. These difficulties were eventually handled through the renormalization procedure, which became and has been ever since a crucial part of the machinery of quantum electrodynamics and quantum field theory. In the case of quantum electrodynamics, renormalization was performed in the later 1940s by S-I. Tomonaga, J. Schwinger, and R. Feynman (which brought them a joint Nobel Prize in 1965), with important contributions by others, especially F. Dyson, and earlier H. Bethe and H. Kramers. The Yang-Mills theory, which grounds the standard model, was eventually shown to be renormalizable as well by M. Veltman and G. 't Hooft in the 1970s (bringing them their Nobel Prize). This allowed a proper development of the standard model of all forces of nature, except for gravity, which has not, thus far, been given its quantum form.

The renormalization procedure is difficult mathematically even in quantum electrodynamics (the mathematics of the electro-weak theory or of quantum chromodynamics, which handles the strong force, is nearly prohibitively difficult). While it is possible to see renormalization in more benign ways (e.g., via the so-called "renormalization group" and "effective quantum field theories") and while it has been vey effective thus fact, its mathematical legitimacy is still not really established. Roughly speaking, the procedure might be seen as manipulating infinite integrals that are divergent and, hence, mathematically illegitimate. At a certain stage of calculation, however, these integrals are replaced by finite integrals through artificial cut-offs that have no proper mathematical justification within the formalism and are performed by putting in, by hand, experimentally obtained numbers that make these integrals finite, which removes the infinities from the final results of calculations. These calculations are experimentally confirmed to a very high degree. Quantum electrodynamics is the best experimentally confirmed theory thus far. I will not address the details of renormalization further here, and will allow myself to refer to (Refs. [38 and 45] and Ref [48, pp 149–168]) and, for a historical account, to (Ref. [46], pp. 595–605). The discussion of the relevant subsequent developments, such as the renormalization group, effective quantum field theories, and so forth is also beyond my scope (but see, again, Refs. [38 and 45]).

One might ask, however, why quantum field theory, say, quantum electrodynamics, contains such infinities, in contrast to quantum mechanics. The subject is still little explored philosophically, although some helpful insights are found in Refs. [38, 45, and 48]. Bohr's argumentation, cited above, arising from the analysis given in his paper with Rosenfeld, suggests one possible reason, which is of particular interest in the epistemological context of these lectures. I hasten to add that other possible and perhaps more plausible reasons have been advanced, but I shall not address here, even though some of them may be connected to Bohr's argumentation. As just explained, the mathematical formalism of quantum electrodynamics is essentially linked to the same idealization of measurement that is use in the low-energy the quantum-mechanical regimes, insofar as both idealization disregard "the atomic structure of the field sources and the measuring instruments." Could this idealization, which poses no problems of this type in quantum mechanics, be responsible for the appearance of the illegitimate infinities in quantum field theory? Dyson thought so, in part under impact of Bohr and Rosenfeld's analysis and his discussions with R. Oppenheim (Ref [46], p. 549). Dyson's argument was not without significant qualifications, and, as will be seen presently, he was not arguing, quite the contrary, that one needs to replace quantum electrodynamics or for that matter even avoid renormalization. As Dyson said:

We interpret the contrast between the divergent Hamiltonian formalism and the finite S-matrix as a contrast between two pictures of the world, seen by two observers having a different choice of measuring equipment at their disposal. The first picture is of a collection of quantized fields with localizable interactions, and is seen by a fictitious observer whose apparatus has no atomic structure and whose measurements are limited in accuracy only by the existence of the fundamental constants c and h. This ["ideal"] observer is able to make with complete freedom on a sub-microscopic scale the kind of observations which

Bohr and Rosenfeld employ ... in their classical discussion of the measurability of field-quantities. The second picture is of collections of observable quantities (in the terminology of Heisenberg) and is the picture seen by a real observer, whose apparatus consists of atoms and elementary particles and whose measurements are limited in accuracy not only by c and h, but also by other constants such as α [the fine-structure constant] and m [the mass of the electron]. (Ref. [49], p. 1755, cited in Ref.'[46], pp. 547–548)

S. Schweber's commentary on this passage further clarifies the situation:

A "real observer" can measure energy levels, and perform experiments involving the scattering of various elementary particles – the observables of S-matrix theory – but cannot measure field strengths in small regions of spacetime. The "ideal" observer, making use of the kind of "ideal" apparatus described by Bohr and Rosenfeld, can make measurements of this last kind, and the commutation relations of the fields can be interpreted in terms of such measurements. The Hamiltonian density will presumably always remain unobservable to the real observer whereas the ideal observer, "using nonatomic apparatus whose location in space and time is known with infinite precision" is presumed to be able to measure the interaction's Hamiltonian density. "In conformity with the Heisenberg uncertainty principle, it can perhaps be considered a physical consequence of the infinitely precise knowledge of location allowed to the ideal observer, that the value obtained by him when he measures Hamiltonian density is infinity." If this analysis is correct, Dyson speculated, *the divergences of QED are directly attributable "to the fact that the Hamiltonian formalism is based upon an idealized conceptualization of measurability"* (Ref. [46] 1994, p. 548, citing Dyson in Ref. [49], p. 1755; emphasis added)

In other words, "*if this analysis is correct*" (it continues to remain conjectural), in our interactions, via measuring instruments, with quantum objects we are, thus far, "ideal observers." The observable strata of these instruments are still described classically, just as they are in quantum mechanics, and specifically by disregarding their atomic constitution, which would compel us to take into account such constants as α and m. The necessity to take these constants into account does not mean that we need to actually describe this atomic constitution and the quantum interaction between quantum objects and the measuring instruments, which may not be not be possible and is not possible, according to the epistemology adopted here, which in principle preclude such a description. However, even if we continue to restrict our theory to probabilistic predictions, in the absence of any description of quantum objects and processes (including, again, those occurring in the measuring instruments while interacting with quantum objects), taking into account, in this nondescriptive way, the atomic structure of the measuring instruments would require a different type of theory. Would such a theory, assuming it is possible, be able to avoid the infinities? Perhaps! The idealized conceptualization of measurability in question is only one of many idealizations found in quantum field theory. Some of these idealizations are shared with quantum mechanics, such as disregarding the atomic structure of measuring instruments, or treating elementary particles as dimensionless points, which is reflected in the Hamiltonian formalism of both (still keeping in mind that in the epistemology adopted here, no claim is made concerning the independent constitution of elementary particles or any quantum objects). Others are going further than those of quantum mechanics. Renormalization may be due to the insufficiency of any of these idealizations, or any given combination of them. And then, is there even a strongly compelling necessity to avoid infinities or get rid of renormalization?

Dyson did not think that his considerations, just cited, implied that quantum electrodynamics should be replaced by an essentially different theory, and in fact he did not think that that there is, in view of his argument, anything especially wrong with the infinities found quantum electrodynamics or renormalization; and Bohr and Rosenfeld's updated (1950) version of their analysis, which addressed renormalizability, appears to confirm this view as well (Ref. [47]). As Schweber explains: "If [Dyson's] notion of measurability is accepted, the correlation between expressions which are unobservable to a real observer and expressions which are infinite 'is a physically intelligible and acceptable feature of the theory. The paradox is the fact that it is necessary ... to start from the infinite expressions in order to deduce finite ones.' What may be therefore looked for in 'a finite theory' is not necessarily a modification of the present theory which will make all infinite quantities finite 'but rather a turning around of the theory so that all the finite quantities shall become primary and the infinite quantities secondary"' (Ref. [46], p. 548, Ref. [49], p. 1755]. I shall comment on one way in which Dyson thought this "turning around" might be possible presently. Dyson concluded that "the purpose of the foregoing remarks is merely to point out that there is no longer [once renormalization is shown to work], as there seemed to be in the past, a compelling necessity for a future theory to abandon some essential features of the present electrodynamics. The present electrodynamics is certainly incomplete, but is [in view of its renormalizablity] no longer certainly incorrect" (Ref. [49], p. 1755; cited in Ref. [46], p. 548). A very elegant turn of a phrase: It might still prove to be incorrect, but it is no longer *certainly* incorrect! No surprise here: Dyson knows how to write well.

Nevertheless, Dyson's conjecture is worth pondering a bit further, for the following reason. The Hamiltonian character of the quantum formalism was initially brought into quantum mechanics from classical mechanics, via Bohr's correspondence principle in its mathematical form, by directly transferring the equations of classical mechanics, while changing their variables to those of quantum mechanics. In addition, this formalism was made purely predictive and, moreover, probabilistically predictive, even ideally, and lost its descriptive capacity, which, along with its ideally exact predictions, defined its use in classical mechanics. The formalism was then "transferred" by Dirac, again, as a probabilistically predictive formalism, into

quantum electrodynamics, via the adjusted version of the mathematical correspondence principle, adjusted because Dirac's equation must convert into Schrödinger's
equation at the quantum-mechanical limit, and the same type of conversion is found
elsewhere in quantum electrodynamics and quantum field theory. Accordingly, unlike in quantum mechanics, the formalism was changed formally as well (hence my
quotation marks around "transferred"). Nevertheless, it was still ultimately a de-
scendent of the Hamiltonian mathematical formalism of classical mechanics. Neither
classical nor quantum mechanics need to take into account the atomic structure of
measuring instruments or sources of fields, classical mechanics naturally (and it can
in fact disregard the role of measuring instruments altogether) and quantum mechanics, it appears, without any real physical justification, but luckily still allowing
us to make correct predictions concerning the outcomes of quantum experiments.
With quantum electrodynamics we run out of luck as concerns the finite nature of
the formalism, although not out of luck as concerns physics, because renormalization works. It follows that, while the correspondence principle, which was given
a mathematical form by Heisenberg, was crucial for the development of quantum
mechanics and then quantum electrodynamics, it may have also been primarily
responsible for the divergent nature of the mathematical formalism of quantum
field theory. Indeed, the correspondence principle and a possible inadequacy of the
Hamiltonian character of the quantum-field-theoretical formalism may be responsible for the divergencies found in quantum field theory even if they are, physically,
due to other factors than our neglect of the atomic structure of measuring instruments, insofar as these factors are reflected in the Hamiltonian formalism of the
theory. The Hamiltonian framework may also be responsible for us not being able
to turn it around the way Dyson suggested – by making primary the finite rather
than the infinite quantities in the theory. Indeed, as Schweber notes, Dyson (and a
few others, for example, at one point Dirac) believed that the future quantum-field
theories "would probably be based on a modified Lagrangian approach" (Ref. [46],
p. 549). That did not really proved to be the case, but in any event, it would still
be too close to classical physics, for better or worth. Thus far, though, remaining
connected in this and other ways to classical physics was mostly for the better, thus
both justifying classical physics itself beyond its limits and, through it, giving all
modern physics its continuity, amidst its several revolutions, radical as the latter
may have been in turn. We might need both.

At least, we continue to remain lucky with our physics, because renormalization
works thus far, at least, again, insofar as our quantum theories can disregard gravity
(although there are indication such theory, if developed, are in fact likely to be
finite). Many predictions of quantum field theory have been spectacular. The
discovery of Higgs boson, an essential component of the Standard Model, is the latest
example, although some of the earlier discoveries, such as those of the electroweak
(W^+, W^-, and Z) bosons, the top quark, and the tau-lepton, are hardly less
significant. The Higgs discovery may still require further confirmation, and the data

involved may require advancing and perhaps adjusting the standard model itself, which is not entirely complete in any event (Ref. [50]). However, the discovery was deemed confirmed enough (that two independent detectors indicated a likely presence of the Higgs particle helped) to award the 2013 Nobel Prize to F. Englert and P. Higgs, who were among those involved in development of the mathematical theory of the Higgs field. Our future fundamental theories might prove to be finite (some versions of string and brane theory appear to hold such a promise), thus proving that the necessity of renormalization is merely the result of the limited reach of our quantum theories at present. (Effective quantum field theories are based on this view.) The emergence of other finite alternatives, proceeding along entirely different trajectories, may not be excluded either. While, however, a finite theory may be preferable, renormalization may not be a very big price to pay for the theory's extraordinary capacity to predict the increasingly complex manifold of quantum phenomena that physics has confronted throughout the history of quantum field theory. But then, a finite theory may also not be possible, in which case renormalization will continue to be our main hope.

Quantum field theory continues to provide an extraordinarily effective mathematical technology to handle ever-new types of phenomena observed in ever-new configurations of measuring instruments. The Linear Hadron Collider is the latest example of the machine (in either sense) for creating such configurations, one of which (a very complex one) enabled us to confirm the existence of the Higgs boson. Other discoveries are surely to follow. Of course, as all technologies, the mathematical technology of quantum field theory and the experimental technology of the type found in LHC might, and even one day must, become obsolete and be replaced by other technologies. Thus far, however, both have worked marvelously, even if not always easily, beginning with Dirac's introduction of quantum electrodynamics and his famous equation for the relativistic electron, which I shall now consider.

3.4. *The symmetry of space and time, spinors, and Dirac's equation*

The three key elements were, in Dirac's view, especially required for a *relativistic quantum* equation for a free electron, that is, an equation dealing with the higher levels of energy of the electron at which the effect of special relativity theory cannot be neglected in the way they can be in quantum mechanics. The first, *relativistic*, element is that time and space must enter symmetrically, which is required by relativity but which is not the case in Schrödinger's equation, which contains the first derivative of time and the second derivatives of coordinates. The second, *quantum-theoretical*, element is that it contains the first order derivative in time – an element required by quantum-theoretical considerations, captured by the quantum-mechanical formalism, specifically by Schrödinger's equation, and related to several other key features of the formalism. Among these features are the noncommutativity of certain quantum variables, linear superposition, and the conservation of the probability current (which entails positive definite probability density) and, correlatively, the probabilistic character of the predictions enabled by the formalism. The third key element was that, by the mathematical correspondence principle (as applied in quantum electrodynamics) the nonrelativistic limit of a relativistic equation for the electron needed to be Schrödinger's equation. It follows, then, that to be both relativistic and quantum, a relativistic equation for the electron must be a first-order linear differential equation in both space and time, since quantum theory requires the first-order derivative in time, and relativity requires that space and time must enter symmetrically, and indeed that space and time must be interchangeable.

Although this seems simple enough in retrospect, it appears that at the time only Dirac thought of the situation in this way. This thinking was greatly helped by the transformation theory (his "darling," as he called it), especially as concerned the first order derivative in time, $\partial/\partial t$, and positive definite probability density, both central to the transformation theory (Ref. [51]). His famous conversation on the subject with Bohr at the time is revealing: "Bohr: What are you working on? Dirac: I am trying to get a relativistic theory of the electron. Bohr: But Klein already solved that problem" (Ref. [52]). Dirac disagreed, and, for the reasons just explained, it is clear why he did, and why Bohr should have known better. The Klein-Gordon equation is relativistic and symmetrical in space and time, but it is not a first-order differential equation in either, since both enter via their second derivatives. One can derive the continuity equation from it, but the probability density is not positive definite. By the same token, the Klein-Gordon equation does not give us the correct equation, Schrödinger's equation in the nonrelativistic limit. Schrödinger, who appears to be the first to have written down the Klein-Gordon equation in the process of his discovery of wave quantum mechanics, abandoned it in view of the incorrect predictions it gave in the nonrelativistic limit. (It should be noted that, although not workable in the case of the relativistic electron, the equations of the Klein-Gordon type have been effectively used elsewhere in quantum field theory, including in the case of the Higgs field.) The roles of the first-order derivative in time and of the probability-density considerations were not apparent to Schrödinger (who resisted the probabilistic view of his equation later on); they only came into play in Born's probabilistic interpretation of the wave function. Dirac's equation, on the other hand, does convert into Schrödinger's equation in the nonrelativistic limit, which was, again, a crucial part of Dirac's thinking, following Heisenberg's approach in quantum mechanics. Technically, at its immediate nonrelativistic limit, Dirac's theory converts into Pauli's spin-matrix theory (a major achievement in its own right), while Schrödinger's equation, which does not contain spin, was the limit of Pauli's theory, if one neglects spin (Ref. [42]). To summarize the conditions Dirac's equation had to fulfill:

(1) Relativistic requirements, in particular the symmetry of space and time;

(2) To be a first-order linear differential equation in time, which means the first-order in both space *and* time by (1);

(3) The probability density must be positive definite and the probability current must be conserved;

(4) Schrödinger's equation should be the nonrelativistic limit of the theory.

The conditions (2), (3), and (4) are correlative. None of these three conditions is satisfied by the Klein-Gordon equation. This is not to say that the task of deriving the correct equation becomes simple once these requirements are in place; quite the contrary, this derivation involved highly original and nontrivial moves. Most of these moves were, I argue, parallel to those of Heisenberg in his paper introducing quantum mechanics. However, Dirac's mathematical task was more difficult because conditions (1), (2), (3), (4) require both new variables, as in Heisenberg's scheme, and, in contrast to Heisenberg's scheme (which used the equations of classical mechanics), a new equation, in this respect similarly to Schrödinger's approach. In a way, Dirac's derivation of his equation combined Heisenberg's and Schrödinger's approaches, in the spirit of Dirac's transformation theory. As in Heisenberg, Dirac's new variables proved to be matrix-type variables, but of a more complex character, involving the so-called spinors and the multi-component wave function. The latter was a crucial concept, already discovered by Pauli in his nonrelativistic theory of spin. As Heisenberg's matrices before Heisenberg, Dirac's spinors had never been used in physics previously, although they were introduced in mathematics by W. C. Clifford about fifty years earlier (following the work of Hermann Grassmann on the so-called exterior algebras). But, just as Heisenberg in the case of his matrices, Dirac was unaware of their existence and reinvented them in deriving his equation.

In spite of the elegance of its famous compact form, $i\gamma \cdot \partial\psi = m\psi$, reproduced on the plate in Westminster Abbey commemorating Dirac, Dirac's equation encodes an extremely complex multi-component Hilbert-space machinery. It may also be noted that, unlike that of the Klein-Gordon equation, the Lorentz invariance of Dirac's equation is nontrivial, and it was surprising at the time, given the linear nature of the equation. Mathematically, the problem confronting Dirac may be seen in terms of taking a square root of the Klein-Cordon equation (the solutions of which are, again, complex quantities, a mathematically crucial fact here), which also implies that every solution of Dirac's equation is a solution of the Klein-Gordon equation while the opposite is not true. Dirac used this fact in his derivation of his equation. I shall follow Dirac's paper because it reflects the key aspects of quantum-theoretical thinking that I want to address. First, however, I shall give a general summary. The equation, as introduced by Dirac, is

$$(\beta mc^2 + \sum_{k=1}^{3} \alpha_k p_k c)\psi(x,t) = i\hbar\frac{\partial\psi(x,t)}{\partial t}.$$

The new mathematical elements here, which never previously occurred in physics (quantum mechanics included), are the 4×4 matrices α_k and β and the four component wave function ψ. The Dirac matrices are all Hermitian, satisfy

$$\alpha_i^2 = \beta^2 = I_4,$$

(I_4 is the identity matrix), and mutually anticommute:

$$\alpha_i\alpha_j + \alpha_j\alpha_i = 0,$$

$$\alpha_i\beta + \beta\alpha_i = 0.$$

The above single symbolic equation unfolds into four coupled linear first-order partial differential equations for the four quantities that make up the wave function. The matrices form a basis of the corresponding Clifford algebra. Indeed, following Dirac's work and subsequent developments of quantum theory, one can think of Clifford algebras as *quantizations* of Grassmann's exterior algebras, in the same way that the Weyl algebra is a quantization of symmetric algebra. Here, **p** is the momentum operator in Schrödinger's sense, but in a more complicated Hilbert space than in the standard quantum mechanics. The wave function $\psi(t, \mathbf{x})$ takes value in a Hilbert space $X = C^4$ (Dirac's spinors are elements of X). For each t, $\psi(t, \mathbf{x})$ is an element of $H = L^2(R^3; X) = L^2(R^3) \otimes X = L^2(R^3) \otimes C^4$. I shall comment on the significance of this mathematical architecture below, merely noting now that it allows one to properly predict the probabilities of quantum-electro-dynamical (high-energy) events, which have a greater complexity than quantum-mechanical (low-energy) events. Also, as noted from the outset, following Heisenberg and Dirac, finding new matrix variables or Hilbert-space operators became the uniquely defining mathematical element of any quantum field theory. The discoveries of proper theories of weak forces, electroweak unifications, and strong forces (ultimately quantum chromodynamics) were all established through finding such variables. This is correlative to establishing the transformation group, a Lie group, of the theory and finding representations of this group in the corresponding Hilbert spaces. This is true for Heisenberg's matrix variables as well, as was discovered by Weyl and E. Wigner, for the Heisenberg's group. In modern elementary-particle theory, irreducible representations of such groups correspond to elementary particles. This was essentially how M. Gell-Mann discovered quarks, because at the time there were no particles corresponding to the irreducible representations (initially there were three of those, corresponding to three quarks) of the symmetry group of the theory, the so-called $SO(3)$. It is the group of all rotations around the origin in the three-dimensional space, R,3 rotations represented by all three by three orthogonal matrices with determinant 1. (This group is non-commutative.) The electroweak group that Gell-Mann helped to find as well is $SU(2)$, the group of two by two matrices with the determinant 1. Quarks are part of both theories. As will be seen, the genealogy of this matrix and group theoretical thinking clearly extends from the work of Dirac (four by four spinor matrices) and, earlier, Pauli (two by two spin matrices).

Dirac begins his paper by commenting on previous relativistic treatments of the electron, specifically the Klein-Gordon equation and its insufficiencies. He says in particular:

[The Gordon-Klein approach] appears to be satisfactory as far as emission and absorption of radiation are concerned, but is not so general as the interpretation of the non-relativi[stic] quantum mechanics, which has been developed [specifically in Dirac's and Jordan's transformation theory] sufficiently to enable one to answer the question: What is the probability of any dynamical variables at any specified time having a value laying between any specified limits, when the system is represented by a given wave function ψ_n? The Gordon-Klein interpretation can answer such questions if they refer to the position of the electron ... but not if they refer to its momentum, or angular momentum, or any other dynamic variable. We would expect the interpretation of the relativi[stic] theory to be just as general as that the non-relativi[stic] theory. (Ref. [40], pp. 611–612)

The term "interpretation" means here a mathematical representation of the physical situation (quantum-mechanical or quantum-electrodynamical), rather than a physical or philosophical interpretation of a given quantum formalism cum the phenomena it relates to. Dirac's statement does not mean that a physical description of quantum processes in space and time is provided, as against only predictions, in general, probabilistic, of the outcomes of quantum experiments. Dirac, as is clear from this passage, thought the capacity of a given theory to enable such predictions sufficient, and sufficiently general if such predictions are possible for any dynamic variable. Dirac then argues that the derivative first-order in time, missing in the Klein-Gordon equation, is a proper starting point for the relativistic theory of the electron. He says: "The general interpretation of non-relativi[stic] quantum mechanics is based of the transformation theory, and is made possible by the wave equation being of the form

$$(H - W)\psi = 0 \qquad (3.1)$$

i.e., being linear in W or $\partial/\partial t$, so that the wave function at any time determines the wave function at any later time. The wave function of the relativi[stic] theory must also be linear in W if the general interpretation is to be possible" (Ref. [40], p. 612). Before proceeding to his derivation of his equation, Dirac comments on "the second difficulty" of the Klein-Gordon equation:

[The equation] refers equally well to an electron with charge e as to one with charge $-e$. If one considers for definitiveness the limiting case of large quantum numbers one would find that some of the solutions of the wave equation are wave packets moving in the way a particle of $-e$ would move on the classical theory, while other are wave packets moving in the way a particle with charge e would move classically. For this second class of solutions W has a negative value. One gets over the difficulty on the classical theory by arbitrarily excluding those solutions that have a negative W. One cannot do this on the quantum theory, since in general a perturbation will cause transitions from state with W positive to states with W negative. Such a transition would appear experimentally as the electron suddenly changes its charge from $-e$ to e, a phenomenon which has not been observed. The true relativi[stic] wave equation should thus be such that its solutions split up into two non-combining sets, referring respectively to the charge $-e$ and the charge e. In the present paper we shall only be concerned with the removal of the first of these difficulties. The resulting theory is therefore still only an approximation, but it appears to be good enough to account for all the duplexity phenomena without arbitrary assumptions. (Ref. [40], p. 612)

Dirac's theory, thus, inherits this second problem of the Klein-Gordon theory, because, as I said, mathematically every solution of Dirac's equation is a solution of the Klein-Gordon equation, of which Dirac's equation is essentially a square root (the opposite is, again, not true). The difficulty does not appear in low-energy quantum regimes, and, one might add, it disappears at the low energy limit of Dirac's theory, because his equation converts into Schrödinger's equation. This problem has ultimately proven to be a good thing. Dirac did not know this at the time, but, as he will eventually have learned, the theory is much better because of this difficulty. That "in general a perturbation will cause transitions from state with W positive to states with W negative," and that "such a transition would appear experimentally as the electron suddenly changes its charge from $-e$ to e" is what actually happen, and it will have been experimentally established in a year or so. Antimatter was staring right into Dirac's eyes. It took, however, a few years to realize that it is antimatter and that this type of transition (eventually understood in terms of the creation and annihilation of particles, and virtual particle formation) defines high-energy regimes in quantum physics.

Dirac is now ready to present his derivation of his equation, guided by the two key ideas in question: the invariance under a Lorentz transformation and the equivalence of whatever the new one finds to Schrödinger's equation $(H - W)\psi = 0$ (equation (1) above) in the limit of large quantum numbers (Ref. [40], p. 613). In the case of the absence of the external field, which Dirac considers first and to which I shall restrict myself here, since it is sufficient for my main argument, the Klein-Gordon equation "reduces to

$$(-p_0^2 + \mathbf{p}^2 + m^2c^2)\psi = 0, \tag{3.2}$$

if one puts

$$p_0 = \frac{W}{c} = ih\frac{\partial}{c\,\partial t}. \qquad \text{(Ref. [40], p. 613)} \tag{3.3}$$

Next Dirac uses the symmetry between time, p_0, and space, p_1, p_2, p_3, required by relativity, which implies that because the Hamiltonian one needs is linear in p_0, "it must also be linear in p_1, p_2, and p_3." He then says:

[the necessary] wave equation is therefore in the form

$$(p_0 + \alpha_1 p_1 + \alpha_2 p_2 + \alpha_3 p_3 + \beta)\psi = 0. \tag{3.4}$$

where for the present all that is known about the dynamical variables or operators α_1, α_2, α_3, and β is that they are independent of p_0, p_1, p_2, p_3, i.e., that they commute with t, x_1, x_2, x_3. Since we are considering the case of a particle moving in empty space, so that all points in space are equivalent, we should expect the Hamiltonian not to involve t, x_1, x_2, x_3. This means that α_1, α_2, α_3, and β are independent of t, x_1, x_2, x_3, i.e., that they commute with p_0, p_1, p_2, p_3. We are therefore obliged to have other dynamical variables besides the co-ordinates and momenta of the electron, in order that α_1, α_2, α_3, β may be functions of them. The wave function ψ must then involve more variables than merely x_1, x_2, x_3, t. Equation (4) leads to

$$0 = (-p_0 + \alpha_1 p_1 + \alpha_2 p_2 + \alpha_3 p_3 + \beta)(p_0 + \alpha_1 p_1 + \alpha_2 p_2 + \alpha_3 p_3 + \beta)\psi,$$

$$= [-p_0^2 + S\,\alpha_1^2 p_1^2 + \sum(\alpha_1\alpha_2 + \alpha_2\alpha_1)p_1 p_2 + \beta^2 + S(\alpha_1\beta + \beta\alpha_1)]\psi, \quad (3.5)$$

where S refers to cyclic permutation of the suffixes 1, 2, 3. (Ref. [40], p. 613) Let us pause here to admire Dirac's way of thinking, manifest in this passage and throughout his derivation of his equation. This is "very Dirac" – mathematically elegant and physically profound, for example, in the remarkable and far reaching conclusion that "the wave function ψ must then involve more variables than merely x_1, x_2, x_3, t." Taking advantage of noncommutativity in (5), one of Dirac's fortes, is worth a special notice. Equation (4), a square root of (3), the Klein-Gordon equation (a form of "consistency" Dirac uses throughout the paper), is already Dirac's equation in abstract algebraic terms. One will now need to find α_n and β, to find the actual form of the equation. Physics as much follows mathematics as mathematics physics. While, as I argue, inspired by Heisenberg's thinking, Dirac's more formal approach is different from that adopted by Heisenberg in his discovery of quantum mechanics, an approach equally profound physically but more straightforward and cumbersome; unlike in Dirac, in Heisenberg mathematics nearly always follows physics. This is not to say that mathematics is less important for Heisenberg, but only that one does not find in Heisenberg the same kind of, to use Dirac's word, "play" with abstract structures that one finds in Dirac (Ref. 52). Dirac's earlier work on q-numbers quantum-mechanical formalism manifested the same use of the power of mathematical formalization. Dirac proceeds as follows:

[Equation (5)] agrees with (3) [the Klein-Gordon equation] in the absence of the external field

$$(-p_0^2 + \mathbf{p}^2 + m^2 c^2)\psi = 0.] \tag{3.6}$$

if

$$\alpha_r^2 = 1, \qquad \alpha_r\alpha_s + \alpha_s\alpha_r = 0, \qquad (r \neq s), \qquad r, s = 1, 2, 3.$$

$$\beta^2 = m^2 c^2, \qquad \alpha_r\beta + \beta\alpha_r = 0.$$

If we put $\beta = \alpha_4 mc$, these conditions become

$$\alpha_m^2 = 1, \qquad \alpha_m \alpha_n + \alpha_n \alpha_m = 0, \qquad (m \neq n), \qquad m, n = 1, 2, 3, 4. \quad (3.7)$$

(Ref. [40], p. 613)

Thus, as I said, there is also a partial mathematical correspondence with the Klein-Gordon equation (that between the function of a complex variable and its square root), which allows Dirac to derive certain necessary algebraic conditions upon α_μ and β. Dirac will now state that "we can suppose α_μ's to be expressed in some matrix scheme, the matrix elements of α_μ being, say, $a_m(z, z')$" (p. 613). This supposition is not surprising given both the formal mathematical considerations (such as the anticommuting relations between them) and the preceding history of matrix mechanics, including Dirac's own previous work. We know, or may safely assume from Dirac's account of his work on his equations, that matrix manipulation, "playing with equations," as he called it, was one of his starting points (Ref. [52]). In addition, Pauli's theory, which is about to enter Dirac's argument, provided a handy example of a matrix scheme (Ref. [42]; Ref. [53], pp. 55–56, 60). It is clear that matrix algebra of some sort is a good candidate for α_μ.

Dirac's ways of thinking merits a further reflection. Dirac gets extraordinary mileage from considering the formal properties of the variables involved, even before considering what these variables actually are and as a way of gauging what they should be, which is his next step. He used the same approach – begin with the necessary formal properties and then find the actually variables – earlier in developing his q-number formalism. This approach is, as I said, both analogous to Heisenberg's approach and yet is also more formally oriented. That Heisenberg had to find new variables formally satisfying classical equations may be seen as a partially formal task. However, the primary guidance for finding these variables was provided by certain experimentally established physical conditions (Bohr's energy rules and the Rydberg-Ritz frequency rules). In Dirac, while physical conditions of relativity and quantum mechanics do play a role, the primary driving force is the formal properties of variables and, especially, equations, in particular the linearity and noncommutativity of the formalism. These, along with the role of complex numbers, are the defining mathematical building blocks of quantum theory. Or, to put it another way, the imperative of using them was, for Dirac, one of the fundamental *principles* of quantum theory. Dirac's adherence to these principles, physical (such as the relativistic symmetry of time and space, or the role of probability in quantum physics) and mathematical (such those in question at the moment), defines his thinking in his work on quantum theory, beginning with his q-number formalism of quantum mechanics. Dirac's word "principle" in the title of his celebrated book, *The Principles of Quantum Mechanics*, reflects this adherence, this *principle* of using quantum principles, one might say, and it is not accidental or lightly chosen (Ref. [54]).

Another remarkable consequence of the necessity of the particular matrix variables required by Dirac, appears next. (We still do not know what these variables actually are!) For if we indeed "suppose α_μ's to be expressed in some matrix scheme, the matrix elements of α_μ being, say, $a_m(z, z')$," then "the wave function ψ must be a function of z as well as x_1, x_2, x_3, t. The result of a_m multiplied into ψ will be a function (a_m, ψ) of x_1, x_2, x_3, t, z defined by

$$(a_m, \psi)(x, t, z) = \sum_{z'} a_m(z, z') \psi(x, t, z').\text{"}\qquad \text{(Ref. [40], p. 614)}\qquad (3.8)$$

Dirac is now ready "for finding four matrices α_μ to satisfy the conditions (6)," those de facto forming the Clifford algebra, and for finding the actual form of variables that satisfy formal equation (4) or (5). Dirac considers first the three Pauli spin matrices, which satisfy the conditions (6), but not the equations (4) or (5), which needs four matrices. He says:

We make use of the matrices

$$\sigma_1 = \begin{pmatrix} 0 & 1 \\ 1 & 0 \end{pmatrix}, \qquad \sigma_2 = \begin{pmatrix} 0 & -i \\ i & 0 \end{pmatrix}, \qquad \sigma_3 = \begin{pmatrix} 1 & 0 \\ 0 & -1 \end{pmatrix}, \qquad (3.9)$$

which Pauli introduced to describe the three components of the spin angular momentum. These matrices have just the properties

$$\sigma_r^2 = 1, \qquad \sigma_r \sigma_s + \sigma_s \sigma_r = 0, \qquad (r \neq s), \qquad (3.10)$$

that we require for our α's. We cannot, however, just take the σ's to be thereof our α's, because then it would not be possible to find the fourth. We must extend the σ's in a diagonal matter to bring in tow more rows and columns, so that we can introduced three more matrices ρ_1, ρ_2, ρ_3 of the same form as σ_1, σ_2, σ_3, but referring to different rows and columns, thus:

$$\sigma_1 = \begin{pmatrix} 0 & 1 & 0 & 0 \\ 1 & 0 & 0 & 0 \\ 0 & 0 & 0 & 1 \\ 0 & 0 & 1 & 0 \end{pmatrix}, \qquad \sigma_2 = \begin{pmatrix} 0 & -i & 0 & 0 \\ i & 0 & 0 & 0 \\ 0 & 0 & 0 & -i \\ 0 & 0 & i & 0 \end{pmatrix}, \qquad \sigma_3 = \begin{pmatrix} 1 & 0 & 0 & 0 \\ 0 & -1 & 0 & 0 \\ 0 & 0 & 1 & 0 \\ 0 & 0 & 0 & -1 \end{pmatrix},$$

$$\rho_1 = \begin{pmatrix} 0 & 0 & 1 & 0 \\ 0 & 0 & 0 & 1 \\ 1 & 0 & 0 & 0 \\ 0 & 1 & 0 & 0 \end{pmatrix}, \qquad \rho_2 = \begin{pmatrix} 0 & 0 & -i & 0 \\ 0 & 0 & 0 & -i \\ i & 0 & 0 & 0 \\ 0 & i & 0 & 0 \end{pmatrix}, \qquad \rho_3 = \begin{pmatrix} 1 & 0 & 0 & 0 \\ 0 & 1 & 0 & 0 \\ 0 & 0 & -1 & 0 \\ 0 & 0 & 0 & -1 \end{pmatrix}.$$

The ρ's are obtained from σ's by interchanging the second and the third row, and the second and the third columns. We now have, in addition to equations (3.10),

$$\rho_r^2 = 1, \qquad \rho_r \rho_s + \rho_s \rho_r = 0, \qquad (r \neq s), \qquad (3.10')$$

and also

$$\rho_r \sigma_t = \sigma_t \rho_r. \qquad (3.10'')$$

(Ref. [40], p. 615)

These matrices are Dirac's great mathematical invention, parallel to Heisenberg's invention of his new matrix variables for quantum mechanics. Dirac's matrices form the basis of the corresponding Clifford algebra and define the mathematical architecture, mentioned above, where the multicomponent relativistic wave function for the electron must be defined. The entities they transformed are different from either vectors or tensors and are called spinors, introduced, as mathematical objects, by E. Cartan in 1913. Pauli, again, introduced the two-component non-relativistic wave function, which was necessary to incorporate spin, but he did so phenomenologically, rather than from the first principles, as was done by Dirac.

The rest of the derivation of Dirac's equation is a nearly routine exercise, with a few elegant but easy matrix manipulations. Dirac also needs to prove the relativistic invariance and the conservation of the probability current, and to consider the case of the external field, none of which is automatic, but is the standard textbook material at this point. The most fundamental and profound aspects of Dirac's thinking are contained in the parts of his paper just discussed, and I shall in the remainder of this article consider the implications of Dirac's mathematical architecture for physics. I close this section by summing up my discussion thus far. Dirac uses the key elements of Heisenberg's approach to quantum mechanics:

(1) the mathematical correspondence principle, with Pauli's theory (the immediate limit) and Schrödinger's equation (the far limit) as the nonrelativistic limits of his theory;

(2) the introduction of a new type of matrix variables, and spinors on which these matrices act; while Dirac is, again, helped here by Pauli's theory, he displays a highly original and nontrivial way of mathematical thinking;

(3) the assumption of the probabilistically predictive and not descriptive character of the theory he aims to construct.

The main difference from Heisenberg's discovery of quantum mechanics is:

(4) The appearance of a new equation, and of a *new type of equation.*

A few qualifications may be in order concerning Pauli's theory and Schrödinger's equation as the quantum-mechanical limits of Dirac's theory. Given the preceding discussion, it is easy to surmise that Pauli's theory and, if one further neglects spin, Schrödinger's equation, as itself the limit of Pauli's theory, would appear as the nonrelativistic limits (immediate and far) of Dirac's theory. Pauli's matrices are contained in Dirac's matrices, and one can, accordingly, split Dirac's matrices (or the corresponding spinors) into small and large components. One does need a few calculations to rigorously establish this correspondence, but they are more or less straightforward. What is important is that, while Pauli's argument was phenomenological, Dirac had a theoretical argument from the first principles, which contained spin (also for positrons) and which also suggested that spin might have been the consequence of bringing together quantum mechanics and relativity. Dirac's theory also more rigorously justified the apparently irreducible role of complex numbers and the necessity of a complex wave function in quantum theory (for him, again, part of the mathematical principles of quantum theory and, hence, imperative in any quantum formalism) from the geometry of relativistic spacetime via his spinor algebra. This role appears phenomenologically, almost mysteriously, in quantum mechanics. It is important, however, that splitting the Dirac spinors and matrices into large and small components only applies at a low-energy approximation. Rigorously, one needs all of them. It is the whole composition of Dirac's scheme that reflects new physical phenomena found in the relativistic regime.

3.5. *The architecture of mathematics and the architecture of matter in quantum field theory*

Dirac's equation, thus, encodes a complex mathematical architecture. The Hilbert space associated with given quantum systems in Dirac's theory is a tensor product of the infinite dimensional Hilbert space (encoding the mathematics of continuous variables) and a finite-dimensional Hilbert space over complex numbers, which, in contrast to the two-dimensional Hilbert space of Pauli's theory, C^2, is four-dimensional in Dirac's theory, C^4. (Spin is contained by the theory automatically.) Dirac's wave function $\psi(t, \mathbf{x})$ takes value in a Hilbert space $X = C^4$ (Dirac's spinors are elements of X). For each t, $\psi(t, \mathbf{x})$ is an element of

$$H = L^2(R^3; X) = L^2(R^3) \otimes X = L^2(R^3)C^4. \tag{3.11}$$

Other forms of quantum field theory give this type of architecture an even greater complexity, but, which is one of my main points here, they retains its essential mathematical character, thus defined, and the essential ways in which it relates to quantum experiments. It is difficult to overestimate the significance of this mathematical architecture, which amounted to a very radical view of matter, physically especially manifest in the existence of antimatter. This architecture mathematically responds, and in fact led to a discovery of, the following physical situation, again, keeping in mind that, just as that of quantum mechanics, Dirac's formalism only provides probabilities for the outcomes of quantum events, experimentally registered in the corresponding measuring technology. Suppose that one arranges for an emission of an electron, at a given high energy, from a source and then performs a measurement at a certain distance from that source. Placing a photographic plate at this point would do. The probability of the outcome would be properly predicted by quantum electrodynamics. But what will be the outcome? The answer is not what our classical or even our quantum-mechanical intuition would expect, and this unexpected answer was a revolutionary discovery of quantum electrodynamics. Let us consider, first, what happens if we deal with a classical and then a quantum object in the same type of arrangement.

We can take as a model of the classical situation a small ball that hits a metal plate, which can be considered as either a position or a momentum measurement, or indeed a simultaneous measurement of both, and time t. In classical mechanics we can deal directly with the objects involved, rather than with their effects upon measuring instruments. The place of the collision could, at least in an idealized representation of the situation, be predicted exactly by classical mechanics, and we can repeat the experiment with the same outcome on an identical or even the same object. Most importantly, regardless of where we place the plate, we always find the same object, at least in a well-defined experimental situation, which is shielded from significant outside interferences, such as, for example, those that can deflect or even destroy the ball earlier.

By contrast, if we deal with an electron as a quantum object in the quantum-mechanical regime we cannot predict the place of collision exactly and, correlatively, exactly repeat the experiment on the same electron. In fact, we cannot repeat the experiment on the *same* electron at all, and what helps us here that, on quantum theory, electrons are in principle indistinguishable from each other. Also correlatively, we cannot simultaneously predict, or measure, the position and the momentum of an electron, which makes the situation correlative to the uncertainty relations. Indeed, there is a nonzero probability that we will not observe such a collision at all, or that if we do, that a different electron (coming from somewhere else) is involved. It is also not possible to distinguish two observed traces as belonging to two difference objects of the same type. Unlike in the classical case, in dealing with quantum objects, there is no way to improve the conditions or the precision of the experiment to avoid this situation. Quantum mechanics, however, gives us correct probabilities for such events. Mathematically, this is accomplished by defining the corresponding Hilbert space, with the position and other operators as observables, and writing down Schrödinger's equation for the state vector $|\psi\rangle$ (in the case of a pure state, or a density matrix otherwise), and using Born's or similar rules to obtain the probabilities of possible outcomes.

Once the process occurs at a high energy and is governed by quantum electrodynamics, the situation is still different, even radically different. One might find, in the corresponding region, not only an electron, as in classical physics, or an electron or nothing, as in the quantum-mechanical regime, but also other particles: a positron, a photon, an electron-positron pair. Just as does quantum mechanics, quantum-electrodynamics, beginning with Dirac's equation, rigorously predicts which among such events can occur, and with what probability, and, in the present view it can only predict such probabilities, or statistical correlations between certain quantum events. In order to do so, however, the corresponding Hilbert-space machinery becomes much more complex, essentially making the wave function ψ a four-component Hilbert-space vector, as opposed to a one-component Hilbert-space vector, as in quantum mechanics. This Hilbert space is denoted in Eq. (3.11) and the operators are defined accordingly. This structure naturally allows for a more complex structure of predictions (which are still probabilistic) corresponding to the situation just explained, usually considered in terms of virtual particle formation and Feynman's diagrams.

Once we move to still higher energies or different domains governed by quantum field theory the panoply of possible outcomes becomes much greater. The Hilbert spaces involved would be given a yet more complex structure, in relation to the appropriate Lie groups and their representations, defining (when these representations are irreducible) different elementary particles, as indicated above. In the case of quantum electrodynamics we only have electrons, positrons, and photons. It follows that in quantum field theory an investigation of a particular type of quantum object irreducibly involves not only other particles of the same type but also other types of particles. This qualification is important because the identity of particles within each type is strictly maintained in quantum field theory, as it is in quantum mechanics. In either theory one cannot distinguish different particles of the same type, such as electrons. One can never be certain that one encounters the same electron in the experiment just described even in the quantum-mechanical situation, although the probability that it would in fact be a different electron is low in the quantum-mechanical regime in comparison to that in the regime of quantum electrodynamics. In quantum field theory, it is as if instead of identifiable moving objects and motions of the type studied in classical physics, we encounter a continuous emergence and disappearance, creation and annihilation, of particles from point to point, theoretically governed by the concept of virtual particle formation. The operators used to predict the probability of such events, are the creation and annihilation operators. This view, thus, clearly takes us beyond quantum mechanics. For, while the latter questions the applicability of classical concepts, such as objects (particles or waves) and motion, at the quantum level, it still preserves the identity of quantum objects. It is still possible to speak of this identity, even though, in the present view, these objects themselves remain unthinkable and only manifest themselves and their identity to each other in their effects upon measuring instruments.

The introduction of this new mathematical formalism, involving more complex Hilbert spaces and operator algebras, was a momentous event in the history of quantum physics, comparable to that of Heisenberg's introduction of his matrix variables, as Bohr stated in his 1962 assessment of Dirac's theory, cited at the outset of this article. Heisenberg was even more emphatic. He saw Dirac's theory as an even more radical revolution than quantum mechanics was. In an article, revealingly entitled, "What is an Elementary Particle?" (Still an unanswered question!), written in the early 1970s, Heisenberg spoke of Dirac's discovery of antimatter as

> perhaps the biggest change of all the big changes in physics of our century ... because it changed our whole picture of matter ... It was one of the most spectacular consequences of Dirac's discovery that the old concept of the elementary particle [based on their stable identity] collapsed completely. (Ref. [55], pp. 31–33)

A path to a new understanding of the ultimate constitution of nature became open, however. Following this path, quantum field theory made remarkable progress since its introduction or since Heisenberg's remark, a progress resulting, for example, in the electroweak unification and the quark model of nuclear forces, developments that commenced around the time of these remarks. As already noted, many predictions of the theory, from quarks to electroweak bosons and the concept of confinement and asymptotic freedom, to name just a few, were spectacular, and, since its introduction, the field has garnered arguably the greatest number of Nobel Prizes in physics. It was also quantum field theory that led to string and then brane theories, the current stratosphere of theoretical physics. However, the essential mathematical and experimental architecture of the theory, as considered here, have remained in place. Quantum mechanics and then higher-level quantum theories continue classical physics insofar as it is, just as classical physics, from Galileo on, and then relativity have been, the experimental-mathematical science of nature. On the other hand, quantum theory, at least in the interpretation of the type adopted here, breaks with both classical physics and relativity by establishing radically new relationships between mathematics and physics, or mathematics and nature. The mathematics of quantum theory is able to predict correctly the experimental data in question without offering a description of the physical processes responsible for these data. This is of course remarkable. As I have noted throughout these lectures, and as has been often observed by others, we have been extraordinarily lucky that our mathematics works in the absence of such a description, thus allowing us to bring mathematics and physics together in a new way, vis-à-vis classical physics or relativity (e.g., Ref. [11], p. 11). Many, beginning, again, with Einstein, have found this epistemological situation deeply unsatisfactory and even disturbing; and the debates concerning the epistemology of quantum theory have never subsided or lost any of their intensity. It is not my aim to enter these debates here, and as I stressed from the outset we must always be open to alternative views, physical and philosophical, as two very different ways of thinking, that of Heisenberg and that of Schrödinger, leading to the discovery of quantum mechanics show. It is conceivable that the future development of fundamental physics will bring about more classical (realist and causal) alternatives, as Einstein hoped, although a more radical departure from classical epistemology than that enacted by quantum mechanics and quantum field theory (in the interpretation discussed here) is not inconceivable either. Whatever theory will win the day next time around (this process is unlikely to ever end), physics itself is likely to be remarkable, and, as quantum theory has done for a century now, this new physics is also likely to help us advance the ways we think not only about physics but also about the world and ourselves.

Acknowledgments

I would like to express my sincere gratitude to the organizers of the school on "Quantum Foundations and Open Systems," Joao Pessoa, Brazil (15–28 July 2012), especially Claudia Pombo, Claudio Furtado, Inácio de Almeida Pedrosa, and Theo Nieuwenhuizen, for inviting me, and to the participants of the school for invaluable discussions concerning the subjects considered in these lectures. I am particularly grateful to Amir Caldeira, Giuseppe Falci, Andrei Khrennikov, Kristel Michielsen,

Antonio Murilo, Hans de Raedt, and especially Theo Nieuwenhuizen, who offered most helpful commentaries and criticism concerning my argument here.

References

1. A. Plotnitsky, *Epistemology and Probability: Bohr, Heisenberg, Schrödinger and the Nature of Quantum-Theoretical Thinking* (Springer-Verlag, New York, 2009).
2. A. Plotnitsky, Dark materials to create more worlds: on causality in classical physics, quantum physics, and nanophysics, *J. Comput. Theor. Nanosci.* **8**(6), 983–997 (2011).
3. A. Plotnitsky, *Niels Bohr and Complementarity: An Introduction* (Springer-Verlag, New York, 2012).
4. A. Plotnitsky, Dirac equation and the nature of quantum field theory, *Phys. Scr.* **T151**, 014010 (2012).
5. N. Bohr, *The Philosophical Writings of Niels Bohr, 3 vols* (Ox Bow Press, Woodbridge, CT, 1987).
6. A. Einstein, Physics and reality, *Journal of the Franklin Institute* **221**, 349–382 (1936).
7. W. Heisenberg, Quantum-theoretical re-interpretation of kinematical and mechanical relations (1925). In eds. B. L. Van der Waerden, ed., *Sources of Quantum Mechanics*, pp. 261–277. Dover, New York (1968).
8. E. Schrödinger, The present situation in quantum mechanics (1935). In eds. J. A. Wheeler, and W. H. Zurek, *Quantum Theory and Measurement*, pp. 152–167. Princeton University Press, Princeton, NJ (1983).
9. A. Einstein, B. Podolsky and N. Rosen, Can quantum-mechanical description of physical reality be considered complete? (1935). In eds. J. A. Wheeler and W. H. Zurek, *Quantum Theory and Measurement*, pp. 138–141. Princeton University Press, Princeton, NJ (1983).
10. E. Schrödinger, *The Interpretation of Quantum Mechanics: Dublin Seminars (1949–1955) and Other Unpublished Essays* (Ox Bow Press, Woodbridge, CT, 1995).
11. W. Heisenberg, *The Physical Principles of the Quantum Theory*, tr. K. Eckhart and F. C. Hoyt (Dover, New York, 1930, rpt. 1949).
12. A. E. Allahverdyan, R. Balian and T. M. Nieuwenhuizen, Understanding quantum measurement from the solution of dynamical models, *Physics Reports - Review Section of Physics Letters* **525**, 1–166 (2013).
13. P. A. M. Dirac. The fundamental equations of quantum mechanics (1925). In eds. B. L. van der Warden, *Sources of Quantum Mechanics*, pp. 307–320. Dover, New York (1968).
14. W. Pauli, *Writings on Physics and Philosophy* (Springer, Berlin, 1994).
15. A. Einstein, *Autobiographical Notes*, tr. P. A. Schilpp (Open Court, La Salle, IL, 1991).
16. N. Bohr, Can quantum-mechanical description of physical reality be considered complete? *Phys. Rev.* **48**, 696–702 (1935).
17. J. Mehra and H. Rechenberg, *The Historical Development of Quantum Theory*, 6 vols. (Springer-Verlag, Berlin, 2001).
18. E. Schrödinger, Zur Einsteinschen Gastheorie, *Z. Phys.* **27**, 95–101 (1926).
19. E. Schrödinger, Quantisierung als Eigenwertproblem (Erste Mitteilung), *Ann. Phys.* **79**, 361–376, (1926).
20. *Archive for the History of Quantum Physics (AHQP)*.
21. E. Schrödinger, Quantisierung als Eigenwertproblem (Zweite Mitteilung), *Ann. Phys.* **79**, 489–527 (1926).

22. W. Heisenberg, *Physics and Philosophy: The Revolution in Modern Science* (Harper & Row, New York, 1962).
23. *The Random House Webster's Unabridged Dictionary* (Random House, New York, 2005).
24. N. Born, *Natural Philosophy of Cause and Chance* (Dover Publications, New York, 1949).
25. I. Kant, *Critique of Pure Reason*, tr. P. Guyer and A. W. Wood (Cambridge Uiversity Press, Cambridge, 1997).
26. P. Dowe, *Physical Causation* (Cambridge University Press, Cambridge, 2007).
27. J. C. Maxwell, Thompson and Tait's Natural Philosophy, *Nature* **20**, 213–216 (1879).
28. A. Einstein. Geometry and experience. In A. Einstein, *Ideas and Opinions*, pp. 232–242, Random House, New York (1988).
29. J. Gray, *Plato's Ghost: The Modernist Transformation of Mathematics* (Princeton University Press, Princeton, NJ, 2008).
30. D. Hume, *Treatise on Human Nature*, eds. L. A. Selbe-Bigge and P. H. Nidditch (Clarendon, Oxford, 1978).
31. M. Born, *The Einstein-Born Letters*, tr. I. Born (Walker, New York, 2005).
32. E. T. Jaynes, *Probability Theory: The Logic of Science* (Cambridge University Press, Cambridge, 2003).
33. G. 't Hooft, Entangled quantum states in a local deterministic theory, arXiv:0908.3408v1 [quant-ph] June 2009.
34. R. Feynman, R. B. Leighton and M. Sands, *The Feynman Lectures on Physics*, 3 vols. (Addison-Wesley, Menlo Park, CA, 1977).
35. N. Bohr, *Philosophical Writings of Niels Bohr, Volume 4: Causality and Complementarity, Supplementary Papers* (eds. J. Faye and H. J. Folse (Ox Bow Press, Woodbridge, CT, 1998).
36. A. Yu. Khrennikov, *Interpretation of Probability* (De Gruyter, Berlin, 2009).
37. V. Spicka, Th. M. Nieuwenhuizen and P. D. Keefe, Physics at the FQTM'11 Conference, *Phys. Scr.* **T151**, 014001 (2012).
38. M. Kuhlman, Quantum field theory, in *Stanford Encyclopedia of Philosophy*, http://plato.stanford.edu/entries/quantum-field-theory/
39. N. Bohr and L. Rosenfeld, On the question of the measurability of electromagnetic field quantities (1933). In eds. J. A. Wheeler and W. H. Zurek, *Quantum Theory and Measurement*, 479–522, Princeton University Press, Princeton, NJ (1983).
40. P. A. M. Dirac, The quantum theory of the electron, *Proc. R. Soc. Lond. A* **177**, 610–624 (1928).
41. P. A. M. Dirac, The physical interpretation of the quantum dynamics, *Proc. R. Soc. Lond. A* **113**, 621–641 (1928).
42. W. Pauli, Zur Quantenmechanik des magnetischen Elektrons, *Z. Phys.* **43**, 601–625 (1927).
43. W. Pauli and W. Heisenberg, Quantentheorie der Wellenfelder, 2 mitt, *Z. Phys.* **59**, 168–190 (1930).
44. E. Fermi, Versuch einer Theorie der β-Strahlen. I, *Z. Phys.* **88**, 161–177 (1934).
45. T. Y. Cao, ed., *Conceptual Foundations of Quantum Field Theories* (Cambridge University Press, Cambridge, 1999).
46. S. Schweber, *QED and the Men Who Made It: Dyson, Feynman, Schwinger, and Tomonaga* (Princeton University Press, Princeton, NJ, 1994).
47. N. Bohr and L. Rosenfeld, Field and charge measurements in quantum electrodynamics (1950), in eds. J. A. Wheeler and W. H. Zurek, *Quantum Theory and Measurement*, 523–534, Princeton University Press, Princeton, NJ (1983).

48. P. Teller, *An Interpretive Introduction to Quantum Field Theory* (Princeton University Press, Princeton, NJ, 1995).
49. F. J. Dyson, The S-matrix in quantum electrodynamics, *Phys. Rev.* **75**, 1736–1755 (1949).
50. S. Carroll, *The Particle at the End of the Universe: How the Hunt for the Higgs Boson Leads Us to the Edge of a New World* (Dutton/Penguin USA, New York, 2012).
51. P. A. M. Dirac, Report KFKI-1997-62, Hungarian Academy of Science.
52. P. A. M. Dirac, Interview with T. Kuhn, April 1, 1962, *Niels Bohr Archive* (Copenhagen and American Institute of Physics, College Park, MD); http://www.aip.org/history/ohilist/
53. H. Kragh, *Dirac: A Scientific Biography* (Cambridge University Press, Cambridge, 1990).
54. P. A. M. Dirac, 1958, *The Principles of Quantum Mechanics* (Oxford; Clarendon, rpt. 1995).
55. W. Heisenberg, *Encounters with Einstein, And Other Essays on People, Places, and Particles* (Princeton University Press, Princeton, NJ, 1989).

Printed in the United States
By Bookmasters